More information about this series at http://www.springer.com/series/1244

Nicolas Peltier · Viorica Sofronie-Stokkermans (Eds.)

Automated Reasoning

10th International Joint Conference, IJCAR 2020
Paris, France, July 1–4, 2020
Proceedings, Part I

Springer

Editors
Nicolas Peltier
CNRS, LIG, Université Grenoble Alpes
Saint Martin d'Hères, France

Viorica Sofronie-Stokkermans
University Koblenz-Landau
Koblenz, Germany

ISSN 0302-9743 ISSN 1611-3349 (electronic)
Lecture Notes in Artificial Intelligence
ISBN 978-3-030-51073-2 ISBN 978-3-030-51074-9 (eBook)
https://doi.org/10.1007/978-3-030-51074-9

LNCS Sublibrary: SL7 – Artificial Intelligence

This Springer imprint is published by the registered company Springer Nature Switzerland AG
The registered company address is: Gewerbestrasse 11, 6330 Cham, Switzerland

Preface

These volumes contain the papers presented at the 10th International Joint Conference on Automated Reasoning (IJCAR 2020) initially planned to be held in Paris, but – due to the COVID-19 pandemic – held by remote conferencing during July 1-4, 2020.

IJCAR is the premier international joint conference on all aspects of automated reasoning, including foundations, implementations, and applications, comprising several leading conferences and workshops. IJCAR 2020 united CADE, the Conference on Automated Deduction, TABLEAUX, the International Conference on Automated Reasoning with Analytic Tableaux and Related Methods, FroCoS, the International Symposium on Frontiers of Combining Systems, and ITP, the International Conference on Interactive Theorem Proving. Previous IJCAR conferences were held in Siena (Italy) in 2001, Cork (Ireland) in 2004, Seattle (USA) in 2006, Sydney (Australia) in 2008, Edinburgh (UK) in 2010, Manchester (UK) in 2012, Vienna (Austria) in 2014, Coimbra (Portugal) in 2016, and Oxford (UK) in 2018.

150 papers were submitted to IJCAR: 105 regular papers, 21 system description, and 24 short papers, describing interesting work in progress. Each submission was assigned to three Program Committee (PC) members; in a few cases, a fourth additional review was requested. A rebuttal phase was added for the authors to respond to the reviews before the final deliberation. The PC accepted 62 papers, resulting in an acceptance rate of about 41%: 46 regular papers (43%), 11 system descriptions (52%), and 5 short papers (20%).

In addition, the program included three invited talks, by Clark Barrett, Elaine Pimentel, and Ruzica Piskac, plus two additional invited talks shared with the conference on Formal Structures for Computation and Deduction (FSCD), by John Harrison and René Thiemann (the abstract of the invited talk by René Thiemann is available in the proceedings of FSCD 2020).

The Best Paper Award was shared this year by two papers: "Politeness for The Theory of Algebraic Datatypes" by Ying Sheng, Yoni Zohar, Christophe Ringeissen, Jane Lange, Pascal Fontaine, and Clark Barrett, and "The Resolution of Keller's Conjecture" by Joshua Brakensiek, Marijn Heule, John Mackey, and David Narvaez.

IJCAR acknowledges the generous sponsorship of the CNRS (French National Centre for Scientific Research), Inria (French Institute for Research in Computer Science and Automation), the Northern Paris Computer Science (LIPN: Laboratoire d'Informatique de Paris Nord) at the University of Paris North (Université Sorbonne Paris Nord), and of the Computer Science Laboratory of Ecole Polytechnique (LIX: Laboratoire d'Informatique de l'École Polytechnique) in the École Polytechnique.

The EasyChair system was extremely useful for the reviewing and selection of papers, the organization of the program, and the creation of this proceedings volume. The PC chairs also want to thank Springer for their support of this publication.

We would like to thank the organizers of IJCAR, the members of the IJCAR PC, and the additional reviewers, who provided high-quality reviews, as well as all authors, speakers, and attendees.

The COVID-19 pandemic had a strong impact on the organization of IJCAR and significantly weighted the burden on authors, reviewers, and organizers. We are very grateful to all of them for their hard work under such difficult and unusual circumstances.

April 2020

Nicolas Peltier
Viorica Sofronie-Stokkermans

Organization

Program Committee Chairs

Nicolas Peltier	Université Grenoble Alpes, CNRS, LIG, France
Viorica Sofronie-Stokkermans	University of Koblenz-Landau, Germany

Program Committee

Takahito Aoto	Niigata University, Japan
Carlos Areces	Universidad Nacional de Córdoba (FaMAF), Argentina
Jeremy Avigad	Carnegie Mellon University, USA
Franz Baader	TU Dresden, Germany
Peter Baumgartner	CSIRO, Australia
Christoph Benzmüller	Freie Universität Berlin, Germany
Yves Bertot	Inria, France
Armin Biere	Johannes Kepler University Linz, Austria
Nikolaj Bjørner	Microsoft, USA
Jasmin Blanchette	Vrije Universiteit Amsterdam, The Netherlands
Maria Paola Bonacina	Università degli Studi di Verona, Italy
James Brotherston	University College London, UK
Serenella Cerrito	IBISC, Université d'Évry, Université Paris-Saclay, France
Agata Ciabattoni	Vienna University of Technology, Austria
Koen Claessen	Chalmers University of Technology, Sweden
Leonardo de Moura	Microsoft, USA
Stéphane Demri	LSV, CNRS, ENS Paris-Saclay, France
Gilles Dowek	Inria, ENS Paris-Saclay, France
Marcelo Finger	University of São Paulo, Brazil
Pascal Fontaine	Université de Liège, Belgium
Didier Galmiche	Université de Lorraine, CNRS, LORIA, France
Silvio Ghilardi	Università degli Studi di Milano, Italy
Martin Giese	University of Oslo, Norway
Jürgen Giesl	RWTH Aachen University, Germany
Valentin Goranko	Stockholm University, Sweden
Rajeev Gore	The Australian National University, Australia
Stefan Hetzl	Vienna University of Technology, Austria
Marijn Heule	Carnegie Mellon University, USA
Cezary Kaliszyk	University of Innsbruck, Austria
Deepak Kapur	University of New Mexico, USA
Laura Kovacs	Vienna University of Technology, Austria
Andreas Lochbihler	Digital Asset GmbH, Switzerland

Christopher Lynch	Clarkson University, USA
Assia Mahboubi	Inria, France
Panagiotis Manolios	Northeastern University, USA
Dale Miller	Inria, LIX/Ecole Polytechnique, France
Cláudia Nalon	University of Brasília, Brazil
Tobias Nipkow	Technical University of Munich, Germany
Albert Oliveras	Universitat Politècnica de Catalunya, Spain
Jens Otten	University of Oslo, Norway
Lawrence Paulson	University of Cambridge, UK
Nicolas Peltier	Université Grenoble Alpes, CNRS, LIG, France
Frank Pfenning	Carnegie Mellon University, USA
Andrei Popescu	Middlesex University London, UK
Andrew Reynolds	University of Iowa, USA
Christophe Ringeissen	Inria, France
Christine Rizkallah	The University of New South Wales, Australia
Katsuhiko Sano	Hokkaido University, Japan
Renate A. Schmidt	The University of Manchester, UK
Stephan Schulz	DHBW Stuttgart, Germany
Roberto Sebastiani	University of Trento, Italy
Viorica Sofronie-Stokkermans	University of Koblenz-Landau, Germany
Matthieu Sozeau	Inria, France
Martin Suda	Czech Technical University in Prague, Czech Republic
Geoff Sutcliffe	University of Miami, USA
Sofiène Tahar	Concordia University, Canada
Cesare Tinelli	The University of Iowa, USA
Christian Urban	King's College London, UK
Josef Urban	Czech Technical University in Prague, Czech Republic
Uwe Waldmann	Max Planck Institute for Informatics, Germany
Christoph Weidenbach	Max Planck Institute for Informatics, Germany

Conference Chair

Kaustuv Chaudhuri	Inria, Ecole Polytechnique Paris, France

Workshop Chairs

Giulio Manzonetto	Université Paris-Nord, France
Andrew Reynolds	University of Iowa, USA

IJCAR Steering Committee

Franz Baader	TU Dresden, Germany
Kaustuv Chaudhuri	Inria, Ecole polytechnique Paris, France
Didier Galmiche	Université de Lorraine, CNRS, LORIA, France
Ian Horrocks	University of Oxford, UK

Abstracts of Invited Talks

Domain-Specific Reasoning with Satisfiability Modulo Theories

Clark Barrett

Department of Computer Science, Stanford University

Abstract. General-purpose automated reasoning approaches such as automated theorem proving (ATP) and Boolean satisfiability (SAT) are extremely versatile. Many interesting and practical problems can be reduced to ATP or SAT problems. On the other hand, these techniques cannot easily incorporate customized domain-specific reasoning. In contrast, this is one of the strengths of approaches based on Satisfiabiilty Modulo Theories (SMT). This talk showcases recent work in which domain-specific customization of SMT solvers plays a key role in making otherwise intractable problems tractable. The domains discussed include reasoning about neural networks and reasoning about string constraints.

Adventures in Verifying Arithmetic

John Harrison

Amazon Web Services
jrh013@gmail.com

Abstract. I have focused a lot of the applied side of my work over the last 20 years on formal verification of *arithmetic* in some sense. Originally my main focus was on verification of floating-point algorithms for division, square root, transcendental functions, etc. More recently my interests have shifted to discrete arithmetical primitives, large integer arithmetic, and elliptic curve operations. As well as many contrasts and special problems, there are a number of common themes running through all this work: the challenges of verification at the unstructured machine-code level or indeed even getting adequate specifications for machine instruction sets, the countervailing benefit of generally having clear and incontrovertible specifications of the functions themselves, and the value of special customized decision procedures in making verifications of this kind practical. I will give an overview of some of the highlights of this work, as well as talking in more detail about my current project.

Focusing, Axioms and Synthetic Inference Rules (Extended Abstract)

Elaine Pimentel, Sonia Marin, Dale Miller, Marco Volpe

[1] Departamento de Matemática, UFRN, Brazil
elaine.pimentel@gmail.com
[2] Department of Computer Science, University College London, UK
s.marin@ucl.ac.uk
[3] Inria-Saclay & LIX, Ecole Polytechnique, Palaiseau, France
dale@lix.polytechnique.fr
[4] Fortiss, Munich, Germany
emme.volpe@gmail.com

Proving a sequent in sequent-based systems often involves many choices. For example, at every node of a tree-derivation one could: apply an introduction rule over a non-atomic formula; apply a structural rule; introduce a lemma; apply initial rules, etc. Hence, there is a need for *discipline* in structuring such choices and taming proof-search. One such discipline is *focusing* [1].

Focused proof systems combines two devices: *polarization* and *focused rule application*. In classical/intuitionistic first order systems, polarized formulas are built using atomic formulas and polarized versions of logical connectives and constants. The positive and negative versions of connectives and constants have identical truth conditions but have different inference rules inside the polarized proof systems. For example, left introduction rules for positive connectives are invertible while left introduction rules for negative connectives are not necessarily invertible. The polarity of a non-atomic formula is determined by its top-level connective. Since every polarized formula is classified as positive or negative, a polarity to atomic formulas must also be provided. As it turns out, this assignment of polarity to atomic formulas can be arbitrary [1].

When *focusing* on a formula, the focus is transferred to the active formulas in the premises (focused rule applications). This process goes on in all branches of the derivation, until: an initial rule/introduction rule on constants is applied (and the derivation ends at that branch); either the polarity of the focused formula changes *or* the side (left/right) of the focus flips (but not both). In this case, focus is *released* and the formula is eagerly decomposed into its negative-left, positive-right and/or atomic subformulas, that are *stored* in the context. Reading derivations from the root upwards, this forces a sequent derivation to be organized into *focused phases*, each of them corresponding to an application of a *synthetic inference rule* [2], where the focused formula is rewritten into (some of) its subformulas.

There is a class of formulas corresponding to particularly interesting synthetic rules: the *bipolars*. Bipolars are negative formulas in which polarity can change at most once among its subformulas. This means that left-focusing on a bipolar A gives rise to (possibly many) synthetic inference rules having simple shape, with leaves involving only atomic subformulas of A. We call a synthetic inference rule corresponding to the bipolar A a *bipole* for A.

In this talk, we will present a careful study of bipoles, giving a fresh view to an old problem: how to incorporate inference rules encoding axioms into proof systems for classical and intuitionistic logics.

We start by considering LKF and LJF [6,7] as the basic focused proof systems for classical and intuitionistic logics, respectively. In such systems, leafs of focused phases can be composed of either: (i) a conclusion-sequent of the application of introduction rule on constants; (ii) a (focused) conclusion-sequent of the application of the initial rule; (iii) an (unfocused) sequent after the storage of the remaining formulas. As an example, consider the following first order formula, that relates the subset and membership predicates in set theory:

$$A = \forall yz.(\forall x.x \in y \supset x \in z) \supset y \subseteq z.$$

Assuming that the predicate $\subseteq$ is given *negative* polarity, in the focused phase given by (left) focusing on A

$$\dfrac{\dfrac{\dfrac{\dfrac{\dfrac{\Gamma, x \in y \Uparrow \cdot \vdash \cdot \Uparrow x \in z, \Delta}{\Gamma \Uparrow x \in y \vdash x \in z \Uparrow \Delta}\ \mathsf{store}_l, \mathsf{store}_r}{\Gamma \Uparrow \cdot \vdash \forall x.x \in y \supset x \in z \Uparrow \Delta}\ \forall_r, \supset_r}{\Gamma \vdash \forall x.x \in y \supset x \in z \Downarrow \Delta}\ \mathsf{release}_r \qquad \dfrac{}{\Gamma \Downarrow y \subseteq z \vdash \Delta}\ \mathsf{init}_l}{\Gamma \Downarrow (\forall x.x \in y \supset x \in z) \supset y \subseteq z \vdash \Delta}\ \supset_l}{\Gamma \Downarrow \forall yz.(\forall x.x \in y \supset x \in z) \supset y \subseteq z \vdash \Delta}\ \forall_l$$

the right leaf has shape (ii) while the left one is of the form (iii). The formula between the $\Downarrow$ and $\vdash$ is the *focus* of that sequent.

Observe that it must be the case that $y \subseteq z \in \Delta$ (since $y \subseteq z$ is atomic, negative and under focus), while $x \in y, x \in z$ end-up being stored into contexts. This is not by chance: restricted to bipoles, leaves of the shape (ii) *forces atoms to belong to the context*, while leaves of the shape (iii) *adds atoms to the context*. This implies that principal and active formulas in bipoles for A (if any) are atomic formulas. That is: bipoles can be seen, in a sense, as introduction rules for *atoms*. For example, the bipolar above corresponds to the (unpolarized) synthetic rule

$$\dfrac{x \in y, \Gamma \vdash x \in z, \Delta}{\Gamma \vdash y \subseteq z, \Delta}\ .$$

which introduces $y \subseteq z$ from $x \in y$ and $x \in z$, where x is an eigenvariable.

Using such synthetic inference rules is one method for systematically generating proof systems for axiomatic extensions of classical/intuitionistic logics: focusing on a bipolar axiom yields a bipole.

A key step in transforming a formula into synthetic inference rules involves attaching a polarity to atomic formulas and to some logical connectives. Since there are different choices for assigning polarities, it is possible to produce different synthetic inference rules for the same formula. Indeed, if in our running example the predicate $\subseteq$ is given *positive* polarity, the corresponding (unpolarized) synthetic rule is

$$\frac{x \in y, \Gamma \vdash x \in z, \Delta \qquad y \subseteq z, \Gamma \vdash \Delta}{\Gamma \vdash \Delta}.$$

with x an eigenvariable.

We show that this flexibility allows for the generalization of different approaches for transforming axioms into sequent rules present in the literature, such as the works in [4, 5, 11, 12]. In particular, bipolars correspond to (the first-order version of) the $\mathcal{N}_2$ class presented in [3], which subsumes the class of geometric axioms studied in [11,10].

We finish the talk by showing how to emulate precisely rules for modalities in labeled modal systems as synthetic connectives [8, 9]. Such tight emulation means that proof search/proof checking on the focused version of the translated formulas models exactly proof search/proof checking in the correspondent labeled system. As a result, we are able to show that we can use focused proofs to precisely emulate modal proofs whenever Kripke frames are characterized by bipolar properties.

References

1. Andreoli, J.: Logic programming with focusing proofs in linear logic. J. Log. Comput. **2**(3), 297–347 (1992)
2. Chaudhuri, K.: Focusing Strategies In The Sequent Calculus Of Synthetic Connectives. In: Cervesato I., Veith H., Voronkov A. (eds.) LPAR 2008. LNCS, vol 5330, pp. 467–481. Springer, Heidelberg (2008). https://doi.org/10.1007/978-3-540-89439-1_33
3. Ciabattoni, A., Galatos, N., Terui, K.: From axioms to analytic rules in nonclassical logics. In: LICS, pp. 229–240 (2008)
4. Ciabattoni, A., Maffezioli, P., Spendier, L.: Hypersequent and labelled calculi for intermediate logics. In: Galmiche, D., Larchey-Wendling, D. (eds.) TABLEAUX 2013. LNCS, vol 8123, pp. 81–96. Springer, Heidelberg (2013). https://doi.org/10.1007/978-3-642-40537-2_9
5. Dyckhoff, R., Negri, S.: Geometrisation of first-order logic. Bull. Symb. Logic **21**(2), 123–163 (2015)
6. Liang, C., Miller, D.: Focusing and polarization in intuitionistic logic. In: Duparc, J., Henzinger, T.A. (eds.) CSL 2007. LNCS, vol 4646, pp. 451–465. Springer, Heidelberg (2007). https://doi.org/10.1007/978-3-540-74915-8_34
7. Liang, C., Miller, D.: Focusing and polarization in linear, intuitionistic, and classical logics. Theor. Comput. Sci. **410**(46), 4747–4768 (2009)
8. Marin, S., Miller, D., Volpe, M.: A focused framework for emulating modal proof systems. In: Beklemishev, L.D., Demri, S., Maté, A. (eds.) Advances in Modal Logic, vol. 11, pp. 469–488. College Publications (2016)
9. Miller, D., Volpe, M.: Focused labeled proof systems for modal logic. In: Davis, M., Fehnker, A., McIver, A., Voronkov, A. (eds.) LPAR 2015. LNCS, vol. 9450, pp. 266–280. Springer, Heidelberg (2015). https://doi.org/10.1007/978-3-662-48899-7_19

10. Negri, S.: Contraction-free sequent calculi for geometric theories with an application to Barr's theorem. Arch. Math. Log. **42**(4), 389–401 (2003). https://doi.org/10.1007/s001530100124
11. Simpson, A.K.: The Proof Theory and Semantics of Intuitionistic Modal Logic. Ph.D. thesis, School of Informatics, University of Edinburgh (1994)
12. Viganò, L.: Labelled Non-Classical Logics. Kluwer Academic Publishers (2000)

Efficient Automated Reasoning About Sets and Multisets with Cardinality Constraints

Ruzica Piskac

Yale University
ruzica.piskac@yale.edu

Abstract. When reasoning about container data structures that can hold duplicate elements, multisets are the obvious choice for representing the data structure abstractly. However, the decidability and complexity of constraints on multisets has been much less studied and understood than for constraints on sets. In this presentation, we outline an efficient decision procedure for reasoning about multisets with cardinality constraints. We describe how to translate, in linear time, multisets constraints to constraints in an extension of quantifier-free linear integer arithmetic, which we call LIA*. LIA* extends linear integer arithmetic with unbounded sums over values satisfying a given linear arithmetic formula. We show how to reduce a LIA* formula to an equisatisfiable linear integer arithmetic formula. However, this approach requires an explicit computation of semilinear sets and in practice it scales poorly even on simple benchmarks. We then describe a recent more efficient approach for checking satisfiability of LIA*. The approach is based on the use of under- and over-approximations of LIA* formulas. This way we avoid the space overhead and explicitly computing semilinear sets. Finally, we report on our prototype tool which can efficiently reason about sets and multisets formulas with cardinality constraints.

Keywords: Multisets · Cardinality constraints · Linear interger arithmetic.

This work is partially supported by the National Science Foundation under GrantNo. CCF-1553168 and No. CCF-1715387.

Contents – Part I

Invited Paper

Efficient Automated Reasoning About Sets and Multisets with Cardinality Constraints 3
Ruzica Piskac

SAT, SMT and QBF

An SMT Theory of Fixed-Point Arithmetic 13
Marek Baranowski, Shaobo He, Mathias Lechner, Thanh Son Nguyen, and Zvonimir Rakamarić

Covered Clauses Are Not Propagation Redundant 32
Lee A. Barnett, David Cerna, and Armin Biere

The Resolution of Keller's Conjecture 48
Joshua Brakensiek, Marijn Heule, John Mackey, and David Narváez

How QBF Expansion Makes Strategy Extraction Hard 66
Leroy Chew and Judith Clymo

Removing Algebraic Data Types from Constrained Horn Clauses Using Difference Predicates 83
Emanuele De Angelis, Fabio Fioravanti, Alberto Pettorossi, and Maurizio Proietti

Solving Bitvectors with MCSAT: Explanations from Bits and Pieces 103
Stéphane Graham-Lengrand, Dejan Jovanović, and Bruno Dutertre

Monadic Decomposition in Integer Linear Arithmetic 122
Matthew Hague, Anthony W. Lin, Philipp Rümmer, and Zhilin Wu

Scalable Algorithms for Abduction via Enumerative Syntax-Guided Synthesis 141
Andrew Reynolds, Haniel Barbosa, Daniel Larraz, and Cesare Tinelli

Decision Procedures and Combination of Theories

Deciding the Word Problem for Ground Identities with Commutative and Extensional Symbols 163
Franz Baader and Deepak Kapur

Combined Covers and Beth Definability 181
Diego Calvanese, Silvio Ghilardi, Alessandro Gianola, Marco Montali, and Andrey Rivkin

Deciding Simple Infinity Axiom Sets with One Binary Relation by Means of Superpostulates 201
Timm Lampert and Anderson Nakano

A Decision Procedure for String to Code Point Conversion 218
Andrew Reynolds, Andres Nötzli, Clark Barrett, and Cesare Tinelli

Politeness for the Theory of Algebraic Datatypes 238
Ying Sheng, Yoni Zohar, Christophe Ringeissen, Jane Lange, Pascal Fontaine, and Clark Barrett

Superposition

A Knuth-Bendix-Like Ordering for Orienting Combinator Equations 259
Ahmed Bhayat and Giles Reger

A Combinator-Based Superposition Calculus for Higher-Order Logic 278
Ahmed Bhayat and Giles Reger

Subsumption Demodulation in First-Order Theorem Proving 297
Bernhard Gleiss, Laura Kovács, and Jakob Rath

A Comprehensive Framework for Saturation Theorem Proving 316
Uwe Waldmann, Sophie Tourret, Simon Robillard, and Jasmin Blanchette

Proof Procedures

Possible Models Computation and Revision – A Practical Approach 337
Peter Baumgartner

SGGS Decision Procedures 356
Maria Paola Bonacina and Sarah Winkler

Integrating Induction and Coinduction via Closure Operators and Proof Cycles 375
Liron Cohen and Reuben N. S. Rowe

Logic-Independent Proof Search in Logical Frameworks (Short Paper). 395
Michael Kohlhase, Florian Rabe, Claudio Sacerdoti Coen, and Jan Frederik Schaefer

Layered Clause Selection for Theory Reasoning (Short Paper) 402
Bernhard Gleiss and Martin Suda

Non Classical Logics

Description Logics with Concrete Domains and General Concept Inclusions Revisited. 413
Franz Baader and Jakub Rydval

A Formally Verified, Optimized Monitor for Metric First-Order Dynamic Logic 432
David Basin, Thibault Dardinier, Lukas Heimes, Srđan Krstić, Martin Raszyk, Joshua Schneider, and Dmitriy Traytel

Constructive Hybrid Games 454
Rose Bohrer and André Platzer

Formalizing a Seligman-Style Tableau System for Hybrid Logic (Short Paper) 474
Asta Halkjær From, Patrick Blackburn, and Jørgen Villadsen

NP Reasoning in the Monotone μ-Calculus 482
Daniel Hausmann and Lutz Schröder

Soft Subexponentials and Multiplexing 500
Max Kanovich, Stepan Kuznetsov, Vivek Nigam, and Andre Scedrov

Mechanised Modal Model Theory 518
Yiming Xu and Michael Norrish

Author Index 535

Contents – Part II

Interactive Theorem Proving/HOL

Competing Inheritance Paths in Dependent Type Theory: A Case Study in Functional Analysis . . . 3
Reynald Affeldt, Cyril Cohen, Marie Kerjean, Assia Mahboubi, Damien Rouhling, and Kazuhiko Sakaguchi

A Lean Tactic for Normalising Ring Expressions with Exponents (Short Paper) . . . 21
Anne Baanen

Practical Proof Search for Coq by Type Inhabitation . . . 28
Łukasz Czajka

Quotients of Bounded Natural Functors . . . 58
Basil Fürer, Andreas Lochbihler, Joshua Schneider, and Dmitriy Traytel

Trakhtenbrot's Theorem in Coq: A Constructive Approach to Finite Model Theory . . . 79
Dominik Kirst and Dominique Larchey-Wendling

Deep Generation of Coq Lemma Names Using Elaborated Terms . . . 97
Pengyu Nie, Karl Palmskog, Junyi Jessy Li, and Milos Gligoric

Extensible Extraction of Efficient Imperative Programs with Foreign Functions, Manually Managed Memory, and Proofs . . . 119
Clément Pit-Claudel, Peng Wang, Benjamin Delaware, Jason Gross, and Adam Chlipala

Validating Mathematical Structures . . . 138
Kazuhiko Sakaguchi

Teaching Automated Theorem Proving by Example: PyRes 1.2 (System Description) . . . 158
Stephan Schulz and Adam Pease

Beyond Notations: Hygienic Macro Expansion for Theorem Proving Languages . . . 167
Sebastian Ullrich and Leonardo de Moura

Formalizations

Formalizing the Face Lattice of Polyhedra 185
Xavier Allamigeon, Ricardo D. Katz, and Pierre-Yves Strub

Algebraically Closed Fields in Isabelle/HOL. 204
Paulo Emílio de Vilhena and Lawrence C. Paulson

Formalization of Forcing in Isabelle/ZF 221
Emmanuel Gunther, Miguel Pagano, and Pedro Sánchez Terraf

Reasoning About Algebraic Structures with Implicit Carriers in Isabelle/HOL 236
Walter Guttmann

Formal Proof of the Group Law for Edwards Elliptic Curves 254
Thomas Hales and Rodrigo Raya

Verifying Faradžev-Read Type Isomorph-Free Exhaustive Generation 270
Filip Marić

Verification

Verified Approximation Algorithms. 291
Robin Eßmann, Tobias Nipkow, and Simon Robillard

Efficient Verified Implementation of Introsort and Pdqsort 307
Peter Lammich

A Fast Verified Liveness Analysis in SSA Form. 324
Jean-Christophe Léchenet, Sandrine Blazy, and David Pichardie

Verification of Closest Pair of Points Algorithms 341
Martin Rau and Tobias Nipkow

Reasoning Systems and Tools

A Polymorphic Vampire (Short Paper). 361
Ahmed Bhayat and Giles Reger

N-PAT: A Nested Model-Checker (System Description). 369
Hadrien Bride, Cheng-Hao Cai, Jin Song Dong, Rajeev Gore, Zhé Hóu, Brendan Mahony, and Jim McCarthy

HYPNO: Theorem Proving with Hypersequent Calculi for Non-normal Modal Logics (System Description). 378
Tiziano Dalmonte, Nicola Olivetti, and Gian Luca Pozzato

Implementing Superposition in iProver (System Description) 388
André Duarte and Konstantin Korovin

MOIN: A Nested Sequent Theorem Prover for Intuitionistic Modal Logics (System Description) 398
Marianna Girlando and Lutz Straßburger

Make E Smart Again (Short Paper) 408
Zarathustra Amadeus Goertzel

Automatically Proving and Disproving Feasibility Conditions 416
Raúl Gutiérrez and Salvador Lucas

MU-TERM: Verify Termination Properties Automatically (System Description) 436
Raúl Gutiérrez and Salvador Lucas

ENIGMA Anonymous: Symbol-Independent Inference Guiding Machine (System Description) 448
Jan Jakubův, Karel Chvalovský, Miroslav Olšák, Bartosz Piotrowski, Martin Suda, and Josef Urban

The Imandra Automated Reasoning System (System Description) 464
Grant Passmore, Simon Cruanes, Denis Ignatovich, Dave Aitken, Matt Bray, Elijah Kagan, Kostya Kanishev, Ewen Maclean, and Nicola Mometto

A Programmer's Text Editor for a Logical Theory: The SUMOjEdit Editor (System Description) 472
Adam Pease

Sequoia: A Playground for Logicians (System Description) 480
Giselle Reis, Zan Naeem, and Mohammed Hashim

Prolog Technology Reinforcement Learning Prover (System Description). . . . 489
Zsolt Zombori, Josef Urban, and Chad E. Brown

Author Index 509

Invited Paper

Efficient Automated Reasoning About Sets and Multisets with Cardinality Constraints

Ruzica Piskac(✉)

Yale University, New Haven, USA
ruzica.piskac@yale.edu

Abstract. When reasoning about container data structures that can hold duplicate elements, multisets are the obvious choice for representing the data structure abstractly. However, the decidability and complexity of constraints on multisets has been much less studied and understood than for constraints on sets. In this presentation, we outline an efficient decision procedure for reasoning about multisets with cardinality constraints. We describe how to translate, in linear time, multisets constraints to constraints in an extension of quantifier-free linear integer arithmetic, which we call LIA*. LIA* extends linear integer arithmetic with unbounded sums over values satisfying a given linear arithmetic formula. We show how to reduce a LIA* formula to an equisatisfiable linear integer arithmetic formula. However, this approach requires an explicit computation of semilinear sets and in practice it scales poorly even on simple benchmarks. We then describe a recent more efficient approach for checking satisfiability of LIA*. The approach is based on the use of under- and over-approximations of LIA* formulas. This way we avoid the space overhead and explicitly computing semilinear sets. Finally, we report on our prototype tool which can efficiently reason about sets and multisets formulas with cardinality constraints.

Keywords: Multisets · Cardinality constraints · Linear interger arithmetic

1 Introduction

In the verification of container data structures one often needs to reason about sets of objects – for example, abstracting the content of a container data structure as a set. The need for cardinality constraints naturally arises in order to reason about the number of the elements in the data structure. We have all witnessed to the success of the BAPA logic [4,5] that was, among others, used for verification of distributed algorithms [1].

This work is partially supported by the National Science Foundation under Grant No. CCF-1553168 and No. CCF-1715387.

N. Peltier and V. Sofronie-Stokkermans (Eds.): IJCAR 2020, LNAI 12166, pp. 3–10, 2020.
https://doi.org/10.1007/978-3-030-51074-9_1

Similarly, when reasoning about container data structures that can hold duplicate elements, multisets are the obvious choice of an abstraction. Multisets are collections of objects where an element can occur several times. They can be seen as "sets with counting". Although multisets are interesting mathematical objects that can be widely used in verification, there was no efficient reasoner for multisets and sets with cardinality constraints until recently [6]. Moreover, for a long time it was not known if the logic of multisets with cardinality constraints is even decidable [7]. Nevertheless, researchers have recognized the importance of this logic and they have been studied multisets in combination with other theories.

Zarba [13] investigated decision procedures for quantifier-free multisets but without the cardinality operator. He showed how to reduce a multiset formula to a quantifier-free defining each multiset operation pointwise on the elements of the set. Adding the cardinality operator makes such a reduction impossible.

Lugiez studied multiset constraints in the context of a more general result on multitree automata [7] and proved the decidability of multiset formulas with a weaker form of cardinality operator that counts only distinct elements in a multiset.

1.1 Multisets with Cardinality Constraints

In this paper we revive the first decision procedure for multisets with cardinality constraints [9,10]. We represent multisets (*bags*) with their characteristic functions. A multiset m is a function $\mathbb{E} \to \mathbb{N}$, where $\mathbb{E}$ is the universe used for populating multisets and $\mathbb{N}$ is the set of non-negative integers. The value $m(e)$ is the multiplicity (the number of occurrences) of an element e in a multiset m. We assume that the domain $\mathbb{E}$ is fixed and finite but of an unknown size. We consider the logic of multisets constraints with the cardinality operator (MAPA), given in Fig. 1. An atom in MAPA is either a multiset comparison, or it is a standard quantifier-free linear integer arithmetic atom, or it is a quantified formula ($\forall e.\mathsf{F}^{\mathsf{in}}$), or it is a collecting sum formula. We allow only universal quantification over all elements of $\mathbb{E}$. This way we can express, for example, that for a multiset k it holds $\forall e.k(e) = 0 \lor k(e) = 1$ – in other words, k is a set. A collecting sum atom is used to group several formulas involving sums into a single atom. This is needed for the next step of the decision procedure. The sums are used in the definition of the cardinality operator:

$$|m| = \sum_{e \in \mathbb{E}} m(e)$$

Piskac and Kuncak [9] showed that every MAPA formula can be translated to an equisatisfiable LIA* formula. The translation is linear and described in [9]. This way reasoning about MAPA formulas reduces to reasoning about LIA* formulas.

top-level formulas:

$F ::= A \mid F \wedge F \mid \neg F$

$A ::= M{=}M \mid M \subseteq M \mid \forall e.\mathsf{F}^{\mathsf{in}} \mid \mathsf{A}^{\mathsf{out}}$

outer linear arithmetic formulas:

$\mathsf{F}^{\mathsf{out}} ::= \mathsf{A}^{\mathsf{out}} \mid \mathsf{F}^{\mathsf{out}} \wedge \mathsf{F}^{\mathsf{out}} \mid \neg\mathsf{F}^{\mathsf{out}}$

$\mathsf{A}^{\mathsf{out}} ::= \mathsf{t}^{\mathsf{out}} \le \mathsf{t}^{\mathsf{out}} \mid \mathsf{t}^{\mathsf{out}}{=}\mathsf{t}^{\mathsf{out}} \mid (\mathsf{t}^{\mathsf{out}},\ldots,\mathsf{t}^{\mathsf{out}}){=}\sum_{\mathsf{F}^{\mathsf{in}}}(\mathsf{t}^{\mathsf{in}},\ldots,\mathsf{t}^{\mathsf{in}})$

$\mathsf{t}^{\mathsf{out}} ::= k \mid \ |M|\ \mid C \mid \mathsf{t}^{\mathsf{out}} + \mathsf{t}^{\mathsf{out}} \mid C \cdot \mathsf{t}^{\mathsf{out}} \mid \mathsf{ite}(\mathsf{F}^{\mathsf{out}}, \mathsf{t}^{\mathsf{out}}, \mathsf{t}^{\mathsf{out}})$

inner linear arithmetic formulas:

$\mathsf{F}^{\mathsf{in}} ::= \mathsf{A}^{\mathsf{in}} \mid \mathsf{F}^{\mathsf{in}} \wedge \mathsf{F}^{\mathsf{in}} \mid \neg\mathsf{F}^{\mathsf{in}}$

$\mathsf{A}^{\mathsf{in}} ::= \mathsf{t}^{\mathsf{in}} \le \mathsf{t}^{\mathsf{in}} \mid \mathsf{t}^{\mathsf{in}}{=}\mathsf{t}^{\mathsf{in}}$

$\mathsf{t}^{\mathsf{in}} ::= m(e) \mid P \mid \mathsf{t}^{\mathsf{in}} + \mathsf{t}^{\mathsf{in}} \mid P \cdot \mathsf{t}^{\mathsf{in}} \mid \mathsf{ite}(\mathsf{F}^{\mathsf{in}}, \mathsf{t}^{\mathsf{in}}, \mathsf{t}^{\mathsf{in}})$

multiset expressions:

$M ::= m \mid \emptyset \mid M \cap M \mid M \cup M \mid M \uplus M \mid M \setminus M \mid M \setminus\!\!\setminus M \mid \mathsf{set}(M)$

terminals:

m - multiset variables; e - index variable (fixed)

k - integer variable; C - integer constant; P - non-negative integer constant

Fig. 1. The logic of multiset constraints with Presburger Arithmetic (MAPA)

1.2 Reasoning About LIA* Formulas

The LIA* logic [10] is a standard linear integer arithmetic extended with a new operator: the star operator, which is defined over a set of integer vectors as follows:

$$S^{\star} \triangleq \left\{ \sum_{i=1}^{n} \boldsymbol{s}_i \mid \forall i.1 \le i \le n.\ \boldsymbol{s}_i \in S \right\} \tag{1}$$

The result of the star operator applied to set S is a set if all linear additive combinations of vectors from S. Its syntax is given in Fig. 2.

LIA* formulas: $\varphi ::= F_1 \wedge \boldsymbol{x_1} \in \{\boldsymbol{x_2} \mid F_2\}^{\star}$

such that $dim(\boldsymbol{x_1}) = dim(\boldsymbol{x_2})$ and $\textit{free-vars}(F_2) \subseteq \boldsymbol{x_2}$

LIA formulas:

$F ::= A \mid F_1 \wedge F_2 \mid F_1 \vee F_2 \mid \neg F_1 \mid \exists x.\ F \mid \forall x.\ F$

$A ::= T_1 \le T_2 \mid T_1 = T_2$

$T ::= x \mid C \mid T_1 + T_2 \mid C \cdot T_1 \mid \mathsf{ite}(F, T_1, T_2)$

terminals: x - integer variable; C - integer constant

Fig. 2. Linear integer arithmetic (LIA) and an extension with the Star Operator.

To check a satisfiability of a LIA* formula, we use the semilinear set characterization of solutions of integer linear arithmetic formulas.

Definition 1 (Semilinear sets). *A linear set $LS(\boldsymbol{a}, B)$ is defined by an integer vector $\boldsymbol{a}$ and a finite set of integer vectors $B = \{\boldsymbol{b_1}, \ldots, \boldsymbol{b_n}\}$, all of the same dimension, as follows:*

$$LS(\boldsymbol{a}, B) \quad \triangleq \quad \left\{ \boldsymbol{a} + \sum_{i=1}^{n} \lambda_i \boldsymbol{b_i} \quad | \quad \bigwedge_{i=1}^{n} \lambda_i \geq 0 \right\} \tag{2}$$

Sometimes, as a shorthand, we use $\boldsymbol{\lambda} B = \sum_{i=1}^{n} \lambda_i \boldsymbol{b_i}$.

A semilinear set $SLS(ls_1, \ldots, ls_n)$ is a finite union of linear sets $ls_1, \ldots, ls_n$, i.e., $SLS(ls_1, \ldots, ls_n) = \bigcup_{i=1}^{n} ls_i$.

Ginsburg and Spanier showed (Theorem 1.3 in [3]) that a solution set for every Presburger arithmetic formula is a semilinear set, and we use that result to eliminate the star operator.

Theorem 1 (Lemmas 2 and 3 in [10]). *Given a* LIA* *atom $\boldsymbol{x_1} \in \{\boldsymbol{x_2} \mid F_2\}^\star$, let $SLS(LS(\boldsymbol{a_1}, B_1), \ldots, LS(\boldsymbol{a_k}, B_k))$ be a semilinear set describing the set of the solutions of formula F_2. The atom $\boldsymbol{x_1} \in \{\boldsymbol{x_2} \mid F_2\}^\star$ is equisatisfiable to the following* LIA *formula:*

$$\exists \mu_1 \geq 0, \ldots, \mu_k \geq 0, \boldsymbol{\lambda}_1 \geq 0, \ldots \boldsymbol{\lambda}_k \geq 0.$$
$$\boldsymbol{x_1} = \sum_{i=1}^{k} \mu_i \boldsymbol{a}_i + \boldsymbol{\lambda}_i B_i \wedge \bigwedge_{i=1}^{k} (\mu_i = 0 \rightarrow \boldsymbol{\lambda}_i = 0)$$

By applying Theorem 1, checking satisfiability of a LIA* formula reduces to reasoning about linear integer arithmetic. Note, however, that this approach results in automatically constructing a formula might be really large, depending on the size of a semilinear set. In addition, this approach relies on computing semilinear sets explicitly, both of which make it scale poorly even on simple benchmarks.

2 Illustrating Example

We illustrate now how is a decision procedure for MAPA working on the following simple multiset formula: for two multisets X and Y, the size of their disjoint union is the sum of their respective sizes. In other words, we need to prove the validity of the following formula

$$|X \uplus Y| = |X| + |Y|$$

As usual, we prove the unsatisfiability of the formula $|X \uplus Y| \neq |X| + |Y|$. The first step is to reduce this formula into an equisatisfiable LIA* formula. To do that, we perform a sequence of steps that resemble the purification step in the Nelson-Oppen combination procedure [8]. In a nutshell, we introduce a new variable for every non-terminal expression.

We first introduce a multiset variable M defining multiset expression $X \uplus Y$ and then we introduce integer variables k_1, k_2, k_3 for each of the cardinality expressions. This way the formula becomes:

$$k_1 \neq k_2 + k_3 \wedge k_1 = |M| \wedge k_2 = |X| \wedge k_3 = |Y| \wedge M = X \uplus Y$$

We next apply the point-wise definitions of the cardinality and $\uplus$ operators and we obtain the following formula:

$$k_1 \neq k_2 + k_3 \wedge k_1 = \sum_{e \in \mathbb{E}} M(e) \wedge k_2 = \sum_{e \in \mathbb{E}} X(e) \wedge k_3 = \sum_{e \in \mathbb{E}} Y(e) \\ \wedge \forall e. M(e) = X(e) + Y(e)$$

Grouping all the sum expressions together results in the formula:

$$k_1 \neq k_2 + k_3 \wedge (k_1, k_2, k_3) = \sum_{e \in \mathbb{E}} (M(e), X(e), Y(e)) \wedge \forall e. M(e) = X(e) + Y(e)$$

Piskac and Kuncak have shown in [9] that every multiset formula can be reduced to this form. They call it the *sum normal form*. It consists of three conjuncts. One is a pure LIA formula, the other is the summation and the third part is a universally quantified formula. By applying Theorem 2 from [9], the above MAPA formula is translated into an equisatisfiable LIA* formula, where m, x and y are non-negative integer variables:

$$k_1 \neq k_2 + k_3 \wedge (k_1, k_2, k_3) \in \{(m, x, y) | m = x + y\}^\star$$

To check the satisfiability of this formula, we first need to eliminate the star operator, which is done by computing a semilinear set describing the set of solutions of $m = x + y$. In this particular case, the semilinear set is actually a linear set, consisting of the zero vector and two vectors defining linear combinations:

$$\{(m, x, y) | m = x + y\} = LS((0, 0, 0), \{(1, 1, 0), (1, 0, 1)\})$$

Having the semilinear set representation, we can apply Theorem 1. In particular, only one linear set and the zero vector can significantly simplify the corresponding equisatisfiable formula. As the result of applying Theorem 1, we obtain that formula $(k_1, k_2, k_3) \in \{(m, x, y) | m = x+y\}^\star$ is equisatisfiable to the formula $(k_1, k_2, k_3) = \boldsymbol{\lambda}\{(1, 1, 0), (1, 0, 1)\} \Leftrightarrow (k_1, k_2, k_3) = \lambda_1(1, 1, 0) + \lambda_2(1, 0, 1)$.

This way we have eliminated the star operator from the given LIA* formula. It is now reduced to an equisatisfiable linear integer arithmetic formula:

$$k_1 \neq k_2 + k_3 \wedge k_1 = \lambda_1 + \lambda_2 \wedge k_2 = \lambda_1 \wedge k_3 = \lambda_2$$

The resulting LIA formula is unsatisfiable.

3 Efficient Reasoning About LIA* Formulas

The described decision procedure is sound and complete. However, its crucial component is a computation of semilinear sets. While it is possible to compute Hilbert basis using the z3 [2] SMT solver, to the best of our knowledge there are no efficient tools for computing semilinear sets. Moreover, Pottier [12] showed that a semilinear set might contain an exponential number of vectors. To overcome the explicit computation of semilinear sets, Piskac and Kuncak [10] developed a new decision procedure for LIA* which eliminates the star operator from the atom $\boldsymbol{x_1} \in \{\boldsymbol{x_2} \mid F\}^*$ by showing that $\boldsymbol{x_1}$ is a linear combination of $\mathcal{O}(n^2 \log n)$ solution vectors of F, where n is the size of the input formula. Although this new decision procedure avoids computing semilinear sets, it instantly produces a very large formula that could not be solved in practice by existing tools, not even for the most simple benchmarks.

Levatich et al. [6] used those insights to develop a new efficient and scalable approach for solving LIA* formulas. The approach is based on the use of under- and over-approximations of LIA* formulas. This way one avoids the space overhead and explicitly computing semilinear sets.

The key insight of their approach is to construct a solution or a proof of unsatisfiability "on demand". Given a LIA* formula $F_1(\boldsymbol{x_1}) \wedge \boldsymbol{x_1} \in \{\boldsymbol{x_2} \mid F_2(\boldsymbol{x_2})\}^*$, we first find any solution vector for formula F_2, let us name it $\boldsymbol{u_1}$. We next check if formula $F_1(\boldsymbol{x_1}) \wedge \boldsymbol{x_1} = \lambda_1 * \boldsymbol{u_1}$ is satisfiable. If this is the case, the given LIA* formula is satisfiable as well. However, if this is not the case, we cannot conclude anything about the satisifiability of the given LIA* formula, so we find a new different solution of formula F_2, denoted by $\boldsymbol{u_2}$: $F_2(\boldsymbol{u_2}) \wedge \boldsymbol{u_1} \neq \boldsymbol{u_2}$. Next, we check if the vector $\boldsymbol{x_1}$ is a linear combination of those two solution vectors: $F_1(\boldsymbol{x_1}) \wedge \boldsymbol{x_1} = \lambda_1 * \boldsymbol{u_1} + \lambda_2 * \boldsymbol{u_2}$. If this newly constructed formula is satisfiable, so is the original LIA* formula, otherwise we repeat the process. This way, by finding and checking solution vectors of F_2, we construct underapproximations of the set $\{\boldsymbol{x_2} \mid F_2(\boldsymbol{x_2})\}^*$. Moreover, we know that this process will terminate once we check sufficiently many solution vectors, as shown in [10].

However, if the given LIA* formula is unsatisfiable, this approach will result in an equally large formula as in [10], and again it does not scale. Therefore, in parallel to finding an under-approximation of the set $\{\boldsymbol{x_2} \mid F_2(\boldsymbol{x_2})\}^*$, we are also constructing a sequence of its over-approximation. The properties, that such an overapproximation should have, are encoded as a set of Constraint Horn Clauses and we use existing solvers to compute them. Such an overapproximation, if exists, is an interpolant that separates two conjuncts in the given LIA* formula, proving this way that the formula is unsatisfiable.

Finally, we have implemented the presented decision procedure and the tool is publicly available at https://github.com/mlevatich/sls-reachability. Because there were no MAPA benchmarks available, we had to create our own benchmarks. In addition, we also treated 240 BAPA benchmarks about sets, available in [1], as MAPA benchmarks. While the full report on the empirical results is available in [6], our general assessment is that the presented algorithm is effective on both SAT and UNSAT benchmarks. Our tool solved 83% of benchmarks in less than 50 seconds, and over 75% of those in under 3 seconds. We believe that this tool is the first efficient reasoner for multisets and sets with cardinality constraints.

4 Conclusions

The presented work describes a sequence of decision procedures that has lead towards an efficient reasoner for multisets and sets with cardinality constraints. We noticed that some constraints arising in formal verification of protocols and data structures could have been expressed more succinctly and naturally, were they using multisets as the underlying abstract datatype in the specification. Nevertheless, due to the lack of tool support they use sets, resulting in more complex constraints. While there was an older tool for reasoning about multisets with cardinality constraints [11], that tool served mainly as a proof of concept and was evaluated only on a handful of manually written formulas. We here presented a recent tool for reasoning about sets and multisets and we showed empirically that this tool scales well and can easily reason about complex multiset formulas. We hope that this work will lead to a renewed research interest in multisets and encourage their use in software analysis and verification.

Acknowledgments. This presentation is based on the previously published results on reasoning about multisets with cardinality constraints [6,9–11]. We sincerely thank the collaborators on these projects: Nikolaj Bjørner, Maxwell Levatich, Viktor Kunčak and Sharon Shoham, without whom this work would not be possible.

References

1. Berkovits, I., Lazić, M., Losa, G., Padon, O., Shoham, S.: Verification of threshold-based distributed algorithms by decomposition to decidable logics. In: Dillig, I., Tasiran, S. (eds.) CAV 2019. LNCS, vol. 11562, pp. 245–266. Springer, Cham (2019). https://doi.org/10.1007/978-3-030-25543-5_15
2. de Moura, L.M., Bjørner, N.: Z3: an efficient SMT solver. In: Ramakrishnan, C.R., Rehof, J. (eds.) TACAS 2008. LNCS, vol. 4963, pp. 337–340. Springer, Heidelberg (2008). https://doi.org/10.1007/978-3-540-78800-3_24
3. Ginsburg, S., Spanier, E.H.: Semigroups, Presburger formulas, and languages. Pacific J. Math. **16**(2), 285–296 (1966)
4. Kuncak, V., Nguyen, H.H., Rinard, M.C.: An algorithm for deciding BAPA: Boolean algebra with Presburger arithmetic. In: Nieuwenhuis, R. (ed.) CADE 2005. LNCS (LNAI), vol. 3632, pp. 260–277. Springer, Heidelberg (2005). https://doi.org/10.1007/11532231_20
5. Kuncak, V., Nguyen, H.H., Rinard, M.C.: Deciding Boolean algebra with Presburger arithmetic. J. Autom. Reason. **36**(3), 213–239 (2006)
6. Levatich, M., Bjørner, N., Piskac, R., Shoham, S.: Solving LIA* using approximations. In: Beyer, D., Zufferey, D. (eds.) VMCAI 2020. LNCS, vol. 11990, pp. 360–378. Springer, Cham (2020). https://doi.org/10.1007/978-3-030-39322-9_17
7. Lugiez, D.: Multitree automata that count. Theor. Comput. Sci. **333**(1–2), 225–263 (2005)

8. Nelson, G., Oppen, D.C.: Fast decision procedures based on congruence closure. J. ACM **27**(2), 356–364 (1980)
9. Piskac, R., Kuncak, V.: Decision procedures for multisets with cardinality constraints. In: Logozzo, F., Peled, D.A., Zuck, L.D. (eds.) VMCAI 2008. LNCS, vol. 4905, pp. 218–232. Springer, Heidelberg (2008). https://doi.org/10.1007/978-3-540-78163-9_20
10. Piskac, R., Kuncak, V.: Linear arithmetic with stars. In: Gupta, A., Malik, S. (eds.) CAV 2008. LNCS, vol. 5123, pp. 268–280. Springer, Heidelberg (2008). https://doi.org/10.1007/978-3-540-70545-1_25
11. Piskac, R., Kuncak, V.: MUNCH - automated reasoner for sets and multisets. In: Giesl, J., Hähnle, R. (eds.) IJCAR 2010. LNCS (LNAI), vol. 6173, pp. 149–155. Springer, Heidelberg (2010). https://doi.org/10.1007/978-3-642-14203-1_13
12. Pottier, L.: Minimal solutions of linear diophantine systems: bounds and algorithms. In: Book, R.V. (ed.) RTA 1991. LNCS, vol. 488, pp. 162–173. Springer, Heidelberg (1991). https://doi.org/10.1007/3-540-53904-2_94
13. Zarba, C.G.: Combining multisets with integers. In: Voronkov, A. (ed.) CADE-18. LNCS (LNAI), vol. 2392, pp. 363–376. Springer, Heidelberg (2002). https://doi.org/10.1007/3-540-45620-1_30

SAT, SMT and QBF

An SMT Theory of Fixed-Point Arithmetic

Marek Baranowski[1], Shaobo He[1(✉)], Mathias Lechner[2], Thanh Son Nguyen[1], and Zvonimir Rakamarić[1]

[1] School of Computing, University of Utah, Salt Lake City, UT, USA
{baranows,shaobo,thanhson,zvonimir}@cs.utah.edu

[2] IST Austria, Klosterneuburg, Austria
mathias.lechner@ist.ac.at

Abstract. Fixed-point arithmetic is a popular alternative to floating-point arithmetic on embedded systems. Existing work on the verification of fixed-point programs relies on custom formalizations of fixed-point arithmetic, which makes it hard to compare the described techniques or reuse the implementations. In this paper, we address this issue by proposing and formalizing an SMT theory of fixed-point arithmetic. We present an intuitive yet comprehensive syntax of the fixed-point theory, and provide formal semantics for it based on rational arithmetic. We also describe two decision procedures for this theory: one based on the theory of bit-vectors and the other on the theory of reals. We implement the two decision procedures, and evaluate our implementations using existing mature SMT solvers on a benchmark suite we created. Finally, we perform a case study of using the theory we propose to verify properties of quantized neural networks.

Keywords: SMT · Fixed-point arithmetic · Decision procedure

1 Introduction

Algorithms based on real arithmetic have become prevalent. For example, the mathematical models in machine learning algorithms operate on real numbers. Similarly, signal processing algorithms often implemented on embedded systems (e.g., fast Fourier transform) are almost always defined over real numbers. However, real arithmetic is not implementable on computer systems due to its unlimited precision. Consequently, we use implementable approximations of real arithmetic, such as floating-point and fixed-point arithmetic, to realize these algorithms in practice.

Floating-point arithmetic is the dominant approximation of real arithmetic that has mature hardware support. Although it enjoys the benefits of being able to represent a large spectrum of real numbers and high precision of arithmetic

This work was supported in part by NSF awards CCF 1552975, CCF 1704715, and the Austrian Science Fund (FWF) under grant Z211-N23 (Wittgenstein Award).

N. Peltier and V. Sofronie-Stokkermans (Eds.): IJCAR 2020, LNAI 12166, pp. 13–31, 2020.
https://doi.org/10.1007/978-3-030-51074-9_2

operations over small numbers, floating-point arithmetic, due to its complexity, can be too expensive in terms of speed and power consumption on certain platforms. These platforms are often deployed in embedded systems such as mobile phones, video game consoles, and digital controllers. Recently, the machine learning community revived the interest in fixed-point arithmetic since popular machine learning algorithms and models can be implemented using (even very low bit-width) fixed-points with little accuracy loss [11,27,37]. Therefore, fixed-point arithmetic has been a popular alternative to floating-point arithmetic on such platforms since it can be efficiently realized using integer arithmetic. There are several software implementations of fixed-point arithmetic in different programming languages [22,28,34]; moreover, some programming languages, such as Ada and GNU C, have built-in fixed-point types.

While fixed-point arithmetic is less popular in mainstream applications than floating-point arithmetic, the systems employing the former are often safety-critical. For example, fixed-point arithmetic is often used in medical devices, cars, and robots. Therefore, there is a need for formal methods that can rigorously ensure the correctness of these systems. Although techniques that perform automated verification of fixed-point programs already exist [1,3,15], all of them implement a custom dedicated decision procedure without formalizing the details of fixed-point arithmetic. As a result, it is hard to compare these techniques, or reuse the implemented decision procedures.

On the other hand, ever since the SMT theory of floating-point numbers was formalized [8,44] in SMT-LIB [46], there has been a flurry of research in developing novel and faster decision procedures for the theory [6,7,14,29,35, 50]. Meanwhile, the floating-point theory has also been used by a number of approaches that require rigorous reasoning about floating-point arithmetic [2, 36,39,41]. The published formalization of the theory enables fair comparison between the decision procedures, sharing of benchmarks, and easy integration of decision procedures within tools that need this functionality. In this paper, we propose and formalize an SMT theory of fixed-point arithmetic, in the spirit of the SMT theory of floating-point arithmetic, with the hope that it will lead to similar outcomes and advances.

Contributions. We summarize our main contributions as follows:

- We present an intuitive and comprehensive syntax of fixed-point arithmetic (Sect. 3) that captures common use cases of fixed-point operations.
- We provide formal semantics of the fixed-point theory based on rational arithmetic (Sect. 4).
- We propose and implement two decision procedures for the fixed-point theory: one that leverages the theory of fixed-size bit-vectors and the other the theory of real numbers (Sect. 5).
- We evaluate the two decision procedures on a set of benchmarks using mature SMT solvers (Sect. 6), and perform a case study of verifying quantized neural networks that uses our theory of fixed-point arithmetic (Sect. 7).

2 Background

Fixed-point arithmetic, like floating-point arithmetic, is used as an approximation for computations over the reals. Both fixed-point and floating-point numbers (excluding the special values) can be represented using rational numbers. However, unlike floating-point numbers, fixed-point numbers in a certain format maintain a fixed divisor, hence the name fixed-point. Consequently, fixed-point numbers have a reduced range of values. However, this format allows for custom precision systems to be implemented efficiently in software—fixed-point arithmetic operations can be implemented in a much smaller amount of integer arithmetic operations compared to their floating-point counterparts. For example, a fixed-point addition operation simply amounts to an integer addition instruction provided that wrap-around is the intended behavior when overflows occur. This feature gives rise to the popularity of fixed-point arithmetic on embedded systems where computing resources are fairly constrained.

A fixed-point number is typically interpreted as a fraction whose numerator is an integer with fixed bit-width in its two's complement representation and denominator is a power of 2. Therefore, a fixed-point format is parameterized by two natural numbers—*tb* that defines the bit-width of the numerator and *fb* that defines the power of the denominator. A fixed-point number in this format can be treated as a bit-vector of length *tb* that is the two's complement representation of the numerator integer and has an implicit binary point between the $fb + 1^{th}$ and fb^{th} least significant bits. We focus on the binary format (as opposed to decimal, etc.) of fixed-point arithmetic since it is widely adopted in hardware and software implementations in practice. Moreover, depending on the intended usage, developers leverage both signed and unsigned fixed-point formats. The signed or unsigned format determines whether the bit pattern representing the fixed-point number should be interpreted as a signed or unsigned integer, respectively. Therefore, signed and unsigned fixed-point formats having the same *tb* and *fb* have different ranges ($[\frac{-2^{tb-1}}{2^{fb}}, \frac{2^{tb-1}-1}{2^{fb}}]$ and $[0, \frac{2^{tb}-1}{2^{fb}}]$), respectively.

Fixed-point addition (resp. subtraction) is typically implemented by adding (resp. subtracting) the two bit-vector operands (i.e., two's complements), amounting to a single operation. Because the denominators are the same between the two operands, we do not need to perform rounding. However, we still have to take care of potential overflows that occur when the result exceeds the allowed range of the chosen fixed-point format. Fixed-point libraries typically implement two methods to handle overflows: saturation and wrap-around. Saturation entails fixing overflowed results to either the minimal or maximal representable value. The advantage of this method is that it ensures that the final fixed-point result is the closest to the actual result not limited by finite precision. Wrap-around allows for the overflowing result to wrap according to two's complement arithmetic. The advantage of this method is that it is efficient and can be used to ensure the sum of a set of (signed) numbers has a correct final value despite potential overflows (if it falls within the supported range). Note that addition is commutative under both methods, but only addition using the wrap-around method is associative. The multiplication and division operations are more involved since they have to include the rounding step as well.

3 Syntax

In this section, we describe the syntax of our proposed theory of fixed-point arithmetic. It is inspired by the syntax of the SMT theory of floating-points [8,44] and the ISO/IEC TR 18037 standard [23].

Fixed-Points. We introduce the indexed SMT nullary sorts (_ `SFXP` *tb fb*) to represent signed fixed-point sorts, where *tb* is a natural number specifying the total bit-width of the scaled integer in its two's complement form and *fb* is a natural number specifying the number of fractional bits; *tb* is greater than or equal to *fb*. Similarly, we represent unsigned fixed-point sorts with (_ `UFXP` *tb fb*). Following the SMT-LIB notation, we define the following two functions for constructing fixed-points literals:

((_ `sfxp` *fb*) (_ `BitVec` *tb*) (_ `SFXP` *tb fb*))
((_ `ufxp` *fb*) (_ `BitVec` *tb*) (_ `UFXP` *tb fb*))

where (_ `sfxp` *fb*) (resp. (_ `ufxp` *fb*)) produces a function that takes a bit-vector (_ `BitVec` *tb*) and constructs a fixed-point (_ `SFXP` *tb fb*) (resp. (_ `UFXP` *tb fb*)).

Rounding Modes. Similarly to the theory of floating-point arithmetic, we also introduce the `RoundingMode` sort (abbreviated as `RM`) to represent the rounding mode, which controls the direction of rounding when an arithmetic result cannot be precisely represented by the specified fixed-point format. However, unlike the floating-point theory that specifies five different rounding modes, we only adopt two rounding mode constants, namely `roundUp` and `roundDown`, as they are common in practice.

Overflow Modes. We introduce the nullary sort `OverflowMode` (abbreviated as `OM`) to capture the behaviors of fixed-point arithmetic when the result of an operation is beyond the representable range of the used fixed-point format. We adopt two constants, `saturation` and `wrapAround`, to represent the two common behaviors. The `saturation` mode rounds any out-of-bound results to the maximum or minimum values of the representable range, while the `wrapAround` mode wraps the results around similar to bit-vector addition.

Comparisons. The following operators return a Boolean by comparing two fixed-point numbers:

(sfxp.geq (_ SFXP *tb fb*) (_ SFXP *tb fb*) Bool)
(ufxp.geq (_ UFXP *tb fb*) (_ UFXP *tb fb*) Bool)
(sfxp.gt (_ SFXP *tb fb*) (_ SFXP *tb fb*) Bool)
(ufxp.gt (_ UFXP *tb fb*) (_ UFXP *tb fb*) Bool)
(sfxp.leq (_ SFXP *tb fb*) (_ SFXP *tb fb*) Bool)
(ufxp.leq (_ UFXP *tb fb*) (_ UFXP *tb fb*) Bool)
(sfxp.lt (_ SFXP *tb fb*) (_ SFXP *tb fb*) Bool)
(ufxp.lt (_ UFXP *tb fb*) (_ UFXP *tb fb*) Bool)

Arithmetic. We support the following binary arithmetic operators over fixed-point sorts parameterized by *tb* and *fb*:

(sfxp.add OM (_ SFXP *tb fb*) (_ SFXP *tb fb*) (_ SFXP *tb fb*))
(ufxp.add OM (_ UFXP *tb fb*) (_ UFXP *tb fb*) (_ UFXP *tb fb*))
(sfxp.sub OM (_ SFXP *tb fb*) (_ SFXP *tb fb*) (_ SFXP *tb fb*))
(ufxp.sub OM (_ UFXP *tb fb*) (_ UFXP *tb fb*) (_ UFXP *tb fb*))
(sfxp.mul OM RM (_ SFXP *tb fb*) (_ SFXP *tb fb*) (_ SFXP *tb fb*))
(ufxp.mul OM RM (_ UFXP *tb fb*) (_ UFXP *tb fb*) (_ UFXP *tb fb*))
(sfxp.div OM RM (_ SFXP *tb fb*) (_ SFXP *tb fb*) (_ SFXP *tb fb*))
(ufxp.div OM RM (_ UFXP *tb fb*) (_ UFXP *tb fb*) (_ UFXP *tb fb*))

Note that we force the sorts of operands and return values to be the same. The addition and subtraction operations never introduce error into computation according to our semantics in Sect. 4. Hence, these operators do not take a rounding mode as input like multiplication and division.

Conversions. We introduce two types of conversions between sorts. First, the conversions between different fixed-point sorts:

((_ to_sfxp *tb fb*) OM RM (_ SFXP *tb′ fb′*) (_ SFXP *tb fb*))
((_ to_ufxp *tb fb*) OM RM (_ UFXP *tb′ fb′*) (_ UFXP *tb fb*))

Second, the conversions between the real and fixed-point sorts:

((_ to_sfxp *tb fb*) OM RM Real (_ SFXP *tb fb*))
((_ to_ufxp *tb fb*) OM RM Real (_ UFXP *tb fb*))
(sfxp.to_real (_ SFXP *tb fb*) Real)
(ufxp.to_real (_ UFXP *tb fb*) Real)

4 Semantics

In this section, we formalize the semantics of the fixed-point theory by treating fixed-points as rational numbers. We first define fixed-points as indexed subsets of rationals. Then, we introduce two functions, rounding and overflow, that are crucial for the formalization of the fixed-point arithmetic operations. Finally, we present the formal semantics of the arithmetic operations based on rational arithmetic and the two aforementioned functions.

Let $\mathbb{F}_{fb} = \{\frac{n}{2^{fb}} \mid n \in \mathbb{Z}\}$ be the infinite set of rationals that can be represented as fixed-points using *fb* fractional bits. We interpret a signed fixed-point sort (`_ SFXP` *tb fb*) as the finite subset $\mathbb{S}_{tb,fb} = \{\frac{n}{2^{fb}} \mid -2^{tb-1} \leq n < 2^{tb-1}, n \in \mathbb{Z}\}$ of $\mathbb{F}_{fb}$. We interpret an unsigned fixed-point sort (`_ UFXP` *tb fb*) as the finite subset $\mathbb{U}_{tb,fb} = \{\frac{n}{2^{fb}} \mid 0 \leq n < 2^{tb}, n \in \mathbb{Z}\}$ of $\mathbb{F}_{fb}$. The rational value of an unsigned fixed-point constant constructed using (`ufxp` *bv fb*) is $\frac{bv2nat(bv)}{2^{fb}}$, where function *bv2nat* converts a bit-vector to its unsigned integer value. The rational value of its signed counterpart constructed using (`sfxp` *bv fb*) is $\frac{bv2int(bv)}{2^{fb}}$, where function *bv2int* converts a bit-vector to its signed value. Since we treat fixed-point numbers as subsets of rational numbers, we interpret fixed-point comparison operators, such as =, `fxp.le`, `fxp.leq`, as simply their corresponding rational comparison relations, such as =, <, ≤, respectively. To be able to formalize the semantics of arithmetic operations, we first introduce the round and overflow helper functions.

We interpret the rounding mode sort `RoundingMode` as the set $rmode = \{ru, rd\}$, where $[\![\texttt{roundUp}]\!] = ru$ and $[\![\texttt{roundDown}]\!] = rd$. Let $rnd_{\mathbb{F}_{fb}} : rmode \times \mathbb{R} \mapsto \mathbb{F}_{fb}$ be a family of round functions parameterized by *fb* that map a rounding mode and real number to an element of $\mathbb{F}_{fb}$. Then, we define $rnd_{\mathbb{F}_{fb}}$ as

$$rnd_{\mathbb{F}_{fb}}(ru, r) = \min(\{x \mid x \geq r, x \in \mathbb{F}_{fb}\})$$
$$rnd_{\mathbb{F}_{fb}}(rd, r) = \max(\{x \mid x \leq r, x \in \mathbb{F}_{fb}\})$$

We interpret the overflow mode sort `OverflowMode` as the set $omode = \{sat, wrap\}$, where $[\![\texttt{saturation}]\!] = sat$ and $[\![\texttt{wrapAround}]\!] = wrap$. Let $ovf_{\mathbb{F}} : omode \times \mathbb{F}_{fb} \mapsto \mathbb{F}$ be a family of overflow functions parameterized by $\mathbb{F}$ that map a rounding mode and element of $\mathbb{F}_{fb}$ to an element of $\mathbb{F}$; here, $\mathbb{F}$ is either $\mathbb{S}_{tb,fb}$ or $\mathbb{U}_{tb,fb}$ depending on whether we are using signed or unsigned fixed-point numbers, respectively. Then, we define $ovf_{\mathbb{F}}$ as

$$ovf_{\mathbb{F}}(sat, x) = \begin{cases} x & \text{if } x \in \mathbb{F} \\ \max(\mathbb{F}) & \text{if } x > \max(\mathbb{F}) \\ \min(\mathbb{F}) & \text{if } x < \min(\mathbb{F}) \end{cases}$$
$$ovf_{\mathbb{F}}(wrap, x) = y \;\text{ such that }\; y \cdot 2^{fb} \equiv x \cdot 2^{fb} \pmod{2^{tb}} \;\wedge\; y \in \mathbb{F}$$

Note that $x \cdot 2^{fb}, y \cdot 2^{fb} \in \mathbb{Z}$ according to the definition of $\mathbb{F}$, and also there is always exactly one y satisfying the constraint.

Now that we introduced our helper round and overflow functions, it is easy to define the interpretation of fixed-point arithmetic operations:

$$
\begin{aligned}
[\![\texttt{sfxp.add}]\!](om, x_1, x_2) &= \mathit{ovf}_{\mathbb{S}_{tb,fb}}(om, x_1 + x_2)\\
[\![\texttt{ufxp.add}]\!](om, x_1, x_2) &= \mathit{ovf}_{\mathbb{U}_{tb,fb}}(om, x_1 + x_2)\\
[\![\texttt{sfxp.sub}]\!](om, x_1, x_2) &= \mathit{ovf}_{\mathbb{S}_{tb,fb}}(om, x_1 - x_2)\\
[\![\texttt{ufxp.sub}]\!](om, x_1, x_2) &= \mathit{ovf}_{\mathbb{U}_{tb,fb}}(om, x_1 - x_2)\\
[\![\texttt{sfxp.mul}]\!](om, rm, x_1, x_2) &= \mathit{ovf}_{\mathbb{S}_{tb,fb}}(om, \mathit{rnd}_{\mathbb{F}_{fb}}(rm, x_1 \cdot x_2))\\
[\![\texttt{ufxp.mul}]\!](om, rm, x_1, x_2) &= \mathit{ovf}_{\mathbb{U}_{tb,fb}}(om, \mathit{rnd}_{\mathbb{F}_{fb}}(rm, x_1 \cdot x_2))\\
[\![\texttt{sfxp.div}]\!](om, rm, x_1, x_2) &= \mathit{ovf}_{\mathbb{S}_{tb,fb}}(om, \mathit{rnd}_{\mathbb{F}_{fb}}(rm, x_1 / x_2))\\
[\![\texttt{ufxp.div}]\!](om, rm, x_1, x_2) &= \mathit{ovf}_{\mathbb{U}_{tb,fb}}(om, \mathit{rnd}_{\mathbb{F}_{fb}}(rm, x_1 / x_2))
\end{aligned}
$$

Note that it trivially holds that $\forall x_1, x_2 \in \mathbb{F}_{fb} \cdot x_1 + x_2 \in \mathbb{F}_{fb} \wedge x_1 - x_2 \in \mathbb{F}_{fb}$. Therefore, we do not need to round the results of the addition and subtraction operations. In the case of division by zero, we adopt the semantics of other SMT theories such as reals: (= x (**sfxp.div** *om rm* y 0)) and (= x (**ufxp.div** *om rm* y 0)) are satisfiable for every $x, y \in \mathbb{F}$, $om \in \mathit{omode}$, $rm \in \mathit{rmode}$. Furthermore, for every $x, y \in \mathbb{F}$, $om \in \mathit{omode}$, $rm \in \mathit{rmode}$, if (= x y) then (= (**sfxp.div** *om rm* x 0) (**sfxp.div** *om rm* y 0)) and (= (**ufxp.div** *om rm* x 0) (**ufxp.div** *om rm* y 0)).

Note that the order of applying the *rnd* and *ovf* functions to the results in real arithmetic matters. We choose *rnd* followed by *ovf* since it matches the typical real-world fixed-point semantics. Conversely, reversing the order can lead to out-of-bound results. For example, assume that we extend the signature of the *ovf* function to $\mathit{omode} \times \mathbb{R} \mapsto \mathbb{R}$ while preserving its semantics as a modulo operation over 2^{tb-fb}. Then, $\mathit{ovf}_{\mathbb{U}_{3,2}}(\mathit{wrap}, 7.5)$ evaluates to $\frac{7.5}{4}$, and applying $\mathit{rnd}_{\mathbb{F}_2}$ to it when the rounding mode is *ru* evaluates to $\frac{8}{4}$; this is greater than the maximum number in $\mathbb{U}_{3,2}$, namely $\frac{7}{4}$. On the other hand, evaluating $\mathit{ovf}_{\mathbb{U}_{3,2}}(\mathit{wrap}, \mathit{rnd}_{\mathbb{F}_2}(\mathit{ru}, 7.5))$ produces 0, which is the expected result. We could apply the *ovf* function again to the out-of-bound results, but the current semantics achieves the same without this additional operation.

Let $\mathit{cast}_{\mathbb{F},\mathbb{F}_{fb}} : \mathit{omode} \times \mathit{rmode} \times \mathbb{R} \mapsto \mathbb{F}$ be a family of cast functions parameterized by $\mathbb{F}$ and $\mathbb{F}_{fb}$ that map an overflow mode, rounding mode, and real number to an element of $\mathbb{F}$; as before, $\mathbb{F}$ is either $\mathbb{S}_{tb,fb}$ or $\mathbb{U}_{tb,fb}$ depending on whether we are using signed or unsigned fixed-point numbers, respectively. Then, we define $\mathit{cast}_{\mathbb{F},\mathbb{F}_{fb}}(om, rm, r) = \mathit{ovf}_{\mathbb{F}}(om, \mathit{rnd}_{\mathbb{F}_{fb}}(rm, r))$, and the interpretation of the conversions between reals and fixed-points as

$$
\begin{aligned}
[\![(\_\ \texttt{to\_sfxp}\ tb\ fb)]\!](om, rm, r) &= \mathit{cast}_{\mathbb{S}_{tb,fb},\mathbb{F}_{fb}}(om, rm, r)\\
[\![(\_\ \texttt{to\_ufxp}\ tb\ fb)]\!](om, rm, r) &= \mathit{cast}_{\mathbb{U}_{tb,fb},\mathbb{F}_{fb}}(om, rm, r)\\
[\![\texttt{sfxp.to\_real}]\!](r) &= r\\
[\![\texttt{ufxp.to\_real}]\!](r) &= r
\end{aligned}
$$

5 Decision Procedures

In this section, we propose two decision procedures for the fixed-point theory by leveraging the theory of fixed-size bit-vectors in one and the theory of reals in the other.

Bit-Vector Encoding. The decision procedure based on the theory of fixed-size bit-vectors is akin to the existing software implementations of fixed-point arithmetic that use machine integers. More specifically, a fixed-point sort parameterized by tb is encoded as a bit-vector sort of length tb. Therefore, the encoding of the constructors of fixed-point values simply amounts to identity functions. Similarly, the encoding of the comparison operators uses the corresponding bit-vector relations. For example, the comparison operator `sfxp.lt` is encoded as `bvslt`. The essence of the encoding of the arithmetic operations is expressing the numerator of the result, after rounding and overflow handling, using bit-vector arithmetic. We leverage the following two observations in our encoding. First, rounding a real value v to the value in the set $\mathbb{F}_{fb}$ amounts to rounding $v \cdot 2^{fb}$ to an integer following the same rounding mode. This observation explains why rounding is not necessary for the linear arithmetic operations. Second, we can encode the wrap-around of the rounded result as simply extracting tb bits from the encoded result thanks to the wrap-around nature of the two's complement SMT representation. We model the behavior of division-by-zero using uninterpreted functions of the form (`RoundingMode OverflowMode (_ BitVec` tb`) (_ BitVec` tb`))`, with one such function for each fixed-point sort appearing in the query. The result of division-by-zero is then the result of applying this function to the numerator, conditioned on the denominator being equal to zero. This ensures that all divisions-by-zero with equal numerators produce equal results when the overflow and rounding modes are also equal.

Real Encoding. The decision procedure based on the theory of reals closely mimics the semantics defined in Sect. 4. We encode all fixed-point sorts as the real sort, while we represent fixed-point values as rational numbers. Therefore, we can simply encode fixed-point comparisons as real relations. For example, both `sfxp.lt` and `ufxp.lt` are translated into $<$ relation. We rely on the first observation above to implement the rounding function rnd_{fb} using an SMT real-to-integer conversion. We implement the overflow function $ovf_{tb,fb}$ using the SMT remainder function. Note that the encodings of both functions involve non-linear real functions, such as the real-to-int conversion. Finally, we model division as the rounded, overflow-corrected result of the real theory's division operation. Since the real theory's semantics ensures that equivalent division operations produce equivalent results, this suffices to capture the fixed-point division-by-zero semantics.

Implementation. We implemented the two decision procedures within the pySMT framework [25]: the two encodings are rewriting classes of pySMT. We made our implementations publicly available.[1] We also implemented a random generator of queries in our fixed-point theory, and used it to perform thorough differential testing of our decision procedures.

6 Experiments

We generated the benchmarks we use to evaluate the two encodings described in Sect. 5 by translating the SMT-COMP non-incremental QF_FP benchmarks [45]. The translation accepts benchmarks that contain only basic arithmetic operations defined in both theories. Moreover, we exclude all the benchmarks in the *wintersteiger* folder because they are mostly simple regressions to test the correctness of an implementation of the floating-point theory. In the end, we manage to translate 218 QF_FP benchmarks in total.

We translate each QF_FP benchmark into 4 benchmarks in the fixed-point theory, which differ in the configurations of rounding and overflow modes. We denote a configuration as a (rounding mode, overflow mode) tuple. Note that changing a benchmark configuration alters the semantics of its arithmetic operations, which might affect its satisfiability. Our translation replaces floating-point sorts with fixed-point sorts that have the same total bit-widths; the number of fractional bits is half of the bit-width. This establishes a mapping from single-precision floats to Q16.16 fixed-points implemented by popular software libraries such as libfixmath [34]. It translates arithmetic operations into their corresponding fixed-point counterparts using the chosen configuration uniformly across a benchmark. The translation also replaces floating-point comparison operations with their fixed-point counterparts. Finally, we convert floating-point constants by treating them as reals and performing real-to-fixed-point casts. We made our fixed-point benchmarks publicly available.[2]

The SMT solvers that we use in the evaluation are Boolector [9] (version 3.1.0), CVC4 [4] (version 1.7), MathSAT [13] (version 5.5.1), Yices2 [19] (version 2.6.1), and Z3 [17] (version 4.8.4) for the decision procedure based on the theory of bit-vectors. For the evaluation of the decision procedure based on the theory of reals, we use CVC4, MathSAT, Yices2, and Z3. We ran the experiments on a machine with four Intel E7-4830 sockets, for a total of 32 physical cores, and 512GB of RAM, running Ubuntu 18.04. Each benchmark was limited to 1200 s of wall time and 8GB of memory, and no run of any benchmark exceeded the memory limit. We set processor affinity for each solver instance in order to reduce variability due to cache effects.

Table 1 shows the results of running the SMT solvers on each configuration with both encodings (bit-vector and real). We do not observe any inconsistencies in terms of satisfiability reported among all the solvers and between both encodings. The performance of the solvers on the bit-vector encoding is typically better than on the real encoding since it leads to fewer timeouts and crashes. Moreover, all the solvers demonstrate similar performance for the bit-vector encoding

[1] https://github.com/soarlab/pysmt/tree/fixed-points.

[2] https://github.com/soarlab/QF_FXP.

Table 1. The results of running SMT solvers on the four different configurations of the benchmarks using both encodings. Boolector and MathSAT are denoted by Btor and MSAT, respectively. Column "All" indicates the number of benchmarks for which any solver answered sat or unsat; benchmarks for which no solver gave an answer are counted as unknown.

(a) (RoundUp, Saturation)

Result	Bit-Vector Encoding					Real Encoding				All
	Btor	CVC4	MSAT	Yices2	Z3	CVC4	MSAT	Yices2	Z3	
sat	57	52	47	65	43	15	50	52	22	65
unsat	129	127	127	125	131	125	126	126	124	132
timeout	32	39	44	28	44	75	15	40	37	
unknown	0	0	0	0	0	3	0	0	35	21
error	0	0	0	0	0	0	27	0	0	

(b) (RoundUp, WrapAround)

Result	Bit-Vector Encoding					Real Encoding				All
	Btor	CVC4	MSAT	Yices2	Z3	CVC4	MSAT	Yices2	Z3	
sat	58	51	53	67	48	14	34	52	14	72
unsat	128	128	126	128	134	123	65	124	121	134
timeout	32	39	39	23	36	79	102	42	60	
unknown	0	0	0	0	0	2	0	0	23	12
error	0	0	0	0	0	0	17	0	0	

(c) (RoundDown, Saturation)

Result	Bit-Vector Encoding					Real Encoding				All
	Btor	CVC4	MSAT	Yices2	Z3	CVC4	MSAT	Yices2	Z3	
sat	59	52	54	62	50	29	53	57	22	64
unsat	128	127	127	125	134	127	130	130	124	135
timeout	31	39	37	31	34	58	11	31	46	
unknown	0	0	0	0	0	4	0	0	26	19
error	0	0	0	0	0	0	24	0	0	

(d) (RoundDown, WrapAround)

Result	Bit-Vector Encoding					Real Encoding				All
	Btor	CVC4	MSAT	Yices2	Z3	CVC4	MSAT	Yices2	Z3	
sat	57	65	54	67	50	23	39	55	14	71
unsat	128	128	127	129	133	125	81	90	121	134
timeout	33	25	37	22	35	68	65	73	57	
unknown	0	0	0	0	0	2	0	0	26	13
error	0	0	0	0	0	0	33	0	0	

Table 2. Comparison of the number of benchmarks (considering all configurations) solved by a solver but not solved by another solver. Each row shows the number of benchmarks solved by the row's solver but not solved by the column's solver. We mark the bit-vector (resp. real) encoding with B (resp. R).

	Btor-B	CVC4-B	MSAT-B	Yices2-B	Z3-B	CVC4-R	MSAT-R	Yices2-R	Z3-R
Btor-B	—	33	37	11	52	165	183	86	185
CVC4-B	19	—	46	6	57	154	160	70	170
MSAT-B	8	31	—	4	39	141	174	78	160
Yices2-B	35	44	57	—	79	194	198	95	208
Z3-B	31	50	47	34	—	151	189	103	168
CVC4-R	2	5	7	7	9	—	113	49	41
MSAT-R	17	8	37	8	44	110	—	23	118
Yices2-R	28	26	49	13	66	154	131	—	162
Z3-R	3	2	7	2	7	22	102	38	—

across all the configurations, whereas they generally produce more timeouts for the real encoding when the overflow mode is wrap-around. We believe that this can be attributed to the usage of nonlinear operations (e.g., real to int casts) in the handling of wrap-around behaviors. This hypothesis could also explain the observation that the bit-vector encoding generally outperform the real encoding when the overflow mode is wrap-around since wrap-around comes at almost no cost for the bit-vector encoding (see Sect. 5).

Column "All" captures the performance of the solvers when treated as one portfolio solver. This improves the overall performance since the number of solved benchmarks increases, indicating that each solver has different strengths and weaknesses. Table 2 further analyzes this behavior, and we identify two reasons for it when we consider unique instances solved by each individual solver. First, when the overflow mode is saturation, Yices2 is the only solver to solve unique instances for both encodings. Second, when the overflow mode is wrap-around, the uniquely solved instances come from solvers used on the bit-vector encoding, except one that comes from Yices2 on the real encoding. These results provide further evidence that the saturation configurations are somewhat easier to solve with reals, and that wrap-around is easier with bit-vectors.

Figure 1 uses quantile plots [5] to visualize our experimental results in terms of runtimes. A quantile plot shows the minimum runtime on y-axis within which each of the x-axis benchmarks is solved. Some characteristics of a quantile plot are helpful in analyzing the runtimes. First, the rightmost x coordinate is the number of benchmarks that a solver returns meaningful results for (i.e., sat or unsat). Second, the uppermost y coordinate is the maximum runtime of all the benchmarks. Third, the area under a line approximates the total runtime.

Although the semantics of the benchmarks vary for each configuration, we can observe that the shapes of the bit-vector encoding curves are similar, while those of the real encoding differ based on the chosen overflow mode. More precisely, solvers tend to solve benchmarks faster when their overflow mode is saturation

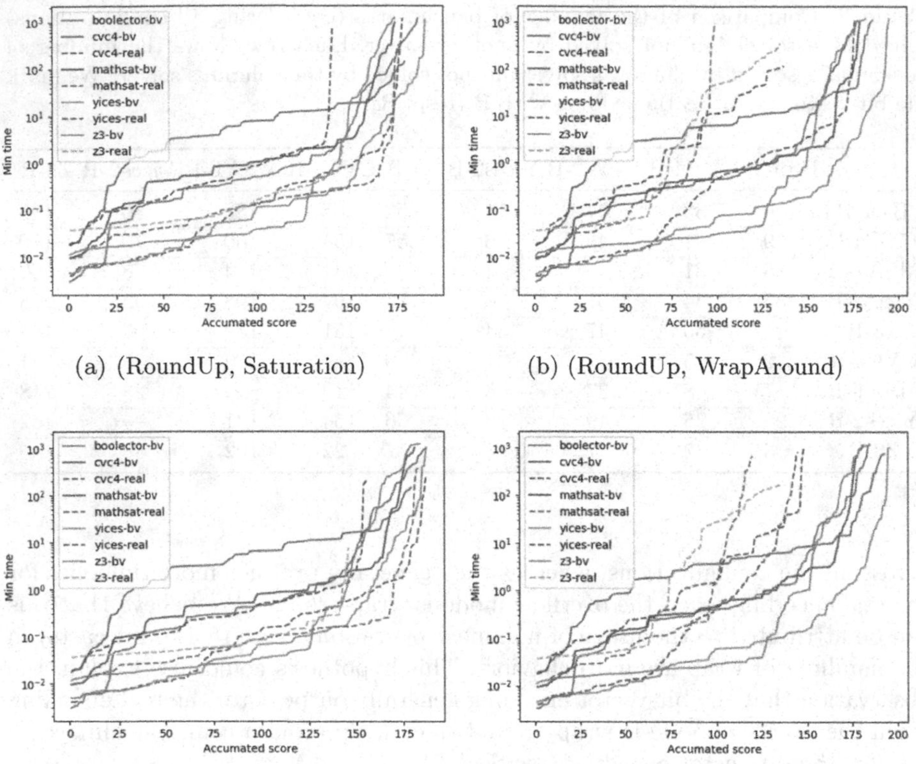

(a) (RoundUp, Saturation)

(b) (RoundUp, WrapAround)

(c) (RoundDown, Saturation)

(d) (RoundDown, WrapAround)

Fig. 1. Quantile plots of our experimental results.

as opposed to wrap-around. We observe the same behavior in Table 1, and it is likely due to the fact that we introduce nonlinear operations to handle wrap-around behaviors when using the real encoding.

7 Case Study: Verification of Quantized Neural Networks

Neural networks have experienced a significant increase in popularity in the past decade. Such networks that are realized by a composition of non-linear layers are able to efficiently solve a large variety of previously unsolved learning tasks. However, neural networks are often viewed as black-boxes, whose causal structure cannot be interpreted easily by humans [40]. This property makes them unfit for applications where guaranteed correctness has a high priority. Advances in formal methods, in particular SMT solvers, leveraging the piece-wise linear structure of neural networks [20,31,47], have made it possible to verify certain formal properties of neural networks of reasonable size. While these successes provide an essential step towards applying neural networks to safety-critical tasks, these

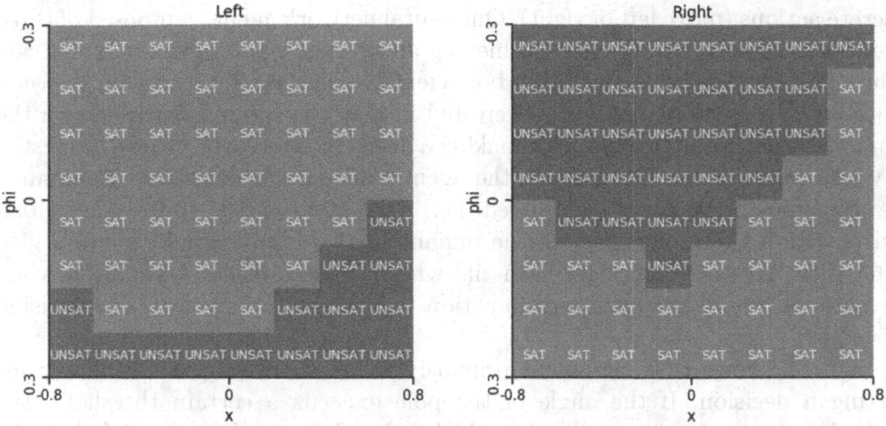

Fig. 2. Satisfiability of specifications of our cart-pole controller.

methods leave out one crucial aspect—neural networks are usually *quantized* before being deployed to production systems [30].

Quantization converts a network that operates over 32-bit floating-point semantics into a fewer-bit fixed-point representation. This process serves two goals: compressing the memory requirement and increasing the computational efficiency of running the network. Quantization introduces additional non-linear rounding operations to the semantics of a neural network. Recently, Giacobbe et al. [26] have shown that, in practice, this can lead to situations where a network that satisfies formal specifications might violate them after the quantization step. Therefore, when checking formal properties of quantized neural networks, we need to take their fixed-point semantics into account.

We derive a set of example fixed-point problem instances based on two machine learning tasks to demonstrate the capabilities of our fixed-point SMT theory on realistic problems. For all tasks, we train multi-layer perceptron modules [43] with ReLU-7 activation function [32] using quantization-aware training [30]. This way we avoid that quantization results in a considerable loss of accuracy. To encode a neural network into an SMT formula, we rely on the Giacobbe et al.'s [26] approach for encoding the summations and activation functions. We quantize all neural networks using the signed fixed-point format with 8 bits total and 4 fractional bits. We are using the bit-vector encoding decision procedure in combination with the Boolector SMT solver.

7.1 Cart-Pole Controller

In our first task, we train a neural network controller using the cart-pole environment of OpenAI's "gym" reinforcement learning suite. In this task, an agent has to balance a pole mounted on a movable cart in an upright position. The cart provides four observation variables x,$\dot{x}$,φ,$\dot{\varphi}$ to the controller, where x is the position of the cart and φ the angle of the pole. The controller then steers the cart by discrete actions (move left or right). Our neural network agent, composed of three layers (4,8,1), solves the task by achieving an average score of the maximal 500 points. We analyze what our black-box agent has learned by using our decision procedure. In particular, we are interested in how much our agent relies on the input variable x compared to φ for making a decision. Moreover, we are interested in which parts of the input space the agent's decision is constant. We assume the dynamics of the cart is bounded, i.e., $-0.3 \leq \dot{x} \leq 0.3, -0.02 \leq \dot{\varphi} \leq 0.2$, and partition the input space of the remaining two input variables into a grid of 64 tiles. We then check for each tile whether there exists a situation when the network would output a certain action (left, right) by invoking our decision procedure.

Figure 2 shows that the agent primarily relies on the input variable φ for making a decision. If the angle of the pole exceeds a certain threshold, the network is guaranteed to make the vehicle move left; on the other hand, if the angle of the pole is below a different threshold, the network moves the vehicle right. Interestingly, this pattern is non-symmetric, despite the task being entirely symmetric.

7.2 Checking Fairness

For our second task, we checked the fairness specification proposed by Giacobbe et al. [26] to evaluate the maximum influence of a single input variable on the decision of a network. We train a neural network on student data to predict the score on a math exam. Among other personal features, the gender of a person is fed into the network for making a decision. As the training data contains a bias in the form of a higher average math score for male participants, the network might learn to underestimate the math score of female students. We employ our decision procedure to compute the maximum influence of the gender of a person to its predicted math score. First, we create encodings of the same network (3 layers of size 6, 16, and 1) that share all input variables except the gender as a single fixed-point theory formula. We then constrain the predicted scores such that the one network outputs a score that is c higher than the score predicted by the other network. Finally, we perform binary search by iteratively invoking our decision procedure to find out at what bias c the formula changes from satisfiable to unsatisfiable.

Table 3. Satisfiability of specifications of our fairness example.

Score Diff	Status	Runtime
11.25	sat	10s
11.5	sat	9s
11.75	unsat	200s
12	unsat	706s

Table 3 shows that there exists a hypothetical person whose predicted math score would drop by 11.5 points out of 100 if the person is female instead of male. Moreover, our results also show that for no person the math score would change by 11.75 points if the gender would be changed.

8 Related Work

Ruemmer and Wahl [44] and Brain et al. [8] propose and formalize the SMT theory of the IEEE-754 floating-point arithmetic. We were inspired by these papers both in terms of the syntax and the formalization of the semantics of our theory. There are several decision procedures for the floating-point theory. In particular, Brain et al. [7] present an efficient and verified reduction from the theory of floating-points to the theory of bit-vectors, while Leeser et al. [33] solve the floating-point theory by reducing it to the theory of reals. These two decision procedures are much more complicated than the ones we describe in Sect. 5 due to the more complex nature of floating-point arithmetic.

In the rest of this section, we introduce related approaches that perform verification or synthesis of programs that use fixed-point arithmetic. Many of these approaches, and in particular the SMT-based ones, could benefit from our unified formalization of the theory of fixed-point arithmetic. For example, they could leverage our decision procedures instead of developing their own from scratch. Moreover, having the same format allows for easier sharing of benchmarks and comparison of results among different decision procedures.

Eldib et al. [21] present an SMT-based method for synthesizing optimized fixed-point computations that satisfy certain acceptance criteria, which they rigorously verify using an SMT solver. Similarly to our paper, their approach encodes fixed-point arithmetic operations using the theory of bit-vectors. Anta et al. [3] tackle the verification problem of the stability of fixed-point controller implementations. They provide a formalization of fixed-point arithmetic semantics using bit-vectors, but unlike our paper they do not formalize rounding and overflows. Furthermore, they encode the fixed-point arithmetic using unbounded integer arithmetic, arguing that unbounded integer arithmetic is a better fit for their symbolic analysis. We could also reduce our bit-vector encoding to unbounded integers following a similar scheme as Anta et al.

Bounded model checker ESMBC [15,24] supports fixed-point arithmetic and has been used to verify safety properties of fixed-point digital controllers [1]. Like us, it also employs a bit-vector encoding. However, it is unclear exactly which fixed-point operations are supported. UppSAT [50] is an approximating SMT solver that leverages fixed-point arithmetic as an approximation theory to floating-point arithmetic. Like the aforementioned work, UppSAT also encodes fixed-point arithmetic using the theory of bit-vectors. Its encoding ignores rounding modes, but adds special values such as infinities.

In addition to SMT-based verification, another important aspect of reasoning about fixed-point computations is error bound analysis, which is often used for the synthesis of fixed-point implementations. Majumdar et al. [38] synthesize Pareto optimal fixed-point implementations of control software in regard to performance criteria and error bounds. They reduce error bound computation to an optimization problem solved by mixed-integer linear programming. Darulova et al. [16] compile real-valued expressions to fixed-point expressions, and rigorously show that the generated expressions satisfy given error bounds. The error bound analysis is static and based on affine arithmetic. Volkova et al. [48,49] propose an approach to determine the fixed-point format that ensures the absence of overflows and minimizes errors; their error analysis is based on Worst-Case Peak Gain measure. TAFFO [12] is an LLVM plugin that performs precision tuning by replacing floating-point computations with their fixed-point counterparts. The quality of precision tuning is determined by a static error propagation analysis.

9 Conclusions and Future Work

In this paper, we propose an SMT theory of fixed-point arithmetic to facilitate SMT-based software verification of fixed-point programs and systems by promoting the development of decision procedures for the proposed theory. We introduce the syntax of fixed-point sorts and operations in the SMT-LIB format similar to that of the SMT floating-point theory. Then, we formalize the semantics of the fixed-point theory, including rounding and overflow, based on the exact rational arithmetic. We develop two decision procedures for the fixed-point theory that encode it into the theory of bit-vectors and reals. Finally, we study the performance of our prototype decision procedures on a set of benchmarks, and perform a realistic case study by proving properties of quantized neural networks.

As future work, we plan to add more complex operations to the fixed-point theory, such as conversions to/from floating-points and the remainder operation. Moreover, we would like to apply the fixed-point theory to verify existing software implementations of fixed-point arithmetic in different programming languages. We plan to do this by integrating it into the Boogie intermediate verification language [18] and the SMACK verification toolchain [10,42].

References

1. Abreu, R.B., Gadelha, M.Y.R., Cordeiro, L.C., de Lima Filho, E.B., da Silva, W.S.: Bounded model checking for fixed-point digital filters. J. Braz. Comput. Soc. **22**(1), 1:1–1:20 (2016). https://doi.org/10.1186/s13173-016-0041-8
2. Andrysco, M., Nötzli, A., Brown, F., Jhala, R., Stefan, D.: Towards verified, constant-time floating point operations. In: Proceedings of the ACM Conference on Computer and Communications Security (CCS), pp. 1369–1382 (2018). https://doi.org/10.1145/3243734.3243766
3. Anta, A., Majumdar, R., Saha, I., Tabuada, P.: Automatic verification of control system implementations. In: Proceedings of the International Conference on Embedded Software (EMSOFT), pp. 9–18 (2010). https://doi.org/10.1145/1879021.1879024
4. Barrett, C., et al.: CVC4. In: Gopalakrishnan, G., Qadeer, S. (eds.) CAV 2011. LNCS, vol. 6806, pp. 171–177. Springer, Heidelberg (2011). https://doi.org/10.1007/978-3-642-22110-1_14
5. Beyer, D.: Software verification and verifiable witnesses. In: Baier, C., Tinelli, C. (eds.) TACAS 2015. LNCS, vol. 9035, pp. 401–416. Springer, Heidelberg (2015). https://doi.org/10.1007/978-3-662-46681-0_31

6. Brain, M., D'Silva, V., Griggio, A., Haller, L., Kroening, D.: Deciding floating-point logic with abstract conflict driven clause learning. Formal Methods Syst. Des. **45**(2), 213–245 (2013). https://doi.org/10.1007/s10703-013-0203-7
7. Brain, M., Schanda, F., Sun, Y.: Building better bit-blasting for floating-point problems. In: Vojnar, T., Zhang, L. (eds.) TACAS 2019. LNCS, vol. 11427, pp. 79–98. Springer, Cham (2019). https://doi.org/10.1007/978-3-030-17462-0_5
8. Brain, M., Tinelli, C., Rümmer, P., Wahl, T.: An automatable formal semantics for IEEE-754 floating-point arithmetic. In: Proceedings of the IEEE International Symposium on Computer Arithmetic (ARITH), pp. 160–167 (2015). https://doi.org/10.1109/ARITH.2015.26
9. Brummayer, R., Biere, A.: Boolector: an efficient SMT solver for bit-vectors and arrays. In: Kowalewski, S., Philippou, A. (eds.) TACAS 2009. LNCS, vol. 5505, pp. 174–177. Springer, Heidelberg (2009). https://doi.org/10.1007/978-3-642-00768-2_16
10. Carter, M., He, S., Whitaker, J., Rakamarić, Z., Emmi, M.: SMACK software verification toolchain. In: Proceedings of the International Conference on Software Engineering (ICSE), pp. 589–592 (2016). https://doi.org/10.1145/2889160.2889163
11. Cherkaev, A., Tai, W., Phillips, J.M., Srikumar, V.: Learning in practice: reasoning about quantization. CoRR abs/1905.11478 (2019). http://arxiv.org/abs/1905.11478
12. Cherubin, S., Cattaneo, D., Chiari, M., Bello, A.D., Agosta, G.: TAFFO: tuning assistant for floating to fixed point optimization. Embed. Syst. Lett. **12**(1), 5–8 (2020). https://doi.org/10.1109/LES.2019.2913774
13. Cimatti, A., Griggio, A., Schaafsma, B.J., Sebastiani, R.: The MathSAT5 SMT solver. In: Piterman, N., Smolka, S.A. (eds.) TACAS 2013. LNCS, vol. 7795, pp. 93–107. Springer, Heidelberg (2013). https://doi.org/10.1007/978-3-642-36742-7_7
14. Conchon, S., Iguernlala, M., Ji, K., Melquiond, G., Fumex, C.: A three-tier strategy for reasoning about floating-point numbers in SMT. In: Majumdar, R., Kunčak, V. (eds.) CAV 2017. LNCS, vol. 10427, pp. 419–435. Springer, Cham (2017). https://doi.org/10.1007/978-3-319-63390-9_22
15. Cordeiro, L., Fischer, B., Marques-Silva, J.: SMT-based bounded model checking for embedded ANSI-C software. In: Proceedings of the International Conference on Automated Software Engineering (ASE), pp. 137–148 (2009). https://doi.org/10.1109/ASE.2009.63
16. Darulova, E., Kuncak, V., Majumdar, R., Saha, I.: Synthesis of fixed-point programs. In: Proceedings of the International Conference on Embedded Software (EMSOFT), pp. 22:1–22:10 (2013). https://doi.org/10.1109/EMSOFT.2013.6658600
17. de Moura, L., Bjørner, N.: Z3: an efficient SMT solver. In: Ramakrishnan, C.R., Rehof, J. (eds.) TACAS 2008. LNCS, vol. 4963, pp. 337–340. Springer, Heidelberg (2008). https://doi.org/10.1007/978-3-540-78800-3_24
18. DeLine, R., Leino, K.R.M.: BoogiePL: a typed procedural language for checking object-oriented programs. Technical report, MSR-TR-2005-70, Microsoft Research (2005)
19. Dutertre, B.: Yices 2.2. In: Biere, A., Bloem, R. (eds.) CAV 2014. LNCS, vol. 8559, pp. 737–744. Springer, Cham (2014). https://doi.org/10.1007/978-3-319-08867-9_49
20. Ehlers, R.: Formal verification of piece-wise linear feed-forward neural networks. In: D'Souza, D., Narayan Kumar, K. (eds.) ATVA 2017. LNCS, vol. 10482, pp. 269–286. Springer, Cham (2017). https://doi.org/10.1007/978-3-319-68167-2_19

21. Eldib, H., Wang, C.: An SMT based method for optimizing arithmetic computations in embedded software code. IEEE Trans. Comput. Aided Des. Integr. Circ. Syst. **33**(11), 1611–1622 (2014). https://doi.org/10.1109/TCAD.2014.2341931
22. The fixed crate. https://gitlab.com/tspiteri/fixed
23. Programming languages – C – extensions to support embedded processors. Standard 18037, ISO/IEC (2008). https://www.iso.org/standard/51126.html
24. Gadelha, M.R., Monteiro, F.R., Morse, J., Cordeiro, L.C., Fischer, B., Nicole, D.A.: ESBMC 5.0: an industrial-strength C model checker. In: Proceedings of the International Conference on Automated Software Engineering (ASE), pp. 888–891. https://doi.org/10.1145/3238147.3240481
25. Gario, M., Micheli, A.: PySMT: a solver-agnostic library for fast prototyping of SMT-based algorithms. In: International Workshop on Satisfiability Modulo Theories (SMT) (2015)
26. Giacobbe, M., Henzinger, T.A., Lechner, M.: How many bits does it take to quantize your neural network? In: Biere, A., Parker, D. (eds.) TACAS 2020. LNCS, vol. 12079, pp. 79–97. Springer, Cham (2020). https://doi.org/10.1007/978-3-030-45237-7_5
27. Gupta, S., Agrawal, A., Gopalakrishnan, K., Narayanan, P.: Deep learning with limited numerical precision. In: Proceedings of the International Conference on Machine Learning (ICML), pp. 1737–1746 (2015)
28. Signed 15.16 precision fixed-point arithmetic. https://github.com/ekmett/fixed
29. He, S., Baranowski, M., Rakamarić, Z.: Stochastic local search for solving floating-point constraints. In: Zamani, M., Zufferey, D. (eds.) NSV 2019. LNCS, vol. 11652, pp. 76–84. Springer, Cham (2019). https://doi.org/10.1007/978-3-030-28423-7_5
30. Jacob, B., et al.: Quantization and training of neural networks for efficient integer-arithmetic-only inference. In: Proceedings of the IEEE Conference on Computer Vision and Pattern Recognition (CVPR), pp. 2704–2713 (2018). https://doi.org/10.1109/CVPR.2018.00286
31. Katz, G., Barrett, C., Dill, D.L., Julian, K., Kochenderfer, M.J.: Reluplex: an efficient SMT solver for verifying deep neural networks. In: Majumdar, R., Kunčak, V. (eds.) CAV 2017. LNCS, vol. 10426, pp. 97–117. Springer, Cham (2017). https://doi.org/10.1007/978-3-319-63387-9_5
32. Krizhevsky, A.: Convolutional deep belief networks on CIFAR-10 (2010, unpublished manuscript)
33. Leeser, M., Mukherjee, S., Ramachandran, J., Wahl, T.: Make it real: effective floating-point reasoning via exact arithmetic. In: Proceedings of the Design, Automation and Test in Europe Conference and Exhibition (DATE), pp. 1–4 (2014). https://doi.org/10.7873/DATE.2014.130
34. Cross platform fixed point maths library. https://github.com/PetteriAimonen/libfixmath
35. Liew, D., Cadar, C., Donaldson, A.F., Stinnett, J.R.: Just fuzz it: solving floating-point constraints using coverage-guided fuzzing. In: Proceedings of the ACM Joint European Software Engineering Conference and Symposium on the Foundations of Software Engineering (ESEC/FSE), pp. 521–532 (2019). https://doi.org/10.1145/3338906.3338921
36. Liew, D., Schemmel, D., Cadar, C., Donaldson, A.F., Zähl, R., Wehrle, K.: Floating-point symbolic execution: a case study in n-version programming. In: Proceedings of the International Conference on Automated Software Engineering (ASE), pp. 601–612 (2017). https://doi.org/10.1109/ASE.2017.8115670

37. Lin, D.D., Talathi, S.S., Annapureddy, V.S.: Fixed point quantization of deep convolutional networks. In: Proceedings of the International Conference on Machine Learning (ICML), pp. 2849–2858 (2016)
38. Majumdar, R., Saha, I., Zamani, M.: Synthesis of minimal-error control software. In: Proceedings of the International Conference on Embedded Software (EMSOFT), pp. 123–132 (2012). https://doi.org/10.1145/2380356.2380380
39. Menendez, D., Nagarakatte, S., Gupta, A.: Alive-FP: automated verification of floating point based peephole optimizations in LLVM. In: Rival, X. (ed.) SAS 2016. LNCS, vol. 9837, pp. 317–337. Springer, Heidelberg (2016). https://doi.org/10.1007/978-3-662-53413-7_16
40. Olah, C., et al.: The building blocks of interpretability. Distill (2018). https://doi.org/10.23915/distill.00010
41. Paganelli, G., Ahrendt, W.: Verifying (in-)stability in floating-point programs by increasing precision, using SMT solving. In: Proceedings of the International Symposium on Symbolic and Numeric Algorithms for Scientific Computing (SYNASC), pp. 209–216 (2013). https://doi.org/10.1109/SYNASC.2013.35
42. Rakamarić, Z., Emmi, M.: SMACK: decoupling source language details from verifier implementations. In: Biere, A., Bloem, R. (eds.) CAV 2014. LNCS, vol. 8559, pp. 106–113. Springer, Cham (2014). https://doi.org/10.1007/978-3-319-08867-9_7
43. Rumelhart, D.E., Hinton, G.E., Williams, R.J.: Learning representations by back-propagating errors. Nature **323**(6088), 533–536 (1986). https://doi.org/10.1038/323533a0
44. Rümmer, P., Wahl, T.: An SMT-LIB theory of binary floating-point arithmetic. In: Informal Proceedings of the International Workshop on Satisfiability Modulo Theories (SMT) (2010)
45. SMT-LIB benchmarks in the QF_FP theory. https://clc-gitlab.cs.uiowa.edu:2443/SMT-LIB-benchmarks/QF_FP
46. SMT-LIB: the satisfiability modulo theories library. http://smtlib.cs.uiowa.edu
47. Tjeng, V., Xiao, K.Y., Tedrake, R.: Evaluating robustness of neural networks with mixed integer programming. In: International Conference on Learning Representations (ICLR) (2019)
48. Volkova, A., Hilaire, T., Lauter, C.: Determining fixed-point formats for a digital filter implementation using the worst-case peak gain measure. In: Proceedings of the Asilomar Conference on Signals, Systems and Computers, pp. 737–741 (2015). https://doi.org/10.1109/ACSSC.2015.7421231
49. Volkova, A., Hilaire, T., Lauter, C.Q.: Arithmetic approaches for rigorous design of reliable fixed-point LTI filters. IEEE Trans. Comput. **69**(4), 489–504 (2020). https://doi.org/10.1109/TC.2019.2950658
50. Zeljić, A., Backeman, P., Wintersteiger, C.M., Rümmer, P.: Exploring approximations for floating-point arithmetic using UppSAT. In: Galmiche, D., Schulz, S., Sebastiani, R. (eds.) IJCAR 2018. LNCS (LNAI), vol. 10900, pp. 246–262. Springer, Cham (2018). https://doi.org/10.1007/978-3-319-94205-6_17

Covered Clauses Are Not Propagation Redundant

Lee A. Barnett(✉), David Cerna, and Armin Biere

Johannes Kepler University Linz, Altenbergerstraße 69, 4040 Linz, Austria
{lee.barnett,david.cerna,armin.biere}@jku.at

Abstract. Propositional proof systems based on recently-developed redundancy properties admit short refutations for many formulas traditionally considered hard. Redundancy properties are also used by procedures which simplify formulas in conjunctive normal form by removing redundant clauses. Revisiting the covered clause elimination procedure, we prove the correctness of an explicit algorithm for identifying covered clauses, as it has previously only been implicitly described. While other elimination procedures produce redundancy witnesses for compactly reconstructing solutions to the original formula, we prove that witnesses for covered clauses are hard to compute. Further, we show that not all covered clauses are propagation redundant, the most general, polynomially-verifiable standard redundancy property. Finally, we close a gap in the literature by demonstrating the complexity of clause redundancy itself.

Keywords: SAT · Clause elimination · Redundancy

1 Introduction

Boolean satisfiability (SAT) solvers have become successful tools for solving reasoning problems in a variety of applications, from formal verification [6] and security [27] to pure mathematics [10,19,24]. Significant recent progress in the design of SAT solvers has come as a result of exploiting the notion of clause redundancy (for instance, [14,16,18]). For a propositional formula F in conjunctive normal form (CNF), a clause C is redundant if it can be added to, or removed from, F without affecting whether F is satisfiable [23].

In particular, redundancy forms a basis for clausal proof systems. These systems refute an unsatisfiable CNF formula F by listing instructions to add or delete clauses to or from F, where the addition of a clause C is permitted only if C meets some criteria ensuring its redundancy. By eventually adding the empty clause, the formula is proven to be unsatisfiable. Crucially, the redundancy

Supported by the Austrian Science Fund (FWF) under project W1255-N23, the LogiCS Doctoral College on Logical Methods in Computer Science, as well as the LIT Artificial Intelligence Lab funded by the State of Upper Austria.

N. Peltier and V. Sofronie-Stokkermans (Eds.): IJCAR 2020, LNAI 12166, pp. 32–47, 2020.
https://doi.org/10.1007/978-3-030-51074-9_3

criteria of a system can also be used as an inference rule by a solver searching for such refutations, or for satisfying assignments.

Proof systems based on the recently introduced PR (Propagation Redundancy) criteria [15] have been shown to admit short refutations of the famous pigeonhole formulas [11,17]. These are known to have only exponential-size refutations in many systems, including resolution [9] and constant-depth Frege systems [1], but have polynomial-size PR refutations. In fact, many problems typically considered hard have short PR refutations, spurring interest in these systems from the viewpoint of proof complexity [4]. Further, systems based on PR are strong even without introducing new variables, and have the potential to afford substantial improvements to SAT solvers (such as in [16,18]).

The PR criteria is very general, encompassing nearly all other established redundancy criteria, and it is NP-complete to decide whether it is met by a given clause [18]. However, when the clause is given alongside a witness, a partial assignment providing additional evidence for the clause's redundancy, the PR criteria can be polynomially verified [15]. SAT solvers producing refutations in the PR system must find and record a witness for each PR clause addition.

Redundancy is also a basis for clause elimination procedures, which simplify a CNF formula by removing redundant clauses [12,14]. These are useful preprocessing and inprocessing techniques that also make use of witnesses, but for the task of solution reconstruction: correcting satisfying assignments found after simplifying to ensure they solve the original formula. A witness for a clause C details how to fix assignments falsifying C without falsifying other clauses in the formula [15,17], so solvers using elimination procedures that do not preserve formula equivalence typically provide a witness for each removed clause.

Covered clause elimination (CCE) [13] is a strong procedure which removes covered clauses, a generalization of blocked clauses [20,25], and has been implemented in various SAT solvers (for example, [2,3,8]) and the CNF preprocessing tool Coprocessor [26]. CCE does not preserve formula equivalence, but provides no witnesses for the clauses it removes. Instead, it uses a complex technique to reconstruct solutions in multiple steps, requiring at times a quadratic amount of space to reconstruct a single clause [14,21]. CCE has so far only been implicitly described, and it is not clear how to produce witnesses for covered clauses.

In this paper we provide an explicit algorithm for identifying covered clauses, and show that their witnesses are difficult to produce. We also demonstrate that although covered clauses are redundant, they do not always meet the criteria required by PR. This suggests it may be beneficial to consider redundancy properties beyond PR which allow alternative types of witnesses. There has already been some work in this direction with the introduction of the SR (Substitution Redundancy) property by Buss and Thapen [4].

The paper is organized as follows. In Sect. 2 we provide necessary background and terminology, while Sect. 3 reviews covered clause elimination, provides the algorithm for identifying covered clauses, and proves that this algorithm and its reconstruction strategy are correct. Section 4 includes proofs about witnesses for covered clauses, and shows that they are not encompassed by PR. In Sect. 5 we consider the complexity of deciding clause redundancy in general, followed by a conclusion and discussion of future work in Sect. 6.

2 Preliminaries

A literal is a boolean variable x or its negation $\neg x$. A clause is a disjunction of literals, and a formula is a conjunction of clauses. We often identify a clause with the set of its literals, and a formula with the set of its clauses. For a set of literals S we write $\neg S$ to refer to the set $\neg S = \{\neg l \mid l \in S\}$. The set of variables occurring in a formula F is written $var(F)$. The empty clause is represented by $\bot$, and the satisfied clause by $\top$.

An *assignment* is a function from a set of variables to the truth values *true* and *false*. An assignment is *total* for a formula F if it assigns a value for every variable in $var(F)$, otherwise it is partial. An assignment is represented by the set of literals it assigns to true. The *composition* of assignments τ and υ is

$$\tau \circ \upsilon(x) = \begin{cases} \tau(x) & \text{if } x, \neg x \notin \upsilon \\ \upsilon(x) & \text{otherwise} \end{cases}$$

for a variable x in the domain of τ or υ. For a literal l, we write τ_l to represent the assignment $\tau \circ \{l\}$. An assignment *satisfies* (resp., *falsifies*) a variable if it assigns that variable true (resp., false). Assignments are lifted to functions assigning literals, clauses, and formulas in the usual way.

Given an assignment τ and a clause C, the *partial application* of τ to C is written $C|_\tau$ and is defined as follows: $C|_\tau = \top$ if C is satisfied by τ, otherwise, $C|_\tau = \{l \mid l \in C \text{ and } \neg l \notin \tau\}$. Likewise, the partial application of the assignment τ to a formula F is written $F|_\tau$ and defined by: $F|_\tau = \top$ if σ satisfies F, otherwise $F|_\tau = \{C|_\tau \mid C \in F \text{ and } C|_\tau \neq \top\}$. *Unit propagation* refers to the iterated application of the *unit clause rule*, replacing F by $F|_{\{l\}}$ for each unit clause $(l) \in F$, until there are no unit clauses left.

We write $F \vDash G$ to indicate that every assignment satisfying F, and which is total for G, satisfies G as well. Further, we write $F \vdash_1 G$ to mean F implies G *by unit propagation*: for every $D \in G$, unit propagation on $\neg D \wedge F$ produces $\bot$.

A clause C is *redundant* with respect to a formula F if the formulas $F \setminus \{C\}$ and $F \cup \{C\}$ are satisfiability-equivalent: both satisfiable, or both unsatisfiable [23]. The following theorem provides a characterization of clause redundancy based on logical implication.

Theorem 1 (Heule, Kiesel, and Biere [17]**).** *A non-empty clause C is redundant with respect to a formula F (with $C \notin F$) if and only if there is a partial assignment ω such that ω satisfies C, and $F|_\alpha \vDash F|_\omega$, where $\alpha = \neg C$.*

As a result, redundancy can be shown by providing a *witnessing assignment* ω (or *witness*) and demonstrating that $F|_\alpha \vDash F|_\omega$. When the logical implication relation "$\vDash$" is replaced with "$\vdash_1$," the result is the definition of a *propagation redundant* or PR *clause*, and ω is called a PR *witness* [15]. Determining whether a clause is PR with respect to a formula is NP-complete [18], but since it can be decided in polynomial time whether $F \vdash_1 G$ for arbitrary formulas F and G, it can be efficiently decided whether a given assignment is a PR witness.

A *clause elimination procedure* iteratively identifies and removes clauses satisfying a particular redundancy property from a formula, until no such clauses remain. A simple example is *subsumption elimination*, which removes any clauses $C \in F$ that are *subsumed* by another clause $D \in F$; that is, $D \subseteq C$. Subsumption elimination is *model-preserving*, as it only removes clauses C such that any assignment satisfying $F \setminus \{C\}$ also satisfies $F \cup \{C\}$.

Some clause elimination procedures are not model-preserving. *Blocked clause elimination* [20,25] iteratively removes from a formula F any clauses C satisfying the following property: C is *blocked* by a literal $l \in C$ if for every clause $D \in F$ containing $\neg l$, there is some other literal $k \in C$ with $\neg k \in D$. For a blocked clause C, there may be assignments satisfying $F \setminus \{C\}$ which falsify $F \cup \{C\}$. However, blocked clauses are redundant, so if $F \cup \{C\}$ is unsatisfiable, then so is $F \setminus \{C\}$, thus blocked clause elimination is still *satisfiability-preserving.*

Clause elimination procedures which are not model-preserving must provide a way to *reconstruct* solutions to the original formula out of solutions to the reduced formula. Witnesses provide a convenient framework for reconstruction: if C is redundant with respect to F, and τ is a total or partial assignment satisfying F but not C, then $\tau \circ \omega$ satisfies $F \cup \{C\}$, for any witness ω for C with respect to F [7,17]. For reconstructing solutions after removing multiple clauses, a sequence σ of *witness-labeled clauses* $(\omega : C)$, called a *reconstruction stack*, can be maintained and used as follows [7,21].

Definition 1. *Given a sequence σ of witness-labeled clauses, the* reconstruction function *(w.r.t. σ) is defined recursively as follows, for an assignment τ:*[1]

$$\mathcal{R}_\epsilon(\tau) = \tau, \qquad \mathcal{R}_{\sigma\cdot(\omega:D)}(\tau) = \begin{cases} \mathcal{R}_\sigma(\tau) & \textit{if } \tau(D) = \top \\ \mathcal{R}_\sigma(\tau \circ \omega) & \textit{otherwise.} \end{cases}$$

For a set of clauses S, a sequence σ of witness-labeled clauses satisfies the *reconstruction property* for S, or is a *reconstruction sequence* for S, with respect to a formula F if $\mathcal{R}_\sigma(\tau)$ satisfies $F \cup S$ for any assignment τ satisfying $F \setminus S$. As long as a witness is recorded for each clause C removed by a non-model-preserving procedure, even combinations of different clause elimination procedures can be used to simplify the same formula. Specifically, $\sigma = (\omega_1 : C_1) \cdots (\omega_n : C_n)$ is a reconstruction sequence for $\{C_1, \ldots, C_n\} \subseteq F$ if ω_i is a witness for C_i with respect to $F \setminus \{C_1, \ldots, C_i\}$, for all $1 \leq i \leq n$ [7].

The following lemma results from the fact that the reconstruction function satisfies $\mathcal{R}_{\sigma\cdot\sigma'}(\tau) = \mathcal{R}_\sigma(\mathcal{R}_{\sigma'}(\tau))$, for any sequences σ, σ' and assignment τ [7].

Lemma 1. *If σ is a reconstruction sequence for a set of clauses S with respect to $F \cup \{C\}$, and σ' is a reconstruction sequence for $\{C\}$ with respect to F, then $\sigma \cdot \sigma'$ is a reconstruction sequence for S with respect to F.*

[1] This improved variant over [7] is due to Christoph Scholl (3rd author of [7]).

3 Covered Clause Elimination

This section reviews covered clause elimination (CCE) and its asymmetric variant (ACCE), introduced by Heule, Järvisalo, and Biere [13], and presents an explicit algorithm implementing the more general ACCE procedure. The definitions as given here differ slightly from the original work, but are generally equivalent. A proof of correctness for the algorithm and its reconstruction sequence are given.

CCE is a clause elimination procedure which iteratively extends a clause by the addition of so-called "covered" literals. If at some point the extended clause becomes blocked, the original clause is redundant and can be eliminated. To make this precise, the set of *resolution candidates* in F of C upon l, written $\mathrm{RC}(F, C, l)$, is defined as the collection of clauses in F with which C has a non-tautological resolvent upon l (where "$\otimes_l$" denotes resolution):

$$\mathrm{RC}(F, C, l) = \{C' \vee \neg l \in F \mid C' \vee \neg l \otimes_l C \not\equiv \top\}.$$

The *resolution intersection* in F of C upon l, written $\mathrm{RI}(F, C, l)$, consists of those literals occurring in each of the resolution candidates, apart from $\neg l$:

$$\mathrm{RI}(F, C, l) = \left(\bigcap \mathrm{RC}(F, C, l)\right) \setminus \{\neg l\}.$$

If $\mathrm{RI}(F, C, l) \neq \emptyset$, its literals are covered by l and can be used to extend C.

Definition 2. *A literal k is* covered *by $l \in C$ with respect to F if $k \in \mathrm{RI}(F, C, l)$. A literal is covered by C if it is covered by some literal in C.*

Covered literals can be added to a clause in a satisfiability-preserving manner, meaning that if the extended clause $C \cup \mathrm{RI}(F, C, l)$ is added to F, then C is redundant. In fact, C is a PR clause.

Proposition 1. *C is* PR *with respect to $F' = F \wedge (C \cup \mathrm{RI}(F, C, l))$ with witness $\omega = \alpha_l$, for $l \in C$ and $\alpha = \neg C$.*

Proof. Consider a clause $D|_\omega \in F'|_\omega$, for some $D \in F'$. We prove that ω is a PR witness by showing that $F'|_\alpha$ implies $D|_\omega$ by unit propagation. First, we know $l \notin D$, since otherwise $D|_\omega = \top$ would vanish in $F|_\omega$. If also $\neg l \notin D$, this means $D|_\omega = D|_\alpha$, and therefore $F'|_\alpha \vdash_1 D|_\omega$. Now, suppose $\neg l \in D$. Notice that D contains no other literal k such that $\neg k \in C$, since otherwise $D|_\omega = \top$ here as well. As a result $D \in \mathrm{RC}(F, C, l)$, so $\mathrm{RI}(F, C, l) \subset D$ and $\mathrm{RI}(F, C, l) \setminus C \subseteq D|_\omega$. Notice $\mathrm{RI}(F, C, l) \setminus C = (C \cup \mathrm{RI}(F, C, l))|_\alpha \in F'|_\alpha$, therefore $F'|_\alpha \vdash_1 D|_\omega$. □

Consequently, C is redundant with respect to $F \cup \{C'\}$ for any $C' \supseteq C$ constructed by iteratively adding covered literals to C. In other words, F and $(F \setminus \{C\}) \cup \{C'\}$ are satisfiability-equivalent, so that C could be replaced by C' in F without affecting the satisfiability of the formula. Thus if some such extension C' would be blocked in F, that C' would be redundant, and therefore C is redundant itself. CCE identifies and removes such clauses from F.

Definition 3. *A clause C is* covered *in F if an extension of C by iteratively adding covered literals is blocked.*

CCE refers to the following procedure: while some clause C in F is covered, remove C (that is, replace F with $F \setminus \{C\}$).

ACCE strengthens this procedure by extending clauses using a combination of covered literals and asymmetric literals. A literal k is *asymmetric* to C with respect to F if there is a clause $C' \vee \neg k \in F$ such that $C' \subseteq C$. The addition of asymmetric literals to a clause is model-preserving, so that the formulas F and $(F \setminus \{C\}) \cup \{C \vee k\}$ are equivalent, for any k which is asymmetric to C [12].

Definition 4. *A clause $C' \supseteq C$ is an* ACC extension *of C with respect to F if C' can be constructed from C by the iterative addition of covered and asymmetric literals. If some* ACC *extension of C is blocked or subsumed in F, then C is an* asymmetric covered clause (ACC).

ACCE performs the following: while some C in F is an ACC, remove C from F. Solvers aiming to eliminate covered clauses more often implement ACCE than plain CCE, since asymmetric literals can easily be found by unit propagation, and ACCE is more powerful than CCE, eliminating more clauses [12,13].

The procedure $\mathsf{ACC}(F, C)$ in Fig. 1 provides an algorithm identifying whether a clause C is an ACC with respect to a formula F. This procedure differs in some ways, and includes optimizations over the original procedure as implicitly given by the definition of ACCE. Notably, two extensions of the original clause C are maintained: E consists of C and any added covered literals, while α tracks C and all added literals, both covered and asymmetric. The literals in α are kept negated, so that $E \subseteq \neg\alpha$, and the clause represented by $\neg\alpha$ is the ACC extension of the original clause C being computed.

The E and α extensions are maintained separately for two purposes. First, the covered literal addition loop (lines 9–16) needs to iterate only over those literals in E, and can ignore those in $(\neg\alpha) \setminus E$, as argued below.

Lemma 2. *If k is covered by $l \in (\neg\alpha) \setminus E$, then $k \in \neg\alpha$ already.*

Proof. If l belongs to $\neg\alpha$ but not to E, then there is some clause $D \vee \neg l$ in F such that $D \subseteq \neg\alpha$. But then $D \vee \neg l$ occurs in $\mathrm{RC}(F, \neg\alpha, l)$, and consequently $\mathrm{RI}(F, \neg\alpha, l) \subseteq D \subseteq \neg\alpha$. Thus $k \in \mathrm{RI}(F, \neg\alpha, l)$ implies $k \in \neg\alpha$. □

Notice that the computation of the literals covered by $l \in E$ also prevents any of these literals already in $\neg\alpha$ from being added again.

The second reason for separating E and α is as follows. When a covered literal is found, or when the extended clause is blocked, the algorithm appends a new witness-labeled clause to the reconstruction sequence σ (lines 11 and 14). Instead of $(\neg\alpha_l : \alpha)$, the procedure adds the shorter witness-labeled clause $(\neg E_l : E)$. The proof of statement (3) in Lemma 3 below shows that this is sufficient.

Certain details are omitted, especially concerning the addition of asymmetric literals (lines 6–7), but notice that it is never necessary to recompute $F|_\alpha$ entirely. Instead the assignment falsifying each u newly added to α can simply be applied

```
ACC(F, C)
1   σ := ε
2   E := C
3   α := ¬C
4   repeat
5       if ⊥ ∈ F|α then return (true, σ)
6       if there are unit clauses in F|α then
7           α := α ∪ {u} for each unit u
8       else
9           for each l ∈ E
10              G := {D|α | (D ∨ ¬l) ∈ F and D|α ≠ ⊤}
11              if G = ∅ then return (true, σ · (¬E_l : E))
12              Φ := ⋂G
13              if Φ ≠ ∅ then
14                  σ := σ · (¬E_l : E)
15                  E := E ∪ Φ
16                  α := α ∪ ¬Φ
17  until no updates to α
18  return (false, ε)
```

Fig. 1. Asymmetric Covered Clause (ACC) Identification. The procedure $\mathsf{ACC}(F, C)$ maintains a sequence σ of witness-labeled clauses, and two sets of literals E and α. The main loop iteratively searches for literals which could be used to extend C and adds their negations to α, so that the clause represented by $\neg\alpha$ is an ACC extension of C. The set E records only those which could be added as covered literals. If C is an ACC, then $\mathsf{ACC}(F, C)$ returns $(\mathbf{true}, \sigma)$: in line 5 if the extension $\neg\alpha$ becomes subsumed in F, or in line 11 if it becomes blocked. In either case, the witness-labeled clauses in σ form a reconstruction sequence for the clause C. Note that lines 5–7 implement Boolean constraint propagation (over the partial assignment α) and can make use of efficient watched clause data structures, while line 10 has to collect all clauses containing $\neg l$, which are still unsatisfied by α, and thus requires full occurrence lists.

to the existing $F|_\alpha$. In contrast, the **for each** loop (lines 9–16) should re-iterate over the entirety of E each time, as added literals may allow new coverings:

Example 1. Let $C = a \vee b \vee c$ and

$$F = (\neg a \vee \neg x_1) \wedge (\neg a \vee x_2) \wedge (\neg b \vee \neg x_1) \wedge (\neg b \vee \neg x_2) \wedge (\neg c \vee x_1)$$

Initially, neither a nor b cover any literals, but c covers x_1, so it can be added to the clause. After extending, a in $C \vee x_1$ covers x_2, and b blocks $C \vee x_1 \vee x_2$.

The following lemma supplies invariants for arguing about $\mathsf{ACC}(F, C)$.

Lemma 3. *After each update to α, for the clauses represented by $\neg\alpha$ and E:*

(1) *$\neg\alpha$ is an* ACC *extension of C,*
(2) *$F \cup \{\neg\alpha\} \models F \cup \{E\}$, and*
(3) *σ is a reconstruction sequence for $\{C\}$ with respect to $F \cup \{\neg\alpha\}$.*

Proof. Let α_i, σ_i, and E_i refer to the values of α, σ, and E, respectively, after $i \geq 0$ updates to α (so that $\alpha_i \subsetneq \alpha_{i+1}$ for each i, but possibly $\sigma_i = \sigma_{i+1}$ and $E_i = E_{i+1}$). Initially, (1) and (2) hold as $E = \neg\alpha_0 = C$. Further, $\sigma_0 = \epsilon$ is a reconstruction sequence for $\{C\}$ with respect to $F \cup \{C\}$, so (3) holds as well. Assuming these claims hold after update i, we show that they hold after $i + 1$.

First suppose update $i + 1$ is the result of executing line 7.

(1) $\alpha_{i+1} = \alpha_i \cup U$, where $u \in U$ implies (u) is a unit clause in $F|_{\alpha_i}$. Then $\neg\alpha_{i+1}$ is the extension of $\neg\alpha_i$ by the addition of asymmetric literals $\neg U$. Assuming $\neg\alpha_i$ is an ACC extension of C, then so is $\neg\alpha_{i+1}$.
(2) Asymmetric literal addition is model-preserving, so $F \cup \{\neg\alpha_{i+1}\} \models F \cup \{\neg\alpha_i\}$. Since E was not updated, $E_{i+1} = E_i$. Assuming $F \cup \{\neg\alpha_i\} \models F \cup \{E_i\}$, we get $F \cup \{\neg\alpha_{i+1}\} \models F \cup \{E_{i+1}\}$.
(3) Again, asymmetric literal addition is model-preserving. Assuming σ_i is a reconstruction sequence for $\{C\}$ with respect to $F \cup \{\alpha_i\}$, then Lemma 1 implies $\sigma_{i+1} = \sigma_i \cdot \varepsilon = \sigma_i$ reconstructs $\{C\}$ with respect to $F \cup \{\neg\alpha_{i+1}\}$.

Now, suppose instead update $i + 1$ is executed in line 16.

(1) $\alpha_{i+1} = \alpha_i \cup \Phi$, for some set of literals $\Phi \neq \emptyset$ constructed for $l \in E \subseteq \neg\alpha$. Notice for $k \in \Phi$ that $k \in \mathrm{RI}(F, \neg\alpha, l)$, so k is covered by $\neg\alpha$. Thus assuming $\neg\alpha_i$ is a ACC extension of C, then $\neg\alpha_{i+1}$ is as well.
(2) Consider an assignment τ satisfying $F \cup \{\neg\alpha_{i+1}\}$. If τ satisfies $\neg\alpha_i \subset \neg\alpha_{i+1}$ then τ satisfies $F \cup \{\neg\alpha_i\}$ and by assumption, $F \cup \{E_i\}$. Since $E_i \subset E_{i+1}$ in this case, τ satisfies $F \cup \{E_{i+1}\}$. If instead τ satisfies some literal in $\neg\alpha_{i+1} \setminus \neg\alpha_i$ then τ satisfies $\Phi \subseteq E_{i+1}$, so τ satisfies $F \cup \{E_{i+1}\}$. Thus $F \cup \{\neg\alpha_{i+1}\} \models F \cup \{E_{i+1}\}$ in this case as well.
(3) Proposition 1 implies $((\alpha_i)_l : \neg\alpha_i)$ is a reconstruction sequence for $\{\neg\alpha_i\}$ with respect to $F \cup \{\neg\alpha_{i+1}\}$. As $E_i \subseteq \neg\alpha_i$, and $F \cup \{\neg\alpha_i\} \models F \cup \{E_i\}$ by assumption, then any τ falsifies $\neg\alpha_i$ if and only if τ falsifies E_i. Since $l \in E_i$ as well, then $((\neg E_i)_l : E_i)$ is, in fact, also a reconstruction sequence for $\{\neg\alpha_i\}$ with respect to $F \cup \{\neg\alpha_{i+1}\}$. Finally, with the assumption σ_i is a reconstruction sequence for C with respect to $F \cup \{\neg\alpha_i\}$ and Lemma 1, then $\sigma_{i+1} = \sigma_i \cdot ((\neg E_i)_l : E_i)$ is a reconstruction sequence for $\{C\}$ in $F \cup \{\neg\alpha_{i+1}\}$.

Thus both updates maintain invariants (1)–(3). □

With the help of this lemma we can now show the correctness of $\mathsf{ACC}(F, C)$:

Theorem 2. *For a formula F and a clause C, the procedure $\mathsf{ACC}(F, C)$ returns $(\textbf{true}, \sigma)$ if and only if C is an* ACC *with respect to F. Further, if $\mathsf{ACC}(F, C)$ returns $(\textbf{true}, \sigma)$, then σ is a reconstruction sequence for $\{C\}$ with respect to F.*

Proof. ($\Rightarrow$) Suppose (**true**, σ) is returned in line 5. Then $\bot \in F|_\alpha$, so there is some $D \in F$ such that $D \subseteq \neg\alpha$; that is, $\neg\alpha$ is subsumed by D. By Lemma 3 then an ACC extension of C is subsumed in F, so C is an ACC with respect to F. Further, subsumption elimination is model-preserving, so that Lemmas 1 and 3 imply σ is a reconstruction sequence for C with respect to F.

Suppose now that (**true**, σ) is returned in line 11. Then for α and some $l \in E$, all clauses in F with $\neg l$ are satisfied by α. Since $E \subseteq \neg\alpha$, then $\neg\alpha$ is blocked by l. By Lemma 3 then C is an ACC with respect to F. Now, α_l is a witness for $\neg\alpha$ with respect to F, and $(\alpha_l : \neg\alpha)$ is a reconstruction sequence for $\{\neg\alpha\}$ in F. Further, $E \subseteq \neg\alpha$, and Lemma 3 gives $F \cup \{\neg\alpha\} \vDash F \cup \{E\}$, therefore $((\neg E_i)_l : E_i)$ is a reconstruction sequence for $\{\neg\alpha\}$ in F as well. Then Lemma 1 implies $\sigma \cdot (\neg E_l : E)$ is a reconstruction sequence for C with respect to F.

($\Leftarrow$) Suppose C is an ACC; that is, some $C' = C \vee k_1 \vee \cdots \vee k_n$ is blocked or subsumed in F, where k_1 is an asymmetric or covered literal for C, and k_i is an asymmetric or covered literal for $C \vee k_1 \vee \cdots \vee k_{i-1}$ for $i > 1$. Towards a contradiction, assume $\mathsf{ACC}(F, C)$ returns (**false**, ε). Then for the final value of α, the clause represented by $\neg\alpha$ is not blocked nor subsumed in F, and hence, $C' \nsubseteq \neg\alpha$. As $C \subseteq \neg\alpha$, there must be some values of i such that $\neg k_i \notin \alpha$.

Let m refer to the least such i; that is, $\neg k_m \notin \alpha$, but $\neg k_i \in \alpha$ for all $1 \leq i < m$. Thus k_m is asymmetric, or covered by, $C_{m-1} = C \vee k_1 \vee \cdots \vee k_{m-1}$.

If k_m is asymmetric to C_{m-1}, there is some clause $D \vee \neg k_m$ in F such that $D \subseteq C_{m-1}$. By assumption, $\neg k_m \notin \alpha$ but $\neg C_{m-1} \subseteq \alpha$. Further, $k_m \notin \alpha$, as otherwise $(D \vee \neg k_m)|_\alpha = \bot$ and $\mathsf{ACC}(F, C)$ would have returned **true**. But $(D \vee \neg k_m)|_\alpha = \neg k_m$ would be a unit in $F|_\alpha$ and added to α by line 7.

If instead k_m is covered by C_{m-1}, then $k_m \in \mathrm{RI}(F, C_{m-1}, l)$ for some literal $l \in C_{m-1} \subseteq \neg\alpha$. In fact $l \in E$, by Lemma 2. During the lth iteration of the **for each** loop, then $k_m \in \Phi$, and $\neg k_m$ would be added to α by line 16. □

$\mathsf{ACC}(F, C)$ produces, for any asymmetric covered clause C in F, a reconstruction sequence σ for C with respect to F. This allows ACCE to be used during preprocessing or inprocessing like other clause elimination procedures, appending this σ to the solver's main reconstruction stack whenever an ACC is removed. However, the algorithm does not produce redundancy witnesses for the clauses it removes. Instead, σ consists of possibly many witness-labeled clauses, starting with the redundant clause C, and reconstructs solutions for C in multiple steps.

In contrast, most clause elimination procedures produce a single witness-labeled clause $(\omega : C)$ for each removed clause C. In practice, only the part of ω which differs from $\neg C$ must be recorded; for most procedures this difference includes only literals in C, so that reconstruction for $\{C\}$ needs only linear space in the size of C. In contrast, the size of σ produced by $\mathsf{ACC}(F, C)$ to reconstruct $\{C\}$ can be quadratic in the length of the extended clause.

Example 2. Consider $C = x_0$ and

$$F_n = (\neg x_{n-2} \vee x_{n-1} \vee x_n) \wedge (\neg x_{n-1} \vee \neg x_n) \wedge \bigwedge_{i=1}^{n-2} (\neg x_{i-1} \vee x_i).$$

The extended clause $\neg\alpha = x_0 \vee x_1 \vee \cdots \vee x_n$ is blocked in F_n by x_{n-1}. Then $\mathsf{ACC}(F_n, C)$ returns the pair with **true** and the reconstruction sequence[2]

$$\sigma = (x_0 \wr x_0) \cdot (x_1 \wr x_0 \vee x_1) \cdots (x_{n-2} \wr x_0 \vee x_1 \vee \cdots \vee x_{n-2}) \cdot (x_{n-1} \wr x_0 \vee x_1 \vee \cdots \vee x_n).$$

The extended clause includes n literals, and the size of σ is $O(n^2)$.

4 Witnesses for Covered Clauses

In this section, we consider the specific problem of finding witnesses for (asymmetric) covered clauses. As these clauses are redundant, such witnesses are guaranteed to exist by Theorem 1, though they are not produced by $\mathsf{ACC}(F, C)$. More precisely, we are interested in the *witness problem* for covered clauses.

Definition 5. *The* witness problem *for a redundancy property P is as follows: given a formula F and a clause C, if P is met by C with respect to F then return a witness for C, or decide that P is not met by C.*

For instance, the witness problem for blocked clauses is solved as follows: test each $l \in C$ to see if l blocks C in F. As soon as a blocking literal l is found then α_l is a witness for C, where $\alpha = \neg C$. If no blocking literal is found, then C is not blocked. For blocked clauses, this polynomial procedure decides whether C is blocked or not and also determines a witness $\omega = \alpha_l$ for C.

Solving the witness problem for covered clauses is not as straightforward, as it is not clear how a witness could be produced when deciding a clause is covered, or from a sequence σ constructed by $\mathsf{ACC}(F, C)$. The following theorem shows that this problem is as difficult as producing a satisfying assignment for an arbitrary formula, if one exists. In particular, we present a polynomial time reduction from the search analog of the SAT problem: given a formula F, return a satisfying assignment of F, or decide that F is unsatisfiable.

Specifically, given a formula G, we construct a pair (F, C) as an instance to the witness problem for covered clauses. In this construction, C is covered in F and has some witness ω. Moreover, any witness ω for this C necessarily provides a satisfying assignment to G, if there is one.

Proposition 2. *Given a formula $G = D_1 \wedge \cdots \wedge D_n$, let $G' = D'_1 \wedge \cdots \wedge D'_n$ refer to a variable-renamed copy of G, containing v' everywhere G contains v, so that $var(G) \cap var(G') = \emptyset$. Further, let $C = k \vee l$ and construct the formula:*

$$\begin{aligned} F = \;& (x \vee \neg k) \wedge (\neg x \vee \neg y) \wedge (y \vee \neg l) \;\wedge \\ & (x \vee D_1) \wedge \quad \cdots \quad \wedge (x \vee D_n) \wedge \\ & (y \vee D'_1) \wedge \quad \cdots \quad \wedge (y \vee D'_n) \end{aligned}$$

for variables $x, y, k, l \notin var(G) \cup var(G)'$. Finally, let ω be a witness for C with respect to F. Either ω satisfies at least one of G or G', or G is unsatisfiable.

[2] In order to simplify the presentation, only the part of the witness differing from the negated clause is written, so that $(l \wr C)$ actually stands for $(\neg C_l : C)$. The former is in essence the original notation used in [21], while set, super or globally blocked, as well as PR clauses [15,22,23] require the more general one used in this paper.

Proof. First notice for C that x is covered by k and y is covered by l, so that the extension $(k \vee l \vee x \vee y)$ is blocked in F (with blocking literal x or y). Thus C is redundant in F, so a witness ω exists.

We show that ω satisfies G or G' if and only if G is satisfiable.

($\Rightarrow$) If ω satisfies G then surely G is satisfiable. If ω satisfies G' but not G then the assignment $\omega_G = \{x \in var(G) \mid x' \in G' \text{ and } x \in \omega\}$ satisfies G.

($\Leftarrow$) Assume G is satisfiable, and without loss of generality[3] further assume $\omega = \{k\} \circ \omega'$ for some ω' not assigning $var(k)$. Then $F|_\alpha \models (F|_{\{k\}})|_{\omega'}$; that is,

$$F|_\alpha \models ((x) \wedge (\neg x \vee \neg y) \wedge (y \vee \neg l) \wedge (x \vee D_1) \wedge \cdots \wedge (y \vee D'_n))|_{\omega'}.$$

G is satisfiable, so there are models of $F|_\alpha$ in which $\neg x$ is true. However, x occurs as a unit clause in $F|_{\{k\}}$, so it must be the case that $x \in \omega'$. Therefore $\omega = \{k, x\} \circ \omega''$ for some ω'' assigning neither $var(k)$ nor $var(x)$ such that

$$F|_\alpha \models ((\neg y) \wedge (y \vee \neg l) \wedge (y \vee D'_1) \wedge \cdots \wedge (y \wedge D'_n))|_{\omega''}.$$

By similar reasoning, ω'' must assign y to false, so now $\omega = \{k, x, \neg y\} \circ \omega'''$ for some ω''', assigning none of $var(k)$, $var(x)$, or $var(y)$, such that

$$F|_\alpha \models ((\neg l) \wedge (D'_1) \wedge \cdots \wedge (D'_n))|_{\omega'''}.$$

Finally, consider any clause $D'_i \in G'$. We show that ω satisfies D'_i. As ω is a witness, $F|_\alpha \models F|_\omega$, so that $(D'_i)|_\omega$ is true in all models of $F|_\alpha$, including models which assign y to true. In particular, let τ be a model of G; then $(D'_i)|_\omega$ is satisfied by $\tau \cup \{\neg x, y\} \cup \nu$, for every assignment ν over $var(G')$. Because $var(D'_i) \subseteq var(G')$, then $(D'_i)|_\omega \equiv \top$. Therefore $G'|_\omega \equiv \top$. □

Proposition 2 suggests there is likely no polynomial procedure for computing witnesses for covered clauses. The existence of witnesses is the basis for solution reconstruction, but witnesses which cannot be efficiently computed make the use of non-model-preserving clause elimination procedures more challenging; that is, we are not aware of any polynomial algorithm for generating a compact (sub-quadratic) reconstruction sequence (see also Example 2).

As PR clauses are defined by witnesses, procedures deciding PR generally solve the witness problem for PR. For example, the PR reduct [16] provides a formula whose satisfying assignments encode PR witnesses, if they exist. However, this does not produce witnesses for covered clauses, which are not encompassed by PR. In other words, although any clause extended by a single covered literal addition is a PR clause by Proposition 1, this is not true for covered clauses.

Theorem 3. *Covered clauses are not all propagation redundant.*

[3] If $k \notin \omega$, then $\omega = \{l\} \circ \omega'$ for some ω' not assigning $var(l)$ and the argument is symmetric, ending with $G|_\omega = \top$. Note that by definition ω satisfies C thus k or l.

Proof. By counterexample. Consider the clause $C = k \vee l$ and the formula

$$\begin{aligned} F = {} & (x \vee \neg k) \wedge (\neg x \vee \neg y) \wedge (y \vee \neg l) \wedge \\ & (x \vee a \vee b) \wedge (x \vee a \vee \neg b) \wedge (x \vee \neg a \vee b) \wedge (x \vee \neg a \vee \neg b) \wedge \\ & (y \vee c \vee d) \wedge (y \vee c \vee \neg d) \wedge (y \vee \neg c \vee d) \wedge (y \vee \neg c \vee \neg d). \end{aligned}$$

The extension $C \vee x \vee y$ is blocked with respect to F, so C is covered. However, C is not PR with respect to F. To see this, suppose to the contrary that ω is a PR witness for C. Similar to the reasoning in the proof of Theorem 2, assume, without loss of generality, that $\omega = \{k\} \circ \omega'$ for some ω' not assigning k. Notice that $(x) \in F|_k$, but unit propagation on $\neg x \wedge F|_\alpha$ stops without producing $\bot$. Therefore $x \in \omega'$, and $\omega = \{k, x\} \circ \omega''$ for some ω'' assigning neither k nor x. By similar reasoning, it must be the case that $\neg y \in \omega''$, so that $\omega = \{k, x, \neg y\} \circ \omega'''$. Now, $(c \vee d) \in F|_{\{k,x,\neg y\}}$, but once more, unit propagation on $F|_\alpha \wedge \neg c \wedge \neg d$ does not produce $\bot$, so either c or d belongs to ω'''. Without loss of generality, assume $c \in \omega'''$ so that $\omega = \{k, x, \neg y, c\} \circ \omega''''$. Finally, both d and $\neg d$ are clauses in $F'|_{\{k,x,\neg y,c\}}$, but neither are implied by $F|_\alpha$ by unit propagation. However, if either d or $\neg d$ belongs to ω, then $\bot \in F|_\omega$. As unit propagation on $F|_\alpha$ alone does not produce $\bot$, this is a contradiction. □

Notice the formula in Theorem 3 can be seen as an instance of the formula in Proposition 2, with G as $(a \vee b) \wedge (a \vee \neg b) \wedge (\neg a \vee b) \wedge (\neg a \vee \neg b)$. In fact, as long as unit propagation on G does not derive $\bot$, then G could be any, arbitrarily hard, unsatisfiable formula (such as an instance of the pigeonhole principle).

5 Complexity of Redundancy

In the previous section we introduced the witness problem for a redundancy property (Definition 5) and showed that it is not trivial, even when the redundancy property itself can be efficiently decided. Further, the witness problem for PR clauses is solvable by encoding it into SAT [16].

Note that PR is considered to be a very general redundancy property. The proof of theorem 1 in [15,17] shows that if F is satisfiable and C redundant, then $F \wedge C$ is satisfiable by definition. In addition, any satisfying assignment τ of $F \wedge C$ is a PR witness for C with respect to F. This yields the following:

Proposition 3. *Let F be a satisfiable formula. A clause C is redundant with respect to F if and only if it is a* PR *clause with respect to F.*

While not all covered clauses are PR, this motivates the question of whether witnesses for all redundancy properties can be encoded as an instance to SAT, and solved similarly. In this section we show that this is likely not the case by demonstrating the complexity of the *redundancy problem*: given a clause C and a formula F, is C redundant with respect to F?

Deciding whether a clause is PR belongs to NP: assignments can be chosen non-deterministically and efficiently verified as PR witnesses, since the relation $\vdash_1$ is polynomially decidable [18]. For clause redundancy in general, it is not clear that this holds, as the corresponding problem is co-NP-complete.

Proposition 4. *Deciding whether an assignment ω is a witness for a clause C with respect to a formula F is complete for* co-NP.

Proof. The problem belongs to co-NP since $F|_\alpha \vDash F|_\omega$ whenever $\neg(F|_\alpha) \vee F|_\omega$ is a tautology. In the following we show a reduction from the tautology problem. Given a formula F, construct the formula F' as below, for $x \notin var(F)$. Further, let $C' = x$, so that $\alpha = \neg x$, and let also $\omega = x$.

$$F' = \bigwedge_{C \in F} (C \vee \neg x)$$

Then $F'|_\alpha = \top$ and $F'|_\omega = F$. Therefore $F'|_\alpha \vDash F'|_\omega$ if and only if $\top \vDash F$. □

Theorem 4 below shows that the *irredundancy problem*, the complement of the redundancy problem, is complete for the class $\mathrm{D^P}$, the class of languages that are the intersection of a language in NP and a language in co-NP [28]:

$$\mathrm{D^P} = \{L_1 \cap L_2 \mid L_1 \in \mathrm{NP} \text{ and } L_2 \in \text{co-NP}\}.$$

This class was originally introduced to classify certain problems which are hard for both NP and co-NP, but do not seem to be complete for either, and it characterizes a variety of optimization problems. It is the second level of the Boolean hierarchy over NP, which is the completion of NP under Boolean operations [5,29]. We provide a reduction from the canonical $\mathrm{D^P}$-complete problem, SAT-UNSAT: given formulas F and G, is F satisfiable and G unsatisfiable?

Theorem 4. *The irredundancy problem is* $\mathrm{D^P}$*-complete.*

Proof. Notice that the irredundancy problem can be expressed as

$$\begin{aligned} \mathrm{IRR} &= \{(F, C) \mid F \text{ is satisfiable, and } F \wedge C \text{ is unsatisfiable}\} \\ &= \{(F, C) \mid F \in \mathrm{SAT}\} \cap \{(F, C) \mid F \wedge C \in \mathrm{UNSAT}\}. \end{aligned}$$

That is, IRR is the intersection of a language in NP and a language in co-NP, and so the irredundancy problem belongs to $\mathrm{D^P}$.

Now, let (F, G) be an instance to SAT-UNSAT. Construct the formula F' as follows, for $x \notin var(F) \cup var(G)$:

$$F' = \bigwedge_{C \in F} (C \vee x) \wedge \bigwedge_{D \in G} (D \vee \neg x).$$

Further, let $C' = x$. We demonstrate that $(F, G) \in$ SAT-UNSAT if and only if C' is irredundant with respect to F'.

($\Leftarrow$) Suppose C' is irredundant with respect to F'. In other words, F' is satisfiable but $F' \wedge C'$ is unsatisfiable. Since $F' \wedge C'$ is unsatisfiable, it must be the case that $F'|_{\{x\}}$ is unsatisfiable; however, F' is satisfiable, therefore $F'|_{\{\neg x\}}$ must be satisfiable. Since $F'|_{\{\neg x\}} = F$ and $F'|_{\{x\}} = G$, then $(F, G) \in$ SAT-UNSAT.

($\Rightarrow$) Now, suppose F is satisfiable and G is unsatisfiable. Then some assignment τ over $var(F)$ satisfies F. As a result, $\tau \cup \{\neg x\}$ satisfies F'. Because G

is unsatisfiable, there is no assignment satisfying $F'|_{\{x\}} = G$. This means there is no σ satisfying both F' and $C' = x$, and so $F' \wedge C'$ is unsatisfiable as well. Therefore C' is irredundant with respect to F'. □

Consequently the redundancy problem is complete for co-D^P. This suggests that sufficient SAT encodings of the clause redundancy problem, and its corresponding witness problem, are not possible.

6 Conclusion

We revisit a strong clause elimination procedure, covered clause elimination, and provide an explicit algorithm for both deciding its redundancy property and reconstructing solutions after its use. Covered clause elimination is unique in that it does not produce redundancy witnesses for clauses it eliminates, and uses a complex, multi-step reconstruction strategy. We prove that while witnesses exist for covered clauses, computing such a witness is as hard as finding a satisfying assignment for an arbitrary formula.

For PR, a very general redundancy property used by strong proof systems, witnesses can be found through encodings into SAT. We show that covered clauses are not described by PR, and SAT encodings for finding general redundancy witnesses likely do not exist, as deciding clause redundancy is hard for the class D^P, the second level of the Boolean hierarchy over NP.

Directions for future work include the development of redundancy properties beyond PR, and investigating their use for solution reconstruction after clause elimination, as well as in proof systems. Extending redundancy notions by using a structure for witnesses other than partial assignments may provide more generality while remaining polynomially verifiable.

We are also interested in developing notions of redundancy for adding or removing more than a single clause at a time, and exploring proof systems and simplification techniques which make use of non-clausal redundancy properties.

References

1. Ajtai, M.: The complexity of the pigeonhole principle. Combinatorica **14**(4), 417–433 (1994)
2. Biere, A.: CaDiCaL, Lingeling, Plingeling, Treengeling and YalSAT entering the SAT competition 2018. In: Heule, M.J.H., Järvisalo, M., Suda, M. (eds.) Proceedings of SAT Competition 2018, pp. 13–14. Department of Computer Science Series of Publications B, University of Helsinki (2018)
3. Biere, A.: CaDiCaL at the SAT race 2019. In: Heule, M.J.H., Järvisalo, M., Suda, M. (eds.) Proceedings of SAT Race 2019, pp. 8–9. Department of Computer Science Series of Publications B, University of Helsinki (2019)
4. Buss, S., Thapen, N.: DRAT proofs, propagation redundancy, and extended resolution. In: Janota, M., Lynce, I. (eds.) SAT 2019. LNCS, vol. 11628, pp. 71–89. Springer, Cham (2019). https://doi.org/10.1007/978-3-030-24258-9_5

5. Cai, J.Y., Hemachandra, L.: The Boolean hierarchy: hardware over NP. In: Selman, A.L. (ed.) Structure in Complexity Theory. LNCS, vol. 223, pp. 105–124. Springer, Heidelberg (1986). https://doi.org/10.1007/3-540-16486-3_93
6. Clarke, E., Biere, A., Raimi, R., Zhu, Y.: Bounded model checking using satisfiability solving. Formal Methods Syst. Des. **19**(1), 7–34 (2001)
7. Fazekas, K., Biere, A., Scholl, C.: Incremental inprocessing in SAT solving. In: Janota, M., Lynce, I. (eds.) SAT 2019. LNCS, vol. 11628, pp. 136–154. Springer, Cham (2019). https://doi.org/10.1007/978-3-030-24258-9_9
8. Gableske, O.: BossLS preprocessing and stochastic local search. In: Balint, A., Belov, A., Diepold, D., Gerber, S., Järvisalo, M., Sinz, C. (eds.) Proceedings of SAT Challenge 2012, pp. 10–11. Department of Computer Science Series of Publications B, University of Helsinki (2012)
9. Haken, A.: The intractability of resolution. Theor. Comput. Sci. **39**(2–3), 297–308 (1985)
10. Heule, M.J.H.: Schur number five. In: Proceedings of the 32nd AAAI Conference on Artificial Intelligence (AAAI 2018), pp. 6598–6606 (2018)
11. Heule, M.J.H., Biere, A.: What a difference a variable makes. In: Beyer, D., Huisman, M. (eds.) TACAS 2018. LNCS, vol. 10806, pp. 75–92. Springer, Cham (2018). https://doi.org/10.1007/978-3-319-89963-3_5
12. Heule, M.J.H., Järvisalo, M., Biere, A.: Clause elimination procedures for CNF formulas. In: Fermüller, C.G., Voronkov, A. (eds.) LPAR 2010. LNCS, vol. 6397, pp. 357–371. Springer, Heidelberg (2010). https://doi.org/10.1007/978-3-642-16242-8_26
13. Heule, M.J.H., Järvisalo, M., Biere, A.: Covered clause elimination. In: Voronkov, A., Sutcliffe, G., Baaz, M., Fermüller, C. (eds.) Logic for Programming, Artificial Intelligence and Reasoning - LPAR 17 (short paper). EPiC Series in Computing, vol. 13, pp. 41–46. EasyChair (2010)
14. Heule, M.J.H., Järvisalo, M., Lonsing, F., Seidl, M., Biere, A.: Clause elimination for SAT and QSAT. J. Artif. Intell. Res. **53**(1), 127–168 (2015)
15. Heule, M.J.H., Kiesl, B., Biere, A.: Short proofs without new variables. In: de Moura, L. (ed.) CADE 2017. LNCS (LNAI), vol. 10395, pp. 130–147. Springer, Cham (2017). https://doi.org/10.1007/978-3-319-63046-5_9
16. Heule, M.J.H., Kiesl, B., Biere, A.: Encoding redundancy for satisfaction-driven clause learning. In: Vojnar, T., Zhang, L. (eds.) TACAS 2019. LNCS, vol. 11427, pp. 41–58. Springer, Cham (2019). https://doi.org/10.1007/978-3-030-17462-0_3
17. Heule, M.J.H., Kiesl, B., Biere, A.: Strong extension-free proof systems. J. Autom. Reason. **64**, 533–554 (2020)
18. Heule, M.J.H., Kiesl, B., Seidl, M., Biere, A.: PRuning through satisfaction. In: Strichman, O., Tzoref-Brill, R. (eds.) HVC 2017. LNCS, vol. 10629, pp. 179–194. Springer, Cham (2017). https://doi.org/10.1007/978-3-319-70389-3_12
19. Heule, M.J.H., Kullmann, O., Marek, V.W.: Solving and verifying the Boolean Pythagorean triples problem via Cube-and-Conquer. In: Creignou, N., Le Berre, D. (eds.) SAT 2016. LNCS, vol. 9710, pp. 228–245. Springer, Cham (2016). https://doi.org/10.1007/978-3-319-40970-2_15
20. Järvisalo, M., Biere, A., Heule, M.J.H.: Blocked clause elimination. In: Esparza, J., Majumdar, R. (eds.) TACAS 2010. LNCS, vol. 6015, pp. 129–144. Springer, Heidelberg (2010). https://doi.org/10.1007/978-3-642-12002-2_10
21. Järvisalo, M., Heule, M.J.H., Biere, A.: Inprocessing rules. In: Gramlich, B., Miller, D., Sattler, U. (eds.) IJCAR 2012. LNCS (LNAI), vol. 7364, pp. 355–370. Springer, Heidelberg (2012). https://doi.org/10.1007/978-3-642-31365-3_28

22. Kiesl, B., Heule, M.J.H., Biere, A.: Truth assignments as conditional autarkies. In: Chen, Y.-F., Cheng, C.-H., Esparza, J. (eds.) ATVA 2019. LNCS, vol. 11781, pp. 48–64. Springer, Cham (2019). https://doi.org/10.1007/978-3-030-31784-3_3
23. Kiesl, B., Seidl, M., Tompits, H., Biere, A.: Super-blocked clauses. In: Olivetti, N., Tiwari, A. (eds.) IJCAR 2016. LNCS (LNAI), vol. 9706, pp. 45–61. Springer, Cham (2016). https://doi.org/10.1007/978-3-319-40229-1_5
24. Konev, B., Lisitsa, A.: Computer-aided proof of Erdős discrepancy properties. Artif. Intell. **224**, 103–118 (2015)
25. Kullmann, O.: On a generalization of extended resolution. Discret. Appl. Math. **96**(97), 149–176 (1999)
26. Manthey, N.: Coprocessor 2.0 – a flexible CNF simplifier. In: Cimatti, A., Sebastiani, R. (eds.) SAT 2012. LNCS, vol. 7317, pp. 436–441. Springer, Heidelberg (2012). https://doi.org/10.1007/978-3-642-31612-8_34
27. Mironov, I., Zhang, L.: Applications of SAT solvers to cryptanalysis of hash functions. In: Biere, A., Gomes, C.P. (eds.) SAT 2006. LNCS, vol. 4121, pp. 102–115. Springer, Heidelberg (2006). https://doi.org/10.1007/11814948_13
28. Papadimitriou, C., Yannakakis, M.: The complexity of facets (and some facets of complexity). J. Comput. Syst. Sci. **28**(2), 244–259 (1984)
29. Wechsung, G.: On the Boolean closure of NP. In: Budach, L. (ed.) FCT 1985. LNCS, vol. 199, pp. 485–493. Springer, Heidelberg (1985). https://doi.org/10.1007/BFb0028832

The Resolution of Keller's Conjecture

Joshua Brakensiek[1], Marijn Heule[2(✉)], John Mackey[2], and David Narváez[3]

[1] Stanford University, Stanford, CA, USA
[2] Carnegie Mellon University, Pittsburgh, PA, USA
marijn@cmu.edu
[3] Rochester Institute of Technology, Rochester, NY, USA

Abstract. We consider three graphs, $G_{7,3}$, $G_{7,4}$, and $G_{7,6}$, related to Keller's conjecture in dimension 7. The conjecture is false for this dimension if and only if at least one of the graphs contains a clique of size $2^7 = 128$. We present an automated method to solve this conjecture by encoding the existence of such a clique as a propositional formula. We apply satisfiability solving combined with symmetry-breaking techniques to determine that no such clique exists. This result implies that every unit cube tiling of $\mathbb{R}^7$ contains a facesharing pair of cubes. Since a faceshare-free unit cube tiling of $\mathbb{R}^8$ exists (which we also verify), this completely resolves Keller's conjecture.

1 Introduction

In 1930, Keller conjectured that any tiling of n-dimensional space by translates of the unit cube must contain a pair of cubes that share a complete $(n-1)$-dimensional face [13]. Figure 1 illustrates this for the plane and the 3-dimensional space. The conjecture generalized a 1907 conjecture of Minkowski [24] in which the centers of the cubes were assumed to form a lattice. Keller's conjecture was proven to be true for $n \leq 6$ by Perron in 1940 [25,26], and in 1942 Hajós [6] showed Minkowski's conjecture to be true in all dimensions.

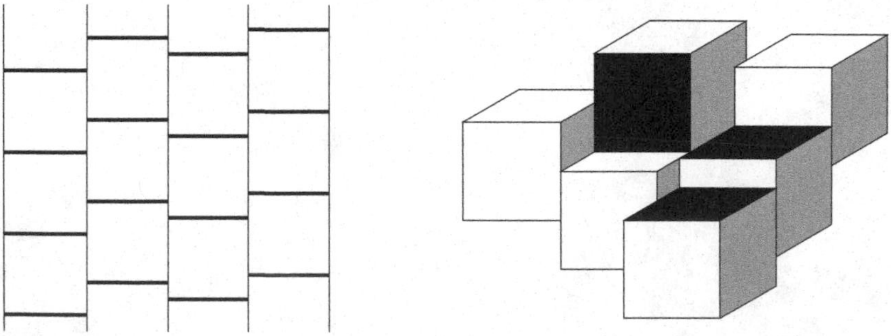

Fig. 1. Left, a tiling of the plane (2-dimensional space) with unit cubes (squares). The bold black edges are fully face-sharing edges. Right, a partial tiling of the 3-dimensional space with unit cubes. The only way to tile the entire space would result in a fully face-sharing square at the position of the black squares.

N. Peltier and V. Sofronie-Stokkermans (Eds.): IJCAR 2020, LNAI 12166, pp. 48–65, 2020.
https://doi.org/10.1007/978-3-030-51074-9_4

In 1986 Szabó [28] reduced Keller's conjecture to the study of periodic tilings. Using this reduction Corrádi and Szabó [3] introduced the Keller graphs: the graph $G_{n,s}$ has vertices $\{0,1,\ldots,2s-1\}^n$ such that a pair are adjacent if and only if they differ by exactly s in at least one coordinate and they differ in at least two coordinates. The size of cliques in $G_{n,s}$ is at most 2^n [5] and the size of the largest clique in $G_{n,s}$ is at most the size of the largest clique in $G_{n,s+1}$.

A clique in $G_{n,s}$ of size 2^n demonstrates that Keller's conjecture is false for dimensions greater than or equal to n. Lagarias and Shor [19] showed that Keller's conjecture is false for $n \geq 10$ in 1992 by exhibiting clique of size 2^{10} in $G_{10,2}$. In 2002, Mackey [22] found a clique of size 2^8 in $G_{8,2}$ to show that Keller's conjecture is false for $n \geq 8$. In 2011, Debroni, Eblen, Langston, Myrvold, Shor, and Weerapurage [5] showed that the largest clique in $G_{7,2}$ has size 124.

In 2015, Kisielewicz and Łysakowska [14,16] made substantial progress on reducing the conjecture in dimension 7. More recently, in 2017, Kisielewicz [15] reduced the conjecture in dimension 7 as follows: Keller's conjecture is true in dimension 7 if and only if there does not exist a clique in $G_{7,3}$ of size 2^7 [21].

The main result of this paper is the following theorem.

Theorem 1. *Neither $G_{7,3}$ nor $G_{7,4}$ nor $G_{7,6}$ contains a clique of size* $2^7 = 128$.

Although proving this property for $G_{7,3}$ suffices to prove Keller's conjecture true in dimension 7, we also show this for $G_{7,4}$ and $G_{7,6}$ to demonstrate that our methods need only depend on prior work of Kisielewicz and Łysakowska [14,16]. In particular, the argument for $G_{7,6}$ [14] predates and is much simpler than the one for $G_{7,4}$ [16] (although the publication dates indicate otherwise). It is not explicitly stated in either that it suffices to prove that $G_{7,4}$ or $G_{7,6}$ lacks a clique of size 128 to prove Keller's conjecture. We show this in the Appendix of the extended version, available at https://arxiv.org/abs/1910.03740.

We present an approach based on satisfiability (SAT) solving to show the absence of a clique of size 128. SAT solving has become a powerful tool in computer-aided mathematics in recent years. For example, it was used to prove the Erdős discrepancy conjecture with discrepancy 2 [17], the Pythagorean triples problem [10], and Schur number five [7]. Modern SAT solvers can also emit proofs of unsatisfiability. There exist formally verified checkers for such proofs as developed in the ACL2, Coq, and Isabelle theorem-proving systems [4,20].

The outline of this paper is as follows. After describing some background concepts in Sect. 2, we present a compact encoding whether $G_{n,s}$ contains a clique of size 2^n as a propositional formula in Sect. 3. Without symmetry breaking, these formulas with $n > 5$ are challenging for state-of-the-art tools. However, the Keller graphs contain many symmetries. We perform some initial symmetry breaking that is hard to express on the propositional level in Sect. 4. This allows us to partially fix three vertices. On top of that we add symmetry-breaking clauses in Sect. 5. The soundness of their addition has been mechanically verified. We prove in Sect. 6 the absence of a clique of size 128 in $G_{7,3}$, $G_{7,4}$ and $G_{7,6}$. We optimize the proofs of unsatisfiability obtained by the SAT solver and certify them using a formally verified checker. Finally we draw some conclusions in Sect. 7 and present directions for future research.

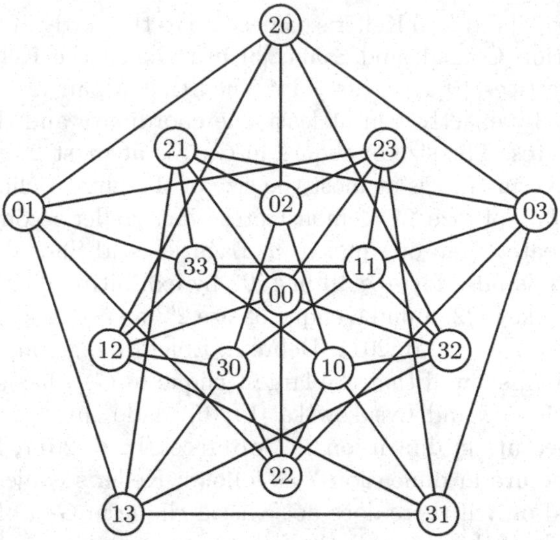

Fig. 2. Illustration of $G_{2,2}$. The coordinates of the vertices are compactly represented by a sequence of the digits.

2 Preliminaries

We present the most important background concepts related to this paper and introduce some properties of $G_{n,s}$. First, for positive integers k, we define two sets: $[k] := \{1, 2, \ldots, k\}$ and $\langle k \rangle := \{0, 1, \ldots, k-1\}$.

Keller Graphs. The Keller graph $G_{n,s}$ consists of the vertices $\langle 2s \rangle^n$. Two vertices are adjacent if and only if they differ by exactly s in at least one coordinate and they differ in at least two coordinates. Figure 2 shows a visualization of $G_{2,2}$.

As noted in [5], $\{sw + \langle s \rangle^n : w \in \{0,1\}^n\}$ is a partition of the vertices of $G_{n,s}$ into 2^n independent sets. Consequently, any clique in $G_{n,s}$ has at most 2^n vertices. For example, $V(G_{2,2})$ is partitioned as follows:

$$\{\{2(0,0) + \{0,1\}^2, 2(0,1) + \{0,1\}^2, 2(1,0) + \{0,1\}^2, 2(1,1) + \{0,1\}^2\} =$$
$$\{\{(0,0), (0,1), (1,0), (1,1)\}, \{(0,2), (0,3), (1,2), (1,3)\},$$
$$\{(2,0), (2,1), (3,0), (3,1)\}, \{(2,2), (2,3), (3,2), (3,3)\}\}.$$

We use the above observation for encoding whether $G_{n,s}$ has a clique of size 2^n. Instead of searching for such a clique on the graph representation of $G_{n,s}$, which consists of $(2s)^n$ vertices, we search for 2^n vertices, one from each $sw + \langle s \rangle^n$, such that every pair is adjacent.

For every $i \in \langle 2^n \rangle$, we let $w(i) = (w_1, w_2, \ldots, w_n) \in \{0,1\}^n$ be defined by $i = \sum_{k=1}^{n} 2^{k-1} \cdot w_k$. Given a clique of size 2^n, we let c_i be its unique element in $sw(i) + \langle s \rangle^n$ and we let $c_{i,j}$ be the jth coordinate of c_i.

Useful Automorphisms of Keller Graphs. Let S_n be the set of permutations of $[n]$ and let H_s be the set of permutations of $\langle 2s \rangle$ generated by the swaps $(i\ i+s)$ composed with any permutation of $\langle s \rangle$ which is identically applied to $s + \langle s \rangle$. The maps

$$(x_1, x_2, \ldots, x_n) \mapsto (\tau_1(x_{\sigma(1)}), \tau_2(x_{\sigma(2)}), \ldots, \tau_n(x_{\sigma(n)})),$$

where $\sigma \in S_n$ and $\tau_1, \tau_2, \ldots, \tau_n \in H_s$ are automorphisms of $G_{n,s}$. Note that applying an automorphism to every vertex of a clique yields another clique of the same size.

Propositional Formulas. We consider formulas in *conjunctive normal form* (CNF), which are defined as follows. A *literal* is either a variable x (a *positive literal*) or the negation $\overline{x}$ of a variable x (a *negative literal*). The *complement* $\overline{l}$ of a literal l is defined as $\overline{l} = \overline{x}$ if $l = x$ and $\overline{l} = x$ if $l = \overline{x}$. For a literal l, $var(l)$ denotes the variable of l. A *clause* is a disjunction of literals and a *formula* is a conjunction of clauses.

An *assignment* is a function from a set of variables to the truth values 1 (*true*) and 0 (*false*). A literal l is *satisfied* by an assignment α if l is positive and $\alpha(var(l)) = 1$ or if it is negative and $\alpha(var(l)) = 0$. A literal is *falsified* by an assignment if its complement is satisfied by the assignment. A clause is satisfied by an assignment α if it contains a literal that is satisfied by α. A formula is satisfied by an assignment α if all its clauses are satisfied by α. A formula is *satisfiable* if there exists an assignment that satisfies it and *unsatisfiable* otherwise.

Clausal Proofs. Our proof that Keller's conjecture is true for dimension 7 is predominantly a clausal proof, including a large part of the symmetry breaking. Informally, a clausal proof system allows us to show the unsatisfiability of a CNF formula by continuously deriving more and more clauses until we obtain the empty clause. Thereby, the addition of a derived clause to the formula and all previously derived clauses must preserve satisfiability. As the empty clause is trivially unsatisfiable, a clausal proof shows the unsatisfiability of the original formula. Moreover, it must be checkable in polynomial time that each derivation step does preserve satisfiability. This requirement ensures that the correctness of proofs can be efficiently verified. In practice, this is achieved by allowing only the derivation of specific clauses that fulfill some efficiently checkable criterion.

Formally, clausal proof systems are based on the notion of *clause redundancy*. A clause C is *redundant* with respect to a formula F if adding C to F preserves satisfiability. Given a formula $F = C_1 \land \cdots \land C_m$, a *clausal proof* of F is a sequence $(C_{m+1}, \omega_{m+1}), \ldots, (C_n, \omega_n)$ of pairs where each C_i is a clause, each ω_i (called the *witness*) is a string, and C_n is the empty clause [9]. Such a sequence gives rise to formulas $F_m, F_{m+1}, \ldots, F_n$, where $F_i = C_1 \land \cdots \land C_i$. A clausal proof is *correct* if every clause C_i $(i > m)$ is redundant with respect to F_{i-1}, and if this redundancy can be checked in polynomial time (with respect to the size of the proof) using the witness ω_i.

An example for a clausal proof system is the resolution proof system, which only allows the derivation of resolvents (with no or empty witnesses). However, the resolution proof system does not allow to compactly express symmetry breaking. Instead we will construct a proof in the resolution asymmetric tautology (RAT) proof system. This proof system is also used to validate the results of the SAT competitions [11]. For the details of RAT, we refer to the original paper [9]. Here, we just note that (1) for RAT clauses, it can be checked efficiently that their addition preserves satisfiability, and (2) every resolvent is a RAT clause but not vice versa.

3 Clique Existence Encoding

Recall that $G_{n,s}$ has a clique of size 2^n if and only if there exist vertices $c_i \in sw(i) + \langle s \rangle^n$ for all $i \in \langle 2^n \rangle$ such that for all $i \neq i'$ there exists at least two $j \in [n]$ such that $c_{i,j} \neq c_{i',j}$ and there exists at least one $j \in [n]$ such that $c_{i,j} = c_{i',j} \pm s$.

Our CNF will encode the coordinates of the c_i. For each $i \in \langle 2^n \rangle$, $j \in [n]$, $k \in \langle s \rangle$, we define Boolean variables $x_{i,j,k}$ which are true if and only if $c_{i,j} = sw(i)_j + k$. We therefore need to encode that exactly one of $x_{i,j,0}$, $x_{i,j,1}$, …, $x_{i,j,s-1}$ is true. We use the following clauses

$$\forall i \in \langle 2^n \rangle, \forall j \in [n], \; (x_{i,j,0} \vee x_{i,j,1} \vee \cdots \vee x_{i,j,s-1}) \wedge \bigwedge_{k<k' \in \langle s \rangle} (\overline{x}_{i,j,k} \vee \overline{x}_{i,j,k'}). \tag{1}$$

Next we enforce that every pair of vertices c_i and $c_{i'}$ in the clique differ in at least two coordinates. For most pairs of vertices, no clauses are required because $w(i)$ and $w(i')$ differ in at least two positions. Hence, a constraint is only required for two vertices if $w(i)$ and $w(i')$ differ in exactly one coordinate.

Let $\oplus$ be the binary XOR operator and e_j be the indicator vector of the jth coordinate. If $w(i) \oplus w(i') = e_j$, then in order to ensure that c_i and $c_{i'}$ differ in at least two coordinates we need to make sure that there is some coordinate $j' \neq j$ for which $c_{i,j'} \neq c_{i',j'}$

$$\forall i \neq i' \in \langle 2^n \rangle \text{ s.t. } w(i) \oplus w(i') = e_j, \quad \bigvee_{j' \in [n] \setminus \{j\}, k \in \langle s \rangle} (x_{i,j',k} \neq x_{i',j',k}). \tag{2}$$

We use the Plaisted-Greenbaum [27] encoding to convert the above constraint into CNF. We refer to the auxiliary variables introduced by the encoding as $y_{i,i',j',k}$, which if true imply $x_{i,j',k} \neq x_{i',j',k}$, or written as an implication

$$y_{i,i',j',k} \rightarrow (x_{i,j',k} \neq x_{i',j',k})$$

The following two clauses express this implication

$$(\overline{y}_{i,i',j',k} \vee x_{i,j',k} \vee x_{i',j',k}) \wedge (\overline{y}_{i,i',j',k} \vee \overline{x}_{i,j',k} \vee \overline{x}_{i',j',k}) \tag{3}$$

Notice that the implication is only in one direction as Plaisted-Greenbaum takes the polarity of constraints into account. The clauses that represent the other direction, i.e., $(y_{i,i',j',k} \vee x_{i,j',k} \vee \overline{x}_{i',j',k})$ and $(y_{i,i',j',k} \vee \overline{x}_{i,j',k} \vee x_{i',j',k})$ are redundant (and more specifically, they are blocked [18]).

Using the auxiliary variables, we can express the constraint (2) using clauses of length $s \cdot (n-1)$

$$\forall i \neq i' \in \langle 2^n \rangle \text{ s.t. } w(i) \oplus w(i') = e_j, \quad \bigvee_{j' \in [n] \setminus \{j\}, k \in \langle s \rangle} y_{i,i',j',k}. \tag{4}$$

The last part of the encoding consists of clauses to ensure that each pair of vertices in the clique have at least one coordinate in which they differ by exactly s. Observe that $c_{i,j} = c_{i',j} \pm s$ implies that $c_{i,j} \neq c_{i',j}$ and $x_{i,j,k} = x_{i',j,k}$ for all $k \in \langle s \rangle$. We use auxiliary variables $z_{i,i',j}$, whose truth implies $c_{i,j} = c_{i',j} \pm s$, or written as an implication

$$\forall i \neq i' \in \langle 2^n \rangle, \forall j \in [n] \text{ s.t. } c_{i,j} \neq c_{i',j},$$
$$z_{i,i',j} \rightarrow \big((x_{i,j,0} = x_{i',j,0}) \wedge \cdots \wedge (x_{i,j,s-1} = x_{i',j,s-1})\big).$$

Notice that the implication is again in one direction only. Below we enforce that some $z_{i,i',j}$ variables must be true, but there are no constraints that enforce $z_{i,i',j}$ variables to be false.

This can be written as a CNF using the following clauses:

$$\bigwedge_{k \in \langle s \rangle} \big((\overline{z}_{i,i',j} \vee x_{i,j,k} \vee \overline{x}_{i',j,k}) \wedge (\overline{z}_{i,i',j} \vee \overline{x}_{i,j,k} \vee x_{i',j,k})\big) \tag{5}$$

Finally, to make sure that $c_{i,j} = c_{i',j} \pm s$ for some $j \in [n]$, we specify

$$\forall i \neq i' \in \langle 2^n \rangle, \quad \bigvee_{j : c_{i,j} \neq c_{i',j}} z_{i,i',j}. \tag{6}$$

The variables and clauses, including precise formulas for their counts, are summarized in Table 1. The sizes of the CNF encodings (before the addition of symmetry breaking clauses) of $G_{7,3}$, $G_{7,4}$, and $G_{7,6}$ are listed in Table 2. Notice that for fixed n, the dependence on s is quadratic, which is better than the s^{2n} dependence one would get in the naive encoding of $G_{n,s}$ as a graph. This compact encoding, when combined with symmetry breaking, is a core reason that we were able to prove Theorem 1.

The instances with $n = 7$ are too hard for state-of-the-art SAT solvers if no symmetry breaking is applied. We experimented with general-purpose symmetry-breaking techniques, similar to the symmetry-breaking predicates produced by shatter [1]. This allows solving the formula for $G_{7,3}$, but the computation takes a few CPU years. The formulas for $G_{7,4}$ and $G_{7,6}$ with these symmetry-breaking predicates are significantly harder.

Instead we employ problem-specific symmetry breaking by making use of the observations in Sects. 4 and 5. This allows solving the clique of size 2^n existence problem for all three graphs in reasonable time.

Table 1. Summary of variable and clause counts in the CNF encoding.

Clauses	New Variable Count	Clause Count
(1)	$2^n \cdot n \cdot s$	$2^n \cdot n \cdot (1 + \binom{s}{2})$
(3)	$2^{n-1} \cdot n \cdot s \cdot (n-1)$	$2^n \cdot n \cdot s \cdot (n-1)$
(4)		$2^{n-1} \cdot n$
(5)	$2^{2n-2} \cdot n$	$2^{2n-1} \cdot n \cdot s$
(6)		$\binom{2^n}{2}$
Total	$2^{n-1} \cdot n \cdot (s(n+1) + 2^{n-1})$	$2^n \cdot n \cdot (\frac{3}{2} + \binom{s}{2} + n \cdot s - s) + 2^{2n-1}ns + \binom{2^n}{2}$

4 Initial Symmetry Breaking

Our goal is to prove that there exists no clique of size 128 in $G_{7,s}$ for $s \in \{3, 4, 6\}$. In this section, and the subsequent, we assume that such a clique exists and adapt some of the arguments of Perron [25,26] to show that it may be assumed to have a canonical form. We will use $\star_i$ to denote an element in $\langle i \rangle$.

Lemma 1. *If there is a clique of size* 128 *in* $G_{7,s}$*, then there is a clique of size* 128 *in* $G_{7,s}$ *containing the vertices* $(0,0,0,0,0,0,0)$ *and* $(s,1,0,0,0,0,0)$.

Proof. Let K be a clique of size 128 in $G_{7,s}$. Consider the following sets of vertices in $G_{6,s}$:

$$K_{<s} := \{(v_2, \ldots, v_7) \mid \exists v_1 \in \langle s \rangle \text{ s.t. } (v_1, \ldots, v_7) \in K\}$$

and

$$K_{\geq s} := \{(v_2, \ldots, v_7) \mid \exists v_1 \in s + \langle s \rangle \text{ s.t. } (v_1, \ldots, v_7) \in K\}.$$

Every pair of vertices in $K_{<s}$ differs by exactly s in at least one coordinate, because the corresponding pair of vertices in K can't differ by exactly s in the first coordinate. Similarly, every pair of vertices in $K_{\geq s}$ differs by exactly s in at least one coordinate. Although $K_{<s}$ and $K_{\geq s}$ are not necessarily cliques in $G_{6,s}$, they satisfy the first condition of the adjacency requirement. The partition of Sect. 2 can thus be applied to deduce that $|K_{<s}| \leq 64$ and $|K_{\geq s}| \leq 64$. Since $|K_{<s}| + |K_{\geq s}| = 128$, we conclude that $|K_{<s}| = 64$ and $|K_{\geq s}| = 64$.

By the truth of Keller's conjecture in dimension 6, $K_{<s}$ is not a clique in $G_{6,s}$. Thus, some pair of vertices in $K_{<s}$ are identical in five of the six coordinates.

Table 2. Summary of variable and clause counts of the CNF encoding for $G_{7,3}$, $G_{7,4}$, and $G_{7,6}$. These counts do not include the clauses introduced by the symmetry breaking.

Keller Graph	Variable Count	Clause Count
$G_{7,3}$	39 424	200 320
$G_{7,4}$	43 008	265 728
$G_{7,6}$	50 176	399 232

After application of an automorphism, we may without loss of generality assume that this pair is $(s,0,0,0,0,0)$ and $(0,0,0,0,0,0)$. Since the pair comes from $K_{<s}$, there exist $v_1 \neq v_1' \in \langle s \rangle$ such that $(v_1, s, 0, 0, 0, 0, 0)$ and $(v_1', 0, 0, 0, 0, 0, 0)$ are in the clique.

After application of an automorphism that moves v_1 to 1 and v_1' to 0, we deduce that without loss of generality $(1, s, 0, 0, 0, 0, 0)$ and $(0, 0, 0, 0, 0, 0, 0)$ are in the clique. Application of the automorphism that interchanges the first two coordinates yields a clique of size 128 containing the vertices $c_0 = (0, 0, 0, 0, 0, 0, 0)$ and $c_1 = (s, 1, 0, 0, 0, 0, 0)$. □

Theorem 2. *If there is a clique of size 128 in $G_{7,s}$, then there is a clique of size 128 in $G_{7,s}$ containing the vertices $(0,0,0,0,0,0,0)$, $(s,1,0,0,0,0,0)$, and $(s, s+1, \star_2, \star_2, 1, 1, 1)$.*

Proof. Using the preceding lemma, we can choose from among all cliques of size 128 that contain $c_0 = (0, 0, 0, 0, 0, 0, 0)$ and $c_1 = (s, 1, 0, 0, 0, 0, 0)$, one in which c_3 has the fewest number of coordinates equal to 0. Let λ be this least number.

Observe that the first two coordinates of c_3 must be $(s, s+1)$ in order for it to be adjacent with both c_0 and c_1. Thus, we have

$$\begin{aligned} c_0 &= (0,\ 0,\ 0,\ 0,\ 0,\ 0,\ 0) \\ c_1 &= (s,\ 1,\ 0,\ 0,\ 0,\ 0,\ 0) \\ c_3 &= (s,\ s+1,\ \star_s,\ \star_s,\ \star_s,\ \star_s,\ \star_s) \end{aligned}$$

In the above, we can apply automorphisms that fix 0 in the last five coordinates to replace $\star_s$ by $\star_2$. We can apply an automorphism that permutes the last five coordinates to assume that the 0's and 1's in c_3 are sorted in increasing order. Notice that not all of the $\star_2$ coordinates in c_3 can be 0, because c_1 and c_3 are adjacent and must therefore differ in at least two coordinates. Hence at least the last coordinate of c_3 is 1.

Case 1) $\lambda = 4$. In this case $c_3 = (s, s+1, 0, 0, 0, 0, 1)$. In order for c_{67} to be adjacent with c_0, c_1, and c_3, it must start with $(s, s+1)$ and end with $s+1$:

$$\begin{aligned} c_0 &= (0,\ 0,\ 0,\ 0,\ 0,\ 0,\ 0) \\ c_1 &= (s,\ 1,\ 0,\ 0,\ 0,\ 0,\ 0) \\ c_3 &= (s,\ s+1,\ 0,\ 0,\ 0,\ 0,\ 1) \\ c_{67} &= (s,\ s+1,\ \star_s,\ \star_s,\ \star_s,\ \star_s,\ s+1) \end{aligned}$$

Not all $\star_s$ elements in c_{67} can be 0, because c_3 and c_{67} differ in at least two coordinates. However, if one of the $\star_s$ elements in c_{67} is nonzero, then we can swap 1 and $s+1$ in the last coordinate to obtain a clique in which c_3 has three or fewer coordinates equal to 0, contradicting $\lambda = 4$. Thus, $\lambda \leq 3$.

Case 2) $\lambda = 3$, in which case $c_3 = (s, s+1, 0, 0, 0, 1, 1)$:

$$\begin{aligned} c_0 &= (0,\ 0,\ 0,\ 0,\ 0,\ 0,\ 0) \\ c_1 &= (s,\ 1,\ 0,\ 0,\ 0,\ 0,\ 0) \\ c_3 &= (s,\ s+1,\ 0,\ 0,\ 0,\ 1,\ 1) \\ c_{35} &= (s,\ s+1,\ \star_s,\ \star_s,\ \star_s,\ s+1,\ \star_s) \\ c_{67} &= (s,\ s+1,\ \star_s,\ \star_s,\ \star_s,\ \star_s,\ s+1) \end{aligned}$$

Since c_{67} is adjacent with c_0, c_1, and c_3, it must start with $(s, s+1)$ and end with $s+1$. Similarly, since c_{35} is adjacent with c_0, c_1, and c_3, it must start with $(s, s+1)$ and have $s+1$ as its penultimate coordinate. Since c_{35} and c_{67} are adjacent, either the last coordinate of c_{35} must be 1, or the penultimate coordinate of c_{67} must be 1. Without loss of generality we can assume that the penultimate coordinate of c_{67} is 1 as we can permute the last two coordinates which would swap c_{35} and c_{67} without involving the other cubes. The remaining three $\star_s$ elements in c_{67} cannot all be 0, since c_3 and c_{67} differ in at least two coordinates. However, if one of the $\star_s$ elements is non-zero, then we can swap 1 and $s+1$ in the last coordinate to obtain a clique in which c_3 has two or fewer coordinates equal to 0, contradicting $\lambda = 3$. Thus, we have $\lambda \leq 2$ and $c_3 = (s, s+1, \star_2, \star_2, 1, 1, 1)$, as desired. □

Notice that most of the symmetry breaking discussed in this section is challenging, if not impossible, to break on the propositional level: The proof of Lemma 1 uses the argument that Keller's conjecture holds for dimension 6, while the proof of Theorem 2 uses the interchangeability of 1 and $s+1$, which is not a symmetry on the propositional level. We will break these symmetries by adding some unit clauses to the encoding. All additional symmetry breaking will be presented in the next section and will be checked mechanically.

5 Clausal Symmetry Breaking

Our symmetry-breaking approach starts with enforcing the initial symmetry breaking: We assume that vertices $c_0 = (0, 0, 0, 0, 0, 0, 0)$, $c_1 = (s, 1, 0, 0, 0, 0, 0)$ and $c_3 = (s, s+1, \star_s, \star_s, 1, 1, 1)$ are in our clique K, which follows from Theorem 2. We will not use the observation that $\star_s$ occurrences in c_3 can be reduced to $\star_2$ and instead add and validate clauses that realize this reduction.

We fix the above initial vertices by adding unit clauses to the CNF encoding. This is the only part of the symmetry breaking that is not checked mechanically. Let $\Phi_{7,s}$ be the formula obtained from our encoding in Sect. 3 together with the unit clauses corresponding to the 19 coordinates fixed among c_0, c_1 and c_3. In this section we will identify several symmetries in $\Phi_{7,s}$ that can be further broken at the CNF level by adding symmetry breaking clauses. The formula ultimately used in Sect. 6 for the experiments is the result of adding these symmetry breaking clauses to $\Phi_{7,s}$. Symmetry breaking clauses are added in an incremental fashion. For each addition, a clausal proof of its validity with respect to $\Phi_{7,s}$ and the clauses added so far is generated, as well. Each of these clausal proofs has been validated using the `drat-trim` proof checker.

Our approach can be described in general terms as identifying groups of coordinates whose assignments exhibit interesting symmetries and calculating the equivalence classes of these assignments. Given a class of symmetric assignments, it holds that one of these assignments can be extended to a clique of size 128 if and only if every assignment in that class can be extended as well. It is then enough to pick a canonical representative for each class, add clauses forbidding every assignment that is not canonical, and finally determine the satisfiability of the formula under the canonical representative of every class of assignments: if no canonical assignment can be extended to a satisfying assignment for the formula, then the formula is unsatisfiable. In order to forbid assignments that are not canonical, we use an approach similar to the one described in [8].

5.1 The Last Three Coordinates of c_{19}, c_{35} and c_{67}

The reasoning in the proof of Theorem 2 leads to the following forced settings, once we assign $c_3 = (s, s+1, \star_s, \star_s, 1, 1, 1)$ and apply unit propagation:

$$(c_{19,1}, c_{19,2}, c_{19,5}) = (s, s+1, s+1),$$
$$(c_{35,1}, c_{35,2}, c_{35,6}) = (s, s+1, s+1),$$
$$(c_{67,1}, c_{67,2}, c_{67,7}) = (s, s+1, s+1).$$

Let's now focus on the 3×3 matrix of the coordinates below and do a case split on all of the s^6 possible assignments of coordinates labeled with $\star_s$.

	5	6	7
c_{19}	$s+1$	$\star_s$	$\star_s$
c_{35}	$\star_s$	$s+1$	$\star_s$
c_{67}	$\star_s$	$\star_s$	$s+1$

Notice, however, that since the only positions in which c_{19} and c_{35} can differ by exactly s are positions 5 and 6, and since $c_{19,5}$ and $c_{35,6}$ are already set to $s+1$, at least one of $c_{19,6}$ and $c_{35,5}$ has to be set to 1. Similarly, it is not possible for both $c_{35,7}$ and $c_{67,6}$ to not be 1 and for both $c_{67,5}$ and $c_{19,7}$ to not be 1. By the inclusion-exclusion principle, this reasoning alone discards $3(s-1)^2s^4 - 3(s-1)^4s^2 + (s-1)^6$ cases. All of these cases can be blocked by adding the binary clauses: $(x_{19,6,1} \vee x_{35,5,1}) \wedge (x_{35,7,1} \vee x_{67,6,1}) \wedge (x_{67,5,1} \vee x_{19,7,1})$. These three clauses are RAT clauses [12] with respect to the formula $\Phi_{7,s}$.

Furthermore, among the remaining $(2s-1)^3$ cases, several assignment pairs are symmetric. For example, the following two assignments are symmetric because one can be obtained from the other by swapping columns and rows:

	5	6	7
c_{19}	$s+1$	1	2
c_{35}	2	$s+1$	2
c_{67}	1	1	$s+1$

	5	6	7
c_{19}	$s+1$	1	1
c_{35}	2	$s+1$	1
c_{67}	2	2	$s+1$

As with many problems related to symmetries, we can encode each assignment as a vertex-colored graph and use canonical labeling algorithms to determine a canonical assignment representing all the symmetric assignments of each equivalence class, and which assignments are symmetric to each canonical form. Our approach is similar to the one by McKay and Piperno for isotopy of matrices [23].

This additional symmetry breaking reduces the number of cases for the last three coordinates of the vertices c_{19}, c_{37}, and c_{67} from the trivial s^6 to 25 cases for $s = 3$ and 28 cases for $s \geq 4$. Figure 3 shows the 25 canonical cases for $s = 3$.

(0, 0, 1, 0, 1, 1)	(0, 0, 1, 1, 1, 1)	(0, 0, 1, 1, 1, 2)	(0, 1, 1, 0, 0, 1)	(0, 1, 1, 0, 1, 1)
(0, 1, 1, 0, 2, 1)	(0, 1, 1, 1, 0, 2)	(0, 1, 1, 1, 1, 0)	(0, 1, 1, 1, 1, 1)	(0, 1, 1, 1, 1, 2)
(0, 1, 1, 1, 2, 0)	(0, 1, 1, 1, 2, 1)	(0, 1, 1, 1, 2, 2)	(0, 1, 1, 2, 1, 1)	(0, 1, 1, 2, 2, 1)
(0, 2, 1, 1, 1, 1)	(0, 2, 1, 1, 1, 2)	(0, 2, 1, 2, 1, 1)	(1, 1, 1, 1, 1, 1)	(1, 1, 1, 1, 1, 2)
(1, 1, 1, 1, 2, 2)	(1, 1, 1, 2, 2, 1)	(1, 1, 2, 1, 2, 1)	(1, 1, 2, 1, 2, 2)	(1, 2, 2, 1, 1, 2)

Fig. 3. The 25 canonical cases for $s = 3$. Each vector corresponds to the values of the coordinates $(c_{19,6}, c_{19,7}, c_{35,5}, c_{35,7}, c_{67,5}, c_{67,6})$.

5.2 Coordinates Three and Four of Vertices c_3, c_{19}, c_{35} and c_{67}

The symmetry breaking in the previous subsection allows us to fix, without loss of generality, the last coordinate of c_{19} to 1. It also constrains the third and fourth coordinates of c_3 to take values in $\langle 2 \rangle$ instead of $\langle s \rangle$.

We break the computation into further cases by enumerating over choices for the third and fourth coordinates of vertices c_3, c_{19}, c_{37}, and c_{67} (Fig. 4). Up to this point, our description of the partial clique is invariant under the permutations of $\langle s - 1 \rangle$ in the third and fourth coordinates as well as swapping the third and fourth coordinates. With respect to these automorphisms, for $s = 3$ there are only 861 equivalence classes for how to fill in the $\star_s$ cases for these four vertices. For $s = 4$ there are 1326 such equivalence classes, and for $s = 6$ there are 1378 such equivalence classes. This gives a total of $25 \times 861 = 21\,525$ cases to check for $s = 3$, $28 \times 1326 = 37\,128$ cases to check for $s = 4$, and $28 \times 1378 = 38\,584$ cases to check for $s = 6$.

5.3 Identifying Hardest Cases

In initial experiments we observed for each $s \in \{3, 4, 6\}$ that out of the many thousands of subformulas (cases), one subformula was significantly harder to solve compared to the other subformulas. Figure 5 shows the coordinates of the key vertices of this subformula for $s \in \{3, 4, 6\}$. Notice that the third and fourth coordinates are all 0 for all the key vertices. We therefore applied additional symmetry breaking in case all of these coordinates are 0. Under this case, the third and the fourth coordinates of vertex c_2 can be restricted to $(0, 0)$, $(0, 1)$,

$$\begin{aligned}
c_0 &= (\mathbf{0}, \ \mathbf{0}, \mathbf{0}, \mathbf{0}, \ \mathbf{0}, \ \mathbf{0}, \ \mathbf{0}) \\
c_1 &= (\boldsymbol{s}, \ \mathbf{1}, \mathbf{0}, \mathbf{0}, \ \mathbf{0}, \ \mathbf{0}, \ \mathbf{0}) \\
c_3 &= (\boldsymbol{s}, \boldsymbol{s}+\mathbf{1}, \star_2, \star_2, \ \mathbf{1}, \ \mathbf{1}, \ \mathbf{1}) \\
c_{19} &= (s, s+1, \star_3, \star_3, s+1, \star_3, \ \mathbf{1}) \\
c_{35} &= (s, s+1, \star_4, \star_4, \star_3, s+1, \star_3) \\
c_{67} &= (s, s+1, \star_5, \star_5, \star_4, \star_4, s+1)
\end{aligned}$$

Fig. 4. Part of the symmetry breaking on the key vertices. The bold coordinates show the (unverified) initial symmetry breaking. The bold s and $s+1$ coordinates in c_1 and c_3 are also implied by unit propagation. The additional symmetry breaking is validated by checking a DRAT proof expressing the symmetry breaking clauses.

c_0	$(0,0,0,0,0,0,0)$	$(0,0,0,0,0,0,0)$	$(0,0,0,0,0,0,0)$
c_1	$(3,1,0,0,0,0,0)$	$(4,1,0,0,0,0,0)$	$(6,1,0,0,0,0,0)$
c_3	$(3,4,0,0,1,1,1)$	$(4,5,0,0,1,1,1)$	$(6,7,0,0,1,1,1)$
c_{19}	$(3,4,0,0,4,0,1)$	$(4,5,0,0,5,0,1)$	$(6,7,0,0,7,0,1)$
c_{35}	$(3,4,0,0,1,4,0)$	$(4,5,0,0,1,5,0)$	$(6,7,0,0,1,7,0)$
c_{67}	$(3,4,0,0,0,1,4)$	$(4,5,0,0,0,1,5)$	$(6,7,0,0,0,1,7)$

Fig. 5. The hardest instance for $s = 3$ (left), $s = 4$ (middle), and $s = 6$ (right).

and $(1, 1)$, and the last three coordinates of c_2 can only take values in $\langle 3 \rangle$. Furthermore, any assignment (a, b, c) to the last three coordinates of c_2 is symmetric to the same assignment "shifted right", i.e. (c, a, b), by swapping columns and rows appropriately. These symmetries define equivalence classes of assignments that can also be broken at the CNF level. Under the case shown in Fig. 5, there are only 33 non-isomorphic assignments remaining for vertex c_2 for $s \geq 3$. We replace the hard case for each $s \in \{3, 4, 6\}$ by the corresponding 33 cases, thereby increasing the total number of cases mentioned above by 32.

5.4 SAT Solving

Each of the cases was solved using a SAT solver, which produced a proof of unsatisfiability that was validated using a formally verified checker (details are described in the following section). To ensure that the combined cases cover the entire search space, we constructed for each $s \in \{3, 4, 6\}$ a tautological formula in disjunctive normal form (DNF). The building blocks of a DNF are conjuctions of literals known as cube. We will use α as a symbol for cubes as they can also be considered variable assignments. For each cube α in the DNF, we prove that the formula after symmetry breaking under α is unsatisfiable. Additionally, we mechanically check that the three DNFs are indeed tautologies.

6 Experiments

We used the CaDiCaL[1] SAT solver developed by Biere [2] and ran the simulations on a cluster of Xeon E5-2690 processors with 24 cores per node. CaDiCaL supports proof logging in the DRAT format. We used DRAT-trim [29] to optimize the emitted proof of unsatisfiability. Afterwards we certified the optimized proofs with ACL2check, a formally verified checker [4]. All of the code that we used is publicly available on GitHub.[2] We have also made the logs of the computation publicly available on Zenodo.[3]

---

1 Commit 92d72896c49b30ad2d50c8e1061ca0681cd23e60 of https://github.com/arminbiere/cadical.

2 https://github.com/marijnheule/Keller-encode.

3 https://doi.org/10.5281/zenodo.3755116.

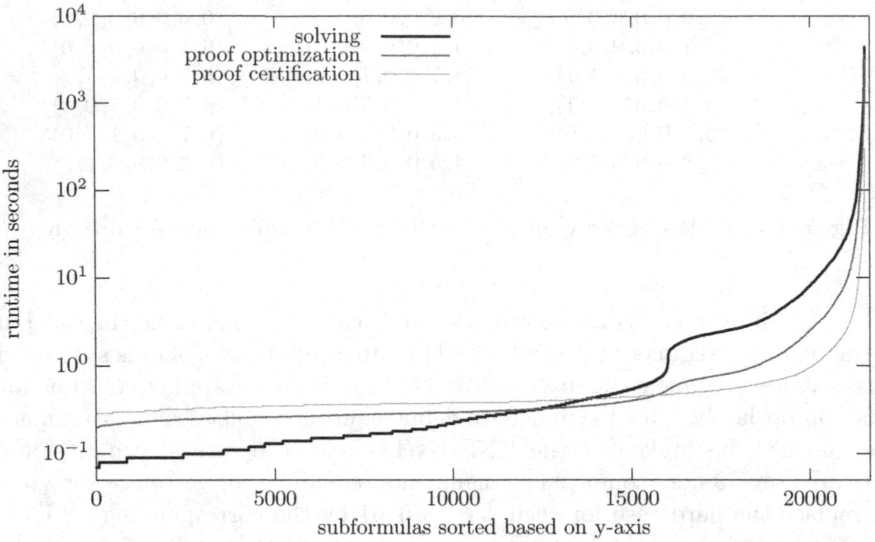

Fig. 6. Cactus plot of the runtime in seconds (logscale) to solve the 21 557 subformulas of $G_{7,3}$ as well as the times to optimize and certify the proofs of unsatisfiability.

6.1 Results for Dimension 7

Table 3 summarizes the running times are for experiment. The subformula-solving runtimes for $s = 3, 4$ and 6 are summarized in cactus plots in Figs. 6, 7 and 8. The combined size of all unsatisfiability proofs of the subformulas of $s = 6$ is 224 gigabyte in the binary DRAT proof format. These proofs contained together $6.18 \cdot 10^9$ proof steps (i.e., additions of redundant clauses). The DRAT-trim proof checker only used $6.39 \cdot 10^8$ proof steps to validate the unsatisfiability of all subformulas. In other words, almost 90% of the clauses generated by CaDiCaL are not required to show unsatisfiability. It is therefore likely that a single DRAT proof for the formula after symmetry breaking can be constructed that is about 20 gigabytes in size. That is significantly smaller compared to other recently solved problems in mathematics that used SAT solvers [7,10].

Table 3. Summary of solve times for $s = 3, 4, 6$. Times without a unit are in CPU hours. "No. Hard" is the number of subformulas which required more than 900 seconds to solve. "Hardest" is the solve time of the hardest subformula in CPU hours.

s	Tot. Solve	Avg. Solve	Proof Opt.	Proof Cert.	No. Hard	Hardest
3	43.27	7.23 s	22.46	4.98	28 form.	≈ 1.2
4	77.00	7.46 s	44.00	9.70	62 form.	≈ 2.7
6	81.85	7.63 s	34.84	14.53	63 form.	≈ 1.25

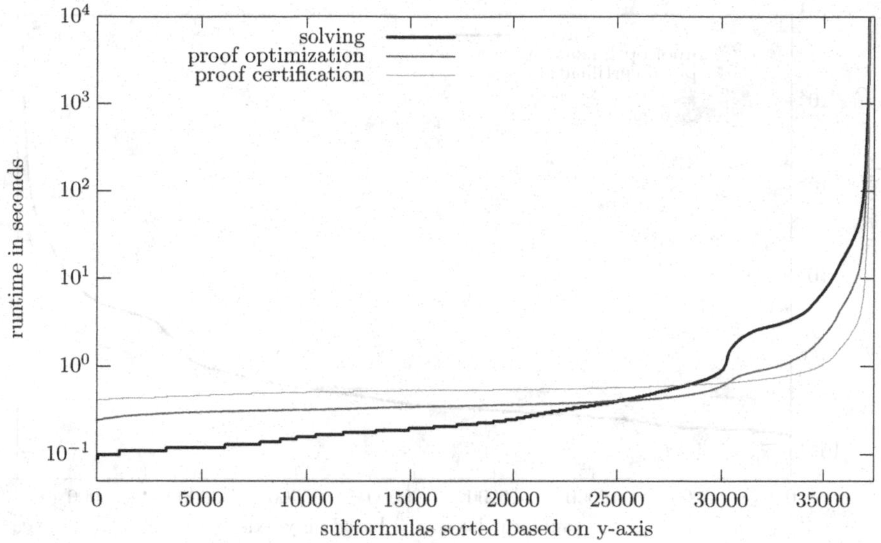

Fig. 7. Cactus plot of the runtime in seconds (logscale) to solve the 37 160 subformulas of $G_{7,4}$ as well as the times to optimize and certify the proofs of unsatisfiability.

We ran all three experiments simultaneously on 20 nodes on the Lonestar 5 cluster and computing on 24 CPUs per node in parallel. All instances were reported unsatisfiable and all proofs of unsatisfiability were certified by the formally verified checker. This proves Theorem 1.

6.2 Refuting Keller's Conjecture in Dimension 8

To check the accuracy of the CNF encoding, we verified that the generated formulas for $G_{8,2}$, $G_{8,3}$, $G_{8,4}$ and $G_{8,6}$ are satisfiable — thereby confirming that Keller's conjecture is false for dimension 8. These instances, by themselves, have too many degrees of freedom for the solver to finish. Instead, we added to the CNF the unit clauses consistent with the original clique found in the paper of Mackey [22] (as suitably embedded for the larger graphs). Specification of the vertices was per the method in Sect. 3 and 4. These experiments were run on Stanford's Sherlock cluster and took less than a second to confirm satisfiability.

Figure 9 shows an illustration of a clique of size 256 in $G_{8,2}$. This is the smallest counterexample for Keller's conjecture, both in the dimension ($n = 8$) as in the number of coordinates ($s = 2$). The illustration uses a black circle, black square, white circle, or white square to represent a coordinate set to 0, 1, 2, or 3, respectively. Notice that for each pair of vertices it holds that they have a complementary (black vs white) circle or square and at least one other different coordinate (black/white or circle/square or both).

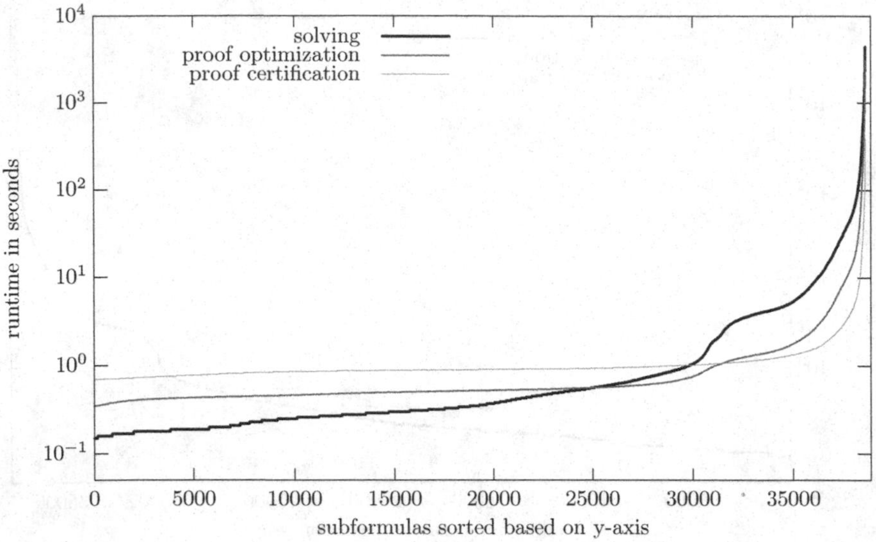

Fig. 8. Cactus plot of the runtime in seconds (logscale) to solve the 38 616 subformulas of $G_{7,6}$ as well as the times to optimize and certify the proofs of unsatisfiability.

7 Conclusions and Future Work

In this paper, we analyzed maximal cliques in the graphs $G_{7,3}$, $G_{7,4}$, and $G_{7,6}$ by combining symmetry-breaking and SAT-solving techniques. For the initial symmetry breaking we adapt some of the arguments of Perron. Additional symmetry breaking is performed on the propositional level and this part is mechanically verified. We partitioned the resulting formulas into thousands of subformulas and used a SAT solver to check that each subformula cannot be extended to a clique of size 128. Additionally, we optimized and certified the resulting proofs of unsatisfiability. As a result, we proved Theorem 1, which resolves Keller's conjecture in dimension 7.

In the future, we hope to construct a formally verified argument for Keller's conjecture, starting with a formalization of Keller's conjecture down to the relation of the existence of cliques of size 2^n in Keller graphs and finally the correctness of the presented encoding. This effort would likely involve formally verifying most of the theory discussed in the Appendix of the extended version of the paper. On top of that, we would like to construct a single proof of unsatisfiability that incorporates all the clausal symmetry breaking and the proof of unsatisfiability of all the subformulas and validate this proof using a formally verified checker.

Furthermore, we would like to extend the analysis to $G_{7,s}$, including computing the size of the largest cliques for various values of s. Another direction to consider is to study the maximal cliques in $G_{8,s}$ in order to have some sort of classification of all maximal cliques.

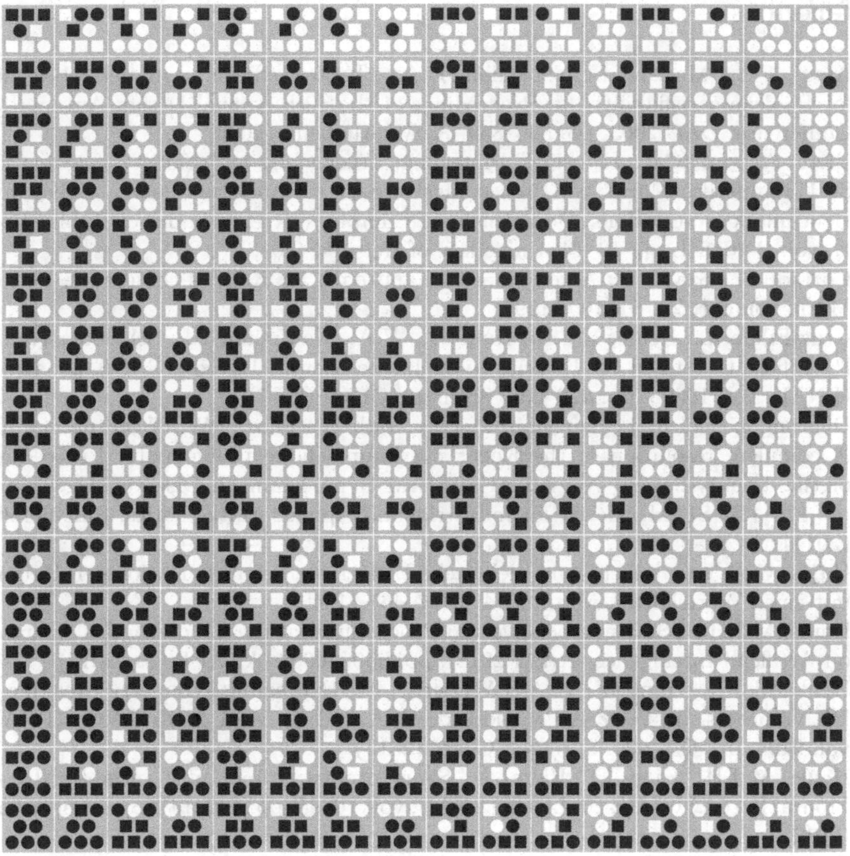

Fig. 9. Illustration of a clique of 256 vertices in $G_{8,2}$. Each "dice" with eight dots represents a vertex, and each dot represents a coordinate. A black circle, black square, white circle, and white square represent a coordinate set to 0, 1, 2, and 3, respectively.

Acknowledgments. The authors acknowledge the Texas Advanced Computing Center (TACC) at The University of Texas at Austin, RIT Research Computing, and the Stanford Research Computing Center for providing HPC resources that have contributed to the research results reported within this paper. Joshua is supported by an NSF graduate research fellowship. Marijn and David are supported by NSF grant CCF-1813993. We thank Andrzej Kisielewicz and Jasmin Blanchette for valuable comments on an earlier version of the manuscript. We thank William Cooperman for helpful discussions on a previous attempt at programming simulations to study the half-integral case. We thank Alex Ozdemir for helpful feedback on both the paper and the codebase. We thank Xinyu Wu for making this collaboration possible.

References

1. Aloul, F.A., Markov, I.L., Sakallah, K.A.: Shatter: efficient symmetry-breaking for Boolean satisfiability. In: Proceedings of the 40th Annual Design Automation Conference, DAC 2003, pp. 836–839. ACM, Anaheim (2003)
2. Biere, A.: CaDiCaL, Lingeling, Plingeling, Treengeling and YalSAT entering the SAT competition 2018. In: Proceedings of SAT Competition 2018 – Solver and Benchmark Descriptions. Department of Computer Science Series of Publications B, vol. B-2018-1, pp. 13–14. University of Helsinki (2018)
3. Corrádi, K., Szabó, S.: A combinatorial approach for Keller's conjecture. Period. Math. Hungar. **21**, 91–100 (1990)
4. Cruz-Filipe, L., Heule, M.J.H., Hunt Jr., W.A., Kaufmann, M., Schneider-Kamp, P.: Efficient certified RAT verification. In: de Moura, L. (ed.) CADE 26. LNCS (LNAI), vol. 10395, pp. 220–236. Springer, Cham (2017)
5. Debroni, J., Eblen, J., Langston, M., Myrvold, W., Shor, P.W., Weerapurage, D.: A complete resolution of the Keller maximum clique problem. In: Proceedings of the Twenty-Second Annual ACM-SIAM Symposium on Discrete Algorithms, pp. 129–135. SIAM, Society for Industrial and Applied Mathematics, Philadelphia (2011)
6. Hajós, G.: Uber einfache und mehrfache Bedeckung des n-dimensionalen Raumes mit einen Wurfelgitter. Math. Z. **47**, 427–467 (1942)
7. Heule, M.J.H.: Schur number five. In: Proceedings of the 32nd AAAI Conference on Artificial Intelligence (AAAI 2018), pp. 6598–6606. AAAI Press (2018)
8. Heule, M.J.H., Hunt Jr., W.A., Wetzler, N.D.: Expressing symmetry breaking in DRAT proofs. In: Felty, A.P., Middeldorp, A. (eds.) CADE 2015. LNCS (LNAI), vol. 9195, pp. 591–606. Springer, Cham (2015)
9. Heule, M.J.H., Kiesl, B., Biere, A.: Short proofs without new variables. In: de Moura, L. (ed.) CADE 26. LNCS (LNAI), vol. 10395, pp. 130–147. Springer, Cham (2017)
10. Heule, M.J.H., Kullmann, O., Marek, V.W.: Solving and verifying the Boolean Pythagorean Triples problem via Cube-and-Conquer. In: Creignou, N., Le Berre, D. (eds.) SAT 2016. LNCS, vol. 9710, pp. 228–245. Springer, Cham (2016)
11. Heule, M.J.H., Schaub, T.: What's hot in the SAT and ASP competition. In: Twenty-Ninth AAAI Conference on Artificial Intelligence 2015, pp. 4322–4323. AAAI Press (2015)
12. Järvisalo, M., Heule, M.J.H., Biere, A.: Inprocessing rules. In: Gramlich, B., Miller, D., Sattler, U. (eds.) IJCAR 2012. LNCS (LNAI), vol. 7364, pp. 355–370. Springer, Heidelberg (2012)
13. Keller, O.H.: Über die lückenlose Erfüllung des Raumes mit Würfeln. Journal für die reine und angewandte Mathematik **163**, 231–248 (1930)
14. Kisielewicz, A.P.: Rigid polyboxes and Keller's conjecture. Adv. Geom. **17**(2), 203–230 (2017)
15. Kisielewicz, A.P.: Towards resolving Keller's cube tiling conjecture in dimension seven. arXiv preprint arXiv:1701.07155 (2017)
16. Kisielewicz, A.P., Łysakowska, M.: On Keller's conjecture in dimension seven. Electron. J. Comb. **22**(1), P1–16 (2015)
17. Konev, B., Lisitsa, A.: Computer-aided proof of Erdős discrepancy properties. Artif. Intell. **224**(C), 103–118 (2015)
18. Kullmann, O.: On a generalization of extended resolution. Discret. Appl. Math. **96–97**, 149–176 (1999)

19. Lagarias, J.C., Shor, P.W.: Keller's cube-tiling conjecture is false in high dimensions. Bull. Am. Math. Soc. **27**(2), 279–283 (1992)
20. Lammich, P.: Efficient verified (UN)SAT certificate checking. In: de Moura, L. (ed.) CADE 26. LNCS (LNAI), vol. 10395, pp. 237–254. Springer, Cham (2017)
21. Łysakowska, M.: Extended Keller graph and its properties. Quaestiones Mathematicae **42**(4), 551–560 (2019)
22. Mackey, J.: A cube tiling of dimension eight with no facesharing. Discret. Comput. Geom. **28**(2), 275–279 (2002)
23. McKay, B.D., Piperno, A.: Nauty and Traces User' Guide (version 2.6). http://users.cecs.anu.edu.au/~bdm/nauty/nug26.pdf
24. Minkowski, H.: Diophantische Approximationen. B.G. Teubner, Leipzig (1907)
25. Perron, O.: Über lückenlose ausfüllung desn-dimensionalen raumes durch kongruente würfel. Math. Z. **46**(1), 1–26 (1940)
26. Perron, O.: Über lückenlose ausfüllung desn-dimensionalen raumes durch kongruente würfel. ii. Math. Z. **46**(1), 161–180 (1940)
27. Plaisted, D.A., Greenbaum, S.: A structure-preserving clause form translation. J. Symb. Comput. **2**(3), 293–304 (1986)
28. Szabó, S.: A reduction of Keller's conjecture. Period. Math. Hung. **17**(4), 265–277 (1986)
29. Wetzler, N.D., Heule, M.J.H., Hunt, W.A.: DRAT-trim: efficient checking and trimming using expressive clausal proofs. In: Sinz, C., Egly, U. (eds.) SAT 2014. LNCS, vol. 8561, pp. 422–429. Springer, Cham (2014)

How QBF Expansion Makes Strategy Extraction Hard

Leroy Chew[1] and Judith Clymo[2(✉)]

[1] Computer Science Department, Carnegie Mellon University, Pittsburgh, PA, USA
`lchew@andrew.cmu.edu`
`http://www.leroychew.wordpress.com`
[2] School of Computing, University of Leeds, Leeds, UK
`scjc@leeds.ac.uk`

Abstract. In this paper we show that the QBF proof checking format QRAT (Quantified Resolution Asymmetric Tautologies) by Heule, Biere and Seidl cannot have polynomial time strategy extraction unless P=PSPACE. In our proof, the crucial property that makes strategy extraction PSPACE-hard for this proof format is universal expansion, even expansion on a single variable.

While expansion reasoning used in other QBF calculi can admit polynomial time strategy extraction, we find this is conditional on a property studied in proof complexity theory. We show that strategy extraction on expansion based systems can only happen when the underlying propositional calculus has the property of feasible interpolation.

Keywords: QBF · Proof complexity · QRAT · Strategy extraction · Quantifier expansion · Feasible interpolation

1 Introduction

Quantified Boolean logic is an extension of propositional logic in which variables may be existentially or universally quantified. This can allow problems to be represented more succinctly than is possible in propositional logic. Deciding the truth of a quantified Boolean formula (QBF) is PSPACE-complete. Propositional proof systems can be lifted to the QBF situation by the addition of rules to handle the universal quantification.

In addition to deciding whether a given QBF is true or false it is desirable that algorithms for solving QBFs can provide verification by outputting a proof. The QRAT proof system [14] is sufficiently strong to simulate the reasoning steps of all current QBF solvers and preprocessors and is a candidate for a standard format for verification of solvers.

In many settings it is not simply desirable to know that a QBF is true or false but also to find functions that witness to this. For example, QBFs may be used to model safety verification so that if the QBF is false then the modelled system is able to reach an unsafe state. It may be important to also know *how* this state

N. Peltier and V. Sofronie-Stokkermans (Eds.): IJCAR 2020, LNAI 12166, pp. 66–82, 2020.
https://doi.org/10.1007/978-3-030-51074-9_5

is reached. If a QBF is true (resp. false) then there must exist Skolem (resp. Herbrand) functions for the existentially (resp. universally) quantified variables that certify this. Substituting the certifying Skolem functions into the original QBF yields a tautology. Equivalently, substituting Herbrand functions results in an unsatisfiable propositional formula. The ability to efficiently extract Skolem or Herbrand functions from the proof output by a QBF solver is called strategy extraction.

There are generally two main paradigms in QBF solving: QCDCL (Conflict Driven Clause Learning) and QBF expansion. Both of these paradigms borrow techniques from propositional satisfiability solving for existential variables, but they differ in how they handle universal variables. The performance and limitations of these solvers can be analysed by studying proof systems that follow the solver steps. QCDCL adds the universal *reduction* rule, such as in the Q-Res proof system [20]. QBF expansion, on the other hand, adds the universal *expansion* rule such as in the proof system ∀Exp+Res [16]. Both Q-Res and ∀Exp+Res are based on the resolution system in propositional logic.

The relationship between the two systems has been studied extensively in both QBF theory and practice. In [4,16] it was shown that Q-Res and ∀Exp+Res are incomparable. However the picture becomes much more nuanced on certain fragments of QBF. Lonsing and Egly ran experiments on QBFs which were parametrised by the number of quantifier alternations and found better performance in the expansion based solvers on formulas with a low number of alternations [23]. This observation was confirmed in proof complexity in [3] where the expansion calculus ∀Exp+Res was shown to polynomially simulate the QCDCL calculus Q-Res on bounded quantifier alternations.

As well as using the calculi Q-Res and ∀Exp+Res to compare the strengths of the two types of QBF solvers, other properties can be studied for each of these systems. One very important property is the aforementioned strategy extraction. Often the strategies are just as important as whether a QBF is true or false. Many QBF proof systems with the universal reduction rule (from QCDCL) have been studied and shown to have polynomial time strategy extraction using a technique from [1] and later generalised in [2,7]. For QBF systems with universal expansion some strategy extraction results are known using a different technique [4,12].

QRAT is a very different kind of proof system, not only can it simulate both the universal reduction and expansion rules but it draws from a stronger form of propositional reasoning than resolution. With this power it has been shown to simulate a number of different QBF proof systems [17,18].

Strategy extraction on a universal checking format like QRAT would have certain benefits for the solving community. One could extract a QRAT proof from a solver and then from that proof separately extract the Skolem/Herbrand functions that give the winning strategy. This would avoid having to extract strategies directly from solvers while they are running which may affect performance.

Conversely, the property of strategy extraction can actually provide a source of weakness in QBF proof systems. In particular Q-Res can always extract strategies as bounded depth circuits. This means that QBFs with winning strategies that cannot be expressed in small bounded depth circuits necessarily have large Q-Res proofs [4]. This is similar to the proof size lower bound technique based on feasible interpolation [21,24] where if a propositional proof system can extract Craig interpolants in polynomial time then super-polynomial interpolant size lower bounds become super-polynomial proof size lower bounds.

It was shown in [13] that Skolem functions to certify that a QBF is true can be extracted in polynomial time from a QRAT proof. In [8] a partial result was presented showing that Herbrand functions may be extracted from proofs in a restricted version of refutational QRAT. Here, we show that it is not possible in general to efficiently extract Herbrand functions certifying falsity from proofs in QRAT. This is due to QRAT providing short proofs to formulas that have PSPACE-hard strategies. Thus we show an asymmetry between the refutation of false QBF and proof of true QBF in the QRAT system. We demonstrate that this is due to the presence of universal expansion steps which manifest from the powerful reduction rules in the full QRAT proof system [14,18].

The universal expansion reasoning technique is present in QBF proof systems other than QRAT, but does not always exhibit the same hardness issues that we demonstrate for QRAT regarding strategy extraction. For example, the proof system ∀Exp+Res [16] uses expansion, but allows polynomial time strategy extraction [4]. In this paper we strengthen the important connection, first explored in [5], between strategy extraction and feasible interpolation.

This paper is organised as follows: Sect. 2 introduces the main concepts used in this paper. We show that strategy extraction in QRAT is PSPACE-hard in Sects. 3 and 4. In Sect. 5 we look at expansion based systems that do have strategy extraction. We show that it is necessary for their underlying proof systems to have feasible interpolation. A sufficient condition with a relationship to feasible interpolation is also shown.

2 Preliminiaries

2.1 Proof Complexity

Let Γ be an alphabet, with Γ^* being the set of strings over Γ. Let $\mathcal{L}$ be a language over Γ, in other words $\mathcal{L} \subset \Gamma^*$. A *proof system* [9] for $\mathcal{L}$ is a polynomial time computable partial function $f : \Gamma^\star \rightharpoonup \Gamma^\star$ with range $\mathcal{L}$. The size $|\pi|$ of a proof π in $\Gamma^\star$ is the number of characters it contains. Proof system f maps a proof to the theorem it proves (or refutes, in the case of a refutational proof system). Soundness of f requires that $rng(f) \subseteq \mathcal{L}$ and completeness that $rng(f) \supseteq \mathcal{L}$. Given a family $\{c_i \mid i \in \mathbb{N}\}$ of formulas $f_i \in \mathcal{L}$, and a family of f-proofs $P = \{p_i \mid, i \in \mathbb{N}\}$ such that $f(p_i) = c_i$ we can say that p_i is *polynomially bounded* if $|p_i| \leq |c_i|^{O(1)}$. An even stronger property is that P is said to be *uniform* if there is a polynomial-time function h such that $h(c_i) = p_i$.

In propositional logic a literal is a Boolean variable x or its negation $\neg x$. A clause is a disjunction of literals. A formula in conjunctive normal form (CNF) is a conjunction of clauses. Let l be a literal. If $l = x$ then $\bar{l} = \neg x$, if $l = \neg x$

then $\bar{l} = x$. A CNF is naturally understood as a set of clauses, and a clause as a set of literals. Where it is convenient to do so we will therefore use set notation $C \in \phi$ and $l \in C$ to state that clause C appears in ϕ and literal l appears in C. It is often convenient to notationally treat clauses as unordered disjunctions and sets simultaneously, so we can use $C \vee l$ to denote the clause that contains all literals of clause C and also the literal l if it is not already included, and $D \cup E$ to denote the disjunction of all literals that appear in either clause D or E. An assignment τ for a formula ϕ over n variables is a partial function from the variables of ϕ to $\{0,1\}^n$. $\tau(C)$ is the result of evaluating clause C under assignment τ, and $\phi|_\tau = \{\tau(C) \mid C \in \phi\}$. For formula (or circuit) ϕ, we define $\phi[b/x]$ so that all instances of variable x in ϕ are replaced with $b \in \{0,1\}$.

2.2 Quantified Boolean Formulas

Quantified Boolean formulas (QBF) extend propositional logic by allowing for Boolean variables to be universally or existentially quantified [19]. $\forall x\, \Psi$ is satisfied by the same truth assignments as $\Psi[0/x] \wedge \Psi[1/x]$ and $\exists x\, \Psi$ is satisfied by the same truth assignments as $\Psi[0/x] \vee \Psi[1/x]$. In a closed QBF all variables must be quantified. A prenex QBF Ψ consists of a prefix Π defining how each variable is quantified and a propositional part ϕ called the matrix. We write $\Psi = \Pi\phi$. The prefix Π has a linearly ordered structure. A PCNF is a QBF in prenex form and with the propositional part in conjunctive normal form. We consider only closed PCNFs.

Starting from the left we can assign each variable x a level, denoted $\text{lv}(x)$. The first variable has level 1. The level is incremented by 1 every time the quantifier type changes and otherwise remains the same. It is often convenient to write quantifiers in a prefix only when the level changes. When QBF are written in this way we can think of entire levels quantifying *blocks* of variables.

2.3 Winning Strategies

A closed prenex QBF is analogous to a two player game with perfect information, in which one player is responsible for assigning values to the existentially quantified variables and the other to the universally quantified variables. The players make assignments according to the quantifier prefix, so each level of the prefix corresponds to one turn in the game, moving from left to right. The existential player wins the game if the formula evaluates to true once all assignments have been made, the universal player wins if the formula evaluates to false.

A strategy for the universal player on QBF $\Pi\phi$ is a set of rules for making the assignments to each universal variable u. The rule for setting u must depend only on variables earlier than (to the left of) u in Π, respecting the idea that when u is being decided the universal player cannot know what choices will be made in future turns. If this strategy ensures the universal player always wins games on $\Pi\phi$ (however the existential player makes assignments), then it is called a winning strategy. A QBF is false if and only if the universal player has a winning strategy. Strategies for the existential player are defined analogously. A refutational proof system is said to admit strategy extraction if and only if it is possible to efficiently (i.e. in polynomial time in the size of the proof) construct a circuit representing a winning strategy for the universal player from a refutation of a QBF.

2.4 Expansion Based Proof Systems

Since QBF includes and extends all propositional formulas, proving (or refuting) QBFs typically involves adapting existing propositional proof systems to deal with variables that are now quantified.

One such approach is to take the semantic definition of the universal quantifier $\forall u \Psi \equiv \Psi[0/u] \wedge \Psi[1/u]$, which can be used as a rule to eliminate universal quantifiers. If Ψ is a QBF then $\Psi[0/u]$ and $\Psi[1/u]$ each contain their own quantifiers, so the variables bound by these quantifiers would have to be renamed to avoid repeating the other's variables. We take a convention of renaming these variables by putting a partial assignment in the superscript such that the variables $X = \{x_i \mid i \in I\}$ bound in Ψ are renamed $X^{0/u} = \{x_i^{0/u} \mid i \in I\}$ in $\Psi[0/u]$ and $X^{1/u} = \{x_i^{1/u} \mid i \in I\}$ in $\Psi[1/u]$. Repeated expansions create a larger superscript e.g.

$$
\begin{aligned}
&\forall u \exists x \forall v \exists y\, (\neg u \vee x \vee v \vee \neg y) \\
\equiv &\forall u \exists x \exists y^{0/v}\, \exists y^{1/v}\, (\neg u \vee x \vee 0 \vee \neg y^{0/v}) \wedge (\neg u \vee x \vee 1 \vee \neg y^{1/v}) \\
\equiv &\forall u \exists x \exists y^{0/v}\, \exists y^{1/v}\, (\neg u \vee x \vee \neg y^{0/v}) \\
\equiv &\exists x^{0/u}\, \exists x^{1/u}\, \exists y^{0/u,0/v}\, \exists y^{0/u,1/v}\, \exists y^{1/u,0/v}\, \exists y^{1/u,1/v} \\
&(1 \vee x^{0/u} \vee \neg y^{0/u,0/v}) \wedge (0 \vee x^{1/u} \vee \neg y^{1/u,0/v}) \\
\equiv &\exists x^{0/u}\, \exists x^{1/u}\, \exists y^{0/u,0/v}\, \exists y^{0/u,1/v}\, \exists y^{1/u,0/v}\, \exists y^{1/u,1/v} \\
&(x^{1/u} \vee \neg y^{1/u,0/u})
\end{aligned}
$$

Note that because we started here with a prenex formula, we can maintain that throughout the expansions. In the end, once we expand all universal variables we have a prenex QBF with only existential quantifiers, this is known as a *full expansion.* Deciding the truth of a closed PCNF with only existential quantifiers is simply a propositional satisfiability problem. If we use a refutation system S we can attempt to refute the expanded formula.

In fact for any refutational propositional proof system S we can create a refutational QBF proof system (that is refutationally complete) by taking the full expansion and showing a contradiction using propositional system S. Such a system would easily have many exponential lower bounds due to the explosion caused by the full expansion on a linear number of universal variables.

In practice we can often do better than this. The full expansion gives a large conjunction and we may only need to use *some* of the conjuncts in order to prove a contradiction. This can be tightened up further when the original QBF is a prenexed conjunction (like a PCNF), checking whether a conjunct is in the full expansion can be decided in polynomial time. We define this formally below.

$\mathsf{S}+\forall\mathsf{Exp}_{0,1}$ We start with a propositional proof system S and prenex QBF $\Psi = \Pi\phi$, where Π is the quantifier prefix and ϕ is a propositional matrix in variables of Π. We treat ϕ as a conjunction of formulas.

Let τ be a full assignment to all universal variables and let l be an existentially quantified literal. We define $\mathsf{restrict}_l(\tau)$ to be the partial assignment of τ for all universal variables whose level (in the prefix) is less than that of the variable of l. Now let us use that to define C^τ, where C is a propositional formula in both existential and universal variables.

C^τ is the same as C except that we replace every existential literal l with the annotated literal $l^{\mathsf{restrict}_l(\tau)}$ and every universal variable u with its value $\tau(u)$.

Definition 1. *The refutational QBF proof system $\mathsf{S}+\forall\mathsf{Exp}_{0,1}$ allows the instantiation of* ***axiom*** *C^τ, whenever C is a conjunct from the matrix and τ is an assignment to all universal variables; it generates an S refutation of the conjunction of the axioms, treating differently annotated variables as different.*

An $\mathsf{S}+\forall\mathsf{Exp}_{0,1}$ proof [2] π of QBF Ψ therefore consists of a propositional S proof of a sub-conjunction of the full expansion. We denote this conjunction as $\mathsf{subexp}_\pi(\Psi)$.

A well-known example of $\mathsf{S}+\forall\mathsf{Exp}_{0,1}$ is $\forall\mathsf{Exp}+\mathsf{Res}$ [16] which is obtained when S is propositional resolution. This is the proof system that underlies the reasoning in the competitive QBF solver RAReQS [15]. Resolution is chosen because of its use in SAT solving.

2.5 QRAT

The QRAT proof system [14] was introduced as a universal proof checking format for QBF. It is able to express many QBF preprocessing techniques and proof systems. QRAT works on a QBF $\Pi\phi$ in PCNF which is modified throughout the proof by satisfiability preserving rules. Clauses may be added, altered or deleted depending on the current status of $\Pi\phi$.

QRAT may be used either to prove that a QBF is true or to refute it. The refutational version of QRAT uses the following rules: Asymmetric Tautology Addition (ATA), Quantified Resolution Asymmetric Tautology Addition (QRATA), Quantified Resolution Asymmetric Tautology Universal (QRATU), Extended Universal Reduction (EUR), and Clause Deletion. We define only the rules that are relevant for this paper, for a full definition of the QRAT system please refer to [14].

If C is a clause, then $\bar{C}$ is the conjunction of the negation of the literals in C. *Unit propagation* is a procedure on a formula ϕ in CNF that builds a partial assignment τ. This assignment is applied to ϕ and then for any literal l that appears in a singleton (unit) clause in the resulting formula, the assignment satisfying l is added to τ. This is repeated until reaching fix-point, which must happen in polynomial time in the number of clauses in ϕ. Unit propagation is used extensively in QRAT for deciding whether a derivation rule may be applied. We denote by $\phi \vdash_1 \perp$ that unit propagation derives the empty clause from ϕ.

In the following definitions, $\Pi\phi$ is a closed PCNF and C a clause not in ϕ. Π' is a prefix including the variables of ϕ and C, Π is identical to Π' except that it contains the variables of ϕ only.

Definition 2 (Asymmetric Tautology Addition (ATA)). *Suppose* $\phi \wedge \bar{C} \vdash_1 \perp$. *Then we can make the following inference*

$$\frac{\Pi\phi}{\Pi'\phi \wedge C}\ (ATA)$$

Definition 3 (Outer Clause). *Suppose C contains a literal l. Consider all clauses D in ϕ with $\bar{l} \in D$. The* outer clause *O^l_D of D is* $\{k \in D \mid \text{lv}(k) \leq_\Pi \text{lv}(l), k \neq \bar{l}\}$.

Definition 4 (Quantified Resolution Asymmetric Tautology Addition (QRATA)). *If C contains an existential literal l such that for every $D \in \phi$ with $\bar{l} \in D$, $\phi \wedge \bar{C} \wedge \bar{O}^l_D \vdash_1 \perp$ then we can derive*

$$\frac{\Pi\phi}{\Pi'\phi \wedge C}\ (QRATA\ w.r.t.\ l)$$

Note that for Definition 4 we have used the original definition of QRATA as it appears in [14], where it is explicitly stated that new variables can appear anywhere in the prefix. In [13] another definition is used where new variables only appear at the end of the prefix and are necessarily existential. This paper is in line with the original paper and recent papers such as [8,17,18].

Definition 5 (Extended Universal Reduction (EUR)). *Given a clause $C \vee u$ with universal literal u, consider extending C by*
$C \leftarrow C \cup \{k \in D \mid \text{lv}(k) >_\Pi \text{lv}(u) \text{ or } k = \bar{u}\}$,
where $D \in \phi$ is any clause with some $p : \text{lv}(p) >_\Pi \text{lv}(u)$, $p \in C$ and $\bar{p} \in D$, until we reach a fix-point denoted C'. If $\bar{u} \notin C'$ then we can perform the following rule.

$$\frac{\Pi\phi \wedge (C \vee u)}{\Pi'\phi \wedge C}\ (EUR)$$

Π' differs from Π only in that it does not contain u if $u \notin \phi \cup C$.

We can also define a weaker version of QRAT, QRAT(UR), which uses universal reduction instead of EUR.

Definition 6 (Universal Reduction (UR)). *Given a clause $C \vee u$ with universal literal u such that $\text{lv}(u) > \text{lv}(x)$ for all existentially quantified variables x in C we can apply the following rule.*

$$\frac{\Pi\phi \wedge (C \vee u)}{\Pi'\phi \wedge C}\ (UR)$$

3 Cheating a QBF Game

It is rumoured that the famous chess players Alekhine and Bogoljubov were once both separately challenged to a game of correspondence chess by an anonymous opportunist. The third player had deviously remembered the moves of each opponent to play Alekhine's and Bogoljubov's moves against each other, effectively removing themselves from the game. The player was guaranteed to win or draw in at least one game, and with the money odds against them, they stood to make a profit.

We see that this devious idea can also be used in the conjunction of QBF two-player games. We will show that these conjunctions have short QRAT proofs. We take a QBF and conjunct it with its negation in new variables. We interleave the prefixes so that the existential player plays first and the universal player is able to copy the moves at the right time. The universal player has to win on only one of the conjuncts and an easy winning strategy is to copy the opponent's move for the other side. The easy winning strategy is essential for the short proofs, but despite the guaranteed win, it is PSPACE-hard to find out which game the universal player wins prior to playing it. In the next section we add an extra universal variable that requires the calculation of who wins in order to make the game hard. However we see that expansion allows us to quickly return to the original easy problem.

In this section we will define these formulas that conjunct a QBF and its negation and show how a short QRAT proof can be uniformly obtained.

3.1 Duality Formulas

Let X be the set of variables $\{x_1, \dots, x_{2n}\}$ and $\phi(X)$ a CNF in the variables of X. Then $\Pi\phi(X)$ with prefix $\Pi = \forall x_1 \exists x_2 \forall x_3 \dots \exists x_{2n}$ is a closed PCNF. We also define a second set of $2n$ variables $X' = \{x'_1, \dots, x'_{2n}\}$ and an alternative prefix $\Pi' = \exists x'_1 \forall x'_2 \exists x'_3 \dots \forall x'_{2n}$. The QBF $\Pi\phi(X) \wedge \Pi'\neg\phi(X')$ is necessarily false. However this QBF is not in PCNF, which many proof systems require.

Firstly we will transform $\neg\phi(X')$ into a CNF $\bar{\phi}(X', T)$ via the use of Tseitin variables $T = \{t_K \mid K \in \phi(X)\}$. We overload the $'$ notation:

- For literal l if $l = x_i$ then $l' = x'_i$ and if $l = \neg x_i$ then $l' = \neg x'_i$.
- For each clause K in $\phi(X)$ we denote the corresponding clause in $\phi(X')$ as K' so that $K' = \bigvee_{l \in K} l'$.

We require that $\bar{\phi}(X', T)$ is true precisely when $\phi(X')$ is false. We will introduce clauses stating that variable t_K is true if and only if clause K' is satisfied. Then $\phi(X')$ is false if and only if at least one t_K is false, so we will also add a clause specifying that this must hold.

$\bar{\phi}(X', T)$ contains the following clauses:

- $(\neg t_K \vee K')$ for each clause K in $\phi(X)$
- $(\neg l' \vee t_K)$ for each literal l in K and each K in $\phi(X)$
- $\left(\bigvee_{K \in \phi(X)} \neg t_K\right)$

The next part is the most important – the prenexing of the QBF. We place every universal variable to the right of its existential counterpart. The auxiliary T variables must be placed at the end of the prefix. Thus, from any PCNF $\Psi = \Pi\phi$ we generate a formula $\mathsf{Duality}(\Pi\phi)$ encoding in PCNF the claim that both Ψ and its negation are true:

$$\mathsf{Duality}(\Pi\phi) = \exists x'_1 \forall x_1 \exists x_2 \forall x'_2 \dots \exists x'_{2n-1} \forall x_{2n-1} \exists x_{2n} \forall x'_{2n} \exists T\ \phi(X) \wedge \bar{\phi}(X', T)$$

3.2 Short Proofs of Duality Formulas

In [6], Beyersdorff et al. showed short $\mathsf{Frege} + \forall\mathsf{red}$ proofs of a family of QBFs that take an input of a graph and state that there is a k-clique (Clique) and dually that there is no k-clique (Co-Clique). The short proofs exploited the fact that the Co-Clique part of the formula was structured in a similar way to the Clique part.

We generalise this approach here for short proofs of the $\mathsf{Duality}$ formulas. First we will give a sketch proof of how this can be done using $\mathsf{Frege} + \forall\mathsf{red}$ rules before we show those short proofs formally in QRAT. $\mathsf{Frege} + \forall\mathsf{red}$ is simply a propositional Frege system augmented with the $\forall\mathsf{red}$ rule for removing universally quantified variables. $\forall\mathsf{red}$ allows to substitute a Boolean value for universally quantified u in a previously derived line, provided that $\mathrm{lv}(u) > \mathrm{lv}(x)$ for all existentially quantified x in the proof line.

The clauses in $\mathsf{Duality}(\Pi\phi)$ state $\bigwedge_K (t_K \leftrightarrow K')$, $\bigvee_K \neg t_K$ and $\bigwedge K$.

Recall that clause K is identical to clause K' with all instances of x'_i replaced with x_i (for all i). From assumption $\bigwedge_{i=1}^{2n} (x_i \leftrightarrow x'_i)$ we would find a contradiction in polynomially many Frege steps. The outline of the derivation is given below:

$$\dfrac{\bigvee_K \neg t_K \qquad \dfrac{\bigwedge K \qquad \dfrac{\bigwedge_K (t_K \leftrightarrow K') \qquad \bigwedge_{i=1}^{2n} (x_i \leftrightarrow x'_i)}{\bigwedge_K (t_K \leftrightarrow K)}}{\bigwedge_K t_K}}{\bot}$$

We therefore conclude that $\bigvee_{i=1}^{2n} \neg(x_i \leftrightarrow x'_i)$.

Now, starting from the variables quantified innermost in the prefix, we perform $\forall\mathsf{red}$ on all universally quantified x'_{2j} and x_{2j+1}:

$$\neg(x_{2n} \leftrightarrow 0) \vee \bigvee_{i=1}^{2n-1} \neg(x_i \leftrightarrow x'_i) = x_{2n} \vee \bigvee_{i=1}^{2n-1} \neg(x_i \leftrightarrow x'_i)$$

Reduction can also be done with $x'_{2j} = 1$

$$\neg(x_{2n} \leftrightarrow 1) \vee \bigvee_{i=1}^{2n-1} \neg(x_i \leftrightarrow x'_i) = \neg x_{2n} \vee \bigvee_{i=1}^{2n-1} \neg(x_i \leftrightarrow x'_i)$$

We can resolve these two disjunctions together and conclude $\bigvee_{i=1}^{2n-1} \neg(x_i \leftrightarrow x'_i)$.

Now x_{2n-1} is the innermost universally quantified variable. The same sequence of steps is applied for each universal variable leading to a contradiction which completes the proof.

This proof idea works for showing short proofs in QRAT. In fact these proofs have a uniform structure.

Theorem 1. *Given a formula* $\mathsf{Duality}(\Pi\phi)$ *we can in polynomial time construct a quadratic size QRAT(UR) refutation of* $\mathsf{Duality}(\Pi\phi)$.

Proof. Let $|K|$ be the number of literals in clause $K \in \phi$, then $|\mathsf{Duality}(\Pi\phi)| \geq |\phi| \geq \Sigma_{K\in\phi}|K|$. Recall $\Pi\phi$ has $2n$ variables.

Extension Variables. The refutation begins by using QRATA to introduce extension variable eq_{x_i} for each $x_i \in X$. Every eq_{x_i} is existentially quantified and is introduced to the prefix so that $\mathrm{lv}(\mathrm{eq}_{x_i}) > \mathrm{lv}(x_i), \mathrm{lv}(x'_i)$ and $\mathrm{lv}(\mathrm{eq}_{x_i}) < \mathrm{lv}(x_j), \mathrm{lv}(x'_j)$ for all $j > i$ (which is possible since $\mathrm{lv}(x_i), \mathrm{lv}(x'_i) < \mathrm{lv}(x_j), \mathrm{lv}(x'_j)$ in $\mathsf{Duality}(\Pi\phi)$ whenever $j > i$). For each $x_i \in X$ we use QRATA to add four clauses:

- $(\neg x_i \vee x'_i \vee \neg \mathrm{eq}_{x_i})$
- $(x_i \vee \neg x'_i \vee \neg \mathrm{eq}_{x_i})$
- $(\neg x_i \vee \neg x'_i \vee \mathrm{eq}_{x_i})$
- $(x_i \vee x'_i \vee \mathrm{eq}_{x_i})$

Recall that adding a clause by QRATA requires that we have an existential literal l in the new clause C such that $\phi \wedge \bar{C} \wedge \bar{O}^l_D \vdash_1 \bot$ for all D with $\bar{l} \in D$. For the first two clauses this is vacuously satisfied with $l = \neg\mathrm{eq}_{x_i}$ since eq_{x_i} does not appear positively anywhere in the formula. To add the latter clauses we have $l = \mathrm{eq}_{x_i}$ and must consider the two outer clauses $(\neg x_i \vee x'_i)$ and $(x_i \vee \neg x'_i)$. The QRATA condition is satisfied for $(\neg x_i \vee \neg x'_i \vee \mathrm{eq}_{x_i})$ because $x_i \wedge x'_i \wedge x_i \wedge \neg x'_i \vdash_1 \bot$ and $x_i \wedge x'_i \wedge \neg x_i \wedge x'_i \vdash_1 \bot$, and similarly for the final clause.

For each of the original $2n$ variables in $\Pi\phi$ we have added four clauses of constant size. Following $O(n)$ steps the formula has increased in length by $O(n)$ characters.

Non-Equivalence of X and X′. The next three ATA steps are equivalent to those in the derivation of $\bigvee_{i=1}^{2n} \neg(x_i \leftrightarrow x'_i)$ in the sketch proof above.

- $(\bigvee_{i=1}^{2n} \neg\mathrm{eq}_{x_i} \vee t_K \vee \bar{l})$ for every $K \in \phi(X)$ and every $l \in K$
- $(\bigvee_{i=1}^{2n} \neg\mathrm{eq}_{x_i} \vee t_K)$ for every $K \in \phi(X)$
- $(\bigvee_{i=1}^{2n} \neg\mathrm{eq}_{x_i})$

Each clause has $O(n)$ literals and there are at most $|\phi|$ clauses of each type. In $O(|\phi|)$ proof steps the formula has increased in length by $O(n|\phi|)$.

Removing the Universal Variables. Finally, we want to derive $A_j = (\bigvee_{i=1}^{j-1} \neg \mathrm{eq}_{x_i})$ for $j = 2n \dots 1$ (thus $j = 1$ means that we have derived the empty clause). Assuming that we already have A_{j+1} we can use ATA to add:

- $(\bigvee_{i=1}^{j-1} \neg \mathrm{eq}_{x_i} \vee x_j \vee x'_j)$
- $(\bigvee_{i=1}^{j-1} \neg \mathrm{eq}_{x_i} \vee \neg x_j \vee \neg x'_j)$

In these clauses whichever of x_j and x'_j is universally quantified is innermost by the construction of $\mathsf{Duality}(\Pi\phi)$ and the decision of where to introduce the variables eq_{x_i} in the prefix. Without loss of generality, assume x'_j is universally quantified so we can use UR to derive clauses $(\bigvee_{i=1}^{j-1} \neg \mathrm{eq}_{x_i} \vee x_j)$ and $(\bigvee_{i=1}^{j-1} \neg \mathrm{eq}_{x_i} \vee \neg x_j)$, then ATA allows to add the resolvent A_j.

For each of the $2n$ variables from ϕ there are five proof steps in this final part of the refutation, each introducing a new clause of size $O(n)$, and in total the formula has increased in length by $O(n^2)$. The whole refutation therefore has size $O(|\,\mathsf{Duality}(\Pi\phi)|^2)$. □

4 Making Strategies Hard

The formulas $\mathsf{Duality}(\Pi\phi)$ have short winning strategies for the universal player, namely to always play so that $x_i = x'_i$. We know also that one of $\Pi\phi(X)$ or $\Pi'\bar{\phi}(X', T)$ is false and so has a winning strategy for the universal player. Deciding which subformula is false is PSPACE-hard and the winning strategy for the false formula could be much more complicated than the strategy for $\mathsf{Duality}(\Pi\phi)$. We introduce formulas exploiting this hardness:

$$\mathsf{Select}(\Pi\phi) = \forall u\ \mathcal{Q}\ \exists T\ (\phi_u(X)) \wedge (\bar{\phi}_{\neg u}(X', T))$$

$$\text{where } \phi_l(X) = \bigwedge_{K \in \phi(X)} (K \vee l)$$

$$\text{and } \mathcal{Q} = \exists x'_1 \forall x_1 \exists x_2 \forall x'_2 \dots \exists x'_{2n-1} \forall x_{2n-1} \exists x_{2n} \forall x'_{2n}$$

4.1 Short Proofs of Select Formulas in QRAT

It was shown in [13] that satisfaction QRAT has strategy extraction, and in [8] that refutational QRAT(UR) has strategy extraction. In this section we use the formulas $\mathsf{Select}(\Pi\phi)$ to show that refutational QRAT does not have strategy extraction under a strong complexity assumption.

Theorem 2. *QRAT has short uniform proofs of* $\mathsf{Select}(\Pi\phi)$ *for any QBF* $\Pi\phi$.

Proof. The first step in the proof is to use Extended Universal Reduction (EUR) to remove u from all clauses in $\phi_u(X)$ and $\neg u$ from all clauses in $\bar{\phi}_{\neg u}(X', T)$. Using EUR to reduce l in C requires that $\bar{l}$ does not appear in C' (the fix-point of the inner expansion as given in Definition 5). In other words, there is no inner resolution path between any clauses containing the removed literal and its negation. We can only add literals to the inner expansion from clauses that share variables in common with the current inner expansion. However u and $\neg u$ appear in sections of the formula that have no other variables in common. Hence we can always reduce u (and $\neg u$) in $\mathsf{Select}(\Pi\phi)$.

Having performed these (polynomially many) EUR steps the formula is identical to $\mathsf{Duality}(\Pi\phi)$, which is uniformly refuted as in Theorem 1. □

Corollary 1. *Refutational QRAT does not have strategy extraction unless P = PSPACE.*

Proof. If QRAT has strategy extraction we can decide the truth of closed QBF in polynomial time – a PSPACE-complete problem.

Given a QBF $\Pi\phi$, with Π a prefix and ϕ a propositional formula in the variables of Π, we create the formula $\mathsf{Select}(\Pi\phi)$ and then in polynomial time we can output the proofs as in Theorem 2. Then from the proof of $\mathsf{Select}(\Pi\phi)$ we can extract the strategy for u. Since u is outermost in the prefix, this strategy must be constant. If the strategy sets $u = 0$ then all clauses in $\bar{\phi}_{\neg u}(X', T)$ are immediately satisfied so we know that the rest of the extracted strategy is a strategy for $\Pi\phi$, showing that $\Pi\phi$ is false. Similarly, if the strategy sets $u = 1$ then it must be the case that $\bar{\phi}_{\neg u}(X', T)$ is false and so, by construction, $\Pi\phi$ is true. Therefore we have a polynomial time decision procedure for an arbitrary QBF. □

In fact, the full power of EUR is not required. QRAT(UR) is capable of refuting the formulas $\mathsf{Duality}(\Pi\phi)$, and the initial EUR step can be replaced by universal expansion of u, producing a formula equivalent to $\mathsf{Duality}(\Pi\phi)$ with renamed variables. Even QBF solvers whose underlying proof system uses universal reduction to handle universally quantified variables often employ a preprocessing stage that includes universal expansion. Our $\mathsf{Select}(\Pi\phi)$ formulas show that a single initial expansion step may be sufficient to prevent strategy extraction.

5 Relation to Feasible Interpolation

The results of the previous section indicate that expansion steps may prevent strategy extraction. However we have seen many proof systems and solvers that admit strategy extraction despite using universal expansion. It is clear that the other rules of the calculus play an important role on whether or not strategy extraction is admissible.

If we wish to guarantee strategy extraction in our proof systems and solvers, it may be important for future work to explore *sufficient* conditions for strategy extraction when using expansion. In this section we instead explore a *necessary* condition for strategy extraction.

In Corollary 1 the necessary condition for strategy extraction in QRAT was that we needed efficient circuits that calculated the truth of $\Pi\phi$ in order to have strategy extraction for $\mathsf{Duality}(\Pi\phi)$. We can think of a strategy for u acting as a circuit deciding between $\Pi\phi$ and $\neg(\Pi\phi)$. In propositional logic the efficient extraction of these deciding circuits (known as interpolating circuits or interpolants) from a proof is a well studied technique known as feasible interpolation. In this section, we will use our lower bound technique to place necessary conditions for strategy extraction on a large class of QBF proof systems.

Given a true propositional implication $A(P,Q) \rightarrow B(P,R)$ (or, equivalently, a false conjunction $A(P,Q) \wedge \neg B(P,R)$) Craig's interpolation theorem [11], states that there is an interpolant $C(P)$ in only the joint variables P. Feasible interpolation is a property of proof systems. A proof system has feasible interpolation [21,24] if and only if there is a polynomial time procedure that takes a proof of $A(P,Q) \rightarrow B(P,R)$ as an input and extracts an interpolating circuit $C(P)$.

In [5] Beyersdorff et al. lifted a version of the feasible interpolation lower bound technique from propositional logic to QBF. In Sect. 5 of [5] feasible interpolation was linked to strategy extraction by adding an extra universal variable with similarities to Sect. 4 and how the Select formulas are created from the Duality formulas.

Theorem 3. *Given any propositional refutation system* S, *if the refutational QBF proof system* $\mathsf{S}+\forall\mathsf{Exp}_{0,1}$ *has strategy extraction then* S *must have feasible interpolation.*[1]

Proof. Suppose $\mathsf{S}+\forall\mathsf{Exp}_{0,1}$ has strategy extraction and we have an S-refutation π of $A(P,Q) \wedge B(P,R)$ with P, Q, R disjoint sets of variables. We will show that we can find an interpolant in polynomial time.

We consider the following QBF

$$\exists P \forall u \exists Q \exists R (A(P,Q) \vee u) \wedge (B(P,R) \vee \bar{u})$$

We can refute this formula in $\mathsf{S}+\forall\mathsf{Exp}_{0,1}$ using π. Expansion gives us

$$(A(P,Q^{0/u}) \vee 0) \wedge (B(P,R^{0/u}) \vee 1) \wedge (A(P,Q^{1/u}) \vee 1) \wedge (B(P,R^{1/u}) \vee 0)$$

but this immediately simplifies to $A(P,Q^{0/u}) \wedge B(P,R^{1/u})$.

We can now refute this using π using $Q^{0/u}$ variables instead of Q, and using $R^{1/u}$ variables instead of R. The provision here is important for this as one could make S a pathological proof system that disallowed steps using variables named as in $R^{0/u}$, but allowed them named as in R.

We can then extract a strategy for u as a circuit in the variables P. However this circuit is also an interpolant for $A(P,Q) \wedge B(P,R)$. □

In regards to making *sufficient* conditions for strategy extraction using feasible interpolation we can look at the results that have come before. We see that ∀Exp+Res has strategy extraction. This is done by a "round-based" strategy extraction. This technique gives a winning response for the universal player by

[1] Provided the refutations of S work independently of the variable names. This is usually the case but one could create a pathological proof proof system where annotated variables are treated differently to normal ones.

taking the outermost block of existentially quantified variables and applying a restriction to the $\forall$Exp+Res proof on that block in correspondence to the existential player's moves. The universal player can then "read off", from the restricted proof, which universal assignment to the next block of variables is useful in the proof. The proof can be restricted by the universal assignment and we repeat until we end up with a complete set of universal responses and a falsified formula.

The "reading off" which clauses actually contribute to the proof is a weak form of feasible interpolation and so we can say we have strategy extraction for S+$\forall$Exp$_{0,1}$ whenever refutational proof system S satisfies two conditions. Note that because of Theorem 3 feasible interpolation is implied by these two conditions (although this can be shown without Theorem 3). The extraction technique is inspired by the one used in [12], instead here we use it for expansion systems.

Theorem 4. S+$\forall$Exp$_{0,1}$ *has strategy extraction whenever:*

1. S *is closed under restriction, meaning that from a refutation* π *of* ϕ *one can extract in polynomial time an* S *refutation* π_ϵ *of* $\phi|_\epsilon$ *for any assignment* ϵ *with* $|\pi_\epsilon| \leq |\pi|$.
2. *From any refutation* ρ_1 *in* S *of* $A(Q) \wedge B(R)$ *where* Q, R *share no common variables another refutation* ρ_2 *of either* $A(Q)$ *or* $B(R)$ *can be extracted in polynomial time with* $|\rho_2| \leq |\rho_1|$.

Resolution would be an example of such a system with both properties. It is not possible to say for certain that any proof system lacks one or both of these properties without a separation of P and NP. The first property is fairly common, even in stronger systems, but we do not expect systems such as bounded-depth Frege to have the second property.

Proof. Suppose we have a closed prenex QBF $\exists X \forall Y \Pi \phi$ where Π is a prefix in variables Z and ϕ is a propositional matrix with variables in X,Y and Z. Now suppose we have an S+$\forall$Exp$_{0,1}$ refutation π of $\exists X \forall Y \Pi \phi$. This gives an S proof π' of $\mathsf{subexp}_\pi(\exists X \forall Y \Pi \phi)$, a subset of the full expansion of $\exists X \forall Y \Pi \phi$ using π.

We will show that under conditions 1 and 2 we have a polynomial time procedure that takes any assignment ϵ to X and outputs a response μ in Y and a $\Pi \phi|_{\epsilon,\mu}$ refutation in S+$\forall$Exp$_{0,1}$.

From π', we can extract π'_ϵ in polynomial time, using condition 1, which provides an S refutation of $\mathsf{subexp}_\pi(\exists X \forall Y \Pi \phi)|_\epsilon$. For $D \in \phi$ we have that $C = D|_\epsilon$ is in $\phi|_\epsilon$, so every conjunct $D^\tau|_\epsilon$ of $\mathsf{subexp}_\pi(\exists X \forall Y \Pi \phi)|_\epsilon$ is also an axiom C^τ of $\forall Y \Pi \phi|_\epsilon$. Therefore, π'_ϵ becomes an S+$\forall$Exp$_{0,1}$ refutation π_ϵ of $\forall Y \Pi \phi|_\epsilon$.

Now we find the universal response in universal variables Y. We separate $Y = \{y_1 \dots y_m\}$ and we can start with a response c to y_1 and then find an S+$\forall$Exp$_{0,1}$ refutation of $\forall y_2 \dots y_m \Pi \phi|_{\epsilon, c/y_1}$. We make sure the proofs do not increase in size. Then we can repeat this for each variable in Y in turn.

Suppose we have an S+$\forall$Exp$_{0,1}$ refutation π_i of the QBF $\forall y_i \dots y_m \Pi \phi|_{\epsilon,\mu_i}$, where μ_i $1 \leq i \leq m$ is a Boolean assignment to variables $\{y_1, \dots, y_{i-1}\}$. The variables of $\mathsf{subexp}_{\pi_i}(\forall y_i \dots y_m \Pi \phi|_{\epsilon,\mu_i})$ can be partitioned into $Z_{0/y_i} = \{z^\alpha \mid z \in$

$Z, \alpha(y_i) = 0\}$ and $Z_{1/y_i} = \{z^\alpha \mid z \in Z, \alpha(y_i) = 1\}$. This completely partitions the variables because y_i is leftmost in the prefix.

Conjunct $C \in \mathsf{subexp}_{\pi_i}(\Pi\phi)|_{\epsilon,\mu_i}$ cannot mix variables Z_{0/y_i} and Z_{1/y_i} since the axiom rule in Definition 1 substitutes one or the other everywhere in the conjunct. Therefore $\mathsf{subexp}_{\pi_i}(\forall y_i \dots y_m \Pi\phi|_{\epsilon,\mu_i})$ can be written as $A(Z_{0/y_i}) \wedge B(Z_{1/y_i})$ with S refutation π'_i (based on the $\mathsf{S}{+}\forall\mathsf{Exp}_{0,1}$ refutation π_i).

We define a new partial assignment μ_{i+1}, which is defined as $\mu_{i+1}(y_j) = \mu_i(y_j)$ for $1 \leq j < i$. Now we can use condition 2 to extract from π'_i an S refutation π'_{i+1} of either $A(Z_{0/y_i})$ or $B(Z_{1/y_i})$ in polynomial time. If it is $A(Z_{0/y_i})$ then we let $\mu_{i+1}(y_i) = 0$ and if it is $B(Z_{1/y_i})$ then we let $\mu_{i+1}(y_i) = 1$. π'_{i+1} can be used as part of an $\mathsf{S}{+}\forall\mathsf{Exp}_{0,1}$ refutation π_{i+1} of $\forall y_{i+1} \dots y_m \Pi\phi|_{\epsilon,\mu_{i+1}}$ as $\mathsf{subexp}_{\pi_{i+1}}(\forall y_{i+1} \dots y_m \Pi\phi|_{\epsilon,\mu_{i+1}})$ is equal to $A(Z_{0/y_i})$ or $B(Z_{1/y_i})$. Condition 2 guarantees $|\pi'_{i+1}| \leq |\pi'_i|$ so $|\pi_{i+1}| \leq |\pi_i|$ as well.

Once we get to μ_m we have a complete assignment to Y and a guarantee that the remaining QBF game on $\Pi\phi|_{\epsilon,\mu_m}$ is false by the $\mathsf{S}{+}\forall\mathsf{Exp}_{0,1}$ refutation π_m, with $|\pi_m| \leq |\pi|$.

We can repeat this procedure for every universal block and we end up with the false proposition $\bot$ and since our proofs are non-increasing in size in each step we guarantee this can be done in a polynomial time procedure. □

6 Conclusion

We have answered an open question in QBF proof complexity by showing that refutational QRAT does not have strategy extraction, and have introduced a family of QBFs witnessing this fact. We have also formalised one condition for strategy extraction to be present in QBF proof systems using universal expansion. This adds to an existing awareness of the trade-off between strength of QBF proof systems and the ability to offer explanation via winning strategies [4].

In current QBF solvers that use inference from propositional SAT solvers implementing CDCL the propositional inference used is Resolution and the feasible interpolation property applies. As such it may be possible (as it is indeed possible in $\forall$Exp+Res) to have both expansion solving and strategy extraction together. There is also hope that, because the Cutting Planes proof system [10] has feasible interpolation [24], strategy extraction may still be compatible with QBF expansion if we were to use pseudo-Boolean solvers as a propositional component in QBF solvers.

However, if SAT solvers are developed with more power than both Resolution and Cutting Planes, then the problem of having both strategy extraction and expansion in QBF solvers becomes more serious. Extended Resolution is strong enough to cause these issues [22], but there is a degree of non determinism required in choosing the correct extension variables in practice. In the case of our Select and Duality formulas this task would be to find the eq extension variables, despite the fact we are linearly increasing the number of variables this way and increasing the search space for the solver.

References

1. Balabanov, V., Jiang, J.-H.R.: Resolution proofs and skolem functions in QBF evaluation and applications. In: Gopalakrishnan, G., Qadeer, S. (eds.) CAV 2011. LNCS, vol. 6806, pp. 149–164. Springer, Heidelberg (2011). https://doi.org/10.1007/978-3-642-22110-1_12
2. Beyersdorff, O., Bonacina, I., Chew, L.: Lower bounds: from circuits to QBF proof systems. In: Proceedings of the ACM Conference on Innovations in Theoretical Computer Science (ITCS 2016), pp. 249–260. ACM (2016)
3. Beyersdorff, O., Chew, L., Clymo, J., Mahajan, M.: Short proofs in QBF expansion. In: Janota, M., Lynce, I. (eds.) SAT 2019. LNCS, vol. 11628, pp. 19–35. Springer, Cham (2019). https://doi.org/10.1007/978-3-030-24258-9_2
4. Beyersdorff, O., Chew, L., Janota, M.: Proof complexity of resolution-based QBF calculi. In: Proceedings of the Symposium on Theoretical Aspects of Computer Science, pp. 76–89. LIPIcs series (2015)
5. Beyersdorff, O., Chew, L., Mahajan, M., Shukla, A.: Feasible interpolation for QBF resolution calculi. Logical Methods Comput. Sci. **13**(2) (2017). https://doi.org/10.23638/LMCS-13(2:7)2017. https://lmcs.episciences.org/3702
6. Beyersdorff, O., Chew, L., Mahajan, M., Shukla, A.: Understanding cutting planes for QBFs. Inf. Comput. **262**, 141–161 (2018). https://doi.org/10.1016/j.ic.2018.08.002. http://www.sciencedirect.com/science/article/pii/S0890540118301184
7. Beyersdorff, O., Hinde, L., Pich, J.: Reasons for hardness in QBF proof systems. In: Electronic Colloquium on Computational Complexity (ECCC) 24, Report no. 44 (2017). https://eccc.weizmann.ac.il/report/2017/044
8. Chew, L., Clymo, J.: The equivalences of refutational QRAT. In: Janota, M., Lynce, I. (eds.) SAT 2019. LNCS, vol. 11628, pp. 100–116. Springer, Cham (2019). https://doi.org/10.1007/978-3-030-24258-9_7
9. Cook, S.A., Reckhow, R.A.: The relative efficiency of propositional proof systems. J. Symb. Logic **44**(1), 36–50 (1979)
10. Cook, W.J., Coullard, C.R., Turán, G.: On the complexity of cutting-plane proofs. Discret. Appl. Math. **18**(1), 25–38 (1987)
11. Craig, W.: Three uses of the Herbrand-Gentzen theorem in relating model theory and proof theory. J. Symb. Logic **22**(3), 269–285 (1957)
12. Goultiaeva, A., Van Gelder, A., Bacchus, F.: A uniform approach for generating proofs and strategies for both true and false QBF formulas. In: Walsh, T. (ed.) International Joint Conference on Artificial Intelligence IJCAI, pp. 546–553. IJCAI/AAAI (2011)
13. Heule, M., Seidl, M., Biere, A.: Efficient extraction of Skolem functions from QRAT proofs. In: Formal Methods in Computer-Aided Design, FMCAD 2014, Lausanne, Switzerland, 21–24 October 2014, pp. 107–114 (2014). https://doi.org/10.1109/FMCAD.2014.6987602
14. Heule, M.J.H., Seidl, M., Biere, A.: A unified proof system for QBF preprocessing. In: Demri, S., Kapur, D., Weidenbach, C. (eds.) IJCAR 2014. LNCS (LNAI), vol. 8562, pp. 91–106. Springer, Cham (2014). https://doi.org/10.1007/978-3-319-08587-6_7
15. Janota, M., Klieber, W., Marques-Silva, J., Clarke, E.: Solving QBF with counterexample guided refinement. In: Cimatti, A., Sebastiani, R. (eds.) SAT 2012. LNCS, vol. 7317, pp. 114–128. Springer, Heidelberg (2012). https://doi.org/10.1007/978-3-642-31612-8_10

16. Janota, M., Marques-Silva, J.: Expansion-based QBF solving versus Q-resolution. Theor. Comput. Sci. **577**, 25–42 (2015)
17. Kiesl, B., Heule, M.J.H., Seidl, M.: A little blocked literal goes a long way. In: Gaspers, S., Walsh, T. (eds.) SAT 2017. LNCS, vol. 10491, pp. 281–297. Springer, Cham (2017). https://doi.org/10.1007/978-3-319-66263-3_18
18. Kiesl, B., Seidl, M.: QRAT polynomially simulates ∀-Exp+Res. In: Janota, M., Lynce, I. (eds.) SAT 2019. LNCS, vol. 11628, pp. 193–202. Springer, Cham (2019). https://doi.org/10.1007/978-3-030-24258-9_13
19. Kleine Büning, H., Bubeck, U.: Theory of quantified Boolean formulas. In: Biere, A., Heule, M., van Maaren, H., Walsh, T. (eds.) Handbook of Satisfiability, Frontiers in Artificial Intelligence and Applications, vol. 185, pp. 735–760. IOS Press (2009)
20. Kleine Büning, H., Karpinski, M., Flögel, A.: Resolution for quantified Boolean formulas. Inf. Comput. **117**(1), 12–18 (1995)
21. Krajíček, J.: Interpolation theorems, lower bounds for proof systems and independence results for bounded arithmetic. J. Symb. Logic **62**(2), 457–486 (1997)
22. Krajíček, J., Pudlák, P.: Some consequences of cryptographical conjectures for S_2^1 and EF. Inf. Comput. **140**(1), 82–94 (1998)
23. Lonsing, F., Egly, U.: QRAT$^+$: generalizing QRAT by a more powerful QBF redundancy property. In: Galmiche, D., Schulz, S., Sebastiani, R. (eds.) IJCAR 2018. LNCS (LNAI), vol. 10900, pp. 161–177. Springer, Cham (2018). https://doi.org/10.1007/978-3-319-94205-6_12
24. Pudlák, P.: Lower bounds for resolution and cutting planes proofs and monotone computations. J. Symb. Logic **62**(3), 981–998 (1997)

Removing Algebraic Data Types from Constrained Horn Clauses Using Difference Predicates

Emanuele De Angelis[1,3(✉)], Fabio Fioravanti[1(✉)], Alberto Pettorossi[2,3], and Maurizio Proietti[3]

[1] DEC, University 'G. d'Annunzio', Chieti-Pescara, Pescara, Italy
fabio.fioravanti@unich.it
[2] DICII, University of Rome 'Tor Vergata', Rome, Italy
pettorossi@info.uniroma2.it
[3] IASI-CNR, Rome, Italy
{emanuele.deangelis,maurizio.proietti}@iasi.cnr.it

Abstract. We address the problem of proving the satisfiability of Constrained Horn Clauses (CHCs) with Algebraic Data Types (ADTs), such as lists and trees. We propose a new technique for transforming CHCs with ADTs into CHCs where predicates are defined over basic types, such as integers and booleans, only. Thus, our technique avoids the explicit use of inductive proof rules during satisfiability proofs. The main extension over previous techniques for ADT removal is a new transformation rule, called *differential replacement*, which allows us to introduce auxiliary predicates corresponding to the lemmas used when making inductive proofs. We present an algorithm that applies the new rule, together with the traditional folding/unfolding rules, for the automatic removal of ADTs. We prove that if the set of the transformed clauses is satisfiable, then so is the set of the original clauses. By an experimental evaluation, we show that the use of the new rule significantly improves the effectiveness of ADT removal, and that our approach is competitive with respect to a state-of-the-art tool that extends the CVC4 solver with induction.

1 Introduction

Constrained Horn Clauses (CHCs) constitute a fragment of the first order predicate calculus, where the Horn clause format is extended by allowing *constraints* on specific domains to occur in clause premises. CHCs have gained popularity as a suitable logical formalism for automatic program verification [3]. Indeed, many verification problems can be reduced to the satisfiability problem for CHCs.

Satisfiability of CHCs is a particular case of *Satisfiability Modulo Theories* (SMT), understood here as the general problem of determining the satisfiability of (possibly quantified) first order formulas where the interpretation of some function and predicate symbols is defined in a given constraint (or *background*)

This work has been partially supported by GNCS-INdAM, Italy.

N. Peltier and V. Sofronie-Stokkermans (Eds.): IJCAR 2020, LNAI 12166, pp. 83–102, 2020.
https://doi.org/10.1007/978-3-030-51074-9_6

theory [2]. Recent advances in the field have led to the development of a number of very powerful SMT (and, in particular, CHC) *solvers*, which aim at solving satisfiability problems with respect to a large variety of constraint theories. Among SMT solvers, we would like to mention CVC4 [1], MathSAT [5], and Z3 [14], and among solvers with specialized engines for CHCs, we recall Eldarica [22], HSF [20], RAHFT [26], and Spacer [29].

Even if SMT algorithms for unrestricted first order formulas suffer from incompleteness limitations due to general undecidability results, most of the above mentioned tools work well in practice when acting on constraint theories, such as Booleans, Uninterpreted Function Symbols, Linear Integer or Real Arithmetic, Bit Vectors, and Arrays. However, when formulas contain universally quantified variables ranging over inductively defined *algebraic data types* (ADTs), such as lists and trees, then the SMT/CHC solvers often show a very poor performance, as they do not incorporate induction principles relative to the ADT in use.

To mitigate this difficulty, some SMT/CHC solvers have been enhanced by incorporating appropriate induction principles [38,43,44], similarly to what has been done in automated theorem provers [4]. The most creative step which is needed when extending SMT solving with induction, is the generation of the auxiliary lemmas that are required for proving the main conjecture.

An alternative approach, proposed in the context of CHCs [10], consists in transforming a given set of clauses into a new set: (i) where all ADT terms are removed (without introducing new function symbols), and (ii) whose satisfiability implies the satisfiability of the original set of clauses. This approach has the advantage of separating the concern of dealing with ADTs (at transformation time) from the concern of dealing with simpler, non-inductive constraint theories (at solving time), thus avoiding the complex interaction between inductive reasoning and constraint solving. It has been shown [10] that the transformational approach compares well with induction-based tools in the case where lemmas are not needed in the proofs. However, in some satisfiability problems, if suitable lemmas are not provided, the transformation fails to remove the ADT terms.

The main contributions of this paper are as follows.

(1) We extend the transformational approach by proposing a new rule, called *differential replacement*, based on the introduction of suitable *difference predicates*, which play a role similar to that of lemmas in inductive proofs. We prove that the combined use of the fold/unfold transformation rules [17] and the differential replacement rule is *sound*, that is, if the transformed set of clauses is satisfiable, then the original set of clauses is satisfiable.
(2) We develop a transformation algorithm that removes ADTs from CHCs by applying the fold/unfold and the differential replacement rules in a fully automated way.
(3) Due to the undecidability of the satisfiability problem for CHCs, our technique for ADT removal is incomplete. Thus, we evaluate its effectiveness from an experimental point of view, and in particular we discuss the results obtained by the implementation of our technique in a tool, called AdtRem.

We consider a set of CHC satisfiability problems on ADTs taken from various benchmarks which are used for evaluating inductive theorem provers. The experiments show that AdtRem is competitive with respect to Reynolds and Kuncak's tool that augments the CVC4 solver with inductive reasoning [38].

The paper is structured as follows. In Sect. 2 we present an introductory, motivating example. In Sect. 3 we recall some basic notions about CHCs. In Sect. 4 we introduce the rules used in our transformation technique and, in particular, the novel differential replacement rule, and we show their soundness. In Sect. 5 we present a transformation algorithm that uses the transformation rules for removing ADTs from sets of CHCs. In Sect. 6 we illustrate the AdtRem tool and we present the experimental results we have obtained. Finally, in Sect. 7 we discuss the related work and make some concluding remarks.

2 A Motivating Example

Let us consider the following functional program *Reverse*, which we write using the OCaml syntax [31]:

```
type list = Nil | Cons of int * list;;
let rec append l ys = match l with
  | Nil -> ys       | Cons(x,xs) -> Cons(x,(append xs ys));;
let rec rev l = match l with
  | Nil -> Nil      | Cons(x,xs) -> append (rev xs) (Cons(x,Nil));;
let rec len l = match l with
  | Nil -> 0        | Cons(x,xs) -> 1 + len xs;;
```

The functions `append`, `rev`, and `len` compute list concatenation, list reversal, and list length, respectively. Suppose we want to prove the following property:

$$\forall\, \texttt{xs}, \texttt{ys}.\ \texttt{len (rev (append xs ys))} \;=\; \texttt{(len xs)} \;+\; \texttt{(len ys)} \qquad (1)$$

Inductive theorem provers construct a proof of this property by induction on the structure of the list `l`, by assuming the knowledge of the following lemma:

$$\forall\, \texttt{x}, \texttt{l}.\ \texttt{len (append l (Cons(x,Nil)))} \;=\; \texttt{(len l)} \;+\; \texttt{1} \qquad (2)$$

The approach we follow in this paper avoids the explicit use of induction principles and also the knowledge of *ad hoc* lemmas. First, we consider the translation of Property (1) into a set of constrained Horn clauses [10,43], as follows[1]:

[1] In the examples, we use Prolog syntax for writing clauses, instead of the more verbose SMT-LIB syntax. The predicates `\=` (different from), `=` (equal to), `<` (less-than), `>=` (greater-than-or-equal-to) denote constraints between integers. The last argument of a Prolog predicate stores the value of the corresponding function.

```
1. false :- N2\=N0+N1, append(Xs,Ys,Zs), rev(Zs,Rs),
            len(Xs,N0), len(Ys,N1), len(Rs,N2).
2. append([],Ys,Ys).    3. append([X|Xs],Ys,[X|Zs]) :- append(Xs,Ys,Zs).
4. rev([],[]).          5. rev([X|Xs],Rs) :- rev(Xs,Ts), append(Ts,[X],Rs).
6. len([],N) :- N=0.    7. len([X|Xs],N1) :- N1=N0+1, len(Xs,N0).
```

The set of clauses 1–7 is satisfiable if and only if Property (1) holds. However, state-of-the-art CHC solvers, such as Z3 or Eldarica, fail to prove the satisfiability of clauses 1–7, because those solvers do not incorporate any induction principle on lists. Moreover, some tools that extend SMT solvers with induction [38,43] fail on this particular example because they are not able to introduce Lemma (2).

To overcome this difficulty, we apply the transformational approach based on the fold/unfold rules [10], whose objective is to transform a given set of clauses into a new set without occurrences of list variables, whose satisfiability can be checked by using CHC solvers based on the theory of Linear Integer Arithmetic only. The soundness of the transformation rules ensures that the satisfiability of the transformed clauses implies the satisfiability of the original ones. We apply the *Elimination Algorithm* [10] as follows. First, we introduce a new clause:

```
8. new1(N0,N1,N2) :- append(Xs,Ys,Zs), rev(Zs,Rs),
                     len(Xs,N0), len(Ys,N1), len(Rs,N2).
```

whose body is made out of the atoms of clause 1 which have at least one list variable, and whose head arguments are the integer variables of the body. By folding, from clause 1 we derive a new clause without occurrences of lists:

```
9. false :- N2\=N0+N1, new1(N0,N1,N2).
```

We proceed by eliminating lists from clause 8. By unfolding clause 8, we replace some predicate calls by their definitions and we derive the two new clauses:

```
10. new1(N0,N1,N2) :- N0=0, rev(Zs,Rs), len(Zs,N1), len(Rs,N2).
11. new1(N01,N1,N21) :- N01=N0+1, append(Xs,Ys,Zs), rev(Zs,Rs),
              len(Xs,N0), len(Ys,N1), append(Rs,[X],R1s), len(R1s,N21).
```

We would like to fold clause 11 using clause 8, so as to derive a recursive definition of `new1` without lists. Unfortunately, this folding step cannot be performed because the body of clause 11 does not contain a variant of the body of clause 8, and hence the Elimination Algorithm fails to eliminate lists in this example.

Thus, now we depart from the Elimination Algorithm and we continue our derivation by observing that the body of clause 11 contains the *subconjunction* '`append(Xs,Ys,Zs), rev(Zs,Rs), len(Xs,N0), len(Ys,N1)`' of the body of clause 8. Then, in order to find a variant of the whole body of clause 8, we may replace in clause 11 the remaining subconjunction '`append(Rs,[X],R1s), len(R1s,N21)`' by the new subconjunction '`len(Rs,N2), diff(N2,X,N21)`', where `diff` is a predicate, called *difference predicate*, defined as follows:

```
12. diff(N2,X,N21) :- append(Rs,[X],R1s), len(R1s,N21), len(Rs,N2).
```

From clause 11, by performing that replacement, we derive the following clause:

```
13. new1(N01,N1,N21) :- N01=N0+1, append(Xs,Ys,Zs), rev(Zs,Rs),
                len(Xs,N0), len(Ys,N1), len(Rs,N2), diff(N2,X,N21).
```

Now, we can fold clause 13 using clause 8 and we derive a new clause without list arguments:

```
14. new1(N01,N1,N21) :- N01=N0+1, new1(N0,N1,N2), diff(N2,X,N21).
```

At this point, we are left with the task of removing list arguments from clauses 10 and 12. As the reader may verify, this can be done by applying the Elimination Algorithm without the need of introducing additional difference predicates. By doing so, we get the following final set of clauses without list arguments:

```
false :- N2\=N0+N1, new1(N0,N1,N2).
new1(N0,N1,N2) :- N0=0, new2(N1,N2).
new1(N0,N1,N2) :- N0=N+1, new1(N,N1,M), diff(M,X,N2).
new2(M,N) :- M=0, N=0.
new2(M1,N1) :- M1=M+1, new2(M,N), diff(N,X,N1).
diff(N0,X,N1) :- N0=0, N1=1.
diff(N0,X,N1) :- N0=N+1, N1=M+1, diff(N,X,M).
```

The Eldarica CHC solver proves the satisfiability of this set of clauses by computing the following model (here we use a Prolog-like syntax):

```
new1(N0,N1,N2) :- N2=N0+N1, N0>=0, N1>=0, N2>=0.
new2(M,N) :- M=N, M>=0, N>=0.
diff(N,X,M) :- M=N+1, N>=0.
```

Finally, we note that if in clause 12 we substitute the atom `diff(N2,X,N21)` by its model computed by Eldarica, namely the constraint '`N21=N2+1, N2>=0`', we get exactly the CHC translation of Lemma (2). Thus, in some cases, the introduction of the difference predicates can be viewed as a way of automatically introducing the lemmas needed when performing inductive proofs.

3 Constrained Horn Clauses

Let *LIA* be the theory of linear integer arithmetic and *Bool* be the theory of boolean values. A *constraint* is a quantifier-free formula of $LIA \cup Bool$. Let $\mathcal{C}$ denote the set of all constraints. Let $\mathcal{L}$ be a typed first order language with equality [16] which includes the language of $LIA \cup Bool$. Let *Pred* be a set of predicate symbols in $\mathcal{L}$ not occurring in the language of $LIA \cup Bool$.

The integer and boolean types are said to be the *basic types*. For reasons of simplicity we do not consider any other basic types, such as real number, arrays, and bit-vectors, which are usually supported by SMT solvers [1,14,22]. We assume that all non-basic types are specified by suitable data-type declarations (such as the `declare-datatypes` declarations adopted by SMT solvers), and are collectively called *algebraic data types* (ADTs).

An *atom* is a formula of the form $p(t_1, \ldots, t_m)$, where p is a typed predicate symbol in *Pred*, and $t_1, \ldots, t_m$ are typed terms constructed out of individual variables, individual constants, and function symbols. A *constrained Horn clause* (or simply, a *clause*, or a CHC) is an implication of the form $A \leftarrow c, B$ (for clauses we use the logic programming notation, where comma denotes conjunction). The conclusion (or *head*) A is either an atom or *false*, the premise (or *body*) is the conjunction of a constraint $c \in \mathcal{C}$, and a (possibly empty) conjunction B of atoms. If A is an atom of the form $p(t_1, \ldots, t_n)$, the predicate p is said to be a *head predicate*. A clause whose head is an atom is called a *definite clause*, and a clause whose head is *false* is called a *goal*.

We assume that all variables in a clause are universally quantified in front, and thus we can freely rename those variables. Clause C is said to be a *variant* of clause D if C can be obtained from D by renaming variables and rearranging the order of the atoms in its body. Given a term t, by $vars(t)$ we denote the set of all variables occurring in t. Similarly for the set of all free variables occurring in a formula. Given a formula φ in $\mathcal{L}$, we denote by $\forall(\varphi)$ its *universal closure*.

Let $\mathbb{D}$ be the usual interpretation for the symbols in $LIA \cup Bool$, and let a $\mathbb{D}$-*interpretation* be an interpretation of $\mathcal{L}$ that, for all symbols occurring in $LIA \cup Bool$, agrees with $\mathbb{D}$.

A set P of CHCs is *satisfiable* if it has a $\mathbb{D}$-model and it is *unsatisfiable*, otherwise. Given two $\mathbb{D}$-interpretations $\mathbb{I}$ and $\mathbb{J}$, we say that $\mathbb{I}$ is *included* in $\mathbb{J}$ if for all ground atoms A, $\mathbb{I} \models A$ implies $\mathbb{J} \models A$. Every set P of definite clauses is satisfiable and has a *least* (with respect to inclusion) $\mathbb{D}$-model, denoted $M(P)$ [24]. Thus, if P is any set of constrained Horn clauses and Q is the set of the goals in P, then we define $Definite(P) = P \setminus Q$. We have that P is satisfiable if and only if $M(Definite(P)) \models Q$.

We will often use tuples of variables as arguments of predicates and write $p(X, Y)$, instead of $p(X_1, \ldots, X_m, Y_1, \ldots, Y_n)$, whenever the values of m (≥ 0) and n (≥ 0) are not relevant. Whenever the order of the variables is not relevant, we will feel free to identify tuples of distinct variables with finite sets, and we will extend to finite tuples the usual operations and relations on finite sets. Given two tuples X and Y of distinct elements, (i) their union $X \cup Y$ is obtained by concatenating them and removing all duplicated occurrences of elements, (ii) their intersection $X \cap Y$ is obtained by removing from X the elements which do not occur in Y, (iii) their difference $X \backslash Y$ is obtained by removing from X the elements which occur in Y, and (iv) $X \subseteq Y$ holds if every element of X occurs in Y. For all $m \geq 0$, $(u_1, \ldots, u_m) = (v_1, \ldots, v_m)$ iff $\bigwedge_{i=1}^{m} u_i = v_i$. The empty tuple $()$ is identified with the empty set $\emptyset$.

By $A(X, Y)$, where X and Y are disjoint tuples of distinct variables, we denote an atom A such that $vars(A) = X \cup Y$. Let P be a set of definite clauses. We say that the atom $A(X, Y)$ is *functional from X to Y with respect to P* if

(F1) $M(P) \models \forall X, Y, Z.\ A(X, Y) \wedge A(X, Z) \rightarrow Y = Z$

The reference to the set P of definite clauses is omitted, when understood from the context. Given a functional atom $A(X, Y)$, we say that X and Y are its *input* and *output* (tuples of) variables, respectively. The atom $A(X, Y)$ is said to be *total from X to Y with respect to P* if

(F2) $M(P) \models \forall X \exists Y.\ A(X, Y)$

If $A(X, Y)$ is a total, functional atom from X to Y, we may write $A(X; Y)$, instead of $A(X, Y)$. For instance, `append(Xs,Ys,Zs)` is a total, functional atom from `(Xs,Ys)` to `Zs` with respect to the set of clauses 1–7 of Sect. 2.

Now we extend the above notions from atoms to conjunctions of atoms. Let F be a conjunction $A_1(X_1; Y_1), \ldots, A_n(X_n; Y_n)$ such that: (i) $X = (\bigcup_{i=1}^{n} X_i) \setminus (\bigcup_{i=1}^{n} Y_i)$, (ii) $Y = (\bigcup_{i=1}^{n} Y_i)$, and (iii) for $i = 1, \ldots, n$, Y_i is disjoint from $(\bigcup_{j=1}^{i} X_j) \cup (\bigcup_{j=1}^{i-1} Y_j)$. Then, the conjunction F is said to be a *total, functional conjunction from X to Y* and it is also written as $F(X; Y)$. For $F(X; Y)$, the above properties (F1) and (F2) hold if we replace A by F. For instance, `append(Xs,Ys,Zs), rev(Zs,Rs)` is a total, functional conjunction from `(Xs,Ys)` to `(Zs,Rs)` with respect to the set of clauses 1–7 of Sect. 2.

4 Transformation Rules for Constrained Horn Clauses

In this section we present the rules that we propose for transforming CHCs, and in particular, for introducing difference predicates, and we prove their soundness. We refer to Sect. 2 for examples of how the rules are applied.

First, we introduce the following notion of a *stratification* for a set of clauses. Let $\mathbb{N}$ denote the set of the natural numbers. A *level mapping* is a function $\ell : \mathit{Pred} \rightarrow \mathbb{N}$. For every predicate p, the natural number $\ell(p)$ is said to be the *level* of p. Level mappings are extended to atoms by stating that the level $\ell(A)$ of an atom A is the level of its predicate symbol. A clause $H \leftarrow c, A_1, \ldots, A_n$ is *stratified with respect to ℓ* if, for $i = 1, \ldots, n$, $\ell(H) \geq \ell(A_i)$. A set P of CHCs is *stratified w.r.t. ℓ* if all clauses in P are stratified w.r.t. ℓ. Clearly, for every set P of CHCs, there exists a level mapping ℓ such that P is stratified w.r.t. ℓ [33].

A *transformation sequence from P_0 to P_n* is a sequence $P_0 \Rightarrow P_1 \Rightarrow \ldots \Rightarrow P_n$ of sets of CHCs such that, for $i = 0, \ldots, n-1$, P_{i+1} is derived from P_i, denoted $P_i \Rightarrow P_{i+1}$, by applying one of the following Rules R1–R7. We assume that P_0 is stratified w.r.t. a given level mapping ℓ.

(R1) *Definition Rule.* Let D be the clause $\mathit{newp}(X_1, \ldots, X_k) \leftarrow c, A_1, \ldots, A_n$, where: (i) newp is a predicate symbol in Pred not occurring in the sequence $P_0 \Rightarrow P_1 \Rightarrow \ldots \Rightarrow P_i$, (ii) c is a constraint, (iii) the predicate symbols of $A_1, \ldots, A_n$ occur in P_0, and (iv) $(X_1, \ldots, X_k) \subseteq \mathit{vars}(c, A_1, \ldots, A_n)$. Then, $P_{i+1} = P_i \cup \{D\}$. We set $\ell(\mathit{newp}) = \mathit{max}\{\ell(A_i) \mid i = 1, \ldots, n\}$.

For $i = 0, \ldots, n$, by Defs_i we denote the set of clauses, called *definitions*, introduced by Rule R1 during the construction of $P_0 \Rightarrow P_1 \Rightarrow \ldots \Rightarrow P_i$. Thus, $\emptyset = \mathit{Defs}_0 \subseteq \mathit{Defs}_1 \subseteq \ldots$. However, by using Rules R2–R7 we can replace a definition in P_i, for $i > 0$, and hence it may happen that $\mathit{Defs}_{i+1} \not\subseteq P_{i+1}$.

(R2) *Unfolding Rule.* Let C: $H \leftarrow c, G_L, A, G_R$ be a clause in P_i, where A is an atom. Without loss of generality, we assume that $\mathit{vars}(C) \cap \mathit{vars}(P_0) = \emptyset$. Let *Cls*: $\{K_1 \leftarrow c_1, B_1,\ \ldots,\ K_m \leftarrow c_m, B_m\}$, $m \geq 0$, be the set of clauses in P_0, such that: for $j = 1, \ldots, m$, (1) there exists a most general unifier ϑ_j of A and K_j, and (2) the conjunction of constraints $(c, c_j)\vartheta_j$ is satisfiable. Let $\mathit{Unf}(C, A, P_0) = \{(H \leftarrow c, c_j, G_L, B_j, G_R)\vartheta_j \mid j = 1, \ldots, m\}$. Then, by *unfolding C w.r.t. A*, we derive the set $\mathit{Unf}(C, A, P_0)$ of clauses and we get $P_{i+1} = (P_i \setminus \{C\}) \cup \mathit{Unf}(C, A, P_0)$.

When we apply Rule R2, we say that, for $j = 1, \ldots, m$, the atoms in the conjunction $B_j\vartheta_j$ are *derived* from A, and the atoms in the conjunction $(G_L, G_R)\vartheta_j$ are *inherited* from the corresponding atoms in the body of C.

(R3) *Folding Rule.* Let C: $H \leftarrow c, G_L, Q, G_R$ be a clause in P_i, and let D: $K \leftarrow d, B$ be a clause in Defs_i. Suppose that: (i) either H is *false* or $\ell(H) \geq \ell(K)$, and (ii) there exists a substitution ϑ such that $Q = B\vartheta$ and $\mathbb{D} \models \forall(c \rightarrow d\vartheta)$. By *folding C using definition D*, we derive clause E: $H \leftarrow c, G_L, K\vartheta, G_R$, and we get $P_{i+1} = (P_i \setminus \{C\}) \cup \{E\}$.

(R4) *Clause Deletion Rule.* Let C: $H \leftarrow c, G$ be a clause in P_i such that the constraint c is unsatisfiable. Then, we get $P_{i+1} = P_i \setminus \{C\}$.

(R5) *Functionality Rule.* Let C: $H \leftarrow c, G_L, F(X;Y), F(X;Z), G_R$ be a clause in P_i, where $F(X;Y)$ is a total, functional conjunction in $\mathit{Definite}(P_0) \cup \mathit{Defs}_i$. By *functionality*, from C we derive clause D: $H \leftarrow c, Y = Z, G_L, F(X;Y), G_R$, and we get $P_{i+1} = (P_i \setminus \{C\}) \cup \{D\}$.

(R6) *Totality Rule.* Let C: $H \leftarrow c, G_L, F(X;Y), G_R$ be a clause in P_i such that $Y \cap \mathit{vars}(H \leftarrow c, G_L, G_R) = \emptyset$ and $F(X;Y)$ is a total, functional conjunction in $\mathit{Definite}(P_0) \cup \mathit{Defs}_i$. By *totality*, from C we derive clause $D: H \leftarrow c, G_L, G_R$ and we get $P_{i+1} = (P_i \setminus \{C\}) \cup \{D\}$.

Since the initial set of clauses is obtained by translating a terminating functional program, the functionality and totality properties hold by construction and we do not need to prove them when we apply Rules R5 and R6.

(R7) *Differential Replacement Rule.* Let C: $H \leftarrow c, G_L, F(X;Y), G_R$ be a clause in P_i, and let D: $\mathit{diff}(Z) \leftarrow d, F(X;Y), R(V;W)$ be a definition clause in Defs_i, where: (i) $F(X;Y)$ and $R(V;W)$ are total, functional conjunctions with respect to $\mathit{Definite}(P_0) \cup \mathit{Defs}_i$, (ii) $W \cap \mathit{vars}(C) = \emptyset$, (iii) $\mathbb{D} \models \forall(c \rightarrow d)$, and (iv) $\ell(H) > \ell(\mathit{diff})$. By *differential replacement*, we derive E: $H \leftarrow c, G_L, R(V;W), \mathit{diff}(Z), G_R$ and we get $P_{i+1} = (P_i \setminus \{C\}) \cup \{E\}$.

Rule R7 has a very general formulation that eases the proof of the Soundness Theorem, which extends to Rules R1–R7 correctness results for transformations of (constraint) logic programs [17,18,39,42] (see [13] for a proof). In the transformation algorithm of Sect. 5, we will use a specific instance of Rule R7 which is sufficient for ADT removal (see, in particular, the Diff-Introduce step).

Theorem 1 (Soundness). *Let $P_0 \Rightarrow P_1 \Rightarrow \ldots \Rightarrow P_n$ be a transformation sequence using Rules* R1–R7. *Suppose that the following condition holds:*

(U) *for $i = 1, \ldots, n-1$, if $P_i \Rightarrow P_{i+1}$ by folding a clause in P_i using a definition $D : H \leftarrow c, B$ in Defs_i, then, for some $j \in \{1, \ldots, i-1, i+1, \ldots, n-1\}$, $P_j \Rightarrow P_{j+1}$ by unfolding D with respect to an atom A such that $\ell(H) = \ell(A)$.*

If P_n is satisfiable, then P_0 is satisfiable.

Thus, to prove the satisfiability of a set P_0 of clauses, it suffices: (i) to construct a transformation sequence $P_0 \Rightarrow \ldots \Rightarrow P_n$, and then (ii) to prove that P_n is satisfiable. Note, however, that if Rule R7 is used, it may happen that P_0 is satisfiable and P_n is unsatisfiable, that is, some false counterexamples to satisfiability, so-called *false positives*, may be generated, as we now show.

Example 1. Let us consider the following set P_1 of clauses derived by adding the definition clause `D` to the initial set $P_0 = \{\mathtt{C}, 1, 2, 3\}$ of clauses:

```
C. false :- X=0, Y>0, a(X,Y).
1. a(X,Y) :- X=<0, Y=0.      2. a(X,Y) :- X>0, Y=1.      3. r(X,W) :- W=1.
D. diff(Y,W) :- a(X,Y), r(X,W).
```

where: (i) `a(X,Y)` is a total, functional atom from `X` to `Y`, (ii) `r(X,W)` is a total, functional atom from `X` to `W`, and (iii) `D` is a definition in $Defs_1$. By applying Rule R7, from P_1 we derive the set $P_2 = \{\mathtt{E}, 1, 2, 3, \mathtt{D}\}$ of clauses where:

```
E. false :- X=0, Y>0, r(X,W), diff(Y,W).
```

Now we have that P_0 is satisfiable, while P_2 is unsatisfiable.

5 An Algorithm for ADT Removal

Now we present Algorithm $\mathcal{R}$ for eliminating ADT terms from CHCs by using the transformation rules presented in Sect. 4 and automatically introducing suitable difference predicates. If $\mathcal{R}$ terminates, it transforms a set *Cls* of clauses into a new set *TransfCls* where the arguments of all predicates have basic type. Theorem 1 guarantees that if *TransfCls* is satisfiable, then also *Cls* is satisfiable.

Algorithm $\mathcal{R}$ (see Fig. 1) removes ADT terms starting from the set *Gs* of goals in *Cls*. $\mathcal{R}$ collects these goals in *InCls* and stores in *Defs* the definitions of new predicates introduced by Rule R1.

Algorithm $\mathcal{R}$
Input: A set *Cls* of clauses;
Output: A set *TransfCls* of clauses that have basic types.

Let *Cls* = *Ds* ∪ *Gs*, where *Ds* is a set of definite clauses and *Gs* is a set of goals;
InCls := *Gs*; *Defs* := ∅; *TransfCls* := ∅;
while *InCls*≠∅ **do**
- ■ *Diff-Define-Fold*(*InCls*, *Defs*, *NewDefs*, *FldCls*);
- ■ *Unfold*(*NewDefs*, *Ds*, *UnfCls*);
- ■ *Replace*(*UnfCls*, *Ds*, *RCls*);
  InCls := *RCls*; *Defs* := *Defs*∪*NewDefs*; *TransfCls* := *TransfCls*∪*FldCls*;

Fig. 1. The ADT removal algorithm $\mathcal{R}$.

Before describing the procedures used by Algorithm $\mathcal{R}$, let us first introduce the following notions.

Given a conjunction G of atoms, *bvars*(G) (or *adt−vars*(G)) denotes the set of variables in G that have a basic type (or an ADT type, respectively). We say that an atom (or clause) *has basic types* if *all* its arguments (or atoms, respectively) have a basic type. An atom (or clause) *has ADTs* if *at least one* of its arguments (or atoms, respectively) has an ADT type.

Given a set (or a conjunction) S of atoms, *SharingBlocks*(S) denotes the partition of S with respect to the reflexive, transitive closure $\Downarrow_S$ of the relation $\downarrow_S$ defined as follows. Given two atoms A_1 and A_2 in S, $A_1 \downarrow_S A_2$ holds iff *adt-vars*$(A_1) \cap$ *adt-vars*$(A_2) \neq \emptyset$. The elements of the partition are called the *sharing blocks* of S.

A *generalization* of a pair (c_1, c_2) of constraints is a constraint $\alpha(c_1, c_2)$ such that $\mathbb{D} \models \forall(c_1 \rightarrow \alpha(c_1, c_2))$ and $\mathbb{D} \models \forall(c_2 \rightarrow \alpha(c_1, c_2))$ [19]. In particular, we consider the following generalization operator based on *widening* [7]. Suppose that c_1 is the conjunction $(a_1, \ldots, a_m)$ of atomic constraints, then $\alpha(c_1, c_2)$ is defined as the conjunction of all a_i's in $(a_1, \ldots, a_m)$ such that $\mathbb{D} \models \forall(c_2 \rightarrow a_i)$. For any constraint c and tuple V of variables, the *projection* of c onto V is a constraint $\pi(c, V)$ such that: (i) $vars(\pi(c, V)) \subseteq V$, and (ii) $\mathbb{D} \models \forall(c \rightarrow \pi(c, V))$. In our implementation, $\pi(c, V)$ is computed from $\exists Y.c$, where $Y = vars(c) \setminus V$, by a quantifier elimination algorithm in the theory of booleans and *rational* numbers. This implementation is safe in our context, and avoids relying on modular arithmetic, as is often done when eliminating quantifiers in LIA [37].

For two conjunctions G_1 and G_2 of atoms, $G_1 \lessapprox G_2$ holds if $G_1 = (A_1, \ldots, A_n)$ and there exists a subconjunction $(B_1, \ldots, B_n)$ of G_2 (modulo reordering) such that, for $i = 1, \ldots, n$, B_i is an instance of A_i. A conjunction G of atoms is *connected* if it consists of a single sharing block.

■ *Procedure Diff-Define-Fold* (see Fig. 2). At each iteration of the body of the **for** loop, the *Diff-Define-Fold* procedure removes the ADT terms occurring in a sharing block B of the body of a clause C: $H \leftarrow c, B, G'$ of *InCls*. This is done by possibly introducing some new definitions (using Rule R1) and applying the Folding Rule R3. To allow folding, some applications of the Differential Replacement Rule R7 may be needed. We have the following four cases.

• (**Fold**). We remove the ADT arguments occurring in B by folding C using a definition D introduced at a previous step. Indeed, the head of each definition introduced by Algorithm $\mathcal{R}$ is by construction a tuple of variables of basic type.

• (**Generalize**). We introduce a new definition *GenD* : *genp*$(V) \leftarrow \alpha(d, c), B$ whose constraint is obtained by generalizing (d, c), where d is the constraint occurring in an already available definition whose body is B. Then, we remove the ADT arguments occurring in B by folding C using *GenD*.

• (**Diff-Introduce**). Suppose that B *partially matches* the body of an available definition D: *newp*$(U) \leftarrow d, B'$, that is, for some substitution ϑ, $B = (M, F(X;Y))$, and $B'\vartheta = (M, R(V;W))$. Then, we introduce a difference predicate through the new definition $\widehat{D}$: *diff*$(Z) \leftarrow \pi(c, X), F(X;Y), R(V;W)$, where $Z =$ *bvars*$(F(X;Y), R(V;W))$ and, by Rule R7, we replace the conjunction $F(X;Y)$ by $(R(V;W),$ *diff*$(Z))$ in the body of C, thereby deriving C'. Finally,

Procedure *Diff-Define-Fold*(*InCls*, *Defs*, *NewDefs*, *FldCls*)
Input: A set *InCls* of clauses and a set *Defs* of definitions;
Output: A set *NewDefs* of definitions and a set *FldCls* of clauses with basic types.

$NewDefs := \emptyset$; $FldCls := \emptyset$;
for each clause C: $H \leftarrow c, G$ in *InCls* **do**
  if C has basic types **then** $InCls := InCls \setminus \{C\}$; $FldCls := FldCls \cup \{C\}$
  else
    let C be $H \leftarrow c, B, G'$ where B is a sharing block in G that contains at least one atom that has ADTs;
    • **(Fold)** **if** in $Defs \cup NewDefs$ there is a (variant of) clause D: $newp(V) \leftarrow d, B$ such that $\mathbb{D} \models \forall(c \rightarrow d)$ **then** fold C using D and derive E: $H \leftarrow c, newp(V), G'$;
    • **(Generalize)** **else if** in $Defs \cup NewDefs$ there is a (variant of a) clause $newp(V) \leftarrow d, B$ and $\mathbb{D} \not\models \forall(c \rightarrow d)$ **then**
      introduce definition $GenD$: $genp(V) \leftarrow \alpha(d, c), B$;
      fold C using $GenD$ and derive E: $H \leftarrow c, genp(V), G'$;
      $NewDefs := NewDefs \cup \{GenD\}$;
    • **(Diff-Introduce)** **else if** in $Defs \cup NewDefs$ there is a (variant of a) clause D: $newp(U) \leftarrow d, B'$ such that: (i) $vars(C) \cap vars(D) = \emptyset$, and (ii) $B' \lessapprox B$ **then**
      take a maximal subconjunction M of B, if any, such that:
        (i) $B = (M, F(X;Y))$, for some connected conjunction M and non-empty conjunction $F(X;Y)$, (ii) $B'\vartheta = (M, R(V;W))$, for some substitution ϑ such that $W \cap vars(C) = \emptyset$, and (iii) for every atom A in $\{F(X;Y), R(V;W)\}$, $\ell(H) > \ell(A)$;
      introduce definition $\widehat{D}$: $diff(Z) \leftarrow \pi(c, X), F(X;Y), R(V;W)$
        where $Z = bvars(F(X;Y), R(V;W))$;
      $NewDefs := NewDefs \cup \{\widehat{D}\}$;
      replace $F(X;Y)$ by $(R(V;W), diff(Z))$ in C, and derive clause
        C': $H \leftarrow c, M, R(V;W), diff(Z), G'$;
      if $\mathbb{D} \models \forall(c \rightarrow d\vartheta)$
        then fold C' using D and derive E: $H \leftarrow c, newp(U\vartheta), diff(Z), G'$;
        else introduce definition $GenD$: $genp(U') \leftarrow \alpha(d\vartheta, c), B'\vartheta$
            where $U' = bvars(B'\vartheta)$;
          fold C' using $GenD$ and derive E: $H \leftarrow c, genp(U'), diff(Z), G'$;
          $NewDefs := NewDefs \cup \{GenD\}$;
    • **(Project)** **else**
      introduce definition $ProjC$: $newp(V) \leftarrow \pi(c, V), B$ where $V = bvars(B)$;
      fold C using $ProjC$ and derive clause E: $H \leftarrow c, newp(V), G'$;
      $NewDefs := NewDefs \cup \{ProjC\}$;
  $InCls := (InCls \setminus \{C\}) \cup \{E\}$;

Fig. 2. The *Diff-Define-Fold* procedure.

we remove the ADT arguments in B by folding C' using either D or a clause *GenD* whose constraint is a generalization of the pair $(d\vartheta, c)$ of constraints.

The example of Sect. 2 allows us to illustrate this (Diff-Introduce) case. With reference to that example, clause C: $H \leftarrow c, G$ that we want to fold is clause `11`, whose body has the single sharing block B: ‘`append(Xs,Ys,Zs),`

`rev(Zs,Rs), len(Xs,N0), len(Ys,N1), append(Rs,[X],R1s), len(R1s,N21)`'. Block B partially matches the body '`append(Xs,Ys,Zs), rev(Zs,Rs), len(Xs,N0), len (Ys,N1), len(Rs,N2)`' of clause 8 of Sect. 2 which plays the role of definition D: $newp(U) \leftarrow d, B'$ in this example. Indeed, we have that:
$M=$ (`append(Xs,Ys,Zs), rev(Zs,Rs), len(Xs,N0), len(Ys,N1)`),
$F(X;Y)=$(`append(Rs,[X],R1s), len(R1s,N21)`), where $X=$`(Rs,X)`, $Y=$`(R1s,N21)`,
$R(V;W)=$`len(Rs,N2)`, where $V=$ `(Rs)`, $Y=$ `(N2)`.
In this example, ϑ is the identity substitution. Morevover, the condition on the level mapping ℓ required in the *Diff-Define-Fold* Procedure of Fig. 2 can be fulfilled by stipulating that ℓ(`new1`) > ℓ(`append`) and ℓ(`new1`) > ℓ(`len`). Thus, the definition $\widehat{D}$ to be introduced is:

```
12. diff(N2,X,N21) :- append(Rs,[X],R1s), len(R1s,N21), len(Rs,N2).
```

Indeed, we have that: (i) the projection $\pi(c,X)$ is π(`N01=N0+1, (Rs,X)`), that is, the empty conjunction, (ii) $F(X;Y)$, $R(V;W)$ is the body of clause `12`, and (iii) the head variables `N2`, `X`, and `N21` are the integer variables in that body. Then, by applying Rule R7, we replace in clause `11` the conjunction '`append(Rs,[X],R1s), len(R1s,N21)`' by the new conjunction '`len(Rs,N2), diff(N2,X,N21)`', hence deriving clause C', which is clause `13` of Sect. 2. Finally, by folding clause `13` using clause `8`, we derive clause `14` of Sect. 2, which has no list arguments.
• (**Project**). If none of the previous three cases apply, then we introduce a new definition *ProjC*: $newp(V) \leftarrow \pi(c,V), B$, where $V = bvars(B)$. Then, we remove the ADT arguments occurring in B by folding C using *ProjC*.

The *Diff-Define-Fold* procedure may introduce new definitions with ADTs in their bodies, which are added to *NewDefs* and processed by the *Unfold* procedure. In order to present this procedure, we need the following notions.

The application of Rule R2 is controlled by marking some atoms in the body of a clause as *unfoldable*. If we unfold w.r.t. atom A clause C: $H \leftarrow c, L, A, R$ the marking of the clauses in $Unf(C,A,Ds)$ is handled as follows: the atoms derived from A are not marked as unfoldable and each atom A'' inherited from an atom A' in the body of C is marked as unfoldable iff A' is marked as unfoldable.

An atom $A(X;Y)$ in a conjunction $F(V;Z)$ of atoms is said to be a *source atom* if $X \subseteq V$. Thus, a source atom corresponds to an innermost function call in a given functional expression. For instance, in clause `1` of Sect. 2, the source atoms are `append(Xs,Ys,Zs)`, `len(Xs,N0)`, and `len(Ys,N1)`. Indeed, the body of clause `1` corresponds to `len(rev(append xs ys)) ≠ (len xs)+(len ys)`.

An atom $A(X;Y)$ in the body of clause C: $H \leftarrow c, L, A(X;Y), R$ is a *head-instance* w.r.t. a set *Ds* of clauses if, for every clause $K \leftarrow d, B$ in *Ds* such that: (1) there exists a most general unifier ϑ of $A(X;Y)$ and K, and (2) the constraint $(c,d)\vartheta$ is satisfiable, we have that $X\vartheta = X$. Thus, the input variables of $A(X;Y)$ are not instantiated by unification. For instance, the atom `append([X|Xs],Ys,Zs)` is a head-instance, while `append(Xs,Ys,Zs)` is not.

In a set *Cls* of clauses, predicate p *immediately depends on* predicate q, if in *Cls* there is a clause of the form $p(\ldots) \leftarrow \ldots, q(\ldots), \ldots$ The *depends on* relation is the transitive closure of the *immediately depends on* relation. Let $\prec$ be a well-founded ordering on tuples of terms such that, for all terms t, u, if $t \prec u$, then, for all substitutions ϑ, $t\vartheta \prec u\vartheta$. A predicate p is *descending* w.r.t. $\prec$ if, for all clauses, $p(t; u) \leftarrow c,\, p_1(t_1; u_1), \ldots, p_n(t_n; u_n)$, for $i = 1, \ldots, n$, if p_i depends on p then $t_i \prec t$. An atom is descending if its predicate is descending. The well-founded ordering $\prec$ we use in our implementation is based on the *subterm* relation and is defined as follows: $(x_1, \ldots, x_k) \prec (y_1, \ldots, y_m)$ if every x_i is a subterm of some y_j and there exists x_i which is a strict subterm of some y_j. For instance, the predicates `append`, `rev`, and `len` in the example of Sect. 2 are all descending.

■ *Procedure Unfold* (see Fig. 3) repeatedly applies Rule R2 in two phases. In Phase 1, the procedure unfolds the clauses in *NewDefs* w.r.t. at least one source atom. Then, in Phase 2, clauses are unfolded w.r.t. head-instance atoms. Unfolding is repeated only w.r.t descending atoms. The termination of the *Unfold* procedure is ensured by the fact that the unfolding w.r.t. a non-descending atom is done at most once in each phase.

Procedure *Unfold(NewDefs, Ds, UnfCls)*
Input: A set *NewDefs* of definitions and a set *Ds* of definite clauses;
Output: A set *UnfCls* of clauses.

UnfCls := *NewDefs*; Mark as unfoldable a nonempty set of source atoms in the body of each clause of *UnfCls*;
- **while** there exists a clause C: $H \leftarrow c, L, A, R$ in *UnfCls*, for some conjunctions L and R, such that A is an unfoldable atom **do**
  UnfCls := (*UnfCls* $-$ $\{C\}$) $\cup$ *Unf*(C, A, Ds);
- Mark as unfoldable all atoms in the body of each clause in *UnfCls*;
- **while** there exists a clause C: $H \leftarrow c, L, A, R$ in *UnfCls*, for some conjunctions L and R, such that A is a head-instance atom w.r.t. *Ds* and A is either unfoldable or descending **do**
  UnfCls := (*UnfCls* $-$ $\{C\}$) $\cup$ *Unf*(C, A, Ds);

Fig. 3. The *Unfold* procedure.

■ *Procedure Replace* simplifies some clauses by applying Rules R5 and R6 as long as possible. *Replace* terminates because each application of either rule decreases the number of atoms.

Thus, each execution of the *Diff-Define-Fold*, *Unfold*, and *Replace* procedures terminates. However, Algorithm $\mathcal{R}$ might not terminate because new predicates may be introduced by *Diff-Define-Fold* at each iteration of the **while-do** of $\mathcal{R}$. Soundness of $\mathcal{R}$ follows from soundness of the transformation rules [13].

Theorem 2 (Soundness of Algorithm $\mathcal{R}$). *Suppose that Algorithm $\mathcal{R}$ terminates for an input set Cls of clauses, and let TransfCls be the output set of clauses. Then, every clause in TransfCls has basic types, and if TransfCls is satisfiable, then Cls is satisfiable.*

Algorithm $\mathcal{R}$ is not complete, in the sense that, even if *Cls* is a satisfiable set of input clauses, then $\mathcal{R}$ may not terminate or, due to the use of Rule R7, it may terminate with an output set *TransfCls* of unsatisfiable clauses, thereby generating a false positive (see Example 1 in Sect. 4). However, due to well-known undecidability results for the satisfiability problem of CHCs, this limitation cannot be overcome, unless we restrict the class of clauses we consider. The study of such restricted classes of clauses is beyond the scope of the present paper and, instead, in the next section, we evaluate the effectiveness of Algorithm $\mathcal{R}$ from an experimental viewpoint.

6 Experimental Evaluation

In this section we present the results of some experiments we have performed for assessing the effectiveness of our transformation-based CHC solving technique. We compare our technique with the one proposed by Reynolds and Kuncak [38], which extends the SMT solver CHC4 with inductive reasoning.

Implementation. We have developed the AdtRem tool for ADT removal, which is based on an implementation of Algorithm $\mathcal{R}$ in the VeriMAP system [8].

Benchmark Suite and Experiments. Our benchmark suite consists of 169 verification problems over inductively defined data structures, such as lists, queues, heaps, and trees, which have been adapted from the benchmark suite considered by Reynolds and Kuncak [38]. These problems come from benchmarks used by various theorem provers: (i) 53 problems come from CLAM [23], (ii) 11 from HipSpec [6], (iii) 63 from IsaPlanner [15,25], and (iv) 42 from Leon [41]. We have performed the following experiments, whose results are summarized in Table 1[2].
(1) We have considered Reynolds and Kuncak's **dtt** encoding of the verification problems, where natural numbers are represented using the built-in SMT type *Int*, and we have discarded: (i) problems that do not use ADTs, and (ii) problems that cannot be directly represented in Horn clause format. Since AdtRem does not support higher order functions, nor user-provided lemmas, in order to make a comparison between the two approaches on a level playing field, we have replaced higher order functions by suitable first order instances and we have removed all auxiliary lemmas from the input verification problems. We have also replaced the basic functions recursively defined over natural numbers, such as the *plus* and *less-or-equal* functions, by LIA constraints.
(2) Then, we have translated each verification problem into a set, call it P, of CHCs in the Prolog-like syntax supported by AdtRem by using a modified version of the SMT-LIB parser of the ProB system [32]. We have run Eldarica and Z3[3], which use no induction-based mechanism for handling ADTs, to check the satisfiability of P. Rows 'Eldarica' and 'Z3' show the number of solved problems, that is, problems whose CHC encoding has been proved satisfiable.

---

[2] The tool and the benchmarks are available at https://fmlab.unich.it/adtrem/.
[3] More specifically, Eldarica v2.0.1 and Z3 v4.8.0 with the Spacer engine [28].

(3) We have run algorithm $\mathcal{R}$ on P to produce a set T of CHCs without ADTs. Row '$\mathcal{R}$' reports the number of problems for which Algorithm $\mathcal{R}$ terminates.
(4) We have converted T into the SMT-LIB format, and then we have run Eldarica and Z3 for checking its satisfiability. Rows 'Eldarica$_{\text{noADT}}$' and 'Z3$_{\text{noADT}}$' report outside round parentheses the number of solved problems. There was only one false positive (that is, a satisfiable set P of clauses transformed into an unsatisfiable set T), which we have classified as an unsolved problem.
(5) In order to assess the improvements due to the use of the differential replacement rule we have applied to P a modified version, call it $\mathcal{R}^{\circ}$, of the ADT removal algorithm $\mathcal{R}$ that *does not* introduce difference predicates, that is, the *Diff-Introduce* case of the *Diff-Define-Fold* Procedure of Fig. 2 is never executed. The number of problems for which $\mathcal{R}^{\circ}$ terminates and the number of solved problems using Eldarica and Z3 are shown within round parentheses in rows '$\mathcal{R}$', 'Eldarica$_{\text{noADT}}$', and 'Z3$_{\text{noADT}}$', respectively.
(6) Finally, we have run the **cvc4+ig** configuration of the CVC4 solver extended with inductive reasoning [38] on the 169 problems in SMT-LIB format obtained at Step (1). Row 'CVC4+Ind' reports the number of solved problems.

Table 1. *Experimental results.* For each problem we have set a timeout limit of 300 seconds. Experiments have been performed on an Intel Xeon CPU E5-2640 2.00 GHz with 64GB RAM under CentOS.

	CLAM	HipSpec	IsaPlanner	Leon	Total
Number of problems	53	11	63	42	169
Eldarica	0	2	4	9	15
Z3	6	0	2	10	18
$\mathcal{R}$	(18) 36	(2) 4	(56) 59	(18) 30	(94) 129
Eldarica$_{\text{noADT}}$	(18) 32	(2) 4	(56) 57	(18) 29	(94) 122
Z3$_{\text{noADT}}$	(18) 29	(2) 3	(55) 56	(18) 26	(93) 114
CVC4+Ind	17	5	37	15	74

Evaluation of Results. The results of our experiments show that ADT removal considerably increases the effectiveness of CHC solvers without inductive reasoning support. For instance, Eldarica is able to solve 15 problems out of 169, while it solves 122 problems after the removal of ADTs. When using Z3, the improvement is slightly lower, but still very considerable. Note also that, when the ADT removal terminates (129 problems out of 169), the solvers are very effective (95% successful verifications for Eldarica). The improvements specifically due to the use of the difference replacement rule are demonstrated by the increase of the number of problems for which the ADT removal algorithm terminates (from 94 to 129), and of the number of problems solved (from 94 to 122, for Eldarica).

AdtRem compares favorably to CVC4 extended with induction (compare rows 'Eldarica$_{\text{noADT}}$' and 'Z3$_{\text{noADT}}$' to row 'CVC4+Ind'). Interestingly, the effectiveness of CVC4 may be increased if one extends the problem formalization with extra lemmas which may be used for proving the main conjecture. Indeed, CVC4 solves 100 problems when auxiliary lemmas are added, and 134 problems when, in addition, it runs on the **dti** encoding, where natural numbers are represented using both the built-in type *Int* and the ADT definition with the zero and successor constructors. Our results show that in most cases AdtRem needs neither those extra axioms nor that sophisticated encoding.

Finally, in Table 2 we report some problems solved by AdtRem with Eldarica that are not solved by CVC4 with induction (using any encoding and auxiliary lemmas), or vice versa. For details, see https://fmlab.unich.it/adtrem/.

Table 2. A comparison between AdtRem with Eldarica and CVC4 with induction.

Problem	*Property proved by* AdtRem *and not by* CVC4
CLAM goal6	$\forall x, y.\ len(rev(append(x, y))) = len(x) + len(y)$
CLAM goal49	$\forall x.\ mem(x, sort(y)) \Rightarrow mem(x, y)$
IsaPlanner goal52	$\forall n, l.\ count(n, l) = count(n, rev(l))$
IsaPlanner goal80	$\forall l.\ sorted(sort(l))$
Leon heap-goal13	$\forall x, l.\ len(qheapsorta(x, l)) = hsize(x) + len(l)$
Problem	*Property proved by* CVC4 *and not by* AdtRem
CLAM goal18	$\forall x, y.\ rev(append(rev(x), y)) = append(rev(y), x)$
HipSpec rev-equiv-goal4	$\forall x, y.\ qreva(qreva(x, y), nil) = qreva(y, x)$
HipSpec rev-equiv-goal6	$\forall x, y.\ append(qreva(x, y), z) = qreva(x, append(y, z))$

7 Related Work and Conclusions

Inductive reasoning is supported, with different degrees of human intervention, by many theorem provers, such as ACL2 [27], CLAM [23], Isabelle [34], HipSpec [6], Zeno [40], and PVS [35]. The combination of inductive reasoning and SMT solving techniques has been exploited by many tools for program verification [30,36,38,41,43,44].

Leino [30] integrates inductive reasoning into the Dafny program verifier by implementing a simple strategy that rewrites user-defined properties that may benefit from induction into proof obligation to be discharged by Z3. The advantage of this technique is that it fully decouples inductive reasoning from SMT solving. Hence, no extensions to the SMT solver are required.

In order to extend CVC4 with induction, Reynolds and Kuncak [38] also consider the rewriting of formulas that may take advantage from inductive reasoning, but this is done dynamically, during the proof search. This approach allows CVC4 to perform the rewritings lazily, whenever new formulas are generated during the proof search, and to use the partially solved conjecture, to generate lemmas that may help in the proof of the initial conjecture.

The issue of generating suitable lemmas during inductive proofs has been also addressed by Yang et al. [44] and implemented in ADTIND. In order to conjecture new lemmas, their algorithm makes use of a syntax-guided synthesis strategy driven by a grammar, which is dynamically generated from user-provided templates and the function and predicate symbols encountered during the proof search. The derived lemma conjectures are then checked by the SMT solver Z3.

In order to take full advantage of the efficiency of SMT solvers in checking satisfiability of quantifier-free formulas over LIA, ADTs, and finite sets, the Leon verification system [41] implements an SMT-based solving algorithm to check the satisfiability of formulas involving recursively defined first-order functions. The algorithm interleaves the unrolling of recursive functions and the SMT solving of the formulas generated by the unrolling. Leon can be used to prove properties of Scala programs with ADTs and integrates with the Scala compiler and the SMT solver Z3. A refined version of that algorithm, restricted to *catamorphisms*, has been implemented into a solver-agnostic tool, called RADA [36].

In the context of CHCs, Unno et al. [43] have proposed a proof system that combines inductive theorem proving with SMT solving. This approach uses Z3-PDR [21] to discharge proof obligations generated by the proof system, and has been applied to prove relational properties of OCaml programs.

The distinctive feature of the technique presented in this paper is that it does not make use of any explicit inductive reasoning, but it follows a transformational approach. First, the problem of verifying the validity of a universally quantified formula over ADTs is reduced to the problem of checking the satisfiability of a set of CHCs. Then, this set of CHCs is transformed with the aim of deriving a set of CHCs over basic types (i.e., integers) only, whose satisfiability implies the satisfiability of the original set. In this way, the reasoning on ADTs is separated from the reasoning on satisfiability, which can be performed by specialized engines for CHCs on basic types (e.g. Eldarica [22] and Z3-Spacer [29]). Some of the ideas presented here have been explored in [11,12], but there neither formal results nor an automated strategy were presented.

A key success factor of our technique is the introduction of difference predicates, which can be viewed as the transformational counterpart of lemma generation. Indeed, as shown in Sect. 6, the use of difference predicates greatly increases the power of CHC solving with respect to previous techniques based on the transformational approach, which do not use difference predicates [10].

As future work, we plan to apply our transformation-based verification technique to more complex program properties, such as relational properties [9].

References

1. Gopalakrishnan, G., Qadeer, S. (eds.): CAV 2011. LNCS, vol. 6806. Springer, Heidelberg (2011). https://doi.org/10.1007/978-3-642-22110-1
2. Barrett, C., Tinelli, C.: Satisfiability modulo theories. In: Clarke, E., Henzinger, T., Veith, H., Bloem, R. (eds.) Handbook of Model Checking, pp. 305–343. Springer, Cham (2018). https://doi.org/10.1007/978-3-319-10575-8_11

3. Bjørner, N., Gurfinkel, A., McMillan, K., Rybalchenko, A.: Horn Clause Solvers for Program Verification. In: Beklemishev, L.D., Blass, A., Dershowitz, N., Finkbeiner, B., Schulte, W. (eds.) Fields of Logic and Computation II. LNCS, vol. 9300, pp. 24–51. Springer, Cham (2015). https://doi.org/10.1007/978-3-319-23534-9_2
4. Bundy, A.: The automation of proof by mathematical induction. In: Robinson, A., Voronkov, A. (eds.) Handbook of Automated Reasoning, vol. I, pp. 845–911. North Holland (2001)
5. Cimatti, A., Griggio, A., Schaafsma, B.J., Sebastiani, R.: The MathSAT5 SMT solver. In: Piterman, N., Smolka, S.A. (eds.) TACAS 2013. LNCS, vol. 7795, pp. 93–107. Springer, Heidelberg (2013). https://doi.org/10.1007/978-3-642-36742-7_7
6. Claessen, K., Johansson, M., Rosén, D., Smallbone, N.: Automating inductive proofs using theory exploration. In: Bonacina, M.P. (ed.) CADE 2013. LNCS (LNAI), vol. 7898, pp. 392–406. Springer, Heidelberg (2013). https://doi.org/10.1007/978-3-642-38574-2_27
7. Cousot, P., Halbwachs, N.: Automatic discovery of linear restraints among variables of a program. In: Proceedings of the Fifth ACM Symposium on Principles of Programming Languages, POPL 1978, pp. 84–96. ACM (1978)
8. De Angelis, E., Fioravanti, F., Pettorossi, A., Proietti, M.: VeriMAP: a tool for verifying programs through transformations. In: Ábrahám, E., Havelund, K. (eds.) TACAS 2014. LNCS, vol. 8413, pp. 568–574. Springer, Heidelberg (2014). https://doi.org/10.1007/978-3-642-54862-8_47. http://www.map.uniroma2.it/VeriMAP
9. De Angelis, E., Fioravanti, F., Pettorossi, A., Proietti, M.: Relational verification through Horn clause transformation. In: Rival, X. (ed.) SAS 2016. LNCS, vol. 9837, pp. 147–169. Springer, Heidelberg (2016). https://doi.org/10.1007/978-3-662-53413-7_8
10. De Angelis, E., Fioravanti, F., Pettorossi, A., Proietti, M.: Solving Horn clauses on inductive data types without induction. Theor. Pract. Logic Program. **18**(3–4), 452–469 (2018). Special Issue on ICLP 2018
11. De Angelis, E., Fioravanti, F., Pettorossi, A., Proietti, M.: Lemma generation for Horn clause satisfiability: a preliminary study. In: Lisitsa, A., Nemytykh, A.P. (eds.) Proceedings Seventh International Workshop on Verification and Program Transformation, VPT@Programming 2019, EPTCS, Genova, Italy, 2nd April 2019, vol. 299, pp. 4–18 (2019)
12. De Angelis, E., Fioravanti, F., Pettorossi, A., Proietti, M.: Proving properties of sorting programs: a case study in Horn clause verification. In: De Angelis, E., Fedyukovich, G., Tzevelekos, N., Ulbrich, M. (eds.) Proceedings of the Sixth Workshop on Horn Clauses for Verification and Synthesis and Third Workshop on Program Equivalence and Relational Reasoning, HCVS/PERR@ETAPS 2019, EPTCS, Prague, Czech Republic, 6–7 April 2019, vol. 296, pp. 48–75 (2019)
13. De Angelis, E., Fioravanti, F., Pettorossi, A., Proietti, M.: Removing algebraic data types from constrained Horn clauses using difference predicates - Preliminary version. CoRR (2020). http://arXiv.org/abs/2004.07749
14. de Moura, L., Bjørner, N.: Z3: an efficient SMT solver. In: Ramakrishnan, C.R., Rehof, J. (eds.) TACAS 2008. LNCS, vol. 4963, pp. 337–340. Springer, Heidelberg (2008). https://doi.org/10.1007/978-3-540-78800-3_24
15. Dixon, L., Fleuriot, J.: IsaPlanner: a prototype proof planner in Isabelle. In: Baader, F. (ed.) CADE 2003. LNCS (LNAI), vol. 2741, pp. 279–283. Springer, Heidelberg (2003). https://doi.org/10.1007/978-3-540-45085-6_22
16. Enderton, H.: A Mathematical Introduction to Logic. Academic Press, Cambridge (1972)

17. Etalle, S., Gabbrielli, M.: Transformations of CLP modules. Theor. Comput. Sci. **166**, 101–146 (1996)
18. Fioravanti, F., Pettorossi, A., Proietti, M.: Transformation rules for locally stratified constraint logic programs. In: Bruynooghe, M., Lau, K.-K. (eds.) Program Development in Computational Logic. LNCS, vol. 3049, pp. 291–339. Springer, Heidelberg (2004). https://doi.org/10.1007/978-3-540-25951-0_10
19. Fioravanti, F., Pettorossi, A., Proietti, M., Senni, V.: Generalization strategies for the verification of infinite state systems. Theor. Pract. Logic Program. **13**(2), 175–199 (2013). Special Issue on the 25th Annual GULP Conference
20. Grebenshchikov, S., Lopes, N.P., Popeea, C., Rybalchenko, A.: Synthesizing software verifiers from proof rules. In: Proceedings of the ACM SIGPLAN Conference on Programming Language Design and Implementation, PLDI 2012, pp. 405–416 (2012)
21. Hoder, K., Bjørner, N.: Generalized property directed reachability. In: Cimatti, A., Sebastiani, R. (eds.) SAT 2012. LNCS, vol. 7317, pp. 157–171. Springer, Heidelberg (2012). https://doi.org/10.1007/978-3-642-31612-8_13
22. Hojjat, H., Rümmer, P.: The ELDARICA Horn solver. In: Bjørner, N., Gurfinkel, A. (eds.) 2018 Formal Methods in Computer Aided Design, FMCAD 2018, Austin, TX, USA, 30 Oct–2 Nov 2018, pp. 1–7. IEEE (2018)
23. Ireland, A., Bundy, A.: Productive use of failure in inductive proof. J. Autom. Reason. **16**(1), 79–111 (1996)
24. Jaffar, J., Maher, M.: Constraint logic programming: a survey. J. Logic Program. **19**(20), 503–581 (1994)
25. Johansson, M., Dixon, L., Bundy, A.: Case-analysis for rippling and inductive proof. In: Kaufmann, M., Paulson, L.C. (eds.) ITP 2010. LNCS, vol. 6172, pp. 291–306. Springer, Heidelberg (2010). https://doi.org/10.1007/978-3-642-14052-5_21
26. Kafle, B., Gallagher, J.P., Morales, J.F.: RAHFT: a tool for verifying Horn clauses using abstract interpretation and finite tree automata. In: Chaudhuri, S., Farzan, A. (eds.) CAV 2016. LNCS, vol. 9779, pp. 261–268. Springer, Cham (2016). https://doi.org/10.1007/978-3-319-41528-4_14
27. Kaufmann, M., Manolios, P., Moore, J.S.: Computer-Aided Reasoning: An Approach. Kluwer Academic Publishers, Berlin (2000)
28. Komuravelli, A., Gurfinkel, A., Chaki, S.: SMT-based model checking for recursive programs. In: Biere, A., Bloem, R. (eds.) CAV 2014. LNCS, vol. 8559, pp. 17–34. Springer, Cham (2014). https://doi.org/10.1007/978-3-319-08867-9_2
29. Komuravelli, A., Gurfinkel, A., Chaki, S., Clarke, E.M.: Automatic abstraction in SMT-based unbounded software model checking. In: Sharygina, N., Veith, H. (eds.) CAV 2013. LNCS, vol. 8044, pp. 846–862. Springer, Heidelberg (2013). https://doi.org/10.1007/978-3-642-39799-8_59
30. Leino, K.R.M.: Automating induction with an SMT solver. In: Kuncak, V., Rybalchenko, A. (eds.) VMCAI 2012. LNCS, vol. 7148, pp. 315–331. Springer, Heidelberg (2012). https://doi.org/10.1007/978-3-642-27940-9_21
31. Leroy, X., Doligez, D., Frisch, A., Garrigue, J., Rémy, D., Vouillon, J.: The OCaml system, Release 4.06. Documentation and user's manual, Institut National de Recherche en Informatique et en Automatique, France (2017)
32. Leuschel, M., Butler, M.: ProB: a model checker for B. In: Araki, K., Gnesi, S., Mandrioli, D. (eds.) FME 2003. LNCS, vol. 2805, pp. 855–874. Springer, Heidelberg (2003). https://doi.org/10.1007/978-3-540-45236-2_46
33. Lloyd, J.W.: Foundations of Logic Programming, 2nd edn. Springer, Berlin (1987). https://doi.org/10.1007/978-3-642-83189-8

34. Nipkow, T., Wenzel, M., Paulson, L.C.: Isabelle/HOL: A Proof Assistant for Higher-Order Logic. Springer, Heidelberg (2002)
35. Owre, S., Rushby, J.M., Shankar, N.: PVS: a prototype verification system. In: Kapur, D. (ed.) CADE 1992. LNCS, vol. 607, pp. 748–752. Springer, Heidelberg (1992). https://doi.org/10.1007/3-540-55602-8_217
36. Pham, T.-H., Gacek, A., Whalen, M.W.: Reasoning about algebraic data types with abstractions. J. Autom. Reason. **57**(4), 281–318 (2016)
37. Rabin, M.O.: Decidable theories. In: Barwise, J. (ed.) Handbook of Mathematical Logic, pp. 595–629. North-Holland, Amsterdam (1977)
38. Reynolds, A., Kuncak, V.: Induction for SMT solvers. In: D'Souza, D., Lal, A., Larsen, K.G. (eds.) VMCAI 2015. LNCS, vol. 8931, pp. 80–98. Springer, Heidelberg (2015). https://doi.org/10.1007/978-3-662-46081-8_5
39. Seki, H.: On Inductive and coinductive proofs via unfold/fold transformations. In: De Schreye, D. (ed.) LOPSTR 2009. LNCS, vol. 6037, pp. 82–96. Springer, Heidelberg (2010). https://doi.org/10.1007/978-3-642-12592-8_7
40. Sonnex, W., Drossopoulou, S., Eisenbach, S.: Zeno: an automated prover for properties of recursive data structures. In: Flanagan, C., König, B. (eds.) TACAS 2012. LNCS, vol. 7214, pp. 407–421. Springer, Heidelberg (2012). https://doi.org/10.1007/978-3-642-28756-5_28
41. Suter, P., Köksal, A.S., Kuncak, V.: Satisfiability modulo recursive programs. In: Yahav, E. (ed.) SAS 2011. LNCS, vol. 6887, pp. 298–315. Springer, Heidelberg (2011). https://doi.org/10.1007/978-3-642-23702-7_23
42. Tamaki, H., Sato, T.: A generalized correctness proof of the unfold/fold logic program transformation. Technical Report 86–4, Ibaraki University, Japan (1986)
43. Unno, H., Torii, S., Sakamoto, H.: Automating induction for solving Horn clauses. In: Majumdar, R., Kunčak, V. (eds.) CAV 2017. LNCS, vol. 10427, pp. 571–591. Springer, Cham (2017). https://doi.org/10.1007/978-3-319-63390-9_30
44. Yang, W., Fedyukovich, G., Gupta, A.: Lemma synthesis for automating induction over algebraic data types. In: Schiex, T., de Givry, S. (eds.) CP 2019. LNCS, vol. 11802, pp. 600–617. Springer, Cham (2019). https://doi.org/10.1007/978-3-030-30048-7_35

Solving Bitvectors with MCSAT: Explanations from Bits and Pieces

Stéphane Graham-Lengrand(✉), Dejan Jovanović, and Bruno Dutertre

SRI International, Menlo Park, USA
stephane.graham-lengrand@csl.sri.com

Abstract. We present a decision procedure for the theory of fixed-sized bitvectors in the MCSAT framework. MCSAT is an alternative to CDCL(T) for SMT solving and can be seen as an extension of CDCL to domains other than the Booleans. Our procedure uses BDDs to record and update the sets of feasible values of bitvector variables. For explaining conflicts and propagations, we develop specialized word-level interpolation for two common fragments of the theory. For full generality, explaining conflicts outside of the covered fragments resorts to local bit-blasting. The approach is implemented in the Yices 2 SMT solver and we present experimental results.

1 Introduction

Model-constructing satisfiability (MCSAT) [19,20,24] is an alternative to the CDCL($\mathcal{T}$) scheme [28] for Satisfiability Modulo Theories (SMT). While CDCL($\mathcal{T}$) interfaces a CDCL SAT solver [23] with black-box decision procedures, MCSAT integrates first-order reasoning into CDCL directly. Like CDCL, MCSAT alternates between search and conflict analysis. In the search phase, MCSAT assigns values to first-order variables and propagates unit consequences of these assignments. If a conflict occurs during search, e.g., when the domain of a first-order variable is empty, MCSAT enters conflict analysis and learns an explanation, which is a symbolic representation of what was wrong with the assignments causing the conflict. As in CDCL, the learned clause triggers backtracking from which search can resume. Decision procedures based on MCSAT have demonstrated strong performance in theories such as non-linear real [24] and integer arithmetic [19]. These theories are relatively well-behaved and provide features such as quantifier elimination and interpolation—the building blocks of conflict resolution in MCSAT.

We describe an MCSAT decision procedure for the theory of bitvectors ($\mathcal{BV}$). In contrast to arithmetic, the complexity of $\mathcal{BV}$ in terms of syntax and semantics, combined with the lack of word-level interpolation and quantifier elimination, makes the development of $\mathcal{BV}$ decision procedures (MCSAT or not) very difficult. The state-of-the art $\mathcal{BV}$ decision procedures are all based on a "preprocess and bitblast" pipeline [12,22,25]: they reduce the $\mathcal{BV}$ problems to a pure SAT problem by reducing the word-level semantics to bit-level semantics. Exceptions

N. Peltier and V. Sofronie-Stokkermans (Eds.): IJCAR 2020, LNAI 12166, pp. 103–121, 2020.
https://doi.org/10.1007/978-3-030-51074-9_7

to the bitblasting approach do exist, such as [4,16], which generally do not perform as well as bitblasting except on small classes of crafted examples, and the MCSAT approach of [30], which we discuss below and in the conclusion.

An MCSAT decision procedure must provide two theory-specific reasoning mechanisms.

First, the procedure must maintain a set of values that are feasible for each variable. This set is updated during the search. It is used to propagate variable values and to detect a conflict when the set becomes empty. Finding a suitable representation for domains is a key step in integrating a theory into MCSAT. We represent variable domains with Binary Decision Diagrams (BDDs) [5]. BDDs can represent any set of bitvector values. By being canonical, they offer a simple mechanism to detect when a domain becomes a singleton—in which case MCSAT can perform a theory propagation—and when a domain becomes empty–in which case MCSAT enters conflict analysis. In short, BDDs offer a generic mechanism for proposing and propagating values, and for detecting conflicts. In contrast, previous work by Zeljić et al. [30] represents bitvector domains using *intervals* and *patterns*, which cannot represent every set of bitvector values precisely; they over-approximate the domains.

Second, once a conflict has been detected, the procedure must construct a symbolic explanation of the conflict. This explanation must rule out the partial assignment that caused the conflict, but it is desirable for explanations to generalize and rule out larger parts of the search space. For this purpose, previous work [30] relied on incomplete abstraction techniques (replace a value by an interval; extend a value into a larger set by leaving some bits unassigned), and left open the idea of using interpolation to produce explanations. Instead of aiming for a uniform, generic explanation mechanism, we take a modular approach. We develop efficient word-level explanation procedures for two useful fragments of $\mathcal{BV}$, based on interpolation. Our first fragment includes bitvector equalities, extractions, and concatenations where word-level explanations can be constructed through model-based variants of classic equality reasoning techniques (e.g., [4,9,10]). Our second fragment is a subset of linear arithmetic where explanations are constructed by interval reasoning in modular arithmetic. When conflicts do not fit into either fragment, we build an explanation by bitblasting and extracting an unsat core. Although this fallback produces theory lemmas expressed at the bit-level, it is used only as a last resort. In addition, this bitblasting-based procedure is local and limited to constraints that are relevant to the current conflict; we do not apply bitblasting to the full problem.

Section 2, is an overview of MCSAT. It also presents the BDD approach and general considerations for conflict explanation. Section 3 describes our interpolation algorithm for equality with concatenation and extraction. Section 4 presents our interpolation method for a fragment of linear bitvector arithmetic. Section 5 presents the normalization technique we apply to conflicts in the hope of expressing them in that bitvector arithmetic fragment. Section 6 presents an evaluation of the approach, which we implemented in the Yices 2 solver [11].[1]

2 A General Scheme for Bitvectors

By $\mathcal{BV}$, we denote the theory of quantifier-free fixed-sized bitvectors, a.k.a. QF_BV in SMT-LIB [1]. A first-order term u of $\mathcal{BV}$ is sorted as either a Boolean or a bitvector of a fixed *length* (a.k.a. *bitwidth*), denoted $|u|$. Its set of variables (a.k.a. uninterpreted constants) is denoted $\mathsf{var}(u)$. This paper only uses a few $\mathcal{BV}$ operators. The concatenation of bitvector terms t and u is denoted $t \circ u$; the binary predicates $<^{\mathsf{u}}$, $\leq^{\mathsf{u}}$ denote unsigned comparisons, and $<^{\mathsf{s}}$, $\leq^{\mathsf{s}}$ denote signed comparisons. In such comparisons, both operands must have the same bitwidth. If n is the bitwidth of u, and l and h are two integer indices such that $0 \leq l < h \leq n$, then $u[h{:}l]$, extracts $h{-}l$ bits of u, namely the bits at indices between l and $h{-}1$ (included). We write $u[{:}l]$, $u[h{:}]$, and $u[l]$ as abbreviations for $u[n{:}l]$, $u[h{:}0]$, and $u[l{+}1{:}l]$, respectively. Our convention is to have bitvector indices start from the right-hand side, so that bit 0 is the right-most bit and $0011[2{:}]$ is 11. We use standard notations for bitvector arithmetic, which coincides with arithmetic modulo 2^w where w is the bitwidth. We sometimes use integer constants e.g., 0, 1, -1 for bitvectors when the bitwidth is clear. We use the standard (quantifier-free) notions of *literal*, *clause*, *cube*, and *formula* [29].

A *model* of a $\mathcal{BV}$ formula Φ is an assignment that gives a bitvector (resp. Boolean) value to all bitvector (resp. Boolean) variables of Φ, in such a way that Φ evaluates to true, under the standard interpretation of Boolean and bitvector symbols. To simplify the presentation, we assume in this paper that there are no Boolean variables, although they are supported in our implementation.

2.1 MCSAT Overview

MCSAT searches for a model of an input quantifier-free formula by building a partial assignment—maintained in a *trail*—and extends the concepts of unit propagation and consistency to first-order terms and literals [19,20,24]. Reasoning is implemented by theory-specific plugins, each of which has a partial view of the trail. In the case of $\mathcal{BV}$, the bitvector plugin sees in the trail an assignment $\mathcal{M}$ of the form $x_1 \mapsto v_1, \ldots, x_n \mapsto v_n$ that gives values to bitvector variables, and a set of bitvector literals $L_1, \ldots, L_t$, called *constraints*, that must be true in the current trail. MCSAT and its bitvector plugin maintain the invariant that none of the literals L_i evaluates to false under $\mathcal{M}$; either L_i is true or some variable of L_i has no value in $\mathcal{M}$. To maintain this invariant, they detect *unit inconsistencies*: We say that literal L_i is *unit in* y if y is the only unassigned variable of L_i, and that a trail is *unit inconsistent* if there is a variable y and a subset $\{C_1, \ldots, C_m\}$ of $\{L_1, \ldots, L_t\}$, called a *conflict*, such that every C_j is unit

[1] This paper extends preliminary results presented at the SMT workshop [13,14] and includes a full implementation and experimental evaluation.

in y and the formula $\exists y \bigwedge_{i=1}^{m} C_i$ evaluates to false under $\mathcal{M}$. In such a case, y is called the *conflict variable* and $C_1, \ldots, C_m$ are called the *conflict literals*.

When such a conflict is detected, the current assignment, or partial model, $\mathcal{M}$ cannot be extended to a full model; some values assigned to $x_1, \ldots, x_n$ must be revised. As in CDCL, MCSAT backtracks and updates the current assignment by learning a new clause that explains the conflict. This new clause must not contain other variables than $x_1, \ldots, x_n$ and it must rule out the current assignment. For some theories, this *conflict explanation* can be built by quantifier elimination. More generally, we can build an explanation from an *interpolant*.

Definition 1 (Interpolant). *A clause I is an* interpolant[2] *for formula F at model $\mathcal{M}$ assigning values to $x_1, \ldots, x_n$, if (1) $F \Rightarrow I$ is valid (in $\mathcal{BV}$), (2) The variables in I are in $\{x_1, \ldots, x_n\} \cap \mathsf{var}(F)$, and (3) I evaluates to false in $\mathcal{M}$.*

Given an interpolant I for the conjunction $\bigwedge_{i=1}^{m} C_i$ of the conflict literals (or equivalently, for $\exists y \bigwedge_{i=1}^{m} C_i$) at the current model $\mathcal{M}$, the conflict explanation is clause $(\bigwedge_{i=1}^{m} C_i) \Rightarrow I$. Our main goal is constructing such interpolants in $\mathcal{BV}$.

2.2 BDD Representation and Conflict Detection

To detect conflicts, we must keep track of the set of feasible values for every unassigned variable y. These sets are frequently updated during search so an efficient representation is critical. The following operations are needed:

- updating the set when a new constraint becomes unit in y,
- detecting when the set becomes empty,
- selecting a value from the set.

For $\mathcal{BV}$, Zeljić et al. [30] represent sets of feasible values using both intervals and bit patterns. For example, the set defined by the interval [0000, 0011] and the pattern ???1 is the pair {0001, 0011} (i.e., all bitvectors in the interval whose low-order bit is 1). This representation is lightweight and efficient but it is not precise. Some sets are not representable exactly. We use Binary Decision Diagrams (BDD) [5] over the bits of y. The major advantage is that BDDs provide an exact implementation of any set of values for y. Updating sets of values amounts to computing the conjunction of BDDs (i.e., set intersection). Checking whether a set is empty and selecting a value in the set (if it is not), can be done efficiently by, respectively, checking whether the BDD is false, and performing a top-down traversal of the BDD data structure. There is a risk that the BDD representation explodes but this risk is reduced in our context since each BDD we build is for a single variable (and most variables do not have too many bits). We use the CUDD package [8] to implement BDDs.

---

[2] This is the same as the usual notion of (reverse) interpolant between formulas if we see $\mathcal{M}$ as the formula $F_{\mathcal{M}}$ defined by $(x_1 \simeq v_1) \wedge \cdots \wedge (x_n \simeq v_n)$: the interpolant is implied by F, it is inconsistent with $F_{\mathcal{M}}$, and its variables occur in both F and $F_{\mathcal{M}}$.

2.3 Baseline Conflict Explanation

Given a conflict as described previously, the clause $(x_1 \not\simeq v_1) \vee \cdots \vee (x_n \not\simeq v_n)$, which is falsified by model $\mathcal{M}$ only, is an interpolant for $\bigwedge_{i=1}^{m} C_i$ at $\mathcal{M}$ according to Definition 1. This gives the following trivial conflict explanation:

$$C_1 \wedge \cdots \wedge C_m \Rightarrow (x_1 \not\simeq v_1) \vee \cdots \vee (x_n \not\simeq v_n)$$

We seek to generalize model $\mathcal{M}$ with a formula that rules out bigger parts of the search space than just $\mathcal{M}$. A first improvement is replacing the constraints by a *core* $\mathcal{C}$, that is, a minimal subset of $\{C_1, \ldots, C_n\}$ that evaluates to false in $\mathcal{M}$.[3]

To produce the interpolant I, we can bitblast the constraints $C_1, \ldots, C_m$ and solve the resulting SAT problem *under the assumptions* that each bit of $x_1, \ldots, x_n$ is true or false as indicated by the values $v_1, \ldots, v_n$. Since the SAT problem encodes a conflict, the SAT solver will return an *unsat core*, from which we can extract bits of $v_1, \ldots, v_n$ that contribute to unsatisfiability. This generalizes $\mathcal{M}$ by leaving some bits unassigned, as in [30].

This method is general. It works whatever the constraints $C_1, \ldots, C_m$, so we use it as a default procedure. The bitblasting step focuses on constraints that are unit in y, which typically leads to a much smaller SAT problem than bitblasting the whole problem from the start. However, the bitblasting approach can still be costly and it may produce weak explanations.

Example 1. Consider the constraints $\{x_1 \not\simeq x_2,\ x_1 \simeq y,\ x_2 \simeq y\}$ and the assignment $x_1 \mapsto 1001, x_2 \mapsto 0101$. The bitblasting approach might produce explanation $(x_1 \simeq y \wedge x_2 \simeq y) \Rightarrow (x_1[3] \Rightarrow x_2[3])$. After backtracking, we might similarly learn that $(x_2[3] \Rightarrow x_1[3])$. In this way, it will take eight iterations to learn enough information to represent the high-level explanation:

$$(x_1 \simeq y \wedge x_2 \simeq y) \Rightarrow x_1 \simeq x_2.$$

A procedure that can produce $(x_1 \simeq x_2)$ directly is much more efficient.

3 Equality, Concatenation, Extraction

Our first specialized interpolation mechanism applies when constraints $\mathcal{C} = \{C_1, \ldots, C_m\}$ belong to the following grammar:

$$\begin{array}{llcl} \text{Constraints} & C & ::= & t \simeq t \mid t \not\simeq t \\ \text{Terms} & t & ::= & e \mid y[h{:}l] \mid t \circ t \end{array}$$

where e ranges over any bitvector terms such that $y \notin \mathsf{var}(e)$. Without loss of generality, we can assume that $\mathcal{C}$ is a core. We split $\mathcal{C}$ into a set of equalities $E = \{a_i \simeq b_i\}_{i \in \mathfrak{E}}$ and a set of disequalities $D = \{a_i \not\simeq b_i\}_{i \in \mathfrak{D}}$.

[3] In our implementation, we construct $\mathcal{C}$ using the QuickXplain algorithm [21].

Algorithm 1. E-graph with value management

function E_GRAPH($E_s, \mathcal{M}$)
  INITIALIZE($\mathcal{G}$) ▷ each evaluable term or slice is its own component
  for $t_1 \simeq t_2 \in E_s$ **do**
    $t_1' \leftarrow$ REP($t_1, \mathcal{G}$) ▷ get representative for t_1's component
    $t_2' \leftarrow$ REP($t_2, \mathcal{G}$) ▷ get representative for t_2's component
    if $y \notin \mathsf{var}(t_1')$ and $y \notin \mathsf{var}(t_2')$ and $[\![t_1']\!]_{\mathcal{M}} \neq [\![t_2']\!]_{\mathcal{M}}$ **then**
      raise_conflict($E \Rightarrow t_1' \simeq t_2'$) ▷ D must be empty
    $t_3 \leftarrow$ SELECT(t_1', t_2') ▷ select representative for merged component
    $\mathcal{G} \leftarrow$ MERGE($t_1, t_2, t_3, \mathcal{G}$) ▷ merge the components with representative t_3
  return $\mathcal{G}$

Slicing. Our first step rewrites $\mathcal{C}$ into an equivalent *sliced* form. This computes the *coarsest-base slicing* [4,9] of equalities and disequalities in $\mathcal{C}$. The goal of this rewriting step is to split the variables into slices that can be treated as independent terms. The terms in coarsest-base slicing are either of the form $y[h{:}l]$ (slices), or are *evaluable terms* e with $y \notin \mathsf{var}(e)$.

Example 2. Consider the constraints $E = \{x_1[4{:}0] \simeq x_1[8{:}4], y[6{:}2] \simeq y[4{:}0]\}$ and $\{y[4{:}0] \not\simeq x_1[8{:}4]\}$ over variables y of length 6, and x_1 of length 8. We cannot treat $y[6{:}2]$ and $y[4{:}0]$ as independent terms because they overlap. To break the overlap, we introduce slices: $y[6{:}4]$, $y[4{:}2]$, and $y[2{:}0]$. Equality $y[6{:}2] \simeq y[4{:}0]$ is rewritten to $(y[6{:}4] \simeq y[4{:}2]) \wedge (y[4{:}2] \simeq y[2{:}0])$. Disequality $y[4{:}0] \not\simeq x_1[8{:}4]$ is rewritten to $(y[4{:}2] \not\simeq x_1[8{:}6]) \vee (y[2{:}0] \not\simeq x_1[6{:}4])$. The final result is

$$E_s = \{\ x_1[4{:}2] \simeq x_1[8{:}6]\ ,\ x_1[2{:}0] \simeq x_1[6{:}4]\ ,\ y[6{:}4] \simeq y[4{:}2]\ ,\ y[4{:}2] \simeq y[2{:}0]\ \},$$
$$D_s = \{\ (y[4{:}2] \not\simeq x_1[8{:}6]) \vee (y[2{:}0] \not\simeq x_1[6{:}4])\ \}.$$

Explanations. After slicing, we obtain a set E_s of equalities and a set D_s that contains disjunctions of disequalities. We can treat each slice as a separate variable, so the problem lies within the theory of equality on a *finite domain.*

We first analyze the conflict with equality reasoning against the model, as shown in Algorithm 1. We construct the E-graph $\mathcal{G}$ from E_s [10], while also taking into account the partial model $\mathcal{M}$ that triggered the conflict. The model can evaluate terms e such that $y \notin \mathsf{var}(e)$ to values $[\![e]\!]_{\mathcal{M}}$, and those can be the source of the conflict. To use the model for evaluating terms, we maintain two invariants during E-graph construction:

1. If a component contains an evaluable term c, then the representative of that component is evaluable.
2. Two evaluable terms c_1 and c_2 in the same component must evaluate to the same value, otherwise this is the source of the conflict.

The E-graph construction can detect and explain basic conflicts between the equalities in E and the current assignment.

Example 3. Let r_1, r_2, r_3 be bit ranges of the same width. Let E be such that $E_s = \{x_1[r_1] \simeq y[r_3],\ x_2[r_2] \simeq y[r_3]\}$, and let $D = \emptyset$. Consider the model $\mathcal{M} :=$

Algorithm 2. Disequality conflict

function DIS_CONFLICT($D_s, \mathcal{M}, \mathcal{G}$)
  $S \leftarrow \emptyset$ ▷ where we collect interface terms
  $C_0 \leftarrow \emptyset$ ▷ where we collect the disequalities that evaluate to false
  for $C \in D_s$ **do**
    $C_{\mathcal{M}}^{\mathsf{rep}} \leftarrow \bigvee\{\text{REP}(t_1, \mathcal{G}) \not\simeq \text{REP}(t_2, \mathcal{G}) \mid (t_1 \not\simeq t_2) \in C_{\mathcal{M}}\}$
    if IS_EMPTY($C_{\mathsf{interface}}$) and IS_EMPTY(C_{free}) **then**
      raise_conflict($E \wedge D \Rightarrow C_{\mathcal{M}}^{\mathsf{rep}}$)
    else
      $C_0 \leftarrow C_0 \vee C_{\mathcal{M}}^{\mathsf{rep}}$ ▷ we collect the disequalities made false in the model
      for $t_1 \not\simeq t_2 \in C_{\mathsf{interface}}$ with $y \notin \mathsf{var}(\text{REP}(t_1, \mathcal{G}))$ **do**
        $S \leftarrow S \cup \{\text{REP}(t_1, \mathcal{G})\}$ ▷ we collect the interface term
  $C_{\neq} \leftarrow \bigvee\{t_1 \simeq t_2 \mid [\![t_1]\!]_{\mathcal{M}} \neq [\![t_2]\!]_{\mathcal{M}},\ t_1, t_2 \in S\}$
  $C_{=} \leftarrow \bigvee\{t_1 \not\simeq t_2 \mid [\![t_1]\!]_{\mathcal{M}} = [\![t_2]\!]_{\mathcal{M}},\ t_1 \neq t_2,\ t_1, t_2 \in S\}$
  return $E \wedge D \Rightarrow C_0 \vee C_{\neq} \vee C_{=}$

$x_1 \mapsto 0\ldots0, x_2 \mapsto 1\ldots1$. Then, E_GRAPH($E_s, \mathcal{M}$) produces the conflict clause $E \Rightarrow x_1[r_1] \simeq x_2[r_2]$.

If the E-graph construction does not raise a conflict, then $\mathcal{M}$ is compatible with the equalities in E_s. Since $\mathcal{C}$ conflicts with $\mathcal{M}$, the conflict explanation must involve D_s. To obtain an explanation, we decompose each disjunct $C \in D_s$ into $(C_{E_s} \vee C_{\mathcal{M}} \vee C_{\mathsf{interface}} \vee C_{\mathsf{free}})$ as follows.

- C_{E_s} contains disequalities $t_1 \not\simeq t_2$ such that t_1 and t_2 have the same E-graph representatives; such disequalities are false because of the equalities in E_s.
- $C_{\mathcal{M}}$ contains disequalities $t_1 \not\simeq t_2$ such that t_1 and t_2 have distinct representatives t_1' and t_2' with $[\![t_1']\!]_{\mathcal{M}} = [\![t_2']\!]_{\mathcal{M}}$; these are false because of $\mathcal{M}$.
- $C_{\mathsf{interface}}$ contains disequalities $t_1 \not\simeq t_2$ such that t_1 and t_2 have distinct representatives t_1' and t_2', t_1' is evaluable and t_2' is a slice; we can still satisfy $t_1 \not\simeq t_2$ by picking a good value for y; we say t_1' is an *interface term.*
- C_{free} contains disequalities $t_1 \not\simeq t_2$ such that t_1 and t_2 have distinct slices as representatives; we can still satisfy $t_1 \not\simeq t_2$ by picking a good value for y.

The disjuncts in D_s take part in the conflict either when (i) one of the clauses in D_s is false because $C_{\mathsf{interface}}$ and C_{free} are both empty; or (ii) the finite domains are too small to satisfy the disequalities in $C_{\mathsf{interface}}$ and C_{free}, given the values assigned in $\mathcal{M}$. In either case, we can produce a conflict explanation with Algorithm 2.

In a type (i) conflict, the algorithm produces an interpolant $C_{\mathcal{M}}^{\mathsf{rep}}$ that is derived from a single element of D_s. Because we assume that $\mathcal{C}$ is a core, a type (i) conflict can happen only if D_s is a singleton. Here is how the algorithm behaves on such a conflict:

Example 4. Let r_1 and r_2 be bit ranges of the same length, let r_3, r_4, r_5 be bit ranges of the same length. Assume E_s contains

$$\{\ x_1[r_1] \simeq y[r_1]\ ,\ x_2[r_2] \simeq y[r_2]\ ,\ y[r_3] \simeq y[r_5]\ ,\ y[r_4] \simeq y[r_5]\ \},$$

and assume D_s is the singleton $\{ (y[r_1] \not\simeq y[r_2] \vee y[r_3] \not\simeq y[r_4]) \}$. Let $\mathcal{M}$ map x_1 and x_2 to $0\ldots0$ and assume $y[r_5]$ is the E-graph representative for component

$$\{ y[r_3], y[r_4], y[r_5] \}.$$

The unique clause of D_s contains two disequalities:

- The first one, $y[r_1] \not\simeq y[r_2]$, belongs to $C_{\mathcal{M}}$ because the representatives of $y[r_1]$ and $y[r_2]$, namely $x_1[r_1]$ and $x_2[r_2]$, both evaluate to $0\ldots0$.
- The second one, $y[r_3] \not\simeq y[r_4]$, belongs to C_{E_s} because the representatives of $y[r_3]$ and $y[r_4]$ are both $y[r_5]$,

As $C_{\mathsf{interface}}$ and C_{free} are empty, Algorithm 2 outputs $E \wedge D \Rightarrow x_1[r_1] \not\simeq x_2[r_2]$.

For a conflict of type (ii), the equalities and disequalities that hold in $\mathcal{M}$ between the interface terms make the slices of y require more values than there exist. So the produced conflict clause includes (the negation of) all such equalities and disequalities. An example can be given as follows:

Example 5. Assume E (and then E_s) is empty and assume D_s is

$$\{ x_2[0] \not\simeq x_2[1] \vee y[0] \not\simeq y[1] \ ,\ x_1[0] \not\simeq y[0] \ ,\ x_1[1] \not\simeq y[1] \}$$

Let $\mathcal{M}$ map x_1 and x_2 to 00. Then DIS_CONFLICT$(D_s, \mathcal{M}, \mathcal{G})$ behaves as follows:

- In the first clause, call it C, the first disequality is in $C_{\mathcal{M}}$, as the two sides are in different components but evaluate to the same value; so C_0 becomes $\{ x_2[0] \not\simeq x_2[1] \}$; the second disequality features two slices and is thus in C_{free}; The clause is potentially satisfiable and we move to the next clause.
- The second clause contains a single disequality that cannot be evaluated (since $y[0]$ is not evaluable in $\mathcal{M}$). Term $x_1[0]$ is added to S. The clause is potentially satisfiable so we move to the next clause.
- The third clause of D_s is similar. It contains a single disequality that cannot be evaluated. The interface term $x_1[1]$ is added to S.

Since all clauses of D_s have been processed, the conflict is of type (ii). Indeed, $y[0]$ must be different from 0 because of the second clause, $y[1]$ must also be different from 0 because of the third clause, but $y[0]$ and $y[1]$ must be different from each other because of the first clause. Since both $y[0]$ and $y[1]$ have only one bit, there are only two possible values for these two slices, so the three constrains are in conflict. Algorithm 2 produces the conflict clause

$$D \Rightarrow (\ x_2[0] \not\simeq x_2[1] \vee x_1[0] \not\simeq x_1[1]\ ).$$

The disequality $x_2[0] \not\simeq x_2[1]$ is necessary because, if it were true in $\mathcal{M}$, we would not have to satisfy $y[0] \not\simeq y[1]$ and therefore $y \leftarrow 11$ would work. Disequality $x_1[0] \not\simeq x_1[1]$ is also necessary because, if it were true in $\mathcal{M}$, say with $x_1 \leftarrow 01$ (resp. $x_1 \leftarrow 10$), then $y \leftarrow 11$ (resp. $y \leftarrow 00$) would work.

Correctness of the method relies on the following lemma, whose proof can be found in [15].

Lemma 1 (The produced clauses are interpolants).

1. *If Algorithm 1 reaches line 7, $t_1' \simeq t_2'$ is an interpolant for $E \wedge D$ at $\mathcal{M}$.*
2. *If Algorithm 2 reaches line 7, $C_{\mathcal{M}}^{\mathsf{rep}}$ is an interpolant for $E \wedge D$ at $\mathcal{M}$.*
3. *If it reaches line 14, $C_0 \vee C_{\neq} \vee C_{=}$ is an interpolant for $E \wedge D$ at $\mathcal{M}$.*

Table 1. Creating the forbidden intervals

Atom a	Forbidden interval that a (resp. $\neg a$) specifies for t			
	I_a	$I_{\neg a}$	Condition $c_a/c_{\neg a}$	
$e_1 + t \leq^{\mathsf{u}} e_2 + t$	$[-e_2\,;\,-e_1[$	$[-e_1\,;\,-e_2[$	$e_1 \not\simeq e_2$	1
	$[0\,;0[$	full	$e_1 \simeq e_2$	2
$e_1 \leq^{\mathsf{u}} e_2 + t$	$[-e_2\,;e_1 - e_2[$	$[e_1 - e_2\,;\,-e_2[$	$e_1 \not\simeq 0$	3
	$[0\,;0[$	full	$e_1 \simeq 0$	4
$e_1 + t \leq^{\mathsf{u}} e_2$	$[e_2 - e_1 + 1\,;\,-e_1[$	$[-e_1\,;e_2 - e_1 + 1[$	$e_2 \not\simeq -1$	5
	$[0\,;0[$	full	$e_2 \simeq -1$	6

4 A Linear Arithmetic Fragment

Our second specialized explanation mechanism applies when constraints $\mathcal{C} = \{C_1, \ldots, C_m\}$ belong to the following grammar:

$$\begin{array}{lll} \text{Constraints} & C & ::= a \mid \neg a \\ \text{Atoms} & a & ::= e_1 + t \leq^{\mathsf{u}} e_2 + t \mid e_1 \leq^{\mathsf{u}} e_2 + t \mid e_1 + t \leq^{\mathsf{u}} e_2 \\ \text{Terms} & t & ::= y[h{:}] \mid t[{:}l] \mid t + e_1 \mid -t \mid 0_k \circ t \mid t \circ 0_k \end{array}$$

where e_1 and e_2 range over *evaluable* bitvector terms (i.e., $y \notin \mathsf{var}(e_1) \cup \mathsf{var}(e_2)$), and 0_k is 0 on k bits. We can represent variable y as the term $y[|y|{:}]$. This fragment of bitvector arithmetic is *linear* in y and there can be only one occurrence of y in terms. Constraints in Sect. 3 are then outside this fragment in general.

Let $\mathcal{A}$ be $\exists y(C_1 \wedge \cdots \wedge C_m)$, and $\mathcal{M}$ be the partial model involved in the conflict. The interpolant for $\mathcal{A}$ at model $\mathcal{M}$ is (roughly) produced as follows:

1. For each constraint C_i, $1 \leq i \leq m$, featuring a (necessarily unique) lower-bits extract $y[w_i{:}]$, we compute a *condition cube* c_i satisfied by $\mathcal{M}$ and a *forbidden interval* I_i of the form $[l_i\,;u_i[$, where l_i and u_i are evaluable terms, such that $c_i \Rightarrow (C_i \Leftrightarrow (y[w_i{:}] \notin I_i))$ is valid.
2. We group the resulting intervals $(I_i)_{1 \leq i \leq m}$ according to their bitwidths: if $\mathcal{S}_w$ is the set of intervals forbidding values for $y[w{:}]$, $1 \leq w \leq |y|$, then under condition $\bigwedge_{i=1}^{m} c_i$ formula $\mathcal{A}$ is equivalent to $\exists y(\bigwedge_{w=1}^{|y|} (\, y[w{:}] \notin \bigcup_{I \in \mathcal{S}_w} I\,))$.
3. We produce a series of constraints $d_1, \ldots,\, d_p$ that are satisfied by $\mathcal{M}$ and that are inconsistent with $\bigwedge_{w=1}^{|y|} (\, y[w{:}] \notin \bigcup_{I \in \mathcal{S}_w} I\,)$. The interpolant will be $(\bigwedge_{i=1}^{m} c_i \wedge \bigwedge_{i=1}^{p} d_i) \Rightarrow \bot$: it is implied by $\mathcal{A}$, and evaluates to false in $\mathcal{M}$.

4.1 Forbidden Intervals

An *interval* takes the form $[l\,;u[$, where the lower bound l and upper bound u are evaluable terms of some bitwidth w, with l included and u excluded. The notion of interval used here is considered modulo 2^w. We do not require $l \leq^{\mathsf{u}} u$ so an interval may "wrap around" in $\mathbb{Z}/2^w\mathbb{Z}$. For instance, the interval $[1111\,;0001[$ contains two bitvector values, namely, 1111 and 0000. If l and u evaluate to the same value, then we consider $[l\,;u[$ to be empty (as opposed to the full domain,

$$
\begin{array}{ll}
\mathsf{forbid}(t,\ [0\,;0[,\ c) := (1,\ [0\,;0[,\ c) & \mathsf{forbid}(0_k \circ t,\ I,\ c) := \mathsf{utrim}_k(t,\ I,\ c)\\
\mathsf{forbid}(t,\ \mathsf{full},\ c) := (1,\ \mathsf{full},\ c) & \mathsf{forbid}(t \circ 0_k,\ I,\ c) := \mathsf{dtrim}_k(t,\ I,\ c)\\
\mathsf{forbid}(y[w{:}],\ I,\ c) := (w,\ I,\ c) & \text{when } I \text{ is not } [0\,;0[\text{ nor full}\\
\mathsf{forbid}(t[{:}w],\ [l\,;u[,\ c) := \mathsf{forbid}(t,\ [l \circ 0_w\,; u \circ 0_w[,\ c) & \\
\mathsf{forbid}(t + c,\ [l\,;u[,\ c) := \mathsf{forbid}(t,\ [l{-}c\,; u{-}c[,\ c) & \\
\mathsf{forbid}(-t,\ [l\,;u[,\ c) := \mathsf{forbid}(t,\ [1{-}u\,; 1{-}l[,\ c) &
\end{array}
$$

$$
\mathsf{utrim}_k(t,\ [l\,;u[,\ c) := \begin{cases}
\mathsf{forbid}(t,\ [l'\,;u'[,\ c \wedge c_l \wedge c_u) & \text{if } [l'\,;u'[\text{ is not } [0\,;0[\\
(1,\ \mathsf{full},\ c \wedge c_l \wedge c_u \wedge c') & \text{if } [l'\,;u'[\text{ is } [0\,;0[\text{ and } [\![c']\!]_{\mathcal{M}} \text{ is true}\\
(1,\ [0\,;0[,\ c \wedge c_l \wedge c_u \wedge \neg c') & \text{if } [l'\,;u'[\text{ is } [0\,;0[\text{ and } [\![c']\!]_{\mathcal{M}} \text{ is false}
\end{cases}
$$

where l' is $l[w{:}]$ (resp. 0_w) and c_l is a_l (resp. $\neg a_l$) if $[\![a_l]\!]_{\mathcal{M}}$ is true (resp. false),
u' is $u[w{:}]$ (resp. 0_w) and c_u is a_u (resp. $\neg a_u$) if $[\![a_u]\!]_{\mathcal{M}}$ is true (resp. false),
a_l is $l[{:}w] \simeq 0_k$, a_u is $u[{:}w] \simeq 0_k$, c' is $(0_{k+w} \in [l\,;u[)$, and w is $|t|$.

$$
\mathsf{dtrim}_k(t,\ [l\,;u[,\ c) := \begin{cases}
\mathsf{forbid}(t,\ [l'\,;u'[,\ p \wedge c_l \wedge c_u) & \text{if } [l'\,;u'[\text{ is not } [0\,;0[\\
(1,\ \mathsf{full},\ c \wedge c_l \wedge c_u \wedge c') & \text{if } [l'\,;u'[\text{ is } [0\,;0[\text{ and } [\![c']\!]_{\mathcal{M}} \text{ is true}\\
(1,\ [0\,;0[,\ c \wedge c_l \wedge c_u \wedge \neg c') & \text{if } [l'\,;u'[\text{ is } [0\,;0[\text{ and } [\![c']\!]_{\mathcal{M}} \text{ is false}
\end{cases}
$$

where l' is $l[{:}k]$ (resp. $l[{:}k]{+}1$) and c_l is a_l (resp. $\neg a_l$) if $[\![a_l]\!]_{\mathcal{M}}$ is true (resp. false),
u' is $u[{:}k]$ (resp. $u[{:}k]{+}1$) and c_u is a_u (resp. $\neg a_u$) if $[\![a_u]\!]_{\mathcal{M}}$ is true (resp. false),
a_l is $l[k{:}] \simeq 0_k$, a_u is $u[k{:}] \simeq 0_k$, c' is $(u' \circ 0_k \in [l\,;u[)$, and w is $|t|$.

Fig. 1. Transforming the forbidden intervals

which we denote by full^w or just full). Notation $t \in I$ stands for literal $\top$ if I is full and literal $t{-}l <^u u{-}l$ if I is $[l\,;u[$. The value in model $\mathcal{M}$ of an evaluable term e (resp. evaluable cube c, interval I) is denoted $[\![e]\!]_{\mathcal{M}}$ (resp. $[\![c]\!]_{\mathcal{M}}$, $[\![I]\!]_{\mathcal{M}}$).

Given a constraint C with unevaluable term t, we produce an interval I_C of forbidden values for t according to the rules of Table 1. A side condition literal c_C identifies when the lower and upper bounds would coincide, in which case the interval produced is either empty or full. For every row of the table, the formula $c_C \Rightarrow (C \Leftrightarrow t \notin I_C)$ is valid in $\mathcal{BV}$. Given a partial model $\mathcal{M}$, we convert C to such an interval by selecting the row where $[\![c_C]\!]_{\mathcal{M}} = \mathsf{true}$.

Example 6.

6.1 Assume C_1 is literal $\neg(x_1 \leq^{\mathsf{u}} y)$ and $\mathcal{M} = \{x_1 \mapsto 0000\}$. Then line 4 of Table 1 applies, and I_{C_1} is interval full with condition $x_1 \simeq 0$.

6.2 Assume C_1 is $\neg(y \simeq x_1)$, C_2 is $(x_1 \leq^{\mathsf{u}} x_3 + y)$, C_3 is $\neg(y - x_2 \leq^{\mathsf{u}} x_3 + y)$, and $\mathcal{M} = \{x_1 \mapsto 1100, x_2 \mapsto 1101, x_3 \mapsto 0000\}$. Then by line 5, $I_{C_1} = [x_1\,; x_1{+}1[$ with trivial condition $(0 \not\simeq -1)$, by line 3, $I_{C_2} = [-x_3\,; x_1{-}x_3[$ with condition $(x_1 \not\simeq 0)$, and by line 1, $I_{C_3} = [x_2\,; -x_3[$ with condition $(-x_2 \not\simeq x_3)$.

Given the supported grammar, term t contains a unique subterm of the form $y[w{:}]$. We transform I_C into an interval of forbidden values for $y[w{:}]$ by applying procedure $\mathsf{forbid}(t,\ I_C,\ c_C)$ shown in Fig. 1, which proceeds by recursion on t. Its specification is given below, and correctness is proved by induction on t.

bitwidth	w_1	>	w_2	> ··· >	w_j
Interval layer	w_1-intervals $\mathcal{S}_1 = \{I_{1.1}, I_{1.2}, \ldots\}$		w_2-intervals $\mathcal{S}_2 = \{I_{2.1}, I_{2.2}, \ldots\}$	... ...	w_j-intervals $\mathcal{S}_j = \{I_{j.1}, I_{j.2}, \ldots\}$
Forbidding values for	$y[w_1{:}]$		$y[w_2{:}]$	...	$y[w_j{:}]$

Fig. 2. Intervals collected from $C_1 \wedge \cdots \wedge C_m$

Lemma 2 (Correctness of forbidden intervals). *Assuming cube c is true in $\mathcal{M}$, then forbid(t , I , c) returns a triple (w, I', c') such that c' is a cube that is true in $\mathcal{M}$, and both $c' \Rightarrow c$ and $c' \Rightarrow (t \notin I \Leftrightarrow y[w{:}] \notin I')$ are valid in $\mathcal{BV}$.*

Running forbid(t_{C_i} , I_{C_i} , c_{C_i}) for all constraints C_i, $1 \le i \le m$, produces a family of triples $(w_i, I'_i, c'_i)_{1 \le i \le m}$ such that, for each i, formula $c'_i \Rightarrow (C_i \Leftrightarrow (y[w_i{:}] \notin I'_i))$ is valid in $\mathcal{BV}$ and c'_i is true in $\mathcal{M}$.

4.2 Interpolant

First, assume that one of the triples obtained above is of the form (w, full, c), coming from constraint C. As the interval forbids the full domain of values for $y[w{:}]$, we produce conflict clause $C \wedge c \Rightarrow \bot$. This formula is an interpolant for $\mathcal{A}$ at $\mathcal{M}$. This is illustrated in Example 7.1.

Example 7.

7.1 In Example 6.1 where C_1 is literal $\neg(x_1 \le^{\mathsf{u}} y)$ and $\mathcal{M} = \{x_1 \mapsto 0000\}$, the interpolant for $\neg(x_1 \le^{\mathsf{u}} y)$ at $\mathcal{M}$ is $(x_1 \simeq 0) \Rightarrow \bot$.

7.2 Example 6.2 does not contain a full interval. Model $\mathcal{M}$ satisfies the three conditions $c_1 := (0 \not\simeq -1)$, $c_2 := (x_1 \not\simeq 0)$ and $c_3 := (-x_2 \not\simeq x_3)$, and the intervals $I_1 = [x_1 ; x_1{+}1[$, $I_2 = [-x_3 ; x_1 - x_3[$, and $I_3 = [x_2 ; -x_3[$, evaluate to $[\![I_1]\!]_{\mathcal{M}} = [1100 ; 1101[$, $[\![I_2]\!]_{\mathcal{M}} = [0000 ; 1100[$, and $[\![I_3]\!]_{\mathcal{M}} = [1101 ; 0000[$, respectively. Note how $\bigcup_{i=1}^{3} [\![I_i]\!]_{\mathcal{M}}$ is the full domain.

Assume now that no interval is full (as in Example 7.2). We group the triples (w, I, c) into different *layers* characterized by their bitwidths w: I will henceforth be called a *w-interval*, restricting the feasible values for $y[w{:}]$, and c_I denotes its associated condition in the triple. Ordering the groups of intervals by decreasing bitwidths $w_1 > w_2 > \cdots > w_j$, as shown in Fig. 2, $\mathcal{S}_j$ denotes the set of produced w_j-intervals. The properties satisfied by the triples entail that

$$\mathcal{A} \wedge (\textstyle\bigwedge_{i=1}^{j} \bigwedge_{I \in \mathcal{S}_i} c_I) \Rightarrow \mathcal{B}$$

is valid, where $\mathcal{B}$ is $\exists y \bigwedge_{i=1}^{j} (y[w_i{:}] \notin \bigcup_{I \in \mathcal{S}_i} I)$. And formula $(\bigwedge_{i=1}^{j} \bigwedge_{I \in \mathcal{S}_i} c_I) \Rightarrow \mathcal{B}$ is false in $\mathcal{M}$. To produce an interpolant, we replace $\mathcal{B}$ by a quantifier-free clause.

The simplest case is when there is only one bitwidth $w = w_1$: the fact that $\mathcal{B}$ is falsified by $\mathcal{M}$ means that $\bigcup_{I \in \mathcal{S}_1} [\![I]\!]_{\mathcal{M}}$ is the full domain $\mathbb{Z}/2^w\mathbb{Z}$. Property "$\bigcup_{I \in \mathcal{S}_1} I$ is the full domain" is then expressed symbolically as a conjunction of constraints in the bitvector language. To compute them, we first extract a sequence $I_1, \ldots, I_q$ of intervals from the set $\mathcal{S}_1$, originating from a subset $\mathcal{C}$ of

Algorithm 3. Producing the interpolant with multiple bitwidths

```
function COVER((S_1, ..., S_j), M)
    output ← ∅                                   ▷ output initialized with the empty set of constraints
    longest ← LONGEST(S_1, M)                                         ▷ longest interval identified
    baseline ← longest.upper                                     ▷ where to extend the coverage from
    while ⟦baseline⟧_M ∉ ⟦longest⟧_M do
        if ∃I ∈ S_1, ⟦baseline⟧_M ∈ ⟦I⟧_M then
            I ← FURTHEST_EXTEND(baseline, S_1, M)
            output ← output ∪ {c_I, baseline ∈ I}      ▷ adding I's condition and linking constraint
            baseline ← I.upper                   ▷ updating the baseline for the next interval pick
        else                          ▷ there is a hole in the coverage of Z/2^{w_1}Z by intervals in S_1
            next ← NEXT_COVERED_POINT(baseline, S_1, M)              ▷ the hole is [baseline ; next[
            if ⟦next⟧_M − ⟦baseline⟧_M <^u 2^{w_2} then
                I ← [next[w_2:] ; baseline[w_2:][        ▷ it is projected on w_2 bits and complemented
                output ← output ∪ {next−baseline <^u 2^{w_2}} ∪ COVER(((S_2 ∪ I), S_3, ..., S_j), M)
                baseline ← next                  ▷ updating the baseline for the next interval pick
            else                   ▷ intervals of bitwidths ≤ w_2 must forbid all values for y[w_2:]
                return COVER((S_2, ..., S_j), M)                          ▷ S_1 was not needed
    return output ∪ {baseline ∈ longest}                         ▷ adding final linking constraint
```

the original constraints $(C_i)_{i=1}^{m}$, and such that the sequence $[\![I_1]\!]_{\mathcal{M}}, \ldots, [\![I_q]\!]_{\mathcal{M}}$ of *concrete* intervals leaves no "hole" between an interval of the sequence and the next, and goes round the full circle of domain $\mathbb{Z}/2^w\mathbb{Z}$: the sequence forms a circular chain of linking intervals. This chain can be produced by a standard coverage extraction algorithm (see, e.g., [15]). Formula $\mathcal{B} := \exists y (y[w:] \notin \bigcup_{I \in \mathcal{S}_1} I)$ is then replaced by $(\bigwedge_{i=1}^{q} u_i \in I_{i+1}) \Rightarrow \bot$, where u_i is the upper bound of I_i and I_{q+1} is I_1. Each interval has its upper bound in the next interval ($u_i \in I_{i+1}$), i.e., intervals do link up with each other. The conflict clause is then

$$(\mathcal{C} \wedge (\bigwedge_{i=1}^{q} c_{I_i}) \wedge (\bigwedge_{i=1}^{q} u_i \in I_{i+1})) \Rightarrow \bot$$

Example 8. For Example 7.2, the coverage-extraction algorithm produces the sequence I_1, I_3, I_2, i.e., $[x_1 ; x_1{+}1[$, $[x_2 ; -x_3[$, $[-x_3 ; x_1{-}x_3[$. The linking constraints are then $d_3 := (x_1{+}1) \in I_3$, $d_2 := (-x_3) \in I_2$, and $d_1 := (x_1{-}x_3) \in I_1$, and the interpolant is $d_3 \wedge d_2 \wedge d_1 \Rightarrow \bot$.[4]

When several bitwidths are involved, the intervals must "complement each other" at different bitwidths so that no value for y is feasible. For a bitwidth w_i, the union of the w_i-intervals in model $\mathcal{M}$ may not necessarily cover the full domain (i.e., $\bigcup_{I \in \mathcal{S}_i} [\![I]\!]_{\mathcal{M}}$ may be different from $\mathbb{Z}/2^{w_i}\mathbb{Z}$). The coverage can leave "holes", and values in that hole are ruled out by constraints of other bitwidths. To produce the interpolant, we adapt the coverage-extraction algorithm into Algorithm 3, which takes as input the sequence of sets $(\mathcal{S}_1, \ldots, \mathcal{S}_j)$ as described in Fig. 2, and produces the interpolant's constraints $d_1, \ldots, d_p$, collected in set output. The algorithm proceeds in decreasing bitwidth order, starting with w_1, and calling itself recursively on smaller bitwidths to cover the holes that the current layer leaves uncovered (termination of that recursion is thus trivial). For every hole that $\bigcup_{I \in \mathcal{S}_1} [\![I]\!]_{\mathcal{M}}$ leaves uncovered, it must determine how intervals of smaller bitwidths can cover it.

[4] We omit c_1, c_2, c_3 here, since they are subsumed by d_1, d_2, d_3, respectively.

$u_1 <^{\mathsf{s}} u_2$	$\rightsquigarrow \neg(u_2 \leq^{\mathsf{s}} u_1)$		$u_1 \leq^{\mathsf{s}} u_2$	$\rightsquigarrow u_1{+}2^{\lvert u_1\rvert-1} \leq^{\mathsf{u}} u_2{+}2^{\lvert u_2\rvert-1}$	
$u_1 <^{\mathsf{u}} u_2$	$\rightsquigarrow \neg(u_2 \leq^{\mathsf{u}} u_1)$		$u_1 \simeq u_2$	$\rightsquigarrow u_1 - u_2 \leq^{\mathsf{u}} 0$	
$u[h{:}l]$	$\rightsquigarrow u[h{:}][{:}l]$		$u[{:}l][h{:}]$	$\rightsquigarrow u[h{+}l{:}][{:}l]$	
$(u_1 {\circ} u_2)[{:}l]$	$\rightsquigarrow u_1[{:}l{-}\lvert u_2\rvert]$	if $\lvert u_2\rvert \leq l$	$(u_1 {\circ} u_2)[h{:}]$	$\rightsquigarrow u_2[h{:}]$	if $h \leq \lvert u_2\rvert$
$(u_1 {\circ} u_2)[{:}l]$	$\rightsquigarrow u_1 \circ u_2[{:}l]$	if not	$(u_1 {\circ} u_2)[h{:}]$	$\rightsquigarrow u_1[h{-}\lvert u_2\rvert{:}] \circ u_2$	if not
$2^n {\times} u$	$\rightsquigarrow u[\lvert u\rvert{-}n{:}] \circ 0_n$	$(n < \lvert u\rvert)$	$(u_1{+}u_2)[h{:}]$	$\rightsquigarrow u_1[h{:}] + u_2[h{:}]$	
$\mathrm{bvnot}(u)$	$\rightsquigarrow -(u+1)$		$(u_1{\times}u_2)[h{:}]$	$\rightsquigarrow u_1[h{:}]{\times}u_2[h{:}]$	
$\pm\text{-ext}_k(u)$	$\rightsquigarrow (0_k{\circ}(u{+}2^{\lvert u\rvert-1})){-}(0_k{\circ}2^{\lvert u\rvert-1})$		$(-u)[h{:}]$	$\rightsquigarrow -u[h{:}]$	
$u_1{\circ}u_2$	$\rightsquigarrow (u_1{\circ}0_{\lvert u_2\rvert}) + (0_{\lvert u_1\rvert}{\circ}u_2)$				

Fig. 3. Rewriting rules

Algorithm 3 relies on the following ingredients:

- LONGEST($\mathcal{S}, \mathcal{M}$) returns an interval among $\mathcal{S}$ whose concrete version $[\![I]\!]_{\mathcal{M}}$ has maximal length;
- I.upper denotes the upper bound of an interval I;
- FURTHEST_EXTEND($a, \mathcal{S}, \mathcal{M}$) returns an interval $I \in \mathcal{S}$ that furthest extends a according to $\mathcal{M}$ (technically, an interval I that $\leq^{\mathsf{u}}$-maximizes $[\![I.\mathsf{upper} - a]\!]_{\mathcal{M}}$ among those intervals I such that $[\![a]\!]_{\mathcal{M}} \in [\![I]\!]_{\mathcal{M}}$).
- If no interval in $\mathcal{S}$ covers a in $\mathcal{M}$, NEXT_COVERED_POINT($a, \mathcal{S}, \mathcal{M}$) outputs the lower bound l of an interval in $\mathcal{S}$ that $\leq^{\mathsf{u}}$-minimizes $[\![l - a]\!]_{\mathcal{M}}$.

Algorithm 3 proceeds by successively moving a concrete bitvector value baseline around the circle $\mathbb{Z}/2^{w_1}\mathbb{Z}$. The baseline is moved when a symbolic reason why it is a forbidden value is found, in a while loop that ends when the baseline has gone round the full circle. If there is at least one interval in $\mathcal{S}_1$ that covers baseline in $\mathcal{M}$ (l. 6), the call to FURTHEST_EXTEND(baseline, $\mathcal{S}_1, \mathcal{M}$) succeeds, and output is extended with condition c_I and (baseline $\subset I$) (l. 8). If not, a hole has been discovered, whose extent is given by NEXT_COVERED_POINT(baseline, $\mathcal{S}_1, \mathcal{M}$) (l. 11). If the hole is bigger than 2^{w_2} (i.e., $2^{w_2} \leq^{\mathsf{u}} [\![\mathsf{next}{-}\mathsf{baseline}]\!]_{\mathcal{M}}$), then the intervals of layers w_2 and smaller must rule out every possible value for $y[w_2{:}]$, and the w_1-intervals were not needed (l. 17). If on the contrary the hole is smaller (i.e., $[\![\mathsf{next}{-}\mathsf{baseline}]\!]_{\mathcal{M}} <^{\mathsf{u}} 2^{w_2}$), then the w_1-interval [baseline ; next[is projected as a w_2-interval $I := [\mathsf{baseline}[w_2{:}]\,;\mathsf{next}[w_2{:}][$ that needs to be covered by the intervals of bitwidth w_2 and smaller. This is performed by a recursive call on bitwidth w_2 (l. 14); the fact that only hole I needs to be covered by the recursive call, rather than the full domain $\mathbb{Z}/2^{w_2}\mathbb{Z}$, is implemented by adding to $\mathcal{S}_2$ in the recursive call the complement $[\mathsf{next}[w_2{:}]\,;\mathsf{baseline}[w_2{:}][$ of I. The result of the recursive call is added to the output variable, as well as the fact that the hole must be small. The final interpolant is $(\bigwedge_{d \in \mathsf{output}} d) \Rightarrow \bot$. An example of run on a variant of Example 6.2 is given in [15].

5 Normalization

As implemented in Yices 2, MCSAT processes a conflict by first computing the conflict core with BDDs, and then normalizing the constraints using the rules of Fig. 3. In the figure, u, u_1 and u_2 stand for any bitvector terms, $\pm$-ext$_k(u)$ is

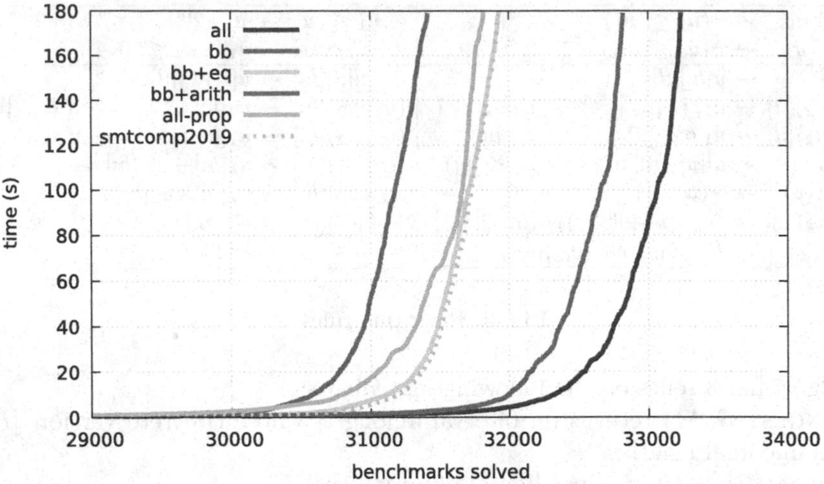

Fig. 4. Evaluation of the MCSAT solver and the effect of different explainer combinations and propagation. Each curve shows the number of benchmarks that the solver variant can solve against the time.

the *sign-extension* of u with k bits, and bvnot(u) is the *bitwise negation* of u. The bottom left rule is applied with lower priority than the others (as upper-bits extraction distributes over $\circ$ but not over $+$) and only if exactly one of $\{u_1, u_2\}$ is evaluable (and not 0). In the implementation, $u[|u|{:}0]$ is identified with u, $\circ$ is associative, and $+, \times$ are subject to ring normalization. This is helped by the internal (flattened) representation of concatenations and bitvector polynomials in Yices 2. Normalization allows the specialized interpolation procedure to apply at least to the following grammar:[5]

$$\begin{array}{lll} \text{Atoms} & a & ::= e_1 + t \lessdot e_2 + t \mid e_1 \lessdot e_2 + t \mid e_1 + t \lessdot e_2 \mid e_1 \lessdot e_2 \\ \text{Terms} & t & ::= t[h{:}l] \mid t + e_1 \mid -t \mid e_1 \circ t \mid t \circ e_1 \mid \pm\text{-ext}_k(t) \end{array}$$

where $\lessdot \in \{\leq^{\mathsf{u}}, <^{\mathsf{u}}, \leq^{\mathsf{s}}, <^{\mathsf{s}}, \simeq\}$. Rewriting can often help further, by eliminating occurrences of the conflict variable (thus making more subterms evaluable) and increasing the chances that two unevaluable terms t_1 and t_2 become syntactically equal in an atom $e_1{+}t_1 \lessdot e_2{+}t_2$.[6] Finally, we cache evaluable terms to avoid recomputing conditions of the form $y \notin \mathsf{var}(e)$. These conditions are needed to determine whether the specialized procedures apply to a given conflict core.

6 Experiments

We implemented our approach in the MCSAT solver within Yices 2 [11]. To evaluate its effectiveness, and the impact of the different modules, we ran the MCSAT solver with different settings on the 41,547 QF_BV benchmarks available

[5] $e_1 \lessdot e_2$ is accepted since it either constitutes the interpolant or it can be ignored.
[6] For this reason we normalize evaluable subterms of, e.g., t_1 and t_2.

in the SMT-LIB library [1]. We used a three-minute timeout per instance. Each curve in Fig. 4 shows the number of solved instances for each solver variant; all: the procedures of Sects. 3 and 4, with the bitblasting baseline when these do not apply; bb: only the bitblasting baseline; bb+eq: procedure of Sect. 3 plus the baseline; bb+arith: procedure of Sect. 4 plus the baseline; all-prop is the same as all but with no propagation of bitvector assignments during search. For reference, we also included the version of the Yices 2 MCSAT solver that entered the 2019 SMT competition[7], marked as smtcomp2019.

The solver combining all explainer modules solved 33,236 benchmarks before timeout, 14,174 of which are solved by pure simplification, and 19,062 of which actually rely on MCSAT explanations. 14,313 of those are solved without ever calling the default bitblasting baseline (only the dedicated explainers of Sects. 3 and 4 are used), while the other 4,749 instances are solved by a combination of the three explainers.

The results show that both equality and arithmetic explainers contribute to the effectiveness of the overall solver, individually and combined. A bit more than half of the problem instances involving MCSAT explanations are fully within the scope of the two dedicated explainers. Of course these explainers are still useful beyond that half, in combination with the bitblasting explainer. The results also show that the eager MCSAT value propagation mechanism introduced in [19] is important for effective solving in practice.

For comparison, we also ran two solvers CDCL($\mathcal{T}$) solvers based on bitblasting on the same benchmarks and with the same timeout. We picked Yices 2 [11] (version 2.6.1) and Boolector [27] (version 3.2.0) and we used the same backend SAT solver for both, namely CaDiCaL [6]. Yices 2 solved 40,962 instances and Boolector solved 40,763 instances. We found 794 instances in the SMTLib benchmarks where our MCSAT solver was faster than Boolector by more than 2 sec. The `pspace/ndist*` and `pspace/shift1add*` instances are trivial for MCSAT (solved in less than 0.25 sec. each), while Boolector hit our 3-minute timeout on all `ndist.a.*` instances and all but 3 `shift1add*` ones. The `brummayerbiere4` instances are trivial for MCSAT (solved in less than 0.03 sec.) while Boolector ran out of memory in our experimentation (except for one instance). Instances with a significant runtime difference in favour of MCSAT are among `spear/openldap_v2.3.35/*` and `brummayerbiere/bitrev*` (MCSAT is systematically better), `float/mult*` (MCSAT is almost systematically better), `float/div*`, `asp/SchurNumbers/*`, `20190311-bv-term-small-rw-Noetzli/*`, and `Sage2/*`. MCSAT is almost systematically faster on `uclid/catchconv/*` and faster on more than half of `spear/samba_v3.0.24/*`.

Using an alternative MCSAT approach to bitvector solving, Zeljić et al. reported that their solver could solve 23704 benchmarks from a larger set of 49971 instances with a larger timeout of 1200 s [30].[8] We have not managed

[7] https://smt-comp.github.io/2019/.

[8] The additional 8424 benchmarks have since been deleted from the SMT-LIB library as duplicates.

to reproduce the results of Zeljić's solver on our Linux server for direct comparison.

To debug the implementation of our explainers, every conflict explanation that is produced when solving in debug mode is sent on-the-fly to (non-MCSAT) Yices 2, which checks the validity of the clause by bitblasting. In debug mode, every normalization we perform with the rules of Sect. 5 is also sent to Yices 2 to prove the equality between the original term and the normalized term. Performance benchmarking was only done after completing, without any red flag, a full run of MCSAT in debug mode on the 41,547 QF_BV benchmarks instances.

7 Discussion and Future Work

The paper presents ongoing work on building an MCSAT solver for the theory of bitvectors. We have presented two main ideas for the treatment of $\mathcal{BV}$ in MCSAT, that go beyond the approach proposed by Zeljić et al. [30].

First, by relying on BDDs for representing feasible sets, our design keeps the main search mechanism of MCSAT generic and leaves fragment-specific mechanisms to conflict explanation. The explanation mechanism is selected based on the constraints involved in the conflict. BDDs are also used to minimize the conflicts, which increases the chances that a dedicated explanation mechanism can be applied. BDDs offer a propagation mechanism that differs from those in [30] in that the justification for a propagated assignment is computed lazily, only when it is needed in conflict analysis. Computing the conflict core at that point effectively recovers justification of the propagations.

Second, we propose explanation mechanisms for two fragments of the theory: the core fragment of $\mathcal{BV}$ that includes equality, concatenation and extraction; and a fragment of linear arithmetic. Compared to previous work on coarsest-base slicing, such as [4], our work applies the slicing on the conflict constraints only, rather than the whole problem. This should in general make the slices coarser, which we expect to positively impact efficiency. Our work on explaining arithmetic constraints is novel, notwithstanding the mechanisms studied by Janota and Wintersteiger [17] that partly inspired our Table 1 but addressed a smaller fragment of arithmetic outside of the context of MCSAT.

We have implemented the overall approach in the Yices 2 SMT solver. Experiments show that the overall approach is effective on practical benchmarks, with all the proposed modules adding to the solver performance. MCSAT is not yet competitive with bitblasting, but we are making progress. The main challenge is devising efficient word-level explanation mechanisms that can handle all or a least a large fragment of $\mathcal{BV}$. Finding high-level interpolants in $\mathcal{BV}$ is still an open problem and our work on MCSAT shows progress for some fragments of the bitvector theory. For MCSAT to truly compete with bitblasting, we will need interpolation methods that cover larger classes of constraints.

A key step in that direction is to extend the bitvector arithmetic explainer so that it handles multiplications by constants, then multiplication by evaluable terms, and, finally, arbitrary multiplications. Deeper integration of fragment-specific explainers could potentially help explaining *hybrid* conflicts that involve constraints from different fragments. To complement the explainers that we are developing, we plan to further explore the connection between interpolant generation and the closely related domain of quantifier elimination, particularly those techniques by John and Chakraborty [18] for the bitvector theory. The techniques by Niemetz et al. [26] for solving quantified bitvector problems using *invertibility conditions* could also be useful for interpolant generation in MCSAT.

Future work also includes relating our approach to the report by Chihani, Bobot, and Bardin [7], which aims at lifting the CDCL mechanisms to the word level of bitvector reasoning, and therefore seems very close to MCSAT. Finally, we plan to explore integrating our MCSAT treatment of bitvectors with other components of SMT-solvers, whether in the context of MCSAT or in different architectures. An approach for this is the recent framework of *Conflict-Driven Satisfiability* (CDSAT) [2,3], which precisely aims at organizing collaboration between generic theory modules.

Acknowledgments. The authors thank Aleksandar Zeljić for fruitful discussions. This material is based upon work supported in part by NSF grants 1528153 and 1816936, and by the Defense Advanced Research Project Agency (DARPA) and Space and Naval Warfare Systems Center, Pacific (SSC Pacific) under Contract No. N66001-18-C-4011. Any opinions, findings and conclusions or recommendations expressed in this material are those of the author(s) and do not necessarily reflect the views of NSF, DARPA, or SSC Pacific.

References

1. Barrett, C., Stump, A., Tinelli, C.: The Satisfiability Modulo Theories Library (SMT-LIB) (2010). www.SMT-LIB.org
2. Bonacina, M.P., Graham-Lengrand, S., Shankar, N.: Conflict-driven satisfiability for theory combination: transition system and completeness. J. Autom. Reasoning **64**(3), 579–609 (2019). https://doi.org/10.1007/s10817-018-09510-y
3. Bonacina, M.P., Graham-Lengrand, S., Shankar, N.: Satisfiability modulo theories and assignments. In: de Moura, L. (ed.) CADE 2017. LNCS (LNAI), vol. 10395, pp. 42–59. Springer, Cham (2017). https://doi.org/10.1007/978-3-319-63046-5_4
4. Bruttomesso, R., Sharygina, N.: A scalable decision procedure for fixed-width bit-vectors. In: Proceedings of the 2009 International Conference on Computer-Aided Design (ICCAD 2009), ICCAD 2009, pp. 13–20. ACM Press (2009). https://doi.org/10.1145/1687399.1687403
5. Bryant, R.E.: Graph-based algorithms for boolean function manipulation. Comput. IEEE Trans. **100**(8), 677–691 (1986)
6. CaDiCaL Simplified Satisfiability Solver. http://fmv.jku.at/cadical/
7. Chihani, Z., Bobot, F., Bardin, S.: CDCL-inspired Word-level Learning for Bit-vector Constraint Solving, June 2017. https://hal.archives-ouvertes.fr/hal-01531336, preprint
8. CUDD: the CU Decision Diagram package. https://github.com/ivmai/cudd
9. Cyrluk, D., Möller, O., Rueß, H.: An efficient decision procedure for the theory of fixed-sized bit-vectors. In: Grumberg, O. (ed.) CAV 1997. LNCS, vol. 1254, pp. 60–71. Springer, Heidelberg (1997). https://doi.org/10.1007/3-540-63166-6_9

10. Detlefs, D., Nelson, G., Saxe, J.B.: Simplify: a theorem prover for program checking. J. ACM (JACM) **52**(3), 365–473 (2005)
11. Dutertre, B.: Yices 2.2. In: Biere, A., Bloem, R. (eds.) CAV 2014. LNCS, vol. 8559, pp. 737–744. Springer, Cham (2014). https://doi.org/10.1007/978-3-319-08867-9_49
12. Damm, W., Hermanns, H. (eds.): CAV 2007. LNCS, vol. 4590. Springer, Heidelberg (2007). https://doi.org/10.1007/978-3-540-73368-3
13. Graham-Lengrand, S., Jovanović, D.: An MCSAT treatment of bit-vectors. In: Brain, M., Hadarean, L. (eds.) 15th International Workshop on Satisfiability Modulo Theories (SMT 2017), July 2017
14. Graham-Lengrand, S., Jovanović, D.: Interpolating bit-vector arithmetic constraints in MCSAT. In: Sharygina, N., Hendrix, J. (eds.) 17th International Workshop on Satisfiability Modulo Theories (SMT 2019), July 2019
15. Graham-Lengrand, S., Jovanović, D., Dutertre, B.: Solving bitvectors with MCSAT: explanations from bits and pieces (long version). Technical report, SRI International (Apr 2020), https://arxiv.org/abs/2004.07940
16. Biere, A., Bloem, R. (eds.): CAV 2014. LNCS, vol. 8559. Springer, Cham (2014). https://doi.org/10.1007/978-3-319-08867-9
17. Janota, M., Wintersteiger, C.M.: On intervals and bounds in bit-vector arithmetic. In: King, T., Piskac, R. (eds.) Proceedings of the 14th International Workshop on Satisfiability Modulo Theories (SMT 2016). CEUR Workshop Proceedings, vol. 1617, pp. 81–84. CEUR-WS.org, July 2016. http://ceur-ws.org/Vol-1617/paper8.pdf
18. John, A.K., Chakraborty, S.: A layered algorithm for quantifier elimination from linear modular constraints. Formal Methods Syst. Des. **49**(3), 272–323 (2016). https://doi.org/10.1007/s10703-016-0260-9
19. Jovanović, D.: Solving nonlinear integer arithmetic with MCSAT. In: Bouajjani, A., Monniaux, D. (eds.) VMCAI 2017. LNCS, vol. 10145, pp. 330–346. Springer, Cham (2017). https://doi.org/10.1007/978-3-319-52234-0_18
20. Jovanović, D., Barrett, C., de Moura, L.: The design and implementation of the model constructing satisfiability calculus. In: Proceedings of the 13th International Conference on Formal Methods In Computer-Aided Design (FMCAD 2013). FMCAD Inc. October 2013
21. Junker, U.: Quickxplain: conflict detection for arbitrary constraint propagation algorithms. In: IJCAI 2001 Workshop on Modelling and Solving Problems with Constraints (2001)
22. Kroening, D., Strichman, O.: Decision Procedures - An Algorithmic Point of View, Second Edition. Texts in Theoretical Computer Science. An EATCS Series. Springer (2016). https://doi.org/10.1007/978-3-662-50497-0
23. Marques Silva, J., Lynce, I., Malik, S.: Conflict-driven clause learning SAT solvers. In: Biere, A., Heule, M., Maaren, H.V., Walsh, T. (eds.) Handbook of Satisfiability, Frontiers in Artificial Intelligence and Applications, vol. 185, pp. 131–153. IOS Press (2009)
24. de Moura, L., Jovanović, D.: A model-constructing satisfiability calculus. In: Giacobazzi, R., Berdine, J., Mastroeni, I. (eds.) VMCAI 2013. LNCS, vol. 7737, pp. 1–12. Springer, Heidelberg (2013). https://doi.org/10.1007/978-3-642-35873-9_1
25. Niemetz, A., Preiner, M., Biere, A.: Boolector 2.0. J. Satisf. Boolean Model. Comput. **9**(1), 53–58 (2014)

26. Niemetz, A., Preiner, M., Reynolds, A., Barrett, C., Tinelli, C.: Solving quantified bit-vectors using invertibility conditions. In: Chockler, H., Weissenbacher, G. (eds.) CAV 2018. LNCS, vol. 10982, pp. 236–255. Springer, Cham (2018). https://doi.org/10.1007/978-3-319-96142-2_16
27. Niemetz, A., Preiner, M., Wolf, C., Biere, A.: Btor2, BtorMC and Boolector 3.0. In: Chockler, H., Weissenbacher, G. (eds.) CAV 2018. LNCS, vol. 10981, pp. 587–595. Springer, Cham (2018). https://doi.org/10.1007/978-3-319-96145-3_32
28. Nieuwenhuis, R., Oliveras, A., Tinelli, C.: Solving SAT and SAT modulo theories: from an abstract Davis-Putnam-Logemann-Loveland procedure to DPLL (T). J. ACM Press **53**(6), 937–977 (2006). https://doi.org/10.1145/1217856.1217859
29. Robinson, J.A., Voronkov, A. (eds.): Handbook of Automated Reasoning (in 2 volumes). Elsevier and The MIT Press, Cambridge (2001)
30. Zeljić, A., Wintersteiger, C.M., Rümmer, P.: Deciding bit-vector formulas with mcSAT. In: Creignou, N., Le Berre, D. (eds.) SAT 2016. LNCS, vol. 9710, pp. 249–266. Springer, Cham (2016). https://doi.org/10.1007/978-3-319-40970-2_16

Monadic Decomposition in Integer Linear Arithmetic

Matthew Hague[1(✉)], Anthony W. Lin[2], Philipp Rümmer[3], and Zhilin Wu[4]

[1] Royal Holloway, University of London, Egham, UK
matthew.hague@rhul.ac.uk
[2] TU Kaiserslautern, Kaiserslautern, Germany
[3] Uppsala University, Uppsala, Sweden
[4] State Key Laboratory of Computer Science, Institute of Software, Chinese Academy of Sciences, Beijing, China

Abstract. Monadic decomposability is a notion of variable independence, which asks whether a given formula in a first-order theory is expressible as a Boolean combination of monadic predicates in the theory. Recently, Veanes et al. showed the usefulness of monadic decomposability in the context of SMT (i.e. the input formula is quantifier-free), and found various interesting applications including string analysis. However, checking monadic decomposability is undecidable in general. Decidability for certain theories is known (e.g. Presburger Arithmetic, Tarski's Real-Closed Field), but there are very few results regarding their computational complexity. In this paper, we study monadic decomposability of integer linear arithmetic in the setting of SMT. We show that this decision problem is coNP-complete and, when monadically decomposable, a formula admits a decomposition of exponential size in the worst case. We provide a new application of our results to string constraint solving with length constraints. We then extend our results to variadic decomposability, where predicates could admit multiple free variables (in contrast to monadic decomposability). Finally, we give an application to quantifier elimination in integer linear arithmetic where the variables in a block of quantifiers, if independent, could be eliminated with an exponential (instead of the standard doubly exponential) blow-up.

1 Introduction

A formula $\phi(\bar{x})$ in some theory $\mathcal{L}$ is *monadically decomposable* if it is $\mathcal{L}$-equivalent to a Boolean combination of monadic predicates in $\mathcal{L}$, i.e., to a *monadic decomposition* of ϕ. Monadic decomposability measures how tightly the free variables in ϕ are coupled. For example, $x = y$ is not monadically decomposable in any (finitary) logic over an infinite domain, but $x + y \geq 2$ can be decomposed, in Presburger arithmetic over natural numbers, since it can be written as $x \geq 2 \vee (x \geq 1 \wedge y \geq 1) \vee y \geq 2$.

Veanes *et al.* [24] initiated the study of monadic decomposability in the setting of Satisfiability Modulo Theories, wherein formulas are required to be

N. Peltier and V. Sofronie-Stokkermans (Eds.): IJCAR 2020, LNAI 12166, pp. 122–140, 2020.
https://doi.org/10.1007/978-3-030-51074-9_8

quantifier-free. Monadic decomposability has many applications, including symbolic transducers [11] and string analysis [24]. Although the problem was shown to be in general undecidable, a generic semi-algorithm for outputting monadic decompositions (if decomposable) was provided. A termination check could in fact be added if the input formula belongs to a theory for which monadic decomposability is decidable, e.g., linear arithmetic, Tarski's Real-Closed Field, and the theory of uninterpreted functions. Hitherto, not much is known about the computational complexity of monadic decomposability problems for many first-order theories (in particular, quantifier-free theories), and about practical algorithms. This was an open problem raised by Veanes *et al.* in [24].

Monadic decomposability is intimately connected to the variable partition problem, first studied by Libkin [19] nearly 20 years ago. In particular, a monadic decomposition gives rise to a partition of the free variables $\bar{x}$ of a formula $\phi(\bar{x})$, wherein each part consists of a single variable. More precisely, take a partition $\Pi = \{Y_1, \ldots, Y_m\}$ of $\bar{x}$ into sets Y_i of variables, with linearizations $\overline{y}_i$. The formula $\phi(\bar{x})$ is Π-decomposable (in some theory $\mathcal{L}$) if it is $\mathcal{L}$-equivalent to a boolean combination of formulas of the form $\Delta(\overline{y}_i)$. As suggested in [19], such *variadic decompositions* of $\phi(\bar{x})$ have potential applications in optimization of database query processing and quantifier elimination. The author gave a general condition for the decidability of variable independence in first-order theories. This result is unfortunately not easily applicable in the SMT setting for at least two reasons: (i) the full first-order theory might be undecidable (e.g. theory of uninterpreted functions), and (ii) even for a first-order theory that admits decidable monadic decompositions, the complexity of the algorithm obtained from [19] could be too prohibitive for the quantifier-free fragment. One example that epitomizes (ii) is the problem of determining whether a given relation $R \subseteq (\Sigma^*)^k$ over strings represented by a regular transducer could be expressed as a boolean combination of monadic predicates. The result of [19] would give a double exponential-time algorithm for monadic decomposability, whereas it was recently shown in [5] to be solvable in polynomial-time (resp. polynomial-space) when the transducer is given as a deterministic (resp. nondeterministic) machine.

Contributions. First, we determine the complexity of deciding monadic decomposability and outputting monadic decompositions (if they exist) for the theory of integer linear arithmetic in the setting of SMT. Our result is summarized in Theorem 1.

Theorem 1 (Monadic Decomposability). *Given a quantifer-free formula ϕ of Presburger Arithmetic, it is coNP-complete to decide if ϕ is monadically decomposable. This is efficiently reducible to unsatisfiability of quantifier-free Presburger formulas. Moreover, if a decomposition exists, it can be constructed in exponential time.*

We show a new application of monadic decomposability in integer linear arithmetic for SMT over strings, which is currently a very active research area, e.g., see [1–3,6,9,12,16,18,20,22,23]. One problem that makes string constraint solving difficult is the presence of additional *length constraints,* which forces the lengths of the strings in the solutions to satisfy certain linear arithmetic constraints. Whereas satisfiability of string equations with regular constraints is PSPACE-complete (e.g. see [13,17]), it is a long-standing open problem [7,14] whether word equations with length constraints are decidable. Length constraints are omnipresent in Kaluza [22], arguably the first serious string constraint benchmarks obtained from real-world JavaScript applications. Using our monadic decomposability solver, we show that 90% of the Kaluza benchmarks are in fact in a decidable fragment of string constraints, since occurring length constraints can be completely removed by means of decomposition.

Next we extend our result to variadic decomposability (cf. [19]).

Theorem 2 (Variadic Decomposability). *It is coNP-complete to decide if ϕ is Π-decomposable, given a quantifer-free formula $\phi(\bar{x})$ of Presburger Arithmetic and a partition $\Pi = \{Y_1, \ldots, Y_n\}$ of $\bar{x}$. This is efficiently reducible to unsatisfiability of quantifier-free Presburger formulas. Moreover, if a decomposition exists, it can be constructed in exponential time.*

We show how this could be applied to quantifier elimination. In particular, we show that if a formula $\phi(\bar{y}) = \exists \bar{x}.\, \psi(\bar{x}, \bar{y})$, where ψ is quantifier-free, is $\{X, Y\}$-decomposable—where $\bar{x}$ and $\bar{y}$ are linearizations of the variables in X and Y—then we can compute in exponential time a formula $\theta(\bar{y})$ such that $\langle \mathbb{N}, + \rangle \models \theta \leftrightarrow \phi$, i.e., avoiding the standard double-exponential blow-up (cf. [25]).

Organization. Preliminaries are in Sect. 2. Results on monadic (resp. variadic) decomposition are in Sect. 3 (resp. Sect. 4) and applications appear in Sect. 5.

2 Preliminaries

2.1 Presburger Syntax

In this paper we study the problem of monadic decomposition for formulas in linear integer arithmetic. All of our results are presented for Presburger arithmetic over natural numbers, but they can be adapted easily to all integers.

Definition 1 (Fragments of Presburger Arithmetic). *A formula ϕ of* Presburger arithmetic *is a formula of the form $Q_1 x_1 \cdots Q_n x_n.\, \psi$ where $Q_i \in \{\forall, \exists\}$ and ψ is a quantifier-free Presburger formula:*

$$\psi := \sum_i a_i x_i \sim b \mid ax \equiv_k by \mid x \equiv_k c \mid \phi_1 \wedge \phi_2 \mid \neg\phi$$

where $a_i, a, b \in \mathbb{Z}$, $k, c \in \mathbb{N}$ with $0 \leq c < k$, variables x_i, x, y range over $\mathbb{N}$, and $\sim \in \{\leq, \geq\}$. The operator $\equiv_k$ denotes equality modulo k, i.e., $s \equiv_k t$ whenever $s - t$ is a multiple of k. Formulas of the shape $\sum_i a_i x_i \sim b$, $ax \equiv_k by$, or $x \equiv_k c$ are called atoms.

Existential Presburger formulas are formulas of the form $\exists x_1, \ldots, x_n.\, \psi$ for some quantifier-free Presburger formula ψ. We let $QF(\mathbb{N})$ (resp. $\exists^*(\mathbb{N})$) denote the set of all quantifier-free (resp., existential) Presburger formulas.

Let $\overline{x} = (x_1, \ldots, x_n)$ be a tuple of integer variables. We write $f(\overline{x}) = \sum_i a_i x_i$ for a linear sum over $\overline{x}$. Let $\overline{y} = (y_1, \ldots, y_m)$. By slight abuse of notation, we may also write $\phi(\overline{x}, \overline{y})$ to denote a QF(ℕ) formula over the variables $x_1, \ldots, x_n, y_1, \ldots, y_m$.

2.2 Monadic Decomposability

A quantifier-free formula ϕ is called *monadic* if every atom in ϕ contains at most one variable, and it is called *monadically decomposable* if ϕ is equivalent to a monadic formula ϕ'. In this case, ϕ' is also called a *decomposition* of ϕ. For our main results we use a slightly refined notion of a formula being decomposable:

Definition 2 (Monadically Decomposable on x). *Fix a logic $\mathcal{L}$ (e.g. QF(ℕ) or $\exists^*(\mathbb{N})$). We say a formula $\phi(x_1, \ldots, x_n)$ in $\mathcal{L}$ is* monadically decomposable on x_i *whenever*

$$\phi(x_1, \ldots, x_n) \equiv \bigvee_j \Delta_j(x_i) \wedge \psi_j(x_1, \ldots, x_{i-1}, x_{i+1}, \ldots, x_n)$$

for some formulas Δ_j and ψ_j in $\mathcal{L}$.

It can be observed that a formula is monadically decomposable if and only if it is monadically decomposable on all variables occurring in the formula (cf. Lemma 1). We expand on this for the variadic case below.

We recall the following characterization of monadic decomposability for formulas $\phi(x, y)$ with two free variables (cf. [5,8,19,24]), which holds regardless of the theory under consideration. This can be extended easily to formulas with k variables, but is not needed in this paper. Given a formula $\phi(x, y)$, define the formula $\sim$ as follows:

$$x \sim x' := \forall y, y'. \; (\phi(x, y) \wedge \phi(x', y') \rightarrow (\phi(x', y) \wedge \phi(x, y')))$$

Proposition 1. *The relation $\sim$ is an equivalence relation. Furthermore, $\phi(x, y)$ is monadically decomposable iff $\sim$ has a finite index (i.e. the number of $\sim$-equivalence classes is finite).*

Using this proposition, it is easy to show that over a structure with an infinite domain (e.g. integer linear arithmetic) the formula $x = y$ is not monadically decomposable. As was noted already in [19], to check monadic decomposability of a formula ϕ in Presburger Arithmetic in general, we may simply check if there is an upper bound B on the smallest representation of every $\sim$-equivalence class, i.e.,

$$\exists B. \forall x. \exists x_s. \; (x_s \leq B \wedge x_s \sim x).$$

However, to derive tight complexity bounds for checking monadic decomposability, this approach is problematic, since the above characterisation has multiple quantifier alternations. Using known results (e.g. [15]), one would only obtain an upper bound in the weak exponential hierarchy [15], which only admits double-exponential time algorithms.

2.3 Variadic Decomposability

The notion of a *variadic decomposition* generalises monadic decomposition by considering partitions of the occurring variables.

Definition 3 (Π-Decomposable). *Fix a logic $\mathcal{L}$ (e.g. $QF(\mathbb{N})$ or $\exists^*(\mathbb{N})$). Take a formula $\phi(x_1, \ldots, x_n)$ in $\mathcal{L}$ and a partition $\Pi = \{Y_1, \ldots, Y_m\}$ of $x_1, \ldots, x_n$. We say ϕ is* Π-decomposable *whenever*

$$\phi(x_1, \ldots, x_n) \equiv \bigvee_i \Delta_i^1(\overline{y}_1) \wedge \cdots \wedge \Delta_i^m(\overline{y}_m)$$

for some formulas Δ_i^j in $\mathcal{L}$ and linearizations $\overline{y}_j$ of Y_j.

Observe that a formula $\phi(x_1, \ldots, x_n)$ is monadically decomposable on x_i iff it is Π-decomposable with $\Pi = \{\{x_i\}, \{x_1, \ldots, x_{i-1}, x_{i+1}, \ldots, x_n\}\}$. Moreover, we say a formula ϕ over the set of variables X is *variadic decomposable on Y* whenever it is Π-decomposable with $\Pi = \{Y, X \setminus Y\}$.

General Π-decompositions can be computed by decomposing on binary partitions $\{Y, X \setminus Y\}$, which is why we focus on this binary case in the rest of the paper. We argue why this is the case below.

Let a formula ϕ and $\Pi = \{Y_1, \ldots, Y_m\}$ be given. We can first decompose separately on each $\{Y_i, Y\}$ where $Y = Y_1 \cup \cdots \cup Y_{i-1} \cup Y_{i+1} \cup \cdots \cup Y_m$. Using the algorithm in Sect. 4 we obtain for each i a decomposition of a specific form:

$$\bigvee_j \Delta_j^i(\overline{y}_i) \wedge \phi(\overline{y}_1, \ldots, \overline{y}_{i-1}, \overline{c}_j^i, \overline{y}_{i+1}, \ldots, \overline{y}_m).$$

Note, these decompositions can be performed independently using the algorithm in Sect. 4 and the second conjunct of each disjunct is ϕ with y_i replaced by fixed constants $\overline{c}_j$. Additionally, each Δ_j^i is polynomial in size and each $\overline{c}_j^i$ can be represented with polynomially many bits. We note also that our algorithm ensures that each Δ_j^i is satisfiable.

Given such decompositions, we can recursively decompose ϕ on Π. We first use the above decomposition for $i = 1$ and obtain

$$\bigvee_j \Delta_j^1(\overline{y}_1) \wedge \phi(\overline{c}_j^1, \overline{y}_2, \ldots, \overline{y}_m).$$

Next, we use the decomposition for $i = 2$ to decompose the copies of ϕ in the decomposition above. We obtain

$$\bigvee_{j_1} \left(\Delta_{j_1}^1(\overline{y}_1) \wedge \bigvee_{j_2} \Delta_{j_2}^2(\overline{y}_2) \wedge \phi(\overline{c}_{j_1}^1, \overline{c}_{j_2}^2, \overline{y}_3, \ldots, \overline{y}_m) \right).$$

This process repeats until all Y_i have been considered. If ϕ is Π-decomposable, we find a decomposition. If ϕ is not Π-decomposable, then it would not be possible to do the independent decompositions for each i. Thus, for $\Pi = \{Y_1, \ldots, Y_m\}$, we can use variadic decompositions on Y_i to compute Π-decompositions.

The above algorithm runs in exponential time due both to the exponential size of the decompositions and the branching caused by the disjuncts. If we are only interested in whether a formula is Π-decomposable, it is enough to ask whether it is decomposable on Y_i for *each* i. In particular, a formula $\phi(\bar{x})$ is monadically decomposable iff ϕ is decomposable for each variable $y \in \bar{x}$. Since the complexity class coNP is closed under intersection, we obtain the following:

Lemma 1. *A coNP upper bound for monadic decomposability on a given variable y implies a coNP upper bound for monadic decomposability. Likewise, a coNP upper bound for variadic decomposability on a given subset Y of variables implies a coNP upper bound for Π-decomposability.*

2.4 Example

Consider the formula $\phi(x, y, z)$ given by $z = x + 2y \wedge z < 5$. This formula is monadically decomposable, which means, it is Π-decomposable for $\Pi = \{\{x\}, \{y\}, \{z\}\}$.

Our algorithm will first take a decomposition on x and might obtain $\bigvee_{i=0}^{4} \Delta_i^1(x) \wedge \phi(i, y, z)$ where $\Delta_i^1(x) = (x = i)$ and $\phi(i, y, z) = (z = i + 2y) \wedge z < 5$. Next, we use a decomposition on y. For each $\phi(i, y, z)$ we substitute $\bigvee_{j=0}^{2-\lceil \frac{i}{2} \rceil} y = j \wedge \phi(i, j, z)$, and as the final decomposition we get

$$\bigvee_{i=0}^{4} \bigvee_{j=0}^{2-\lceil \frac{i}{2} \rceil} x = i \wedge y = j \wedge z = i + 2j.$$

3 Monadic Decomposability

3.1 Lower Bounds

We first show that unsatisfiability of Boolean formulas can be reduced to monadic decomposability of formulas with only two variables, directly implying coNP-hardness:

Lemma 2 (coNP-Hardness). *Deciding whether a formula $\phi(x, y)$ in $QF(\mathbb{N})$ is monadic decomposable is coNP-hard.*

Proof. We reduce from unsatisfiability of propositional formulas to monadic decomposability of $\phi(x, y)$. Take a propositional formula $S(x_1, \ldots, x_n)$. Let $p_1, \ldots, p_n$ be the first n primes. Let $\psi(x)$ be the formula obtained from S by replacing each occurrence of x_i by $x \equiv_{p_i} 0$. Given an assignment $\nu : \{x_1, \ldots, x_n\} \rightarrow \{0, 1\}$, we let

$$H_\nu = \{m \in \mathbb{N} \mid \forall 1 \leq i \leq n.\ (m \equiv_{p_i} 0 \leftrightarrow \nu(x_i) = 1)\}.$$

Thanks to the Chinese Remainder Theorem, H_ν is non-empty and periodic with period $p = \prod_{i=1}^{n} p_i$, which implies that H_ν is infinite for every ν. We also have that $\nu \models S$ iff, for each $n \in H_\nu$, $\psi(n)$ is true.

Now define $\phi(x, y) = (\psi(x) \wedge x = y)$. If S is unsatisfiable, then ψ is unsatisfiable and so it is decomposable. Conversely, if S can be satisfied by some assignment ν, then $\phi(m, m)$ is true for all (infinitely many) $m \in H_\nu$. Since all solutions to $\phi(x, y)$ imply that $x = y$, by Proposition 1 we have that ϕ is not monadically decomposable. □

We next provide exponential lower bounds for decompositions in either *disjunctive normal form* (DNF) or *conjunctive normal form* (CNF). DNF has been frequently used to represent monadic decompositions by previous papers (e.g. [5,8,19]), and it is most suitable for applications in quantifier elimination.

Lemma 3 (Size of Decomposition). *There exists a family* $\{\phi_n(x, y)\}_{n \in \mathbb{N}}$ *of formulas in* $QF(\mathbb{N})$ *such that* ϕ_n *grows linearly in* n*, while the smallest decomposition on* x *in DNF/CNF is exponential in* n.

Proof. Consider the formulas $\phi_n(x, y) = (x + y \leq 2^n)$. Using a binary encoding of constants, the size of the formulas is linear in n. We show that decompositions in DNF/CNF must be exponential in size.

Disjunctive: Suppose $\psi_n(x, y) = \bigvee_i \psi_i^x(x) \wedge \psi_i^y(y)$ is a monadic decomposition in DNF. Each disjunct $\psi_i^x(x) \wedge \psi_i^y(y)$, if it is satisfiable at all, has an upper right corner (x_i, y_i) such that $\psi_i^x(x_i) \wedge \psi_i^y(y_i)$ holds, but $\psi_i^x(x) \wedge \psi_i^y(y) \Rightarrow x \leq x_i \wedge y \leq y_i$. This immediately implies that exponentially many disjuncts are needed to cover the exponentially many points on the line $x + y = 2^n$.

Conjunctive: Suppose $\psi_n(x, y)$ is a succinct monadic decomposition of ϕ_n in CNF. Since $\neg\psi_n(x, y) \equiv 2^n + 1 \leq x + y \equiv (2^n - x + 1) + (2^n - y) \leq 2^n$, it follows that $\neg\psi_n(2^n - x + 1, 2^n - y) \equiv (2^n - (2^n - x + 1) + 1) + (2^n - (2^n - y)) \leq 2^n \equiv x + y \leq 2^n$. Therefore, $\neg\psi_n(2^n - x + 1, 2^n - y)$ is a succinct decomposition of ϕ_n in DNF, contradicting the lower bound for DNFs. □

3.2 Upper Bound

We prove Theorem 1. Following Lemma 1, it suffices to show that testing decomposability on a variable x is in coNP and that a decomposition can be computed in exponential time. Assume without loss of generality that we have $\phi(x, \overline{y})$ where $\overline{y} = (y_1, \ldots, y_n)$, and that we are decomposing on the first variable x.

We claim that ϕ is monadically decomposable on x iff

$$\forall x_1, x_2 \geq B. \forall \overline{y}.\ \text{SameDiv}(x_1, x_2, \overline{y}) \Rightarrow (\phi(x_1, \overline{y}) \iff \phi(x_2, \overline{y}))$$

where B is a bound exponential in the size of ϕ and SameDiv is a formula asserting that x_1 and x_2 satisfy the same divisibility constraints. This bound is computable in polynomial time and is described in Sect. 3.4. To define SameDiv, let Divs be the set of all divisibility constraints $az_1 \equiv_k bz_2$ or $z_1 \equiv_k c$ appearing (syntactically) in ϕ. Assume without loss of generality that x always appears on the left-hand side of a divisibility constraint (i.e., in the z_1 position of $az_1 \equiv_k bz_2$). We then define

$$\mathrm{SameDiv}(x_1, x_2, \overline{y}) = \left(\begin{array}{c} \bigwedge_{ax \equiv_k bz \in \mathrm{Divs}} (ax_1 \equiv_k bz) \iff (ax_2 \equiv_k bz) \\ \wedge \\ \bigwedge_{x \equiv_k c \in \mathrm{Divs}} (x_1 \equiv_k c) \iff (x_2 \equiv_k c) \end{array} \right).$$

We prove the claim in the following sections and simultaneously show how to construct the decomposition. Once we have established the above, we can test non-decomposability on x by checking

$$\exists x_1, x_2 \geq B. \exists \overline{y}.\ \mathrm{SameDiv}(x_1, x_2, \overline{y}) \wedge \phi(x_1, \overline{y}) \wedge \neg\phi(x_2, \overline{y})$$

which is decidable in NP. Thus we obtain a coNP decision procedure because the above formula is polynomial in the size of ϕ.

Example. We consider some examples. First consider the formula $x = y$ that cannot be decomposed on x. Since there are no divisibility constraints, SameDiv is simply *true*. It is straightforward to see that, $\forall B. \exists x_1, x_2 \geq B. \exists y.\ \mathit{true} \wedge x_1 = y \wedge x_2 \neq y$, for example by setting $x_1 = B$, $x_2 = B + 1$, and $y = B$.

Now consider the monadically decomposable formula

$$\phi(x, y, z) = x + 2y \geq 5 \wedge z < 5 \wedge x \equiv_2 y.$$

In this case $\mathrm{SameDiv}(x_1, x_2, y, z) = (x_1 \equiv_2 y \iff x_2 \equiv_2 y)$. We can verify

$$\forall x_1, x_2 \geq B. \forall y, z.\ \mathrm{SameDiv}(x_1, x_2, y, z) \Rightarrow (\phi(x_1, y, z) \iff \phi(x_2, y, z))$$

holds, as it will be the case that $5 < B$ and for all $x > 5$ the formula ϕ will hold whenever $x \equiv_2 y$ holds and $z < 5$. The precondition SameDiv ensures that the if and only if holds. We will construct the decomposition in the next section.

Expanded Divisibility Constraints. Observe that divisibility constraints are always decomposable. In particular, $az_1 \equiv_k bz_2$ is equivalent to a finite disjunction of clauses $z_1 \equiv_{k'} c \wedge z_2 \equiv_{k'} c$ where k' and c are bounded by a multiple of a, b and k. The expansion is exponential in size, since the values up to k' have to be enumerated explicitly.

We define XDivs be the set of all constraints of the form $x \equiv_k c'$ where $0 \leq c' < k$ and $x \equiv_k c$ appears directly in ϕ or in the expansion of the divisibility constraints of ϕ. This set will be used in the next sections.

3.3 Soundness

We show that if

$$\forall x_1, x_2 \geq B. \forall \overline{y}.\ \mathrm{SameDiv}(x_1, x_2, \overline{y}) \Rightarrow (\phi(x_1, \overline{y}) \iff \phi(x_2, \overline{y}))$$

then ϕ is decomposable on x. We do this by constructing the decomposition.

Although there are doubly exponentially many subsets $D \subseteq$ XDivs, there are only exponentially many *maximal consistent* subsets. We implicitly restrict D to such subsets. This is because, for any k, there is no value of x such that $x \equiv_k c$ and $x \equiv_k c'$ both hold with $c \neq c'$ but $c, c' \in \{0, \ldots, k-1\}$. For any maximal consistent set $D \subseteq$ XDivs, let c_D be the smallest integer greater than or equal to B satisfying all constraints in D. Note, since D is maximal, a value that satisfies all constraints in D also does not satisfy an constraints not in D. The number c_D can be represented using polynomially many bits.

We can now decompose ϕ into

$$\left(\begin{array}{c}(x = 0 \land \phi(0, \overline{y})) \\ \lor \cdots \lor \\ (x = B - 1 \land \phi(B-1, \overline{y}))\end{array}\right) \lor \bigvee_{D \subseteq \text{XDivs}} \left(x \geq B \land \bigwedge_{x \equiv_k c \in D} x \equiv_k c \land \phi(c_D, \overline{y})\right).$$

This formula is exponential in the size of ϕ if D only ranges over the maximal consistent subsets of XDivs. For values of x less than B, equivalence with the original formula is immediate. For larger values, we use the fact that, from our original assumption, for any values x_1 and x_2 that satisfy the same divisibility constraints, we have $\phi(x_1, \overline{y})$ iff $\phi(x_2, \overline{y})$. Hence, we can substitute the values c_D in these cases.

Example. We return to $\phi(x, y, z) = x + 2y \geq 5 \land z < 5 \land x \equiv_2 y$ and compute the decomposition on x. Assuming B is odd, the decomposition will be as follows. In our presentation we slightly simplify the formula. Strictly speaking $x \equiv_2 y$ should be expanded to $(x \equiv_2 0 \land y \equiv_2 0) \lor (x \equiv_2 1 \land y \equiv_2 1)$. We simplify these to $y \equiv_2 0$ and $y \equiv_2 1$, respectively, when instantiated with concrete values of x.

$$\begin{array}{c}(x = 0 \land (0 + 2y \geq 5 \land z < 5 \land y \equiv_2 0)) \lor \\ (x = 1 \land (1 + 2y \geq 5 \land z < 5 \land y \equiv_2 1)) \\ \lor \cdots \lor \\ (x = B - 1 \land (B - 1 + 2y \geq 5 \land z < 5 \land y \equiv_2 0)) \lor \\ ((x \equiv_2 0 \land x \geq B) \land (B + 1 + 2y \geq 5 \land z < 5 \land y \equiv_2 0)) \lor \\ ((x \equiv_2 1 \land x \geq B) \land (B + 2y \geq 5 \land z < 5 \land y \equiv_2 1))\end{array}$$

3.4 Completeness

We now show that every formula ϕ decomposable on x satisfies

$$\forall x_1, x_2 \geq B. \forall \overline{y}.\ \text{SameDiv}(x_1, x_2, \overline{y}) \Rightarrow (\phi(x_1, \overline{y}) \iff \phi(x_2, \overline{y})).$$

We first show that some B must exist. Once the existence has been established, we can argue that it must be at most exponential in ϕ.

Existence of the Bound. If $\phi(x, \overline{y})$ is decomposable on x, then there is an equivalent formula $\bigvee_i \Delta_i(x) \land \psi_i(\overline{y})$. It is known that every formula $\Delta(x)$ is satisfied by a finite union of arithmetic progressions $a+jb$. Let B be larger than the largest value of a in the arithmetic progressions satisfying the $\Delta_i(x)$.

We show when $\text{SameDiv}(x_1, x_2, \overline{y})$ then $\phi(x_1, \overline{y})$ iff $\phi(x_2, \overline{y})$ for all values $x_1, x_2 \geq B$ and $\overline{y}$. Assume towards a contradiction that we have values x_1, x_2 and a tuple of values $\overline{y}$ such that $\text{SameDiv}(x_1, x_2, \overline{y})$ and $\phi(x_1, \overline{y})$, but not $\phi(x_2, \overline{y})$.

Let k be the product of all k' appearing in some divisibility constraint $x \equiv_{k'} c$ in XDivs. We know that there is some disjunct of the monadic decomposition such that $\Delta(x_1) \wedge \psi(\overline{y})$ holds. Moreover, let x_1 belong to the arithmetic progression $a + jb$. Since $x_1 \geq B > a$ we know that $\Delta(x_1') \wedge \psi(\overline{y})$ also holds for any $x_1' = x_1 + j'bk$. That is, we can pump x_1 by adding a multiple of bk, while staying in the same arithmetic progression and satisfying the same divisibility constraints.

Similarly, let d be the product of all b appearing in the (finite number of) arithmetic progressions that define the monadic decomposition of ϕ, limited to disjuncts such that $\psi_i(\overline{y})$ holds. Since $\phi(x_2, \overline{y})$ does not hold, then $\phi(x_2', \overline{y})$ also does not hold for any $x_2' = x_2 + jdk$. This means that we can pump x_2 staying outside of the arithmetic progressions defining permissible values of x for the given values $\overline{y}$, whilst additionally satisfying the same divisibility constraints.

Now, for each value of x_1' satisfying $\phi(x_1', \overline{y})$ we can consider the disjunctive normal form of ϕ. By expanding the divisibility constraints, a disjunct becomes a conjunction of terms of the form, where f represents some linear function on $\overline{y}$,

1. $ax + f(\overline{y}) \leq c$ or $ax + f(\overline{y}) \geq c$, or
2. $y_i \equiv_{k'} c$ or $x \equiv_{k'} c$.

Since there are infinitely many x_1', we can choose one disjunct satisfied by infinitely many x_1'. This means that for constraints of the form $ax + f(\overline{y}) \leq c$ or $ax + f(\overline{y}) \geq c$ with a non-zero a, then a must be negative or positive respectively (or zero). Otherwise, only a finite number of values of x would be permitted.

We know that x_2' and $\overline{y}$ do not satisfy the disjunct. We argue that this is a contradiction by considering each term in turn. Since there are infinitely many x_2' we can assume without loss of generality that $x_2' > x_1'$.

1. If $ax + f(\overline{y}) \leq c$ (resp. $ax + f(\overline{y}) \geq c$) appears and is satisfied by x_1', then a must be negative or zero (resp. positive or zero) and x_2' will also satisfy the atom.
2. Atoms of the form $y_i \equiv_{k'} c$ do not distinguish values of x and thus are satisfied for both x_1' and x_2'. We cannot have $x_1' \equiv_{k'} c$ but not $x_2' \equiv_{k'} c$ since x_1' and x_2' satisfy the same divisibility constraints.

Thus, it cannot be the case that x_1' satisfies the disjunct, while x_2' does not. This is our required contradiction. Hence, for all $x_1, x_2 \geq B$ and $\overline{y}$ such that $\text{SameDiv}(x_1, x_2, \overline{y})$ it must be the case that $\phi(x_1, \overline{y})$ iff $\phi(x_2, \overline{y})$. We have thus established the existence of a bound B.

Size of the Bound. We now argue that this bound is exponential in the size of ϕ, and can thus be encoded in a polynomial number of bits.

Consider the formula that is essentially the negation of our property.

$$\chi(x_1, x_2, \overline{y}) = \text{SameDiv}(x_1, x_2) \wedge \phi(x_1, \overline{y}) \wedge \neg\phi(x_2, \overline{y}).$$

There is some computable bound B' exponential in the size of χ (and thus ϕ) such that, if there exists $x_1, x_2 \geq B'$ and some $\overline{y}$ such that $\chi(x_1, x_2, \overline{y})$ holds, then there are infinitely many x_1' and x_2' such that for some $\overline{y}'$ we have that $\chi(x_1', x_2', \overline{y}')$ holds. An argument for the existence of this bound is given in the full version In short, we first convert the formula above into a disjunction of conjunctions of linear equalities, using a linear number of slack variables to encode inequalities and divisibility constraints. Then, using a result of Chistikov and Haase [10], we set $B' = 2^{dnm+3}$ where d is the number of bits needed to encode the largest constant in the converted formula (polynomially related to the size of the formula above), n is the maximum number of linear equalities in any disjunct, and m is the number of variables (including slack variables).

Now, assume that the smallest B is larger than B'. That is

$$\forall x_1, x_2 \geq B. \forall \overline{y}.\ \text{SameDiv}(x_1, x_2, \overline{y}) \Rightarrow (\phi(x_1, \overline{y}) \iff \phi(x_2, \overline{y}))$$

holds, but it does not hold that

$$\forall x_1, x_2 \geq B'. \forall \overline{y}.\ \text{SameDiv}(x_1, x_2, \overline{y}) \Rightarrow (\phi(x_1, \overline{y}) \iff \phi(x_2, \overline{y}))$$

This implies there exists some $x_1, x_2 \geq B'$ and $\overline{y}$ such that $\chi(x_1, x_2, \overline{y})$ holds. Thus, there are infinitely many such x_1' and x_2', contradicting the fact that all $x_1', x_2' \geq B$ do not satisfy the property. Thus, we take B' as the value of B. It is computable in polynomial time, exponential in size, and representable in a polynomial number of bits.

4 Variadic Decomposability

We consider decomposition along several variables instead of just one. In this section, we assume without loss of generality that ϕ is given in positive normal form and all (in)equalities rearranged into the form $\sum_i a_i x_i \geq b$. We may use negation $\neg\phi$ as a shorthand. We require this form because later we use the set of all linear equations in the DNF of a formula. Since negation alters the linear equations, it is more convenient to assume that negation has already been eliminated.

4.1 *Π*-Decomposability

As described in Sect. 2.3, we refine the notion of Π-decomposability to separate only a single set Y_i in $\Pi = \{Y_1, \ldots, Y_n\}$. Without loss of generality, we assume we are given a formula $\phi(\overline{x}, \overline{y})$ and we separate the variables in $\overline{x}$ from $\overline{y}$.

In particular, given a formula $\phi(\overline{x}, \overline{y})$ we aim to decompose the formula into $\phi(\overline{x}, \overline{y}) \equiv \bigvee_j \Delta_j(\overline{x}) \wedge \psi_j(\overline{y})$ for some QF($\mathbb{N}$) formulas Δ_i and ψ_i.

4.2 Decomposition

We show that testing whether a given formula ϕ is variadic decomposable on $\overline{x}$ is in coNP. This proves Theorem 2 as the coNP lower bound follows from the monadic case.

Lemma 4 (Decomposing on $\overline{x}$). *Given a QF($\mathbb{N}$) formula $\phi(\overline{x}, \overline{y})$ there is a coNP algorithm to decide if ϕ is variadic decomposable on $\overline{x}$. Moreover, if a decomposition exists, it can be constructed in exponential-time and is exponential in size.*

Let F be the set of all f such that $f(\overline{x}) + g(\overline{y}) \geq b$ is a linear inequality appearing in ϕ. Our approach will divide the points of $\overline{x}$ into regions where all points within a region can be paired with the same values of $\overline{y}$ to satisfy the formula. These regions are given by a bound B. If $f(\overline{x})$ is within the bound, then two points $\overline{x}_1$ and $\overline{x}_2$ are in the same region if $f(\overline{x}_1) = f(\overline{x}_2)$. If two points are outside the bound, then by a pumping argument we can show that we have $\phi(\overline{x}_1, \overline{y})$ iff $\phi(\overline{x}_2, \overline{y})$.

Let $\hat{r} = (\mathrm{UB}, \mathrm{EQ})$ be a partition of F into unbounded and bounded functions (where EQ refers to equality being asserted over bounded functions as shown below). Define for each $\hat{r} = (\mathrm{UB}, \mathrm{EQ})$

$$\mathrm{Region}_{\hat{r}}(\overline{x}_1, \overline{x}_2) \triangleq \left(\bigwedge_{f \in \mathrm{EQ}} f(\overline{x}_1) = f(\overline{x}_2) \right).$$

Note, this formula intentionally does not say anything about the unbounded functions. This is important when we need to derive a bound—such a derivation cannot use a pre-existing bound.

We also need to extend SameDiv to account for $\overline{x}_1$ and $\overline{x}_2$ being vectors. This is a straightforward extension asserting that each variable in $\overline{x}_1$ satisfies the same divisibility constraints as its counterpart in $\overline{x}_2$. Again, let Divs be the set of all divisibility constraints $az_1 \equiv_k bz_2$ appearing (syntactically) in ϕ. Let x_i, x_i^1 and x_i^2 denote the ith variable of $\overline{x}$, $\overline{x}_1$, and $\overline{x}_2$ respectively. Assume without loss of generality that variables in $\overline{x}$ always either appear on the left-hand side of a divisibility constraint (i.e. in the z_1 position) or on both sides. Define

$$\mathrm{SameDiv}(x_1, x_2, \overline{y}) =$$
$$\bigwedge_{\substack{ax_i \equiv_k bz \in \mathrm{Divs}, \\ z \neq x_j}} \begin{pmatrix} (ax_i^1 \equiv_k bz) \\ \Longleftrightarrow \\ (ax_i^2 \equiv_k bz) \end{pmatrix} \wedge \bigwedge_{\substack{ax_i \equiv_k bx_j \\ \in \\ \mathrm{Divs}}} \begin{pmatrix} (ax_i^1 \equiv_k bx_j^1) \\ \Longleftrightarrow \\ (ax_i^2 \equiv_k bx_j^2) \end{pmatrix} \wedge \bigwedge_{\substack{x_i \equiv_k c \\ \in \\ \mathrm{Divs}}} \begin{pmatrix} (x_i^1 \equiv_k c) \\ \Longleftrightarrow \\ (x_i^2 \equiv_k c) \end{pmatrix}.$$

Next, we introduce an operator for comparing a vector of variables with a bound. For $a \in \mathbb{Z}$ let $\mathrm{abs}(a)$ denote the absolute value of a. Given a bound B and some $\hat{r} = (\mathrm{UB}, \mathrm{EQ})$ let

$$(\overline{x} \geq_{\hat{r}} B) \triangleq \bigwedge_{f \in \mathrm{UB}} \mathrm{abs}(f(\overline{x})) \geq B \wedge \bigwedge_{f \in \mathrm{EQ}} \mathrm{abs}(f(\overline{x})) < B.$$

We claim there is an exponential bound B such that ϕ is variadic decomposable iff for all $\hat{r}$ we have

$$\forall \overline{x}_1, \overline{x}_2 \geq_{\hat{r}} B. \forall \overline{y}. \begin{pmatrix} \text{Region}_{\hat{r}}(\overline{x}_1, \overline{x}_2) \\ \wedge \\ \text{SameDiv}(\overline{x}_1, \overline{x}_2, \overline{y}) \end{pmatrix} \Rightarrow \begin{pmatrix} \phi(\overline{x}_1, \overline{y}) \\ \Longleftrightarrow \\ \phi(\overline{x}_2, \overline{y}) \end{pmatrix} \qquad \text{(DC-}\hat{r}\text{)}$$

Note, unsatisfiability can be tested in NP. First guess $\hat{r}$, then guess $\overline{x}_1, \overline{x}_2, \overline{y}$.

We prove soundness of the claim in the next section. Completeness is an extension of the argument for the monadic case and is given in the full version. In the monadic case, we were able to take some values of $x_1, x_2 > B$ such that both satisfied the same divisibility constraints, but one value satisfied the formula while the other did not. Since these values were large, we derived an infinite number of such value pairs with increasing values. We then used these growing solutions to show that it was impossible for the value of x_1 to satisfy the formula, while the value of x_2 does not, as they were both beyond the distinguishing power of the linear inequalities. The argument for the variadic case is similar, with the values of x_1 and x_2 being replaced by the values of $f(\overline{x}_1)$ and $f(\overline{x}_2)$.

4.3 Soundness

Assume there is an exponential bound B such that for each $\hat{r}$, Equation DC-$\hat{r}$ holds. We show how to produce a decomposition.

As in the monadic case (Sect. 3.3), let XDivs be the set of all constraints of the form $\overline{x}_i \equiv_k c$ in the expansion of the divisibility constraints of ϕ. Observe again that there are only exponentially many maximal consistent subsets $D \subseteq$ XDivs. For each D fix a vector of values $\overline{c}_D$ that satisfies all constraints in D and is encodable in a polynomial number of bits. Furthermore, we define

$$\text{Div}_D(\overline{z}) \triangleq \bigwedge_{x_i \equiv_k c \in D} z_i \equiv_k c.$$

For each $\hat{r}$ and D we can define an equivalence relation over values of $\overline{x}$ such that $\overline{x} \geq_{\hat{r}} B$ and $\text{Div}_D(\overline{x})$.

$$(\overline{x}_1 =^D_{\hat{r}} \overline{x}_2) \triangleq (\overline{x}_1 \geq_{\hat{r}} B \wedge \overline{x}_2 \geq_{\hat{r}} B \wedge \text{Region}_{\hat{r}}(\overline{x}_1, \overline{x}_2) \wedge \text{Div}_D(\overline{x}_1) \wedge \text{Div}_D(\overline{x}_2)).$$

Observe each equivalence relation has an exponential number of equivalence classes depending on the values of the bounded f. Let $C^D_{\hat{r}}$ be a set of minimal representatives from each equivalence class such that each representative is representable in a polynomial number of bits. These can be computed by solving an existential Presburger constraint for each set of values of the bounded f. In particular, for each $\hat{r} = (\text{UB}, \text{EQ})$ and assignments $\text{abs}(c_f) < B$ for each $f \in \text{EQ}$, we select a solution to the equation

$$\overline{x} \geq_{\hat{r}} B \wedge \bigwedge_{f \in \text{EQ}} f(\overline{x}) = c_f \wedge \text{Div}_D(\overline{x})$$

if such a solution exists. If no such solution exists, the assignment can be ignored.

The decomposition is

$$\bigvee_{\hat{r}} \bigvee_{D} \bigvee_{\overline{c} \in C_{\hat{r}}^{D}} (\overline{x} \geq_{\hat{r}} B \wedge \mathrm{Region}_{\hat{r}}(\overline{x}, \overline{c}) \wedge \mathrm{Div}_D(\overline{x}) \wedge \phi(\overline{c}, \overline{y}))\,.$$

The correctness of this decomposition follows from the Equations DC-$\hat{r}$. For any values $\overline{c}_{\overline{x}}$ and $\overline{c}_{\overline{y}}$ of $\overline{x}$ and $\overline{y}$, first assume $\phi(\overline{c}_{\overline{x}}, \overline{c}_{\overline{y}})$ holds. Since there is some disjunct in the decomposition for which it holds that $\overline{c}_{\overline{x}} \geq_{\hat{r}} B \wedge \mathrm{Region}_{\hat{r}}(\overline{c}_{\overline{x}}, \overline{c}) \wedge \mathrm{Div}_D(\overline{x})$ then, by applying Equation DC-$\hat{r}$ we get $\phi(\overline{c}, \overline{c}_{\overline{y}})$ as required. Conversely, if some disjunct of the decomposition holds, we can apply Equation DC-$\hat{r}$ and obtain $\phi(\overline{c}_{\overline{x}}, \overline{c}_{\overline{y}})$.

5 Applications of Decomposition

5.1 Monadic Decomposition in String Solving

The development of effective techniques for solving string constraints has received a lot of attention over the last years, motivated by applications ranging from program verification [2,16] and security analysis [22,23] to the analysis of access policies of cloud services [4]. Strings give rise to a rich theory that may combine, depending on the studied fragment, (i) word equations, i.e., equations over the free monoid generated by some finite (but often large) alphabet, (ii) regular expression constraints, (iii) transduction, i.e., constraints described by finite-state automata with multiple tracks, (iv) conversion functions, e.g. between integer variables and strings encoding numbers in binary or decimal notation, (v) length constraints, i.e., arithmetic constraints on the length of strings.

Table 1. Statistics about the Kaluza benchmarks [22]. It should be noted (and is well-known [18]) that the categories "sat" and "unsat" do not (always) imply the status of the benchmarks, they only represent the way the benchmarks were organised by the Kaluza authors.

Folder	#Benchmarks	Benchmarks with `str.len`	Decomposition checks	Decomposition checks succeeded
sat/small	19804	2185	2183	2155
sat/big	1741	1318	1317	56
unsat/small	11365	3910	2919	2919
unsat/big	14374	13813	6786	3362
Total	47284	21226	13205	8492

The handling of length constraints has turned out to be particularly challenging in this context, both practically and theoretically. Even for the combination of word equations (or even just quadratic word equations) with length constraints, decidability of the (quantifier-free) theory is a long-standing open problem [21]. At the same time, length constraints are quite frequently used in applications; they are needed, for instance, when encoding operations like indexof or substring, or also when splitting a string into the parts separated by some delimiter. In standard benchmark libraries for string constraints, like the Kaluza set [22], benchmarks with length constraints occur in large numbers.

The notion of monadic decomposition is in this setting important, since any *monadic* length constraint (in Presburger arithmetic) can be reduced to a Boolean combination of regular expression constraints, and is therefore easier to handle than the general case.

Proposition 2. *Satisfiability of a quantifier-free formula* $\phi = \phi_{eq} \wedge \phi_{regex} \wedge \phi_{len}$ *consisting of word equations, regular expression constraints, and* monadically decomposable *length constraints is decidable.*

Proof. Suppose $w_1, \ldots, w_n$ are the string variables occurring in ϕ, and $|w_1|, \ldots, |w_n|$ the terms representing their length. A decision procedure can first compute a monadic representation ϕ'_{len} of ϕ_{len} over lengths $|w_1|, \ldots, |w_n|$, and then turn each atom $\Delta(|w_i|)$ in ϕ'_{len} into an equivalent regular membership constraint $w_i \in \mathcal{L}_\Delta$. This is possible because the Presburger formula Δ can be represented as a semi-linear set, which can directly be translated to a regular expression. Decidability follows from the decidability of word equations combined with regular expression constraints [13]. □

This motivates the use of monadic decomposition as a standard pre-processing step in string solvers, transforming away those length constraints that can be turned into monadic form. To evaluate the effectiveness of such an optimisation, we implemented the decomposition check defined in Sect. 3.2, and used it within the string SMT solver OSTRICH [9] to determine the number of Kaluza benchmarks with monadic decomposable length constraints.[1] The results are summarised in Table 1:

- Of altogether 47 284 benchmarks, 21 226 contain the `str.len` function, and therefore length constraints. This number was determined by a simple textual analysis of the benchmarks.
- Running our decomposition check in OSTRICH, in 13 205 of the 21 226 cases length constraints were found that could be analysed. The remaining 8 021 problems were proven unsatisfiable without ever reaching the string theory solver in OSTRICH, i.e., as a result of pre-processing the input formula, or because Boolean reasoning discovered obvious inconsistencies in the problems.
- In 8 492 of the 13 205 cases, all analysed length constraints were found to be monadically decomposable; 4 713 of the benchmarks contained length constraints that could not be decomposed.

[1] Branch "modec" of https://github.com/uuverifiers/ostrich, which also contains detailed logs of the experiments.

This means that 42 571 of the Kaluza benchmarks (slightly more than 90%) do in principle not require support for length constraints in a string solver, either because there are no length constraints, or because length constraints can be decomposed and then turned into regular expression constraints.

Even with a largely unoptimised implementation, the time required to check whether length constraints can be decomposed was negligible in case of the Kaluza benchmarks, with the longest check requiring 2.1 s (on an AMD Opteron 2220 SE machine). The maximum number of variables in a length constraint was 140.

5.2 Variadic Decomposition in Quantifier Elimination

A second natural application of decomposition is *quantifier elimination*, i.e., the problem of deriving an equivalent quantifier-free formula ϕ' for a given formula ϕ with quantifiers. In Presburger arithmetic, for a formula $\phi = \exists x_1, \ldots, x_n. \psi$ with n quantifiers but no quantifier alternations, quantifier elimination in the worst case causes a doubly-exponential increase in formula size [25].

Variadic decomposition can be used to eliminate quantifiers with a smaller worst-case increase in size, provided that the matrix of a quantifier formula can be decomposed. Suppose $\phi = \exists \bar{x}. \psi(\bar{x}, \bar{y})$ is given and ψ is variadic decomposable on $\bar{x}$, i.e.,

$$\psi(\bar{x}, \bar{y}) \equiv \bigvee_j \Delta_j(\bar{x}) \wedge \psi_j(\bar{y})$$

This means that the existential quantifiers can be distributed over the disjunction, and their elimination turns into a simpler satisfiability check:

$$\exists \bar{x}. \psi(\bar{x}, \bar{y}) \equiv \bigvee_j \exists \bar{x}. \Delta_j(\bar{x}) \wedge \psi_j(\bar{y}) \equiv \bigvee_{j:\ \Delta_j(\bar{x}) \text{ is sat}} \psi_j(\bar{y})$$

Universal quantifiers can be handled in a similar way by negating the matrix first.

Proposition 3. *Take a formula $\phi(\bar{y}) = \exists \bar{x}. \psi(\bar{x}, \bar{y})$ in Presburger arithmetic in which ψ is quantifier-free and variadic decomposable on $\bar{x}$. Then there is a quantifier-free formula $\phi'(\bar{y})$ that is equivalent to ϕ and at most singly-exponentially bigger than ϕ.*

Checking whether a formula can be decomposed is therefore a simple optimisation that can be added to any quantifier elimination procedure for Presburger arithmetic.

6 Conclusion and Future Work

We have shown that the monadic and variadic decomposability problem for QF(ℕ) is coNP-complete. Moreover, when a decomposition exists, it is at most exponential in size and can be computed in exponential time. This formula size is tight for decompositions presented in either disjunctive or conjunctive normal form.

We gave two applications of our results. The first was in string constraint solving. In program analysis, string constraints are often mixed with numerical constraints on the lengths of the strings (for example, via the indexOf function). Length constraints significantly complicate the analysis of strings. However, if the string constraints permit a monadic decomposition, they may be reduced to regular constraints and thus eliminated. We analysed the well-known Kaluza benchmarks and showed that less than 10% of the benchmarks contained length constraints that could not be decomposed.

For the second application, we showed that the doubly exponential blow-up caused by quantifier elimination can be limited to a singly exponential blow up whenever the formula is decomposable on the quantified variables. Thus, variadic decomposition can form an optimisation step in a quantifier elimination algorithm.

Interesting problems are opened up by our results. It would be interesting to study lower bounds for general boolean formulas. If smaller decompositions are possible, they would be useful for applications in string solving.

Second, we may consider variadic decomposition where a partition Π is not given as part of the input. Instead, one must check whether a Π-decomposition exists for some non-trivial Π. This variant of the problem has a simple Σ_2^P algorithm that first guesses some Π and then verifies Π-decomposability. However, the only known lower bound is coNP, which follows the same argument as monadic decomposability. A better algorithm would not improve the worst-case complexity for our quantifier elimination application, but it might provide a way to quickly identify a subset of a block of quantifiers that can be eliminated quickly with Π-decompositions.

Acknowledgments. We thank Christoph Haase, Leonid Libkin, and Pascal Bergsträßer for their help during the preparation of this work. Matthew Hague is supported by EPSRC [EP/T00021X/1]. Anthony Lin is supported by the European Research Council (ERC) under the European Union's Horizon 2020 research and innovation programme (grant agreement no 759969), and by Max-Planck Fellowship. Philipp Rümmer is supported by the Swedish Research Council (VR) under grant 2018-04727, and by the Swedish Foundation for Strategic Research (SSF) under the project WebSec (Ref. RIT17-0011). Zhilin Wu is partially supported by the NSFC grant No. 61872340, Guangdong Science and Technology Department grant (No. 2018B010107004), and the INRIA-CAS joint research project VIP.

References

1. Abdulla, P.A., et al.: TRAU: SMT solver for string constraints. In: Formal Methods in Computer Aided Design, FMCAD 2018 (2018)
2. Abdulla, P.A., et al.: String constraints for verification. In: Biere, A., Bloem, R. (eds.) CAV 2014. LNCS, vol. 8559, pp. 150–166. Springer, Cham (2014). https://doi.org/10.1007/978-3-319-08867-9_10

3. Amadini, R., Gange, G., Stuckey, P.J.: Sweep-based propagation for string constraint solving. In: Proceedings of the Thirty-Second AAAI Conference on Artificial Intelligence (AAAI 2018), the 30th Innovative Applications of Artificial Intelligence (IAAI 2018), and the 8th AAAI Symposium on Educational Advances in Artificial Intelligence (EAAI 2018), New Orleans, Louisiana, USA, 2–7 February 2018, pp. 6557–6564 (2018). https://www.aaai.org/ocs/index.php/AAAI/AAAI18/paper/view/16223
4. Backes, J., et al.: Semantic-based automated reasoning for AWS access policies using SMT. In: Bjørner, N., Gurfinkel, A. (eds.) 2018 Formal Methods in Computer Aided Design, FMCAD 2018, Austin, TX, USA, 30 October–2 November 2018, pp. 1–9. IEEE (2018). https://doi.org/10.23919/FMCAD.2018.8602994
5. Barceló, P., Hong, C., Le, X.B., Lin, A.W., Niskanen, R.: Monadic decomposability of regular relations. In: 46th International Colloquium on Automata, Languages, and Programming, ICALP 2019, Patras, Greece, pp. 103:1–103:14 (2019). https://doi.org/10.4230/LIPIcs.ICALP.2019.103
6. Berzish, M., Ganesh, V., Zheng, Y.: Z3str3: a string solver with theory-aware heuristics. In: 2017 Formal Methods in Computer Aided Design, FMCAD 2017, Vienna, Austria, 2–6 October 2017, pp. 55–59. IEEE (2017). https://doi.org/10.23919/FMCAD.2017.8102241
7. Büchi, J.R., Senger, S.: Definability in the existential theory of concatenation and undecidable extensions of this theory. In: Mac, L.S., Siefkes, D. (eds.) The Collected Works of J. Richard Büchi, pp. 671–683. Springer, Heidelberg (1990). https://doi.org/10.1007/978-1-4613-8928-6_37
8. Carton, O., Choffrut, C., Grigorieff, S.: Decision problems among the main subfamilies of rational relations. ITA **40**(2), 255–275 (2006). https://doi.org/10.1051/ita:2006005
9. Chen, T., Hague, M., Lin, A.W., Rümmer, P., Wu, Z.: Decision procedures for path feasibility of string-manipulating programs with complex operations. CoRR abs/1811.03167 (2018). https://arxiv.org/abs/1811.03167
10. Chistikov, D., Haase, C.: The taming of the semi-linear set. In: Chatzigiannakis, I., Mitzenmacher, M., Rabani, Y., Sangiorgi, D. (eds.) 43rd International Colloquium on Automata, Languages, and Programming (ICALP 2016). Leibniz International Proceedings in Informatics (LIPIcs), vol. 55, pp. 128:1–128:13. Schloss Dagstuhl-Leibniz-Zentrum fuer Informatik, Dagstuhl, Germany (2016). https://doi.org/10.4230/LIPIcs.ICALP.2016.128. http://drops.dagstuhl.de/opus/volltexte/2016/6263
11. D'Antoni, L., Veanes, M.: The power of symbolic automata and transducers. In: Majumdar, R., Kunčak, V. (eds.) CAV 2017, Part 1. LNCS, vol. 10426, pp. 47–67. Springer, Cham (2017). https://doi.org/10.1007/978-3-319-63387-9_3
12. Day, J.D., Ehlers, T., Kulczynski, M., Manea, F., Nowotka, D., Poulsen, D.B.: On solving word equations using SAT. In: Filiot, E., Jungers, R., Potapov, I. (eds.) RP 2019. LNCS, vol. 11674, pp. 93–106. Springer, Cham (2019). https://doi.org/10.1007/978-3-030-30806-3_8
13. Diekert, V.: Makanin's algorithm. In: Lothaire, M. (ed.) Algebraic Combinatorics on Words, Encyclopedia of Mathematics and its Applications, vol. 90, chap. 12, pp. 387–442. Cambridge University Press (2002). https://doi.org/10.1017/CBO9781107326019.013
14. Ganesh, V., Minnes, M., Solar-Lezama, A., Rinard, M.: Word equations with length constraints: what's decidable? In: Biere, A., Nahir, A., Vos, T. (eds.) HVC 2012. LNCS, vol. 7857, pp. 209–226. Springer, Heidelberg (2013). https://doi.org/10.1007/978-3-642-39611-3_21

15. Haase, C.: Subclasses of presburger arithmetic and the weak EXP hierarchy. In: Joint Meeting of the Twenty-Third EACSL Annual Conference on Computer Science Logic (CSL) and the Twenty-Ninth Annual ACM/IEEE Symposium on Logic in Computer Science (LICS), CSL-LICS 2014, Vienna, Austria, 14–18 July 2014, pp. 47:1–47:10 (2014). https://doi.org/10.1145/2603088.2603092
16. Hojjat, H., Rümmer, P., Shamakhi, A.: On strings in software model checking. In: Lin, A.W. (ed.) APLAS 2019. LNCS, vol. 11893, pp. 19–30. Springer, Cham (2019). https://doi.org/10.1007/978-3-030-34175-6_2
17. Jez, A.: Word equations in linear space. CoRR abs/1702.00736 (2017). http://arxiv.org/abs/1702.00736
18. Liang, T., Reynolds, A., Tinelli, C., Barrett, C., Deters, M.: A DPLL(T) theory solver for a theory of strings and regular expressions. In: Biere, A., Bloem, R. (eds.) CAV 2014. LNCS, vol. 8559, pp. 646–662. Springer, Cham (2014). https://doi.org/10.1007/978-3-319-08867-9_43
19. Libkin, L.: Variable independence for first-order definable constraints. ACM Trans. Comput. Log. **4**(4), 431–451 (2003). https://doi.org/10.1145/937555.937557
20. Lin, A.W., Barceló, P.: String solving with word equations and transducers: towards a logic for analysing mutation XSS. In: Proceedings of the 43rd Annual ACM SIGPLAN-SIGACT Symposium on Principles of Programming Languages, POPL 2016, pp. 123–136. Springer (2016). https://doi.org/10.1145/2837614.2837641
21. Lin, A.W., Majumdar, R.: Quadratic word equations with length constraints, counter systems, and presburger arithmetic with divisibility. In: Lahiri, S.K., Wang, C. (eds.) ATVA 2018. LNCS, vol. 11138, pp. 352–369. Springer, Cham (2018). https://doi.org/10.1007/978-3-030-01090-4_21
22. Saxena, P., Akhawe, D., Hanna, S., Mao, F., McCamant, S., Song, D.: A symbolic execution framework for JavaScript. In: 31st IEEE Symposium on Security and Privacy, S&P 2010, Berleley/Oakland, California, USA, 16–19 May 2010, pp. 513–528. IEEE (2010). https://doi.org/10.1109/SP.2010.38
23. Trinh, M., Chu, D., Jaffar, J.: S3: a symbolic string solver for vulnerability detection in web applications. In: Proceedings of the 2014 ACM SIGSAC Conference on Computer and Communications Security, CCS 2014, pp. 1232–1243. ACM (2014). https://doi.org/10.1145/2660267.2660372
24. Veanes, M., Bjørner, N., Nachmanson, L., Bereg, S.: Monadic decomposition. J. ACM **64**(2), 14:1–14:28 (2017). https://doi.org/10.1145/3040488
25. Weispfenning, V.: Complexity and uniformity of elimination in presburger arithmetic. In: Proceedings of the 1997 International Symposium on Symbolic and Algebraic Computation, ISSAC 1997, Maui, Hawaii, USA, 21–23 July 1997, pp. 48–53 (1997)

Scalable Algorithms for Abduction via Enumerative Syntax-Guided Synthesis

Andrew Reynolds[1], Haniel Barbosa[2], Daniel Larraz[1], and Cesare Tinelli[1](✉)

[1] Department of Computer Science, The University of Iowa, Iowa City, USA
cesare-tinelli@uiowa.edu

[2] Department of Computer Science, Universidade Federal de Minas Gerais (UFMG), Belo Horizonte, Brazil

Abstract. The abduction problem in logic asks whether there exists a formula that is consistent with a given set of axioms and, together with these axioms, suffices to entail a given goal. We propose an approach for solving this problem that is based on syntax-guided enumeration. For scalability, we use a novel procedure that incrementally constructs a solution in disjunctive normal form that is built from enumerated formulas. The procedure can be configured to generate progressively weaker and simpler solutions over the course of a run of the procedure. Our approach is fully general and can be applied over any background logic that is handled by the underlying SMT solver in our approach. Our experiments show our approach compares favorably with other tools for abductive reasoning.

1 Introduction

The abduction problem for theory T, a set of axioms A and goal G asks whether there exists a formula φ such that: (i) $\mathsf{A} \wedge \varphi$ is T-satisfiable and (ii) $\mathsf{A} \wedge \varphi \models_T \mathsf{G}$. In other words, it asks for a formula φ that is consistent with the axioms and when added to it allows the goal to be proven. Ideally, φ should be as weak as possible and typically, it is expected to satisfy additional syntactic restrictions, such as, for instance, on its quantifier prefix. Abductive reasoning has gained a variety of applications recently, including extending knowledge bases for failed verification conditions [16] and invariant generation [17,20]. Despite the usefulness of abductive reasoning, and the recent development of a few abductive reasoners, such as GPiD [19] and Explain [15], general tools for automatic abductive inference are not yet mainstream.

Independently from the research on abduction, many high-performance general-purpose solvers for syntax-guided synthesis (SyGuS) have also been developed in the past decade. These solvers have been applied successfully in a number of domains, including the implementation of network protocols [36],

This work was partially supported by NSF grant #1656926 and DARPA grants #N66001-18-C-4006 and #N66001-18-C-4012.

N. Peltier and V. Sofronie-Stokkermans (Eds.): IJCAR 2020, LNAI 12166, pp. 141–160, 2020.
https://doi.org/10.1007/978-3-030-51074-9_9

data processing [22], and code optimization [29]. The performance and scalability of SyGuS solvers has made considerable progress in recently years, as demonstrated by an annual competition [4].

In this paper, we investigate scalable approaches to solving the abduction problem using (enumerative) syntax-guided synthesis techniques. We impose no requirements on the background theory T other than it must be supported by an existing SMT solver and amenable to syntax-guided synthesis, as we explain in more detail later. Our immediate goal is to leverage the power of syntax-guided synthesis solvers. Our longer term goal is to standardize the interface for these solvers for abduction problems and make them available to users of program analysis and automated reasoning who would benefit from high performance automated reasoning systems for abduction.

Contributions

- We introduce a novel procedure for solving abduction problems using enumerative syntax-guided synthesis.
- We give an extension of the procedure that is capable of generating progressively weaker solutions to a given abduction problem.
- We provide an implementation of these techniques in CVC4SY [31], a state-of-the-art SyGuS solver implemented within the SMT solver CVC4 [8], and discuss several experiments we designed to test its effectiveness. We show that it has compelling advantages with respect to to other approaches for abduction including those implemented in EXPLAIN [15] and GPID [19].

2 Preliminaries

We work in the context of many-sorted first-order logic with equality ($\simeq$) and assume the reader is familiar with the notions of signature, terms, and so on (see, e.g., [21]). A *theory* is a pair $T = (\Sigma, I)$ where Σ is a signature and I is a non-empty class of Σ-interpretations, the *models of* T, that is closed under variable reassignment (i.e., every Σ-interpretation that differs from one in I only in how it interprets the variables is also in I) and isomorphism. A Σ-formula φ is T-*satisfiable* (respectively, T-*unsatisfiable*) if it is satisfied by some (resp., no) interpretation in I. A satisfying interpretation for φ is a *model* φ. A formula φ is *valid in* T (or T-*valid*), written $\models_T \varphi$, if every model of T is a model of φ. We write $\varphi[\boldsymbol{x}]$ for a tuple $\boldsymbol{x}$ of distinct variables to indicate that the free variables of φ occur in $\boldsymbol{x}$. Given $\varphi[\boldsymbol{x}]$, we write $\varphi[\boldsymbol{t}]$ to denote the result of replacing every occurrence of every variable of $\boldsymbol{x}$ in φ with the corresponding term in the tuple $\boldsymbol{t}$. We write conjunctions of formulas as sets.

Syntax-Guided Synthesis (SyGuS). Syntax-guided synthesis [2] is a recent paradigm for automated synthesis that combines semantic and syntactic restrictions on the space of solutions. Specifically, a SyGuS problem for a function f in a theory T consists of

1. *semantic restrictions*, a specification given by a (second-order) T-formula of the form $\exists f. \forall \boldsymbol{x}. \varphi[f, \boldsymbol{x}]$, and
2. *syntactic restrictions* on the solutions for f, given by a context-free grammar $\mathcal{R}$.

The grammar $\mathcal{R}$ is a triple (s_0, S, R) where s_0 is an initial symbol, S is a set of symbols with $s_0 \in S$, and R is a set of *production rules* of the form $s \to t$, where $s \in S$ and t is a term built from the symbols in the signature of theory T, free variables, and symbols from S. The rules define a rewrite relation over such terms, also denoted by $\to$, as expected. We say a term t is *generated* by $\mathcal{R}$ if $s_0 \to^* t$ where $\to^*$ is the reflexive-transitive closure of $\to$ and t does not contain symbols from S. For example, the terms x, $(x+x)$ and $((1+x)+1)$ are all generated by the grammar $\mathcal{R} = (\mathsf{I}, \{\mathsf{I}\}, \{\mathsf{I} \to x, \mathsf{I} \to 1, \mathsf{I} \to (\mathsf{I} + \mathsf{I})\})$. A *solution for the SyGuS problem for* f is a lambda term $\lambda \boldsymbol{x}.e$ of the same type as f such that (*i*) $\forall \boldsymbol{x}. \varphi[\lambda \boldsymbol{x}.e, \boldsymbol{x}]$ is T-valid and (*ii*) e is generated by $\mathcal{R}$.

A number of recent approaches for the syntax-guided synthesis problem exist that target specific classes of semantic and syntactic restrictions, including programming-by-examples [22], single invocation conjectures [32], and pointwise specifications [5,27]. General purpose methods for solving the syntax-guided synthesis problem are generally based on *enumerative counterexample-guided inductive synthesis* (CEGIS) [34,35]. Enumerative approach uses a grammar to generate candidate solutions systematically based on some term ordering, typically term size (e.g., the number of non-nullary function applications in the term). The generated candidate solutions are then tested for correctness using a verification oracle (typically an SMT solver). This process is accelerated by the use of *counterexamples* for previously discarded candidates, i.e., valuations for the input variables $\boldsymbol{x}$, or *points*, that witness the failure of those candidates to satisfy the specification. Despite its simplicity, enumerative CEGIS is the de-facto approach for solving the general class of SyGuS problems, as implemented in a several recent tools, notably CVC4SY [31] and the enumerative solver ESOLVER [3]. Its main downside remains scalability to cases where the required solution is very large. As we will show in Sect. 4, we present a more scalable procedure for the abduction problem that builds on top of enumerative CEGIS and is capable of quickly finding (conjunctive) solutions.

3 The Abduction Problem

In general, the abduction problem for a set A of axioms and a goal G is the problem of finding a formula S that is consistent with A and, together with A, entails the goal. We refine the problem by restricting it to first-order logic and to a given background theory T, and also considering syntactic restrictions on the solution S. We refer to this as the *syntax-restricted abduction problem*, which we formalize in the following definition.

Definition 1 (Abduction Problem). *The (syntax-restricted) abduction problem for a theory* T*, a conjunction* $\mathsf{A}[\boldsymbol{x}]$ *of axioms, a goal* $\mathsf{G}[\boldsymbol{x}]$ *and a grammar*

R, *where axioms and goal are first-order formulas, is that of finding a first-order formula* $\mathsf{S}[\boldsymbol{x}]$ *such that:*

1. $\mathsf{A} \wedge \mathsf{S} \models_T \mathsf{G}$,
2. $\mathsf{A} \wedge \mathsf{S}$ *is* T*-satisfiable, and*
3. S *is generated by grammar* R.

In practice, as in SyGuS, syntactic restrictions on the solution space may be used to capture user-requirements on the desired shape of a solution. They can also be used as a mechanism for narrowing the search space to one where one believes the solver is likely to find a solution. Observe that the formulation of the problem includes the case with no syntactic restriction as a trivial case of a grammar that accepts all formulas in the signature of the theory T. In the abduction solver we have developed for this work, the syntax restriction is optional. When it is missing, a grammar generating the full language is constructed internally automatically.

Syntax-restricted abduction bears a strong similarity to SyGuS.[1] We exploit this similarity by leveraging much of the technology we developed for SyGuS, with the goal of achieving generality and scalability.

Normally, an abduction problem admits many solutions. Thus, it may be useful to look for solutions that optimize certain criteria, such as generality with respect to entailment in T, or minimality with respect to size or number of free variables. Our evaluation contains several case studies where we explore this aspect in detail.

Recent Applications. Abduction has a long history in logic and automatic reasoning (see, e.g., [24]). More recently, it has found many useful applications in program analysis. It has been used for identifying the possible facts a verification tool is missing to either discharge or validate a verification condition [16], inferring library specifications that are needed for verifying a client program [37], and synthesizing specifications for multiple unknown procedures called from a main program [1]. Other applications of abduction includes loop invariant generation [17,20], where it is used to iteratively strengthen candidate solutions until they are inductive and strong enough to verify a program, and compositional program verification [25], where it is used for inferring not only loop invariants but also preconditions required for the invariants to hold. Abductive inference has also been applied to modular heap reasoning [12], and the synthesis of missing guards for memory safety [18].

4 Abduction via Enumerative Syntax-Guided Synthesis

In this section, we fix a theory T and describe our approach for solving the abduction problem in T using enumerative syntax-guided synthesis. We first present a basic procedure for abduction in the following section, and then extend it to generate (conjunctive) solutions in a highly scalable manner. We then describe how

---

[1] In fact, it could be readily recast as SyGuS, if one ignored Condition 2 in Definition 1.

```
GetAbductBasic(axioms A[x], goal G[x], grammar R)
 1: Let P = ∅   // set of points
 2: loop
 3:   Let c[x] = NextEnum(R)
 4:   if Eval(c, p) = ⊥ for all p ∈ P then
 5:     if c ∧ A ∧ ¬G is T-satisfiable then
 6:       P := P ∪ {p} with p such that Eval(c ∧ A ∧ ¬G, p) = ⊤
 7:     else if c ∧ A is T-satisfiable then
 8:       return c
 9:     end if
10:   end if
11: end loop
```

Fig. 1. Basic procedure for the abduction problem for axioms A, goal G and grammar R.

either approach can be extended to be incremental so that it constructs progressively logically weaker solutions over time. For simplicity, *we restrict ourselves to abduction problems where axioms, goals, and solutions are quantifier-free.* Note, however, that the procedure can be used for abduction problems where these components are quantified, as long the restrictions below (lifted to quantified formulas) are satisfied.

Requirements on T. We assume that the T-satisfiability of quantifier-free formulas is decidable. For each sort of T, we also assume a distinguished set of variable-free terms of that sort which we call *values* (e.g., numerals and negated numerals in the case of integer arithmetic) such that every T-satisfiable formula is satisfied by a valuation of its free variables to sort elements denoted by values. Finally, we require the availability of a computable function Eval that takes a first-order formula $\varphi[\boldsymbol{x}]$ and a tuple $\boldsymbol{p}$ of values of the same length as $\boldsymbol{x}$, and returns $\top$ if $\varphi[\boldsymbol{p}]$ is T-satisfiable and $\bot$ otherwise. These restrictions are met by most theories used in Satisfiability Modulo Theories (SMT).

4.1 Enumerative Counterexample-Guided Inductive Synthesis for Abduction

We start with a basic CEGIS-style synthesis procedure for solving the syntax-restriction abduction problem where points that represent counterexamples for candidate solutions are cached and used to discard subsequent candidates. The procedure is presented in Fig. 1. It takes as input: axioms A, goal G and grammar R, and maintains an internally set P of points that satisfy the axioms and falsify the goal. On line 3, it invokes the stateful sub-procedure NextEnum(R) which enumerates the formulas generated by grammar R based on enumerative techniques used in SyGuS solvers. We will refer to the return formula c as the current *candidate solution.* Then, using the (fast) evaluation function Eval, it checks at line 4 that c is satisfied by none of the counterexample points in P. If the check fails, the procedure discards c and loops back to line 3 because adding

c to A would definitely be not enough to entail G. If the check succeeds, it also checks, at line 5, whether $c \wedge \mathsf{A} \wedge \neg \mathsf{G}$ is T-satisfiable. If so, it obtains a witness point $\boldsymbol{p}$ for the satisfiability, adds it to current set of points P on line 6, and discards c; otherwise, it checks that c is consistent with A before returning it as a possible solution.

Example 1. Let T be the theory of linear integer arithmetic with the usual signature. Let A be the set $\{y \geqslant 0\}$, let G be the set $\{x + y + z \geqslant 0\}$, and assume R is a grammar generating all linear arithmetic atomic formulas over the variables x, y, z. The results of the procedure are summarized in the table below. We provide, for each iteration, the candidate c generated by syntax-guided enumeration on line 3, the Boolean value of the conditions on lines 4, 5 and 7 of the procedure when applicable, and the point (x, y, z) added to P in when the condition on line 5 evaluates to true. The last column specifies the solution returned on that iteration if any.

#	c	line 4	line 5	$p \in \mathsf{P}$	line 7	return
1	$x \geqslant 0$	true	true	$(0, 0, -1)$		
2	$x < 0$	true	true	$(-1, 0, 0)$		
3	$y \geqslant 0$	false				
4	$y < 0$	true	false		false	
5	$z \geqslant 0$	false				
6	$z < 0$	false				
7	$x + y \geqslant 0$	false				
8	$x + y < 0$	false				
9	$x + z \geqslant 0$	true	false		true	$x + z \geqslant 0$

On the first iteration, the syntax-guided enumeration generates the formula $x \geqslant 0$ as the candidate solution c. This fails to imply the goal, specifically, with $(x, y, z) = (0, 0, -1)$ the axioms and c are satisfied and the goal is falsified. The second candidate fails for similar reasons for point $(-1, 0, 0)$. The check on line 4 fails for five of the next six candidates, with the exception of the candidate $y < 0$. This candidate is falsified by both points in P but it must be discarded since it is inconsistent with the axioms (line 7). Finally, the candidate $x + z \geqslant 0$ generated on the ninth iteration passes all the tests and is returned as a solution for this abduction problem. □

4.2 A Procedure for Abduction Based on Unsat Core Learning

This section extends the procedure from Fig. 1 with techniques that make it scalable when the intended solution to the abduction problem is a conjunction of formulas. The procedure is applicable to cases where the language generated by grammar R is closed under conjunction. In essence, the procedure in this section applies when $s_0 \to s_0 \wedge s_0$ is a production rule in R where s_0 is the start symbol of R. However, it avoids enumerating conjunctive formulas directly, preferring instead to generate them as sets of (non-conjunctive) formulas.

```
GetAbductUCL(axioms A, goal G, grammar R)
Let E, P, U = ∅
loop
    E := E ∪ {NextEnum(R)}
    Let C = ∅
    while EnsureCexFalsify(C, E, P, U) do
        if C ∧ A ∧ ¬G is T-unsatisfiable then
            Let C_min ⊆ C such that C_min ∧ A ∧ ¬G is T-unsatisfiable
            if C_min ∧ A is T-satisfiable then
                return C_min
            else
                Let u ⊆ C_min such that u ∧ A is T-unsatisfiable
                U := U ∪ {u}; C := C \ {e} for some e ∈ u
            end if
        else
            P := P ∪ {p} where Eval((C ∧ A ∧ ¬G), p) = ⊤
        end if
    end while
end loop

EnsureCexFalsify(candidate C, formulas E, points P, cores U)
while C = ∅ or Eval(C, p) = ⊤ for some p ∈ P do
    if Eval(e, p) = ⊥ for some e ∈ E and u ⊈ C ∪ {e} for all u ∈ U then
        C := C ∪ {e}
    else
        return false
    end if
end while
return true
```

Fig. 2. Procedure for the abduction problem for A, G and R based on unsat core learning.

This procedure is presented in Fig. 2. Similarly to the basic procedure from the previous section, it maintains a set of points P that satisfy the axioms and falsify the goal. Additionally, the new procedure maintains a set E of enumerated formulas, and a set U of subsets of E that are inconsistent with the axioms. The procedure modifies to each of these three sets during the course of its run. Each loop iteration attempts to construct a set C of formulas whose conjunction is a solution to the abduction problem. This is in contrast to the basic procedure from Fig. 1 which considers only individual formulas as candidate solutions.

To construct the candidate set C, the procedure uses a helper function EnsureCexFalsify which ensures that (*i*) C is non-empty, (*ii*) the conjunction of the formulas in C is falsified by each point in P and (*iii*) no subset of C occurs in U. The first condition is to ensure that the candidate is generated by the grammar. The second condition ensures that C along with our axioms suffices to prove the goal. The third condition ensures that C is consistent with the axioms. If we are able to successfully construct a candidate solution set C, then line 6 checks whether that candidate indeed suffices when added to the axioms to show the goal. If it does not, we add a counterexample point to P; otherwise, we construct a (ideally minimal) subset of C_{min} of C that also suffices to show the goal. This information can be readily computed by an SMT solver [10] with support respectively for model generation and for unsatisfiable core generation [13], two features common to most modern solvers, including CVC4. We then check whether C_{min} is consistent with our axioms. If it is consistent, we return it as a solution to the abduction problem; if it is not, we add some subset of it to U that is also inconsistent with the axioms, where again the subset can be computed by an SMT solver with support for unsatisfiable cores. Adding such subset amounts to learning that subset should never be included in future candidate solutions. To maintain the invariant that no subset of C occurs in U, we remove one enumerated formula $e \in u$ from C on line 12. In the case where a point is added to P (line 15) or when an unsat core is added to U (line 12), we run the method EnsureCexFalsify starting from the current resultant set C. This will force the procedure to try to construct a new candidate solution based on the set E. When this strategy fails to construct a candidate, the inner loop terminates and the next formula is added to E based on syntax-guided enumeration.

We now revisit Example 1. As demonstrated in this example, GetAbductUCL is often capable of generating solutions to the abduction problem faster than the one from Fig. 1, albeit those solutions may be logically stronger.

Example 2. We revisit Example 1, where A is the set $\{y \geqslant 0\}$ and G is $\{x+y+z \geqslant 0\}$. A run of the procedure from Fig. 2 is summarized in the table below. We list iterations of the outer loop of the procedure (lines 2–18) in the first column of this table. For each iteration, we provide the formula that is added to our pool E (line 3), and the considered candidate set C upon a successful call to EnsureCexFalsify. Notice that the inner loop of the procedure may consider multiple candidates C for a single iteration of the outer loop. For each candidate, when applicable, we give the result of the evaluation of the condition on line 6, the point p added to P if that condition is false (line 15), the minimal candidate set C_{min} constructed on line 7, the evaluation of the condition on line 8, the set of formulas added to our set of unsatisfiable cores if that condition is false (line 12), and finally the formula (if any) returned as a solution (line 9).

#	$e \in \mathsf{E}$	C	line 6	$p \in \mathsf{P}$	C_{min}	line 8	$u \in \mathsf{U}$	return
1	$x \geqslant 0$	$\{x \geqslant 0\}$	false	$(0, 0, -1)$				
2	$x < 0$	$\{x < 0\}$	false	$(-1, 0, 0)$				
		$\{x < 0, x \geqslant 0\}$	true		C	false	$\{x < 0, x \geqslant 0\}$	
3	$y \geqslant 0$							
4	$y < 0$	$\{y < 0\}$	true		C	false	$\{y < 0\}$	
5	$z \geqslant 0$	$\{x \geqslant 0, z \geqslant 0\}$	true		C	true		$x \geqslant 0 \wedge z \geqslant 0$

We assume the same ordered list of formulas enumerated from Fig. 1. On the first iteration, we add $x \geqslant 0$ to our pool of enumerated formulas E. The helper function EnsureCexFalsify constructs the set $\mathsf{C} = \{x \geqslant 0\}$ since (vacuously) it is true for all points in P. Similar to the first iteration of Fig. 1, on line 6 we learn that $x \geqslant 0$ does not suffice with our axioms to show the goal; a counterexample point is $(x, y, z) = (0, 0, -1)$ which is added to P. Afterwards, EnsureCexFalsify is not capable of constructing another C since there are no other formulas in E. In contrast to Fig. 1 which discards the formula $x \geqslant 0$ at this point, here it remains in E and can be added as part of C in future iterations.

On the second iteration, we add $x < 0$ to our pool. We check the candidate set $\mathsf{C} = \{x < 0\}$, which fails to imply the goal for counterexample point $(x, y, z) = (-1, 0, 0)$. To construct the next candidate set C, we must find an additional formula from E that evaluates to false on this point (or otherwise we again would fail to imply our goal). Indeed, $x \geqslant 0 \in \mathsf{E}$ evaluates to false on this point, and thus EnsureCexFalsify returns the set $\{x < 0, x \geqslant 0\}$. This set suffices to prove the goal given the axioms, that is, the condition on line 6 succeeds; the unsatisfiable core C_{min} computed for this query is the same as C. However, on line 8, we learn that this set is inconsistent with our axioms (in fact, the set by itself is equivalent to false). On line 12, we add $\{x < 0, x \geqslant 0\}$ to U. In other words, we learn that *any* solution that contains both these formulas is inconsistent with our axioms. Learning this subset will help prune later candidate solutions. The procedure on this iteration proceeds by removing one of these formulas from our candidate solution set C. Subsequently the helper function EnsureCexFalsify cannot construct a new candidate subset due to $\{x < 0, x \geqslant 0\} \in \mathsf{U}$ and since no other formulas occur in E.

On the third iteration, $y \geqslant 0$ is added to our pool. However, no candidate solution can be constructed, where notice that $y \geqslant 0$ evaluates to $\top$ on both points in P. On the fourth iteration, $y < 0$ is added to our pool and the candidate solution set $\{y < 0\}$ is constructed, where notice that this formula evaluates to $\bot$ on both points in P. This formula suffices to show the goal from the axioms, but is however inconsistent with our axioms. Thus, $\{y < 0\}$ is added to our set of unsatisfiable cores U. In other words, we have learned that no solution C should include the formula $y < 0$ since it is alone inconsistent with our axioms.

On the fifth iteration, $z \geqslant 0$ is added to our pool. The only viable candidate that falsifies all points in P and does not contain a subset from U is $\{x \geqslant 0, z \geqslant 0\}$. This set is a solution to the abduction problem and so the formula $x \geqslant 0 \wedge z \geqslant 0$ is returned. Due to our assumption that R admits conjunctions, this formula meets the syntax restrictions of our grammar. A run of this procedure required the enumeration of only 5 formulas before finding a solution whereas the basic one in Fig. 1 required 9. □

While the solution in the previous example $x \geqslant 0 \wedge z \geqslant 0$ was found in fewer iterations, notice that it is logically stronger than the solution $x + z \geqslant 0$ produced in Example 1, since $x \geqslant 0 \wedge z \geqslant 0$ entails $x + z \geqslant 0$ but not vice versa. We remark that the main advantage of procedure Fig. 2 is that is typically capable of generating *any* feasible solution to the abduction problem faster than the procedure from Fig. 1. This is especially the case if the only solutions to the abduction problem consist of a large conjunction of literals of small term size $\ell_1 \wedge \ldots \wedge \ell_n$. The basic procedure does not scale to this case, if its enumeration is by formula size, since it will have to wait until the conjunction above is enumerated as an individual formula.

Furthermore, we remark that procedure in this section can be configured to have the same solution completeness guarantees as the basic procedure from Fig. 1. In particular, our choice of e in the EnsureCexFalsify method chooses the most recently enumerated formula when the candidate pool C is empty. Since a single loop of the procedure is terminating and due to the above policy for selection, the procedure will terminate in the worst case when the enumerated pool E contains a formula that by itself is the solution to the synthesis conjecture.

4.3 Incremental Weakening for Abduction

The user may be interested in obtaining an abduction problem solution that maximizes some criteria and is not necessarily the first one discovered by (either of) the procedures we have described so far. In this section, we describe an extension to our approach for abduction that maintains the advantage of returning solutions quickly while still seeking to generate the best solution in the long run according to metric such as logical weakness.

We observe that it is straightforward to extend our enumerative syntax-guided approach to generate *multiple* solutions. We are interested, however, in generating increasingly better solutions over time. We briefly give an overview of how the procedures of Fig. 1 and Fig. 2 can be extended in this way and discuss a few relevant details of the extension. We focus on the problem of generating the *logically weakest* solution to the abduction problem in this section.

Figure 3 presents an incremental procedure for generating (multiple) solutions to a given abduction problem. The procedure requires that the language restriction R admit disjunctive formulas which is the case, for instance, if $s_0 \rightarrow s_0 \vee s_0$ is a production rule in R where s_0 is again the start symbol. It maintains a formula S that, when not $\bot$, represents the logically weakest solution to the abduction problem known so far. In its main loop, on line 3, the procedure calls one of the previous procedures for generating single solutions to the abduction problem (written GetAbduct*). Line 4 then checks whether a new solution can be constructed that is logically weaker with respect to the axioms than the current one. In particular, this is the case if $\mathsf{C} \wedge \mathsf{A} \wedge \neg\mathsf{S}$ is T-satisfiable, which means that there is at least one point that satisfies the current candidate but not the current solution S. In that case, the current solution S is updated to $\mathsf{S} \vee \mathsf{C}$, which is by construction guaranteed to also be a solution to the abduction problem. If no such point can be found, then C is redundant with respect to the current candidate solution since it does not generalize it. Optionally, the procedure may learn a subset u of C that is also redundant with respect to the current candidate solution. This subset can be learned as an unsatisfiable core when using the procedure GetAbductUCL as the sub-procedure on line 3.

```
GetAbductInc(axioms A, goal G, grammar R)
Let S = ⊥
loop
   Let C = GetAbduct*(A, G, R).
   if C ∧ A ∧ ¬S is T-satisfiable then
      S := S ∨ C
      print Weakest solution so far is S
   else
      // Exclude {u} for some u ⊆ C such that u ∧ A ∧ ¬S is T-unsatisfiable
   end if
end loop
```

Fig. 3. Incremental abduction procedure for axioms A, goal G and grammar R.

4.4 Implementation Details

We implemented the procedures above in the state-of-the-art SMT solver CVC4 [8]. CVC4 incorporates a SyGuS solver, CVC4SY, implementing several strategies for enumerative syntax-guided synthesis [31]. It accepts as input both SMT problems written in the SMT-LIB version 2.6 format [9], and synthesis problems written in the SyGuS version 2.0 format [30]. SMT-LIB version 2.6 is a scripting language that allows one to assert a formula F to the solver with a command of the form (assert F). The solver checks the satisfiability the formulas asserted so far in response to the command (check-sat). We extended CVC4's SMT-LIB parser to support also commands of the form (get-abduct p G R) where p is a symbol, the identifier of the expected solution formula; G is a formula, the goal of the abduction problem; and the optional R is a grammar expressed in the SyGuS version 2.0 format. This command asks the solver to find a formula that is a solution to the abduction problem (A, G), where A, standing for the set of axioms, consists of the conjunction of the currently asserted formulas. The expected response from the solver is a definition of the form (define-fun p () Bool S) where p is the identifier provided in the first argument of get-abduct and S is a formula that solves the abduction problem.

Internally, invoking a get-abduct command causes a synthesis conjecture to be constructed and passed to CVC4SY. The latter normally accepts conjectures of the form $\exists f. \forall \boldsymbol{x}. \varphi[f, \boldsymbol{x}]$ where φ is quantifier-free. Thus, we must pass the abduction problem in two parts: (i) the *synthesis conjecture* $\exists P. \forall \boldsymbol{x}. \neg(P(\boldsymbol{x}) \wedge A \wedge \neg G)$ where $\boldsymbol{x}$ collects the free variables of A and of G,[2] stating that the expected solution P along with the axioms A must entail the goal G, and (ii) a *side condition* $\exists \boldsymbol{x}. P(\boldsymbol{x}) \wedge A$ stating that P must be consistent with the axioms. The synthesis conjecture is of a form that can be readily handled by CVC4SY and processed using its current techniques. We have modified it so that it considers the side condition as well during solving, as described in Figs. 1 and 2.

[2] We assume that all free symbols in A and G are variables.

The procedure in Fig. 2 is implemented as a strategy on top of the basic enumerative CEGIS loop of CVC4SY. We give some noteworthy implementation details here. Firstly, we use a data structure for efficiently checking whether any subset of C occurs in our set of unsatisfiable cores U, which keeps the sets in U in an index and is traversed dynamically as formulas are added to C. We chose enumerated formulas on line 2 of EnsureCexFalsify by selecting first the most recently generated formula, and then a random one amongst those that meet the criteria to be included in C. Finally, since the number of candidate solutions can be exponential in the worst case for a given iteration of the inner loop of this procedure, we use a heuristic where formulas cannot be added to C more than once in the same iteration of the loop, making the number of candidate sets tried on a given iteration linear in the size of E in the worst case.

5 Evaluation

We evaluated our approach[3] in comparison with CVC4SY's enumerative CEGIS, a general purpose synthesis approach, as well as with GPID [19] and EXPLAIN [15], state-of-the-art solvers for similar abduction problems as the one defined here. In the comparison below, we refer to the basic procedure from Fig. 1 as CVC4SY+B and the one from Fig. 2 as CVC4SY+U. Experiments ran on a cluster with Intel E5-2637 v4 CPUs, Ubuntu 16.04. Each execution of a solver on a benchmark was provisioned one core, 300 s and 8 GB RAM.

5.1 Benchmarks

Since abduction tools are generally focused on specific application domains, there is no standard language or benchmark library for evaluation. Moreover, these tools use abduction as part of a larger verification toolchain. As here we did not target a specific application but rather the abduction problem as a whole, an evaluation with their benchmarks would require integrating our solver in the tools as an alternative abduction engine. This was not feasible due to either the source code not being available or the verification and abduction engines being too tightly coupled for us to use our solver as an alternative. Thus we had to generate our own abduction benchmark sets. We did so using benchmarks relevant for verification from SMT-LIB [9], the standard test suite for SMT solvers. We chose as a basis the SMT-LIB logics QF_LIA, QF_NIA, and QF_SLIA due to their relevance for verification. For QF_NIA, we focus on the benchmark family VeryMax and on kaluza for QF_SLIA. In QF_LIA we excluded benchmark families whose benchmarks explode in size without the let operator. This was necessary to allow a comparison with EXPLAIN, whose parser does not fully support let, on let-free benchmarks. We considered both benchmarks that were (annotated as) satisfiable and unsatisfiable for generating abduction problems, according to the following methodology.

[3] Full material at http://cvc4.cs.stanford.edu/papers/abduction-sygus/.

Given a *satisfiable* SMT-LIB problem $\varphi = \psi_1 \wedge \cdots \wedge \psi_n$[4] in the theory T, we see it as an encoding of a validity problem $\psi_1 \wedge \cdots \wedge \psi_{n-1} \models \neg\psi_n$ that could not be proven. We consider the abduction problem where G is $\neg\psi_n$, A is $\psi_1 \wedge \cdots \wedge \psi_{n-1}$, and R is a grammar that generates any quantifier-free formula in the language of T over the free variables of G and A. A solution S to this problem allows the validity of φ to be proven, since $\varphi \wedge \mathsf{S}$ is unsatisfiable.

Given an *unsatisfiable* SMT-LIB problem φ, let $U = \{\psi_1, \ldots, \psi_n\}$ be a *minimal* unsatisfiable core for this formula, i.e. any conjunctive set $U \setminus \{\psi\}$, for some $\psi \in U$, is satisfiable. Let $\psi_{\max}$ be U's component with maximal size. We will call $\psi_{\max}$ the *reference* to the abduction problem. We consider the abduction problem whose G is $\neg\psi_{\mathsf{G}}$, for some $\psi_{\mathsf{G}} \in U$ and $\psi_{\mathsf{G}} \neq \psi_{\max}$, whose axioms A are $U \setminus \{\psi_{\mathsf{G}}, \psi_{\max}\}$ and R as before is a grammar that generates any formula in the language of T over the free variables of G and A. A solution S to this problem allows proving the validity of $U \setminus \{\psi_{\mathsf{G}}, \psi_{\max}\} \models \psi_{\mathsf{G}}$, since $U \setminus \{\psi_{\max}\} \cup \{\mathsf{S}\}$ is unsatisfiable. Solving this abduction problem amounts to "completing" the original unsatisfiable core with the further restriction that this completion is at least as weak as the reference, as well as consistent with all but one of the other core components, seen as axioms for the abduction problem.

From satisfiable SMT-LIB benchmarks we generated 2025 abduction problems in QF_LIA, 12214 in QF_NIA and 11954 in QF_SLIA. For unsatisfiable benchmarks we were limited not only by the benchmark annotations but also by being able to find minimal unsatisfiable cores. We used the Z3 SMT solver [14] to generate minimal unsatisfiable cores with a 120s timeout. Excluding benchmarks whose cores had less than three assertions (so we could have axioms, a goal and a reference), we ended up with 97 problems in QF_LIA, 781 in QF_NIA and 2546 in QF_SLIA. We chose the reference as the component of the unsatisfiable core with maximal size and the goal as the last formula in the core (viewed as a list) after the reference was removed.

Table 1. Comparison of abduction problems from originally SAT SMT-LIB benchmarks.

Logic	#	CVC4SY+B			CVC4SY+U		
		Solved	Unique	Weaker	Solved	Unique	Weaker
QF_LIA	2025	721	261	183	594	134	2
QF_SLIA	11954	10902	3	466	10980	81	0
QF_NIA	12214	1492	171	671	1712	391	45
Total	26593	13329	435	1320	13628	606	47

5.2 Finding Missing Assumptions in SAT Benchmarks

In this section we evaluate how effective CVC4SY+B and CVC4SY+U are in (i) finding any solution to the abduction problem and (ii) finding logically weak

[4] SMT-LIB problems are represented as sequences of assertions. Here we considered each ψ_i as one of these assertions.

solutions. The evaluation is done on the abduction problems produced from satisfiable SMT-LIB benchmarks as above. Results are summarized in Table 1. The number of solved problems corresponds to the problems for which a given CVC4 configuration could find a solution within 300s. CVC4SY+U solves a significant number of problems more than CVC4SY+B in all logics but QF_LIA. In both QF_LIA and QF_NIA we can see a significant orthogonality between both approaches. We attribute these both to the fragility of integer arithmetic reasoning, where the underlying ground solver checking the consistency of candidate solutions is greatly impacted by the shape of the problems it is given. Overall, the procedure in CVC4SY+U leads to a better success rate than the basic procedure in CVC4SY+B. Solution strength was evaluated on commonly solved problems considering the solutions produced according to the incremental procedures shown in Sect. 4.3, in which the overall solution is a disjunction of individual solutions found over time. As expected, CVC4SY+U is able to solve more problems but at the cost of often producing stronger (and bigger) solutions than CVC4SY+B. This is particularly the case in QF_SLIA and QF_NIA, in which CVC4SY+U both solves many more problems and often finds stronger solutions.

5.3 Completing UNSAT Cores

Here we evaluate how effective CVC4SY+B and CVC4SY+U are in solving the abduction problem with the extra restriction of finding a solution that is at least as weak as a given reference formula. We use the abduction problems produced from unsatisfiable SMT-LIB benchmarks following the methodology of Sect. 5.1 as the basis for this evaluation.

Table 2. Comparison of abduction problems from originally UNSAT SMT-LIB benchmarks.

Logic	#	CVC4SY+B		CVC4SY+U	
		Solved	Unique	Solved	Unique
QF_LIA	97	6	0	6	0
QF_SLIA	2546	2546	32	2514	0
QF_NIA	781	86	49	41	4
Total	3424	2638	81	2561	4

The results are summarized in Table 2. CVC4SY+B significantly outperforms CVC4SY+U in QF_SLIA, in which the references are very simple formulas (generally with size below 3), for which the specialized procedure of CVC4SY+U is not necessary. Overall, as in the previous section when checking who finds the weakest solution, CVC4SY+B has as advantage over CVC4SY+U for finding solutions as weak as the reference.

5.4 Comparison with Explain

EXPLAIN [15] is a tool for abductive inference based on quantifier elimination. It accepts as input a subset of SMT-LIB and we extended it to support abduction problems as generated in Sect. 5.1. However, EXPLAIN imposes more restrictions to their solutions, only producing those with a *minimal* number of variables and for which every other solution with those variables is not stronger than it. Their rationale is finding "simple" solutions, according to the above criteria, which are more interesting to their applications. Since we do not apply these restrictions in CVC4, nor is in the scope of this paper incorporating them into our procedure, it should be noted that comparing CVC4 and EXPLAIN puts the latter at a disadvantage. We considered satisfiable SMT-LIB problems in the QF_LIA logic for our evaluation, as QF_LIA is better supported by EXPLAIN (Table 3).

Table 3. Comparison with EXPLAIN in 2025 abduction problems in QF_LIA

	Solved	Unique	Total time
CVC4SY+B	721	261	418849 s
CVC4SY+U	594	125	449424 s
EXPLAIN	33	0	532839 s

All problems solved by EXPLAIN are solved by CVC4SY+U. Of these 33 problems, CVC4SY+U, in incremental mode, finds a solution with the same minimal number of variables as EXPLAIN for 25 of them. Of the 8 problems to which it only finds solutions with more variables, in 4 of them the difference is of a single variable. All other 4 are in the slacks benchmark family, which contains crafted problems. A similar comparison occurs with CVC4SY+B. This shows that even though CVC4 is not optimized to minimize the number of variables it its solutions, it can still often finds solutions that are optimal (or close to optimal) according to EXPLAIN's criteria, while solving a much larger number of problems with a fully general approach.

5.5 Comparison with GPiD

We also compared CVC4 with GPID [19], a framework for generating implicates, i.e. logical consequences of formulas. As Echenim et al. say in their paper, negating the implicate of a satisfiable formula φ yields the "missing hypothesis" for making φ unsatisfiable. Therefore GPID solves a similar problem to that of Sect. 5.2, differing by they always considering an empty set of axioms and the whole original formula as the goal. Given this similarity, we compare the performance of GPID in generating implicates for satisfiable benchmarks and of CVC4SY+B and CVC4SY+U in solving abduction problems generated from those same benchmarks. We did not consider the benchmarks from the previous sections because we were not able to produce *abduces*, which are the syntactic components GPiD uses to find implicates, for other logics using the tools in GPiD public repository[5]. Thus we restricted our analysis to 400 abduction problems produced, as per the methodology of Sect. 5.1, from satisfiable QF_UFLIA benchmarks that were used in [19]. Note however that the CVC4 configurations will require solutions to be consistent with all but the last assertion in the problems (which are the axioms in the respective abduction problem). Since that, as far as we know, this is not a requirement in GPiD, effectively CVC4SY+B and CVC4SY+U are solving a harder problem than GPiD. We formulated the abduction problem this way, rather than as with all assertions as goals, to avoid trivializing the abduction problem, for which the negation of the goal would always be a solution. Also note that the presence of uninterpreted functions in the abduction problem requires solutions to be generated in a higher-order background logic, which CVC4 supports after a recent extension [7]. As in [19], we used GPiD's version with the Z3 backend. We present their results with (GPiD-1) and without (GPiD) the restriction to limit the set of abduces to size 1.

Table 4. Comparison with GPiD on 400 abduction problems in the QF_UFLIA logic.

	Solved	Unique	Total time
CVC4SY+B	214	0	57290 s
CVC4SY+U	342	0	18735 s
GPiD	193	0	69 s
GPiD-1	398	54	1188 s

Results are summarized in Table 4. CVC4SY+U significantly outperforms CVC4SY+B, both in the number of problems solved and in total time, besides being almost 20% faster on commonly solved problems. We also see that solution finding in GPiD is heavily dependent on which abduces are considered when building solutions, as it solves almost all benchmarks when limited to abduces of size 1 but barely half when unrestricted. It should also be noted that GPiD takes *pre-computed* abduces, whose production time is not accounted for in the evaluation. Despite this, CVC4SY+U is only on average 30% slower on commonly solved problems than GPiD-1 and solves many more problems than GPiD. The big variation of GPiD results in terms of what pre-determined set of candidates can be used in the computation is a severe limitation of their tool. Similarly, while the method proposed in [19] is theory agnostic, their tooling for producing abduces imposes strong limitations on the usage of GPiD for theories other than QF_UFLIA.

[5] At https://github.com/sellamiy/GPiD-Framework.

6 Related Work

The procedure introduced in Sect. 4.2 based on unsat core learning follows a recent trend in enumerative syntax-guided synthesis solving that aims to improve scalability by applying divide-and-conquer techniques, where candidate solutions are built from smaller enumerated pieces rather than being directly enumerated. While previous approaches, both for pointwise [5,27] and for unrestricted specifications [6], have targeted general-purpose function synthesis, we specialize divide and conquer for solving the abduction problem with a lean (see Sect. 4.4) and effective (see Sect. 5) procedure.

Abductive inference tools for the propositional case include the AbHS and AbHS+ tools [26,33], based on SAT solvers [11] and hitting set procedures, and the Hyper [23] tool, that includes a series of algorithmic improvements over the former, and uses a MaxSAT solver for computing the hitting set. Like AbHS, GetAbductUCL checks entailment and consistency using two separate calls to the underlying solver, and uses its failures for the selection of new candidates. In contrast, GetAbductUCL keeps this information in two dedicated data structures rather than encoding it with an implicit hitting set. Another significant difference is that the set of hypotheses in the propositional case is fixed and finite, whereas in our setting it is generated dynamically from a grammar. More general approaches, to which our work bears more resemblance and to which we provided an experimental comparison in Sect. 5, are GPiD [19] and Explain [15]. GPiD uses an off-the-shelf SMT solver as a black box to generate ground implicates. It can be used with any theory supported by the underlying SMT solver, similarly to our SyGuS-based approach. While we enumerate formulas that compose the solution for the abduction problem GPiD's authors use *abducibles*, which are equalities and disequalities over the variables in the problem. They similarly build candidates in a refinement loop by combining abducibles according to consistency checks performed by an underlying SMT solver. They use an order on abducibles to guide the search, which is analogous to the enumeration order in enumerative synthesis. Explain on the other hand is built on top of an SMT solver for the theories of linear integer arithmetic and of equality with uninterpreted functions, although its abduction inference procedure in principle can work with any theory that admits quantifier elimination. The method implemented in Explain is based on first determining a subset of the variables in the abduction problem and trying to build the weakest solution over these variables via quantifier elimination, while computing minimal satisfying assignments to ensure that a found solution covers a minimal subset. This method, however, is not complete, as it can miss solutions. The tool also allows the user to specify costs for each variable, so that a given minimal set may be favored.

7 Conclusion

We have described approaches for solving the abduction problem using a modern enumerative solver for syntax-guided synthesis. Our evaluation shows that procedures based on enumerative CEGIS scale for several non-trivial abduction tasks, and have several compelling advantages with respect to other approaches like those used in EXPLAIN and GPID. In several cases, it suffices to use a basic procedure for enumerative CEGIS to generate solutions to abduction problems that are optimal according to certain metrics. Moreover, the generation of feasible solutions can be complemented and accelerated via a procedure for generating conjunctions of enumerated formulas as shown in Fig. 2.

We believe that new abduction capabilities presented in this paper and implemented in CVC4 will be useful in all the applications of abduction we describe in Sect. 3. In addition, we see a number of promising applications in the context of SMT itself. For example, we plan to use abduction to generate useful *conditional rewrite rules* for SMT solvers. Many such rules are used internally by SMT solvers to simplify their input formulas by (equivalence-preserving) term rewriting. The manual identification and selection of good rewrite rules is a tedious and error-prone process. Abduction can be used to generalize a recent approach for the semi-automated development of rewrite rules [28] by synthesizing (most general) conditions under which two terms are equivalent. This in turn can be used to develop new solving strategies in the SMT solver based on those rewrite rules.

References

1. Albarghouthi, A., Dillig, I., Gurfinkel, A.: Maximal specification synthesis. ACM SIGPLAN Not. **51**(1), 789–801 (2016)
2. Alur, R., et al.: Syntax-guided synthesis. In: Formal Methods in Computer-Aided Design (FMCAD), pp. 1–8. IEEE (2013)
3. Alur, R., Černý, P., Radhakrishna, A.: Synthesis through unification. In: Kroening, D., Păsăreanu, C.S. (eds.) CAV 2015. LNCS, vol. 9207, pp. 163–179. Springer, Cham (2015). https://doi.org/10.1007/978-3-319-21668-3_10
4. Alur, R., Fisman, D., Singh, R., Solar-Lezama, A.: SyGuS-comp 2017: results and analysis. In: Proceedings Sixth Workshop on Synthesis, SYNT@CAV 2017, Heidelberg, Germany, 22 July 2017, pp. 97–115 (2017)
5. Alur, R., Radhakrishna, A., Udupa, A.: Scaling enumerative program synthesis via divide and conquer. In: Legay, A., Margaria, T. (eds.) TACAS 2017. LNCS, vol. 10205, pp. 319–336. Springer, Heidelberg (2017). https://doi.org/10.1007/978-3-662-54577-5_18
6. Barbosa, H., Reynolds, A., Larraz, D., Tinelli, C.: Extending enumerative function synthesis via SMT-driven classification. In: Barrett, C.W., Yang, J. (eds.) Formal Methods in Computer-Aided Design (FMCAD), pp. 212–220. IEEE (2019)
7. Barbosa, H., Reynolds, A., El Ouraoui, D., Tinelli, C., Barrett, C.: Extending SMT solvers to higher-order logic. In: Fontaine, P. (ed.) CADE 2019. LNCS (LNAI), vol. 11716, pp. 35–54. Springer, Cham (2019). https://doi.org/10.1007/978-3-030-29436-6_3
8. Barrett, C., et al.: CVC4. In: Gopalakrishnan, G., Qadeer, S. (eds.) CAV 2011. LNCS, vol. 6806, pp. 171–177. Springer, Heidelberg (2011). https://doi.org/10.1007/978-3-642-22110-1_14
9. Barrett, C., Fontaine, P., Tinelli, C.: The SMT-LIB Standard: Version 2.6. Technical report, Department of Computer Science, The University of Iowa (2017). www.SMT-LIB.org

10. Barrett, C., Sebastiani, R., Seshia, S., Tinelli, C.: Satisfiability modulo theories (chap. 26). In Biere, A., Heule, M.J.H., van Maaren, H., Walsh, T. (eds.) Handbook of Satisfiability. Frontiers in Artificial Intelligence and Applications, vol. 185, pp. 825–885. IOS Press (2009)
11. Biere, A., Heule, M., van Maaren, H., Walsh, T.: Handbook of Satisfiability: Volume 185 Frontiers in Artificial Intelligence and Applications. IOS Press, Amsterdam (2009)
12. Calcagno, C., Distefano, D., O'Hearn, P., Yang, H: Compositional shape analysis by means of bi-abduction. In: Proceedings of the 36th Annual ACM SIGPLAN-SIGACT Symposium on Principles of Programming Languages, pp. 289–300 (2009)
13. Cimatti, A., Griggio, A., Sebastiani, R.: Computing small unsatisfiable cores in satisfiability modulo theories. J. Artif. Intell. Res. (JAIR) **40**, 701–728 (2011)
14. de Moura, L., Bjørner, N.: Z3: an efficient SMT solver. In: Ramakrishnan, C.R., Rehof, J. (eds.) TACAS 2008. LNCS, vol. 4963, pp. 337–340. Springer, Heidelberg (2008). https://doi.org/10.1007/978-3-540-78800-3_24
15. Dillig, I., Dillig, T.: EXPLAIN: a tool for performing abductive inference. In: Sharygina, N., Veith, H. (eds.) CAV 2013. LNCS, vol. 8044, pp. 684–689. Springer, Heidelberg (2013). https://doi.org/10.1007/978-3-642-39799-8_46
16. Dillig, I., Dillig, T., Aiken, A.: Automated error diagnosis using abductive inference. ACM SIGPLAN Not. **47**(6), 181–192 (2012)
17. Dillig, I., Dillig, T., Li, B., McMillan, K.: Inductive invariant generation via abductive inference. ACM SIGPLAN Not. **48**(10), 443–456 (2013)
18. Dillig, T., Dillig, I., Chaudhuri, S.: Optimal guard synthesis for memory safety. In: Biere, A., Bloem, R. (eds.) CAV 2014. LNCS, vol. 8559, pp. 491–507. Springer, Cham (2014). https://doi.org/10.1007/978-3-319-08867-9_32
19. Echenim, M., Peltier, N., Sellami, Y.: A generic framework for implicate generation modulo theories. In: Galmiche, D., Schulz, S., Sebastiani, R. (eds.) IJCAR 2018. LNCS (LNAI), vol. 10900, pp. 279–294. Springer, Cham (2018). https://doi.org/10.1007/978-3-319-94205-6_19
20. Echenim, M., Peltier, N., Sellami, Y.: Ilinva: using abduction to generate loop invariants. In: Herzig, A., Popescu, A. (eds.) FroCoS 2019. LNCS (LNAI), vol. 11715, pp. 77–93. Springer, Cham (2019). https://doi.org/10.1007/978-3-030-29007-8_5
21. Enderton, H.B.: A Mathematical Introduction to Logic, 2nd edn. Academic Press, Cambridge (2001)
22. Gulwani, S.: Programming by examples: applications, algorithms, and ambiguity resolution. In: Olivetti, N., Tiwari, A. (eds.) IJCAR 2016. LNCS (LNAI), vol. 9706, pp. 9–14. Springer, Cham (2016). https://doi.org/10.1007/978-3-319-40229-1_2
23. Ignatiev, A., Morgado, A., Marques-Silva, J.: Propositional abduction with implicit hitting sets. In: ECAI 2016–22nd European Conference on Artificial Intelligence, 29 August–2 September 2016, The Hague, The Netherlands - Including Prestigious Applications of Artificial Intelligence, PAIS 2016, pp. 1327–1335 (2016)
24. Josephson, J.R., Josephson, S.G. (eds.): Abductive Inference: Computation, Philosophy, Technology. Cambridge University Press, Cambridge (1994)
25. Li, B., Dillig, I., Dillig, T., McMillan, K., Sagiv, M.: Synthesis of circular compositional program proofs via abduction. In: Piterman, N., Smolka, S.A. (eds.) TACAS 2013. LNCS, vol. 7795, pp. 370–384. Springer, Heidelberg (2013). https://doi.org/10.1007/978-3-642-36742-7_26
26. Moreno-Centeno, E., Karp, R.M.: The implicit hitting set approach to solve combinatorial optimization problems with an application to multigenome alignment. Oper. Res. **61**(2), 453–468 (2013)

27. Neider, D., Saha, S., Madhusudan, P.: Synthesizing piece-wise functions by learning classifiers. In: Chechik, M., Raskin, J.-F. (eds.) TACAS 2016. LNCS, vol. 9636, pp. 186–203. Springer, Heidelberg (2016). https://doi.org/10.1007/978-3-662-49674-9_11
28. Nötzli, A., et al.: Syntax-guided rewrite rule enumeration for SMT solvers. In: Janota, M., Lynce, I. (eds.) SAT 2019. LNCS, vol. 11628, pp. 279–297. Springer, Cham (2019). https://doi.org/10.1007/978-3-030-24258-9_20
29. Phothilimthana, P.M., Thakur, A., Bodík, R., Dhurjati, D.: Scaling up superoptimization. In: Proceedings of the Twenty-First International Conference on Architectural Support for Programming Languages and Operating Systems, ASPLOS 2016, Atlanta, GA, USA, 2–6 April 2016, pp. 297–310 (2016)
30. Raghothaman, M., Reynolds, A., Udupa, A.: The SyGuS language standard version 2.0 (2019)
31. Reynolds, A., Barbosa, H., Nötzli, A., Barrett, C., Tinelli, C.: CVC4SY: smart and fast term enumeration for syntax-guided synthesis. In: Dillig, I., Tasiran, S. (eds.) CAV 2019. LNCS, vol. 11562, pp. 74–83. Springer, Cham (2019). https://doi.org/10.1007/978-3-030-25543-5_5
32. Reynolds, A., Deters, M., Kuncak, V., Tinelli, C., Barrett, C.: Counterexample-guided quantifier instantiation for synthesis in SMT. In: Kroening, D., Păsăreanu, C.S. (eds.) CAV 2015. LNCS, vol. 9207, pp. 198–216. Springer, Cham (2015). https://doi.org/10.1007/978-3-319-21668-3_12
33. Saikko, P., Wallner, J.P., Järvisalo, M.: Implicit hitting set algorithms for reasoning beyond NP. In: Principles of Knowledge Representation and Reasoning: Proceedings of the Fifteenth International Conference, KR 2016, Cape Town, South Africa, 25–29 April 2016, pp. 104–113 (2016)
34. Solar-Lezama, A., Rabbah, R.M., Bodík, R., Ebcioglu, K.: Programming by sketching for bit-streaming programs. In: Sarkar, V., Hall, M.W. (eds.) Conference on Programming Language Design and Implementation (PLDI), pp. 281–294. ACM (2005)
35. Solar-Lezama, A., Tancau, L., Bodík, R., Seshia, S.A., Saraswat, V.A.: Combinatorial sketching for finite programs. In: Shen, J.P., Martonosi, M. (eds.) Architectural Support for Programming Languages and Operating Systems (ASPLOS), pp. 404–415. ACM (2006)
36. Udupa, A., Raghavan, A., Deshmukh, J.V., Mador-Haim, S., Martin, M.M.K., Alur, R.: TRANSIT: specifying protocols with concolic snippets. In: ACM SIGPLAN Conference on Programming Language Design and Implementation, PLDI 2013, Seattle, WA, USA, 16–19 June 2013, pp. 287–296 (2013)
37. Zhu, H., Dillig, T., Dillig, I.: Automated inference of library specifications for source-sink property verification. In: Shan, C. (ed.) APLAS 2013. LNCS, vol. 8301, pp. 290–306. Springer, Cham (2013). https://doi.org/10.1007/978-3-319-03542-0_21

Decision Procedures and Combination of Theories

Deciding the Word Problem for Ground Identities with Commutative and Extensional Symbols

Franz Baader[1(✉)] and Deepak Kapur[2]

[1] Theoretical Computer Science, TU Dresden, Dresden, Germany
franz.baader@tu-dresden.de
[2] Department of Computer Science, University of New Mexico, Albuquerque, USA
kapur@cs.unm.edu

Abstract. The word problem for a finite set of ground identities is known to be decidable in polynomial time, and this is also the case if some of the function symbols are assumed to be commutative. We show that decidability in P is preserved if we also assume that certain function symbols f are *extensional* in the sense that $f(s_1, \ldots, s_n) \approx f(t_1, \ldots, t_n)$ implies $s_1 \approx t_1, \ldots, s_n \approx t_n$. In addition, we investigate a variant of extensionality that is more appropriate for commutative function symbols, but which raises the complexity of the word problem to coNP.

1 Introduction

One motivation for this work stems from Description Logic (DL) [1], where constant symbols (called individual names) are used within knowledge bases to denote objects or individuals in an application domain. If such objects are composed of other objects, it makes sense to represent them as (ground) terms rather than constants. For example, the couple consisting of individual a in the first component and individual b in the second component is more reasonably represented by the term $f(a, b)$ (where f is a binary function symbol denoting the couple constructor) than by a third constant c that is unrelated to a and b. In fact, if we have two couples, one consisting of a and b and the other of a' and b', and we learn (by DL reasoning or from external sources) that a is equal to a' and b is equal to b', then this automatically implies that $f(a, b)$ is equal to $f(a', b')$, i.e., that this is one and the same couple, whereas we would not obtain such a consequence if we had introduced constants c and c' for the two couples.

If we use terms to represent objects, and can learn (e.g., by DL reasoning) that two terms are supposed to be equal, we need to be able to decide which other identities between terms can be derived from the given ones. Fortunately, this problem (usually called the *word problem for ground identities*) is decidable in polynomial time. The standard approach for deciding this word problem is *congruence closure* [3,5,10,12]. Basically, congruence closure starts with the

---

F. Baader—Partially supported by DFG, Grant 389792660, within TRR 248.

N. Peltier and V. Sofronie-Stokkermans (Eds.): IJCAR 2020, LNAI 12166, pp. 163–180, 2020.
https://doi.org/10.1007/978-3-030-51074-9_10

given set of ground identities E, and then extends it using closure under reflexivity, symmetry, transitivity, and congruence. The set $CC(E)$ obtained this way is usually infinite, and the main observation that yields decidability in polynomial time is that one can restrict it to the subterms of E and the subterms of the terms for which one wants to decide the word problem. An alternative approach for deciding the word problem for ground identities is based on *term rewriting*. Basically, in this approach one generates an appropriate canonical term rewriting system from E, and then decides whether two terms are equal modulo the theory E by computing their canonical forms and checking whether they are syntactically equal. This was implicit in [15], and made explicit in [7] (see also [6,16] for other rewriting-based approaches).

In the motivating example from DL, but also in other settings where congruence closure is employed (such as SMT [13,17]), it sometimes makes sense to assume that certain function symbols satisfy additional properties that are not expressible by finitely many ground identities. For example, one may want to considered couples where the order of the components is irrelevant, which means that the couple constructor function is commutative. Another interesting property for (ordered) couples is extensionality: if two couples are equal then they must have the same first and second components, i.e., the couple constructor f must satisfy the extensionality rule $f(x, y) \approx f(x', y') \Rightarrow x \approx x' \wedge y \approx y'$. While it is known that adding commutativity does not increase the complexity (see, e.g., [5,8]), extensionality has, to the best of our knowledge, not been considered in this context before. The problem with extensionality is that it allows us to derive "small" identities from larger ones. Consequently, it is conceivable that one first needs to generate such large identities using congruence and applying other rules, before one can get back to a smaller one through the application of extensionality. Thus, it is not obvious that also with extensionality one can restrict congruence closure to a finite set of terms determined by the input. Here, we will tackle this problem using a rewriting-based approach. Our proofs imply that, also with extensional symbols, proofs of identities that detour through "large" terms can be replaced by proofs using only "small" terms, but it is not clear how this could be shown directly without the rewriting-based approach.

In the next section, we show how the rewriting-based approach of [7] can be extended such that it can also handle commutative symbols. In contrast to approaches that deal with associative-commutative (AC) symbols [4,11] using rewriting modulo AC, we treat commutativity by introducing an additional rewrite system consisting of appropriately ordered ground instances of commutativity. This sets the stage for our rewriting-based approach that works in the presence of commutative symbols and extensional symbols presented in Sect. 4. In this section, we do not consider symbols f that are both commutative and extensional since extensionality as defined until now is not appropriate for commutative symbols: for arbitrary terms s, t, commutativity yields $f(s, t) \approx f(t, s)$, and thus extensionality implies $s \approx t$, which shows that the equational theory becomes trivial. In Sect. 5, we introduce the notion of c-extensionality, which is more appropriate for commutative symbols. Whereas the approaches developed in Sects. 3 and 4 yield polynomial time decision procedures for the word problem, c-extensionality makes the word problem coNP-complete.

Due to space constraints not all proofs can be given here. Detailed proofs can be found in [2].

2 Preliminaries on Equational Theories and Term Rewriting

We assume that the reader is familiar with basic notions and results regarding equational theories, universal algebra, and term rewriting. Here, we briefly recall the most important notions and results and refer the readers to [3] for details. We will keep as close as possible to the notation introduced in [3]. In particular, we use $\approx$ to denote identities between terms and $=$ to denote syntactic equality.

Terms are built as usual from variables, constants, and function symbols. An *identity* is a pair of terms (s, t), which we usually write as $s \approx t$. A *ground term* is a term not containing variables and a *ground identity* is a pair of ground terms. Given a set of identities E, the equational theory induced by E is defined (semantically) as $\approx_E := \{s \approx t \mid \text{every models of } E \text{ is a model of } s \approx t\}$. The notion of *model* used here is the usual one from first-order logic, where we assume that identities are (implicitly) universally quantified. Since we consider signatures consisting only of constant and function symbols, we call first-order interpretations *algebras*.

Birkhoff's theorem provides us with an alternative characterization of $\approx_E$ that is based on rewriting. A given set of identities E induces a binary relation $\rightarrow_E$ on terms. Basically, we have $s \rightarrow_E t$ if there is an identity $\ell \approx r$ in E such that s contains a substitution instance $\sigma(\ell)$ of ℓ as subterm, and t is obtained from s by replacing this subterm with $\sigma(r)$. Birkhoff's theorem says that $\approx_E$ is identical to $\stackrel{*}{\leftrightarrow}_E$, where $\stackrel{*}{\leftrightarrow}_E$ denotes the reflexive, transitive, and symmetric closure of $\rightarrow_E$.

If $\rightarrow_E$ is canonical (i.e., terminating and confluent), then we have $s \stackrel{*}{\leftrightarrow}_E t$ iff s and t have the same canonical forms. The *canonical form* of a term s is an irreducible term $\widehat{s}$ such that $s \stackrel{*}{\rightarrow}_E \widehat{s}$, where $\stackrel{*}{\rightarrow}_E$ denotes the reflexive and transitive closure of $\rightarrow_E$ and $\widehat{s}$ is *irreducible* if there is no s' with $\widehat{s} \rightarrow_E s'$. Termination ensures that the canonical form exists and confluence that it is unique. The relation $\rightarrow_E$ is *confluent* if $s \stackrel{*}{\rightarrow}_E t_1$ and $s \stackrel{*}{\rightarrow}_E t_2$ imply that there is a term t such that $t_1 \stackrel{*}{\rightarrow}_E t$ and $t_2 \stackrel{*}{\rightarrow}_E t$. It is *terminating* if there is no infinite chain $t_0 \rightarrow_E t_1 \rightarrow_E t_2 \rightarrow_E \cdots$.

Termination can be proved using a so-called *reduction order*, which is a well-founded order $>$ on terms such that $\ell > r$ for all $\ell \approx r \in E$ implies $s > t$ for all terms s, t with $s \rightarrow_E t$. Since $>$ is well-founded this then implies termination. If $\rightarrow_E$ is terminating, then confluence can be tested by checking whether all critical pairs of E are joinable. Basically, *critical pairs* (t_1, t_2) consider the most general forks of the form $s \rightarrow_E t_1$ and $s \rightarrow_E t_2$ that are due to overlapping left-hand sides of identities. Such a pair is *joinable* if there is a term t such that $t_1 \stackrel{*}{\rightarrow}_E t$ and $t_2 \stackrel{*}{\rightarrow}_E t$.

Usually, when considering the relation $\rightarrow_E$, one calls E a *term rewriting system* rather than a set of identities, and writes its elements (called rewrite rules) as $\ell \rightarrow r$ rather than $\ell \approx r$. From a formal point of view, however, both rewrite rules and identities are pairs of terms. Given a set of such pairs, we can view it as a set of identities or a term rewriting system, and thus the notions introduced above apply to both.

3 Commutative Congruence Closure Based on Rewriting

Let Σ be a finite set of function symbols of arity ≥ 1 and C_0 a finite set of constant symbols. We denote the set of ground terms built using symbols from Σ and C_0 with $G(\Sigma, C_0)$. In the following, let E be a finite set of identities $s \approx t$ between terms $s, t \in G(\Sigma, C_0)$, and $\approx_E$ the equational theory induced by E on $G(\Sigma, C_0)$, defined either semantically using algebras or (equivalently) syntactically through rewriting [3].

It is well-known (see, e.g., [3], Lemma 4.3.3) that $\approx_E$ (viewed as a subset of $G(\Sigma, C_0) \times G(\Sigma, C_0)$) can be generated using congruence closure, i.e., by exhaustively applying reflexivity, transitivity, symmetry, and congruence to E. To be more precise, $CC(E)$ is the smallest subset of $G(\Sigma, C_0) \times G(\Sigma, C_0)$ that contains E and is closed under the following rules:

- if $s \in G(\Sigma, C_0)$, then $s \approx s \in CC(E)$ (reflexivity);
- if $s_1 \approx s_2, s_2 \approx s_3 \in CC(E)$, then $s_1 \approx s_3 \in CC(E)$ (transitivity);
- if $s_1 \approx s_2 \in CC(E)$, then $s_2 \approx s_1 \in CC(E)$ (symmetry);
- if f is an n-ary function symbol and $s_1 \approx t_1, \ldots, s_n \approx t_n \in CC(E)$, then $f(s_1, \ldots, s_n) \approx f(t_1, \ldots, t_n) \in CC(E)$ (congruence).

The set $CC(E)$ is usually infinite. To obtain a decision procedure for the word problem, one can show that it is sufficient to restrict the application of the above rules to a finite subset of $G(\Sigma, C_0)$, which consists of the subterms of terms occurring in E and of the subterms of the terms s_0, t_0 for which one wants to decide whether $s_0 \approx_E t_0$ holds or not (see, e.g., [3], Theorem 4.3.5).

This actually also works if one adds commutativity of some binary function symbols to the theory. To be more precise, we assume that some of the binary function symbols in Σ are commutative, i.e., there is a set of binary function symbols $\Sigma_c \subseteq \Sigma$ whose elements we call *commutative* symbols. In addition to the identities in E, we assume that the identities $f(x, y) \approx f(y, x)$ are satisfied for all function symbols $f \in \Sigma_c$. From a semantic point of view, this means that we consider algebras $\mathcal{A}$ that satisfy not only the identities in E, but also *commutativity* for the symbols in Σ_c, i.e., for all $f \in \Sigma_c$, and all elements a, b of $\mathcal{A}$ we have that $f^{\mathcal{A}}(a, b) = f^{\mathcal{A}}(b, a)$. Given $s, t \in G(\Sigma, C_0)$, we say that $s \approx t$ *follows from E w.r.t. the commutative symbols in Σ_c* (written $s \approx_E^{\Sigma_c} t$) if $s^{\mathcal{A}} = t^{\mathcal{A}}$ holds in all algebras that satisfy the identities in E and commutativity for the symbols in Σ_c. The relation $\approx_E^{\Sigma_c} \subseteq G(\Sigma, C_0) \times G(\Sigma, C_0)$ can also be generated by extending congruence closure by a commutativity rule.

To be more precise, $CC^{\Sigma_c}(E)$ is the smallest subset of $G(\Sigma, C_0) \times G(\Sigma, C_0)$ that contains E and is closed under reflexivity, transitivity, symmetry, congruence, and the following commutativity rule:

- if $f \in \Sigma_c$ and $s, t \in G(\Sigma, C_0)$, then $f(s,t) \approx f(t,s) \in CC^{\Sigma_c}(E)$ (commutativity).

We call $CC^{\Sigma_c}(E)$ the *commutative congruence closure* of E. Using Birkhoff's theorem, it is easy to see that $CC^{\Sigma_c}(E)$ coincides with $\approx_E^{\Sigma_c}$ in the sense that $s \approx t \in CC^{\Sigma_c}(E)$ iff $s \approx_E^{\Sigma_c} t$ (see Lemma 3.5.13 and Theorem 3.5.14 in [3]). Again, it is not hard to show that the restriction of the commutative congruence closure to a polynomially large set of terms determined by the input E, s_0, t_0 is complete, which yields decidability of $\approx_E^{\Sigma_c}$ [5].

Here, we follow a different approach, which is based on rewriting [7,8]. Let $S(E)$ denote the set of subterms of the terms occurring in E. In a first step, we introduce a new constant c_s for every term $s \in S(E) \setminus C_0$. To simplify notation, for a constant $a \in C_0$ we sometimes use c_a to denote a. Let C_1 be the set of new constants introduced this way and $C := C_0 \cup C_1$. Given a term $u \in G(\Sigma, C)$, we denote with $\widehat{u}$ the term in $G(\Sigma, C_0)$ obtained from u by replacing the occurrences of the constants $c_s \in C_1$ in u with the corresponding terms $s \in S(E)$.

We fix an arbitrary linear order $>$ on C, which will be used to orient identities between constants into rewrite rules. Note that this order does not take into account which terms the constants correspond to, and thus we may well have $c_s > c_{f(s)}$.

The initial rewrite system $R(E)$ induced by E consists of the following rules:

- If $s \in S(E) \setminus C_0$, then s is of the form $f(s_1, \ldots, s_n)$ for an n-ary function symbol f and terms $s_1, \ldots, s_n$ for some $n \geq 1$. For every such s we add the rule $f(c_{s_1}, \ldots, c_{s_n}) \to c_s$ to $R(E)$.
- For every identity $s \approx t \in E$ we add $c_s \to c_t$ to $R(E)$ if $c_s > c_t$, and $c_t \to c_s$ if $c_t > c_s$.

Obviously, the cardinality of C_1 is linear in the size of E, and $R(E)$ can be constructed in time linear in the size of E. From the above construction, it follows that $R(E)$ has two types of rules: *constant rules* of the form $c \to d$ for $c > d$ and *function rules* of the form $f(c_1, \ldots, c_n) \to d$.

Example 1. Consider $E = \{f(a, g(a)) \approx c, g(b) \approx h(a), a \approx b\}$ with $\Sigma_c = \{f\}$. It is easy to see that we have $f(h(a), b) \approx_E^{\Sigma_c} c$. Using our construction, we first introduce the new constants $C_1 = \{c_{f(a,g(a))}, c_{g(a)}, c_{g(b)}, c_{h(a)}\}$. If we fix the linear order on C as $c_{f(a,g(a))} > c_{g(a)} > c_{g(b)} > c_{h(a)} > a > b > c$, then we obtain the following rewrite system: $R(E) = \{f(a, c_{g(a)}) \to c_{f(a,g(a))}, g(a) \to c_{g(a)}, g(b) \to c_{g(b)}, h(a) \to c_{h(a)}, c_{f(a,g(a))} \to c, c_{g(b)} \to c_{h(a)}, a \to b\}$.

The following lemma is an easy consequence of the definition of $R(E)$. The first part can be shown by a simple induction on the structure of s.

Lemma 1. *For all terms $s \in S(E)$ we have $s \approx_{R(E)} c_s$. Consequently, $u \approx_{R(E)} \widehat{u}$ and thus also $u \approx_{R(E)}^{\Sigma_c} \widehat{u}$ for all terms $u \in G(\Sigma, C)$.*

Using this lemma, we can show that the construction of $R(E)$ is correct for consequence w.r.t. commutative symbols in the following sense:

Lemma 2. *Viewed as a set of identities, $R(E)$ is a conservative extension of E w.r.t. the commutative symbols in Σ_c, i.e., for all terms $s_0, t_0 \in G(\Sigma, C_0)$ we have $s_0 \approx_E^{\Sigma_c} t_0$ iff $s_0 \approx_{R(E)}^{\Sigma_c} t_0$.*

In this lemma, we use commutativity of the elements of Σ_c as additional identities. Our goal is, however, to deal both with the ground identities in E and with commutativity by rewriting. For this reason, we consider the rewrite system[1]

$$R(\Sigma_c) := \{f(s,t) \to f(t,s) \mid f \in \Sigma_c, s,t \in G(\Sigma, C), \text{ and } s >_{lpo} t\}, \tag{1}$$

where $>_{lpo}$ denotes the lexicographic path order (see Definition 5.4.12 in [3]) induced by a linear order on $\Sigma \cup C$ that extends $>$ on C, makes each function symbol in Σ greater than each constant symbol in C, and linearly orders the function symbols in an arbitrary way. Note that $>_{lpo}$ is then a linear order on $G(\Sigma, C)$ (see Exercise 5.20 in [3]). Consequently, for every pair of distinct terms $s, t \in G(\Sigma, C)$, we have $f(s,t) \to f(t,s) \in R(\Sigma_c)$ or $f(t,s) \to f(s,t) \in R(\Sigma_c)$.

The term rewriting system $R(E) \cup R(\Sigma_c)$ can easily be shown to terminate using this order. In fact, $>_{lpo}$ is a reduction order, and we have $\ell >_{lpo} r$ for all rules $\ell \to r \in R(E) \cup R(\Sigma_c)$. However, in general $R(E) \cup R(\Sigma_c)$ need not be confluent. For instance, in Example 1 we have the two rewrite sequences $g(a) \to g(b) \to c_{g(b)} \to c_{h(a)}$ and $g(a) \to c_{g(a)}$ w.r.t. $R(E) \cup R(\Sigma_c)$, and the two constants $c_{h(a)}$ and $c_{g(a)}$ are irreducible w.r.t. $R(E) \cup R(\Sigma_c)$, but not equal.

We turn $R(E) \cup R(\Sigma_c)$ into a confluent and terminating system by modifying $R(E)$ appropriately. We start with $R_0^{\Sigma_c}(E) := R(E)$ and $i := 0$:

(a) Let $R_i^{\Sigma_c}(E)|_{con}$ consist of the constant rules in $R_i^{\Sigma_c}(E)$. For every constant $c \in C$, consider

$$[c]_i := \{d \in C \mid c \approx_{R_i^{\Sigma_c}(E)|_{con}} d\},$$

and let e be the least element in $[c]_i$ w.r.t. the order $>$. We call e the *representative* of c w.r.t. $R_i^{\Sigma_c}(E)$ and $>$. If $c \neq e$, then add $c \to e$ to $R_{i+1}^{\Sigma_c}(E)$.

(b) In all function rules in $R_i^{\Sigma_c}(E)$, replace each constant by its representative w.r.t. $R_i^{\Sigma_c}(E)$ and $>$, and call the resulting set of function rules $F_i^{\Sigma_c}(E)$. Then, we distinguish two cases, depending on whether the function symbol occurring in the rule is commutative or not.

(b1) Let f be an n-ary function symbol not belonging to Σ_c. For every term $f(c_1, \ldots, c_n)$ occurring as the left-hand side of a rule in $F_i^{\Sigma_c}(E)$, consider all the rules $f(c_1, \ldots, c_n) \to d_1, \ldots, f(c_1, \ldots, c_n) \to d_k$ in $F_i^{\Sigma_c}(E)$ with this left-hand side. Let d be the least element w.r.t. $>$ in $\{d_1, \ldots, d_k\}$. Add $f(c_1, \ldots, c_n) \to d$ and $d_j \to d$ for all j with $d_j \neq d$ to $R_{i+1}^{\Sigma_c}(E)$.

[1] Since this system is in general infinite, we do not generate it explicitly. But we can nevertheless apply the appropriate rule when encountering a commutative symbol during rewriting by just ordering its arguments according to $>_{lpo}$.

(b2) Let f be a binary function symbol belonging to Σ_c. For all pairs of constant symbols c_1, c_2 such that $f(c_1, c_2)$ or $f(c_2, c_1)$ is the left-hand side of a rule in $F_i^{\Sigma_c}(E)$, consider the set of constant symbols $\{d_1, \ldots, d_k\}$ occurring as right-hand sides of such rules, and let d be the least element w.r.t. $>$ in this set. Add $d_j \rightarrow d$ for all j with $d_j \neq d$ to $R_{i+1}^{\Sigma_c}(E)$. In addition, if $c_2 >_{lpo} c_1$, then add $f(c_1, c_2) \rightarrow d$ to $R_{i+1}^{\Sigma_c}(E)$, and otherwise $f(c_2, c_1) \rightarrow d$.

If at least one constant rule has been added in this step, then set $i := i+1$ and continue with step (a). Otherwise, terminate with output $\widehat{R}^{\Sigma_c}(E) := R_{i+1}^{\Sigma_c}(E)$.

Let us illustrate the construction of $\widehat{R}^{\Sigma_c}(E)$ using Example 1. In step (a), the non-trivial equivalence classes are $[a]_0 = \{a, b\}$ with representative b, $[c_{f(a,g(a))}] = \{c_{f(a,g(a))}, c\}$ with representative c, and $[c_{g(b)}] = \{c_{g(b)}, c_{h(a)}\}$ with representative $c_{h(a)}$. Thus, $a \rightarrow b, c_{f(a,g(a))} \rightarrow c, c_{g(b)} \rightarrow c_{h(a)}$ are the constant rule added to $R_1^{\Sigma_c}(E)$. The function rules in $F_0^{\Sigma_c}(E)$ are then $f(b, c_{g(a)}) \rightarrow c, g(b) \rightarrow c_{g(a)}, g(b) \rightarrow c_{h(a)}, h(b) \rightarrow c_{h(a)}$. For the two rules with left-hand side $g(b)$, we add $c_{g(a)} \rightarrow c_{h(a)}$ and $g(b) \rightarrow c_{h(a)}$ to $R_1^{\Sigma_c}(E)$. The rules with left-hand sides different from $g(b)$ are moved unchanged from $F_0^{\Sigma_c}(E)$ to $R_1^{\Sigma_c}(E)$ since their left-hand sides are unique. Thus, $R_1^{\Sigma_c}(E) = \{a \rightarrow b, c_{f(a,g(a))} \rightarrow c, c_{g(b)} \rightarrow c_{h(a)}, c_{g(a)} \rightarrow c_{h(a)}, f(b, c_{g(a)}) \rightarrow c, g(b) \rightarrow c_{h(a)}, h(b) \rightarrow c_{h(a)}\}$.

In the second iteration step, we now have the new non-trivial equivalence class $[c_{g(b)}]_1 = \{c_{g(b)}, c_{h(a)}, c_{g(a)}\}$ with representative $c_{h(a)}$. The net effect of step (a) is, however, that the constant rules are moved unchanged from $R_1^{\Sigma_c}(E)$ to $R_2^{\Sigma_c}(E)$. The function rules in $F_1^{\Sigma_c}(E)$ are then $f(b, c_{h(a)}) \rightarrow c, g(b) \rightarrow c_{h(a)}, h(b) \rightarrow c_{h(a)}$. Consequently, no constant rules are added in step (b), and the construction terminates with output $\widehat{R}^{\Sigma_c}(E) = \{a \rightarrow b, c_{f(a,g(a))} \rightarrow c, c_{g(b)} \rightarrow c_{h(a)}, c_{g(a)} \rightarrow c_{h(a)}, f(b, c_{h(a)}) \rightarrow c, g(b) \rightarrow c_{h(a)}, h(b) \rightarrow c_{h(a)}\}$.

Our goal is now to show that $\widehat{R}^{\Sigma_c}(E) \cup R(\Sigma_c)$ provides us with a polynomial-time decision procedure for the commutative word problem in E.

Lemma 3. *The system $\widehat{R}^{\Sigma_c}(E)$ can be computed from $R(E)$ in polynomial time, and its construction is correct in the following sense: viewed as a set of identities, $\widehat{R}^{\Sigma_c}(E) \cup R(\Sigma_c)$ is equivalent to $R(E)$ with commutativity, i.e., for all terms $s, t \in G(\Sigma, C)$ we have $s \approx_{R(E)}^{\Sigma_c} t$ iff $s \approx_{\widehat{R}^{\Sigma_c}(E) \cup R(\Sigma_c)} t$.*

If we view $\widehat{R}^{\Sigma_c}(E) \cup R(\Sigma_c)$ as a term rewriting system, then we obtain the following result.

Lemma 4. *$\widehat{R}^{\Sigma_c}(E) \cup R(\Sigma_c)$ is canonical, i.e., terminating and confluent.*

Proof. Termination of the term rewriting system $\widehat{R}^{\Sigma_c}(E) \cup R(\Sigma_c)$ can be shown as for $R(E) \cup R(\Sigma_c)$, using the reduction order $>_{lpo}$ introduced in the definition of $R(\Sigma_c)$. Confluence can thus be proved by showing that all non-trivial critical pairs of this system can be joined (see [2] for details). □

Since $\widehat{R}^{\Sigma_c}(E) \cup R(\Sigma_c)$ is canonical, each term $s \in G(\Sigma, C)$ has a unique normal form (i.e., irreducible term reachable from s) w.r.t. $\widehat{R}^{\Sigma_c}(E) \cup R(\Sigma_c)$, which we call the *canonical form* of s. We can thus use the system $\widehat{R}^{\Sigma_c}(E) \cup R(\Sigma_c)$ to decide whether terms s, t are equivalent w.r.t. E and commutativity of the symbols in Σ_c, i.e., whether $s \approx t \in CC^{\Sigma_c}(E)$, by computing the canonical forms of the terms s and t.

Theorem 1. *Let $s_0, t_0 \in G(\Sigma, C_0)$. Then we have $s_0 \approx t_0 \in CC^{\Sigma_c}(E)$ iff s_0 and t_0 have the same canonical form w.r.t. $\widehat{R}^{\Sigma_c}(E) \cup R(\Sigma_c)$.*

Consider the rewrite system $\widehat{R}^{\Sigma_c}(E)$ that we have computed (above Lemma 3) from the set of ground identities E in Example 1, and recall that $f(h(a), b) \approx_E^{\Sigma_c} c$. The canonical form of c is clearly c, and the canonical form of $f(h(a), b)$ can be computed by the following rewrite sequence:

$$f(h(a), b) \rightarrow_{R(\Sigma_c)} f(b, h(a)) \rightarrow_{\widehat{R}^{\Sigma_c}(E)} f(b, h(b)) \rightarrow_{\widehat{R}^{\Sigma_c}(E)} f(b, c_{h(a)}) \rightarrow_{\widehat{R}^{\Sigma_c}(E)} c.$$

Note that the construction of $\widehat{R}^{\Sigma_c}(E)$ is actually independent of the terms s_0, t_0 for which we want to decide the word problem in E. This is in contrast to approaches that restrict the construction of the congruence closure to the subterms of E and the subterms of the terms s_0, t_0 for which one wants to decide the word problem. This fact will turn out to be useful in the next section.

Since it is easy to show that reduction to canonical forms requires only a polynomial number of rewrite steps, Theorem 1 thus yields the following complexity result.

Corollary 1. *The commutative word problem for finite sets of ground identities is decidable in polynomial time, i.e., given a finite set of ground identities $E \subseteq G(\Sigma, C_0) \times G(\Sigma, C_0)$, a set $\Sigma_c \subseteq \Sigma$ of commutative symbols, and terms $s_0, t_0 \in G(\Sigma, C_0)$, we can decide in polynomial time whether $s_0 \approx_E^{\Sigma_c} t_0$ holds or not.*

This complexity result has been shown before in [5] and [8], but note that, in these papers, detailed proofs are given for the case without commutativity, and then it is only sketched how the respective approach can be extended to accommodate commutativity. Like the approach in this paper, the one employed in [8] is rewriting-based, but in contrast to ours it does not explicitly use the rewrite system $R(\Sigma_c)$.

4 Commutative Congruence Closure with Extensionality

Here, we additionally assume that some of the *non-commutative*[2] function symbols are extensional, i.e., there is a set of function symbols $\Sigma^e \subseteq \Sigma \setminus \Sigma_c$ whose elements we call *extensional* symbols. In addition to the identities in E and

[2] We will explain in the next section why the notion of extensionality introduced in (2) below is not appropriate for commutative symbols.

commutativity for the symbols in Σ_c, we now assume that also the following conditional identities are satisfied for every n-ary function symbol $f \in \Sigma^e$:

$$f(x_1, \ldots, x_n) \approx f(y_1, \ldots, y_n) \Rightarrow x_i \approx y_i \text{ for all } i, 1 \leq i \leq n. \tag{2}$$

From a semantic point of view, this means that we now consider algebras $\mathcal{A}$ that satisfy not only the identities in E and commutativity for the symbols in Σ_c, but also *extensionality* for the symbols in Σ^e, i.e., for all $f \in \Sigma^e$, all $i, 1 \leq i \leq n$, and all elements $a_1, \ldots, a_n, b_1, \ldots, b_n$ of $\mathcal{A}$ we have that $f^{\mathcal{A}}(a_1, \ldots, a_n) = f^{\mathcal{A}}(b_1, \ldots, b_n)$ implies $a_i = b_i$ for all $i, 1 \leq i \leq n$. Let $\Sigma_c^e = (\Sigma_c, \Sigma^e)$ and $s, t \in G(\Sigma, C_0)$. We say that $s \approx t$ follows from E w.r.t. the commutative symbols in Σ_c and the extensional symbols in Σ^e (written $s \approx_E^{\Sigma_c^e} t$) if $s^{\mathcal{A}} = t^{\mathcal{A}}$ holds in all algebras that satisfy the identities in E, commutativity for the symbols in Σ_c, and extensionality for the symbols in Σ^e.

The relation $\approx_E^{\Sigma_c^e} \subseteq G(\Sigma, C_0) \times G(\Sigma, C_0)$ can also be generated using the following extension of congruence closure by an extensionality rule. To be more precise, $CC^{\Sigma_c^e}(E)$ is the smallest subset of $G(\Sigma, C_0) \times G(\Sigma, C_0)$ that contains E and is closed under reflexivity, transitivity, symmetry, congruence, commutativity, and the following extensionality rule:

- if $f \in \Sigma^e$ is an n-ary function symbol, $1 \leq i \leq n$, and $f(s_1, \ldots, s_n) \approx f(t_1, \ldots, t_n) \in CC^{\Sigma_c^e}(E)$, then $s_i \approx t_i \in CC^{\Sigma_c^e}(E)$ (extensionality).

Proposition 1. *For all terms $s, t \in G(\Sigma, C_0)$ we have $s \approx_E^{\Sigma_c^e} t$ iff $s \approx t \in CC^{\Sigma_c^e}(E)$.*

Proof. This proposition is an easy consequence of Theorem 54 in [18], which (adapted to our setting) says that $\approx_E^{\Sigma_c^e}$ is the least congruence containing E that is invariant under applying commutativity and extensionality. Clearly, this is exactly $CC^{\Sigma_c^e}(E)$. □

To obtain a decision procedure for $\approx_E^{\Sigma_c^e}$, we extend the rewriting-based approach from the previous section. Let the term rewriting system $R(E)$ be defined as in Sect. 3.

Example 2. Consider $E' = \{f(a, g(a)) \approx c, g(b) \approx h(a), g(a) \approx g(b)\}$ with $\Sigma_c = \{f\}$ and $\Sigma^e = \{g\}$. It is easy to see that we have $f(h(a), b) \approx_{E'}^{\Sigma_c^e} c$. Let the set C_1 of new constants and the linear order on all constants be defined as in Example 1. Now, we obtain the following rewrite system: $R(E') = \{f(a, c_{g(a)}) \rightarrow c_{f(a,g(a))}, g(a) \rightarrow c_{g(a)}, g(b) \rightarrow c_{g(b)}, h(a) \rightarrow c_{h(a)}, c_{f(a,g(a))} \rightarrow c, c_{g(b)} \rightarrow c_{h(a)}, c_{g(a)} \rightarrow c_{g(b)}\}$.

Lemma 5. *The system $R(E)$ is a conservative extension of E also w.r.t. the commutative symbols in Σ_c and the extensional symbols in Σ^e, i.e., for all terms $s_0, t_0 \in G(\Sigma, C_0)$ we have $s_0 \approx_E^{\Sigma_c^e} t_0$ iff $s_0 \approx_{R(E)}^{\Sigma_c^e} t_0$.*

We extend the construction of the confluent and terminating rewrite system corresponding to $R(E)$ by adding a third step that takes care of extensionality. To be more precise, $\widehat{R}^{\Sigma_c^e}(E)$ is constructed by performing the following steps, starting with $R_0^{\Sigma_c^e}(E) := R(E)$ and $i := 0$:

(a) Let $R_i^{\Sigma_c^e}(E)|_{con}$ consist of the constant rules in $R_i^{\Sigma_c^e}(E)$. For every constant $c \in C$, consider

$$[c]_i := \{d \in C \mid c \approx_{R_i^{\Sigma_c^e}(E)|_{con}} d\},$$

and let e be the least element in $[c]_i$ w.r.t. the order $>$. We call e the *representative* of c w.r.t. $R_i^{\Sigma_c^e}(E)$ and $>$. If $c \neq e$, then add $c \rightarrow e$ to $R_{i+1}^{\Sigma_c^e}(E)$.

(b) In all function rules in $R_i^{\Sigma_c^e}(E)$, replace each constant by its representative w.r.t. $R_i^{\Sigma_c^e}(E)$ and $>$, and call the resulting set of function rules $F_i^{\Sigma_c^e}(E)$. Then, we distinguish two cases, depending on whether the function symbol occurring in the rule is commutative or not.

(b1) Let f be an n-ary function symbol not belonging to Σ_c. For every term $f(c_1, \ldots, c_n)$ occurring as the left-hand side of a rule in $F_i^{\Sigma_c^e}(E)$, consider all the rules $f(c_1, \ldots, c_n) \rightarrow d_1, \ldots, f(c_1, \ldots, c_n) \rightarrow d_k$ in $F_i^{\Sigma_c^e}(E)$ with this left-hand side. Let d be the least element w.r.t. $>$ in $\{d_1, \ldots, d_k\}$. Add $f(c_1, \ldots, c_n) \rightarrow d$ and $d_j \rightarrow d$ for all j with $d_j \neq d$ to $R_{i+1}^{\Sigma_c^e}(E)$.

(b2) Let f be a binary function symbol belonging to Σ_c. For all pairs of constant symbols c_1, c_2 such that $f(c_1, c_2)$ or $f(c_2, c_1)$ is the left-hand side of a rule in $F_i^{\Sigma_c^e}(E)$, consider the set of constant symbols $\{d_1, \ldots, d_k\}$ occurring as right-hand sides of such rules, and let d be the least element w.r.t. $>$ in this set. Add $d_j \rightarrow d$ for all j with $d_j \neq d$ to $R_{i+1}^{\Sigma_c^e}(E)$. In addition, if $c_2 >_{lpo} c_1$, then add $f(c_1, c_2) \rightarrow d$ to $R_{i+1}^{\Sigma_c^e}(E)$, and otherwise $f(c_2, c_1) \rightarrow d$.

If at least one constant rule has been added in this step, then set $i := i + 1$ and continue with step (a). Otherwise, continue with step (c).

(c) For all $f \in \Sigma^e$, all pairs of distinct rules $f(c_1, \ldots, c_n) \rightarrow d, f(c'_1, \ldots, c'_n) \rightarrow d$ in $F_i^{\Sigma_c^e}(E)$, and all $i, 1 \leq i \leq n$ such that $c_i \neq c'_i$, add $c_i \rightarrow c'_i$ to $R_{i+1}^{\Sigma_c^e}(E)$ if $c_i > c'_i$ and otherwise add $c'_i \rightarrow c_i$ to $R_{i+1}^{\Sigma_c^e}(E)$. If at least one constant rule has been added in this step, then set $i := i + 1$ and continue with step (a). Otherwise, terminate with output $\widehat{R}^{\Sigma_c^e}(E) := R_{i+1}^{\Sigma_c^e}(E)$.

We illustrate the above construction using Example 2. In step (a), the non-trivial equivalence classes are $[c_{f(a,g(a))}] = \{c_{f(a,g(a))}, c\}$ with representative c and $[c_{g(b)}] = \{c_{g(a)}, c_{g(b)}, c_{h(a)}\}$ with representative $c_{h(a)}$. Thus, $c_{f(a,g(a))} \rightarrow c, c_{g(a)} \rightarrow c_{h(a)}, c_{g(b)} \rightarrow c_{h(a)}$ are the constant rules added to $R_1^{\Sigma_c^e}(E')$. The function rules in $F_0^{\Sigma_c^e}(E')$ are then $f(a, c_{h(a)}) \rightarrow c, g(a) \rightarrow c_{h(a)}, g(b) \rightarrow c_{h(a)}, h(a) \rightarrow c_{h(a)}$. Since these rules have unique left-hand sides, no constant rule is added in step (b). Consequently, we proceed with step (c). Since $g \in \Sigma^e$, the presence of the rules $g(a) \to c_{h(a)}$ and $g(b) \longrightarrow c_{h(a)}$ triggers the addition of $a \longrightarrow b$ to $R_1^{\Sigma_c^e}(E')$. The function rules in $R_1^{\Sigma_c^e}(E')$ are the ones in $F_0^{\Sigma_c^e}(E')$.

In the second iteration step, we now have the new non-trivial equivalence class $[a]_1 = \{a, b\}$ with representative b. The net effect of step (a) is, again, that the constant rules are moved unchanged from $R_1^{\Sigma_c^e}(E')$ to $R_2^{\Sigma_c^e}(E')$. The function rules in $F_1^{\Sigma_c^e}(E')$ are then $f(b, c_{h(a)}) \to c, g(b) \to c_{h(a)}, h(b) \to c_{h(a)}$. Consequently, no new constant rules are added in steps (b) and (c), and the construction terminates with output $\widehat{R}^{\Sigma_c^e}(E') = \{a \to b, c_{f(a,g(a))} \to c, c_{g(a)} \to c_{h(a)}, c_{g(b)} \to c_{h(a)}, f(b, c_{h(a)}) \to c, g(b) \to c_{h(a)}, h(b) \to c_{h(a)}\}$, which is identical to the system $\widehat{R}^{\Sigma_c}(E)$ computed for the set of identity E of Example 1.

Our goal is now to show that $\widehat{R}^{\Sigma_c^e}(E)$ provides us with a polynomial-time decision procedure for the extensional word problem in E, i.e., it allows us to decide the relation $\approx_E^{\Sigma_c^e}$. Let $R(\Sigma_c)$ and $>_{lpo}$ be defined as in (1).

Lemma 6. *The system $\widehat{R}^{\Sigma_c^e}(E)$ can be computed from $R(E)$ in polynomial time. Viewed as a set of identities, $\widehat{R}^{\Sigma_c^e}(E) \cup R(\Sigma_c)$ is*

- *sound for commutative and extensional reasoning, i.e., for all rules $s \to t$ in $\widehat{R}^{\Sigma_c^e}(E) \cup R(\Sigma_c)$ we have $s \approx_{R(E)}^{\Sigma_c^e} t$, and*
- *complete for commutative reasoning, i.e., or all terms $s, t \in G(\Sigma, C)$ we have that $s \approx_{R(E)}^{\Sigma_c} t$ implies $s \approx_{\widehat{R}^{\Sigma_c^e}(E) \cup R(\Sigma_c)} t$.*

Lemma 7. *Viewed as a term rewriting system, $\widehat{R}^{\Sigma_c^e}(E) \cup R(\Sigma_c)$ is canonical, i.e., terminating and confluent.*

Intuitively, $\widehat{R}^{\Sigma_c^e}(E)$ extends $\widehat{R}^{\Sigma_c}(E)$ by additional rules relating constants that are equated due to extensionality. However, to keep the system confluent, we need to re-apply the other steps once two constants have been equated.

Lemma 8. *If $s, t \in G(\Sigma, C)$ have the same canonical forms w.r.t. $\widehat{R}^{\Sigma_c}(E) \cup R(\Sigma_c)$, then they also have the same canonical forms w.r.t. $\widehat{R}^{\Sigma_c^e}(E) \cup R(\Sigma_c)$.*

We are now ready to prove our main technical result, from which decidability of the commutative and extensional word problem immediately follows.

Theorem 2. *Let $s, t \in G(\Sigma, C_0)$. Then we have $s \approx t \in CC^{\Sigma_c^e}(E)$ iff s and t have the same canonical form w.r.t. $\widehat{R}^{\Sigma_c^e}(E) \cup R(\Sigma_c)$.*

Proof. Since the if-direction is easy to show, we concentrate here on the only-if-direction. If $s, t \in G(\Sigma, C_0)$ are such that $s \approx t \in CC^{\Sigma_c^e}(E)$, then there is a sequence of identities $s_1 \approx t_1, s_2 \approx t_2, \ldots, s_k \approx t_k$ such that $s_k = s, t_k = t$, and for all $i, 1 \leq i \leq k$, the identity $s_i \approx t_i$ belongs to E or can be derived from some of the identities $s_j \approx t_j$ with $j < i$ by apply reflexivity, transitivity, symmetry, congruence, commutativity, or extensionality. We prove that s and t

have the same canonical form w.r.t. $\widehat{R}^{\Sigma_c^e}(E) \cup R(\Sigma_c)$ by induction on the number of applications of the extensionality rule used when creating this sequence.

In the *base case*, no extensionality rule is used, and thus $s \approx t \in CC^{\Sigma_c}(E)$. By Theorem 1, s and t have the same canonical form w.r.t. $\widehat{R}^{\Sigma_c}(E) \cup R(\Sigma_c)$, and thus by Lemma 8 also w.r.t. $\widehat{R}^{\Sigma_c^e}(E) \cup R(\Sigma_c)$.

In the *step case*, we consider the last identity $s_m \approx t_m$ obtained by an application of the extensionality rule. Then, by induction, we know that, for each $i, 1 \leq i < m$, the terms s_i and t_i have the same canonical form w.r.t. $\widehat{R}^{\Sigma_c^e}(E) \cup R(\Sigma_c)$.

Now, consider the application of extensionality to an identity $s_\ell \approx t_\ell$ ($\ell < m$) that produced $s_m \approx t_m$. Thus, we have $s_\ell = f(g_1, \ldots, g_n)$ and $t_\ell = f(h_1, \ldots, h_n)$ for some n-ary function symbol $f \in \Sigma^e$, and extensionality generates the new identity $g_\mu \approx h_\mu$ for some $\mu, 1 \leq \mu \leq n$, such that $s_m = g_\mu$ and $t_m = h_\mu$. For $\nu = 1, \ldots, n$, let g'_ν be the canonical form of g_ν w.r.t. $\widehat{R}^{\Sigma_c^e}(E) \cup R(\Sigma_c)$, and h'_ν the canonical form of h_ν w.r.t. $\widehat{R}^{\Sigma_c^e}(E) \cup R(\Sigma_c)$. We know that the canonical forms of s_ℓ and t_ℓ w.r.t. $\widehat{R}^{\Sigma_c^e}(E) \cup R(\Sigma_c)$ are identical, and these canonical forms can be obtained by normalizing $f(g'_1, \ldots, g'_n)$ and $f(h'_1, \ldots, h'_n)$. Since the rules of $R(\Sigma_c)$ are not applicable to these terms due to the fact that $f \notin \Sigma_c$, there are two possible cases for how the canonical forms of s_ℓ and t_ℓ can look like:

1. s_ℓ and t_ℓ respectively have the canonical forms $f(g'_1, \ldots, g'_n)$ and $f(h'_1, \ldots, h'_n)$, and thus the corresponding arguments are syntactically equal, i.e., $g'_\nu = h'_\nu$ for $\nu = 1, \ldots, n$. In this case, the identity $s_m \approx t_m$ added by the application of the extensionality rule satisfies $s_m \approx_{\widehat{R}^{\Sigma_c^e}(E) \cup R(\Sigma_c)} t_m$ since we have $s_m = g_\mu \approx_{\widehat{R}^{\Sigma_c^e}(E) \cup R(\Sigma_c)} g'_\mu = h'_\mu \approx_{\widehat{R}^{\Sigma_c^e}(E) \cup R(\Sigma_c)} h_\mu = t_m$.
2. s_ℓ and t_ℓ reduce to the same constant d. Then $\widehat{R}^{\Sigma_c^e}(E)$ must contain the rules $f(g'_1, \ldots, g'_n) \to d$ and $f(h'_1, \ldots, h'_n) \to d$. By the construction of $\widehat{R}^{\Sigma_c^e}(E)$, we again have that $g'_\mu = h'_\mu$, i.e., the two terms are syntactically equal. In fact, otherwise a new constant rule $g'_\mu \to h'_\mu$ or $h'_\mu \to g'_\mu$ would have been added, and the construction would not have terminated yet. We thus have again $s_m = g_\mu \approx_{\widehat{R}^{\Sigma_c^e}(E) \cup R(\Sigma_c)} g'_\mu = h'_\mu \approx_{\widehat{R}^{\Sigma_c^e}(E) \cup R(\Sigma_c)} h_\mu = t_m$.

Summing up, we have seen that we have $s_i \approx_{\widehat{R}^{\Sigma_c^e}(E) \cup R(\Sigma_c)} t_i$ for all $i, 1 \leq i \leq m$. Since the identities $s_j \approx t_j$ for $m < j \leq k$ are generated from the identities $s_i \approx t_i$ for $i = 1, \ldots, m$ and E using only reflexivity, transitivity, symmetry, commutativity, and congruence, this implies that also these identities satisfy $s_j \approx_{\widehat{R}^{\Sigma_c^e}(E) \cup R(\Sigma_c)} t_j$. In particular, we thus have $s_k \approx_{\widehat{R}^{\Sigma_c^e}(E) \cup R(\Sigma_c)} t_k$. Since $\widehat{R}^{\Sigma_c^e}(E) \cup R(\Sigma_c)$ is canonical, this implies that $s_k = s$ and $t_k = t$ have the same canonical form w.r.t. $\widehat{R}^{\Sigma_c^e}(E) \cup R(\Sigma_c)$. □

Recall that we have $f(h(a), b) \approx_{E'}^{\Sigma_c^e} c$ for the set of identities E' of Example 2. We have already seen that these two terms rewrite to the same canonical form w.r.t. $\widehat{R}^{\Sigma_c}(E) \cup R(\Sigma_c) = \widehat{R}^{\Sigma_c^e}(E') \cup R(\Sigma_c)$.

Again, it is easy to show that the decision procedure obtained by applying Theorem 2 requires only polynomial time.

Corollary 2. *The commutative and extensional word problem for finite sets of ground identities is decidable in polynomial time, i.e., given a finite set of ground identities* $E \subseteq G(\Sigma, C_0) \times G(\Sigma, C_0)$*, finite sets* $\Sigma_c \subseteq \Sigma$ *of commutative and* $\Sigma^e \subseteq \Sigma \setminus \Sigma_c$ *of non-commutative extensional symbols, and terms* $s_0, t_0 \in G(\Sigma, C_0)$*, we can decide in polynomial time whether* $s_0 \approx_E^{\Sigma_c^e} t_0$ *holds or not.*

We have mentioned in the introduction that it is unclear how this polynomiality result could be obtained by a simple adaptation of the usual approach that restricts congruence closure to a polynomially large set of subterms determined by the input (informally called "small" terms in the following). The main problem is that one might have to generate identities between "large" terms before one can get back to a desired identity between "small" terms using extensionality. The question is now where our rewriting-based approach actually deals with this problem. The answer is: in Case 1 of the case distinction in the proof of Theorem 2. In fact, there we consider a derived identity $s_\ell \approx t_\ell$ such that the (syntactically identical) canonical forms of $s_\ell = f(g_1, \ldots, g_n)$ and $t_\ell = f(h_1, \ldots, h_n)$ are not a constant from C, but of the form $f(g_1', \ldots, g_n') = f(h_1', \ldots, h_n')$. Basically, this means that s_ℓ and t_ℓ are terms that are not equivalent modulo E to subterms of terms occurring in E, since the latter terms have a constant representing them. Thus, s_ℓ, t_ℓ are "large" terms that potentially could cause a problem: an identity between them has been derived, and now extensionality applied to this identity yields a new identity $g_\mu \approx h_\mu$ between smaller terms. Our induction proof shows that this identity can nevertheless be derived from $\widehat{R}^{\Sigma_c^e}(E) \cup R(\Sigma_c)$, and thus does not cause a problem.

5 Symbols that Are Commutative and Extensional

In the previous section, we have made the assumptions that the sets Σ_c and Σ^e are disjoint, i.e., we did not consider extensionality for commutative symbols. The reason is that the presence of a commutative *and* extensional symbol would trivialize the equational theory. In fact, as already mentioned in the introduction, if f is assumed to be commutative and extensional, then commutativity yields $f(s,t) \approx f(t,s)$ for all terms $s, t \in G(\Sigma, C_0)$, and extensionality then $s \approx t$. This shows that, in this case, the commutative and extensional congruence closure would be $G(\Sigma, C_0) \times G(\Sigma, C_0)$, independently of E, and thus even for $E = \emptyset$.

In this section, we consider the following variant of extensionality for commutative function symbols f, which we call *c-extensionality*:

$$f(x_1, x_2) \approx f(y_1, y_2) \Rightarrow (x_1 \approx y_1 \land x_2 \approx y_2) \lor (x_1 \approx y_2 \land x_2 \approx y_1). \qquad (3)$$

For example, if f is a commutative couple constructor, and two couples turn out to be equal, then we want to infer that they consist of the same two persons, independently of the order in which they were put into the constructor.

Unfortunately, adding such a rule makes the word problem coNP-hard, which can be shown by a reduction from validity of propositional formulae.

Proposition 2. *In the presence of at least one commutative and c-extensional symbol, the word problem for finite sets of ground identities is coNP-hard.*

We prove this proposition by a reduction from validity of propositional formulae. Thus, consider a propositional formula ϕ, and let $p_1, \ldots, p_n$ be the propositional variables occurring in ϕ. We take the constants 0 and 1, and for every $i, 1 \leq i \leq n$, we view p_i as a constant symbol, and add a second constant symbol $\overline{p}_i$. In addition, we consider the function symbols $f_\vee, f_\wedge, f_\neg, f$, and assume that f is commutative and satisfies (3). We then consider ground identities that axiomatize the truth tables for $\vee, \wedge, \neg$, i.e.,

$$\begin{aligned} &f_\vee(0,0) \approx 0,\ f_\vee(1,0) \approx 1,\ f_\vee(0,1) \approx 1,\ f_\vee(1,1) \approx 1,\\ &f_\wedge(0,0) \approx 0,\ f_\wedge(1,0) \approx 0,\ f_\wedge(0,1) \approx 0,\ f_\wedge(1,1) \approx 1,\\ &f_\neg(0) \approx 1, \quad f_\neg(1) \approx 0. \end{aligned} \tag{4}$$

In addition, we consider, for every $i, 1 \leq i \leq n$, the identities $f(p_i, \overline{p}_i) \approx f(0,1)$. Let E_ϕ be the set of these ground identities, and let t_ϕ be the term obtained from ϕ by replacing the Boolean operations $\vee, \wedge$, and $\neg$ by the corresponding function symbols $f_\vee, f_\wedge$, and $f_\neg$.

Proposition 2 is now an immediate consequence of the following lemma.

Lemma 9. *The identity $t_\phi \approx 1$ holds in every algebra satisfying E_ϕ together with* (3) *and commutativity of f iff ϕ is valid.*

To prove a *complexity upper bound* that matches the lower bound stated in Proposition 2, we consider a finite signature Σ, a finite set of ground identities $E \subseteq G(\Sigma, C_0) \times G(\Sigma, C_0)$ as well as sets $\Sigma_c \subseteq \Sigma$ and $\Sigma^e \subseteq \Sigma$ of commutative and extensional symbols, respectively, and assume that the non-commutative extensional symbols in $\Sigma^e \setminus \Sigma_c$ satisfy extensionality (2), whereas the commutative extensional symbols in $\Sigma^e \cap \Sigma_c$ satisfy c-extensionality (3). We want to show that, in this setting, the problem of deciding, for given terms $s_0, t_0 \in G(\Sigma, C_0)$, whether s_0 is *not* equivalent to t_0 is in NP.

For this purpose, we employ a *nondeterministic variant* of our construction of $\widehat{R}^{\Sigma^e_c}(E)$. In steps (a) and (b), this procedure works as described in the previous section. For extensional symbols $f \in \Sigma^e \setminus \Sigma_c$, step (c) is also performed as in the previous section. For an extensional symbol $f \in \Sigma^e \cap \Sigma_c$, step (c) is modified as follows: for all pairs of distinct rules $f(c_1, c_2) \to d, f(c_1', c_2') \to d$ in $F_i^{\Sigma^e_c}(E)$, nondeterministically choose whether

- c_1 and c_1' as well as c_2 and c_2' are to be identified, or
- c_1 and c_2' as well as c_2 and c_1' are to be identified,

and then add the corresponding constant rules to $R_{i+1}^{\Sigma^e_c}(E)$ unless the respective constants are already syntactically equal.

This nondeterministic algorithm has different runs, depending on the choices made in the nondeterministic part of step (c). But each run r produces a rewrite system $\widehat{R}_r^{\Sigma^e_c}(E)$.

Example 3. We illustrate the nondeterministic construction using the identities E_ϕ for $\phi = p \vee \neg p$ from our coNP-hardness proof. Then E_ϕ consists of the identities in (4) together with the identity $f(p, \overline{p}) \approx f(0,1)$. Assuming an appropriate order on the constants, the system $R(E_\phi)$ contains, among others, the rules

$$\begin{array}{llll} f_\vee(1,0) \to c_{f_\vee(1,0)}, & c_{f_\vee(1,0)} \to 1, & f_\vee(0,1) \to c_{f_\vee(0,1)}, & c_{f_\vee(0,1)} \to 1 \\ f_\neg(0) \to c_{f_\neg(0)}, & c_{f_\neg(0)} \to 1, & f_\neg(1) \to c_{f_\neg(1)}, & c_{f_\neg(1)} \to 0 \\ f(p,\overline{p}) \to c_{f(p,\overline{p})}, & f(1,0) \to c_{f(1,0)}, & c_{f(p,\overline{p})} \to c_{f(1,0)}. & \end{array}$$

In step (a) and (b) of the construction, these rules are transformed into the form

$$\begin{array}{llll} f_\vee(1,0) \to 1, & c_{f_\vee(1,0)} \to 1, & f_\vee(0,1) \to 1, & c_{f_\vee(0,1)} \to 1 \\ f_\neg(0) \to 1, & c_{f_\neg(0)} \to 1, & f_\neg(1) \to 0, & c_{f_\neg(1)} \to 0 \\ f(p,\overline{p}) \to c_{f(1,0)}, & f(1,0) \to c_{f(1,0)}, & c_{f(p,\overline{p})} \to c_{f(1,0)}. & \end{array} \qquad (5)$$

Since no new constant rule is added, the construction proceeds with step (c). Due to the presence of the rules $f(p, \overline{p}) \to c_{f(1,0)}$ and $f(1,0) \to c_{f(1,0)}$ for $f \in \Sigma_c \cap \Sigma^e$, it now nondeterministically chooses between identifying p with 1 or with 0. In the first case, the constant rules $p \to 1, \overline{p} \to 0$ are added, and in the second $p \to 0, \overline{p} \to 1$ are added. In the next iteration, no new constant rules are added, and thus the construction terminates. It has two runs r_1 and r_2. The generated rewrite systems $\widehat{R}_{r_1}^{\Sigma_c^e}(E)$ and $\widehat{R}_{r_2}^{\Sigma_c^e}(E)$ share the rules in (5), but the first contains $p \to 1$ whereas the second contains $p \to 0$.

Coming back to the general case, as in the proofs of Lemma 6 and Lemma 7, we can show the following for the rewrite systems $\widehat{R}_r^{\Sigma_c^e}(E)$.

Lemma 10. *For every run r, the term rewriting system $\widehat{R}_r^{\Sigma_c^e}(E)$ is produced in polynomial time, and the system $\widehat{R}_r^{\Sigma_c^e}(E) \cup R(\Sigma_c)$ is canonical.*

Using the canonical rewrite systems $\widehat{R}_r^{\Sigma_c^e}(E) \cup R(\Sigma_c)$, we can now characterize when an identity follows from E w.r.t. commutativity of the symbols in Σ_c, extensionality of the symbols in $\Sigma^e \setminus \Sigma_c$, and c-extensionality of the symbols in $\Sigma^e \cap \Sigma_c$ as follows.

Theorem 3. *Let $s_0, t_0 \in G(\Sigma, C_0)$. The identity $s_0 \approx t_0$ holds in every algebra that satisfies E, commutativity for every $f \in \Sigma_c$, extensionality for every $f \in \Sigma^e \setminus \Sigma_c$, and c-extensionality for every $f \in \Sigma^e \cap \Sigma_c$ iff s_0, t_0 have the same canonical forms w.r.t. $\widehat{R}_r^{\Sigma_c^e}(E) \cup R(\Sigma_c)$ for every run r of the nondeterministic construction.*

The main ideas for how to deal with extensionality and c-extensionality in the proof of this theorem are very similar to how extensionality was dealt with in the proof of Theorem 2. As for all the other results stated without proof here, a detailed proof can be found in [2]. Together with Proposition 2, Theorem 3 yields the following complexity results.

Corollary 3. *Consider a finite set of ground identities* $E \subseteq G(\Sigma, C_0) \times G(\Sigma, C_0)$ *as well as sets* $\Sigma_c \subseteq \Sigma$ *and* $\Sigma^e \subseteq \Sigma$ *of commutative and extensional symbols, respectively, and two terms* $s_0, t_0 \in G(\Sigma, C_0)$. *The problem of deciding whether the identity* $s_0 \approx t_0$ *holds in every algebra that satisfies* E, *commutativity for every* $f \in \Sigma_c$, *extensionality for every* $f \in \Sigma^e \setminus \Sigma_c$, *and c-extensionality for every* $f \in \Sigma^e \cap \Sigma_c$ *is coNP-complete.*

Coming back to Example 3, we note that $\phi = p \vee \neg p$ is valid, and thus (by Lemma 9), the identity $f_\vee(p, f_\neg(p)) \approx 1$ holds in all algebra that satisfy E_ϕ and interpret f as a commutative and c-extensional symbol. Using the rewrite system generated by the run r_1, we obtain the following rewrite sequence: $f_\vee(p, f_\neg(p)) \to f_\vee(1, f_\neg(p)) \to f_\vee(1, f_\neg(1)) \to f_\vee(1, 0) \to 1$. For the run r_2, we obtain the sequence $f_\vee(p, f_\neg(p)) \to f_\vee(0, f_\neg(p)) \to f_\vee(0, f_\neg(0)) \to f_\vee(0, 1) \to 1$. Thus, for both runs the terms $f_\vee(p, f_\neg(p))$ and 1 have the same canonical form 1.

6 Conclusion

We have shown, using a rewriting-based approach, that adding commutativity and extensionality of certain function symbols to a finite set of ground identities leaves the complexity of the word problem in P. In contrast, adding c-extensionality for commutative function symbols raises the complexity to coNP. For classical congruence closure, it is well-known that it can actually be computed in $O(n \log n)$ [12,13]. Since this complexity upper bound can also be achieved using a rewriting-based approach [8,16], we believe that the approach developed here can also be used to obtain an $O(n \log n)$ upper bound for the word problem for ground identities in the presence of commutativity and extensionality, as in Sect. 4, but this question was not in the focus here.

The rules specifying extensionality are simple kinds of Horn rules whose atoms are identities. The question arises which other such Horn rules can be added without increasing the complexity of the word problem. It is known that allowing for associative-commutative (AC) symbols leaves the word problem for finite sets of ground identities decidable [4,11]. It would be interesting to see what happens if additionally (non-AC) extensional symbols are added. The approaches employed in [4,11] are rewriting-based, but in contrast to our treatment of commutativity, they use rewriting modulo AC. It is thus not clear whether the approach developed in the present paper can be adapted to deal with AC symbols.

Regarding the application motivation from DL, it should be easy to extend tableau-based algorithms for DLs to deal with individuals named by ground terms and identities between these terms. Basically, the tableau algorithm then works with the canonical forms of such terms, and if it identifies two terms (e.g., when applying a tableau-rule dealing with number restrictions), then the rewrite system and the canonical forms need to be updated. More challenging would be a setting where rules are added to the knowledge base that generate new terms if they find a certain constellation in the knowledge base (e.g., a married couple, for which the rule introduces a ground term denoting the couple and assertions that link the couple with its components). In the context of first-order logic and modal logics, the combination of tableau-based reasoning and congruence closure has respectively been investigated in [9] and [14].

Acknowledgements. The authors would like to thank the reviewers for their careful reading of the paper and their useful comments, which helped to improve the presentation of the paper. The first author thanks Barbara Morawska for helpful discussions.

References

1. Baader, F., et al.: An Introduction to Description Logic. Cambridge University Press, Cambridge (2017)
2. Baader, F., Kapur, D.: Deciding the word problem for ground identities with commutative and extensional symbols. LTCS-Report 20-02, Chair of Automata Theory, Institute of Theoretical Computer Science, Technische Universität Dresden, Dresden (2020). https://tu-dresden.de/inf/lat/reports#BaKa-LTCS-20-02
3. Baader, F., Nipkow, T.: Term Rewriting and All That. Cambridge University Press, Cambridge (1998)
4. Bachmair, L., Ramakrishnan, I.V., Tiwari, A., Vigneron, L.: Congruence closure modulo associativity and commutativity. In: Kirchner, H., Ringeissen, C. (eds.) FroCoS 2000. LNCS (LNAI), vol. 1794, pp. 245–259. Springer, Heidelberg (2000). https://doi.org/10.1007/10720084_16
5. Downey, P., Sethi, R., Tarjan, R.E.: Variations on the common subexpression problem. J. ACM **27**(4), 758–771 (1980)
6. Gallier, J.: An algorithm for finding canonical sets of ground rewrite rules in polynomial time. J. ACM **40**(1), 1–16 (1993)
7. Kapur, D.: Shostak's congruence closure as completion. In: Comon, H. (ed.) RTA 1997. LNCS, vol. 1232, pp. 23–37. Springer, Heidelberg (1997). https://doi.org/10.1007/3-540-62950-5_59
8. Kapur, D.: Conditional congruence closure over uninterpreted and interpreted symbols. J. Syst. Sci. Complex. **32**(1), 317–355 (2019). https://doi.org/10.1007/s11424-019-8377-8
9. Käufl, T., Zabel, N.: The theorem prover of the program verifier *Tatzelwurm*. In: Stickel, M.E. (ed.) CADE 1990. LNCS, vol. 449, pp. 657–658. Springer, Heidelberg (1990). https://doi.org/10.1007/3-540-52885-7_128
10. Kozen, D.: Complexity of finitely presented algebras. In: Proceedings of the 9th ACM Symposium on Theory of Computing, pp. 164–177. ACM (1977)
11. Narendran, P., Rusinowitch, M.: Any ground associative-commutative theory has a finite canonical system. In: Book, R.V. (ed.) RTA 1991. LNCS, vol. 488, pp. 423–434. Springer, Heidelberg (1991). https://doi.org/10.1007/3-540-53904-2_115
12. Nelson, G., Oppen, D.C.: Fast decision procedures based on congruence closure. J. ACM **27**(4), 356–364 (1980)
13. Nieuwenhuis, R., Oliveras, A.: Fast congruence closure and extensions. Inf. Comput. **205**(4), 557–580 (2007)
14. Schmidt, R.A., Waldmann, U.: Modal Tableau systems with blocking and congruence closure. In: De Nivelle, H. (ed.) TABLEAUX 2015. LNCS (LNAI), vol. 9323, pp. 38–53. Springer, Cham (2015). https://doi.org/10.1007/978-3-319-24312-2_4
15. Shostak, R.E.: An algorithm for reasoning about equality. Commun. ACM **21**(7), 583–585 (1978)

16. Snyder, W.: A fast algorithm for generating reduced ground rewriting systems from a set of ground equations. J. Symb. Comput. **15**(4), 415–450 (1993)
17. Stump, A., Barrett, C.W., Dill, D.L., Levitt, J.R.: A decision procedure for an extensional theory of arrays. In: Proceedings of 16th Annual IEEE Symposium on Logic in Computer Science (LICS 2001), pp. 29–37. IEEE Computer Society (2001)
18. Wechler, W.: Universal Algebra for Computer Scientists. EATCS Monographs on Theoretical Computer Science, vol. 25. Springer, Cham (1992). https://doi.org/10.1007/978-3-642-76771-5

Combined Covers and Beth Definability

Diego Calvanese[1], Silvio Ghilardi[2], Alessandro Gianola[1,3](✉), Marco Montali[1], and Andrey Rivkin[1]

[1] Faculty of Computer Science, Free University of Bozen-Bolzano, Bolzano, Italy
{calvanese,gianola,montali,rivkin}@inf.unibz.it

[2] Dipartimento di Matematica, Università degli Studi di Milano, Milan, Italy
silvio.ghilardi@unimi.it

[3] CSE Department, University of California San Diego, San Diego, CA, USA
agianola@eng.ucsd.edu

Abstract. Uniform interpolants were largely studied in non-classical propositional logics since the nineties, and their connection to model completeness was pointed out in the literature. A successive parallel research line inside the automated reasoning community investigated uniform quantifier-free interpolants (sometimes referred to as "covers") in first-order theories. In this paper, we investigate cover transfer to theory combinations in the disjoint signatures case. We prove that, for convex theories, cover algorithms can be transferred to theory combinations under the same hypothesis needed to transfer quantifier-free interpolation (i.e., the equality interpolating property, aka strong amalgamation property). The key feature of our algorithm relies on the extensive usage of the Beth definability property for primitive fragments to convert implicitly defined variables into their explicitly defining terms. In the non-convex case, we show by a counterexample that cover may not exist in the combined theories, even in case combined quantifier-free interpolants do exist.

1 Introduction

Uniform interpolants were originally studied in the context of non-classical logics, starting from the pioneering work by Pitts [26]. We briefly recall what uniform interpolants are; we fix a logic or a theory T and a suitable fragment (propositional, first-order quantifier-free, etc.) of its language L. Given an L-formula $\phi(\underline{x}, \underline{y})$ (here $\underline{x}, \underline{y}$ are the variables occurring free in ϕ), a *uniform interpolant* of ϕ (w.r.t. $\underline{y}$) is a formula $\phi'(\underline{x})$ where only the $\underline{x}$ occur free, and that satisfies the following two properties: *(i)* $\phi(\underline{x}, \underline{y}) \vdash_T \phi'(\underline{x})$; *(ii)* for any further L-formula $\psi(\underline{x}, \underline{z})$ such that $\phi(\underline{x}, \underline{y}) \vdash_T \psi(\underline{x}, \underline{z})$, we have $\phi'(\underline{x}) \vdash_T \psi(\underline{x}, \underline{z})$. Whenever uniform interpolants exist, one can compute an interpolant for an entailment like $\phi(\underline{x}, \underline{y}) \vdash_T \psi(\underline{x}, \underline{z})$ in a way that is *independent* of ψ.

The existence of uniform interpolants is an exceptional phenomenon, which is however not so infrequent; it has been extensively studied in non-classical logics starting from the nineties, as witnessed by a large literature (a non-exhaustive

N. Peltier and V. Sofronie-Stokkermans (Eds.): IJCAR 2020, LNAI 12166, pp. 181–200, 2020.
https://doi.org/10.1007/978-3-030-51074-9_11

list includes [1,11,17–19,23,28,31,32]). The main results from the above papers are that uniform interpolants exist for intuitionistic logic and for some modal systems (like the Gödel-Löb system and the S4.Grz system); they do not exist for instance in $S4$ and $K4$, whereas for the basic modal system K they exist for the local consequence relation but not for the global consequence relation. The connection between uniform interpolants and model completions (for equational theories axiomatizing the varieties corresponding to propositional logics) was first stated in [20] and further developed in [17,23,31].

In the last decade, also the automated reasoning community developed an increasing interest in uniform interpolants, with particular focus on quantifier-free fragments of first-order theories. This is witnessed by various talks and drafts by D. Kapur presented in many conferences and workshops (FloC 2010, ISCAS 2013-14, SCS 2017 [22]), as well as by the paper [21] by Gulwani and Musuvathi in ESOP 2008. In this last paper uniform interpolants were renamed as *covers*, a terminology we shall adopt in this paper too. In these contributions, examples of cover computations were supplied and also some algorithms were sketched. The first formal *proof* about existence of covers in $\mathcal{EUF}$ was however published by the present authors only in [7]; such a proof was equipped with powerful semantic tools (the Cover-by-Extensions Lemma 1 below) coming from the connection to model-completeness, as well as with an algorithm relying on a constrained variant of the Superposition Calculus (two simpler algorithms are studied in [15]). The usefulness of covers in model checking was already stressed in [21] and further motivated by our recent line of research on the verification of data-aware processes [5,6,8,9]. Notably, it is also operationally mirrored in the MCMT [16] implementation since version 2.8. Covers (via quantifier elimination in model completions and hierarchical reasoning) play an important role in symbol elimination problems in theory extensions, as witnessed in the comprehensive paper [29] and in related papers (e.g., [25]) studying invariant synthesis in model checking applications.

An important question suggested by the applications is the cover transfer problem for combined theories: for instance, when modeling and verifying data-aware processes, it is natural to consider the combination of different theories, such as the theories accounting for the read-write and read-only data storage of the process as well as those for the elements stored therein [6–8]. Formally, the cover transfer problem can be stated as follows: *by supposing that covers exist in theories T_1, T_2, under which conditions do they exist also in the combined theory $T_1 \cup T_2$?* In this paper we show that the answer is affirmative in the disjoint signatures convex case, using the same hypothesis (that is, the equality interpolating condition) under which quantifier-free interpolation transfers. Thus, for convex theories we essentially obtain a necessary and sufficient condition, in the precise sense captured by Theorem 6 below. We also prove that if convexity fails, the non-convex equality interpolating property [2] may not be sufficient to ensure the cover transfer property. As a witness for this, we show that $\mathcal{EUF}$ combined with integer difference logic or with linear integer arithmetics constitutes a counterexample.

The main tool employed in our combination result is the *Beth definability theorem for primitive formulae* (this theorem has been shown to be equivalent to the equality interpolating condition in [2]). In order to design a combined cover algorithm, we exploit the equivalence between implicit and explicit definability that is supplied by the Beth theorem. Implicit definability is reformulated, via covers for input theories, at the quantifier-free level. Thus, the combined cover algorithm guesses the implicitly definable variables, then eliminates them via explicit definability, and finally uses the component-wise input cover algorithms to eliminate the remaining (non implicitly definable) variables. The identification and the elimination of the implicitly defined variables via explicitly defining terms is an essential step towards the correctness of the combined cover algorithm: when computing a cover of a formula $\phi(\underline{x}, \underline{y})$ (w.r.t. $\underline{y}$), the variables $\underline{x}$ are (non-eliminable) parameters, and those variables among the $\underline{y}$ that are implicitly definable *need to be discovered and treated in the same way as the parameters* $\underline{x}$. Only after this preliminary step (Lemma 5 below), the input cover algorithms can be suitably exploited (Proposition 1 below).

The combination result we obtain is quite strong, as it is a typical 'black box' combination result: it applies not only to theories used in verification (like the combination of real arithmetics with $\mathcal{EUF}$), but also in other contexts. For instance, since the theory $\mathcal{B}$ of Boolean algebras satisfies our hypotheses (being model completable and strongly amalgamable [14]), we get that uniform interpolants exist in the combination of $\mathcal{B}$ with $\mathcal{EUF}$. The latter is the equational theory algebraizing the basic non-normal classical modal logic system **E** from [27] (extended to n-ary modalities). Notice that this result must be contrasted with the case of many systems of Boolean algebras with operators where existence of uniform interpolation fails [23] (recall that operators on a Boolean algebra are not just arbitrary functions, but are required to be monotonic and also to preserve either joins or meets in each coordinate).

As a last important comment on related work, it is worth mentioning that Gulwani and Musuvathi in [21] also have a combined cover algorithm for convex, signature disjoint theories. Their algorithm looks quite different from ours; apart from the fact that a full correctness and completeness proof for such an algorithm has never been published, we underline that our algorithm is rooted on different hypotheses. In fact, we only need the equality interpolating condition and we show that this hypothesis is not only sufficient, but also necessary for cover transfer in convex theories; consequently, our result is formally stronger. The equality interpolating condition was known to the authors of [21] (but not even mentioned in their paper [21]): in fact, it was introduced by one of them some years before [33]. The equality interpolating condition was then extended to the non convex case in [2], where it was also semantically characterized via the strong amalgamation property.

The paper is organized as follows: after some preliminaries in Sect. 2, the crucial Covers-by-Extensions Lemma and the relationship between covers and model completions from [7] are recalled in Sect. 3. In Sect. 4, we present some preliminary results on interpolation and Beth definability that are instrumental to our machinery. After some useful facts about convex theories in Sect. 5, we introduce the combined cover algorithms for the convex case and we prove its correctness in Sect. 6; we also present a detailed example of application of the combined algorithm in case of the combination of $\mathcal{EUF}$ with linear real arithmetic, and we show that the equality interpolating condition is necessary (in some sense) for combining covers. In Sect. 7 we exhibit a counteraxample to the existence of combined covers in the non-convex case. Section 8 is devoted to the conclusions and discussion of future work. The extended version of the current paper with full proofs and details is available online in [4].

2 Preliminaries

We adopt the usual first-order syntactic notions of signature, term, atom, (ground) formula, and so on; our signatures are always *finite* or *countable* and include equality. To avoid considering limit cases, we assume that signatures always contain at least an individual constant. We compactly represent a tuple $\langle x_1, \dots, x_n \rangle$ of variables as $\underline{x}$. The notation $t(\underline{x}), \phi(\underline{x})$ means that the term t, the formula ϕ has free variables included in the tuple $\underline{x}$. This tuple is assumed to be formed by *distinct variables*, thus we underline that when we write e.g. $\phi(\underline{x}, \underline{y})$, we mean that the tuples $\underline{x}, \underline{y}$ are made of distinct variables that are also disjoint from each other.

A formula is said to be *universal* (resp., *existential*) if it has the form $\forall \underline{x}(\phi(\underline{x}))$ (resp., $\exists \underline{x}(\phi(\underline{x}))$), where ϕ is quantifier-free. Formulae with no free variables are called *sentences*. On the semantic side, we use the standard notion of Σ-structure $\mathcal{M}$ and of truth of a formula in a Σ-structure under a free variables assignment. The support of $\mathcal{M}$ is denoted as $|\mathcal{M}|$. The interpretation of a (function, predicate) symbol σ in $\mathcal{M}$ is denoted $\sigma^{\mathcal{M}}$.

A *Σ-theory* T is a set of Σ-sentences; a *model* of T is a Σ-structure $\mathcal{M}$ where all sentences in T are true. We use the standard notation $T \models \phi$ to say that ϕ is true in all models of T for every assignment to the variables occurring free in ϕ. We say that ϕ is *T-satisfiable* iff there is a model $\mathcal{M}$ of T and an assignment to the variables occurring free in ϕ making ϕ true in $\mathcal{M}$.

We now focus on the constraint satisfiability problem and quantifier elimination for a theory T. A Σ-formula ϕ is a *Σ-constraint* (or just a constraint) iff it is a conjunction of literals. The *constraint satisfiability problem* for T is the following: we are given a constraint $\phi(\underline{x})$ and we are asked whether there exist a model $\mathcal{M}$ of T and an assignment $\mathcal{I}$ to the free variables $\underline{x}$ such that $\mathcal{M}, \mathcal{I} \models \phi(\underline{x})$. A theory T has *quantifier elimination* iff for every formula $\phi(\underline{x})$ in the signature of T there is a quantifier-free formula $\phi'(\underline{x})$ such that $T \models \phi(\underline{x}) \leftrightarrow \phi'(\underline{x})$. Since we are in a computational logic context, when we speak of quantifier elimination, we assume that it is effective, namely that it comes with an algorithm for computing ϕ' out of ϕ. It is well-known that quantifier elimination holds in case we can eliminate quantifiers from *primitive* formulae, i.e., formulae of the kind $\exists \underline{y}\, \phi(\underline{x}, \underline{y})$, with ϕ a constraint.

We recall also some further basic notions. Let Σ be a first-order signature. The signature obtained from Σ by adding to it a set $\underline{a}$ of new constants (i.e.,

0-ary function symbols) is denoted by $\Sigma^{\underline{a}}$. Analogously, given a Σ-structure $\mathcal{M}$, the signature Σ can be expanded to a new signature $\Sigma^{|\mathcal{M}|} := \Sigma \cup \{\bar{a} \mid a \in |\mathcal{M}|\}$ by adding a set of new constants $\bar{a}$ (the *name* for a), one for each element a in the support of $\mathcal{M}$, with the convention that two distinct elements are denoted by different "name" constants. $\mathcal{M}$ can be expanded to a $\Sigma^{|\mathcal{M}|}$-structure $\overline{\mathcal{M}} := (\mathcal{M}, a)_{a \in |\mathcal{M}|}$ just interpreting the additional constants over the corresponding elements. From now on, when the meaning is clear from the context, we will freely use the notation $\mathcal{M}$ and $\overline{\mathcal{M}}$ interchangeably: in particular, given a Σ-structure $\mathcal{M}$ and a Σ-formula $\phi(\underline{x})$ with free variables that are all in $\underline{x}$, we will write, by abuse of notation, $\mathcal{M} \models \phi(\underline{a})$ instead of $\overline{\mathcal{M}} \models \phi(\underline{\bar{a}})$.

A *Σ-homomorphism* (or, simply, a homomorphism) between two Σ-structures $\mathcal{M}$ and $\mathcal{N}$ is a map $\mu : |\mathcal{M}| \longrightarrow |\mathcal{N}|$ among the support sets $|\mathcal{M}|$ of $\mathcal{M}$ and $|\mathcal{N}|$ of $\mathcal{N}$ satisfying the condition ($\mathcal{M} \models \varphi \quad \Rightarrow \quad \mathcal{N} \models \varphi$) for all $\Sigma^{|\mathcal{M}|}$-atoms φ ($\mathcal{M}$ is regarded as a $\Sigma^{|\mathcal{M}|}$-structure, by interpreting each additional constant $a \in |\mathcal{M}|$ into itself and $\mathcal{N}$ is regarded as a $\Sigma^{|\mathcal{M}|}$-structure by interpreting each additional constant $a \in |\mathcal{M}|$ into $\mu(a)$). In case the last condition holds for all $\Sigma^{|\mathcal{M}|}$-literals, the homomorphism μ is said to be an *embedding* and if it holds for all first order formulae, the embedding μ is said to be *elementary*. If $\mu : \mathcal{M} \longrightarrow \mathcal{N}$ is an embedding which is just the identity inclusion $|\mathcal{M}| \subseteq |\mathcal{N}|$, we say that $\mathcal{M}$ is a *substructure* of $\mathcal{N}$ or that $\mathcal{N}$ is an *extension* of $\mathcal{M}$. Universal theories can be characterized as those theories T having the property that if $\mathcal{M} \models T$ and $\mathcal{N}$ is a substructure of $\mathcal{M}$, then $\mathcal{N} \models T$ (see [10]). If $\mathcal{M}$ is a structure and $X \subseteq |\mathcal{M}|$, then there is the smallest substructure of $\mathcal{M}$ including X in its support; this is called the substructure *generated by* X. If X is the set of elements of a finite tuple $\underline{a}$, then the substructure generated by X has in its support precisely the $b \in |\mathcal{M}|$ such that $\mathcal{M} \models b = t(\underline{a})$ for some term t.

Let $\mathcal{M}$ be a Σ-structure. The *diagram* of $\mathcal{M}$, written $\Delta_{\Sigma}(\mathcal{M})$ (or just $\Delta(\mathcal{M})$), is the set of ground $\Sigma^{|\mathcal{M}|}$-literals that are true in $\mathcal{M}$. An easy but important result, called *Robinson Diagram Lemma* [10], says that, given any Σ-structure $\mathcal{N}$, the embeddings $\mu : \mathcal{M} \longrightarrow \mathcal{N}$ are in bijective correspondence with expansions of $\mathcal{N}$ to $\Sigma^{|\mathcal{M}|}$-structures which are models of $\Delta_{\Sigma}(\mathcal{M})$. The expansions and the embeddings are related in the obvious way: $\bar{a}$ is interpreted as $\mu(a)$.

3 Covers and Model Completions

We report the notion of *cover* taken from [21] and also the basic results proved in [7]. Fix a theory T and an existential formula $\exists \underline{e}\, \phi(\underline{e}, \underline{y})$; call a *residue* of $\exists \underline{e}\, \phi(\underline{e}, \underline{y})$ any quantifier-free formula belonging to the set of quantifier-free formulae $Res(\exists \underline{e}\, \phi) = \{\theta(\underline{y}, \underline{z}) \mid T \models \phi(\underline{e}, \underline{y}) \to \theta(\underline{y}, \underline{z})\}$. A quantifier-free formula $\psi(\underline{y})$ is said to be a *T-cover* (or, simply, a *cover*) of $\exists \underline{e}\, \phi(\underline{e}, \underline{y})$ iff $\psi(\underline{y}) \in Res(\exists \underline{e}\, \phi)$ and $\psi(\underline{y})$ implies (modulo T) all the other formulae in $Res(\exists \underline{e}\, \phi)$. The following "cover-by-extensions" Lemma [7] (to be widely used throughout the paper) supplies a semantic counterpart to the notion of a cover:

Lemma 1 (Cover-by-Extensions). *A formula $\psi(\underline{y})$ is a T-cover of $\exists \underline{e}\, \phi(\underline{e}, \underline{y})$ iff it satisfies the following two conditions:(i) $T \models \forall \underline{y}\, (\exists \underline{e}\, \phi(\underline{e}, \underline{y}) \to \psi(\underline{y}))$; (ii)* *for every model $\mathcal{M}$ of T, for every tuple of elements $\underline{a}$ from the support of $\mathcal{M}$ such that $\mathcal{M} \models \psi(\underline{a})$ it is possible to find another model $\mathcal{N}$ of T such that $\mathcal{M}$ embeds into $\mathcal{N}$ and $\mathcal{N} \models \exists \underline{e}\, \phi(\underline{e}, \underline{a})$.* ◁

We underline that, since our language is at most countable, we can assume that the models $\mathcal{M}$, $\mathcal{N}$ from (ii) above are at most countable too, by a Löwenheim-Skolem argument.

We say that a theory T has *uniform quantifier-free interpolation* iff every existential formula $\exists \underline{e}\, \phi(\underline{e}, \underline{y})$ (equivalently, every primitive formula $\exists \underline{e}\, \phi(\underline{e}, \underline{y})$) has a T-cover.

It is clear that if T has uniform quantifier-free interpolation, then it has ordinary *quantifier-free interpolation* [2], in the sense that if we have $T \models \phi(\underline{e}, \underline{y}) \rightarrow \phi'(\underline{y}, \underline{z})$ (for quantifier-free formulae ϕ, ϕ'), then there is a quantifier-free formula $\theta(\underline{y})$ such that $T \models \phi(\underline{e}, \underline{y}) \rightarrow \theta(\underline{y})$ and $T \models \theta(\underline{y}) \rightarrow \phi'(\underline{y}, \underline{z})$. In fact, if T has uniform quantifier-free interpolation, then the interpolant θ is independent on ϕ' (the same $\theta(\underline{y})$ can be used as interpolant for all entailments $T \models \phi(\underline{e}, \underline{y}) \rightarrow \phi'(\underline{y}, \underline{z})$, varying ϕ').

We say that a *universal* theory T has a *model completion* iff there is a stronger theory $T^* \supseteq T$ (still within the same signature Σ of T) such that (i) every Σ-constraint that is satisfiable in a model of T is satisfiable in a model of T^*; (ii) T^* eliminates quantifiers. Other equivalent definitions are possible [10]: for instance, (i) is equivalent to the fact that T and T^* prove the same universal formulae or again to the fact that every model of T can be embedded into a model of T^*. We recall that the model completion, if it exists, is unique and that its existence implies the quantifier-free interpolation property for T [10] (the latter can be seen directly or via the correspondence between quantifier-free interpolation and amalgamability, see [2]).

A close relationship between model completion and uniform interpolation emerged in the area of propositional logic (see the book [17]) and can be formulated roughly as follows. It is well-known that most propositional calculi, via Lindembaum constructions, can be algebraized: the algebraic analogue of classical logic are Boolean algebras, the algebraic analogue of intuitionistic logic are Heyting algebras, the algebraic analogue of modal calculi are suitable variaties of modal agebras, etc. Under suitable hypotheses, it turns out that a propositional logic has uniform interpolation (for the global consequence relation) iff the equational theory axiomatizing the corresponding variety of algebras has a model completion [17]. In the context of first order theories, we prove an even more direct connection:

Theorem 1. *Suppose that T is a universal theory. Then T has a model completion T^* iff T has uniform quantifier-free interpolation. If this happens, T^* is axiomatized by the infinitely many sentences $\forall \underline{y}\, (\psi(\underline{y}) \rightarrow \exists \underline{e}\, \phi(\underline{e}, \underline{y}))$, where $\exists \underline{e}\, \phi(\underline{e}, \underline{y})$ is a primitive formula and ψ is a cover of it.* ◁

The proof (via Lemma 1, by iterating a chain construction) is in [9] (see also [3]).

4 Equality Interpolating Condition and Beth Definability

We report here some definitions and results we need concerning combined quantifier-free interpolation. Most definitions and result come from [2], but are simplified here because we restrict them to the case of universal convex theories. Further information on the semantic side is supplied in the extended version of this paper [4].

A theory T is *stably infinite* iff every T-satisfiable constraint is satisfiable in an infinite model of T. The following Lemma comes from a compactness argument (see [4] for a proof):

Lemma 2. *If T is stably infinite, then every finite or countable model $\mathcal{M}$ of T can be embedded in a model $\mathcal{N}$ of T such that $|\mathcal{N}| \setminus |\mathcal{M}|$ is countable.* ◁

A theory T is *convex* iff for every constraint δ, if $T \vdash \delta \to \bigvee_{i=1}^{n} x_i = y_i$ then $T \vdash \delta \to x_i = y_i$ holds for some $i \in \{1, ..., n\}$. A convex theory T is 'almost' stably infinite in the sense that it can be shown that every constraint which is T-satisfiable in a T-model whose support has at least two elements is satisfiable also in an infinite T-model. The one-element model can be used to build counterexamples, though: e.g., the theory of Boolean algebras is convex (like any other universal Horn theory) but the constraint $x = 0 \wedge x = 1$ is only satisfiable in the degenerate one-element Boolean algebra. Since we take into account these limit cases, we do not assume that convexity implies stable infiniteness.

Definition 1. *A convex universal theory T is* equality interpolating *iff for every pair y_1, y_2 of variables and for every pair of* constraints $\delta_1(\underline{x}, \underline{z}_1, y_1), \delta_2(\underline{x}, \underline{z}_2, y_2)$ *such that*

$$T \vdash \delta_1(\underline{x}, \underline{z}_1, y_1) \wedge \delta_2(\underline{x}, \underline{z}_2, y_2) \to y_1 = y_2 \tag{1}$$

there exists a term $t(\underline{x})$ such that

$$T \vdash \delta_1(\underline{x}, \underline{z}_1, y_1) \wedge \delta_2(\underline{x}, \underline{z}_2, y_2) \to y_1 = t(\underline{x}) \wedge y_2 = t(\underline{x}). \tag{2}$$

◁

Theorem 2 [2,33]. *Let T_1 and T_2 be two universal, convex, stably infinite theories over disjoint signatures Σ_1 and Σ_2. If both T_1 and T_2 are equality interpolating and have quantifier-free interpolation property, then so does $T_1 \cup T_2$.* ◁

There is a converse of the previous result; for a signature Σ, let us call $\mathcal{EUF}(\Sigma)$ the pure equality theory over the signature Σ (this theory is equality interpolating and has the quantifier-free interpolation property).

Theorem 3 [2]. *Let T be a stably infinite, universal, convex theory admitting quantifier-free interpolation and let Σ be a signature disjoint from the signature of T containing at least a unary predicate symbol. Then, $T \cup \mathcal{EUF}(\Sigma)$ has quantifier-free interpolation iff T is equality interpolating.* ◁

In [2] the above definitions and results are extended to the non-convex case and a long list of universal quantifier-free interpolating and equality interpolating theories is given. The list includes $\mathcal{EUF}(\Sigma)$, recursive data theories, as well as linear arithmetics. For linear arithmetics (and fragments of its), it is essential to make a very careful choice of the signature, see again [2] (especially Subsection 4.1) for details. All the above theories admit a model completion (which coincides with the theory itself in case the theory admits quantifier elimination).

The equality interpolating property in a theory T can be equivalently characterized using Beth definability as follows. Consider a primitive formula $\exists \underline{z}\phi(\underline{x}, \underline{z}, y)$ (here ϕ is a conjunction of literals); we say that $\exists \underline{z}\, \phi(\underline{x}, \underline{z}, y)$ *implicitly defines* y *in* T iff the formula

$$\forall y \forall y' \, (\exists \underline{z}\phi(\underline{x}, \underline{z}, y) \wedge \exists \underline{z}\phi(\underline{x}, \underline{z}, y') \to y = y') \tag{3}$$

is T-valid. We say that $\exists \underline{z}\phi(\underline{x}, \underline{z}, y)$ *explicitly defines* y *in* T iff there is a term $t(\underline{x})$ such that the formula

$$\forall y \, (\exists \underline{z}\phi(\underline{x}, \underline{z}, y) \to y = t(\underline{x})) \tag{4}$$

is T-valid.

For future use, we notice that, by trivial logical manipulations, the formulae (3) and (4) are logically equivalent to

$$\forall y \forall \underline{z} \forall y' \forall \underline{z}' (\phi(\underline{x}, \underline{z}, y) \wedge \phi(\underline{x}, \underline{z}', y') \to y = y'). \tag{5}$$

and to

$$\forall y \forall \underline{z} (\phi(\underline{x}, \underline{z}, y) \to y = t(\underline{x})) \tag{6}$$

respectively (we shall use such equivalences without explicit mention).

We say that a theory T has the *Beth definability property for primitive formulae* iff whenever a primitive formula $\exists \underline{z}\, \phi(\underline{x}, \underline{z}, y)$ implicitly defines the variable y then it also explicitly defines it.

Theorem 4 [2]. *A convex theory T having quantifier-free interpolation is equality interpolating iff it has the Beth definability property for primitive formulae.* $\triangleleft$

Proof. We recall the easy proof of the left-to-right side (this is the only side we need in this paper). Suppose that T is equality interpolating and that

$$T \vdash \phi(\underline{x}, \underline{z}, y) \wedge \phi(\underline{x}, \underline{z}', y') \to y = y';$$

then there is a term $t(\underline{x})$ such that

$$T \vdash \phi(\underline{x}, \underline{z}, y) \wedge \phi(\underline{x}, \underline{z}', y') \to y = t(\underline{x}) \wedge y' = t(\underline{x}).$$

Replacing $\underline{z}', y'$ by $\underline{z}, y$ via a substitution, we get precisely (6). $\dashv$

5 Convex Theories

We now collect some useful facts concerning convex theories. We fix for this section a *convex, stably infinite, equality interpolating universal theory T admitting a model completion T^**. We let Σ be the signature of T. We fix also *a Σ-constraint $\phi(\underline{x}, \underline{y})$*, where we assume that $\underline{y} = y_1, \ldots, y_n$ (recall that the tuple $\underline{x}$ is disjoint from the tuple $\underline{y}$ according to our conventions from Sect. 2).

For $i = 1, \ldots, n$, we let the formula $\mathtt{ImplDef}^T_{\phi,y_i}(\underline{x})$ be the quantifier-free formula equivalent in T^* to the formula

$$\forall \underline{y} \forall \underline{y}' (\phi(\underline{x}, \underline{y}) \wedge \phi(\underline{x}, \underline{y}') \rightarrow y_i = y'_i) \tag{7}$$

where the $\underline{y}'$ are renamed copies of the $\underline{y}$. Notice that the variables occurring free in ϕ are $\underline{x}, \underline{y}$, whereas only the $\underline{x}$ occur free in $\mathtt{ImplDef}^T_{\phi,y_i}(\underline{x})$ (the variable y_i is among the $\underline{y}$ and does not occur free in $\mathtt{ImplDef}^T_{\phi,y_i}(\underline{x})$): these facts coming from our notational conventions are crucial and should be kept in mind when reading this and next section. The following semantic technical lemma is proved in the extended version of this paper [4]:

Lemma 3. *Suppose that we are given a model $\mathcal{M}$ of T and elements $\underline{a}$ from the support of $\mathcal{M}$ such that $\mathcal{M} \not\models \mathtt{ImplDef}^T_{\phi,y_i}(\underline{a})$ for all $i = 1, \ldots, n$. Then there exists an extension $\mathcal{N}$ of $\mathcal{M}$ such that for some $\underline{b} \in |\mathcal{N}| \setminus |\mathcal{M}|$ we have $\mathcal{N} \models \phi(\underline{a}, \underline{b})$.* ◁

The following Lemma supplies terms which will be used as ingredients in our combined covers algorithm:

Lemma 4. *Let $L_{i1}(\underline{x}) \vee \cdots \vee L_{ik_i}(\underline{x})$ be the disjunctive normal form (DNF) of $\mathtt{ImplDef}^T_{\phi,y_i}(\underline{x})$. Then, for every $j = 1, \ldots, k_i$, there is a $\Sigma(\underline{x})$-term $t_{ij}(\underline{x})$ such that*

$$T \vdash L_{ij}(\underline{x}) \wedge \phi(\underline{x}, \underline{y}) \rightarrow y_i = t_{ij}. \tag{8}$$

As a consequence, a formula of the kind $\mathtt{ImplDef}^T_{\phi,y_i}(\underline{x}) \wedge \exists \underline{y}\, (\phi(\underline{x}, \underline{y}) \wedge \psi)$ is equivalent (modulo T) to the formula

$$\bigvee_{j=1}^{k_i} \exists \underline{y}\, (y_i = t_{ij} \wedge L_{ij}(\underline{x}) \wedge \phi(\underline{x}, \underline{y}) \wedge \psi). \tag{9}$$

◁

Proof. We have that $(\bigvee_j L_{ij}) \leftrightarrow \mathtt{ImplDef}^T_{\phi,y_i}(\underline{x})$ is a tautology, hence from the definition of $\mathtt{ImplDef}^T_{\phi,y_i}(\underline{x})$, we have that

$$T^* \vdash L_{ij}(\underline{x}) \rightarrow \forall \underline{y} \forall \underline{y}' (\phi(\underline{x}, \underline{y}) \wedge \phi(\underline{x}, \underline{y}') \rightarrow y_i = y'_i);$$

however this formula is trivially equivalent to a universal formula (L_{ij} does not depend on $\underline{y}, \underline{y}'$), hence since T and T^* prove the same universal formulae, we get

$$T \vdash L_{ij}(\underline{x}) \wedge \phi(\underline{x}, \underline{y}) \wedge \phi(\underline{x}, \underline{y}') \rightarrow y_i = y'_i.$$

Using Beth definability property (Theorem 4), we get (8), as required, for some terms $t_{ij}(\underline{x})$. Finally, the second claim of the lemma follows from (8) by trivial logical manipulations. ⊣

In all our concrete examples, the theory T has decidable quantifier-free fragment (namely it is decidable whether a quantifier-free formula is a logical consequence of T or not), thus the terms t_{ij} mentioned in Lemma 4 can be computed just by enumerating all possible $\Sigma(\underline{x})$-terms: the computation terminates, because the above proof shows that the appropriate terms always exist. However, this is terribly inefficient and, from a practical point of view, one needs to have at disposal dedicated algorithms to find the required equality interpolating terms. For some common theories ($\mathcal{EUF}$, Lisp-structures, linear real arithmetic), such algorithms are designed in [33]; in [2] [Lemma 4.3 and Theorem 4.4], the algorithms for computing equality interpolating terms are connected to quantifier elimination algorithms in the case of universal theories admitting quantifier elimination. Still, an extensive investigation on te topic seems to be missed in the SMT literature.

6 The Convex Combined Cover Algorithm

Let us now fix two theories T_1, T_2 over disjoint signatures Σ_1, Σ_2. We assume that both of them satisfy the assumptions from the previous section, meaning that they are convex, stably infinite, equality interpolating, universal and admit model completions T_1^*, T_2^* respectively. We shall supply a cover algorithm for $T_1 \cup T_2$ (thus proving that $T_1 \cup T_2$ has a model completion too).

We need to compute a cover for $\exists \underline{e}\, \phi(\underline{x}, \underline{e})$, where ϕ is a conjunction of $\Sigma_1 \cup \Sigma_2$-literals. By applying rewriting purification steps like

$$\phi \Longrightarrow \exists d\, (d = t \wedge \phi(d/t))$$

(where d is a fresh variable and t is a pure term, i.e. it is either a Σ_1- or a Σ_2-term), we can assume that our formula ϕ is of the kind $\phi_1 \wedge \phi_2$, where ϕ_1 is a Σ_1-formula and ϕ_2 is a Σ_2-formula. Thus we need to compute a cover for a formula of the kind

$$\exists \underline{e}\, (\phi_1(\underline{x}, \underline{e}) \wedge \phi_2(\underline{x}, \underline{e})), \tag{10}$$

where ϕ_i is a conjunction of Σ_i-literals ($i = 1, 2$). We also assume that both ϕ_1 and ϕ_2 contain the literals $e_i \neq e_j$ (for $i \neq j$) as a conjunct: this can be achieved by guessing a partition of the $\underline{e}$ and by replacing each e_i with the representative element of its equivalence class.

Remark 1. It is not clear whether this preliminary guessing step can be avoided. In fact, Nelson-Oppen [24] combined satisfiability for *convex* theories does not need it; however, combining covers algorithms is a more complicated problem than combining mere satisfiability algorithms and for technical reasons related to the correctness and completeness proofs below, we were forced to introduce guessing at this step. ◁

To manipulate formulae, our algorithm employs acyclic explicit definitions as follows. When we write $\texttt{ExplDef}(\underline{z}, \underline{x})$ (where $\underline{z}, \underline{x}$ are tuples of distinct variables), we mean any formula of the kind (let $\underline{z} := z_1 \ldots, z_m$)

$$\bigwedge_{i=1}^{m} z_i = t_i(z_1, \ldots, z_{i-1}, \underline{x})$$

where the term t_i is pure (i.e. it is a Σ_i-term) and only the variables $z_1, \ldots, z_{i-1}, \underline{x}$ can occur in it. When we assert a formula like $\exists \underline{z}\, (\texttt{ExplDef}(\underline{z}, \underline{x}) \wedge \psi(\underline{z}, \underline{x}))$, we are in fact in the condition of recursively eliminating the variables $\underline{z}$ from it via terms containing only the parameters $\underline{x}$ (the 'explicit definitions' $z_i = t_i$ are in fact arranged acyclically).

A *working formula* is a formula of the kind

$$\exists \underline{z}\, (\texttt{ExplDef}(\underline{z}, \underline{x}) \wedge \exists \underline{e}\, (\psi_1(\underline{x}, \underline{z}, \underline{e}) \wedge \psi_2(\underline{x}, \underline{z}, \underline{e}))), \tag{11}$$

where ψ_1 is a conjunction of Σ_1-literals and ψ_2 is a conjunction of Σ_2-literals. The variables $\underline{x}$ are called *parameters*, the variables $\underline{z}$ are called *defined variables* and the variables $\underline{e}$ *(truly) existential variables.* The parameters do not change during the execution of the algorithm. We assume that ψ_1, ψ_2 in a working formula (11) always contain the literals $e_i \neq e_j$ (for distinct e_i, e_j from $\underline{e}$) as a conjunct.

In our starting formula (10), there are no defined variables. However, if via some syntactic check it happens that some of the existential variables can be recognized as defined, then it is useful to display them as such (this observation may avoid redundant cases - leading to inconsistent disjuncts - in the computations below).

A working formula like (11) is said to be *terminal* iff for every existential variable $e_i \in \underline{e}$ we have that

$$T_1 \vdash \psi_1 \to \neg\texttt{ImplDef}^{T_1}_{\psi_1, e_i}(\underline{x}, \underline{z}) \quad \text{and} \quad T_2 \vdash \psi_2 \to \neg\texttt{ImplDef}^{T_2}_{\psi_2, e_i}(\underline{x}, \underline{z}). \tag{12}$$

Roughly speaking, we can say that in a terminal working formula, all variables which are not parameters are either explicitly definable or recognized as not implicitly definable by both theories; of course, a working formula with no existential variables is terminal.

Lemma 5. *Every working formula is equivalent (modulo $T_1 \cup T_2$) to a disjunction of terminal working formulae.* ◁

Proof. We only sketch the proof of this Lemma (see the extended version [4] for full details), by describing the algorithm underlying it. To compute the required terminal working formulae, it is sufficient to apply the following non-deterministic procedure (the output is the disjunction of all possible outcomes). The non-deterministic procedure applies one of the following alternatives.

(1) Update ψ_1 by adding it a disjunct from the DNF of $\bigwedge_{e_i \in \underline{e}} \neg\texttt{ImplDef}^{T_1}_{\psi_1, e_i}(\underline{x}, \underline{z})$ and ψ_2 by adding to it a disjunct from the DNF of $\bigwedge_{e_i \in \underline{e}} \neg \texttt{ImplDef}^{T_2}_{\psi_1, e_i}(\underline{x}, \underline{z})$;

(2.i) Select $e_i \in \underline{e}$ and $h \in \{1, 2\}$; then update ψ_h by adding to it a disjunct L_{ij} from the DNF of $\mathtt{ImplDef}^{T_h}_{\psi_h, e_i}(\underline{x}, \underline{z})$; the equality $e_i = t_{ij}$ (where t_{ij} is the term mentioned in Lemma 4)[1] is added to $\mathtt{ExplDef}(\underline{z}, \underline{x})$; the variable e_i becomes in this way part of the defined variables.

If alternative (1) is chosen, the procedure stops, otherwise it is recursively applied again and again (we have one truly existential variable less after applying alternative (2.i), so we eventually terminate). ⊣

Thus we are left to the problem of computing a cover of a terminal working formula; this problem is solved in the following proposition:

Proposition 1. *A cover of a terminal working formula* (11) *can be obtained just by unravelling the explicit definitions of the variables* $\underline{z}$ *from the formula*

$$\exists \underline{z}\, (\mathtt{ExplDef}(\underline{z}, \underline{x}) \wedge \theta_1(\underline{x}, \underline{z}) \wedge \theta_2(\underline{x}, \underline{z})) \tag{13}$$

where $\theta_1(\underline{x}, \underline{z})$ *is the* T_1*-cover of* $\exists \underline{e} \psi_1(\underline{x}, \underline{z}, \underline{e})$ *and* $\theta_2(\underline{x}, \underline{z})$ *is the* T_2*-cover of* $\exists \underline{e} \psi_2(\underline{x}, \underline{z}, \underline{e})$. ◁

Proof. In order to show that Formula (13) is the $T_1 \cup T_2$-cover of a terminal working formula (11), we prove, by using the Cover-by-Extensions Lemma 1, that, for every $T_1 \cup T_2$-model $\mathcal{M}$, for every tuple $\underline{a}, \underline{c}$ from $|\mathcal{M}|$ such that $\mathcal{M} \models \theta_1(\underline{a}, \underline{c}) \wedge \theta_2(\underline{a}, \underline{c})$ there is an extension $\mathcal{N}$ of $\mathcal{M}$ such that $\mathcal{N}$ is still a model of $T_1 \cup T_2$ and $\mathcal{N} \models \exists \underline{e}(\psi_1(\underline{a}, \underline{c}, \underline{e}) \wedge \psi_2(\underline{a}, \underline{c}, \underline{e}))$. By a Löwenheim-Skolem argument, since our languages are countable, we can suppose that $\mathcal{M}$ is at most countable and actually that it is countable by stable infiniteness of our theories, see Lemma 2 (the fact that $T_1 \cup T_2$ is stably infinite in case both T_1, T_2 are such, comes from the proof of Nelson-Oppen combination result, see [12,24,30]).

According to the conditions (12) and the definition of a cover (notice that the formulae $\neg\mathtt{ImplDef}^{T_h}_{\psi_h, e_i}(\underline{x}, \underline{z})$ do not contain the $\underline{e}$ and are quantifier-free) we have that

$$T_1 \vdash \theta_1 \to \neg\mathtt{ImplDef}^{T_1}_{\psi_1, e_i}(\underline{x}, \underline{z}) \quad \text{and} \quad T_2 \vdash \theta_2 \to \neg\mathtt{ImplDef}^{T_2}_{\psi_2, e_i}(\underline{x}, \underline{z})$$

(for every $e_i \in \underline{e}$). Thus, since $\mathcal{M} \not\models \mathtt{ImplDef}^{T_1}_{\psi_1, e_i}(\underline{a}, \underline{c})$ and $\mathcal{M} \not\models \mathtt{ImplDef}^{T_2}_{\psi_2, e_i}(\underline{a}, \underline{c})$ holds for every $e_i \in \underline{e}$, we can apply Lemma 3 and conclude that there exist a T_1-model $\mathcal{N}_1$ and a T_2-model $\mathcal{N}_2$ such that $\mathcal{N}_1 \models \psi_1(\underline{a}, \underline{c}, \underline{b}_1)$ and $\mathcal{N}_2 \models \psi_2(\underline{a}, \underline{c}, \underline{b}_2)$ for tuples $\underline{b}_1 \in |\mathcal{N}_1|$ and $\underline{b}_2 \in |\mathcal{N}_2|$, both disjoint from $|\mathcal{M}|$. By a Löwenheim-Skolem argument, we can suppose that $\mathcal{N}_1, \mathcal{N}_2$ are countable and by Lemma 2 even that they are both countable extensions of $\mathcal{M}$.

The tuples $\underline{b}_1$ and $\underline{b}_2$ have equal length because the ψ_1, ψ_2 from our working formulae entail $e_i \neq e_j$, where e_i, e_j are different existential variables. Thus there is a bijection $\iota : |\mathcal{N}_1| \to |\mathcal{N}_2|$ fixing all elements in $\mathcal{M}$ and mapping componentwise the $\underline{b}_1$ onto the $\underline{b}_2$. But this means that, exactly as it happens in the proof of

[1] Lemma 4 is used taking as $\underline{y}$ the tuple $\underline{e}$, as $\underline{x}$ the tuple $\underline{x}, \underline{z}$, as $\phi(\underline{x}, \underline{y})$ the formula $\psi_h(\underline{x}, \underline{z}, \underline{e})$ and as ψ the formula ψ_{3-h}.

the completeness of the Nelson-Oppen combination procedure, the Σ_2-structure on $\mathcal{N}_2$ can be moved back via ι^{-1} to $|\mathcal{N}_1|$ in such a way that the Σ_2-substructure from $\mathcal{M}$ is fixed and in such a way that the tuple $\underline{b}_2$ is mapped to the tuple $\underline{b}_1$. In this way, $\mathcal{N}_1$ becomes a $\Sigma_1 \cup \Sigma_2$-structure which is a model of $T_1 \cup T_2$ and which is such that $\mathcal{N}_1 \models \psi_1(\underline{a}, \underline{c}, \underline{b}_1) \wedge \psi_2(\underline{a}, \underline{c}, \underline{b}_1)$, as required. ⊣

From Lemma 5, Proposition 1 and Theorem 1, we immediately get

Theorem 5. *Let T_1, T_2 be convex, stably infinite, equality interpolating, universal theories over disjoint signatures admitting a model completion. Then $T_1 \cup T_2$ admits a model completion too. Covers in $T_1 \cup T_2$ can be effectively computed as shown above.* ◁

Notice that the input cover algorithms in the above combined cover computation algorithm are used not only in the final step described in Proposition 1, but also every time we need to compute a formula $\texttt{ImplDef}^{T_h}_{\psi_h, e_i}(\underline{x}, \underline{z})$: according to its definition, this formula is obtained by eliminating quantifiers in T_i^* from (7) (this is done via a cover computation, reading $\forall$ as $\neg\exists\neg$). In practice, implicit definability is not very frequent, so that in many concrete cases $\texttt{ImplDef}^{T_h}_{\psi_h, e_i}(\underline{x}, \underline{z})$ is trivially equivalent to $\bot$ (in such cases, Step (2.i) above can obviously be disregarded).

An Example. We now analyze an example in detail. Our results apply for instance to the case where T_1 is $\mathcal{EUF}(\Sigma)$ and T_2 is linear real arithmetic. We recall that covers are computed in real arithmetic by quantifier elimination, whereas for $\mathcal{EUF}(\Sigma)$ one can apply the superposition-based algorithm from [7]. Let us show that the cover of

$$\exists e_1 \cdots \exists e_4 \begin{pmatrix} e_1 = f(x_1) \ \wedge \ e_2 = f(x_2) \ \wedge \\ \wedge \ f(e_3) = e_3 \ \wedge \ \ f(e_4) = x_1 \ \wedge \\ \wedge \ x_1 + e_1 \leq e_3 \ \wedge \ e_3 \leq x_2 + e_2 \ \wedge \ e_4 = x_2 + e_3 \end{pmatrix} \tag{14}$$

is the following formula

$$\begin{array}{l} [x_2 = 0 \ \wedge \ f(x_1) = x_1 \ \wedge \ x_1 \leq 0 \ \wedge \ x_1 \leq f(0)] \ \vee \\ \vee \ [x_1 + f(x_1) < x_2 + f(x_2) \ \wedge \ x_2 \neq 0] \ \vee \\ \vee \ \begin{bmatrix} x_2 \neq 0 \ \wedge \ x_1 + f(x_1) = x_2 + f(x_2) \ \wedge \ f(2x_2 + f(x_2)) = x_1 \ \wedge \\ \wedge \ f(x_1 + f(x_1)) = x_1 + f(x_1) \end{bmatrix} \end{array} \tag{15}$$

Formula (14) is already purified. Notice also that the variables e_1, e_2 are in fact already explicitly defined (only e_3, e_4 are truly existential variables).

We first make the partition guessing. There is no need to involve defined variables into the partition guessing, hence we need to consider only two partitions; they are described by the following formulae:

$$\begin{aligned} P_1(e_3, e_4) &\equiv e_3 \neq e_4 \\ P_2(e_3, e_4) &\equiv e_3 = e_4 \end{aligned}$$

We first analyze **the case of** P_1. The formulae ψ_1 and ψ_2 to which we need to apply exhaustively Step (1) and Step (2.i) of our algorithm are:

$$\psi_1 \equiv f(e_3) = e_3 \wedge f(e_4) = x_1 \wedge e_3 \neq e_4$$
$$\psi_2 \equiv x_1 + e_1 \leq e_3 \wedge e_3 \leq x_2 + e_2 \wedge e_4 = x_2 + e_3 \wedge e_3 \neq e_4$$

We first compute the implicit definability formulae for the truly existential variables with respect to both T_1 and T_2.

- We first consider $\mathtt{ImplDef}^{T_1}_{\psi_1,e_3}(\underline{x},\underline{z})$. Here we show that the cover of the negation of formula (7) is equivalent to $\top$ (so that $\mathtt{ImplDef}^{T_1}_{\psi_1,e_3}(\underline{x},\underline{z})$ is equivalent to $\bot$). We must quantify over truly existential variables and their duplications, thus we need to compute the cover of

$$f(e_3') = e_3' \wedge f(e_3) = e_3 \wedge f(e_4') = x_1 \wedge f(e_4) = x_1 \wedge e_3 \neq e_4 \wedge e_3' \neq e_4' \wedge e_3' \neq e_3$$

  This is a saturated set according to the superposition based procedure of [7], hence the result is $\top$, as claimed.
- The formula $\mathtt{ImplDef}^{T_1}_{\psi_1,e_4}(\underline{x},\underline{z})$ is also equivalent to $\bot$, by the same argument as above.
- To compute $\mathtt{ImplDef}^{T_2}_{\psi_2,e_3}(\underline{x},\underline{z})$ we use Fourier-Motzkin quantifier elimination. We need to eliminate the variables e_3, e_3', e_4, e_4' (intended as existentially quantified variables) from

$$x_1 + e_1 \leq e_3' \leq x_2 + e_2 \wedge x_1 + e_1 \leq e_3 \leq x_2 + e_2 \wedge e_4' = x_2 + e_3' \wedge$$
$$\wedge e_4 = x_2 + e_3 \wedge e_3 \neq e_4 \wedge e_3' \neq e_4' \wedge e_3' \neq e_3.$$

  This gives $x_1 + e_1 \neq x_2 + e_2 \wedge x_2 \neq 0$, so that $\mathtt{ImplDef}^{T_2}_{\psi_2,e_3}(\underline{x},\underline{z})$ is $x_1 + e_1 = x_2 + e_2 \wedge x_2 \neq 0$. The corresponding equality interpolating term for e_3 is $x_1 + e_1$.
- The formula $\mathtt{ImplDef}^{T_2}_{\psi_2,e_4}(\underline{x},\underline{z})$ is also equivalent to $x_1 + e_1 = x_2 + e_2 \wedge x_2 \neq 0$ and the equality interpolating term for e_4 is $x_1 + e_1 + x_2$.

So, if we apply Step 1 we get

$$\exists e_1 \cdots \exists e_4 \begin{pmatrix} e_1 = f(x_1) \wedge e_2 = f(x_2) \wedge \\ \wedge f(e_3) = e_3 \wedge f(e_4) = x_1 \wedge e_3 \neq e_4 \wedge \\ \wedge x_1 + e_1 \leq e_3 \wedge e_3 \leq x_2 + e_2 \wedge e_4 = x_2 + e_3 \wedge x_1 + e_1 \neq x_2 + e_2 \end{pmatrix} \tag{16}$$

(notice that the literal $x_2 \neq 0$ is entailed by ψ_2, so we can simplify it to $\top$ in $\mathtt{ImplDef}^{T_2}_{\psi_2,e_3}(\underline{x},\underline{z})$ and $\mathtt{ImplDef}^{T_2}_{\psi_2,e_4}(\underline{x},\underline{z})$). If we apply Step (2.i) (for i = 3), we get (after removing implied equalities)

$$\exists e_1 \cdots \exists e_4 \begin{pmatrix} e_1 = f(x_1) \wedge e_2 = f(x_2) \wedge e_3 = x_1 + e_1 \wedge \\ \wedge f(e_3) = e_3 \wedge f(e_4) = x_1 \wedge e_3 \neq e_4 \wedge \\ \wedge e_4 = x_2 + e_3 \wedge x_1 + e_1 = x_2 + e_2 \end{pmatrix} \tag{17}$$

Step (2.i) (for i = 4) gives a formula logically equivalent to (17). Notice that (17) is terminal too, because all existential variables are now explicitly defined (this is a lucky side-effect of the fact that e_3 has been moved to the defined variables). Thus the exhaustive application of Steps (1) and (2.i) is concluded.

Applying the final step of Proposition 1 to (17) is quite easy: it is sufficient to unravel the acyclic definitions. The result, after little simplification, is

$$\begin{aligned} &x_2 \neq 0 \wedge x_1 + f(x_1) = x_2 + f(x_2) \wedge \\ &\wedge f(x_2 + f(x_1 + f(x_1))) = x_1 \wedge f(x_1 + f(x_1)) = x_1 + f(x_1); \end{aligned}$$

this can be further simplified to

$$\begin{aligned} &x_2 \neq 0 \wedge x_1 + f(x_1) = x_2 + f(x_2) \wedge \\ &\wedge f(2x_2 + f(x_2)) = x_1 \wedge f(x_1 + f(x_1)) = x_1 + f(x_1); \end{aligned} \tag{18}$$

As to formula (16), we need to apply the final cover computations mentioned in Proposition 1. The formulae ψ_1 and ψ_2 are now

$$\begin{aligned} \psi_1' &\equiv f(e_3) = e_3 \wedge f(e_4) = x_1 \wedge e_3 \neq e_4 \\ \psi_2' &\equiv x_1 + e_1 \leq e_3 \leq x_2 + e_2 \wedge e_4 = x_2 + e_3 \wedge x_1 + e_1 \neq x_2 + e_2 \wedge e_3 \neq e_4 \end{aligned}$$

The T_1-cover of ψ_1' is $\top$. For the T_2-cover of ψ_2', eliminating with Fourier-Motzkin the variables e_4 and e_3, we get

$$x_1 + e_1 < x_2 + e_2 \wedge x_2 \neq 0$$

which becomes

$$x_1 + f(x_1) < x_2 + f(x_2) \wedge x_2 \neq 0 \tag{19}$$

after unravelling the explicit definitions of e_1, e_2. Thus, *the analysis of the case of the partition P_1 gives, as a result, the disjunction of* (18) *and* (19).

We now analyze **the case of P_2**. Before proceeding, we replace e_4 with e_3 (since P_2 precisely asserts that these two variables coincide); our formulae ψ_1 and ψ_2 become

$$\begin{aligned} \psi_1'' &\equiv f(e_3) = e_3 \wedge f(e_3) = x_1 \\ \psi_2'' &\equiv x_1 + e_1 \leq e_3 \wedge e_3 \leq x_2 + e_2 \wedge 0 = x_2 \end{aligned}$$

From ψ_1'' we deduce $e_3 = x_1$, thus we can move e_3 to the explicitly defined variables (this avoids useless calculations: the implicit definability condition for variables having an entailed explicit definition is obviously $\top$, so making case split on it produces either tautological consequences or inconsistencies). In this way we get the terminal working formula

$$\exists e_1 \cdots \exists e_3 \begin{pmatrix} e_1 = f(x_1) \wedge e_2 = f(x_2) \wedge e_3 = x_1 \\ \wedge f(e_3) = e_3 \wedge f(e_3) = x_1 \wedge \\ \wedge x_1 + e_1 \leq e_3 \wedge e_3 \leq x_2 + e_2 \wedge 0 = x_2 \end{pmatrix} \tag{20}$$

Unravelling the explicit definitions, we get (after exhaustive simplifications)

$$x_2 = 0 \wedge f(x_1) = x_1 \wedge x_1 \leq 0 \wedge x_1 \leq f(0) \tag{21}$$

Now, the disjunction of (18),(19) and (21) is precisely the final result (15) claimed above. This concludes our detailed analysis of our example.

Notice that the example shows that combined cover computations may introduce terms with arbitrary alternations of symbols from both theories (like $f(x_2+f(x_1+f(x_1)))$ above). The point is that when a variable becomes explicitly definable via a term in one of the theories, then using such additional variable may in turn cause some other variables to become explicitly definable via terms from the other theory, and so on and so forth; when ultimately the explicit definitions are unraveled, highly nested terms arise with many symbol alternations from both theories.

The Necessity of the Equality Interpolating Condition. The following result shows that equality interpolating is a necessary condition for a transfer result, in the sense that it is already required for minimal combinations with signatures adding uninterpreted symbols:

Theorem 6. *Let T be a convex, stably infinite, universal theory admitting a model completion and let Σ be a signature disjoint from the signature of T containing at least a unary predicate symbol. Then $T \cup \mathcal{EUF}(\Sigma)$ admits a model completion iff T is equality interpolating.* ◁

Proof. The necessity can be shown by using the following argument. By Theorem 1, $T \cup \mathcal{EUF}(\Sigma)$ has uniform quantifier-free interpolation, hence also ordinary quantifier-free interpolation. We can now apply Theorem 3 and get that T must be equality interpolating. Conversely, the sufficiency comes from Theorem 5 together with the fact that $\mathcal{EUF}(\Sigma)$ is trivially universal, convex, stably infinite, has a model completion [7] and is equality interpolating [2,33]. ⊣

7 The Non-convex Case: A Counterexample

In this section, we show by giving a suitable counterexample that the convexity hypothesis cannot be dropped from Theorems 5, 6. We make use of basic facts about ultrapowers (see [10] for the essential information we need). We take as T_1 integer difference logic $\mathcal{IDL}$, i.e. the theory of integer numbers under the unary operations of successor and predecessor, the constant 0 and the strict order relation $<$. This is stably infinite, universal and has quantifier elimination (thus it coincides with its own model completion). It is not convex, but it satisfies the equality interpolating condition, once the latter is suitably adjusted to non-convex theories, see [2] for the related definition and all the above mentioned facts.

As T_2, we take $\mathcal{EUF}(\Sigma_f)$, where Σ_f has just one unary free function symbol f (this f is supposed not to belong to the signature of T_1).

Proposition 2. *Let T_1, T_2 be as above; the formula*

$$\exists e\,(0 < e \wedge e < x \wedge f(e) = 0) \tag{22}$$

does not have a cover in $T_1 \cup T_2$. ◁

Proof. Suppose that (22) has a cover $\phi(x)$. This means (according to Cover-by-Extensions Lemma 1) that for every model $\mathcal{M}$ of $T_1 \cup T_2$ and for every element $a \in |\mathcal{M}|$ such that $\mathcal{M} \models \phi(a)$, there is an extension $\mathcal{N}$ of $\mathcal{M}$ such that $\mathcal{N} \models \exists e\,(0 < e \wedge e < a \wedge f(e) = 0)$.

Consider the model $\mathcal{M}$, so specified: the support of $\mathcal{M}$ is the set of the integers, the symbols from the signature of T_1 are interpreted in the standard way and the symbol f is interpreted so that 0 is not in the image of f. Let a_k be the number $k > 0$ (it is an element from the support of $\mathcal{M}$). Clearly it is not possible to extend $\mathcal{M}$ so that $\exists e\,(0 < e \wedge e < a_k \wedge f(e) = 0)$ becomes true: indeed, we know that all the elements in the interval $(0, k)$ are definable as iterated successors of 0 and, by using the axioms of $\mathcal{IDL}$, no element can be added between a number and its successor, hence this interval cannot be enlarged in a superstructure. We conclude that $\mathcal{M} \models \neg\phi(a_k)$ for every k.

Consider now an ultrapower $\Pi_D \mathcal{M}$ of $\mathcal{M}$ modulo a non-principal ultrafilter D and let a be the equivalence class of the tuple $\langle a_k \rangle_{k \in \mathbb{N}}$; by the fundamental Los theorem [10], $\Pi_D \mathcal{M} \models \neg\phi(a)$. We claim that it is possible to extend $\Pi_D \mathcal{M}$ to a superstructure $\mathcal{N}$ such that $\mathcal{N} \models \exists e\,(0 < e \wedge e < a \wedge f(e) = 0)$: this would entail, by definition of cover, that $\Pi_D \mathcal{M} \models \phi(a)$, contradiction. We now show why the claim is true. Indeed, since $\langle a_k \rangle_{k \in \mathbb{N}}$ has arbitrarily big numbers as its components, we have that, in $\Pi_D \mathcal{M}$, a is bigger than all standard numbers. Thus, if we take a further non-principal ultrapower $\mathcal{N}$ of $\Pi_D \mathcal{M}$, it becomes possible to change in it the evaluation of $f(b)$ for some $b < a$ and set it to 0 (in fact, as it can be easily seen, there are elements $b \in |\mathcal{N}|$ less than a but not in the support of $\Pi_D \mathcal{M}$). ⊣

The counterexample still applies when replacing integer difference logic with linear integer arithmetics.

8 Conclusions and Future Work

In this paper we showed that covers (aka uniform interpolants) exist in the combination of two convex universal theories over disjoint signatures in case they exist in the component theories and in case the component theories also satisfy the equality interpolating condition - this further condition is nevertheless needed in order to transfer the existence of (ordinary) quantifier-free interpolants. In order to prove that, Beth definability property for primitive fragments turned out to be the crucial ingredient to extensively employ. In case convexity fails, we showed by a counterexample that covers might not exist anymore in the combined theory. The last result raises the following research problem. Even if in general covers do not exist for combination of non-convex theories, it would be interesting to see under what conditions one can decide whether a given cover exists and, in the affirmative case, to compute it.

Applications suggest a different line of investigations. In database-driven verification [5,6,9] one uses only 'tame' theory combinations. A tame combination of two multi-sorted theories T_1, T_2 in their respective signatures Σ_1, Σ_2 is characterized as follows: the shared sorts *can only be the codomain sort* (and not a domain sort) of a symbol from Σ_1 other than an equality predicate. In other words, if a relation or a function symbol has among its domain sorts a sort from $\Sigma_1 \cap \Sigma_2$, then this symbol is from Σ_2 (and not from Σ_1, unless it is the equality predicate). The key result for tame combination is that covers existence transfers to a tame combination $T_1 \cup T_2$ in case covers exist in the two component theories (the two component theories are only assumed to be stably infinite with respect to all shared sorts). Moreover, the transfer algorithm, in the case relevant for applications to database driven verification (where T_2 is linear arithmetics and T_1 is a multi-sorted version of $\mathcal{EUF}(\Sigma)$ in a signature Σ containing only unary function symbols and relations of any arity), is relatively well-behaved also from the complexity viewpoint because its cost is dominated by the cost of the covers computation in T_2. These results on tame combinations are included in the larger ArXiv version [4] of the present paper; an implementation in the beta version 2.9 of the model-checker MCMT is already available from the web site http://users.mat.unimi.it/users/ghilardi/mcmt.

A final future research line could consider cover transfer properties to non-disjoint signatures combinations, analogously to similar results obtained in [13, 14] for the transfer of quantifier-free interpolation.

References

1. Bílková, M.: Uniform interpolation and propositional quantifiers in modal logics. Stud. Logica **85**(1), 1–31 (2007)
2. Bruttomesso, R., Ghilardi, S., Ranise, S.: Quantifier-free interpolation in combinations of equality interpolating theories. ACM Trans. Comput. Log. **15**(1), 5:1–5:34 (2014)
3. Calvanese, D., Ghilardi, S., Gianola, A., Montali, M., Rivkin, A.: Quantifier elimination for database driven verification. Technical report arXiv:1806.09686, arXiv.org (2018)
4. Calvanese, D., Ghilardi, S., Gianola, A., Montali, M., Rivkin, A.: Combined covers and Beth definability (extended version). Technical report arXiv:1911.07774, arXiv.org (2019)
5. Calvanese, D., Ghilardi, S., Gianola, A., Montali, M., Rivkin, A.: Formal modeling and SMT-based parameterized verification of data-aware BPMN. In: Hildebrandt, T., van Dongen, B.F., Röglinger, M., Mendling, J. (eds.) BPM 2019. LNCS, vol. 11675, pp. 157–175. Springer, Cham (2019). https://doi.org/10.1007/978-3-030-26619-6_12
6. Calvanese, D., Ghilardi, S., Gianola, A., Montali, M., Rivkin, A.: From model completeness to verification of data aware processes. In: Lutz, C., Sattler, U., Tinelli, C., Turhan, A.-Y., Wolter, F. (eds.) Description Logic, Theory Combination, and All That. LNCS, vol. 11560, pp. 212–239. Springer, Cham (2019). https://doi.org/10.1007/978-3-030-22102-7_10

7. Calvanese, D., Ghilardi, S., Gianola, A., Montali, M., Rivkin, A.: Model completeness, covers and superposition. In: Fontaine, P. (ed.) CADE 2019. LNCS (LNAI), vol. 11716, pp. 142–160. Springer, Cham (2019). https://doi.org/10.1007/978-3-030-29436-6_9
8. Calvanese, D., Ghilardi, S., Gianola, A., Montali, M., Rivkin, A.: Verification of data-aware processes: challenges and opportunities for automated reasoning. In: Proceedings of ARCADE, EPTCS, vol. 311 (2019)
9. Calvanese, D., Ghilardi, S., Gianola, A., Montali, M., Rivkin, A.: SMT-based verification of data-aware processes: a model-theoretic approach. Math. Struct. Comput. Sci. **30**(3), 271–313 (2020)
10. Chang, C.-C., Keisler, J.H.: Model Theory, 3rd edn. North-Holland Publishing Co., Amsterdam (1990)
11. Ghilardi, S.: An algebraic theory of normal forms. Ann. Pure Appl. Logic **71**(3), 189–245 (1995)
12. Ghilardi, S.: Model theoretic methods in combined constraint satisfiability. J. Autom. Reason. **33**(3–4), 221–249 (2004)
13. Ghilardi, S., Gianola, A.: Interpolation, amalgamation and combination (The non-disjoint signatures case). In: Dixon, C., Finger, M. (eds.) FroCoS 2017. LNCS (LNAI), vol. 10483, pp. 316–332. Springer, Cham (2017). https://doi.org/10.1007/978-3-319-66167-4_18
14. Ghilardi, S., Gianola, A.: Modularity results for interpolation, amalgamation and superamalgamation. Ann. Pure Appl. Log. **169**(8), 731–754 (2018)
15. Ghilardi, S., Gianola, A., Kapur, D.: Compactly representing uniform interpolants for EUF using (conditional) DAGS. Technical report arXiv:2002.09784, arXiv.org (2020)
16. Ghilardi, S., Ranise, S.: MCMT: a model checker modulo theories. In: Giesl, J., Hähnle, R. (eds.) IJCAR 2010. LNCS (LNAI), vol. 6173, pp. 22–29. Springer, Heidelberg (2010). https://doi.org/10.1007/978-3-642-14203-1_3
17. Ghilardi, S., Zawadowski, M.: Sheaves, Games, and Model Completions: A Categorical Approach to Nonclassical Propositional Logics. Trends in Logic-Studia Logica Library, vol. 14. Kluwer Academic Publishers, Dordrecht (2002)
18. Ghilardi, S., Zawadowski, M.W.: A sheaf representation and duality for finitely presenting heyting algebras. J. Symb. Log. **60**(3), 911–939 (1995)
19. Ghilardi, S., Zawadowski, M.W.: Undefinability of propositional quantifiers in the modal system S4. Stud. Logica **55**(2), 259–271 (1995)
20. Ghilardi, S., Zawadowski, M.W.: Model completions, r-Heyting categories. Ann. Pure Appl. Log. **88**(1), 27–46 (1997)
21. Gulwani, S., Musuvathi, M.: Cover algorithms and their combination. In: Drossopoulou, S. (ed.) ESOP 2008. LNCS, vol. 4960, pp. 193–207. Springer, Heidelberg (2008). https://doi.org/10.1007/978-3-540-78739-6_16
22. Kapur, D.: Nonlinear polynomials, interpolants and invariant generation for system analysis. In: Proceedings of the 2nd International Workshop on Satisfiability Checking and Symbolic Computation Co-Located with ISSAC (2017)
23. Kowalski, T., Metcalfe, G.: Uniform interpolation and coherence. Ann. Pure Appl. Log. **170**(7), 825–841 (2019)
24. Nelson, G., Oppen, D.C.: Simplification by cooperating decision procedures. ACM Trans. Program. Lang. Syst. **1**(2), 245–257 (1979)
25. Peuter, D., Sofronie-Stokkermans, V.: On invariant synthesis for parametric systems. In: Fontaine, P. (ed.) CADE 2019. LNCS (LNAI), vol. 11716, pp. 385–405. Springer, Cham (2019). https://doi.org/10.1007/978-3-030-29436-6_23

26. Pitts, A.M.: On an interpretation of second order quantification in first order intuitionistic propositional logic. J. Symb. Log. **57**(1), 33–52 (1992)
27. Segerberg, K.: An Essay in Classical Modal Logic. Filosofiska Studier, vol. 13. Uppsala Universitet (1971)
28. Shavrukov, V.: Subalgebras of diagonalizable algebras of theories containing arithmetic. Dissertationes Mathematicae, CCCXXIII (1993)
29. Sofronie-Stokkermans, V.: On interpolation and symbol elimination in theory extensions. Log. Methods Comput. Sci. **14**(3), 1–41 (2018)
30. Tinelli, C., Harandi, M.: A new correctness proof of the Nelson-Oppen combination procedure. In: Baader, F., Schulz, K.U. (eds.) Frontiers of Combining Systems. ALS, vol. 3, pp. 103–119. Springer, Dordrecht (1996). https://doi.org/10.1007/978-94-009-0349-4_5
31. van Gool, S.J., Metcalfe, G., Tsinakis, C.: Uniform interpolation and compact congruences. Ann. Pure Appl. Log. **168**(10), 1927–1948 (2017)
32. Visser, A.: Uniform interpolation and layered bisimulation. In Hájek, P. (ed.) Gödel 1996. Logical Foundations on Mathematics, Computer Science and Physics – Kurt Gödel's Legacy. Springer, Heidelberg (1996)
33. Yorsh, G., Musuvathi, M.: A combination method for generating interpolants. In: Nieuwenhuis, R. (ed.) CADE 2005. LNCS (LNAI), vol. 3632, pp. 353–368. Springer, Heidelberg (2005). https://doi.org/10.1007/11532231_26

Deciding Simple Infinity Axiom Sets with One Binary Relation by Means of Superpostulates

Timm Lampert[1](✉) and Anderson Nakano[2]

[1] Humboldt University Berlin, Unter den Linden 6, 10099 Berlin, Germany
lampertt@staff.hu-berlin.de

[2] Pontifícia Universidade Católica, R. Monte Alegre, 984, São Paulo 05014-901, Brazil
alnakano@pucsp.br

Abstract. Modern logic engines widely fail to decide axiom sets that are satisfiable only in an infinite domain. This paper specifies an algorithm that automatically generates a database of independent infinity axiom sets with fewer than 1000 characters. It starts with complete theories of pure first-order logic with only one binary relation (FOL_R) and generates further infinity axiom sets S of FOL_R with fewer than 1000 characters such that no other infinity axiom set with fewer than 1000 characters exists in the database that implies S. We call the generated infinity axiom sets S "superpostulates". Any formula that is derivable from (satisfiable) superpostulates is also satisfiable. Thus far, we have generated a database with 2346 infinity superpostulates by running our algorithm. This paper ends by identifying three practical uses of the algorithmic generation of such a database: (i) for systematic investigations of infinity axiom sets, (ii) for deciding infinity axiom sets and (iii) for the development of saturation algorithms.

Keywords: First-order logic · Decision problem · Complete theories · Infinity axioms · Reduction classes · Dyadic logic

1 Introduction

Modern logic engines for first-order formulas (FOL), such as Vampire, CVC4, Spass or i-Prover, are very powerful in proving theorems (or refutations) or in deciding formulas with finite models (counter-models). For formulas that have only infinite models, however, these engines widely fail.[1] One exception is

[1] Exceptions include infinity axiom sets that can be decided by saturation, such as the theory of immediate successors as stated in FOL with identity (cf. (14) on p. 15). However, in regard to FOL without identity, infinity axiom sets are only rarely solved, and only a few of them are included in the Thousands of Problems for Theorem Provers (TPTP) library, namely, problems SYO635+1 to SYO638+1. Among these problems, only SYO638+1 can be solved by SPASS due to chaining rules, which directly apply to the transitivity axioms. Furthermore, Infinox decides the finite unsatisfiability of SY0638+1 only by virtue of the plain properties of the relation involved.

N. Peltier and V. Sofronie-Stokkermans (Eds.): IJCAR 2020, LNAI 12166, pp. 201–217, 2020.
https://doi.org/10.1007/978-3-030-51074-9_12

Infinox, which decides the finite unsatisfiability of some relevant TPTP problems due to specific model-theoretic principles (cf. [3]).[2] In contrast to Infinox, we investigate how to *generate* hitherto unknown and intricate infinity axiom sets (*superpostulates*) of FOL_R to make it possible to decide an infinite number of infinity axiom sets related to these superpostulates. No engine is currently able to solve most of the superpostulates we generate.

In this paper, we refer only to FOL_R, i.e., FOL without function symbols, without identity and with only one binary relation. Based on Herbrand's reduction of FOL to dyadic FOL, Kalmar has proven that FOL_R is a reduction class (cf. [7]). Boolos, Jeffrey and Burgess have defined and proven the algorithm for reducing FOL to FOL_R in modern terms (cf. [2], Sect. 21.3). Börger, Grädel and Gurevich have specified prenex normal forms of FOL_R-formulas with prenexes of the form $\forall\exists\forall$ (= $[\forall\exists\forall, (0,1)]$-class) or of the form $\forall\forall\forall\exists$ (= $[\forall^3\exists, (0,1)]$-class) as the two *minimal* classes of FOL_R with infinity axioms (cf. Theorem 6.5.4, p. 309) and have presented the following two examples for the two classes (cf. [1], p. 307):

$$\forall x \exists y \forall z(\neg Rxx \wedge Rxy \wedge (Ryz \rightarrow Rxz)) \tag{1}$$

$$\forall x \forall y \forall z \exists u(\neg Rxx \wedge Rxu \wedge (Rxy \wedge Ryz \rightarrow Rxz)) \tag{2}$$

They also specify another formula with the prenex $\forall\exists\forall$ (cf. [1], p. 33):

$$\forall x \exists u \forall y(\neg Rxx \wedge Rxu \wedge (Ryx \rightarrow Ryu)) \tag{3}$$

(1) and (3) are both derivable from (2). Thus, to decide that the latter is satisfiable, it is sufficient to decide that the former are likewise satisfiable. This method of deciding infinity axioms can be extended by specifying infinity axioms that imply other infinity axioms.

However, prenex normal forms do not provide a syntax that is suitable for this task. Different prenex normal forms can be converted into each other by equivalence transformation. For example, formula (2), which is a member of the minimal class $[\forall^3\exists, (0,1)]$, can easily be converted into a member of the minimal class $[\forall\exists\forall, (0,1)]$, as shown in Table 1.

Table 1. Converting different prenex normal forms into each other

No.	Formula	Strategy
(i)	$\forall x \forall y \forall z \exists u(\neg Rxx \wedge Rxu \wedge (Rxy \wedge Ryz \rightarrow Rxz))$	(2)
(ii)	$\forall x \neg Rxx \wedge \forall x \exists u Rxu \wedge \forall x \forall y(\neg Rxy \vee \forall z(\neg Ryz \vee Rxz))$	Miniscoping
(iii)	$\forall x_2 \neg Rx_2x_2 \wedge \forall x_1 \exists y_1 Rx_1y_1 \wedge \forall x_3 \forall x_4(\neg Rx_3x_4 \vee \forall x_5(\neg Rx_4x_5 \vee Rx_3x_5))$	Renaming
(iv)	$\forall x_1 \exists y_1 \forall x_2 \forall x_3 \forall x_4 \forall x_5(\neg Rx_2x_2 \wedge Rx_1y_1 \wedge (\neg Rx_3x_4 \vee (\neg Rx_4x_5 \vee Rx_3x_5)))$	Prenexing

[2] A further exception with respect to the engines on the TPTP site is Decider. This engine is able to identify infinity axiom sets of pure FOL without identity due to the implementation of a strong saturation algorithm that is not based on resolution. However, the complexity of the implemented algorithm is exponential, and the Decider engine is not optimized for rapid decision making. Thus, it fails to decide complex formulas within reasonable time and memory limits.

Likewise, it may well be that a formula of the minimal class $[\forall\exists\forall, (0,1)]$ can be converted into a formula of a decidable class, e.g., with prenexes starting with existential quantifiers followed by universal quantifiers ($=\exists^*\forall^*, (0,1)]$-class). Similarly, the distribution of bound variables in the matrix is not standardized in prenex normal forms. Thus, many equivalent variations are possible. Therefore, the syntax of prenex normal forms is not well suited for investigating the internal properties and relations of infinity axioms.

This is why we do not refer to *infinity axioms* (and, thus, to prenex normal forms) in the following. Instead, we refer to the opposite of prenex normal forms, namely, anti-prenex normal forms, in which the scopes of quantifiers are minimized. We refer to *infinity axiom sets* in terms of sets of *anti-prenex normal forms.* To do so, we define *primary formulas* via negation normal forms (NNFs) as follows:

Definition 1. *A first-order formula ϕ is* primary *(= in* anti-prenex form*) if ϕ is an NNF and either*

1. *does not contain $\wedge$ or $\vee$ or*
2. *contains $\wedge$ or $\vee$ iff either*
   (a) *any conjunction of n conjuncts ($n > 1$) is preceded by a sequence of existential quantifiers of minimal length 1, where all n conjuncts contain each variable of the existential quantifiers in that sequence, or*
   (b) *any disjunction of n disjuncts ($n > 1$) is preceded by a sequence of universal quantifiers of minimal length 1, where all n disjuncts contain each variable of the universal quantifiers in that sequence.*

Thus, lines (ii) and (iii) of Table 1 are conjunctions of three primary formulas, while lines (i) and (iv) do not contain primary formulas. Any first-order formula can be converted into a disjunction of conjunctions of primary formulas (=FOLDNF); cf. [8] for an algorithm for doing so. Since the satisfiability of a disjunction can be decided by deciding the satisfiability of each disjunct, we consider only conjunctions of primary formulas or, analogously, (finite) sets of axioms (= theories) in anti-prenex form.

Definition 2. *An* axiom set *is a finite set of primary formulas.*

Definition 3. *An* infinity axiom set *is a finite set of primary formulas that is satisfiable only within an infinite domain.*

From here on, we will express all axiom sets of FOL_R in standardized notation.

Definition 4. *A set of primary formulas (axioms) of FOL_R is in* standardized notation *if*

1. *the binary predicate used is R and*

2. for each primary formula (axiom) A,
- *the m universal quantifiers of A are binding variables x_1 to x_m from left to right and*
- *the n existential quantifiers of A are binding variables y_1 to y_n from left to right.*

Example 1. In standardized notation, formula (2) is converted into the following infinity axiom set:

$$\forall x_1 \neg Rx_1x_1, \forall x_1 \exists y_1 Rx_1y_1, \forall x_1 \forall x_2(\neg Rx_1x_2 \lor \forall x_3(\neg Rx_2x_3 \lor Rx_1x_3)) \quad (4)$$

This axiom set was the example used by Hilbert and Bernays to motivate proof theory by showing that the consistency of (4) cannot be proven by finite interpretations ("Methode der Aufweisung"; cf. [6], p. 10). The third axiom expresses transitivity.

Formula (3) is equivalent to the following infinity axiom set from [11], p. 183:

$$\forall x_1 \neg Rx_1x_1, \forall x_1 \exists y_1(Rx_1y_1 \land \forall x_2(\neg Rx_2x_1 \lor Rx_2y_1)) \quad (5)$$

Like (1), (3) and (5) are derivable from (4). The following infinity axiom set (6) is taken from [2], p. 138:

$$\forall x_1 \forall x_2(\neg Rx_1x_2 \lor \neg Rx_2x_1), \forall x_1 \exists y_1 Rx_1y_1, \forall x_1 \forall x_2(\neg Rx_1x_2 \lor \forall x_3(\neg Rx_2x_3 \lor Rx_1x_3)) \quad (6)$$

Like (2), this infinity axiom set is equivalent to (4).

The formulas mentioned up to this point are examples of infinity axiom sets of FOL_R given in the standard literature. They are all satisfiable by interpreting Rxy as $x < y$ over $\mathbb{Q}$. Thus, they are all derivable from the complete dense linear order (DLO) axiom set without endpoints. In the following sections, we will generate a system of DLO variants without identity, to which all of the mentioned standard examples belong. We use the term "*system* of formulas" to refer to a recursive set of formulas generated by rules that can be implemented by a computer program. Our investigation may serve as a case study for investigating systems of infinity axiom sets of FOL_R.

Our goal is the automated generation of a system of consistent infinity axiom sets S of a limited length L such that no other infinity axiom sets $S2$ of length $\leq L$ exist such that $S2$ implies S. We call these infinity axiom sets S "superpostulates". Superpostulates enable the reduction of a large number of infinity axiom sets to a small number of infinity axiom sets of limited length. The term and the general idea of studying systems of superpostulates originated in the work of Sheffer (cf. [12]). Sheffer, however, neither published his work nor developed his logic project to a full extent. His student Langford made use of Sheffer's idea in [9]. Langford intended to prove that the axiom set for linear orderings is a complete superpostulate. Urquhart (cf. [14], p. 44), however, objected to the correctness of his proof and showed how to correct it. Langford proved similar results for the theories of betweenness and cyclic order. All of his examples are equivalent to formulas of decidable classes without existential quantifiers and with identity, which have the finite model property (cf. [1], Theorem 6.5.1, p.

307). To our knowledge, neither Langford nor Sheffer studied superpostulates in terms of infinity axiom sets.

The following sections investigate infinity axiom sets related to the complete DLO axiom set without endpoints. Section 2 introduces further terminology and specifies a superpostulate DLO_R for FOL_R, which is a DLO variant without identity and without endpoints from which all of the mentioned infinity axiom sets follow. After having specified a complete infinity axiom set, Sect. 3 specifies an infinite sequence of infinity axiom sets, each strictly implying the next and all implied by (5). Section 4 then specifies a system of superpostulates based on DLO_R; Sect. 5 refers to a system THS_R based on the complete theory of immediate successors without identity. Finally, we conclude by identifying several practical uses of our method of algorithmic generation of systems of superpostulates in Sect. 6.

2 Superpostulates

We first introduce some terminology as a basis for defining the term *superpostulate*.

Definition 5. *An axiom A is* independent *of a set of axioms S if A is not implied by S and $S \wedge A$ is consistent.*

Definition 6. *An axiom set A is* complete *in FOL_R if for any FOL_R-formula B, either B or $\neg B$ follows from A.*

Remark 1. Throughout this paper, we consider FOL_R. This means that an axiom set A is complete iff no FOL_R-formula B is independent of A.

We measure the length of an axiom set by the number of characters it contains in standardized notation.

Definition 7. *An axiom set A of length $\leq L$ is L-*complete *in FOL_R if for any FOL_R-formula B of length $\leq L$, either B or $\neg B$ follows from A.*

Remark 2. Definition 7 implies that no formula B of length $\leq L$ exists such that A of length $\leq L$ is strictly implied by B if A is L-complete.

Definition 8. *An axiom set A is* minimal *if A does not contain any redundant part that can be eliminated to result in a logically equivalent axiom set.*

Definition 9. *A* superpostulate *(SP) is a minimal and consistent set of independent axioms that is L-complete.*

It is desirable to define superpostulates as complete (and not merely L-complete) theories. However, we abstain from doing so because we intend to define a *practical* algorithm for generating infinity axiom sets. The standard of preserving completeness, however, faces both practical and theoretical problems, as we will show in Sect. 4. Thus, we merely presume that the generation of a system of superpostulates is based on complete theories and aims for completeness relative to a length L of up to 1000 characters (cf. p. 11).

Definition 10. *An* infinity superpostulate *is a satisfiable superpostulate that has only infinite models.*

Any semi-decider is able to prove that a formula is implied by or inconsistent with another formula or finite axiom set. Since these are the only two options in the case of *complete* superpostulates, it is decidable whether a given formula is implied by a complete superpostulate. Furthermore, all (finite) infinity axiom sets that are implied by a set of infinity superpostulates are decidable.

From the following axiom set DLO_R, all infinity axiom sets of the standard literature mentioned in Sect. 1 follow:

$$\begin{aligned} &\forall x_1 \neg Rx_1x_1, \forall x_1 \exists y_1 Rx_1y_1, \forall x_1 \exists y_1 Ry_1x_1, \\ &\forall x_1 \forall x_2 (\neg Rx_1x_2 \vee \forall x_3 (\neg Rx_2x_3 \vee Rx_1x_3)), \\ &\forall x_1 \forall x_2 (Rx_1x_2 \vee \forall x_3 (Rx_2x_3 \vee \neg Rx_1x_3)), \\ &\forall x_1 \forall x_2 (\neg Rx_1x_2 \vee \exists y_1 (Rx_1y_1 \wedge Ry_1x_2)) \end{aligned} \tag{7}$$

Axiom 1 (irreflexivity), Axioms 2 and 3 (no right endpoint and no left endpoint), Axiom 4 (transitivity) and Axiom 6 (density) are identical to axioms of DLO without endpoints. However, DLO also contains trichotomy, i.e., the axiom $\forall x \forall y (Rxy \vee Ryx \vee x = y)$, which involves identity. The trichotomy axiom of DLO is replaced by Axiom 5 (transitivity of $\neg R$) in DLO_R.

Formula (7) contains all of the axioms of the infinity axiom set given in formula (4). Additionally, Axioms 3, 5 and 6 are added to those in (4). These axioms are not essential for satisfiability in an infinite domain.

Definition 11. *An axiom A of a superpostulate SP is* essential *if SP without A is no longer an infinity axiom.*

The set of the essential axioms plus the negation of an inessential axiom is also an infinity axiom set, but it is not necessarily a complete one.

In the following, we prove that (7) is a complete superpostulate by proving the relevant properties.

Theorem 1. *(7) is an infinity axiom set.*

Proof. If (7) is satisfiable, then it is satisfiable only in an infinite domain; this is because it implies (4), which is known to be an infinity axiom set. Thus, it remains to be shown that (7) is satisfiable. This is done by providing the following model over $\mathbb{Q}$ for (7): $\Im(R) = \{(x, y) \mid x < y\}$. It can be proven that this interpretation is indeed a model by paraphrasing each axiom. We use the common forms of the transitivity axioms (Axiom 4 and Axiom 5) and the density axiom (Axiom 6) for convenience:

Axiom 1	$\forall x \neg Rxx$
$\Im$(Axiom 1)	All rational numbers (r.n.) are no less than (<) themselves (irreflexivity of <)
Axiom 2	$\forall x_1 \exists y_1 Rx_1y_1$
$\Im$(Axiom 2)	For all r.n. x_1, some r.n. y_1 exists such that $x_1 < y_1$ (no right endpoint)
Axiom 3	$\forall x_1 \exists y_1 Ry_1x_1$
$\Im$(Axiom 3)	For all r.n. x_1, some r.n. y_1 exists such that $y_1 < x_1$ (no left endpoint)
Axiom 4	$\forall x_1 \forall x_2 \forall x_3 (Rx_1x_2 \wedge Rx_2x_3 \rightarrow Rx_1x_3)$
$\Im$(Axiom 4)	For all r.n. x_1, x_2, and x_3, if $x_1 < x_2$ and $x_2 < x_3$, then $x_1 < x_3$ (transitivity of <)
Axiom 5	$\forall x_1 \forall x_2 \forall x_3 (\neg Rx_1x_2 \wedge \neg Rx_2x_3 \rightarrow \neg Rx_1x_3)$
$\Im$(Axiom 5)	For all r.n. x_1, x_2, and x_3, if $x_1 \not< x_2$ and $x_2 \not< x_3$, then $x_1 \not< x_3$ (transitivity of $\not<$)
Axiom 6	$\forall x_1 \forall x_2 (Rx_1x_2 \rightarrow \exists y_1 (Rx_1y_1 \wedge Ry_1x_2))$
$\Im$(Axiom 6)	For all r.n. x_1 and x_2, if $x_1 < x_2$, then there exists an r.n. y_1 such that $x_1 < y_1$ and $y_1 < x_2$ (density)

Therefore, (7) is satisfiable; this is so only in an infinite domain, which means that (7) is an infinity axiom set. □

Theorem 2. *(7) is* minimal.

Proof. This is ensured by systematically producing the following:

1. all possible axiom sets A' from (7) with one axiom eliminated (case 1a) or with the conjunct in Axiom 6 eliminated (case 1b),
2. all possible axiom sets A'' with one disjunct in Axioms 4 to 6 eliminated (case 2), and
3. all possible axiom sets A''' obtained by (α) reducing the universal quantifiers in Axioms 4 to 6 and (β) replacing Axiom 2 and Axiom 3 with $\forall x_1 Rx_1x_1$ or $\exists y_1 Ry_1y_1$ (case 3).

In case 2 and case 3 (β), the resulting sets A'' and A''' have been proven to be inconsistent by Vampire in any arbitrary case; in case 3 (α), the resulting axioms in A''' have been proven to be redundant by Vampire. In case 1a, the results of eliminating Axiom 1 or Axiom 4 have been proven to be finite axiom sets by Vampire. Therefore, the resulting axiom sets are not equivalent. Axioms 2, 3, 5 and 6 are not redundant (and, thus, cannot be eliminated) because they can be proven to be independent axioms by replacing them with their negations, which, in turn, results in infinity axiom sets. This can be seen by referring to the following interpretations:

Axiom 1	Axiom 2	Axiom 3	Axiom 4	Axiom 5	Axiom 6	$\Im$
+	¬	+	+	+	+	$x < y$ over $\mathbb{Q}^+$
+	+	¬	+	+	+	$x < y$ over $\mathbb{Q}^-$
+	+	+	+	¬	+	$x < y - 1$ over $\mathbb{Q}$
+	+	+	+	+	¬	$x < y$ over $\mathbb{Z}$

The conjunct in Axiom 6 is not redundant because eliminating it results in a redundant axiom, whereas Axiom 6 itself is not redundant. □

Theorem 3. *(7) is complete.*

Proof. We extend (7) by the following defining axiom of $x_1 = x_2$:

$$\forall x_1 \forall x_2 (x_1 = x_2 \leftrightarrow \forall x_3 (R x_1 x_3 \leftrightarrow R x_2 x_3)) \tag{8}$$

The union of (7) and (8) is equivalent to DLO without endpoints (as proven by Vampire). Thus, DLO without endpoints is a conservative extension of (7) and, by conservativeness, (7) is complete if DLO is complete.[3] Since DLO without endpoints is known to be complete, (7) is also complete. □

We say that an infinity axiom set B is *weaker* than A if B is strictly implied by A, that is, $A \vdash B$, but $B \vdash A$ does not hold. Before we consider how to generate further superpostulates, let us first consider the opposite: infinity axiom sets that are as weak as possible.

3 Weak Infinity Axiom Sets

The set given in formula (5) is the weakest infinity axiom set mentioned in the standard literature. The following infinite axiom sets, however, strictly follow from (5):

$$\forall x_1 \exists y_1 (R x_1 y_1 \wedge \forall x_2 (\neg R x_2 x_1 \vee R x_2 y_1) \wedge \neg R y_1 y_1) \tag{9}$$

$$\forall x_1 \exists y_1 (R x_1 y_1 \wedge \neg R y_1 x_1 \wedge \forall x_2 (\neg R x_2 x_1 \vee R x_2 y_1)) \tag{10}$$

Formula (9) strictly follows from (5), and (10), in turn, strictly follows from (9). We first prove that (10) is an infinity axiom. We then show that it can still be systematically weakened without losing the property of being an infinity axiom set.

Theorem 4. *(10) is an infinity axiom.*

Proof. We first show that there exists a denumerable model of (10) and then show that (10) has no finite model.

We assume the natural numbers as the domain and interpret Rxy as $x < y$. Then, it is true that for all natural numbers x_1, there exists a number y_1 such

[3] For extensions by definitions, cf., e.g., [13], section 4.6.

that $x_1 < y_1$ and $y_1 \not< x_1$, and that for all x_2, if $x_2 < x_1$, then $x_2 < y_1$. Thus, a denumerable model of (10) exists.

We prove that (10) has no finite model by contradiction. Suppose that there is a finite model M of (10), where the elements of its domain are listed as $m_1, \ldots, m_k$. Let $n_1 = m_1$. In the following, we consider how to satisfy the three conjuncts in the scope of $\forall x_1 \exists y_1$. By the first conjunct, there must be some n in the domain such that (n_1, n) is in $\Im(R)$. Let n_2 be the first element for which this is true. Thus, we have $(n_1, n_2) \in \Im(R)$. By the second conjunct, we have that $(n_2, n_1) \notin \Im(R)$. It follows that $n_1 \neq n_2$ is in $\Im(R)$. Again, by the first conjunct, there must be some n such that (n_2, n). Let n_3 be the first element for which this is the case; thus, we have $(n_2, n_3) \in \Im(R)$ and, by the second conjunct, $(n_3, n_2) \notin \Im(R)$, and therefore, $n_2 \neq n_3$. By the third conjunct, we have either $(n_1, n_2) \notin \Im(R)$ or $(n_1, n_3) \in \Im(R)$. Since we already have $(n_1, n_2) \in \Im(R)$, the second of these statements, i.e., $(n_1, n_3) \in \Im(R)$, must be true. Then, by the second conjunct, we have $(n_3, n_1) \notin \Im(R)$. Thus, we also have $n_1 \neq n_3$ (in addition to $n_2 \neq n_3$) since we have $(n_3, n_1) \notin \Im(R)$ and $(n_1, n_3) \in \Im(R)$. Continuing in this manner, we obtain n_4, which is different from all of n_1, n_2, and n_3; then, we obtain n_5, which is different from n_1 to n_4; and so on. However, by the time we reach n_{k+1}, we will have exceeded the number of elements of the domain. Thus, our supposition that the domain of M is finite leads to a contradiction. Therefore, (10) has no finite model. □

Formula (10) is a simple and weak infinity axiom that no TPTP engine is able to decide.[4]

Theorem 5. *From (10), an infinite number of strictly weaker infinity axiom sets follow.*

Proof. We consider the iterative derivation of strictly weaker axioms from (10) by applying the rule $A \vdash B \vee A$ ($\vee$I), as shown in Table 2.

Table 2. Infinity axioms implied by (10)

No.	Axiom	Rule
(i)	$\forall x_1 \exists y_1 (Rx_1y_1 \wedge \neg Ry_1x_1 \wedge \forall x_2(\neg Rx_2x_1 \vee Rx_2y_1))$	(10)
(ii)	$\forall x_1 \exists y_1 (Rx_1y_1 \wedge \neg Ry_1x_1 \wedge \forall x_2(\forall x_3 \neg Rx_3x_2 \vee \neg Rx_2x_1 \vee Rx_2y_1))$	$\vee$I
(iii)	$\forall x_1 \exists y_1 (Rx_1y_1 \wedge \neg Ry_1x_1 \wedge \forall x_2(\forall x_3(\forall x_4 \neg Rx_4x_3 \vee \neg Rx_3x_2) \vee \neg Rx_2x_1 \vee Rx_2y_1))$	$\vee$I
(iv)	$\forall x_1 \exists y_1 (Rx_1y_1 \wedge \neg Ry_1x_1 \wedge \forall x_2(\forall x_3(\forall x_4(\forall x_5 \neg Rx_5x_4 \vee \neg Rx_4x_3) \vee \neg Rx_3x_2) \vee \neg Rx_2x_1 \vee Rx_2y_1))$	$\vee$I
⋮	⋮	⋮

Lines (ii), (iii), etc., follow from (10) with the application of nothing but $\vee$I, which is a valid derivation rule. Each result is, in turn, a primary formula. To verify that iteratively applying $\vee$I generates a sequence of weaker *infinity* axiom

[4] This can be checked on the site http://tptp.org/cgi-bin/SystemOnTPTP.

sets, one must determine how the proof of Theorem 4 still applies. Formula (10) allows one to directly conclude that $n_1 \neq n_3$ once $n_1 \neq n_2$ and $n_2 \neq n_3$ are established. Applying $\vee$I once (cf. line (ii)), however, allows one to conclude only that either $(n_1, n_3) \in \Im(R)$ or $(n_4, n_2) \notin \Im(R)$ (where n_4 is the element introduced by $\forall x_3$, which cannot be presumed to be either identical to or different from n_1, n_2, or n_3). Due to the first two conjuncts, $n_1 \neq n_3$ follows from $(n_1, n_3) \in \Im(R)$, and $n_2 \neq n_4$ follows from $(n_4, n_2) \notin \Im(R)$. Considering further objects of the domain introduces further objects that must be different from all objects in one of the established alternative sets of objects. With each further application of $\vee$I, further alternative sets arise. Thus, each axiom resulting from $\vee$I establishes a wider range of objects that cannot be identical. Therefore, each application allows for less powerful inferences concerning the non-identity of the elements in pairs satisfying $\Im(R)$, and thus, each resulting axiom is strictly weaker. Nevertheless, the resulting axioms still necessitate that at least one of the newly considered objects (e.g., n_3 and n_4 in step 2 of the argument after $n_1 \neq n_2$ is established in step 1) in each step of the argument must be different from at least one of the already considered objects (e.g., n_1 and n_2 in step 1). Therefore, the resulting axioms still cannot be satisfied under the assumption of only a finite number of objects. Consequently, we generate an endless sequence of increasingly weaker infinity axioms. □

There are many further infinite sequences of non-equivalent infinity axiom sets that are implied by (7). They all can be proven to be satisfiable by proving that they follow from (7).

4 A System of Superpostulates: The DLO$_R$-System

We are searching for a system of infinity superpostulates that is automatically generated by transformation rules from a complete theory such as (7). The rules specified in this section are a starting point for the general project of generating systems of infinity superpostulates from a given complete axiom set. We do not claim that our third rule preserves completeness.

Definition 12. *A DLO$_R$-system is a system of infinity superpostulates generated from (7).*

Definition 13. *An infinity axiom set is a DLO$_R$-variant if it is derivable from a superpostulate of a DLO$_R$-system.*

The following two rules yield structurally similar infinity axiom sets of FOL$_R$. They trivially preserve the property of being a complete infinity axiom set in the case of FOL$_R$:

Rule 1: Exchange all negated literals for non-negated literals and all non-negated literals for negated literals.

Rule 2: Exchange the variables in positions 1 and 2 in all atomic propositional functions.

These rules do not necessarily yield new (non-equivalent) infinity axiom sets. Rule 1, for example, converts Axiom 4 of superpostulate (7) into Axiom 5 and vice versa. Rule 2 does not change Axiom 1, does convert Axiom 2 into Axiom 3 and vice versa, and yields equivalent axioms in the cases of Axioms 4 and 5. In fact, applying these rules to our superpostulate (7) results in only one further complete superpostulate:

$$\begin{gathered}\forall x_1 Rx_1x_1, \forall x_1\exists y_1\neg Rx_1y_1, \forall x_1\exists y_1\neg Ry_1x_1,\\ \forall x_1\forall x_2(Rx_1x_2 \vee \forall x_3(Rx_2x_3 \vee \neg Rx_1x_3)),\\ \forall x_1\forall x_2(\neg Rx_1x_2 \vee \forall x_3(\neg Rx_2x_3 \vee Rx_1x_3)),\\ \forall x_1\forall x_2(Rx_1x_2 \vee \exists y_1(\neg Rx_1y_1 \wedge \neg Ry_1x_2))\end{gathered} \tag{11}$$

The crucial problem in considering systems of superpostulates is specifying rules beyond Rules 1 and 2. Replacing Axiom 2 or Axiom 3 with its negation results in a complete variant of DLO_R with endpoints.[5] Thus, one might consider a rule that replaces axioms with their negations. However, the case in which completeness is preserved by simply replacing an axiom with its negation is an exceptional one. This is true only if one cannot replace an axiom A of a complete theory S with a strictly weaker axiom that is still independent of S without A. For example, there is no primary formula X that is strictly implied by $\forall x_1\exists y_1 Rx_1y_1$ (=Axiom 2) and is not implied by Axiom 1 and Axioms 3–6 of (7). The same does not hold, e.g., for Axiom 1. Therefore, simply replacing Axiom 1 of (7) with its negation does not preserve completeness since $\exists y_1\neg Ry_1y_1$ could be added as an independent axiom.

To preserve completeness when replacing an axiom A of an axiom set S with its negation, one must generate an axiom set S' that is *minimally weaker* than S and does not imply A.

Definition 14. *An axiom set $S2$ is* minimally weaker *than an axiom set $S1$ if $S2$ is strictly weaker than $S1$ and no intermediate axiom set $S3$ exists such that $S3$ is strictly weaker than $S1$ and $S2$ is strictly weaker than $S3$.*

Only adding the negation of an axiom A of S to S', which is minimally weaker than S, preserves completeness. However, the generation of a minimally weaker axiom set S' from a given axiom set S gives rise to intricate practical and theoretical problems. We cannot even prove whether a minimal weaker axiom set S' exists for any axiom set S. The main theoretical and practical problem is that one must specify some upper bound for an axiom set that preserves completeness if one axiom is replaced with its negation. We leave it as an open

[5] The same is not true in the case of negating Axiom 2 *and* Axiom 3. DLO with both endpoints is complete only if one adds the axiom $\exists y_1\exists y_2 y_1 \neq y_2$. Without this further axiom, the remaining axioms have a trivial finite model in a domain with only one object. We abstain from generating a DLO_R-variant with left and right endpoints.

question whether this problem is solvable and, if so, how it can be solved. Instead, we confine our DLO_R-system to superpostulates of a certain length.

To do so, we adopt the following two restrictions:

Restriction 1: We set an upper bound for any considered axiom set with a length L of 1000 characters.

Restriction 2: We make use of an incomplete calculus that does not significantly increase the length of formulas when generating implied formulas.

These restrictions are justified by the *practical* aim of generating a system of infinity superpostulates.

Due to our restrictions, the considered superpostulates are simple enough to make use of the following assumption:

Assumption 1. If one of the following cases holds, this is proven by the standard casc-mode of Vampire within a time limit of 300 s:

1. a considered part of a considered axiom set is redundant,
2. a considered minimal axiom set is inconsistent, or
3. the considered axiom set is implied by another considered axiom set or implies either (10) or (12).

In fact, Vampire is able to almost immediately identify these cases for the simple axiom sets that we consider here in nearly all cases. By Assumption 1, we *assume* that Vampire can solve the semi-decidable problems 1 to 3 for axiom sets with fewer than 1000 characters to obtain a positive result within 300 s. In other words, our algorithm runs Vampire with a time limit of 300 s, and if the problem is not solved within this time limit, then we assume a negative solution. The reliability of Assumption 1 is based on extensive experience.

To generate strictly weaker axioms A' from an axiom A by applying an inference rule, we make use of an incomplete calculus that does not include rules for the introduction of disjuncts ($\vee I$) and conjuncts ($\wedge I$), which would increase the length of axioms. For this reason, we cannot ensure the generation of *minimally* weaker axiom sets and, thus, do not necessarily preserve completeness. Instead, we consider only rules concerning quantifiers: one rule for changing the order of quantifiers (QEx), two for universal quantifier elimination ($\forall$E1 and $\forall$E2) and two for existential quantifier introduction ($\exists$I1 and $\exists$I2); cf. Table 3. The abbreviated notations for the rules specify only the relevant syntactic changes. For example, $\exists\mu\forall\nu \vdash \forall\nu\exists\mu$ means that the order of two quantifiers is changed in a primary formula. We tacitly presume that a sequence of universal (existential) quantifiers is orderless; each of the quantifiers of such a sequence can be considered as either the leftmost or rightmost quantifier of the sequence. μ/ν means that μ is replaced with ν. In the case of $\forall$E1, ν may also be replaced with the variable of a new existential quantifier preceding the resulting axiom. $\forall\nu\varphi(\nu,\nu)$ indicates that ν occurs more than once in the scope of $\forall\nu$. $\varphi(\mu/\nu,\mu)$ means that at least one occurrence of μ is replaced with ν. We tacitly presume that new variables are used if a new quantifier is introduced.

Table 3. Quantifier rules

$$\text{QEx: } \exists\mu\forall\nu \vdash \forall\nu\exists\mu$$

$$\forall\text{E1: } \exists\mu\forall\nu\varphi(\mu,\nu) \vdash \exists\mu\varphi(\mu,\nu/\mu) \qquad \exists\text{I1: } \forall\nu\varphi(\nu,\nu) \vdash \forall\nu\exists\mu\varphi(\nu,\nu/\mu)$$

$$\exists\text{I2: } \exists\mu\varphi(\mu,\mu) \vdash \exists\mu\exists\nu\varphi(\mu/\nu,\mu) \quad \forall\text{E2: } \forall\mu\forall\nu\varphi(\mu,\nu,\nu) \vdash \forall\mu\varphi(\mu,\nu/\mu)$$

We establish the following rules for the application of the quantifier rules. ∀E1 can be applied such that $\exists\mu$ is taken from an axiom that differs from A after having variables renamed and PN laws applied to the effect that $\forall\nu$ is directly to the right of $\exists\mu$. Before ∀E1 can be applied to quantifiers of one and the same axiom, PN laws must be applied to the effect that existential quantifiers that are separated by $\wedge$ from other existential quantifiers are pulled outwards. The same holds for universal quantifiers that are separated by $\vee$ from other universal quantifiers. Thus, the numbers of quantifiers in a sequence of existential quantifiers and a sequence of universal quantifiers from which $\exists\mu$ and $\forall\nu$ are to be chosen are increased. ∀E2 and QEx can also be applied to universal quantifiers that are separated by $\vee$; QEx additionally can be applied to existential quantifiers that are separated by $\wedge$ from the universal quantifier. In these cases, PN laws are applied to exchange the order of the universal quantifiers and $\wedge$ or $\vee$. The results of the application of these rules are converted into a minimized FOLDNF in standardized notation (cf. p. 3). *One* application of a quantifier rule results in conversion from a set of primary formulas to a set of primary formulas. Thus, each application comprises the application of equivalence rules to prepare for the application of QEx, ∀E1 or ∀E2 and to convert the result into an FOLDNF. In the case in which the resulting FOLDNF is, in fact, a disjunction, we consider only disjuncts as axiom sets that are strictly implied by the axiom set prior to the application of a quantifier rule. Furthermore, we consider each conjunct of a disjunct of an FOLDNF as a separate axiom.

Given an axiom set S that includes axiom A, we consider the totality of all possible applications of the 5 quantifier rules to A. In the case of ∀E1, this may involve different orders of the prior application of PN laws in addition to the elimination of the universal quantified variable by variables bound by different existential quantifiers. QEx can be applied to all $\exists\mu$ and all $\forall\nu$ such that $\exists\mu$ occurs in a sequence of existential quantifiers directly to the left of a sequence of universal quantifiers containing $\forall\nu$. ∃I1 and ∃I2 may replace variables in different positions. The totality of possible applications of QEx, ∀E1, ∀E2, ∃I1, and ∃I2 to A realizes all different applications of the five rules.

Our method of generating a DLO_R-system is based on all possible *non-redundant* applications of the quantifier rules starting from (7).

Definition 15. *An application of a quantifier rule to an axiom A of an axiom set S is* non-redundant *if it results in a strictly weaker axiom set S' and the axiom A' that replaces A in S is not implied by S' without A'.*

If, however, S' implies S or if A' is not independent in S', then the application of a quantifier rule is redundant.

To ensure that all superpostulates are, in fact, *infinity* axiom sets, we consider only superpostulates that imply the very weak infinity axiom set given in (10) or its counterpart given in (12), as generated by Rule 2:

$$\forall x_1 \exists y_1 (Ry_1x_1 \wedge \neg Rx_1y_1 \wedge \forall x_2(\neg Rx_1x_2 \vee Ry_1x_2)) \tag{12}$$

DLO_R without endpoints, with a left endpoint, and with a right endpoint all imply either (10) or (12).

By Assumption 1 and our quantifier rules, it is possible to apply the following algorithm to generate a set S^* of further superpostulates from a superpostulate S.

Algorithm 1. Generate S^* from S, which includes A, as follows:

1. Set $S^* = \{\}$.
2. Specify the set A' of all non-redundant applications of the quantifier rules to A.
3. Generate the powerset A'' of A'.
4. Traverse the members $A_1, A_2, \ldots, A_k$ of A'', beginning with the largest set A'. Replace A in S with the members of A_i and denote the result by S'; if $A_i = \{\}$, then $S' = S$ without A.
   (a) Delete redundant axioms and redundant subexpressions in S'.
   (b) If $S', \neg A$ is fewer than 1000 characters and consistent, then delete all proper subsets of A_i in A''.
   (c) If $\neg A$ is additionally independent of S', then denote the conversion of the resulting axiom set $S' \cup \{\neg A\}$ set into standard notation by S''.
   (d) If S'' implies (10) or (12), then append S'' to S^*.

Remark 3. For simplicity, our implemented algorithm in fact considers only A' (instead of its powerset A'') and merely deletes redundant axioms (and no other subexpressions) in S'.

Example 2. Applying Algorithm 1 to Axiom 1 of (7) results in a set S^* with the following superpostulate as its only member:

$$\begin{gathered} \exists y_1 Ry_1y_1, \forall x_1 \exists y_1 \neg Rx_1y_1, \forall x_1 \exists y_1 \neg Ry_1x_1, \\ \forall x_1 \exists y_1 (Rx_1y_1 \wedge \neg Ry_1y_1), \forall x_1 \exists y_1 (Ry_1x_1 \wedge \neg Ry_1y_1), \\ \text{Axioms 4, 5, and 6 of (7)} \end{gathered} \tag{13}$$

Note that the application of $\forall$E1 to replace x_1 in Axiom 1 of (7) with y_1 in Axiom 6 is redundant. Set (13) can, in turn, serve as a starting point for the application of Algorithm 1. Applying Algorithm 1 to Axiom 1 of (13) results in (7) once again (consider the implied minimization strategies!).

Let us use S^{**} to denote the result of applying Algorithm 1 to all of the axioms of S; S^{**} is the union of all results S^*. Furthermore, let us use S_T to denote the totality of the generated superpostulates. Finally, we use S^{***} to denote the subset of S^{**} that contains only superpostulates that are not implied

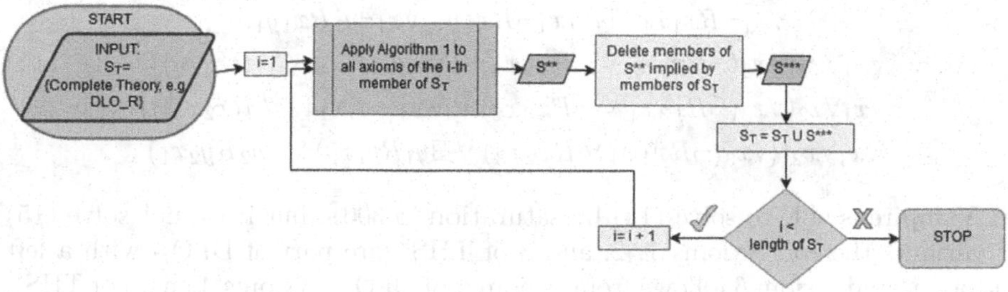

Fig. 1. Flowchart of Rule 3

by superpostulates in S_T. Figure 1 provides a flowchart of our general algorithm (Rule 3).

Based on Algorithm 1, we generate a DLO_R-system by means of the following rule:

Rule 3: Starting from (7), generate infinity superpostulates by iteratively applying Algorithm 1 to all axioms of all generated superpostulates until no further superpostulates are generated. Restriction: Delete superpostulates if they are implied by previously generated superpostulates.

The restriction ensures that no other infinity axiom set of S_T strictly implies S; this is the best we can do to generate L-complete infinity axiom sets up to a length of 1000 characters. Furthermore, the restriction avoids repetitions and loops (cf. the application of Algorithm 1 to Axiom 1 of (13) in Example 2).

Remark 4. With our implementation of Rule 3, we have generated 732 superpostulates in approximately 1000 h of run time. We have generated a database of 2196 infinity superpostulates in TPTP format by applying Rules 1 to 3 starting with DLO_R. None of these superpostulates can be currently solved by any of the TPTP-engines. Additionally, we have generated a database of 60 weak infinity DLO_R-variants. 8 of them can be solved by SPASS within 60 s.

5 The THS_R-System

A prominent alternative complete theory to start with is the theory of immediate successors (THS), which can be defined as follows in familiar notation:

$$\begin{gathered}\forall x \neg Rxx,\\ \forall x \exists y \forall z (Rxz \leftrightarrow z = y),\\ \exists x \forall y (\forall z \neg Rzy \leftrightarrow y = x),\\ \forall x (\exists y Ryx \to \exists y \forall z (Rzx \to z = y))\end{gathered} \tag{14}$$

The following reduction THS_R of THS to FOL_R can be proven to be complete by adding the defining axiom $\forall x \forall y (x = y \leftrightarrow \forall z (Rxz \leftrightarrow Ryz) \wedge \forall z (Rzx \leftrightarrow Rzy))$ and proving the equivalence to (14):

$$\begin{gathered}\forall x_1 \neg Rx_1x_1, \exists y_1 \forall x_1 \neg Rx_1y_1, \forall x_1 \exists y_1 Rx_1y_1,\\ \forall x_1 \forall x_2(\forall x_3(\neg Rx_1x_3 \vee \neg Rx_2x_3) \vee \forall x_4(\neg Rx_4x_1 \vee Rx_4x_2)),\\ \forall x_1 \forall x_2(\forall x_3(\neg Rx_3x_1 \vee \neg Rx_3x_2) \vee \forall x_4(\neg Rx_1x_4 \vee Rx_2x_4)),\\ \forall x_1 \forall x_2((\forall x_3(\neg Rx_1x_3 \vee Rx_2x_3) \vee \exists y_1 Ry_1x_2) \vee \exists y_2 Ry_2x_1)\end{gathered} \tag{15}$$

Vampire is able to solve (14) by saturation in 300 s, but it cannot solve (15) or variants thereof. Axioms 1, 2, and 3 of THS_R are part of DLO_R with a left endpoint, and Axiom 6 follows from Axiom 5 of DLO_R. Axioms 4 and 5 of THS_R contradict Axioms 1, 2, and 6 of DLO_R. Axioms 4, 5, and 6 of DLO_R contradict Axioms 1, 2, and 4 of THS_R.

Remark 5. Our algorithm has generated a database of 156 superpostulates from (15) and terminated after approximately 200 h of run time. 32 of them are solved by CVC4. None of the 156 superpostulates implies any of the superpostulates of the DLO_R-system or any of the weak DLO_R-variants.

6 Conclusion

From a theoretical point of view, it would be desirable to generate a database of complete infinity axiom sets. However, this would involve the problem of generating minimally weaker infinity axiom sets that may be of arbitrary length. Here, our intent is to provide a *pragmatic* algorithm that, in fact, generates significant infinity axiom sets. Thus, we restrict the generated infinity axiom sets to a certain length. This restriction does not imply that the generated sets are not complete theories; rather, it means only that we are unable to *guarantee* that they are. Our algorithm starts with a complete theory of FOL_R, such as DLO_R or THS_R, and automatically generates further infinity axiom sets S *with fewer than 1000 characters* such that, roughly speaking, no other infinity axiom sets *with fewer than 1000 characters* exist that strictly imply infinity axiom sets S. We say "roughly speaking" here because in addition to limiting the set length, we use an incomplete calculus without inference rules that would significantly increase complexity, and we test the restriction only on the basis of already generated entries in our database. We believe that the theoretical question concerning an (unrestricted) algorithm that does, in fact, preserve completeness is an important one, but we do not aim to answer it in this paper. Instead, our final objective is a practical algorithm, and we maintain that we have achieved this aim. Thus far, we have generated more than 2250 superpostulates in 10 weeks of run time. All of these superpostulates are hard problems.

Note that it is not trivial to automatically generate infinity axiom sets. Infinity axiom sets (not to say *systems* of infinity axiom sets) (i) are rarely encountered, (ii) can seldom be solved by logic engines, (iii) are hard problems even when they are short in length, and (iv) have not been systematically studied; moreover, (v) only a few of them exist in the TPTP library, and no systematic archive of infinity axiom sets is available. Therefore, we believe that our database is of practical importance for (i) systematically studying infinity axiom sets, (ii) deciding the satisfiability of additional infinity axiom sets that follow from our database, and (iii) improving saturation algorithms by enabling the decidability of infinity axiom sets. In regard to (i), we are investigating the question of whether starting our algorithm with different complete FOL_R theories will yield identical superpostulates. MacPherson conjectured that any ω-categorical axiom set has the strict order property (cf. [10] and [4] for detailed discussions). Proofs of the property of being ω-categorical widely depend on identity. In light of MacPherson's conjecture, it would be interesting to study the limits of the DLO_R-system. We have also reduced Goldbach's conjecture to FOL_R and intend to study its internal relations with other infinity axiom sets in our still-growing database. In regard to (ii), we have proven in Sect. 3 that an infinite number of additional infinity axiom sets follow from our superpostulates. A database of superpostulates will make it possible to decide the satisfiability of infinity axiom sets that are implied by one of the entries of the database. In regard to (iii), we are currently working on saturation algorithms that can cope with infinity axiom sets. We know from our work how important it is to test such algorithms on a variety of intricate examples during development.

Acknowledgement. We thank the three anonymous referees for their comments and Geoff Suttcliffe for checking our database with CVC4, E, Infinox, SPASS and Vampire.

References

1. Börger, E., Grädel, E., Gurevich, Y.: The classical decision problem. J. Logic Lang. Inform. **8**, 478–481 (2001). https://doi.org/10.1023/A:1008334715902
2. Boolos, G.S., Burgess, J.P., Jeffrey, R.C.: Computability and Logic, 4th edn. Cambridge University Press, Cambridge (2003)
3. Claessen, K., Lilleström, A.: Automated inference of finite unsatisfiability. J. Autom. Reason. **47**, 111–132 (2011)
4. Djordjevic, M.: On first-order sentences without finite models. J. Symb. Log. **69**(2), 329–339 (2004)
5. Dreben, B., Goldfarb, W.D.: The Decision Problem. Solvable Classes of Quantificational Formulas. Addison-Wesley, London (1979)
6. Hilbert, D., Bernays, P.: Grundlagen der Mathematik, vol. 1. Springer, Berlin (1970)
7. Kalmár, L.: Zurückfürung des Entscheidungsproblems auf den Fall von Formeln mit einer einzigen, binären Funktions variablen. Compos. Math. **4**, 137–144 (1936)
8. Lampert, T.: Minimizing disjunctive normal forms of pure first-order logic. Log. J. IGPL **25**(3), 325–347 (2017)
9. Langford, C.: Analytic completeness of sets of postulates. Proc. London Math. Soc. **25**, 115–142 (1926)
10. Macpherson, D.: Finite axiomatizability and theories with trivial algebraic closure. Notre Dame J. Form. Log. **32**(2), 188–192 (1991)
11. Quine, W.V.O.: Methods of Logic, 4th edn. Harvard UP, Cambridge (1982)
12. Sheffer, H.M.: The general theory of notational relativity. Typewritten manuscript of 61 pp. Harvard Widener Library (1921)
13. Shoenfield, J.R.: Mathematical Logic. Addison-Wesley, Reading (1967)
14. Urquhart, A.: Henry M. Sheffer and notational relativity. Hist. Philos. Log. **33**(1), 33–47 (2012)

A Decision Procedure for String to Code Point Conversion

Andrew Reynolds[1], Andres Nötzli[2(✉)], Clark Barrett[2], and Cesare Tinelli[1]

[1] Department of Computer Science, The University of Iowa, Iowa City, USA
[2] Department of Computer Science, Stanford University, Stanford, USA
noetzli@cs.stanford.edu

Abstract. In text encoding standards such as Unicode, text strings are sequences of *code points*, each of which can be represented as a natural number. We present a decision procedure for a concatenation-free theory of strings that includes length and a conversion function from strings to integer code points. Furthermore, we show how many common string operations, such as conversions between lowercase and uppercase, can be naturally encoded using this conversion function. We describe our implementation of this approach in the SMT solver CVC4, which contains a high-performance string subsolver, and show that the use of a native procedure for code points significantly improves its performance with respect to other state-of-the-art string solvers.

1 Introduction

String processing is an important part of many kinds of software. In particular, strings often serve as a common representation for the exchange of data at interfaces between different programs, between different programming languages, and between programs and users. At such interfaces, strings often represent values of types other than strings, and developers have to be careful to sanitize and parse those strings correctly. This is a challenging task, making the ability to automatically reason about such software and interfaces appealing. Applications of automated reasoning about strings include finding or proving the absence of SQL injections and XSS vulnerabilities in web applications [27,30,33], reasoning about access policies in cloud infrastructure [7], and generating database tables from SQL queries for unit testing [31]. To make this type of automated reasoning scalable, several approaches for reasoning natively about string constraints have been proposed [3,4,11,20,21].

To reason about complex string operations such as conversions between strings and numeric values, string solvers typically reduce these operations to operations in some basic fragment of the theory of strings which they support natively. The scalability of a string solver thus depends on the efficiency of the

This work was partially funded by Amazon Web Services.

N. Peltier and V. Sofronie-Stokkermans (Eds.): IJCAR 2020, LNAI 12166, pp. 218–237, 2020.
https://doi.org/10.1007/978-3-030-51074-9_13

reductions as well as the performance of the solver over the basic constraints. In such approaches, the set of operations in the basic fragment of strings has to be chosen carefully. If the set is too extensive, the implementation becomes complex and its performance as well as its correctness may suffer as a result. On the other hand, if the set is too restrictive, the reductions may become too verbose or only approximate, also leading to suboptimal performance. In current string solvers, basic constraints typically include only *word equations* (i.e., equalities between concatenations of variables and constants) and length constraints. Certain operations, however, such as conversions between strings and numeric values, cannot be represented efficiently in this fragment because the encoding requires reasoning by cases on the concrete characters that may occur in the string values assigned to a string variable.

In this work, we investigate extending the set of basic operators supported in a modern string solver to bridge the gap between character and integer domains. We assume a finite character domain of some cardinality n and, similarly to the Unicode standard, we assume a bijective mapping between its character set and the first n natural numbers which associates each character with a unique *code point*. We introduce then a new string operator, code, from characters strings to integers which can be used to encode the code point value of strings of length one and, more generally, reason about the code point of any character in a string. We propose an approach that involves extending a previous decision procedure with native support for this operator, obtaining a new decision procedure which avoids splitting on character values. Using the code operator, we can succinctly represent string operations including common string transducers, conversion between strings and integers, lexicographic ordering on strings, and regular expression membership constraints involving character ranges. We have implemented our proposed decision procedure in the state-of-the-art SMT solver CVC4 as an extension of its decision procedure for word equations by Liang et al. [21]. We have modified CVC4's reductions to take advantage of code. Using benchmarks generated by the concolic execution of Python code, we show that our technique provides significant benefits compared to doing case splitting on values.

To summarize, our contributions are as follows:

- We provide a decision procedure for a simple set of string operations containing length and a code point conversion function code, and prove its correctness. We describe how it can be combined with existing procedures for other string operators.
- We demonstrate how the code operator can be used in the reductions of several classes of useful string constraints.
- We implement and evaluate our approach in CVC4, showing that it leads to significant performance gains with respect to the state of the art.

In the following, we discuss related work. We then describe a fragment of the theory of strings in Sect. 2 that includes code. In Sect. 3, we provide a decision procedure for this fragment, prove its correctness, and describe how it can be integrated with existing decision procedures. Finally, we discuss applications

Σ_{A}	$n : \mathsf{Int}$ for all $n \in \mathbb{N}$	$+ : \mathsf{Int} \times \mathsf{Int} \rightarrow \mathsf{Int}$	$- : \mathsf{Int} \rightarrow \mathsf{Int}$	$\geqslant : \mathsf{Int} \times \mathsf{Int} \rightarrow \mathsf{Bool}$
Σ_{S}	$l : \mathsf{Str}$ for $l \in \mathcal{A}^*$	$\mathsf{len} : \mathsf{Str} \rightarrow \mathsf{Int}$	$\mathsf{code} : \mathsf{Str} \rightarrow \mathsf{Int}$	

Fig. 1. Functions in signature Σ_{AS}. Str and Int denote strings and integers, respectively.

of reasoning about code points in Sect. 4 and evaluate our implementation in Sect. 5.

Related Work. The study of the decidability of different fragments of string constraints has a long history. We know that solvability of word equations over unbounded strings is decidable [23], whereas the addition of quantifiers makes the problem undecidable [25]. The boundary between decidable and undecidable fragments, however, remains unclear—a long-standing open question is whether word equations combined with equalities over string lengths are decidable [14]. Adding extended string operators such as replace [13] or conversions between strings and integers [18] leads to undecidability. Weakly chaining string constraints make up one decidable fragment. This fragment requires that the graph of relational constraints appearing in the constraints only contains limited types of cycles. It generalizes the straight-line fragment [22], which disallows equalities between initialized string variables, and the acyclic fragment [5], which disallows equalities involving multiple occurrences of a string variable and does not include transducers.

In practice, string solvers have to deal with undecidable fragments or fragments of unknown decidability, so current solvers for strings such as CVC4 [9], Z3 [16], Z3STR3 [11] and TRAU [3] implement efficient semi-decision procedures. In this work, we present a decision procedure that can be combined modularly with those procedures.

2 Preliminaries

We work in the context of many-sorted first-order logic with equality and assume the reader is familiar with the notions of signature, term, literal, (quantified) formula, and free variable (see, e.g., [17]). We consider many-sorted signatures Σ that contain an (infix) logical symbol $\approx$ for equality—which has type $\sigma \times \sigma$ for all sorts σ in Σ and is always interpreted as the identity relation. A *theory* is a pair $T = (\Sigma, \mathbf{I})$ where Σ is a signature and $\mathbf{I}$ is a class of Σ-interpretations, the *models* of T. A Σ-formula φ is *satisfiable* (resp., *unsatisfiable*) *in* T if it is satisfied by some (resp., no) interpretation in $\mathbf{I}$. Given a (set of) terms S, we write $\mathcal{T}(S)$ to denote the set of subterms of S. By convention and unless otherwise stated, we use letters x, y, z to denote variables and s, t to denote terms.

We consider a theory T_{AS} of strings with length and code point functions, with a signature Σ_{AS} given in Fig. 1. We *fix a finite totally ordered set* $\mathcal{A}$ *of characters as our alphabet* and define T_{AS} as a set of Σ_{AS}-structures with universe $\mathcal{A}^*$ (the set of all words over $\mathcal{A}$) which differ only on the value they assign to variables. The signature includes the sorts Str and Int, interpreted as $\mathcal{A}^*$ and $\mathbb{Z}$, respectively. Figure 1 partitions the signature Σ_{AS} into the subsignatures Σ_{A} and Σ_{S}, as indicated. The first includes the usual symbols of linear integer arithmetic, interpreted as expected. We will write $t_1 \bowtie t_2$, with $\bowtie \in \{>, <, \leqslant\}$, as syntactic sugar for the equivalent inequality between t_1 and t_2 expressed using only $\geqslant$. The subsignature Σ_{S} includes: all the words of $\mathcal{A}^*$ (including the empty word ϵ) as constant symbols, or *string constants*, each interpreted as itself; a function symbol $\mathsf{len} : \mathsf{Str} \to \mathsf{Int}$, interpreted as the word length function; and a code point function whose semantics is defined as follows.

Definition 1. *Given alphabet $\mathcal{A}$ and its associated total order $<$, let $(c_0, \ldots, c_{n-1})$ be the enumeration of $\mathcal{A}$ induced by $<$ (with $c_i < c_{i+1}$ for all $i = 0 \ldots, n-2$). For each character c_i in the enumeration, we refer to i as its code point. The function symbol* code : Str → Int *is interpreted in T_{AS} as the unique code point function* code *such that:*

1. *for all words $w \in \mathcal{A}^1$, $\text{code}(w)$ is the code point of the (single) character of w, and*
2. *for all other words $w \in \mathcal{A}^*$, $\text{code}(w)$ is -1.*

The code point function can be used in practice to reason about the *code point* values of Unicode strings.[1] We will see in Sect. 4 that this operator is very useful for encoding constraints that occur in applications. We stress, however, that the procedure presented in this paper is agnostic with respect to the concrete alphabet $\mathcal{A}$ and its character ordering.

Note that we do *not* consider string concatenation in the signature above. This omission is for the sake of modularity; also, procedures for word equations have been addressed in a number of recent works [4,21]. In practice, our procedure for string constraints involving code can be naturally combined with existing procedures for a signature that includes string concatenation, as we discuss in Sect. 3.1.

An *atomic term* is either a constant or a variable. A *string term* is either a constant or one that contains function symbols from Σ_{S} only. Notice that integer constants are string terms. A *string constraint* is a (dis)equality between string terms. An *arithmetic constraint* is an inequality or (dis)equality between linear combinations of atomic and/or string terms with integer sort. Notice that the equality $\mathsf{code}(x) \approx \mathsf{code}(y)$ with variables x and y is both a string constraint and an arithmetic constraint.

3 A Decision Procedure for String to Code Point Conversion

In this section, we introduce a decision procedure for a fragment of string constraints involving code but not containing string concatenation. In particular, we introduce a decision procedure for sets of (quantifier-free) Σ_{AS}-constraints

[1] For technical details on Unicode see [28].

for the signature introduced in Fig. 1. A key property of this procedure is that it is able to reason about terms of the form $\mathsf{code}(x)$ without having to do case splitting on concrete values for string x.

Following Liang et al. [21], we describe this procedure as a set of derivation rules that modify configurations of the form $\langle \mathsf{A}, \mathsf{S} \rangle$, where A is a set of arithmetic constraints, and S is a set of string constraints. At a high level, the procedure can be understood as a cooperation between two subsolvers, an *arithmetic subsolver* and a *string subsolver*, which handle these two sets respectively. Our procedure assumes the following preconditions on $\langle \mathsf{A}, \mathsf{S} \rangle$ and maintains them as an invariant for all derived configurations:

1. $\mathsf{A} \cup \mathsf{S}$ contains no terms of the form $\mathsf{len}(l)$ or $\mathsf{code}(l)$ for any string literal l.
2. For every string literal $l \in \mathcal{T}(\mathsf{A} \cup \mathsf{S})$, the set S contains $x \approx l$ for some variable x.

The above restrictions come with no loss of generality since terms of the form $\mathsf{len}(l)$ and $\mathsf{code}(l)$ can be replaced by an equivalent (constant) integer, and fresh variables can be introduced as necessary for the second requirement.

We present the rules of the procedure in two parts, given in Figs. 2 and 3. The rules are given in *guarded assignment form*, where the top of the rule describes the conditions under which the rule can be applied, and the bottom of the rule either is unsat, or otherwise describes the resulting modifications to the components of our configuration. A rule may have multiple, alternative conclusions separated by $\|$. In the premises of the rules, we write $\mathsf{S} \models \varphi$ to denote that S entails formula φ in the empty theory. This can be checked using a standard algorithm for congruence closure, where string literals are treated as distinct values; thus $\mathsf{S} \models l_1 \not\approx l_2$ for any S, $l_1 \neq l_2$. Observe that, for $f \in \{\mathsf{len}, \mathsf{code}\}$, $\mathsf{S} \models f(x) \approx f(y)$ iff $\mathsf{S} \models x \approx y$.

An application of a rule is *redundant* if it has a conclusion where each component in the derived configuration is a subset of the corresponding component in the premise configuration. A configuration other than unsat is *saturated* if every possible application of a derivation rule to it is redundant. A *derivation tree* is a tree where each node is a configuration whose children, if any, are obtained by a non-redundant application of a rule of the calculus. A derivation tree is *closed* if all of its leaves are unsat. We show later that a closed derivation tree with root node $\langle \mathsf{A}, \mathsf{S} \rangle$ is a proof that $\mathsf{A} \cup \mathsf{S}$ is unsatisfiable in T_{AS}. In contrast, a derivation tree with root node $\langle \mathsf{A}, \mathsf{S} \rangle$ and a saturated leaf is a witness that $\mathsf{A} \cup \mathsf{S}$ is satisfiable in T_{AS}.

Figure 2 presents rules adapted from previous work [21,26] that model the interaction between the string and arithmetic subsolvers. First, either subsolver can report that the current set of constraints is unsatisfiable by the rules A-Conf or S-Conf. For the former, the entailment $\models_{\mathsf{LIA}}$ can be checked by a standard procedure for linear integer arithmetic. The rules A-Prop and S-Prop correspond to a form of Nelson-Oppen-style theory combination between the two subsolvers. In particular, each theory solver propagates entailed equalities between terms of type Int. The next two rules ensure that length constraints are satisfied. In particular, L-Intro ensures that the length of a term x is equal to the length of

$$\textsf{A-Conf}\ \frac{\mathsf{A} \models_{\mathsf{LIA}} \bot}{\mathsf{unsat}} \qquad \textsf{A-Prop}\ \frac{\mathsf{A} \models_{\mathsf{LIA}} s \approx t \quad s,t \in \mathcal{T}(\mathsf{A} \cup \mathsf{S})}{\mathsf{S} := \mathsf{S}, s \approx t}$$

$$\textsf{S-Conf}\ \frac{\mathsf{S} \models \bot}{\mathsf{unsat}} \qquad \textsf{S-Prop}\ \frac{\mathsf{S} \models s \approx t \quad s,t : \mathsf{Int} \quad s,t \in \mathcal{T}(\mathsf{A} \cup \mathsf{S})}{\mathsf{A} := \mathsf{A}, s \approx t}$$

$$\textsf{L-Intro}\ \frac{\mathsf{S} \models x \approx l \quad x : \mathsf{Str}}{\mathsf{S} := \mathsf{S},\ \mathsf{len}(x) \approx (\mathsf{len}(l))\!\downarrow} \qquad \textsf{L-Valid}\ \frac{x \in \mathcal{T}(\mathsf{A} \cup \mathsf{S}) \quad x : \mathsf{Str}}{\mathsf{S} := \mathsf{S}, x \approx \epsilon \quad \| \quad \mathsf{A} := \mathsf{A}, \mathsf{len}(x) > 0}$$

$$\textsf{Card}\ \frac{\mathsf{S} \models \mathsf{len}(x_1) \approx \ldots \approx \mathsf{len}(x_n) \quad n > 1}{\|_{1 \leqslant i < j \leqslant n}\ \mathsf{S} := \mathsf{S}, x_i \approx x_j \quad \| \quad \mathsf{A} := \mathsf{A}, \mathsf{len}(x_1) > \lfloor \log_{|\mathcal{A}|}(n-1) \rfloor}$$

Fig. 2. Core derivation rules.

$$\textsf{C-Intro}\ \frac{\mathsf{S} \models x \approx l \quad l \in \mathcal{A}^1}{\mathsf{A} := \mathsf{A}, \mathsf{code}(x) \approx (\mathsf{code}(l))\!\downarrow} \qquad \textsf{C-Collapse}\ \frac{\mathsf{S} \models x \approx l \quad \mathsf{code}(x) \in \mathcal{T}(\mathsf{S})}{\mathsf{A} := \mathsf{A}, \mathsf{code}(x) \approx (\mathsf{code}(l))\!\downarrow}$$

$$\textsf{C-Valid}\ \frac{\mathsf{code}(x) \in \mathcal{T}(\mathsf{S})}{\mathsf{A} := \mathsf{A}, \mathsf{len}(x) \not\approx 1, \mathsf{code}(x) \approx -1 \quad \| \quad \mathsf{A} := \mathsf{A}, \mathsf{len}(x) \approx 1, 0 \leqslant \mathsf{code}(x) < |\mathcal{A}|}$$

$$\textsf{C-Inj}\ \frac{\mathsf{code}(x), \mathsf{code}(y) \in \mathcal{T}(\mathsf{S}) \quad x, y \text{ distinct}}{\mathsf{A} := \mathsf{A}, \mathsf{code}(x) \approx -1 \quad \| \quad \mathsf{A} := \mathsf{A}, \mathsf{code}(x) \not\approx \mathsf{code}(y) \quad \| \quad \mathsf{S} := \mathsf{S}, x \approx y}$$

Fig. 3. Code point derivation rules.

string literals x is equated to in S. We write $(\mathsf{len}(l))\!\downarrow$ to denote the constant integer corresponding to the result of evaluating the expression $\mathsf{len}(l)$. The rule L-Valid has two conclusions. It ensures that either x is the empty string or the value assigned to $\mathsf{len}(x)$ is positive. Finally, since our alphabet is finite, the rule Card is used to determine when a length constraint is implied due to the number of distinct terms of a given length. In particular, if there are n distinct variables $x_1, \ldots, x_n$ whose length is the same, then either x_i is equal to x_j for some $i \neq j$, or their length must be large enough so that they each can be assigned a unique string value. The lower bound on their length is determined by taking the floor of the logarithm of $n-1$ base the cardinality of the alphabet, where this expression denotes an integer constant.[2]

Figure 3 lists rules for reasoning about the code point function. In C-Intro, if a string variable x is equal to a string literal of length one, we add to S an equality between $\mathsf{code}(x)$ and the concrete value of the code point of l. An equality of this form is also added via the rule C-Collapse if a string term x is equated in S to a string literal and, in addition, $\mathsf{code}(x)$ occurs in S. Rule C-Valid splits on whether an instance of $\mathsf{code}(x)$ from S is equal to a valid code point. The left conclusion considers the case where the code point is -1, which means that x must have a length different from 1. The right conclusion considers the case where the code point is between 0 and $|\mathcal{A}|-1$, meaning that x is a one-character string. Finally, rule C-Inj reflects the fact that code denotes an injective function over the domain of strings of length 1. More precisely, it captures the fact that

[2] In the degenerate case where the cardinality of the alphabet is one, we assume this branch is omitted from the conclusion since logarithm base one is undefined.

for any pair of string values l_x and l_y for x and y respectively, one of the following (non-necessarily disjoint) cases always holds: (i) l_x has a length different from 1, (ii) l_x has length 1 and differs from l_y, and has a different code point from that of l_y, or (iii) l_x and l_y are the same.

We now demonstrate the procedure with a few simple examples. Recall that we assume a fixed alphabet $\mathcal{A}$ and write c_n to denote a character from this alphabet whose code point is some n between 0 and the cardinality of $\mathcal{A}$ minus one.

Example 1. Let A_0 be $\{\mathsf{len}(x) > \mathsf{len}(y),\ \mathsf{code}(x) \approx \mathsf{code}(y),\ \mathsf{code}(x) \geqslant 0\}$ and let S_0 be $\varnothing$. We can generate the following closed derivation tree with root $\langle \mathsf{A}_0, \mathsf{S}_0\rangle$. At each node, we list the new constraint that is added to the configuration at that node. All the leaf nodes are derived by A-Conf (not shown in the tree).

$$\dfrac{\dfrac{\langle \mathsf{A},\mathsf{S}\rangle := \langle \mathsf{A}_0,\mathsf{S}_0\rangle}{\mathsf{code}(x)\approx\mathsf{code}(y)\in\mathsf{S}}\ \textsf{A-Prop}}{\dfrac{\mathsf{code}(x)\approx -1\in\mathsf{A}}{\mathsf{unsat}} \quad \dfrac{\mathsf{code}(x)\not\approx\mathsf{code}(y)\in\mathsf{A}}{\mathsf{unsat}} \quad \dfrac{\dfrac{x\approx y\in\mathsf{S}}{\mathsf{len}(x)\approx\mathsf{len}(y)\in\mathsf{A}}\ \textsf{S-Prop}}{\mathsf{unsat}}}\ \textsf{C-Inj}$$

First, since $\mathsf{A} \models_{\mathsf{LIA}} \mathsf{code}(x) \approx \mathsf{code}(y)$, we apply A-Prop which adds the equality $\mathsf{code}(x) \approx \mathsf{code}(y)$ to S. Subsequently, since $\mathsf{code}(x), \mathsf{code}(y) \in \mathcal{T}(\mathsf{S})$, we apply C-Inj which considers three cases. The first two branches result in the arithmetic component of our configuration A being unsatisfiable, and thus unsat may be derived by A-Conf. In the third branch, we consider the case where x is equal to y. We have that S entails that $\mathsf{len}(x) \approx \mathsf{len}(y)$, and hence, by S-Prop, this equality is added to A. Since $\mathsf{len}(x) > \mathsf{len}(y)$ is already in A, we can derive unsat in this branch by A-Conf as well. Since there is a closed derivation tree with root $\langle \mathsf{A}_0, \mathsf{S}_0\rangle$, $\mathsf{A}_0 \cup \mathsf{S}_0$ is unsatisfiable in T_{AS}. □

Example 2. Let A_0 be $\{97 \leqslant \mathsf{code}(x) \leqslant 106\}$ and let S_0 be $\{x \not\approx y,\ x \not\approx z,\ y \approx c_{97},\ z \approx c_{106}\}$. We may obtain a derivation tree with root $\langle \mathsf{A}_0, \mathsf{S}_0\rangle$ and a saturated configuration $\langle \mathsf{A}, \mathsf{S}\rangle$ where A extends A_0 with the constraints:

$$\{\mathsf{code}(x) \not\approx \mathsf{code}(y),\ \mathsf{code}(x) \not\approx \mathsf{code}(z),\ \mathsf{code}(y) \approx 97,\ \mathsf{code}(z) \approx 106\}$$

The constraints $\mathsf{code}(x) \not\approx \mathsf{code}(y)$ and $\mathsf{code}(x) \not\approx \mathsf{code}(z)$ may be obtained by C-Inj, and $\mathsf{code}(y) \approx 97$ and $\mathsf{code}(z) \approx 106$ may be obtained by C-Intro. Since a saturated configuration exists in a derivation tree with root node $\langle \mathsf{A}_0, \mathsf{S}_0\rangle$, we have that $\mathsf{A}_0 \cup \mathsf{S}_0$ is satisfiable in T_{AS}. As we show in Theorem 1 (below), a model for $\mathsf{A}_0 \cup \mathsf{S}_0$ can be obtained by constructing an arbitrary model for $\mathsf{A} \cup \mathsf{S}$. In particular, notice that due to our derived constraints, it must be the case that $\mathsf{code}(x)$ is assigned a value in the range $[98 \ldots 105]$. Indeed, a model $\mathcal{M}$ exists for $\mathsf{A}_0 \cup \mathsf{S}_0$, where $\mathcal{M}(x) = c_k, \mathcal{M}(y) = c_{97}$, and $\mathcal{M}(z) = c_{106}$, for any k in the range $[98 \ldots 105]$. Note that we do not explicitly case split on the value of x. Instead, as we later describe in Definition 2, our procedure assigns a value to x based on the value that the arithmetic subsolver gives to $\mathsf{code}(x)$. □

Example 3. Let A_0 be $\{48 \leqslant \mathsf{code}(x) < 58, \mathsf{len}(x) < 1\}$ and let S_0 be $\varnothing$. We may obtain the following closed derivation tree with root $\langle \mathsf{A}_0, \mathsf{S}_0 \rangle$.

$$\dfrac{\langle \mathsf{A}, \mathsf{S}\rangle := \langle \mathsf{A}_0, \mathsf{S}_0\rangle}{\dfrac{\dfrac{x \approx \epsilon \in \mathsf{S}}{\mathsf{code}(x) \approx -1 \in \mathsf{A}}\ \textsf{C-Collapse}}{\mathsf{unsat}}\ \textsf{A-Conf} \qquad \dfrac{\mathsf{len}(x) > 0 \in \mathsf{A}}{\mathsf{unsat}}\ \textsf{A-Conf}}\ \textsf{L-Valid}$$

Since x is a string term from $\mathcal{T}(\mathsf{A} \cup \mathsf{S})$, we apply $\textsf{L-Valid}$. The left branch considers the case where x is empty. Since $\mathsf{S} \models x \approx \epsilon$, we apply $\textsf{C-Collapse}$ which adds $\mathsf{code}(x) \approx \mathsf{code}(\epsilon){\downarrow} = -1$ to A. This makes A unsatisfiable, and we can derive unsat by $\textsf{A-Conf}$. In the right branch, we consider the case that $\mathsf{len}(x) > 0$, which results in a case where A is unsatisfiable since $\mathsf{len}(x) < 1 \in \mathsf{A}$. Thus, $\mathsf{A}_0 \cup \mathsf{S}_0$ is unsatisfiable in T_{AS}. □

Example 4. Let $\mathsf{A}_0 = \{0 \leqslant \mathsf{code}(x) < \mathsf{len}(x)\}$ and $\mathsf{S}_0 = \varnothing$. We may obtain a saturated configuration $\langle \mathsf{A}, \mathsf{S}\rangle$ where A extends A_0 with $\{\mathsf{len}(x) \approx 1, 0 \leqslant \mathsf{code}(x) < |\mathcal{A}|\}$. These constraints are obtained by considering the right branch of an application of $\textsf{C-Valid}$ since $\mathsf{code}(x) \in \mathcal{T}(\mathsf{S})$ (after the trivial propagation $\mathsf{code}(x) \approx \mathsf{code}(x)$ by $\textsf{A-Prop}$). The only models for $\mathsf{A} \cup \mathsf{S}$ are those where $\mathsf{code}(x)$ is assigned the value for 0; hence the only models $\mathcal{M}$ for $\mathsf{A} \cup \mathsf{S}$ (and hence $\mathsf{A}_0 \cup \mathsf{S}_0$) are where $\mathcal{M}(x) = c_0$. □

We now discuss the formal properties of our calculus, proving that it is refutation-sound, model-sound, and terminating for any set of Σ_{AS}-constraints, and thus yields a decision procedure. We also show that, for any saturated configuration, it is possible to construct a model for the input constraints based on the procedure given in the following definition. In each step, we argue the well-formedness of this construction. In the subsequent theorem, we show that the constructed model indeed satisfies our input constraints.

Definition 2 (Model Construction). *Let $\langle \mathsf{A}, \mathsf{S}\rangle$ be a saturated configuration. Construct a model $\mathcal{M}$ for $\mathsf{A} \cup \mathsf{S}$ based on the following steps.*

1. *Let U be the set of terms of the form $\mathsf{len}(x)$ or $\mathsf{code}(x)$ that occur in A. Let $\mathcal{Z}$ be a model of A', where A' is the result of replacing in A each of its subterms $t \in U$ with a fresh integer variable u_t. Notice that $\mathcal{Z}$ exists, since $\textsf{A-Conf}$ does not apply to our configuration, meaning that A (and hence A') is satisfiable in LIA.*
2. *Construct $\mathcal{M}$ by assigning values to the variables in $\mathsf{A} \cup \mathsf{S}$ in the following order. Below, let $\widehat{\mathsf{S}}$ denote the congruence closure of S.*[3]
   (a) For all integer variables x, set $\mathcal{M}(x) = \mathcal{Z}(x)$.
   (b) For all string equivalence classes $e \in \widehat{\mathsf{S}}$ that contain a string constant l (including the case where $l = \epsilon$), set $\mathcal{M}(y) = l$ for all variables $y \in e$. Notice that l is unique since $\textsf{S-Conf}$ does not apply to our configuration.

[3] That is, the equivalence relation over $\mathcal{T}(\mathsf{S})$ such that s, t are in the same equivalence class if and only if $\mathsf{S} \models s \approx t$.

(c) *For all string equivalence classes* $e \in \widehat{\mathsf{S}}$, *such that* $\mathcal{Z}(u_{\mathsf{len}(z)}) = 1$ *and* $\mathsf{code}(z) \in \mathcal{T}(\mathsf{S})$ *for some* $z \in e$, *we let* $\mathcal{M}(y) = c_k$ *for each variable* $y \in e$, *where* $k = \mathcal{Z}(u_{\mathsf{code}(z)})$. *Since C-Valid cannot be applied to our configuration, it must be the case that* A' *contains the constraint* $0 \leqslant u_{\mathsf{code}(z)} < |\mathcal{A}|$. *Since* $\mathcal{Z}$ *satisfies* A', *the value of* $\mathcal{Z}(u_{\mathsf{code}(z)})$ *is guaranteed to be a valid code point and thus* c_k *is indeed a character in* $\mathcal{A}$.

(d) *For all remaining unassigned string equivalence classes* $e \in \widehat{\mathsf{S}}$, *we have that* $\mathsf{len}(z) \in \mathcal{T}(\mathsf{A})$ *for all variables* $z \in e$, *since L-Valid cannot be applied to our configuration. We choose some* l *of length* $\mathcal{Z}(u_{\mathsf{len}(z)})$, *such that* l *is not already assigned to any other string variable in* $\mathcal{M}$, *and set* $\mathcal{M}(y) = l$ *for all variables* $y \in e$. *Since our configuration is saturated with respect to Card, we know that at least one such string literal exists: if the set of string literals of length* $\mathcal{Z}(u_{\mathsf{len}(z)})$ *were each in the range of* $\mathcal{M}$, *it would imply that there are* $|\mathcal{A}|^{\mathcal{Z}(u_{\mathsf{len}(z)})} + 1$ *distinct terms whose length is* $\mathcal{Z}(u_{\mathsf{len}(z)})$, *in which case Card would require* $\mathsf{len}(z)$ *to be greater than the value of* $\lfloor \log_{|\mathcal{A}|}(|\mathcal{A}|^{\mathcal{Z}(u_{\mathsf{len}(z)})} + 1 - 1) \rfloor = \mathcal{Z}(u_{\mathsf{len}(z)})$. *However, this is not the case since* A *is satisfiable in LIA.*

Theorem 1. *Let* $\mathsf{M} = \mathsf{A}_0 \cup \mathsf{S}_0$ *be a set of* Σ_{AS}*-constraints where* A_0 *are arithmetic constraints and* S_0 *are non-arithmetic constraints. The following statements hold.*

1. *There is a closed derivation tree with root* $\langle \mathsf{A}_0, \mathsf{S}_0 \rangle$ *only if* M *is unsatisfiable in* T_{AS}.
2. *There is a derivation tree with root* $\langle \mathsf{A}_0, \mathsf{S}_0 \rangle$ *containing a saturated configuration only if* M *is satisfiable in* T_{AS}.
3. *All derivation trees with root* $\langle \mathsf{A}_0, \mathsf{S}_0 \rangle$ *are finite.*

Proof. To show (1), assume there exists a model $\mathcal{M}$ of $\mathsf{A}_0 \cup \mathsf{S}_0$. It is straightforward to show that for every rule of the calculus, applying that rule to any node $\langle \mathsf{A}, \mathsf{S} \rangle$ results in a tree where at least one child $\langle \mathsf{A}', \mathsf{S}' \rangle$ is such that $\mathcal{M}$ also satisfies $\mathsf{A}' \cup \mathsf{S}'$. Thus, by induction on the size of the derivation tree, there exists at least one terminal node that is not closed. Thus, if there exists a closed derivation tree with root node $\langle \mathsf{A}_0, \mathsf{S}_0 \rangle$, then it must be the case that no model exists for $\mathsf{A}_0 \cup \mathsf{S}_0$, so M is unsatisfiable in T_{AS}.

To show (2), assume there exists a derivation tree with a saturated configuration $\langle \mathsf{A}, \mathsf{S} \rangle$. Let $\mathcal{M}$ be the model constructed based on the procedure in Definition 2. Below, we argue that $\mathcal{M}$ is a model for $\mathsf{A} \cup \mathsf{S}$, which is a superset of $\mathsf{A}_0 \cup \mathsf{S}_0$ and thus satisfies M. Let U and $\mathcal{Z}$ respectively be the set of terms and model as computed in Step 1 of Definition 2. Below, we show that $\mathcal{M}$ is a model for each constraint in $\mathsf{A} \cup \mathsf{S}$.

- To show $\mathcal{M}$ satisfies each constraint in A, we show $\mathcal{Z}(x) = \mathcal{M}(x \cdot \sigma)$ for all integer variables x, where σ is the substitution $\{u_t \mapsto t \mid u_t \in U\}$.
  - Consider the case $x = u_{\mathsf{len}(y)}$ for some y, that is, x is a variable introduced in Step 1 of Definition 2 for an application of a length term. If y was assigned a value in Step 2(b) of Definition 2, then $\mathcal{M}(y) = l$ for some l

such that $\mathsf{S} \models y \approx l$. Since L-Intro cannot be applied to our configuration, we have that $\mathsf{len}(y) \approx (\mathsf{len}(l))\!\downarrow\, \in \mathsf{A}$, and hence $\mathcal{Z}(u_{\mathsf{len}(y)}) = \mathcal{Z}(\mathsf{len}(l)\!\downarrow) = \mathcal{M}(\mathsf{len}(y))$. If y was assigned in Step 2(c) or 2(d), we have that $\mathcal{M}(y) = l$ for some l whose length is $\mathcal{Z}(u_{\mathsf{len}(y)})$, and hence $\mathcal{Z}(u_{\mathsf{len}(y)}) = \mathcal{M}(\mathsf{len}(y))$.

- Consider the case $x = u_{\mathsf{code}(y)}$ for some y. If y was assigned in Step 2(b) of Definition 2, then $\mathcal{M}(y) = l$ for some l such that $\mathsf{S} \models y \approx l$. Since C-Collapse cannot be applied to our configuration, we have that $\mathsf{code}(y) \approx (\mathsf{code}(l))\!\downarrow\, \in \mathsf{A}$ and hence $\mathcal{Z}(u_{\mathsf{code}(y)}) = \mathcal{M}(\mathsf{code}(l)\!\downarrow) = \mathcal{M}(\mathsf{code}(y))$. If y was assigned in Step 2(c) or 2(d), we have that $\mathcal{M}(y) = l$ for some l whose length is $\mathcal{Z}(u_{\mathsf{len}(y)})$. If it was assigned in Step 2(c), we have that $l = c_k$ for $k = \mathcal{Z}(u_{\mathsf{code}(y)})$ and hence $\mathcal{Z}(u_{\mathsf{code}(y)}) = \mathcal{M}(\mathsf{code}(y))$. If y was assigned in Step 2(d), we have that $\mathsf{code}(y) \in \mathcal{T}(\mathsf{S})$. Since y was not assigned in Step 2(c), it must be the case that $\mathcal{Z}(u_{\mathsf{len}(y)}) \neq 1$. Since C-Valid cannot be applied, and since $\mathsf{code}(y) \in \mathcal{T}(\mathsf{S})$, we have, by its left conclusion, that $\mathsf{len}(y) \not\approx 1$ and $\mathsf{code}(y) \approx -1$ are in A. Due to the former constraint, $\mathcal{Z}(u_{\mathsf{len}(y)}) \neq 1$ and the length of l is not one, and thus $\mathcal{M}(\mathsf{code}(y)) = -1$. Due to the latter constraint, $\mathcal{Z}(u_{\mathsf{code}(y)}) = -1$ as well.
- For all other $x : \mathsf{Int}$, we have that $\mathcal{Z}(x) = \mathcal{M}(x)$ by Step 2(a) of Definition 2.

In all cases above, we have shown that $\mathcal{Z}(x) = \mathcal{M}(t)$ where $t = x \cdot \sigma$. Since all free variables of A' are of integer type and since $\mathcal{Z}$ is a model for A', we have that $\mathcal{M}$ satisfies $\mathsf{A}' \cdot \sigma = \mathsf{A}$.

- To show $\mathcal{M}$ satisfies the equalities between terms of type Int in S, since our configuration is saturated with respect to S-Prop, equalities between integer terms are a subset of those in A, and since $\mathcal{M}$ satisfies A, it satisfies these equalities as well. Furthermore, S' does not contain disequalities between terms of type Int by construction.
- To show $\mathcal{M}$ satisfies the equalities between terms of type Str in S, notice that $s \approx t \in \mathsf{S}'$ implies that s and t reside in the same equivalence class of $e \in \widehat{\mathsf{S}}$. By construction of $\mathcal{M}$ every variable in e is assigned the same value and that value is the same value as the string literal in e if one exists. Thus $\mathcal{M}(s) = \mathcal{M}(t)$ for all terms s, t of type Str that reside in the same equivalence class, and thus $\mathcal{M}$ satisfies $s \approx t$.
- To show that $\mathcal{M}$ satisfies the disequalities $s \not\approx t$ between terms of string type in S, it suffices to show that distinct values are assigned to variables in each distinct equivalence class of $\widehat{\mathsf{S}}$. Moreover, by assumption of the configurations, each equivalence class of terms of type string has at least one variable in it. Let x and y be variables residing in two distinct equivalence classes of $\widehat{\mathsf{S}}$, and without loss of generality, assume y was assigned after x in the construction of $\mathcal{M}$. We show $\mathcal{M}(x) \neq \mathcal{M}(y)$ in the following. If y was assigned in Step 2(d) of Definition 2, then the statement holds since by construction, its value was chosen to be distinct from the value of string variables in previous equivalence classes, including the one containing x. If both x and y were assigned in Step 2(b), the statement holds since S-Conf does not apply. Otherwise, y must have been assigned in Step 2(c) to a string literal of length one. If x was assigned in Step 2(b) and $\mathsf{S} \models x \approx l$ for some string literal l not of length one, then x

and y are assigned different values trivially. Otherwise, x is assigned (either by Step 2(b) or Step 2(c)) to a string of length one. Moreover, $\mathsf{code}(z)$ is a term in S for some z such that $\mathsf{S} \models x \approx z$: if x was assigned in Step 2(b), since C-Intro cannot be applied we have $\mathsf{code}(x) \approx \mathsf{code}(l){\downarrow} \in \mathsf{S}$; if x was assigned in Step 2(c) it holds by construction. Since C-Inj cannot be applied, either $\mathsf{code}(y) \approx -1 \in \mathsf{A}$, $\mathsf{code}(z) \not\approx \mathsf{code}(y) \in \mathsf{A}$, or $z \approx y \in \mathsf{S}$. The first case cannot hold since $\mathcal{M}$ satisfies A, and thus $\mathcal{M}(\mathsf{code}(y))$ is not equal to -1. In the second case, since $\mathcal{M}$ satisfies A, we have that $\mathcal{M}(\mathsf{code}(z)) \neq \mathcal{M}(\mathsf{code}(y))$, and hence, since code is injective over the domain of strings of length one, we have that $\mathcal{M}(z) \neq \mathcal{M}(y)$. Since $\mathcal{M}(z) = \mathcal{M}(x)$, it then follows that $\mathcal{M}(x) \neq \mathcal{M}(y)$. The third case cannot hold since z and y are in distinct equivalence classes. Thus, variables in distinct equivalence classes are assigned distinct values. All disequalities $s \not\approx t \in \mathsf{S}$ are such that s and t are in different equivalence classes since S-Conf cannot be applied. Thus, $\mathcal{M}$ satisfies $s \not\approx t$.

Thus, $\mathcal{M}$ satisfies all constraints in $\mathsf{A} \cup \mathsf{S}$ and the part (2) of the theorem holds.

To show (3), it is enough to show that only finitely many constraints can be generated by the rules of the calculus. Let T^* be the (finite) set of terms that includes $\mathcal{T}(\mathsf{A}_0 \cup \mathsf{S}_0) \cup \{\epsilon, -1\}$ and contains $\mathsf{len}(x)$ and $\mathsf{code}(x)$ for all variables $x \in \mathcal{T}(\mathsf{A}_0 \cup \mathsf{S}_0)$ of type Str, and $(\mathsf{len}(l)){\downarrow}$ and $(\mathsf{code}(l)){\downarrow}$ for all string literals $l \in \mathcal{T}(\mathsf{A}_0 \cup \mathsf{S}_0)$. Let A^* be the set containing A_0, equalities between terms from T^* of type Int, literals of the form $\mathsf{len}(x) > 0, 0 \leqslant \mathsf{code}(x) < |\mathcal{A}|$, $\mathsf{len}(x) \not\approx 1$ for all variables $x \in \mathcal{T}(\mathsf{A}_0 \cup \mathsf{S}_0)$ of type Str, and inequalities of the form $\mathsf{len}(x) > \log_{|\mathcal{A}|}(n-1)$ where n is any positive integer less than or equal to the number of terms of type Str in $\mathcal{T}(\mathsf{A}_0 \cup \mathsf{S}_0)$. Let S^* be the set containing S_0 and equalities between string terms from T^*. Notice that both A^* and S^* are finite. By definition of the rules of our calculus, and by induction on the size of the derivation tree, one can show that all derived configurations $\langle \mathsf{A}, \mathsf{S} \rangle$ are such that $\mathsf{A} \cup \mathsf{S}$ is a subset of $\mathsf{A}^* \cup \mathsf{S}^*$. Since no application of a derivation rule in a tree is redundant, each node in the derivation tree contains at least one more constraint from this set than its parent. Thus, the depth of any tree is bounded by the cardinality of $\mathsf{A}^* \cup \mathsf{S}^*$, and the statement holds. □

An immediate consequence of Theorem 1 is that any strategy for applying the derivation rules in Figs. 2 and 3 is a decision procedure for Σ_{AS}-constraints. We stress that, thanks to the constructiveness of the proof of Part 2, the procedure can also compute a satisfying assignment for the free variables of M when it halts with a saturated configuration.

3.1 Implementation in an SMT Solver

The procedure in this section can be integrated into the DPLL(T) solving architecture [24] used by modern SMT solvers such as CVC4. In the most basic version of this architecture, given an arbitrary quantifier-free Σ_{AS}-formula, an incremental propositional SAT solver first searches for a truth assignment for the literals of this formula that satisfies the formula at the propositional level. If none can

substr :	Str × Int × Int → Str	to_int :	Str → Int	to_lower :	Str → Str
$\preceq$:	Str × Str → Bool	from_int :	Int → Str	to_upper :	Str → Str

Fig. 4. A sample of the extended string functions.

be found, the input is declared unsatisfiable. Otherwise, the found assignment is given as a set of Σ_{AS}-literals to a theory solver that implements the calculus above. If the solver finds a saturated configuration, then the input is declared satisfiable. Otherwise, either a *conflict clause* or a *lemma* is asserted to the SAT solver in the form of additional T_{AS}-valid constraints and the process restarts with the extended set of formulas.

We have integrated the procedure in CVC4. CVC4's linear arithmetic subsolver acts as the arithmetic subsolver of our procedure and reports a conflict clause when the rule A-Conf is applied. Similarly, the string subsolver reports conflict clauses when S-Conf is applied. The rules A-Prop and S-Prop are implemented using the standard Nelson-Oppen theory combination mechanism. Rules with multiple conclusions are implemented via the splitting-on-demand paradigm [10], where the conclusions of the rule are sent as a disjunctive lemma to the SAT solver. The remaining rules are implemented using a solver whose core data structure implements congruence closure, where additional (dis)equalities are added to this structure based on the specific rules of the calculus.

We remark that the procedure presented in this section can be naturally combined with procedures for other kinds of string constraints. While the rules we presented had premises of the form $\mathsf{S} \models s \approx t$ denoting entailment in the empty theory, the procedure can be applied in the same manner for premises $\mathsf{S} \models_{T_S} s \approx t$ for any extension T_S of the core theory of strings. In practice, our theory solver interleaves reasoning about code points with reasoning about other string operators, e.g., string concatenation and regular expressions operators, via the procedure by Liang et al. [21].

The derivation rules of the calculus are applied with consideration to combinations with the other subsolvers of CVC4. For the rules in Fig. 2, we follow the strategy used by Liang et al., which applies L-Intro and L-Valid eagerly and Card only after a configuration is saturated with respect to all other rules. Moreover, since Card is very expensive, we split on equalities between string terms $(x_1, \ldots, x_n$ in the premise of this rule) if some x_i, x_j such that neither $x_i \approx x_j$ or $x_i \not\approx x_j$ is in our current set of assertions. Among the rules in Fig. 3, C-Valid and C-Collapse are applied eagerly, the former when a term $\mathsf{code}(x)$ is registered with the string subsolver, and the latter as soon as our congruence closure procedure puts that term in the same equivalence class as a string literal. Rules C-Intro and C-Inj are applied lazily, only after the arithmetic subsolver determines A is satisfiable in LIA and the string subsolver is finished computing the set of equalities that are entailed by S.

4 Applications

In this section, we describe how a number of common string functions can be implemented efficiently using reductions involving the code function. Previous work has focused on efficient techniques for handling *extended* string functions, which include operators like substring (substr) and string replace (replace), among others [26]. Here we consider the alphabet $\mathcal{A}$ to be the set of all Unicode characters and interpret code as mapping one-character strings to the character's Unicode code point.

A few commonly used extended functions are listed in Fig. 4. In the following, we say a string l is *numeric* if it is non-empty, all of its characters are in the range "0" ... "9", and it its has no leading zeroes, that is, it starts with "0" only if it has length 1.[4] At a high level, the semantics of the operators in Fig. 4 is the following. First, $\mathsf{substr}(x, n, m)$ is interpreted as the maximal substring of x starting at position n with length at most m, or the empty string if n is outside the interval $[0, |x| - 1]$ or m is negative; $\mathsf{to\_int}(x)$ is the non-negative integer represented by x in decimal notation if x is numeric, and is -1 otherwise; $\mathsf{from\_int}(n)$ is the result of converting the value of n to its decimal notation if n is non-negative, and is ϵ otherwise; $x \preceq y$ holds if x is equal to y or precedes it lexicographically, that is, in the lexicographic extension of the character ordering $<$ introduced in Definition 1; $\mathsf{to\_upper}(x)$ maps each lower case letter character from the Basic Latin Unicode block (code points 97 to 122) in x to its uppercase version and all the other characters to themselves. The inverse function $\mathsf{to\_lower}(x)$ is similar except that it maps upper case letters (code points 65 to 90) to their lower case version.

Note that our restriction to the Latin alphabet $\mathsf{to\_upper}(x)$ and to_lower is only for simplicity since case conversions for the entire Unicode alphabet depend on the locale and follow complex rules. However, our definition and reduction can be extended as needed depending on the application.

Generally speaking, current string solvers handle the additional functions above using lazy reductions to a core language of string constraints. We say ρ is a *reduction predicate* for an extended function f if ρ does not contain f and is equivalent to $\lambda \boldsymbol{x}, y.\, f(\boldsymbol{x}) \approx y$ where $\boldsymbol{x}, y$ consist of distinct variables. All applications of f can be eliminated from a quantifier-free formula φ by replacing their occurrences with fresh variables y and conjoining φ with the appropriate applications of the reduction predicate. Reduction predicates are chosen so that their dependencies are not circular (for instance, we do not use reduction predicates for two functions that each introduce applications of the other). In practice, reduction predicates often may contain universally quantified formulas over (finite) integer ranges, which can be handled via a finite model finding strategy that incrementally sets upper bounds on the lengths of strings [26]. These reductions often generate constraints that are both large and hard to reason about. Further-

[4] Treatment of leading zeroes is slightly different in the SMT-LIB theory of strings [29]; our implementation actually conforms to the SMT-LIB semantics. Here, we provide an alternative semantics for simplicity since it admits a simpler reduction.

more, the reduction of certain extended functions cannot be expressed concisely. For example, a reduction for the $\mathsf{to\_upper}(s)$ function naively requires splitting on 26 cases to ensure that "a" is converted to "A", "b" to "B", and so on, for each character in s. As part of this work, we have revisited these reductions and incorporated the use of code. The new reduction predicates are more concise and lead to significant performance gains in practice as we demonstrate in Sect. 5.

Conversions to Lower/Upper Case. The equality $\mathsf{to\_lower}(s) \approx r$ is equivalent to:

$$\mathsf{len}(r) \approx \mathsf{len}(s) \wedge \forall_{0 \leqslant i < \mathsf{len}(s)}.\, \mathsf{code}(r_i) \approx \mathsf{code}(s_i) - \mathsf{ite}(97 \leqslant \mathsf{code}(s_i) \leqslant 122, 32, 0)$$

where r_i is $\mathsf{substr}(r, i, 1)$, s_i is $\mathsf{substr}(s, i, 1)$ and ite is the if-then-else operator. Intuitively, the formula above states that the result of $\mathsf{to\_upper}(s)$ is a string r of the same length as s such that for all positions i in s, the character at that position has a code point that is 32 less than the character at the same position in s if that character is a lowercase character; otherwise it has the same code point. Similarly, the equality $\mathsf{to\_lower}(s) \approx r$ is equivalent to:

$$\mathsf{len}(r) \approx \mathsf{len}(s) \wedge \forall_{0 \leqslant i < \mathsf{len}(s)}.\, \mathsf{code}(r_i) \approx \mathsf{code}(s_i) + \mathsf{ite}(65 \leqslant \mathsf{code}(s_i) \leqslant 90, 32, 0)$$

More generally, the code operator allows us to concisely encode many common string transducers, which have been studied in a number of recent works [6,19, 22].

String to Integer Conversion. The equality $\mathsf{to\_int}(s) \approx r$ is equivalent to:

$$(\neg\varphi_s^{\mathsf{is\_num}} \Rightarrow r \approx -1) \wedge (\varphi_s^{\mathsf{is\_num}} \Rightarrow (r \approx \mathsf{sti}_s(\mathsf{len}(s)) \wedge \varphi_s^{\mathsf{sti}}))$$

where sti_s is an (uninterpreted) function of type $\mathsf{Int} \rightarrow \mathsf{Int}$, $\varphi_s^{\mathsf{is\_num}}$ is:

$$s \not\approx \epsilon \wedge \forall_{0 \leqslant i < \mathsf{len}(s)}.\, \mathsf{ite}(\mathsf{len}(s) > 1 \wedge i \approx 0, 49, 48) \leqslant \mathsf{code}(s_i) \leqslant 57$$

s_i is $\mathsf{substr}(s, i, 1)$, and φ_s^{sti} is:

$$\mathsf{sti}_s(0) \approx 0 \wedge \forall_{0 \leqslant i < \mathsf{len}(s)}.\, \mathsf{sti}_s(i+1) \approx 10 * \mathsf{sti}_s(i) + \mathsf{code}(s_i) - 48$$

In the above reduction, the formula $\varphi_s^{\mathsf{is\_num}}$ states that s is numeric. It must be non-empty, and each of its characters must have a code point in the interval [48, 57], which corresponds to the characters for digits "0" through "9". The term $\mathsf{ite}(\mathsf{len}(s) > 1 \wedge i \approx 0, 49, 48)$ insists that the code point of the first index of s be at least 49 to exclude the possibility that its first character is "0" if the string has length greater than 1.

For a numeric string s, the formula φ_s^{sti} ensures that for each non-zero position i in s, the value of $\mathsf{sti}_s(i)$ is the result of converting the first i characters in s to the integer it denotes. The definition of φ_s^{sti} first constrains that $\mathsf{sti}_s(0)$ is zero. Then, for each $i \geqslant 0$, the value of $\mathsf{sti}_s(i+1)$ is determined by shifting the previously considered characters to the left by a digits place ($10 * \mathsf{sti}_s(i)$) and adding the integer interpretation of the current character ($\mathsf{code}(s_i) - 48$). In the end, the above formula ensures that the value of $\mathsf{sti}_s(\mathsf{len}(s))$ is equivalent to the overall value of $\mathsf{to\_int}(s)$, which is constrained to be equal to the result r in the above reduction.

Given these definitions, it is straightforward to define the opposite reduction from integers to strings. The equality $\mathsf{from\_int}(n) \approx r$ is equivalent to the following:

$$(n < 0 \Rightarrow r \approx \epsilon) \wedge (n \geqslant 0 \Rightarrow (\varphi_r^{\mathsf{is\_num}} \wedge n \approx \mathsf{sti}_r(\mathsf{len}(r)))) \wedge \varphi_r^{\mathsf{sti}}$$

By definition, $\mathsf{from\_int}$ maps negative integers to the empty string. For non-negative integers, the above reduction states that the result of converting integer n to a string is a string r that is a string representation of an integer (due to $\varphi_r^{\mathsf{is\_num}}$), and moreover is such that sti_r for this string results in n. We additionally insist that the formula constraining the semantics of this conversion (φ_r^{sti}) holds.

In practice, these reductions are implemented by introducing a fresh uninterpreted function of type $\mathsf{Int} \Rightarrow \mathsf{Int}$ to represent sti_s for each string s. The functions above are introduced during solving as needed for strings that occur as arguments to $\mathsf{to\_int}$ or those that represent the result of $\mathsf{from\_int}$ according to the above reduction.

Lexicographic Ordering. The (Boolean) equality $(x \preceq y) \approx r$ is equivalent to:

$$(x \approx y \Rightarrow r \approx \top) \wedge (x \not\approx y \Rightarrow \exists k.\, \varphi_{k,x,y}^{\mathsf{diff}} \wedge r \approx \mathsf{code}(x_k) < \mathsf{code}(y_k))$$

where x_k is $\mathsf{substr}(x, k, 1)$, y_k is $\mathsf{substr}(y, k, 1)$, and $\varphi_{k,x,y}^{\mathsf{diff}}$ is:

$$x_k \not\approx y_k \wedge \mathsf{substr}(x, 0, k) \approx \mathsf{substr}(y, 0, k)$$

Above, $\varphi_{k,x,y}^{\mathsf{diff}}$ states that x and y are different and k is the first position at which they differ. If x is a prefix of y or vice versa, then k is the length of the shorter of the two.

The reduction above considers two cases. First, if x and y are the same string, then $x \preceq y$ is trivially true. If x and y are different, then they must differ at some smallest position k. The value of r is equivalent to the comparison of $\mathsf{code}(x_k)$ and $\mathsf{code}(y_k)$. This definition correctly handles cases when k refers to the end position of x or y. If x is a strict prefix of y, k must be $\mathsf{len}(x)$, x_k is the empty string and hence $\mathsf{code}(x_k)$ must be -1. In this case, r must be true since y_k is non-empty and hence the value of $\mathsf{code}(y_k)$ is non-negative; indeed $x \preceq y$ is true when x is a prefix of y. Similarly, r must be false if y is a strict prefix of x; indeed $x \preceq y$ is false when y is a strict prefix of x.

Regular Expression Ranges. In practice, the theory of strings is often extended with memberships constraints of the form $x \in R$, where $\in$ is an infix binary predicate whose first argument is a string and whose second argument R is a regular expression denoting a sublanguage $\mathcal{L}(R)$ of $\mathcal{A}^*$. This constraint holds if x is a member of $\mathcal{L}(R)$.

The code operator can be used for regular expressions that occur often in applications. In particular, the constraint $x \in \mathsf{range}(c_m, c_n)$, where $m \leqslant n$ and c_i is is singleton string constant with code point i, states that x consists of one character whose code point is in the interval $[m, n]$. This is equivalent to $n \leqslant \mathsf{code}(x) \wedge \mathsf{code}(x) \leqslant m$. Our implementation of regular expressions in CVC4 utilizes this as a rewrite rule on membership constraints since it can eliminate the expensive computation of certain regular expression intersections. For example, consider the following equivalent formulas:

$$x \in \mathsf{range}(\texttt{"A"}, \texttt{"M"}) \wedge x \in \mathsf{range}(\texttt{"J"}, \texttt{"Z"}) \tag{1}$$

$$65 \leqslant \mathsf{code}(x) \leqslant 77 \wedge 74 \leqslant \mathsf{code}(x) \leqslant 90 \tag{2}$$

A naive approach to regular expression solving may compute the intersection of the two regular expressions above by explicitly splitting on characters in the ranges of (1). Our approach instead reasons about the arithmetic constraints in (2) and infers the constraint $74 \leqslant \mathsf{code}(x) \leqslant 77$ without expensive case splits. If the latter constraint persists in a saturated configuration, our procedure will then assign x a character in $\mathsf{range}(\texttt{"J"}, \texttt{"M"})$.

5 Evaluation

In this section, we evaluate whether our approach is practical and whether code can enable more efficient implementations of common string functions.[5] As outlined in Sect. 3.1, we have implemented our approach in CVC4, which has a state-of-the-art subsolver for the theory of strings with length and regular expressions. We evaluated it on 21,573 benchmarks [1] originating from the concolic execution of Python code involving `int()` using Py-Conbyte [8,32]. The benchmarks make extensive use of to_int, from_int and regular expression ranges. They are divided into four sets, one for each solver used to generate the benchmarks (CVC4, TRAU [3], Z3 [16], and Z3STR3).

We compare two configurations of CVC4 to show the impact of our approach: A configuration (**cvc4+c**) that uses the reductions from Sect. 4 and a configuration (**cvc4**) that disables all code-derivations and uses reductions without code. For regular expression ranges, **cvc4** disables the rewrite to inequalities involving code and uses its regular expression solver to process them. The reductions in **cvc4** use nested ite terms of the form $\mathsf{ite}(c = \texttt{"9"}, 9, \mathsf{ite}(c = \texttt{"8"}, 8, \ldots))$, i.e., do case splitting on the 10 concrete string values that correspond to valid digits, instead of the code operator but keep the reductions the same otherwise. As a point of reference, we also compare against Z3 version 4.8.7, another state-of-the-art string solver. We omit a comparison against Z3STR3 4.8.7 and Z3-TRAU 1.0 [2] (the new version of TRAU) because our experiments have shown that the current versions are unsound.[6]

[5] The implementation, the benchmarks, and the results are available at https://cvc4.github.io/papers/ijcar2020-strings.

[6] CVC4 and Z3STR3 disagreed on 498 benchmarks whereas CVC4 and Z3-TRAU disagreed on 9. In all instances, Z3STR3 and Z3-TRAU answered that the benchmark is unsatisfiable but accepted CVC4's model when we incorporated it as an additional constraint to the benchmark.

Benchmark Set		cvc4+c	cvc4	z3
	sat	**1344**	1104	1187
py-conbyte_cvc4	unsat	**8576**	8547	8482
	×	13	282	264
	sat	**1009**	929	697
py-conbyte_trauc	unsat	1424	1407	**1428**
	×	13	110	321
	sat	**1354**	1126	1343
py-conbyte_z3seq	unsat	**5864**	5797	5719
	×	35	330	191
	sat	**711**	652	692
py-conbyte_z3str	unsat	**1227**	1223	1223
	×	3	66	26
	sat	**4418**	3811	3919
Total	unsat	**17091**	16974	16852
	×	64	788	802

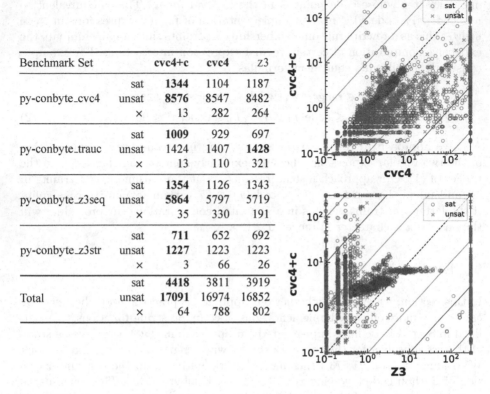

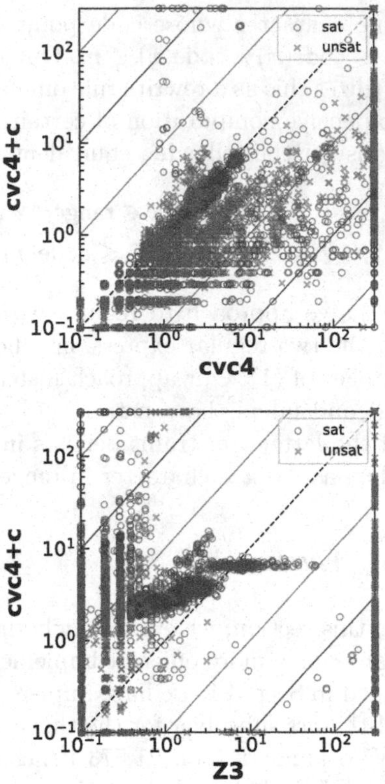

Fig. 5. Number of solved problems per benchmark set and scatter plots comparing the different solvers and configurations on a log-log scale. Best results are in bold. All benchmarks ran with a timeout of 300 s.

We ran our experiments on a cluster with Intel Xeon E5-2637 v4 CPUs running Ubuntu 16.04 and allocated one CPU core, 8 GB of RAM, and 300 s for each job.

Figure 5 summarizes the results of our experiments. The table lists the number of satisfiable and unsatisfiable answers as well as timeouts/memouts (×). z3 ran out of memory on a benchmark but had no other memouts. The figure shows two scatter plots comparing the performance of **cvc4+c** and **cvc4** and comparing **cvc4+c** and z3. Configuration **cvc4** solves more unsatisfiable benchmarks than z3 and fewer satisfiable ones, which suggests that **cvc4** is a reasonable baseline. Our new approach performs significantly better than both **cvc4** and z3. Compared to **cvc4**, configuration **cvc4+c** times out on an order of magnitude fewer benchmarks (64 versus 788) and also improves performance on commonly solved benchmarks, as the scatter plot indicates. While **cvc4** performs worse than z3 on satisfiable benchmarks, **cvc4+c** performs significantly better than both on those benchmarks. The scatter plot indicates that z3 manages to solve a subset of the benchmarks quickly. However, when z3 is not able to solve a benchmark quickly, it is unlikely that it solves it within our timeout. This results in **cvc4+c** having significantly fewer timeouts overall. The results indicate that our new approach is practical and capable of improving the performance of state-of-the-art solvers by enabling more efficient encodings.

6 Conclusion

We have presented a decision procedure for a fragment of strings that includes a string to code point conversion function. We have shown that models can be generated for satisfiable inputs, and that existing techniques for handling strings in SMT solvers can be extended with this procedure. Due to its use for encoding extended string functions, our implementation in CVC4 significantly improves on the state of the art for benchmarks involving string-to-integer conversions and regular expression ranges.

In future work, we plan to extend CVC4 to solve new constraints of interest to user applications. This includes instrumenting our string solver to be capable of generating proofs based on the procedure described in this paper. Further directions such as configuring the solver to generate interpolants for constraints in the theory of strings combined with linear arithmetic could also be explored. Finally we conjecture that efficient support for reasoning about string-to-code conversions can be leveraged for further extensions, such as handling user-defined string transducers.

References

1. str_int_benchmarks (2019). https://github.com/plfm-iis/str_int_benchmarks
2. z3-TRAU (2020). https://github.com/guluchen/z3/releases/tag/z3-trau
3. Abdulla, P.A., et al.: Flatten and conquer: a framework for efficient analysis of string constraints. In: Cohen and Vechev [15], pp. 602–617 (2017)
4. Abdulla, P.A., et al.: String constraints for verification. In: Biere and Bloem [12], pp. 150–166 (2014)
5. Abdulla, P.A., et al.: Norn: an SMT solver for string constraints. In: Kroening, D., Păsăreanu, C.S. (eds.) CAV 2015. LNCS, vol. 9206, pp. 462–469. Springer, Cham (2015). https://doi.org/10.1007/978-3-319-21690-4_29
6. Abdulla, P.A., Atig, M.F., Diep, B.P., Holík, L., Janků, P.: Chain-free string constraints. In: Chen, Y.-F., Cheng, C.-H., Esparza, J. (eds.) ATVA 2019. LNCS, vol. 11781, pp. 277–293. Springer, Cham (2019). https://doi.org/10.1007/978-3-030-31784-3_16
7. Backes, J., et al.: Semantic-based automated reasoning for AWS access policies using SMT. In: Bjørner, N., Gurfinkel, A. (eds.) 2018 Formal Methods in Computer Aided Design, FMCAD 2018, Austin, TX, USA, 30 October–2 November 2018, pp. 1–9. IEEE (2018)
8. Ball, T., Daniel, J.: Deconstructing dynamic symbolic execution. In: Irlbeck, M., Peled, D.A., Pretschner, A. (eds.) Dependable Software Systems Engineering, volume 40 of NATO Science for Peace and Security Series, D: Information and Communication Security, pp. 26–41. IOS Press (2015)

9. Barrett, C., et al.: CVC4. In: Gopalakrishnan, G., Qadeer, S. (eds.) CAV 2011. LNCS, vol. 6806, pp. 171–177. Springer, Heidelberg (2011). https://doi.org/10.1007/978-3-642-22110-1_14
10. Barrett, C., Nieuwenhuis, R., Oliveras, A., Tinelli, C.: Splitting on demand in SAT modulo theories. In: Hermann, M., Voronkov, A. (eds.) LPAR 2006. LNCS (LNAI), vol. 4246, pp. 512–526. Springer, Heidelberg (2006). https://doi.org/10.1007/11916277_35
11. Berzish, M., Ganesh, V., Zheng, Y.: Z3str3: a string solver with theory-aware heuristics. In: Stewart, D., Weissenbacher, G. (eds.) 2017 Formal Methods in Computer Aided Design, FMCAD 2017, Vienna, Austria, 2–6 October 2017, pp. 55–59. IEEE (2017)
12. Biere, A., Bloem, R. (eds.): CAV 2014. LNCS, vol. 8559. Springer, Cham (2014). https://doi.org/10.1007/978-3-319-08867-9
13. Bjørner, N., Tillmann, N., Voronkov, A.: Path feasibility analysis for string-manipulating programs. In: Kowalewski, S., Philippou, A. (eds.) TACAS 2009. LNCS, vol. 5505, pp. 307–321. Springer, Heidelberg (2009). https://doi.org/10.1007/978-3-642-00768-2_27
14. Büchi, J.R., Senger, S.: Definability in the existential theory of concatenation and undecidable extensions of this theory. Math. Log. Q. **34**(4), 337–342 (1988)
15. Cohen, A., Vechev, M.T. (eds.): Proceedings of the 38th ACM SIGPLAN Conference on Programming Language Design and Implementation, PLDI 2017, Barcelona, Spain, 18–23 June 2017. ACM (2017)
16. de Moura, L., Bjørner, N.: Z3: an efficient SMT solver. In: Ramakrishnan, C.R., Rehof, J. (eds.) TACAS 2008. LNCS, vol. 4963, pp. 337–340. Springer, Heidelberg (2008). https://doi.org/10.1007/978-3-540-78800-3_24
17. Enderton, H.B.: A mathematical Introduction to Logic, 2nd edn. Academic Press (2001)
18. Ganesh, V., Berzish, M.: Undecidability of a theory of strings, linear arithmetic over length, and string-number conversion. CoRR, abs/1605.09442 (2016)
19. Hu, Q., D'Antoni, L.: Automatic program inversion using symbolic transducers. In: Cohen and Vechev [15], pp. 376–389 (2017)
20. Kiezun, A., Ganesh, V., Artzi, S., Guo, P.J., Hooimeijer, P., Ernst, M.D.: HAMPI: a solver for word equations over strings, regular expressions, and context-free grammars. ACM Trans. Softw. Eng. Methodol. **21**(4), 25:1–25:28 (2012)
21. Liang, T., Reynolds, A., Tinelli, C., Barrett, C., Deters, M.: A DPLL(T) theory solver for a theory of strings and regular expressions. In: Biere and Bloem [12], pp. 646–662 (2014)
22. Lin, A.W., Barceló, P.: String solving with word equations and transducers: towards a logic for analysing mutation XSS. In: Bodík, R., Majumdar, R. (eds.) Proceedings of the 43rd Annual ACM SIGPLAN-SIGACT Symposium on Principles of Programming Languages, POPL 2016, St. Petersburg, FL, USA, 20–22 January 2016, pp. 123–136. ACM (2016)
23. Makanin, G.S.: The problem of solvability of equations in a free semigroup. Matematicheskii Sbornik **145**(2), 147–236 (1977)
24. Nieuwenhuis, R., Oliveras, A., Tinelli, C.: Solving SAT and SAT modulo theories: from an abstract Davis-Putnam-Logemann-Loveland Procedure to DPLL(T). J. ACM **53**(6), 937–977 (2006)
25. Quine, W.V.O.: Concatenation as a basis for arithmetic. J. Symb. Log. **11**(4), 105–114 (1946)

26. Reynolds, A., Woo, M., Barrett, C., Brumley, D., Liang, T., Tinelli, C.: Scaling up DPLL(T) string solvers using context-dependent simplification. In: Majumdar, R., Kunčak, V. (eds.) CAV 2017. LNCS, vol. 10427, pp. 453–474. Springer, Cham (2017). https://doi.org/10.1007/978-3-319-63390-9_24
27. Saxena, P., Akhawe, D., Hanna, S., Mao, F., McCamant, S., Song, D.: A symbolic execution framework for Javascript. In: 31st IEEE Symposium on Security and Privacy, S&P 2010, 16–19 May 2010, Berleley/Oakland, California, USA, pp. 513–528. IEEE Computer Society (2010)
28. The Unicode Consortium. The Unicode Standard, Version 12.1.0 (2019). http://www.unicode.org/versions/Unicode12.1.0/
29. Tinelli, C., Barrett, C., Fontaine, P.: Unicode Strings (2020). http://smtlib.cs.uiowa.edu/theories-UnicodeStrings.shtml
30. Trinh, M., Chu, D., Jaffar, J.: S3: a symbolic string solver for vulnerability detection in web applications. In: Ahn, G., Yung, M., Li, N. (eds.) Proceedings of the 2014 ACM SIGSAC Conference on Computer and Communications Security, Scottsdale, AZ, USA, 3–7 November 2014, pp. 1232–1243. ACM (2014)
31. Veanes, M., Tillmann, N., de Halleux, J.: Qex: symbolic SQL query explorer. In: Clarke, E.M., Voronkov, A. (eds.) LPAR 2010. LNCS (LNAI), vol. 6355, pp. 425–446. Springer, Heidelberg (2010). https://doi.org/10.1007/978-3-642-17511-4_24
32. Wu, W.-C.: Py-Conbyte (2019). https://github.com/spencerwuwu/py-conbyte
33. Yu, F., Alkhalaf, M., Bultan, T.: STRANGER: an automata-based string analysis tool for PHP. In: Esparza, J., Majumdar, R. (eds.) TACAS 2010. LNCS, vol. 6015, pp. 154–157. Springer, Heidelberg (2010). https://doi.org/10.1007/978-3-642-12002-2_13

Politeness for the Theory of Algebraic Datatypes

Ying Sheng[1], Yoni Zohar[1(✉)], Christophe Ringeissen[2], Jane Lange[1], Pascal Fontaine[2,3], and Clark Barrett[1]

[1] Stanford University, Stanford, USA
yoni206@gmail.com
[2] Université de Lorraine, CNRS, Inria, LORIA, 54000 Nancy, France
[3] Université de Liège, Liège, Belgium

Abstract. Algebraic datatypes, and among them lists and trees, have attracted a lot of interest in automated reasoning and Satisfiability Modulo Theories (SMT). Since its latest stable version, the SMT-LIB standard defines a theory of algebraic datatypes, which is currently supported by several mainstream SMT solvers. In this paper, we study this particular theory of datatypes and prove that it is strongly polite, showing also how it can be combined with other arbitrary disjoint theories using polite combination. Our results cover both inductive and finite datatypes, as well as their union. The combination method uses a new, simple, and natural notion of additivity, that enables deducing strong politeness from (weak) politeness.

1 Introduction

Algebraic datatypes such as lists and trees are extremely common in many programming languages. Reasoning about them is therefore crucial for modeling and verifying programs. For this reason, various decision procedures for algebraic datatypes have been, and continue to be developed and employed by formal reasoning tools such as theorem provers and Satisfiability Modulo Theories (SMT) solvers. For example, the general algorithm of [4] describes a decision procedure for datatypes suitable for SMT solvers. Consistently with the SMT paradigm, [4] leaves the combination of datatypes with other theories to general combination methods, and focuses on parametric datatypes (or *generic* datatypes as they are called in the programming languages community).

The traditional combination method of Nelson and Oppen [20] is applicable for the combination of this theory with many other theories, as long as the other theory is *stably infinite* (a technical condition that intuitively amounts to the ability to extend every model to an infinite one). Some theories of interest, however, are not stably infinite, the most notable one being the theory of

This project was partially supported by a grant from the Defense Advanced Research Projects Agency (N66001-18-C-4012), the Stanford CURIS program, and Jasmin Blanchette's European Research Council (ERC) starting grant Matryoshka (713999).

N. Peltier and V. Sofronie-Stokkermans (Eds.): IJCAR 2020, LNAI 12166, pp. 238–255, 2020.
https://doi.org/10.1007/978-3-030-51074-9_14

fixed-width bit-vectors, which is commonly used for modeling and verifying both hardware and software. To be able to perform combinations with such theories, a more general combination method was designed [21], which relies on *polite theories*. Roughly speaking, a theory is polite if: (*i*) every model can be arbitrarily enlarged; and (*ii*) there is a *witness*, a function that transforms any quantifier-free formula to an equivalent quantifier-free formula such that if the original formula is satisfiable, the new formula is satisfiable in a "minimal" interpretation. This notion was later strengthened to *strongly polite theories* [14], which also account for possible arrangements of the variables in the formula. Strongly polite theories can be combined with any other disjoint decidable theory, even if that other theory is not stably infinite. While strong politeness was already proven for several useful theories (such as equality, arrays, sets, multisets [21]), strong politeness of algebraic datatypes remained an unanswered question.

The main contribution of this paper is an affirmative answer to this question. We introduce a *witness* function that essentially "guesses" the right constructors of variables without an explicit constructor in the formula. We show how to "shrink" any model of a formula that is the output of this function into a minimal model. The witness function, as well as the model-construction, can be used by any SMT solver for the theory of datatypes that implements polite theory combination. We introduce and use the notion of additive witnesses, which allows us to prove politeness and conclude strong politeness. We further study the theory of datatypes beyond politeness and extend a decision procedure for a subset of this theory presented in [9] to support the full theory.

Related Work

The theory investigated in this paper is that of algebraic datatypes, as defined by the SMT-LIB 2 standard [3]. Detailed information on this theory, including a decision procedure and related work, can be found in [4]. Later work extends this procedure to handle shared selectors [23] and co-datatypes [22]. More recent approaches for solving formulas about datatypes use, e.g., theorem provers [15], variant satisfiability [12,19], and reduction-based decision procedures [1,6,13].

In this paper, we focus on polite theory combination. Other combination methods for non stably infinite theories include shiny theories [27], gentle theories [11], and parametric theories [17]. The politeness property was introduced in [21], and extends the stable infiniteness assumption initially used by Nelson and Oppen. Polite theories can be combined à la Nelson-Oppen with any arbitrary decidable theory. Later, a flaw in the original definition of politeness was found [14], and a corrected definition (here called *strong politeness*) was introduced. Strongly polite theories were further studied in [8], where the authors proved their equivalence with shiny theories.

More recently, it was proved [9] that a general family of datatype theories extended with bridging functions is strongly polite. This includes the theories of lists/trees with length/size functions. The authors also proved that a class of axiomatizations of datatypes is strongly polite. In contrast, in this paper we focus on standard interpretations, as defined by the SMT-LIB 2 standard, without any size function, but including selectors and testers. One can notice that the theory of standard lists without the length function, and more generally the theory of finite trees without the size function, were not mentioned as polite in a recent survey [7]. Actually, it was unclear to the authors of [7] whether these theories are strongly polite. This is now clarified in the current paper.

Outline

The paper is organized as follows. Section 2 provides the necessary notions from first-order logic and polite theories, and it introduces our working definition of the theory of datatypes, which is based on SMT-LIB 2. Section 3 discusses the difference between politeness and strong politeness, and introduces a useful condition for their equivalence. Section 4 contains the main result of this paper, namely that the theory of algebraic datatypes is strongly polite. Section 5 studies various axiomatizations of the theory of datatypes, and relates them to politeness. Section 6 concludes with directions for further research.

2 Preliminaries

2.1 Signatures and Structures

We briefly review usual definitions of many-sorted first-order logic with equality (see [10,26] for more details). For any set S, an *S-sorted set* A is a function from S to $\mathcal{P}(X) \setminus \{\emptyset\}$ for some set X (i.e., A assigns a non-empty set to every element of S), such that $A(s) \cap A(s') = \emptyset$ whenever $s \neq s'$. We use A_s to denote $A(s)$ for every $s \in S$, and call the elements of S *sorts*. When there is no ambiguity, we sometimes treat sorted sets as sets (e.g., when writing expressions like $x \in A$). Given a set S (of sorts), the *canonical S-sorted set*, denoted $[[S]]$, satisfies $[[S]]_s = \{s\}$ for every $s \in S$. A *many-sorted signature* Σ consists of a set $\mathcal{S}_\Sigma$ (of *sorts*), a set $\mathcal{F}_\Sigma$ of function symbols, and a set $\mathcal{P}_\Sigma$ of predicate symbols. Function symbols have arities of the form $\sigma_1 \times \ldots \times \sigma_n \to \sigma$, and predicate symbols have arities of the form $\sigma_1 \times \ldots \times \sigma_n$, with $\sigma_1, \ldots, \sigma_n, \sigma \in \mathcal{S}_\Sigma$. For each sort $\sigma \in \mathcal{S}_\Sigma$, $\mathcal{P}_\Sigma$ includes an *equality symbol* $=_\sigma$ of arity $\sigma \times \sigma$. We denote it by $=$ when σ is clear from context. Σ is called *finite* if $\mathcal{S}_\Sigma$, $\mathcal{F}_\Sigma$, and $\mathcal{P}_\Sigma$ are finite.

We assume an underlying $\mathcal{S}_\Sigma$-sorted set of *variables*. Terms, formulas, and literals are defined in the usual way. For a Σ-formula ϕ and a sort σ, we denote the set of free variables in ϕ of sort σ by $vars_\sigma(\phi)$. This notation naturally extends to $vars_S(\phi)$ when S is a set of sorts. A sentence is a formula without free variables. We denote by $QF(\Sigma)$ the set of quantifier-free formulas of Σ. A Σ-literal is called *flat* if it has one of the following forms: $x = y$, $x \neq y$, $x = f(x_1, \ldots, x_n)$, $P(x_1, \ldots, x_n)$, or $\neg P(x_1, \ldots, x_n)$ for some variables $x, y, x_1, \ldots, x_n$ and function and predicate symbols f and P from Σ.

A *Σ-structure* is a many-sorted structure for Σ, without interpretation of variables. It consists of a $\mathcal{S}_\Sigma$-sorted set A, and interpretations to the function and predicate symbols of Σ. We further require that $=_\sigma$ is interpreted as the identity relation over A_σ for every $\sigma \in \mathcal{S}_\Sigma$. A *$\Sigma$-interpretation* $\mathcal{A}$ is an extension of a Σ-structure with interpretations to some set of variables. For any Σ-term α, $\alpha^{\mathcal{A}}$ denotes the interpretation of α in $\mathcal{A}$. When α is a set of Σ-terms, $\alpha^{\mathcal{A}} = \{x^{\mathcal{A}} \mid x \in \alpha\}$. Similarly, $\sigma^{\mathcal{A}}$, $f^{\mathcal{A}}$ and $P^{\mathcal{A}}$ denote the interpretation of σ, f and P in $\mathcal{A}$. Satisfaction is defined as usual. $\mathcal{A} \models \varphi$ denotes that $\mathcal{A}$ satisfies φ.

A Σ-theory T is a class of Σ-structures. A Σ-interpretation whose variable-free part is in T is called a T-interpretation. A Σ-formula ϕ is T-satisfiable if $\mathcal{A} \models \phi$ for some T-interpretation $\mathcal{A}$. Two formulas ϕ and ψ are *T-equivalent* if they are satisfied by the same class of T-interpretations. Let Σ_1 and Σ_2 be signatures, T_1 a Σ_1-theory, and T_2 a Σ_2-theory. The combination of T_1 and T_2, denoted $T_1 \oplus T_2$, is the class of $\Sigma_1 \cup \Sigma_2$-structures $\mathcal{A}$ such that $\mathcal{A}^{\Sigma_1}$ is in T_1 and $\mathcal{A}^{\Sigma_2}$ is in T_2, where $\mathcal{A}^{\Sigma_i}$ is the restriction of $\mathcal{A}$ to Σ_i for $i \in \{1, 2\}$.

2.2 The SMT-LIB 2 Theory of Datatypes

In this section we formally define the SMT-LIB 2 theory of algebraic datatypes. The formalization is based on [3], but is adjusted to suit our investigation of politeness.

Definition 1. *Given a signature Σ, a set $S \subseteq \mathcal{S}_\Sigma$ and an S-sorted set A, the set of Σ-*trees *over A of sort $\sigma \in \mathcal{S}_\Sigma$ is denoted by $T_\sigma(\Sigma, A)$ and is inductively defined as follows:*

- $T_{\sigma,0}(\Sigma, A) = A_\sigma$ *if* $\sigma \in S$ *and* $\emptyset$ *otherwise.*
- $T_{\sigma,i+1}(\Sigma, A) = T_{\sigma,i}(\Sigma, A) \cup \{c(t_1, \ldots, t_n) \mid c : \sigma_1 \times \ldots \times \sigma_n \to \sigma \in \mathcal{F}_\Sigma, t_j \in T_{\sigma_j,i}(\Sigma, A)$ *for* $j = 1, \ldots, n\}$ *for each* $i \geq 0$.

Then $T_\sigma(\Sigma, A) = \bigcup_{i \geq 0} T_{\sigma,i}(\Sigma, A)$. The depth *of a Σ-tree over A is inductively defined by $depth(a) = 0$ for every $a \in A$, $depth(c) = 1$ for every 0-ary function symbol $c \in \mathcal{F}_\Sigma$, and $depth(c(t_1, \ldots, t_n)) = 1 + max(depth(t_1), \ldots, depth(t_n))$ for every n-ary function symbol c of Σ.*

The idea behind Definition 1 is that $T_\sigma(\Sigma, A)$ contains all ground σ-sorted terms constructed from the elements of A (considered as constant symbols) and the function symbols of Σ.

Example 1. Let Σ be a signature with two sorts, **elem** and **struct**, and whose function symbols are b of arity **struct**, and c of arity (**elem** $\times$ **struct** $\times$ **struct**) $\to$ **struct**. Consider the {**elem**}-sorted set $A = \{a\}$. For the **elem** sort, $T_{\mathbf{elem}}(\Sigma, A)$ is the singleton $A = \{a\}$ and the Σ-tree a is of depth 0. For the **struct** sort, $T_{\mathbf{struct}}(\Sigma, A)$ includes infinitely many Σ-trees, such as b of depth 1, $c(a, b, b)$ of depth 2, and $c(a, c(a, b, b), b)$ of depth 3.

Definition 2. *A finite signature Σ is called a* datatypes signature *if $\mathcal{S}_\Sigma$ is the disjoint union of two sets of sorts $\mathcal{S}_\Sigma = \mathbf{Elem}_\Sigma \uplus \mathbf{Struct}_\Sigma$ and $\mathcal{F}_\Sigma$ is the disjoint union of two sets of function symbols $\mathcal{F}_\Sigma = \mathcal{CO}_\Sigma \uplus \mathcal{SE}_\Sigma$, such that $\mathcal{SE}_\Sigma = \{s_{c,i} : \sigma \to \sigma_i \mid c \in \mathcal{CO}_\Sigma, c : \sigma_1, \ldots, \sigma_n \to \sigma, 1 \leq i \leq n\}$ and $\mathcal{P}_\Sigma = \{is_c : \sigma \mid c \in \mathcal{CO}_\Sigma, c : \sigma_1, \ldots, \sigma_n \to \sigma\}$. We denote by $\Sigma_{|\mathcal{CO}}$ the signature with* *the same sorts as Σ, no predicate symbols (except $=_\sigma$ for $\sigma \in \mathcal{S}_\Sigma$), and whose function symbols are $\mathcal{CO}_\Sigma$. We further require the following well-foundedness requirement: $T_\sigma(\Sigma_{|\mathcal{CO}}, [[\mathbf{Elem}_\Sigma]]) \neq \emptyset$ for any $\sigma \in \mathbf{Struct}_\Sigma$.*

From now on, we omit the subscript Σ from the above notations (e.g., when writing $[[\mathbf{Elem}]]$ rather than $[[\mathbf{Elem}_\Sigma]]$, $\mathcal{CO}$ rather than $\mathcal{CO}_\Sigma$) whenever Σ is clear from the context. Notice that Definition 2 remains equivalent if we replace $[[\mathbf{Elem}]]$ by any (non-empty) **Elem**-sorted set A. The set $[[\mathbf{Elem}]]$ has been chosen since this minimal **Elem**-sorted set is sufficient.

In accordance with SMT-LIB 2, we call the elements of $\mathcal{CO}$ *constructors*, the elements of $\mathcal{SE}$ *selectors*, and the elements of $\mathcal{P}$ *testers*. 0-ary constructors are called *nullary*. In what follows, Σ denotes an arbitrary datatypes signature.

In the next example we review some common datatypes signatures.

Example 2. The signature Σ_{list} has two sorts, **elem** and **list**. Its function symbols are *cons* of arity $(\mathbf{elem} \times \mathbf{list}) \to \mathbf{list}$, *nil* of arity **list**, *car* of arity $\mathbf{list} \to \mathbf{elem}$ and *cdr* of arity $\mathbf{list} \to \mathbf{list}$. Its predicate symbols are is_{nil} and is_{cons}, both of arity **list**. It is a datatypes signature, with $\mathbf{Elem} = \{\mathbf{elem}\}$, $\mathbf{Struct} = \{\mathbf{list}\}$, $\mathcal{CO} = \{nil, cons\}$ and $\mathcal{SE} = \{car, cdr\}$. It is often used to model lisp-style linked lists. *car* represents the head of the list and *cdr* represents its tail. *nil* represents the empty list. Σ_{list} is well-founded as $T_{\mathbf{list}}(\Sigma_{list|\mathcal{CO}}, [[\mathbf{Elem}]])$ includes *nil*.

The signature Σ_{pair} also has two sorts, **elem** and **pair**. Its function symbols are *pair* of arity $(\mathbf{elem} \times \mathbf{elem}) \to \mathbf{pair}$ and *first* and *second* of arity $\mathbf{pair} \to \mathbf{elem}$. Its predicate symbol is is_{pair} of arity **pair**. It is a datatypes signature, with $\mathbf{Elem} = \{\mathbf{elem}\}$, $\mathbf{Struct} = \{\mathbf{pair}\}$, $\mathcal{CO} = \{pair\}$, and $\mathcal{SE} = \{first, second\}$. It can be used to model ordered pairs, together with projection functions. It is well-founded as $T_{\mathbf{pair}}(\Sigma_{pair|\mathcal{CO}}, [[\mathbf{Elem}]])$ is not empty (as $[[\mathbf{Elem}]]$ is not empty).

The signature Σ_{lp} has three sorts, **elem**, **pair** and **list**. Its function symbols are *cons* of arity $(\mathbf{pair} \times \mathbf{list}) \to \mathbf{list}$, *car* of arity $\mathbf{list} \to \mathbf{pair}$, as well as *nil*, *cdr*, *first*, *second* with arities as above. Its predicate symbols are is_{pair}, is_{cons} and *nil*, with arities as above. It can be used to model lists of ordered pairs. Similarly to the above signatures, it is a datatypes signature.

Next, we distinguish between finite datatypes (e.g., records) and inductive datatypes (e.g., lists).

Definition 3. *A sort $\sigma \in$ **Struct** is* finite *if $T_\sigma(\Sigma_{|\mathcal{CO}}, [[\mathbf{Elem}]])$ is finite, and is called* inductive *otherwise.*

We denote the set of inductive sorts in Σ by $Ind(\Sigma)$ and the set of its finite sorts by $Fin(\Sigma)$. Note that if σ is inductive, then according to Definitions 1 and 3 we have that for any natural number i there exists a natural number $i' > i$ such that $T_{\sigma,i'}(\Sigma_{|\mathcal{CO}}, [[\mathbf{Elem}]]) \neq T_{\sigma,i}(\Sigma_{|\mathcal{CO}}, [[\mathbf{Elem}]])$. Further, for any natural number d and every **Elem**-sorted set D there exists a natural number i' such that $T_{\sigma,i'}(\Sigma_{|\mathcal{CO}}, D)$ contains an element whose depth is greater than d.

Example 3. **list** is inductive in Σ_{list} and Σ_{lp}. **pair** is finite in Σ_{pair} and Σ_{lp}.

Finally, we define datatypes structures and the theory of algebraic datatypes.

Definition 4. *Let Σ be a datatypes signature and D an* **Elem**-*sorted set. A Σ-structure $\mathcal{A}$ is said to be a* datatypes Σ-structure generated by D *if:*

- $\sigma^{\mathcal{A}} = T_{\sigma}(\Sigma_{|\mathcal{CO}}, D)$ *for every sort* $\sigma \in \mathcal{S}_{\Sigma}$,
- $c^{\mathcal{A}}(t_1, \ldots, t_n) = c(t_1, \ldots, t_n)$ *for every* $c \in \mathcal{CO}$ *of arity* $(\sigma_1 \times \ldots \times \sigma_n) \rightarrow \sigma$ *and* $t_1 \in \sigma_1^{\mathcal{A}}, \ldots, t_n \in \sigma_n^{\mathcal{A}}$,
- $s_{c,i}^{\mathcal{A}}(c(t_1, \ldots, t_n)) = t_i$ *for every* $c \in \mathcal{CO}$ *of arity* $(\sigma_1 \times \ldots \times \sigma_n) \rightarrow \sigma$, $t_1 \in \sigma_1^{\mathcal{A}}, \ldots, t_n \in \sigma_n^{\mathcal{A}}$ *and* $1 \leq i \leq n$,
- $is_c^{\mathcal{A}} = \{c(t_1, \ldots, t_n) \mid t_1 \in \sigma_1^{\mathcal{A}}, \ldots, t_n \in \sigma_n^{\mathcal{A}}\}$ *for every* $c \in \mathcal{CO}$ *of arity* $(\sigma_1 \times \ldots \times \sigma_n) \rightarrow \sigma$.

$\mathcal{A}$ is said to be a datatypes Σ-structure *if it is a datatypes Σ-structure generated by D for some* **Elem**-*sorted set D. The* Σ-theory of datatypes, *denoted $\mathcal{T}_{\Sigma}$ is the class of datatypes Σ-structures.*

Notice that the interpretation of selector functions $s_{c,i}$ when applied to terms that are constructed using a constructor different than c is not fixed and can be set arbitrarily in datatypes structures, consistently with SMT-LIB 2.

Example 4. If $\mathcal{A}$ is a datatypes Σ_{list}-structure then $\mathbf{list}^{\mathcal{A}}$ is the set of terms constructed from $\mathbf{elem}^{\mathcal{A}}$ and *cons*, plus *nil*. If $\mathbf{elem}^{\mathcal{A}}$ is the set of natural numbers, then $\mathbf{list}^{\mathcal{A}}$ contains, e.g., *nil*, $cons(1, nil)$, and $cons(1, cons(1, cons(2, nil)))$. These correspond to the lists [] (the empty list), [1] and [1, 1, 2], respectively.

If $\mathcal{A}$ is a datatypes Σ_{pair}-structure then $\mathbf{pair}^{\mathcal{A}}$ is the set of terms of the form $pair(a, b)$ with $a, b \in \mathbf{elem}^{\mathcal{A}}$. If $\mathbf{elem}^{\mathcal{A}}$ is again interpreted as the set of natural numbers, $\mathbf{pair}^{\mathcal{A}}$ includes, for example, the terms $pair(1, 1)$ and $pair(1, 2)$, that correspond to (1,1) and (1, 2), respectively. Notice that in this case, $\mathbf{pair}^{\mathcal{A}}$ is an infinite set even though **pair** is a finite sort (in terms of Definition 3).

Datatypes Σ_{lp}-structures with the same interpretation for **elem** include the terms *nil*, $cons(pair(1, 1), nil)$, and $cons(pair(1, 1), cons(pair(1, 2), nil))$ in the interpretation for **list**, that correspond to [], [(1, 1)] and [(1, 1), (1, 2)], respectively. If we rename **elem** in the definition of Σ_{list} to **pair**, we get that $\mathcal{T}_{\Sigma_{lp}} = \mathcal{T}_{\Sigma_{list}} \oplus \mathcal{T}_{\Sigma_{pair}}$.

2.3 Polite Theories

Given two theories T_1 and T_2, a combination method à la Nelson-Oppen provides a modular way to decide $T_1 \cup T_2$-satisfiability problems using the satisfiability procedures known for T_1 and T_2. Assuming that T_1 and T_2 have disjoint signatures is not sufficient to get a complete combination method for the satisfiability problem. The reason is that T_1 and T_2 may share sorts, and the equality symbol on these shared sorts. To be complete, T_1 and T_2 must agree on the cardinality of their respective models, and there must be an agreement between T_1 and T_2 on the interpretation of shared formulas built over the equality symbol. These two requirements can be easily fulfilled, based on the following definitions:

Definition 5 (Stable Infiniteness). *Given a signature Σ and a set $S \subseteq \mathcal{S}_\Sigma$, we say that a Σ-theory T is* stably infinite with respect to S *if every quantifier-free Σ-formula that is T-satisfiable is also T-satisfiable by a T-interpretation $\mathcal{A}$ in which $\sigma^{\mathcal{A}}$ is infinite for every $\sigma \in S$.*

Definition 6 (Arrangement). *Let V be a finite set of variables whose sorts are in S and $\{V_\sigma \mid \sigma \in S\}$ a partition of V such that V_σ is the set of variables of sort σ in V. We say that a formula δ is an* arrangement of V *if $\delta = \bigwedge_{\sigma \in S}(\bigwedge_{(x,y)\in E_\sigma}(x = y) \wedge \bigwedge_{(x,y)\notin E_\sigma}(x \neq y))$, where E_σ is some equivalence relation over V_σ for each $\sigma \in S$.*

Assume that both T_1 and T_2 are stably infinite with disjoint signatures, and let V be the finite set of variables shared by T_1 and T_2. Under this assumption, T_1 and T_2 can agree on an infinite cardinality, and guessing an arrangement of V suffices to get an agreement on the interpretation of shared formulas.

In this paper we are interested in an asymmetric disjoint combination where T_1 and T_2 are not both stably infinite. In this scenario, one theory can be arbitrary. As a counterpart, the other theory must be more than stably infinite: it must be polite, meaning that it is always possible to increase the cardinality of a model and to have a model whose cardinality is finite.

In the following we decompose the politeness definition from [14,21] in order to distinguish between politeness and strong politeness (in terms of [8]) in various levels of the definition. In what follows, Σ is an arbitrary (many-sorted) signature, $S \subseteq \mathcal{S}_\Sigma$, and T is a Σ-theory.

Definition 7 (Smooth). *The theory T is* smooth *w.r.t. S if for every quantifier-free formula ϕ, T-interpretation $\mathcal{A}$ that satisfies ϕ, and function κ from S to the class of cardinals such that $\kappa(\sigma) \geq |\sigma^{\mathcal{A}}|$ for every $\sigma \in S$ there exists a Σ-interpretation $\mathcal{A}'$ that satisfies ϕ with $|\sigma^{\mathcal{A}'}| = \kappa(\sigma)$ for every $\sigma \in S$.*

In definitions introduced above, as well as below, we often identify singletons with their single elements when there is no ambiguity (e.g., when saying that a theory is smooth w.r.t. a sort σ).

We now introduce some concepts in order to define finite witnessability. Let ϕ be a quantifier-free Σ-formula and $\mathcal{A}$ a Σ-interpretation. We say that $\mathcal{A}$ *finitely witnesses ϕ for T w.r.t. S* (or, is a *finite witness of ϕ for T w.r.t. S*), if $\mathcal{A}$ is a T-interpretation, $\mathcal{A} \models \phi$, and $\sigma^{\mathcal{A}} = vars_\sigma(\phi)^{\mathcal{A}}$ for every $\sigma \in S$. We say that ϕ is *finitely witnessed for T w.r.t. S* if it is either T-unsatisfiable or it has a finite witness for T w.r.t. S. ϕ is *strongly finitely witnessed for T w.r.t. S* if $\phi \wedge \delta_V$ is finitely witnessed for T w.r.t. S for every arrangement δ_V of V, where V is any set of variables whose sorts are in S. We say that a function $wtn : QF(\Sigma) \to QF(\Sigma)$ is a *(strong) witness for T w.r.t. S* if for every $\phi \in QF(\Sigma)$ we have that: 1. ϕ and $\exists \overrightarrow{w}.\, wtn(\phi)$ are T-equivalent for $\overrightarrow{w} = vars(wtn(\phi)) \setminus vars(\phi)$; and 2. $wtn(\phi)$ is (strongly) finitely witnessed for T w.r.t. S.[1]

[1] We note that in practice, the new variables in $wtn(\phi)$ are assumed to be fresh not only with respect to ϕ, but also with respect to the formula from the second theory being combined.

Definition 8 (Finitely Witnessable). *The theory* T *is* (strongly) finitely witnessable *w.r.t.* S *if there exists a (strong) witness for* T *w.r.t.* S *which is computable.*

Definition 9 (Polite). T *is called* (strongly) polite w.r.t. S *if it is smooth and (strongly) finitely witnessable w.r.t.* S*.*

Finally, we recall the following theorem from [14].

Theorem 1 ([14]). *Let* Σ_1 *and* Σ_2 *be signatures and let* $S = \mathcal{S}_{\Sigma_1} \cap \mathcal{S}_{\Sigma_2}$*. If* T_1 *is a* Σ_1*-theory strongly polite w.r.t.* $S_1 \subseteq \mathcal{S}_{\Sigma_1}$*,* T_2 *is a* Σ_2*-theory strongly polite w.r.t.* $S_2 \subseteq \mathcal{S}_{\Sigma_2}$*, and* $S \subseteq S_2$*, then* $T_1 \oplus T_2$ *is strongly polite w.r.t.* $S_1 \cup (S_2 \setminus S)$*.*

3 Additive Witnesses

It was shown in [14] that politeness is not sufficient for the proof of the polite combination method from [21]. Strong politeness was introduced to fix the problem. It is unknown, however, whether there are theories that are polite but not strongly polite. In this section we offer a simple (yet useful) criterion for the equivalence of the two notions. Throughout this section, unless stated otherwise, Σ and S denote an arbitrary signature and a subset of its set of sorts, and T, T_1, T_2 denote arbitrary Σ-theories.

The following example, which is based on [14] using notions of the current paper, shows that the strong and non-strong witnesses are different. Let Σ_0 be a signature with a single sort σ and no function or predicate symbols (except $=_\sigma$), T_0 the Σ_0-theory consisting of all Σ_0-structures $\mathcal{A}$ with $|\sigma^{\mathcal{A}}| \geq 2$, ϕ the formula $x = x \wedge w = w$, and δ the arrangement $(x = w)$ of $\{x, w\}$. Then $\phi \wedge \delta$ is T_0-satisfiable, but every interpretation $\mathcal{A}$ with $\sigma^{\mathcal{A}} = \{x, w\}^{\mathcal{A}}$ that satisfies it has only one element in $\sigma^{\mathcal{A}}$ and so ϕ is not strongly finitely witnessed for T_0 w.r.t. σ. It is straightforward to show, however, that ϕ is finitely witnessed for T_0 w.r.t. σ. Moreover, the function wtn defined by $wtn(\phi) = (\phi \wedge w_1 = w_1 \wedge w_2 = w_2)$ for fresh w_1, w_2 is a witness for T_0 w.r.t. σ, but not a strong one. This does not show, however, that T_0 is not strongly polite. In fact, it is indeed strongly polite since the function $wtn'(\phi) = \phi \wedge w_1 \neq w_2$ for fresh w_1, w_2 is a strong witness for T_0 w.r.t. σ.

We introduce the notion of additivity, which ensures that the witness is able to "absorb" arrangements and thus lift politeness to strong politeness.

Definition 10 (Additivity). *Let* $f : QF(\Sigma) \to QF(\Sigma)$*. We say that* f *is* S*-additive for* T *if* $f(f(\phi) \wedge \varphi)$ *and* $f(\phi) \wedge \varphi$ *are* T*-equivalent and have the same set of* S*-sorted variables for every* $\phi, \varphi \in QF(\Sigma)$*, provided that* φ *is a conjunction of flat literals such that every term in* φ *is a variable whose sort is in* S*. When* T *is clear from the context, we just say that* f *is* S*-additive. We say that* T *is* additively *finitely witnessable w.r.t.* S *if there exists a witness for* T *w.r.t.* S *which is both computable and* S*-additive.* T *is said to be* additively polite w.r.t. S *if it is smooth and additively finitely witnessable w.r.t.* S*.*

Proposition 1. *Let wtn be a witness for T w.r.t. S. If wtn is S-additive then it is a strong witness for T w.r.t. S.*[2]

Corollary 1. *Suppose T is additively polite w.r.t. S. Then it is strongly polite w.r.t. S.*

The theory T_0 from the example above is additively finitely witnessable w.r.t. σ, even though wtn' is not σ-additive. Indeed, it is possible to define a new witness for T_0 w.r.t. σ, say wtn'', which is σ-additive. This function wtn'' is defined by: $wtn''(\phi) = wtn'(\phi)$ if ϕ is a conjunction that includes some disequality $x \neq y$ for some x, y. Otherwise, $wtn''(\phi) = \phi$.

T_0 is an *existential theory*: it consists of all the structures that satisfy an existential sentence (in this case, $\exists x, y \,.\, x \neq y$). The construction of wtn'' can be generalized to any existential theory. Such theories are also smooth w.r.t. any set of sorts and so existential theories are additively polite.

The notion of additive witnesses is useful for proving that a polite theory is strongly polite. In particular, the witnesses for the theories of equality, arrays, sets and multisets from [21] are all additive, and so strong politeness of these theories follows from their politeness. The same will hold later, when we conclude strong politeness of theories of algebraic datatypes from their politeness.

4 Politeness for the SMT-LIB 2 Theory of Datatypes

Let Σ be a datatypes signature with $\mathcal{S}_\Sigma = \mathbf{Elem} \uplus \mathbf{Struct}$ and $\mathcal{F}_\Sigma = \mathcal{CO} \uplus \mathcal{SE}$. In this section, we prove that $\mathcal{T}_\Sigma$ is strongly polite with respect to **Elem**. In Sect. 4.1, we consider theories with only inductive sorts, and consider theories with only finite sorts in Sect. 4.2. We combine them in Sect. 4.3, where arbitrary theories of datatypes are considered. This separation is only needed for finite witnessability. For smoothness, however, it is straightforward to show that the **Elem** domain of a given interpretation can always be augmented without changing satisfiability of quantifier-free formulas.

Lemma 1. *$\mathcal{T}_\Sigma$ is smooth w.r.t.* **Elem**.

Lemma 1 holds for any datatypes signature.

4.1 Inductive Datatypes

In this section, we assume that all sorts in **Struct** are inductive.

To prove finite witnessability, we now introduce an additive witness function. Following arguments from [21], it suffices to define the witness only for conjunctions of flat literals. A complete witness can then use the restricted one by first transforming the input formula to flat DNF form and then creating a

[2] Due to lack of space, some proofs have been omitted. They can be found in an extended version at https://arxiv.org/abs/2004.04854.

disjunction where each disjunct is the result of applying the witness on the corresponding disjunct. Similarly, it suffices to show that $wtn(\phi)$ is finitely witnessed for ϕ which is a conjunction of flat literals. Essentially, our witness guesses possible constructors for variables whose constructors are not explicit in the input formula.

Definition 11 (A Witness for $\mathcal{T}_\Sigma$). *Let ϕ be a quantifier-free conjunction of flat Σ-literals. $wtn_i(\phi)$ is obtained from ϕ by performing the following steps:*

1. *For any literal of the form $y = s_{c,i}(x)$ such that $x = c(\overrightarrow{u_1}, y, \overrightarrow{u_2})$ does not occur in ϕ and $x = d(\overrightarrow{u_d})$ does not occur in ϕ for any $\overrightarrow{u_1}, \overrightarrow{u_2}, \overrightarrow{u_d}$, we conjunctively add $x = c(\overrightarrow{u_1}, y, \overrightarrow{u_2}) \vee (\bigvee_{d \neq c} x = d(\overrightarrow{u_d}))$ with fresh $\overrightarrow{u_1}, \overrightarrow{u_2}, \overrightarrow{u_d}$, where c and d range over $\mathcal{CO}$.*
2. *For any literal of the form $is_c(x)$ such that $x = c(\overrightarrow{u})$ does not occur in ϕ for any $\overrightarrow{u}$, we conjunctively add $x = c(\overrightarrow{u})$ with fresh $\overrightarrow{u}$.*
3. *For any literal of the form $\neg is_c(x)$ such that $x = d(\overrightarrow{u_d})$ does not occur in ϕ for any $d \neq c$ and $\overrightarrow{u_d}$, we conjunctively add $\bigvee_{d \neq c} x = d(\overrightarrow{u_d})$, with fresh $\overrightarrow{u_d}$.*
4. *For any sort $\sigma \in$ **Elem** such that ϕ does not include a variable of sort σ we conjunctively add a literal $x = x$ for a fresh variable x of sort σ.*

Example 5. Let ϕ be the Σ_{list}-formula $y = cdr(x) \wedge y' = cdr(x) \wedge is_{cons}(y)$. $wtn_i(\phi)$ is $\phi \wedge (x = nil \vee x = cons(e, y)) \wedge (x = nil \vee x = cons(e', y')) \wedge y = cons(e'', z) \wedge e''' = e'''$ where e, e', e'', e''', z are fresh.

In Definition 11, Item 1 guesses the constructor of the argument for the selector. Items 2 and 3 correspond to the semantics of testers. Item 4 is meant to ensure that we can construct a finite witness with non-empty domains. The requirement for absence of literals before adding literals or disjunctions to ϕ is used to ensure additivity of wtn_i. And indeed:

Lemma 2. *wtn_i is* **Elem***-additive.*

Further, it can be verified that:

Lemma 3. *Let ϕ be a conjunction of flat literals. ϕ and $\exists \overrightarrow{w}\,.\,\Gamma$ are $\mathcal{T}_\Sigma$-equivalent, where $\Gamma = wtn_i(\phi)$ and $\overrightarrow{w} = vars(\Gamma) \setminus vars(\phi)$.*

The remainder of this section is dedicated to the proof of the following lemma:

Lemma 4 (Finite Witnessability). *Let ϕ be a conjunction of flat literals. Then, $\Gamma = wtn_i(\phi)$ is finitely witnessed for $\mathcal{T}_\Sigma$ with respect to* **Elem**.

Suppose that Γ is $\mathcal{T}_\Sigma$-satisfiable, and let $\mathcal{A}$ be a satisfying $\mathcal{T}_\Sigma$-interpretation. We define a $\mathcal{T}_\Sigma$-interpretation $\mathcal{B}$ as follows, and then show that $\mathcal{B}$ is a finite witness of Γ for $\mathcal{T}_\Sigma$ w.r.t. **Elem**. First for every $\sigma \in$ **Elem** we set $\sigma^{\mathcal{B}} = vars_\sigma(\Gamma)^{\mathcal{A}}$, and for every variable $e \in vars_\sigma(\Gamma)$, we set $e^{\mathcal{B}} = e^{\mathcal{A}}$. The interpretations of **Struct**-sorts, testers and constructors are uniquely determined by the theory. It is left to define the interpretation of **Struct**-variables in $\mathcal{B}$, as well as the interpretation of the selectors (the interpretation of selectors is fixed by the theory only when applied to the "right" constructor). We do this in several steps:

Step 1 – Simplifying Γ: since ϕ is a conjunction of flat literals, Γ is a conjunction whose conjuncts are either flat literals or disjunctions of flat literals (introduced in Items 1 and 3 of Definition 11). Since $\mathcal{A} \models \Gamma$, $\mathcal{A}$ satisfies exactly one disjunct of each such disjunction. We can thus obtain a formula Γ_1 from Γ by replacing every disjunction with the disjunct that is satisfied by $\mathcal{A}$. Notice that $\mathcal{A} \models \Gamma_1$ and that it is a conjunction of flat literals. Let Γ_2 be obtained from Γ_1 by removing any literal of the form $is_c(x)$ and any literal of the form $\neg is_c(x)$. Let Γ_3 be obtained from Γ_2 by removing any literal of the form $x = s_{c,i}(y)$. For convenience, we denote Γ_3 by Γ'. Obviously, $\mathcal{A} \models \Gamma'$, and Γ' is a conjunction of flat literals without selectors and testers.
Step 2 – Working with Equivalence Classes: We would like to preserve equalities between **Struct**-variables from $\mathcal{A}$. To this end, we group all variables in $vars(\Gamma)$ to equivalence classes according to their interpretation in $\mathcal{A}$. Let $\equiv_{\mathcal{A}}$ denote an equivalence relation over $vars(\Gamma)$ such that $x \equiv_{\mathcal{A}} y$ iff $x^{\mathcal{A}} = y^{\mathcal{A}}$. We denote by $[x]$ the equivalence class of x. Let α be an equivalence class, thus $\alpha^{\mathcal{A}} = \{x^{\mathcal{A}} \mid x \in \alpha\}$ is a singleton. Identifying this singleton with its only element, we have that $\alpha^{\mathcal{A}}$ denotes $a^{\mathcal{A}}$ for an arbitrary element a of the equivalence class α.
Step 3 – Ordering Equivalence Classes: We would also like to preserve disequalities between **Struct**-variables from $\mathcal{A}$. Thus we introduce a relation $\prec$ over the equivalence classes, such that $\alpha \prec \beta$ if $y = c(w_1, \ldots, w_n)$ occurs as one of the conjuncts in Γ' for some $w_1, \ldots, w_n$ and c such that $w_k \in \alpha$ for some $y \in \beta$, $c \in \mathcal{CO}$, and k. Call an equivalence class α *nullary* if $\mathcal{A} \models is_c(x)$ for some $x \in \alpha$ and nullary constructor c. Call an equivalence class α *minimal* if $\beta \not\prec \alpha$ for every β. Notice that each nullary equivalence class is minimal. The relation $\prec$ induces a directed acyclic graph (DAG), denoted G. The vertices are the equivalence classes. Whenever $\alpha \prec \beta$, we draw an edge from vertex α to β.
Step 4 – Interpretation of Equivalence Classes: We define $\alpha^{\mathcal{B}}$ for every equivalence class α. Then, $x^{\mathcal{B}}$ is simply defined as $[x]^{\mathcal{B}}$, for every **Struct**-variable x. The idea goes as follows. Nullary classes are assigned according to $\mathcal{A}$. Other minimal classes are assigned arbitrarily, but it is important to assign different classes to terms whose depths are far enough from each other to ensure that the disequalities in $\mathcal{A}$ are preserved. Non-minimal classes are uniquely determined after minimal ones are assigned. Formally, let m be the number of equivalence classes, l the number of minimal equivalence classes, r the number of nullary equivalence classes, and $\alpha_1, \ldots, \alpha_m$ a topological sort of G, such that all minimal classes occur before all others, and the first r classes are nullary. Let d be the length of the longest path in G. We define $\alpha_i^{\mathcal{B}}$ by induction on i. In the definition, we use $\mathcal{B}_{\mathbf{Elem}}$ to denote the **Elem**-sorted set assigning $\sigma^{\mathcal{B}}$ to every $\sigma \in$ **Elem**.

1. If $0 < r$ and $i \leq r$ then α_i is a nullary class and so we set $\alpha_i^{\mathcal{B}} = \alpha_i^{\mathcal{A}}$.
2. If $r < i \leq l$ then α_i is minimal and not nullary. Let σ be the sort of variables in α_i. If $\sigma \in$ **Elem**, then all variables in the class have already been defined. Otherwise, $\sigma \in$ **Struct**. In this case, we define $\alpha_i^{\mathcal{B}}$ to be an arbitrary element of $T_\sigma(\Sigma_{|\mathcal{CO}}, \mathcal{B}_{\mathbf{Elem}})$ that has depth strictly greater than $\max\{depth(\alpha_j^{\mathcal{B}}) \mid 0 < j < i\} + d$ (here $\max \emptyset = 0$).

3. If $i > l$ then we set $\alpha_i^{\mathcal{B}} = c(\beta_1^{\mathcal{B}}, \ldots, \beta_n^{\mathcal{B}})$ for the unique equivalence classes $\beta_1, \ldots, \beta_n \subseteq \{\alpha_1, \ldots, \alpha_{i-1}\}$ and c such that $y = c(x_1, \ldots, x_n)$ occurs in Γ' for some $y \in \alpha_i$ and $x_1 \in \beta_1, \ldots, x_n \in \beta_n$.

Since Σ is a datatypes signature in which all **Struct**-sorts are inductive, the second case of the definition is well-defined. Further, the topological sort ensures $\beta_1, \ldots, \beta_n$ exist, and the partition to equivalence classes ensures that they are unique. Hence:

Lemma 5. *$\alpha_i^{\mathcal{B}}$ is well-defined.*

Step 5 – Interpretation of Selectors: Let $s_{c,i} \in \mathcal{SE}$ for $c : \sigma_1 \times \ldots \times \sigma_n \to \sigma$, $1 \leq i \leq n$ and $a \in \sigma^{\mathcal{B}}$. If $a \in is_c^{\mathcal{B}}$, we must have $a = c(a_1, \ldots, a_n)$ for some $a_1 \in \sigma_1^{\mathcal{B}}, \ldots, a_n \in \sigma_n^{\mathcal{B}}$. We then set $s_{c,i}^{\mathcal{B}}(a) = a_i$. Otherwise, we consider two cases. If $x^{\mathcal{B}} = a$ for some $x \in vars(\Gamma)$ such that $y = s_{c,i}(x)$ occurs in Γ_2 for some y, we set $s_{c,i}^{\mathcal{B}}(a) = y^{\mathcal{B}}$. Otherwise, $s_{c,i}^{\mathcal{B}}(a)$ is set arbitrarily.

Example 6. Let Γ be the following Σ_{list}-formula: $x_1 = cons(e_1, x_2) \wedge x_3 = cons(e_2, x_4) \wedge x_2 \neq x_4$. Then $\Gamma' = \Gamma$. We have the following satisfying interpretation $\mathcal{A}$: $\mathbf{elem}^{\mathcal{A}} = \{1, 2, 3, 4\}$, $e_1^{\mathcal{A}} = 1, {e_2}^{\mathcal{A}} = 2, {x_1}^{\mathcal{A}} = [1, 2, 3], {x_2}^{\mathcal{A}} = [2, 3], {x_3}^{\mathcal{A}} = [2, 2, 4], {x_4}^{\mathcal{A}} = [2, 4]$. The construction above yields the following interpretation $\mathcal{B}$: $\mathbf{elem}^{\mathcal{B}} = \{1, 2\}$, ${e_1}^{\mathcal{B}} = 1, {e_2}^{\mathcal{B}} = 2$. For **list**-variables, we proceed as follows. The equivalence classes of **list**-variables are $[x_1], [x_2], [x_3], [x_4]$, with $[x_2] \prec [x_1]$ and $[x_4] \prec [x_3]$. The length of the longest path in G is 1. Assuming $[x_2]$ comes before $[x_4]$ in the topological sort, $x_2^{\mathcal{B}}$ will get an arbitrary list over $\{1, 2\}$ with length greater than 1 (the depth of $e_2^{\mathcal{B}}$ plus the length of the longest path), say, $[1, 1, 1]$. $x_4^{\mathcal{B}}$ will then get an arbitrary list of length greater than 4 (the depth of $x_2^{\mathcal{B}}$ plus the length of the longest path). Thus we could have $x_4^{\mathcal{B}} = [1, 1, 1, 1, 1]$. Then, $x_1^{\mathcal{B}} = [1, 1, 1, 1]$ and $x_3^{\mathcal{B}} = [2, 1, 1, 1, 1, 1]$.

Now that $\mathcal{B}$ is defined, it is left to show that it is a finite witness of Γ for $\mathcal{T}_\Sigma$ w.r.t. **Elem**. By construction, $\sigma^{\mathcal{B}} = vars_\sigma(\Gamma)^{\mathcal{B}}$ for every $\sigma \in$ **Elem**. $\mathcal{B}$ also preserves the equalities and disequalities in $\mathcal{A}$, and by considering every shape of a literal in Γ' we can prove that $\mathcal{B} \models \Gamma'$. Our interpretation of the selectors then ensures that:

Lemma 6. *$\mathcal{B} \models \Gamma$.*

Lemma 6, together with the definition of the domains of $\mathcal{B}$, gives us that $\mathcal{B}$ is a finite witness of Γ for $\mathcal{T}_\Sigma$ w.r.t. **Elem**, and so Lemma 4 is proven. As a corollary of Lemmas 1, 2 and 4, strong politeness is obtained.

Theorem 2. *If Σ is a datatypes signature and all sorts in* $\mathbf{Struct}_\Sigma$ *are inductive, then $\mathcal{T}_\Sigma$ is strongly polite w.r.t.* $\mathbf{Elem}_\Sigma$.

4.2 Finite Datatypes

In this section, we assume that all sorts in **Struct** are finite.

For finite witnessability, we define the following witness, that guesses the construction of each **Struct**-variables until a fixpoint is reached. For every quantifier-free conjunction of flat Σ-literals ϕ, define the sequence $\phi_0, \phi_1, \ldots$, such that $\phi_0 = \phi$, and for every $i \geq 0$, ϕ_{i+1} is obtained from ϕ_i by conjuncting it with a disjunction $\bigvee_{c \in \mathcal{CO}} x = c(w_1^c, \ldots, w_{n_c}^c)$ for fresh $w_1^c, \ldots, w_{n_c}^c$, where x is some arbitrary **Struct**-variable in ϕ_i such that there is no literal of the form $x = c(y_1, \ldots, y_n)$ in ϕ_i for any constructor c and variables $y_1, \ldots, y_n$, if such x exists. Since **Struct** only has finite sorts, this sequence becomes constant at some ϕ_k.

Definition 12 (A Witness for $\mathcal{T}_\Sigma$). *$wtn_f(\phi)$ is ϕ_k for the minimal k such that $\phi_k = \phi_{k+1}$.*

Example 7. Let ϕ be the Σ_{pair}-formula $x = first(y) \wedge x' = first(y') \wedge x \neq x'$. $wtn_f(\phi)$ is $\phi \wedge y = pair(e_1, e_2) \wedge y' = pair(e_3, e_4)$.

Similarly to Sect. 4.1, we have:

Lemma 7. *wtn_f is* **Elem**-*additive.*

Lemma 8. *ϕ and $\exists \overrightarrow{w} \, . \, wtn_f(\phi)$ are $\mathcal{T}_\Sigma$-equivalent, where $\overrightarrow{w} = vars(wtn_f(\phi)) \setminus vars(\phi)$.*

We now prove the following lemma:

Lemma 9 (Finite Witnessability). *Let ϕ be a conjunction of flat literals. Then, $wtn_f(\phi)$ is finitely witnessed for $\mathcal{T}_\Sigma$ with respect to* **Elem**.

Suppose $\Gamma = wtn_f(\phi)$ is $\mathcal{T}_\Sigma$-satisfiable, and let $\mathcal{A}$ be a satisfying $\mathcal{T}_\Sigma$-interpretation. We define a $\mathcal{T}_\Sigma$-interpretation $\mathcal{B}$ which is a finite witness of Γ for $\mathcal{T}_\Sigma$ w.r.t. **Elem**. We set $\sigma^\mathcal{B} = vars_\sigma(\Gamma)^\mathcal{A}$ for every $\sigma \in$ **Elem**, $e^\mathcal{B} = e^\mathcal{A}$, for every variable $e \in vars_{\mathbf{Elem}}(\Gamma)$ and $x^\mathcal{B} = x^\mathcal{A}$ for every variable $x \in vars_{\mathbf{Struct}}(\Gamma)$. Selectors are also interpreted as they are interpreted in $\mathcal{A}$. This is well-defined: for any **Struct**-variable x, every element in $\sigma^\mathcal{A}$ for $\sigma \in$ **Elem** that occurs in $x^\mathcal{A}$ has a corresponding variable e in Γ such that $e^\mathcal{A}$ is that element. This holds by the finiteness of the sorts in **Struct** and the definition of wtn_f. Further, for any **Struct**-variable x such that $s_{c,i}(x)$ occurs in Γ, we must have that it occurs in some literal of the form $y = s_{c,i}(x)$ of Γ. Similarly to the above, all elements that occur in $y^\mathcal{A}$ and $x^\mathcal{A}$ have corresponding variables in Γ. Therefore, $\mathcal{B} \models \Gamma$ is a trivial consequence of $\mathcal{A} \models \Gamma$. By the definition of its domains, $\mathcal{B}$ is a finite witness of Γ for $\mathcal{T}_\Sigma$ w.r.t. **Elem**, and so Lemma 9 is proven. Then, by Lemmas 1 7 and 9 , strong politeness is obtained.

Theorem 3. *If Σ is a datatypes signature and all sorts in $\mathbf{Struct}_\Sigma$ are finite, then $\mathcal{T}_\Sigma$ is strongly polite w.r.t. $\mathbf{Elem}_\Sigma$.*

4.3 Combining Finite and Inductive Datatypes

Now we consider the general case. Let Σ be a datatypes signature. We prove that $\mathcal{T}_\Sigma$ is strongly polite w.r.t. **Elem**. We show that there are datatypes signatures $\Sigma_1, \Sigma_2 \subseteq \Sigma$ such that $\mathcal{T}_\Sigma = \mathcal{T}_{\Sigma_1} \oplus \mathcal{T}_{\Sigma_2}$, and then use Theorem 1. In Σ_1, inductive sorts are excluded, while in Σ_2, finite sorts are considered to be element sorts.

Formally, we set Σ_1 as follows: where $\mathbf{Elem}_{\Sigma_1} = \mathbf{Elem}_\Sigma$ and $\mathbf{Struct}_{\Sigma_1} = \mathit{Fin}(\Sigma)$. $\mathcal{F}_{\Sigma_1} = \mathcal{CO}_{\Sigma_1} \uplus \mathcal{SE}_{\Sigma_1}$, where $\mathcal{CO}_{\Sigma_1} = \{c : \sigma_1 \times \ldots \times \sigma_n \rightarrow \sigma \mid c \in \mathcal{CO}_\Sigma, \sigma \in \mathbf{Struct}_{\Sigma_1}\}$ and $\mathcal{SE}_{\Sigma_1}$ and $\mathcal{P}_{\Sigma_1}$ are the corresponding selectors and testers. Notice that if σ is finite and $c : \sigma_1 \times \ldots \times \sigma_n \rightarrow \sigma$ is in $\mathcal{CO}_\Sigma$, then σ_i must be finite or in $\mathbf{Elem}_\Sigma$ for every $1 \leq i \leq n$. Next, we set Σ_2 as follows: $\mathcal{S}_{\Sigma_2} = \mathbf{Elem}_{\Sigma_2} \uplus \mathbf{Struct}_{\Sigma_2}$, where $\mathbf{Elem}_{\Sigma_2} = \mathbf{Elem}_\Sigma \cup \mathit{Fin}(\Sigma)$ and $\mathbf{Struct}_{\Sigma_2} = \mathit{Ind}(\Sigma)$. $\mathcal{F}_{\Sigma_2} = \mathcal{CO}_{\Sigma_2} \uplus \mathcal{SE}_{\Sigma_2}$, where $\mathcal{CO}_{\Sigma_2} = \{c : \sigma_2 \times \ldots \times \sigma_n \rightarrow \sigma \mid c \in \mathcal{CO}_\Sigma, \sigma \in \mathbf{Struct}_{\Sigma_2}\}$ and $\mathcal{SE}_{\Sigma_2}$ and $\mathcal{P}\Sigma_2$ are the corresponding selectors and testers. Thus, $\mathcal{T}_\Sigma = \mathcal{T}_{\Sigma_1} \oplus \mathcal{T}_{\Sigma_2}$. Now set $S = \mathbf{Elem}_\Sigma \cup \mathit{Fin}(\Sigma)$, $S_1 = \mathbf{Elem}_\Sigma$, $S_2 = \mathbf{Elem}_\Sigma \cup \mathit{Fin}(\Sigma)$, $T_1 = \mathcal{T}_{\Sigma_1}$, and $T_2 = \mathcal{T}_{\Sigma_2}$.

By Theorem 3, T_1 is strongly polite w.r.t. S_1 and by Theorem 2, T_2 is strongly polite w.r.t. S_2. By Theorem 1 we have:

Theorem 4. *If Σ is a datatypes signature then $\mathcal{T}_\Sigma$ is strongly polite w.r.t.* $\mathbf{Elem}_\Sigma$.

Remark 1. A concrete witness for $\mathcal{T}_\Sigma$ in the general case, that we call wtn_Σ, is obtained by first applying the witness from Definition 11 and then applying the witness from Definition 12 on the literals that involve finite sorts. A direct finite witnessability proof can be obtained by using the same arguments from the proofs of Lemmas 4 and 9. This witness is simpler than the one produced in the proof from [14] of Theorem 1, that involves purification and arrangements. In our case, we do not consider arrangements, but instead notice that the resulting function is additive, and hence ensures strong finite witnessability.

5 Axiomatizations

In this section, we discuss the possible connections between the politeness of $\mathcal{T}_\Sigma$ and some axiomatizations of trees. We show how to get a reduction of any $\mathcal{T}_\Sigma$-satisfiability problem into a satisfiability problem modulo an axiomatized theory of trees. The latter can be decided using syntactic unification.

Let Σ be a datatypes signature. The set TREE_Σ^* of axioms is defined as the union of all the sets of axioms in Fig. 1 (where upper case letters denote implicitly universally quantified variables). Let TREE_Σ be the set obtained from TREE_Σ^* by dismissing Ext_1 and Ext_2. Note that because of Acyc, we have that TREE_Σ is infinite (that is, consists of infinitely many axioms) unless all sorts in **Struct** are finite. TREE_Σ is a generalization of the theory of Absolutely Free Data Structures (AFDS) from [9] to many-sorted signatures with selectors and testers. In what follows we identify TREE_Σ (and TREE_Σ^*) with the class of structures that satisfy them when there is no ambiguity.

(*Inj*)	$\{c(X_1,\ldots,X_n)=c(Y_1,\ldots,Y_n)\rightarrow\bigwedge_{i=1}^n X_i=Y_i \mid c\in\mathcal{CO}\}$
(*Dis*)	$\{c(X_1,\ldots,X_n)\neq d(Y_1,\ldots,Y_m) \mid c,d\in\mathcal{CO}, c\neq d\}$
(*Proj*)	$\{s_{c,i}(c(X_1,\ldots,X_n))=X_i \mid c\in\mathcal{CO}, i\in[1,n]\}$
(*Is*$_1$)	$\{is_c(c(X_1,\ldots,X_n)) \mid c\in\mathcal{CO}\}$
(*Is*$_2$)	$\{\neg is_c(d(X_1,\ldots,X_n)) \mid c,d\in\mathcal{CO}, c\neq d\}$
(*Acyc*)	$\{X\neq t[X] \mid t$ is a non-variable $\Sigma_{\mid\mathcal{CO}}$-term that contains $X\}$
(*Ext*$_1$)	$\{\bigvee_{c:\sigma_1\times\ldots\times\sigma_n\rightarrow\sigma\in\mathcal{CO}} is_c(X) \mid \sigma\in\mathbf{Struct}\}$
(*Ext*$_2$)	$\{\exists\overrightarrow{y}\,.\,is_c(X)\rightarrow X=c(\overrightarrow{y}) \mid c\in\mathcal{CO}\}$

Fig. 1. Axioms for $TREE_\Sigma$ and $TREE^*_\Sigma$

Proposition 2. *Every $TREE^*_\Sigma$-unsatisfiable formula is $\mathcal{T}_\Sigma$-unsatisfiable.*

Remark 2. Along the lines of [1], a superposition calculus can be applied to get a $TREE_\Sigma$-satisfiability procedure. Such a calculus has been used in [6,9] for a theory of trees with selectors but no testers. To handle testers, one can use a classical encoding of predicates into first-order logic with equality, by representing an atom $is_c(x)$ as a flat equality $Is_c(x)=\mathbb{T}$ where Is_c is now a unary function symbol and $\mathbb{T}$ is a constant. Then, a superposition calculus dedicated to $TREE_\Sigma$ can be obtained by extending the standard superposition calculus [1] with some expansion rules, one for each axiom of $TREE_\Sigma$ [9]. For the axioms Is_1 and Is_2, the corresponding expansion rules are respectively $x=c(x_1,\ldots,x_n)\vdash Is_c(x)=\mathbb{T}$ if $c\in\mathcal{CO}$, and $x=d(x_1,\ldots,x_n)\vdash Is_c(x)\neq\mathbb{T}$ if $c,d\in\mathcal{CO}, c\neq d$. Further, consider the theory of finite trees defined from $TREE_\Sigma$ by dismissing *Proj*, Is_1 and Is_2. Being defined by Horn clauses, it is convex. Further, it is a Shostak theory [16,18,24] admitting a solver and a canonizer [9]. The solver is given by a syntactic unification algorithm [2] and the canonizer is the identity function. The satisfiability procedure built using the solver and the canonizer can be applied to decide $TREE_\Sigma$-satisfiability problems containing $\Sigma_{\mid\mathcal{CO}}$-atoms.

The following result shows that any $\mathcal{T}_\Sigma$-satisfiability problem can be reduced to a $TREE_\Sigma$-satisfiability problem. This leads to a $\mathcal{T}_\Sigma$-satisfiability procedure.

Proposition 3. *Let Σ be a finite datatypes signature and φ any conjunction of flat Σ-literals including an arrangement over the variables in φ. Then, there exists a Σ-formula φ' such that:*

1. *φ and $\exists\overrightarrow{w}\,.\,\varphi'$ are $\mathcal{T}_\Sigma$-equivalent, where $\overrightarrow{w}=vars(\varphi')\backslash vars(\varphi)$.*
2. *φ' is $\mathcal{T}_\Sigma$-satisfiable iff φ' is $TREE_\Sigma$-satisfiable.*

Proposition 3 can be easily lifted to any conjunction of Σ-literals φ by flattening and then guessing all possible arrangements over the variables. Further, $\exists\overrightarrow{w}\,.\,\varphi'$ and φ are not only $\mathcal{T}_\Sigma$-equivalent but also $TREE^*_\Sigma$-equivalent. As a consequence, Proposition 3 also holds when stated using $TREE^*_\Sigma$ instead of $\mathcal{T}_\Sigma$.

We conclude this section with a short discussion on the connection to Sect. 4. Both the current section and Sect. 4 rely on two constructions: (*i*) A formula transformation (wtn_Σ in Sect. 4, $\varphi \mapsto \varphi'$ in the current section); and (*ii*) A small model construction (finite witnessability in Sect. 4, equisatisfiability between $\mathcal{T}_\Sigma$ and *TREE* in Proposition 3). While these constructions are similar in both sections, they are not the same. A nice feature of the constructions of Sect. 4 is that they clearly separate between steps (*i*) and (*ii*). The witness is very simple, and amounts to adding to the input formula literals and disjunctions that trivially follow from the original formula in $\mathcal{T}_\Sigma$. Then, the resulting formula is post-processed in step (*ii*), according to a given satisfying interpretation. Having a satisfying interpretation allows us to greatly simplify the formula, and the simplified formula is useful for the model construction. In contrast, the satisfying $TREE_\Sigma$-interpretation that we start with in step (*ii*) of the current section is not necessarily a $\mathcal{T}_\Sigma$-interpretation, which makes the approach of Sect. 4 incompatible, compared to the syntactic unification approach that we employ here. For that, some of the post-processing steps of Sect. 4 are employed in step (*i*) itself, in order to eliminate all testers and as much selectors as possible. In addition, a pre-processing is applied in order to include an arrangement. The constructed interpretation finitely witnesses φ' and so this technique can be used to produce an alternative proof of strong politeness.

6 Conclusion

In this paper we have studied the theory of algebraic datatypes, as it is defined by the SMT-LIB 2 standard. Our investigation included both finite and inductive datatypes. For this theory, we have proved that it is strongly polite, making it amenable for combination with other theories by the polite combination method. Our proofs used the notion of additive witnesses, also introduced in this paper. We concluded by extending existing axiomatizations and a decision procedure of trees to support this theory of datatypes.

There are several directions for further research that we plan to explore. First, we plan to continue to prove that more important theories are strongly polite, with an eye to recent extensions of the datatypes theory, namely datatypes with shared selectors [23] and co-datatypes [22]. Second, we envision to further investigate the possibility to prove politeness using superposition-based satisfiability procedures. Third, we plan to study extensions of the theory of datatypes corresponding to finite trees including function symbols with some equational properties such as associativity and commutativity to model data structures such as multisets [25]. We want to focus on the politeness of such extensions. Initial work in that direction has been done in [5], that we plan to build on.

Acknowledgments. We are thankful to the anonymous reviewers for their comments.

References

1. Armando, A., Bonacina, M.P., Ranise, S., Schulz, S.: New results on rewrite-based satisfiability procedures. ACM Trans. Comput. Log. **10**(1), 4:1–4:51 (2009)
2. Baader, F., Snyder, W., Narendran, P., Schmidt-Schauß, M., Schulz, K.U.: Unification theory. In: Robinson, J.A., Voronkov, A. (eds.) Handbook of Automated Reasoning (in 2 volumes), pp. 445–532. Elsevier and MIT Press (2001)
3. Barrett, C., Fontaine, P., Tinelli, C.: The SMT-LIB Standard: Version 2.6. Technical report, Department of Computer Science, The University of Iowa (2017). www.SMT-LIB.org
4. Barrett, C.W., Shikanian, I., Tinelli, C.: An abstract decision procedure for a theory of inductive data types. J. Satisfiability Boolean Model. Comput. **3**(1–2), 21–46 (2007)
5. Berthon, R., Ringeissen, C.: Satisfiability modulo free data structures combined with bridging functions. In: King, T., Piskac, R. (eds.) Proceedings of SMT@IJCAR 2016. CEUR Workshop Proceedings, vol. 1617, pp. 71–80. CEUR-WS.org (2016)
6. Bonacina, M.P., Echenim, M.: Rewrite-based satisfiability procedures for recursive data structures. Electron. Notes Theor. Comput. Sci. **174**(8), 55–70 (2007)
7. Bonacina, M.P., Fontaine, P., Ringeissen, C., Tinelli, C.: Theory combination: beyond equality sharing. In: Lutz, C., Sattler, U., Tinelli, C., Turhan, A.-Y., Wolter, F. (eds.) Description Logic, Theory Combination, and All That. LNCS, vol. 11560, pp. 57–89. Springer, Cham (2019). https://doi.org/10.1007/978-3-030-22102-7_3
8. Casal, F., Rasga, J.: Many-sorted equivalence of shiny and strongly polite theories. J. Autom. Reasoning **60**(2), 221–236 (2018)
9. Chocron, P., Fontaine, P., Ringeissen, C.: Politeness and combination methods for theories with bridging functions. J. Autom. Reasoning **64**(1), 97–134 (2020)
10. Enderton, H.B.: A Mathematical Introduction to Logic. Academic Press (2001)
11. Fontaine, P.: Combinations of theories for decidable fragments of first-order logic. In: Ghilardi, S., Sebastiani, R. (eds.) FroCoS 2009. LNCS (LNAI), vol. 5749, pp. 263–278. Springer, Heidelberg (2009). https://doi.org/10.1007/978-3-642-04222-5_16
12. Gutiérrez, R., Meseguer, J.: Variant-based decidable satisfiability in initial algebras with predicates. In: Fioravanti, F., Gallagher, J.P. (eds.) LOPSTR 2017. LNCS, vol. 10855, pp. 306–322. Springer, Cham (2018). https://doi.org/10.1007/978-3-319-94460-9_18
13. Hojjat, H., Rümmer, P.: Deciding and interpolating algebraic data types by reduction. In: Jebelean, T., Negru, V., Petcu, D., Zaharie, D., Ida, T., Watt, S.M. (eds.) 19th International Symposium on Symbolic and Numeric Algorithms for Scientific Computing, SYNASC 2017, Timisoara, Romania, 21–24 September 2017, pp. 145–152. IEEE Computer Society (2017)
14. Jovanović, D., Barrett, C.: Polite theories revisited. In: Fermüller, C.G., Voronkov, A. (eds.) LPAR 2010. LNCS, vol. 6397, pp. 402–416. Springer, Heidelberg (2010). https://doi.org/10.1007/978-3-642-16242-8_29
15. Kovács, L., Robillard, S., Voronkov, A.: Coming to terms with quantified reasoning. In: Castagna, G., Gordon, A.D. (eds.) Proceedings of the 44th ACM SIGPLAN Symposium on Principles of Programming Languages, POPL 2017, Paris, France, 18–20 January 2017, pp. 260–270. ACM (2017)
16. Krstic, S., Conchon, S.: Canonization for disjoint unions of theories. Inf. Comput. **199**(1–2), 87–106 (2005)

17. Krstić, S., Goel, A., Grundy, J., Tinelli, C.: Combined satisfiability modulo parametric theories. In: Grumberg, O., Huth, M. (eds.) TACAS 2007. LNCS, vol. 4424, pp. 602–617. Springer, Heidelberg (2007). https://doi.org/10.1007/978-3-540-71209-1_47
18. Manna, Z., Zarba, C.G.: Combining decision procedures. In: Aichernig, B.K., Maibaum, T. (eds.) Formal Methods at the Crossroads. From Panacea to Foundational Support. LNCS, vol. 2757, pp. 381–422. Springer, Heidelberg (2003). https://doi.org/10.1007/978-3-540-40007-3_24
19. Meseguer, J.: Variant-based satisfiability in initial algebras. Sci. Comput. Program. **154**, 3–41 (2018)
20. Nelson, G., Oppen, D.C.: Simplification by cooperating decision procedures. ACM Trans. Program. Lang. Syst. **1**(2), 245–257 (1979)
21. Ranise, S., Ringeissen, C., Zarba, C.G.: combining data structures with nonstably infinite theories using many-sorted logic. In: Gramlich, B. (ed.) FroCoS 2005. LNCS, vol. 3717, pp. 48–64. Springer, Heidelberg (2005). https://doi.org/10.1007/11559306_3. extended technical report is available at https://hal.inria.fr/inria-00070335/
22. Reynolds, A., Blanchette, J.C.: A decision procedure for (co)datatypes in SMT solvers. J. Autom. Reasoning **58**(3), 341–362 (2017)
23. Reynolds, A., Viswanathan, A., Barbosa, H., Tinelli, C., Barrett, C.: Datatypes with shared selectors. In: Galmiche, D., Schulz, S., Sebastiani, R. (eds.) IJCAR 2018. LNCS (LNAI), vol. 10900, pp. 591–608. Springer, Cham (2018). https://doi.org/10.1007/978-3-319-94205-6_39
24. Shostak, R.E.: A practical decision procedure for arithmetic with function symbols. J. ACM **26**(2), 351–360 (1979)
25. Sofronie-Stokkermans, V.: Locality results for certain extensions of theories with bridging functions. In: Schmidt, R.A. (ed.) CADE 2009. LNCS (LNAI), vol. 5663, pp. 67–83. Springer, Heidelberg (2009). https://doi.org/10.1007/978-3-642-02959-2_5
26. Tinelli, C., Zarba, C.G.: Combining decision procedures for sorted theories. In: Alferes, J.J., Leite, J. (eds.) JELIA 2004. LNCS (LNAI), vol. 3229, pp. 641–653. Springer, Heidelberg (2004). https://doi.org/10.1007/978-3-540-30227-8_53
27. Tinelli, C., Zarba, C.G.: Combining nonstably infinite theories. J. Autom. Reasoning **34**(3), 209–238 (2005)

Superposition

A Knuth-Bendix-Like Ordering for Orienting Combinator Equations

Ahmed Bhayat(✉) and Giles Reger(✉)

University of Manchester, Manchester, UK
giles.reger@manchester.ac.uk

Abstract. We extend the graceful higher-order basic Knuth-Bendix order (KBO) of Becker et al. to an ordering that orients combinator equations left-to-right. The resultant ordering is highly suited to parameterising the first-order superposition calculus when dealing with the theory of higher-order logic, as it prevents inferences between the combinator axioms. We prove a number of desirable properties about the ordering including it having the subterm property for ground terms, being transitive and being well-founded. The ordering fails to be a reduction ordering as it lacks compatibility with certain contexts. We provide an intuition of why this need not be an obstacle when using it to parameterise superposition.

1 Introduction

There exists a wide range of methods for automated theorem proving in higher-order logic. Some provers such as AgsyHOL [17], Satallax [10] and Leo-II [4] implement dedicated higher-order proof calculi. A common approach, followed by the Leo-III prover [21], is to use a co-operative architecture with a dedicated higher-order prover working in conjunction with a first-order prover. It has long been part of theorem proving folklore that sound and complete translations from higher-order to first-order logic exist. Kerber [15] proves this result for a higher-order logic that does not assume comprehension axioms (otherwise known as applicative first-order logic). Thus, translating higher-order problems to first-order logic and running first-order provers on the translations is another method of automated higher-order theorem proving. Variations of this method are widely utilised by interactive theorem provers and their hammers such as Sledgehammer [18] and the CoqHammer [11].

Almost all translations to first-order logic translate λ-expressions using combinators. It is well known that the set of combinators S, K and I is sufficient to translate any λ-expression. For purposes of completeness, these combinators must be axiomatised: $\mathsf{S}\langle\tau_1, \tau_2, \tau_3\rangle\, x\, y\, z = x\, z\, (y\, z)$, $\mathsf{K}\langle\tau_1, \tau_2\rangle\, x\, y = x$ and $\mathsf{I}\langle\tau\rangle\, x = x$. If translating to a monomorphic logic a finite set of axioms cannot achieve completeness.

However, till now, translation based methods have proven disappointing and only achieved decent results with interactive theorem provers when the problems

N. Peltier and V. Sofronie-Stokkermans (Eds.): IJCAR 2020, LNAI 12166, pp. 259–277, 2020.
https://doi.org/10.1007/978-3-030-51074-9_15

are first-order or nearly first-order [22]. One major reason for this is that inferences between combinator axioms can be hugely explosive. A common first-order proof calculus is superposition [19]. Consider a superposition inference from the **K** axiom onto the right-hand of the **S** axiom. The result is $\mathsf{S}\,\mathsf{K}\,y\,z = z$. There is little to restrict such inferences.

Superposition is parameterised by a simplification ordering and inferences are only carried out on the larger side of literals with respect to this ordering. Inferences are not carried out at variables. Consider the **S**-, **K**- and **I**-axioms given above. There can clearly be no unifiers between a subterm of the left side of one axiom and the left side of another except at a variable. Thus, if a simplification ordering exists that orients the axioms left-to-right, inferences amongst the axioms would be impossible.

Currently, no such simplification ordering is known to exist and the authors suspect that no such ordering can exist. Whilst there is a large body of work on higher-order orderings, all either lack some property required for them to be simplification orderings or are unsuitable for orienting the combinator axioms. Jouannaud and Rubio introduced a higher-order version of the recursive path order called HORPO [14]. HORPO is compatible with β-reduction which suggests that without much difficulty it could be modified to be compatible with weak reduction. However, the ordering does not enjoy the subterm property, nor is it transitive. Likewise, is the case for orderings based on HORPO such as the computability path ordering [8] and the iterative HOIPO of Kop and Van Raamsdonk [16]. More recently, a pair of orderings for λ-free higher-order terms have been developed [2,7]. These orderings lack a specific monotonicity property, but this does not prevent their use in superposition [3]. However, neither ordering orients combinator axioms directly.

We investigate an extension of the graceful higher-order basic KBO $>_{\mathsf{hb}}$ introduced by Becker et al. [2]. Our new ordering, $>_{\mathsf{ski}}$, orients combinator equations left-to-right. Thus, if it is used to parameterise a superposition calculus, there can be no inferences among the axioms. The $>_{\mathsf{ski}}$ ordering lacks full compatibility with contexts which is normally a requirement for an ordering to parameterise superposition. In particular, the ordering is not compatible with the so-called unstable contexts. In separate work we show that this is not an obstacle to achieving completeness [5].

A complete superposition calculus for HOL already exists [3]. This calculus has the λ-calculus rather than combinatory logic as its underlying logic. It also employs higher-order unification. There appear to be two potential benefits to using a slightly modified first-order superposition calculus parameterised by our new ordering $>_{\mathsf{ski}}$ over lambda superposition as developed in [3].

- A superposition calculus parameterised by $>_{\mathsf{ski}}$ is far closer to standard first-order superposition than lambda superposition. Unification is first-order and there is no need to deal with binders and bound variables. This allows the re-use of the well-studied data-structures and algorithms used in first-order superposition [12,20].

- As discussed further in the conclusion (Sect. 6), the $>_{\mathsf{ski}}$ ordering allows the comparison of a larger class of non-ground terms than the ordering used in [3]. This results in fewer superposition inferences.

In Sect. 2, we provide the necessary preliminaries and then move on to the main contributions of this paper which are:

- Two approaches extending the $>_{\mathsf{hb}}$ ordering by first comparing terms by the length of the longest weak reduction from them. The approaches differ in the manner in which they compare non-ground terms. A useful trait for an ordering that parameterises superposition is to be able to compare a large class of non-ground terms since this reduces the number of inferences carried out. The most powerful method of defining a non-ground ordering $\succ$ is to *semantically lift* a ground ordering, i.e., to define $t \succ s$ to hold iff $t\theta \succ s\theta$ for all grounding substitutions θ. Such an ordering in non-computable and both our methods attempt to approximate it (Sect. 3).
- A set of proofs that the introduced $>_{\mathsf{ski}}$ ordering enjoys the necessary properties required for its use within the superposition calculus (Sect. 4) and a set of examples demonstrating how the ordering applies to certain terms (Sect. 5).

2 Preliminaries

Syntax of Types and Terms: We work in a polymorphic applicative first-order logic. Let $\mathcal{V}_{\mathsf{ty}}$ be a set of type variables and Σ_{ty} be a set of type constructors with fixed arities. It is assumed that a binary type constructor $\rightarrow$ is present in Σ_{ty} which is written infix. The set of types is defined:

$$\textbf{Polymorphic Types} \quad \tau ::= \kappa(\overline{\tau_n}) \mid \alpha \mid \tau \rightarrow \tau \quad \text{where } \alpha \in \mathcal{V}_{\mathsf{ty}} \text{ and } \kappa \in \Sigma_{\mathsf{ty}}$$

The notation $\overline{t_n}$ is used to denote a tuple or list of types or terms depending on the context. A type declaration is of the form $\Pi\,\overline{\alpha}.\sigma$ where σ is a type and all type variables in σ appear in $\overline{\alpha}$. Let Σ be a set of typed function symbols and $\mathcal{V}$ a set of variables with associated types. It is assumed that Σ contains the following function symbols, known as *basic combinators*:

$$\begin{array}{ll} \mathsf{S} : \Pi\alpha\tau\gamma.(\alpha \rightarrow \tau \rightarrow \gamma) \rightarrow (\alpha \rightarrow \tau) \rightarrow \alpha \rightarrow \gamma & \mathsf{I} : \Pi\alpha.\alpha \rightarrow \alpha \\ \mathsf{C} : \Pi\alpha\tau\gamma.(\alpha \rightarrow \tau \rightarrow \gamma) \rightarrow \tau \rightarrow \alpha \rightarrow \gamma & \mathsf{K} : \Pi\alpha\gamma.\alpha \rightarrow \gamma \rightarrow \alpha \\ \mathsf{B} : \Pi\alpha\tau\gamma.(\alpha \rightarrow \gamma) \rightarrow (\tau \rightarrow \alpha) \rightarrow \tau \rightarrow \gamma & \end{array}$$

The set of terms over Σ and $\mathcal{V}$ is defined below. In what follows, type subscripts, and at times even type arguments, are omitted.

$$\textbf{Terms} \quad \mathcal{T} ::= x \mid \mathsf{f}\langle \overline{\tau_n} \rangle \mid t_{1\tau_1 \rightarrow \tau_2} t_{2\tau_1}$$
$$\text{where } x \in \mathcal{V},\ t_1, t_2 \in \mathcal{T},\ \mathsf{f} \in \Sigma,\ \mathsf{f} : \Pi\,\overline{\alpha_n}.\sigma \text{ and } \overline{\tau_n} \text{ are types}$$

The type of the term $\mathsf{f}\langle \overline{\tau_n} \rangle$ is $\sigma\{\overline{\alpha_n} \to \overline{\tau_n}\}$. Following [2], terms of the form $t_1\,t_2$ are called applications. Non-application terms are called heads. A term can uniquely be decomposed into a head and n arguments. Let $t = \zeta\ \overline{t'_n}$. Then $head(t) = \zeta$ where ζ could be a variable or constant applied to possibly zero type arguments. The symbol $\mathcal{C}_{\mathsf{any}}$ denotes a member of $\{\mathbf{S}, \mathbf{C}, \mathbf{B}, \mathbf{K}, \mathbf{I}\}$, whilst $\mathcal{C}_3$ denotes a member of $\{\mathbf{S}, \mathbf{C}, \mathbf{B}\}$. These symbols are only used when the combinator is assumed to have a full complement of arguments. Thus, in $\mathcal{C}_3\ \overline{t_n}$, $n \geq 3$ is assumed. The symbols $x, y, z \ldots$ are reserved for variables, $\mathsf{c}, \mathsf{d}, \mathsf{f} \ldots$ for non-combinator constants and ζ, ξ range over arbitrary symbols and, by an abuse of notation, at times even terms. A term is *ground* if it contains no variables and *term ground* if it contains no term variables.

Positions over Terms: For a term t, if $t \in \mathcal{V}$ or $t = \mathsf{f}\langle \overline{\tau} \rangle$, then $pos(t) = \{\epsilon\}$ (type arguments have no position). If $t = t_1 t_2$ then $pos(t) = \{\epsilon\} \cup \{i.p \mid 1 \leq i \leq 2,\ p \in pos(t_i)\}$. Subterms at positions of the form $p.1$ are called *prefix* subterms and subterms at positions of the form $p.2$ are known as *first-order* subterms. A position p is *strictly above* a position p' (denoted $p < p'$) if $\exists p''.p'' \neq \epsilon \wedge p' = p.p''$. Positions p and p' are *incomparable* (denoted $p \parallel p'$) if neither $p < p'$ nor $p' < p$, nor $p = p'$. By $|t|$, the number of symbols occurring in t is denoted. By $vars_{\#}(t)$ the multiset of variables in t is denoted. The expression $A \subseteq B$ means that either A is a subset of B or A is a submultiset of B depending on whether A and B are sets or multisets.

Stable Subterms: We define a subset of first-order subterms called *stable* subterms. Let $\mathsf{LPP}(t,p)$ (LPP stands for Longest Proper Prefix) be a partial function that takes a term t and a position p and returns the longest proper prefix p' of p such that $head(t|_{p'})$ is not a partially applied combinator if such a position exists. For a position $p \in pos(t)$, p is a stable position in t if $\mathsf{LPP}(t,p)$ is not defined or $head(t|_{\mathsf{LPP}(t,p)})$ is not a combinator. A *stable subterm* is a subterm occurring at a stable position and is denoted $t\langle\!\langle u \rangle\!\rangle_p$. We call $t\langle\!\langle\rangle\!\rangle_p$ a *stable context* and drop the position where it is not relevant. For example, the subterm a is not stable in $\mathsf{f}\,(\mathbf{S}\,\mathsf{a}\,\mathsf{b}\,\mathsf{c})$, $\mathbf{S}\,(\mathbf{S}\,\mathsf{a})\,\mathsf{b}\,\mathsf{c}$ (in both cases, $head(t|_{\mathsf{LPP}(t,p)}) = \mathbf{S}$) and $\mathsf{a}\,\mathsf{c}$ (a is not a first-order subterm), but is in $\mathsf{g}\,\mathsf{a}\,\mathsf{b}$ and $\mathsf{f}\,(\mathbf{S}\,\mathsf{a})\,\mathsf{b}$. A subterm that is not stable is known as an *unstable* subterm.

The notation $t[u]_p$ denotes an arbitrary subterm u of t that occurs at position p and may be unstable. The notation $t[u_1, \ldots, u_n]$ (or $t[\overline{u_n}]$) denotes the term t containing n *non-overlapping* subterms u_1 to u_n. By $u[]_n$, we refer to a context with n non-overlapping holes. Whilst this resembles the notation for a term at position n, ambiguity is avoided by never using n to denote a position or p to denote a natural number.

Weak Reduction: Each combinator is defined by its characteristic equation; $\mathbf{S}\,x\,y\,z = x\,z\,(y\,z)$, $\mathbf{C}\,x\,y\,z = x\,z\,y$, $\mathbf{B}\,x\,y\,z = x\,(y\,z)$, $\mathbf{K}\,x\,y = x$ and $\mathbf{I}\,x = x$. A term t *weak-reduces* to a term t' in one step (denoted $t \longrightarrow_w t'$) if $t = u[s]_p$ and there exists a combinator axiom $l = r$ and substitution σ such that $l\sigma = s$ and $t' = u[r\sigma]_p$. The term $l\sigma$ in t is called a *weak redex* or just redex. By $\longrightarrow_w^*$, the reflexive transitive closure of $\longrightarrow_w$ is denoted. If term t weak-reduces to term

t' in n steps, we write $t \longrightarrow_w^n t'$. Further, if there exists a weak-reduction path from a term t of length n, we say that $t \in n_w$. Weak-reduction is terminating and confluent as proved in [13]. By $(t)\downarrow^w$, we denote the term formed from t by contracting its leftmost redex.

The length of the longest weak reduction from a term t is denoted $\|t\|$. This measure is one of the crucial features of the ordering investigated in this paper.

2.1 A Maximal Weak-Reduction Strategy

To show that the measure $\|\,\|$ is computable we provide a maximal weak-reduction strategy and prove its maximality. The strategy is used in a number of proofs later in the paper. It is in a sense equivalent to Barendregt's 'perpetual strategy' in the λ-calculus [1]. Our proof of its maximality follows the style of Van Raamsdonk et al. [23] in their proof of the maximality of a particular β-reduction strategy. We begin by proving the fundamental lemma of maximality for combinatory terms.

Lemma 1 (Fundamental Lemma of Maximality). $\|\mathcal{C}_{\mathsf{any}}\ \overline{t_n}\,\| = \|(\mathcal{C}_{\mathsf{any}}\ \overline{t_n}\,)\downarrow^w\| + 1 + isK(\mathcal{C}_{\mathsf{any}}) \times \|t_2\|$ *where* $isK(\mathcal{C}_{\mathsf{any}}) = 1$ *if* $\mathcal{C}_{\mathsf{any}} = \mathsf{K}$ *and is* 0 *otherwise. The lemma holds for* $n \geq 3$ *if* $\mathcal{C}_{\mathsf{any}} \in \{\mathsf{S}, \mathsf{C}, \mathsf{B}\}$, $n \geq 2$ *if* $\mathcal{C}_{\mathsf{any}} = \mathsf{K}$ *and* $n \geq 1$ *otherwise.*

Proof. Assume that $\mathcal{C}_{\mathsf{any}} = \mathsf{K}$. Then any maximal reduction from $\mathsf{K}\,\overline{t_n}$ is of the form: $\mathsf{K}\,t_1\,t_2 \ldots t_n \longrightarrow_w^m \mathsf{K}\,t_1'\,t_2' \ldots t_n' \longrightarrow_w t_1'\,t_3' \ldots t_n' \longrightarrow_w^{m'} s$ where $\|s\| = 0$, $t_1 \longrightarrow_w^{m_1} t_1' \ldots t_n \longrightarrow_w^{m_n} t_n'$, $\|t_2\| = m_2$ and $m = m_1 + \cdots + m_n$. Thus, $\|\mathsf{K}\,\overline{t_n}\,\| = \sum_{i=1}^n m_i + 1 + m'$. There is another method of reducing $\mathsf{K}\,\overline{t_n}$ to s:

$$\begin{array}{llll} \mathsf{K}\,t_1\,t_2 \ldots t_n \longrightarrow_w^{m_2} & \mathsf{K}\,t_1\,t_2' \ldots t_n & \longrightarrow_w & t_1\,t_3 \ldots t_n \\ \longrightarrow_w^{m-m_2} & t_1'\,t_3' \ldots t_n' & \longrightarrow_w^{m'} & s \end{array}$$

As the length of this reduction is the same as the previous reduction, it must be a maximal reduction as well. Therefore we have that: $\|\mathsf{K}\,t_1\,t_2 \ldots t_n\| = m+m'+1 = (m - m_2 + m') + m_2 + 1 = \|t_1\,t_3 \ldots t_n\| + \|t_2\| + 1$

Conversely, assume that $\mathcal{C}_{\mathsf{any}}$ is not K. We prove that the formula holds if $\mathcal{C}_{\mathsf{any}} = \mathsf{S}$. The other cases are similar. If $\mathcal{C}_{\mathsf{any}} = \mathsf{S}$, any maximal reduction from $\mathsf{S}\,\overline{t_n}$ must be of the form: $\mathsf{S}\,t_1 \ldots t_n \longrightarrow_w^m \mathsf{S}\,t_1' \ldots t_n' \longrightarrow_w t_1'\,t_3'\,(t_2'\,t_3')\,t_4' \ldots t_n' \longrightarrow_w^{m'} s$ where $\|s\| = 0$, $t_1 \longrightarrow_w^{m_1} t_1' \ldots t_n \longrightarrow_w^{m_n} t_n'$ and $m = m_1 + \cdots + m_n$. There is another method of reducing $\mathsf{S}\,\overline{t_n}$ to s:

$$\begin{array}{ll} \mathsf{S}\,t_1 \ldots t_n \longrightarrow_w & t_1\,t_3\,(t_2\,t_3)\,t_4 \ldots t_n \\ \longrightarrow_w^{m+m_3} & t_1'\,t_3'\,(t_2'\,t_3')\,t_4' \ldots t_n \\ \longrightarrow_w^{m'} & s \end{array}$$

Thus, we have that $\|\mathsf{S}\,\overline{t_n}\,\| = m+m'+1 \leq m+m_3+m'+1 = \|(\mathsf{S}\,\overline{t_n}\,)\downarrow^w\|+1$. Since $m + m' + 1$ is the length of the maximal reduction, equality must hold.

Lemma 2. *Define a map F_∞ from $\mathcal{T}$ to $\mathcal{T}$ as follows:*

$$F_\infty(t) = t \quad \textit{if } \|t\| = 0 \qquad F_\infty(\zeta\,\overline{t_n}) = \zeta\, t_1 \ldots t_{i-1}\, F_\infty(t_i)\, t_{i+1} \ldots t_n$$
where $\|t_j\| = 0$ for $1 \leq j < i$ and ζ is not a fully applied combinator

$$F_\infty(\mathbf{C_3}\,\overline{t_n}) = (\mathbf{C_3}\,\overline{t_n})\!\downarrow^w \qquad F_\infty(\mathbf{I}\, t_1\, t_2 \ldots t_n) = t_1\, t_2 \ldots t_n$$

$$F_\infty(\mathbf{K}\, t_1\, t_2 \ldots t_n) = \begin{cases} t_1\, t_3 \ldots t_n & \textit{if } \|t_2\| = 0 \\ \mathbf{K}\, t_1\, F_\infty(t_2) \ldots t_n & \textit{otherwise} \end{cases}$$

The reduction strategy F_∞ is maximal.

Proof. As the Lemma is not of direct relevance to the paper, its proof is relegated to the accompanying technical report [6].

3 Term Order

First, Becker et al.'s [2] graceful higher-order basic KBO is presented as it is utilised within our ordering. The presentation here differs slightly from that in [2] because we do not allow ordinal weightings and all function symbols have finite arities. Furthermore, we do not allow the use of different operators for the comparison of tuples, but rather restrict the comparison of tuples to use only the length-lexicographic extension of the base order. This is denoted $\gg_{\mathsf{hb}}^{\mathsf{length\_lex}}$. The length-lexicographic extension first compares the lengths of tuples and if these are equal, carries out a lexicographic comparison. For this section, terms are assumed to be untyped following the original presentation.

3.1 Graceful Higher-Order Basic KBO

Standard first-order KBO first compares the weights of terms, then compares their head-symbols and finally compares arguments recursively. When working with higher-order terms, the head symbol may be a variable. To allow the comparison of variable heads, a mapping *ghd* is introduced that maps variable heads to members of Σ that could possibly instantiate the head. This mapping *respects arities* if for any variable x, all members of $ghd(x)$ have arities greater or equal to that of x. The mapping can be extended to constant heads by taking $ghd(\mathsf{f}) = \{\mathsf{f}\}$. A substitution σ *respects* the mapping *ghd*, if for all variables x, $ghd(x\sigma) \subseteq ghd(x)$.

Let $\succ$ be a total well-founded ordering or *precedence* on Σ. The precedence $\succ$ is extended to arbitrary heads by defining $\zeta \succ \xi$ iff $\forall f \in ghd(\zeta)$ and $\forall g \in ghd(\xi), f \succ g$. Let w be a function from Σ to $\mathbb{N}$ that denotes the weight of a function symbol and $\mathcal{W}$ a function from $\mathcal{T}$ to $\mathbb{N}$ denoting the weight of a term. Let $\varepsilon \in \mathbb{N}_{>0}$. For all constants c, $w(\mathsf{c}) \geq \varepsilon$. The weight of a term is defined recursively:

$$\mathcal{W}(\mathsf{f}) = w(\mathsf{f}) \qquad \mathcal{W}(x) = \varepsilon \qquad \mathcal{W}(s\,t) = \mathcal{W}(s) + \mathcal{W}(t)$$

The graceful higher-order basic Knuth-Bendix order $>_{\mathsf{hb}}$ is defined inductively as follows. Let $t = \zeta\,\bar{t}$ and $s = \xi\,\bar{s}$. Then $t >_{\mathsf{hb}} s$ if $vars_{\#}(s) \subseteq vars_{\#}(t)$ and any of the following are satisfied:

Z1 $\mathcal{W}(t) > \mathcal{W}(s)$
Z2 $\mathcal{W}(t) = \mathcal{W}(s)$ and $\zeta \succ \xi$
Z3 $\mathcal{W}(t) = \mathcal{W}(s), \zeta = \xi$ and $\bar{t} \gg_{\mathsf{hb}}^{\mathsf{length\_lex}} \bar{s}$

3.2 Combinator Orienting KBO

The combinator orienting KBO is the focus of this paper. It has the property that all ground instances of combinator axioms are oriented by it left-to-right. This is achieved by first comparing terms by the length of the longest weak reduction from the term and then using $>_{\mathsf{hb}}$. This simple approach runs into problems with regards to stability under substitution, a crucial feature for any ordering used in superposition.

Consider the terms $t = \mathsf{f}\,x\,\mathsf{a}$ and $s = x\,\mathsf{b}$. As the length of the maximum reduction from both terms is 0, the terms would be compared using $>_{\mathsf{hb}}$ resulting in $t \succ s$ as $\mathcal{W}(t) > \mathcal{W}(s)$. Now, consider the substitution $\theta = \{x \to \mathsf{I}\}$. Then, $\|s\theta\| = 1$ whilst $\|t\theta\| = 0$ resulting in $s\theta \succ t\theta$.

The easiest and most general way of obtaining an order which is stable under substitution would be to restrict the definition of the combinator orienting KBO to ground terms and then *semantically lift* it to non-ground terms as mentioned in the introduction. However, the semantic lifting of the ground order is non-computable and therefore useless for practical purposes. We therefore provide two approaches to achieving an ordering that can compare non-ground terms and is stable under substitution both of which approximate the semantic lifting. Both require some conditions on the forms of terms that can be compared. The first is simpler, but more conservative than the second.

First, in the spirit of Bentkamp et al. [3], we provide a translation that replaces "problematic" subterms of the terms to be compared with fresh variables. With this approach, the simple variable condition of the standard KBO, $vars_{\#}(s) \subseteq vars_{\#}(t)$, ensures stability. However, this approach is over-constrained and prevents the comparison of terms such as $t = x\,\mathsf{a}$ and $s = x\,\mathsf{b}$ despite the fact that for all substitutions θ, $\|t\theta\| = \|s\theta\|$. Therefore, we present a second approach wherein no replacement of subterms occurs. This comes at the expense of a far more complex variable condition. Roughly, the condition stipulates that two terms are comparable if and only if the variables and relevant combinators are in identical positions in each.

Approach 1. Because the $>_{\mathsf{hb}}$ ordering is not defined over typed terms, type arguments are replaced by equivalent term arguments before comparison. The translation $(\!|\,|\!)$ from $\mathcal{T}$ to untyped terms is given below. First we define precisely the subterms that require replacing by variables.

Definition 1 (Type-1 term). *Consider a term* t *of the form* $\mathsf{C}_{\mathsf{any}}\ \overline{t_n}$. *If there exists a position* p *such* $t|_p$ *is a variable, then* t *is a type-1 term.*

Definition 2 (Type-2 term). *A term* $x\ \overline{t_n}$ *where* $n > 0$ *is a type-2 term.*

The translation to untyped terms is defined as follows. If t is a type variable τ, then $(\![t]\!) = \tau$. If $t = \kappa(\overline{\sigma_n})$, then $(\![t]\!) = \kappa\,(\![\overline{\sigma_n}]\!)$. If t is a term variable x, then $(\![t]\!) = x$. If t is a type-1 or type-2 term, then $(\![t]\!)$ is a fresh variable x_t. If $t = \mathsf{f}\langle\overline{\tau_n}\rangle$, then $(\![t]\!) = \mathsf{f}\,(\![\overline{\tau_n}]\!)$. Finally, if $t = t_1\,t_2$, then $(\![t]\!) = (\![t_1]\!)(\![t_2]\!)$.

An untyped term t weak reduces to an untyped term t' in one step if $t = u[s]_p$ and there exists a combinator axiom $l = r$ and substitution σ such that $(\![l]\!)\sigma = s$ and $t' = u[(\![r]\!)\sigma]_p$. The aim of the ordering presented here is to parametrise the superposition calculus. For this purposc, the property that for terms t and t', $t \longrightarrow_w t' \implies t \succ t'$, is desired. To this end, the following lemma is proved.

Lemma 3. *For all term ground polymorphic terms* t *and* t', *it is the case that* $t \longrightarrow_w t' \iff (\![t]\!) \longrightarrow_w (\![t']\!)$.

Proof. The $\implies$ direction can be proved by a straightforward induction on the t. The opposite direction is proved by an induction on $(\![t]\!)$.

Corollary 1. *A straightforward corollary of the above lemma is that for all term-ground polymorphic terms* t, $\|t\| = \|(\![t]\!)\|$.

The combinator orienting Knuth-Bendix order (approach 1) $>_{\mathsf{ski1}}$ is defined as follows. For terms t and s, let $t' = (\![t]\!)$ and $s' = (\![s]\!)$. Then $t >_{\mathsf{ski1}} s$ if $vars_{\#}(s') \subseteq vars_{\#}(t')$ and:

R1 $\|t'\| > \|s'\|$ or,
R2 $\|t'\| = \|s'\|$ and $t' >_{\mathsf{hb}} s'$.

Approach 2. Using approach 1, terms $t = y\,\mathsf{a}$ and $s = y\,\mathsf{b}$ are incomparable. Both are type-2 terms and therefore $(\![t]\!) = x_t$ and $(\![s]\!) = x_s$. The variable condition obviously fails to hold between x_t and x_s. Therefore, we consider another approach which does not replace subterms with fresh variables. We introduce a new translation $[\![]\!]$ from $\mathcal{T}$ to untyped terms that merely replaces type arguments with equivalent term arguments and does not affect term arguments at all. The simpler translation comes at the cost of a more complex variable condition. Before the revised variable definition can be provided, some further terminology requires introduction.

Definition 3 (Safe Combinator). *Let* $\mathsf{C}_{\mathsf{any}}$ *occur in* t *at position* p *and let* p' *be the shortest prefix of* p *such that* $head(t|_{p'})$ *is a combinator and for all positions* p'' *between* p *and* p', $head(t|_{p''})$ *is a combinator. Let* p'' *be a prefix of* p *of length one shorter than* p' *if such a position exists and* ϵ *otherwise. Then* $\mathsf{C}_{\mathsf{any}}$ *is* safe *in* t *if* $t|_{p'}$ *is ground and* $head(t|_{p''}) \notin \mathcal{V}$ *and* unsafe *otherwise.*

Intuitively, unsafe combinators are those that could affect a variable on a longest reduction path or could become applied to a subterm of a substitution. For example, all combinators in the term $\mathsf{S}\,(\mathsf{K}\,\mathsf{I})\,\mathsf{a}\,x$ are unsafe because they affect x, whilst the combinator in $\mathsf{f}\,(\mathsf{I}\,\mathsf{b})\,y$ is safe. The combinators in $x\,(\mathsf{S}\,\mathsf{I})\,\mathsf{a}$ are unsafe because they could potentially interact with a term substituted for x.

Definition 4. *We say a subterm is* top-level *in a term t if it doesn't appear beneath an applied variable or fully applied combinator head in t.*

Definition 5 (Safe). *Let t_1 and t_2 be untyped terms. The predicate $safe(t_1, t_2)$ holds if for every position p in t_2 such that $t_2|_p = \mathcal{C}_{\mathsf{any}}\ \overline{t_n}$ and $\mathcal{C}_{\mathsf{any}}$ (not necessarily fully applied) is unsafe, then $t_1|_p = \mathcal{C}_{\mathsf{any}}\ \overline{s_n}$ and for $1 \leq i \leq n$, $\|s_i\| \geq \|t_i\|$. Further, for all p in $pos(t_2)$ such that $t_2|_p = x\ \overline{t_n}$, then $t_1|_p = x\ \overline{s_n}$ and for $1 \leq i \leq n$, $\|s_i\| \geq \|t_i\|$.*

The definition of $safe$ ensures that if $safe(t, s)$ and $\|t\| \geq \|s\|$, then $\|t\sigma\| \geq \|s\sigma\|$ for any substitution σ a result we prove in Lemma 13. Consider terms $t = x\,(\mathsf{I}(\mathsf{I}\,(\mathsf{I}\,\mathsf{a})))\,\mathsf{b}$ and $s = x\,\mathsf{a}\,(\mathsf{I}\,(\mathsf{I}\,\mathsf{b}))$. We have that $\|t\| = 3 > \|s\| = 2$. However, it is not the case that $safe(t, s)$ because the condition that $\|t_i\| \geq \|s_i\|$ for all i is not met. $\|t_2\| = \|b\| = 0 < 2 = \|\mathsf{I}\,(\mathsf{I}\,\mathsf{b})\| = \|s_2\|$. Now consider the substitution $\sigma = \{x \to \mathsf{S}\,\mathsf{c}\}$. Because this substitution duplicates the second argument in s and t, $\|t\sigma\| = 4 < \|s\sigma\| = 5$ showing the importance of the $safe$ predicate in ensuring stability.

We draw out some obvious consequences of the definition of safety. Firstly, the predicate enjoys the subterm property in the following sense. If p is a position defined in terms t_1 and t_2, then $safe(t_1, t_2) \implies safe(t_1|_p, t_2|_p)$. Secondly, the predicate is transitive; $safe(t_1, t_2) \wedge safe(t_2, t_3) \implies safe(t_1, t_3)$.

There is a useful property that holds for non-ground terms t and s such that $safe(t, s)$.

Definition 6 (Semisafe). *Let t and s be untyped terms. Let $\mathcal{C}_{\mathsf{any}}\ \overline{s_n}$ be a term that occurs in s at p such that all head symbols above $\mathcal{C}_{\mathsf{any}}$ in s are combinators. Then $semisafe(t, s)$ if $t|_p = \mathcal{C}_{\mathsf{any}}\ \overline{t_n}$ and for $1 \leq i \leq n$, $\|t_i\| \geq \|s_i\|$.*

It is clearly the case that $(t \text{ not ground }) \wedge (s \text{ not ground }) \wedge safe(t, s) \implies semisafe(t, s)$. The implication does not hold in the other direction. A useful property of $semisafe$ is that it is stable under head reduction. If for terms t and s that reduce at their heads to t' and s' respective, we have $semisafe(t, s)$, then we have $semisafe(t', s')$.

Variable Condition:
Let $t' = [\![t]\!]$ and $s' = [\![s]\!]$ for polymorphic terms t and s. Let A be the multiset of all top-level, non-ground, first-order subterms in s' of the form $x\ \overline{s_n}$ (n may be 0) or $\mathcal{C}_{\mathsf{any}}\ \overline{t_n}$. Let B be a similarly defined multiset of subterms of t'. Then, $var\_cond(t', s')$ holds if there exists an injective total function f from A to B such that f only associates terms t_1 and t_2 if $safe(t_1, t_2)$.

For example *var_cond*(t, s) holds where $t = \mathsf{f}\, y\, (x\, \mathsf{a})$ and $s = \mathsf{g}\, (x\, b)$. In this case $A = \{x\, \mathsf{b}\}$ and $B = \{y, x\, \mathsf{a}\}$. There exists and injective total function from A to B that matches the requirements by relating $x\, \mathsf{b}$ to $x\, \mathsf{a}$. However, the variable condition does not hold in either direction if $t = \mathsf{f}\, y\, (x\, \mathsf{a})$ and $s = \mathsf{g}\, (x\, (\mathsf{I}\, \mathsf{b}))$. In this case, $x\, (\mathsf{I}\, \mathsf{b})$ cannot be related to $x\, a$ since the condition that $\|a\| \geq \|\mathsf{I}\, \mathsf{b}\|$ is not fulfilled.

We now define the combinator orienting Knuth-Bendix order (approach 2) $>_{\mathsf{ski}}$. For terms t and s, let $t' = [\![t]\!]$ and $s' = [\![s']\!]$. Then $t >_{\mathsf{ski}} s$ if *var_cond*(t', s') and:

R1 $\|t'\| > \|s'\|$ or,
R2 $\|t'\| = \|s'\|$ and $t' >_{\mathsf{hb}} s'$.

Lemma 4. *For all ground instances of combinator axioms* $l \approx r$*, we have* $l >_{\mathsf{ski}} r$.

Proof. Since for all ground instances of the axioms $l \approx r$, we have $\|l\| > \|r\|$, the theorem follows by an application of R1.

It should be noted that for non-ground instances of an axiom $l \approx r$, we do **not** necessarily have $l >_{\mathsf{ski}} r$ since l and r may be incomparable. This is no problem since the definition of $>_{\mathsf{ski}}$ could easily be amended to have $l >_{\mathsf{ski}} r$ by definition if $l \approx r$ is an instance of an axiom. Lemma 4 ensures that stability under substitution would not be affected by such an amendment.

4 Properties

Various properties of the order $>_{\mathsf{ski}}$ are proved here. The proofs provided here lack detail, the full proofs can be found in our report [6]. The proofs can easily be modified to hold for the less powerful $>_{\mathsf{ski1}}$ ordering. In general, for an ordering to parameterise a superposition calculus, it needs to be a *simplification* ordering [19]. That is, superposition is parameterised by an irreflexive, transitive, total on ground-terms, compatible with contexts, stable under substitution and well-founded binary relation. Compatibility with contexts can be relaxed at the cost of extra inferences [3,5,9]. A desirable property to have in our case is coincidence with first-order KBO, since without this, the calculus would not behave on first-order problems as standard first-order superposition would.

Theorem 1 (Irreflexivity). *For all terms* s*, it is not the case that* $s >_{\mathsf{ski}} s$.

Proof. Let $s' = [\![s]\!]$. It is obvious that $\|s'\| = \|s'\|$. Therefore $s >_{\mathsf{ski}} s$ can only be derived by rule R2. However, this is precluded by the irreflexivity of $>_{\mathsf{hb}}$.

Theorem 2 (Transitivity). *For terms* s*,* t *and* u*, if* $s >_{\mathsf{ski}} t$ *and* $t >_{\mathsf{ski}} u$ *then* $s >_{\mathsf{ski}} u$.

Proof. Let $s' = [\![s]\!]$, $t' = [\![t]\!]$ and $u' = [\![u]\!]$. From $var\_cond(s', t')$ and $var\_cond(t', u')$, $var\_cond(s', u')$ by the definition of var_cond and the application of the transitivity of $safe$. If $\|s'\| > \|t'\|$ or $\|t'\| > \|u'\|$ then $\|s'\| > \|u'\|$ and $s >_{\mathsf{ski}} u$ follows by an application of rule R1. Therefore, suppose that $\|s'\| = \|t'\| = \|u'\|$. Then it must be the case that $s' >_{\mathsf{hb}} t'$ and $t' >_{\mathsf{hb}} u'$. It follows from the transitivity of $>_{\mathsf{hb}}$ that $s' >_{\mathsf{hb}} u'$ and thus $s >_{\mathsf{ski}} u$.

Theorem 3 (Ground Totality). *Let s and t be ground terms that are not syntactically equal. Then either $s >_{\mathsf{ski}} t$ or $t >_{\mathsf{ski}} s$.*

Proof. Let $s' = [\![s]\!]$ and $t' = [\![t]\!]$. If $\|s'\| \neq \|t'\|$ then by R1 either $s >_{\mathsf{ski}} t$ or $t >_{\mathsf{ski}} s$. Otherwise, s' and t' are compared using $>_{\mathsf{hb}}$ and either $t' >_{\mathsf{hb}} s'$ or $s' >_{\mathsf{hb}} t'$ holds by the ground totality of $>_{\mathsf{hb}}$ and the injectivity of $[\![]\!]$.

Theorem 4 (Subterm Property for Ground Terms). *If t and s are ground and t is a proper subterm of s then $s >_{\mathsf{ski}} t$.*

Proof. Let $s' = [\![s]\!]$ and $t' = [\![t]\!]$. Since t is a subterm of s, t' is a subterm of s' and $\|s'\| \geq \|t'\|$ because any weak reduction in t' is also a weak reduction in s'. If $\|s'\| > \|t'\|$, the theorem follows by an application of R1. Otherwise s' and t' are compared using $>_{\mathsf{hb}}$ and $s' >_{\mathsf{hb}} t'$ holds by the subterm property of $>_{\mathsf{hb}}$. Thus $s >_{\mathsf{ski}} t$.

Next, a series of lemmas are proved that are utilised in the proof of the ordering's compatibility with contexts and stability under substitution. We prove two monotonicity properties Theorems 5 and 6. Both hold for non-ground terms, but to show this, it is required to show that the variable condition holds between terms $u[t]$ and $u[s]$ for t and s such that $t >_{\mathsf{ski}} s$. To avoid this complication, we prove the Lemmas for ground terms which suffices for our purposes. To avoid clutter, assume that terms mentioned in the statement of Lemmas 5–16 are all untyped, formed by translating polymorphic terms.

Lemma 5. $\|\zeta\,\overline{t_n}\,\| = \sum_{i=1}^{n} \|t_i\|$ *if ζ is not a fully applied combinator.*

Lemma 6. *Let $t = \zeta\,\overline{t_n}$. Then $\|t\| > \sum_{i=1}^{n} \|t_i\|$ if ζ is a fully applied combinator.*

Lemma 7. *Let $\overline{t_n}$ be terms such that for each t_i, $head(t_i) \notin \{\mathsf{I}, \mathsf{K}, \mathsf{B}, \mathsf{C}, \mathsf{S}\}$. Let $\overline{t'_n}$ be terms with the same property. Moreover, let $\|t_i\| \geq \|t'_i\|$ for $1 \leq i \leq n$. Let $s = u[\,\overline{t_n}\,]$ and $s' = u[\,\overline{t'_n}\,]$ where each t_i and t'_i is at position p_i in s and s'. If the F_∞ redex in s is within t_i for some i, then the F_∞ redex in s' is within t'_i unless t'_i is in normal form.*

Proof. Proof is by induction on $|s| + |s'|$. If u has a hole at head position, then $s = \mathsf{f}\,\overline{r_m}\;\overline{s_{m''}}$ and $s' = \mathsf{g}\,\overline{v_{m'}}\;\overline{s'_{m''}}$ where $t_1 = \mathsf{f}\,\overline{r_m}$ and $t'_1 = \mathsf{g}\,\overline{v_{m'}}$. Assume that the F_∞ redex of s is in t_1. Further, assume that $\|t'_1\| > 0$. Then, for some i in $\{1 \ldots m'\}$, it must be the case that $\|v_i\| > 0$. Let j be the smallest index such that $\|v_j\| > 0$. Then by the definition of F_∞, $F_\infty(s') = \mathsf{g}\,v_1 \ldots v_{j-1}\,F_\infty(v_j)\,v_{j+1} \ldots v_{m'}\;\overline{s'_{m''}}$ and the F_∞ redex of s' is in t'_1.

Suppose that the F_∞ redex of s is not in t_1. This can only be the case if $\|t_1\| = 0$ in which case $\|t'_1\| = 0$ as well. In this case, by the definition of F_∞, $F_\infty(s) = \mathsf{f}\ \overline{r_m}\ s_1 \ldots s_{i-1}\, F_\infty(s_i)\, s_{i+1} \ldots s''_m$ where $\|s_j\| = 0$ for $1 \leq j < i$. Without loss of generality, assume that the F_∞ redex of s_i occurs inside t_i. Then t'_i must be a subterm of s'_i. Assume that $\|t'_i\| > 0$ and thus $\|s'_i\| > 0$. Since for all i, s_i and s'_i only differ at positions where one contains a t_j and the other contains a t'_j and $\|t_i\| \geq \|t'_i\|$ for $1 \leq i \leq m''$, we have that $\|s_j\| = 0$ implies $\|s'_j\| = 0$. Thus, using the definition of F_∞, $F_\infty(s') = \mathsf{g}\ \overline{v_{m'}}\ s'_1 \ldots s'_{i-1}\, F_\infty(s'_i)\, s'_{i+1} \ldots s'_{m''}$. The induction hypothesis can be applied to s_i and s'_i to conclude that the F_∞ redex of s'_i occurs inside t'_i. The lemma follows immediately.

If u does not have a hole at its head, then $s = \zeta\ \overline{s_n}$ and $s' = \zeta\ \overline{s'_n}$ where ζ is not a fully applied combinator other than $\mathbf{K}$ (if it was, the F_∞ redex would be at the head).

If ζ is not a combinator, the proof follows by a similar induction to above. Therefore, assume that $\zeta = \mathbf{K}$. It must be the case that $\|s_2\| > 0$ otherwise the F_∞ redex in s would be at the head and not within a t_i. By the definition of F_∞, $F_\infty(s) = \mathbf{K}\, s_1\, F_\infty(s_2)\, s_3 \ldots s_n$. Let the F_∞ redex of s_2 occur inside t_j. Then t'_j is a subterm of s'_2. If $\|t'_j\| > 0$ then $\|s'_2\| > 0$ and $F_\infty(s') = \mathbf{K}\, s'_1\, F_\infty(s'_2)\, s'_3 \ldots s'_n$. By the induction hypothesis, the F_∞ redex of s'_2 occurs in t'_j.

Lemma 8. *Let $\overline{t_n}$ be terms such that for $1 \leq i \leq n$, $head(t_i) \notin \{\mathbf{I}, \mathbf{K}, \mathbf{B}, \mathbf{C}, \mathbf{S}\}$. Then for all contexts $u[]_n$, if $u[\overline{t_n}] \longrightarrow_w u'$ then either:*

1. *$\exists i. u' = u[t_1, \ldots, \hat{t_i}, \ldots, t_n]$ where $t_i \longrightarrow_w \hat{t_i}$ or*
2. *$u' = \hat{u}\{x_1 \to t_1, \ldots, x_n \to t_n\}$ where $u[x_1, \ldots, x_n] \longrightarrow_w \hat{u}$*

Proof. Let $s = u[\overline{t_n}]$ and let $p_1, \ldots p_n$ be the positions of $\overline{t_n}$ in s. Since s is reducible, there must exist a p such that $s|_p$ is a redex.

If $p > p_i$ for some i, there exists a $p' \neq \epsilon$ such that $p = p_i p'$. Then, $u[t_1, \ldots, t_i, \ldots, t_n]|_{p_i} = t_i[\mathcal{C}_{\mathsf{any}}\ \overline{r_n}\,]_{p'} \longrightarrow_w t_i[(\mathcal{C}_{\mathsf{any}}\ \overline{r_n}\,)\downarrow^w]_{p'}$. Let $\hat{t_i} = t_i[(\mathcal{C}_{\mathsf{any}}\ \overline{r_n}\,)\downarrow^w]_{p'}$. We thus have that $t_i \longrightarrow_w \hat{t_i}$ and thus $u[t_1, \ldots, t_i, \ldots, t_n] \longrightarrow_w u[t_1, \ldots, \hat{t_i}, \ldots, t_n]$.

It cannot be the case that $p = p_i$ for any i because $head(t_i)$ is not a combinator for any t_i. In the case where $p < p_i$ or $p \parallel p_i$ for all i, we have that $u[\overline{t_n}\,] = (u[\overline{x_n}\,])\sigma$ and $u[\overline{x_n}\,]|_p$ is a redex where $\sigma = \{\overline{x_n} \to \overline{t_n}\,\}$. Let $\hat{u}$ be formed from $u[\overline{x_n}\,]$ by reducing its redex at p. Then , $s = u[\overline{t_n}\,] = (u[\overline{x_n}\,])\sigma \longrightarrow_w \hat{u}\sigma = \hat{u}\{x_1 \to t_1 \ldots x_n \to t_n\}$

Lemma 9. *Let $\overline{t_n}$ be terms such that for each t_i, $head(t_i) \notin \{\mathbf{I}, \mathbf{K}, \mathbf{B}, \mathbf{C}, \mathbf{S}\}$. Let $\overline{t'_n}$ be terms with the same property. Then:*

1. *If $\|t_i\| = \|t'_i\|$ for all i in $\{1, \ldots, n\}$, then $\|u[\overline{t_n}\,]\| = \|u[\overline{t'_n}\,]\|$ for all n holed contexts u.*
2. *If $\|t_j\| > \|t'_j\|$ for some $j \in \{1, \ldots, n\}$ and $\|t_i\| \geq \|t'_i\|$ for $i \neq j$, then $\|u[\overline{t_n}\,]\| > \|u[\overline{t'_n}\,]\|$ for all n holed contexts u.*

Proof. Let $p_1, \ldots, p_n$ be the positions of the holes in u and let $s = u[\overline{t_n}\,]$ and $s' = u[\overline{t'_n}\,]$. Proof is by induction on $\|s\| + \|s'\|$. We prove part (1) first:

Assume that $\|u[\overline{t_n}]\| = 0$. Then $\|t_i\| = 0$ for $1 \leq i \leq n$. Now assume that $\|u[\overline{t'_n}]\| \neq 0$. Then there must exist some position p such that $s'|_p$ is a redex. We have that $p \neq p_i$ for all p_i as $head(t'_i) \notin \{\mathsf{I}, \mathsf{K}, \mathsf{B}, \mathsf{C}, \mathsf{S}\}$. Assume $p > p_i$ for some p_i. But then, $\|t'_i\| > 0$ which contradicts the fact that $\|t_i\| = \|t'_i\|$ for all i. Therefore, for all p_i either $p < p_i$ or $p \,\|\, p_i$. But then, if $s'|_p$ is a redex, so must $s|_p$ be, contradicting the fact that $\|u[\overline{t_n}]\| = 0$. Thus, we conclude that $\|u[\overline{t'_n}]\| = 0$.

Assume that $\|u[\overline{t_n}]\| > 0$. Let $u' = F_\infty(s)$. By Lemma 8 either $u' = u[t_1, \ldots, \hat{t_i}, \ldots, t_n]$ where $t_i \longrightarrow_w \hat{t_i}$ for $1 \leq i \leq n$ or $u' = \hat{u}\{\overline{x_n} \to \overline{t_n}\}$ where $u[\overline{x_n}] \longrightarrow_w \hat{u}$. In the first case, by Lemma 7 and $\|t_i\| = \|t'_i\|$ we have $F_\infty(s') = u'' = u[t'_1, \ldots, \hat{t'_i}, \ldots, t'_n]$ where $t'_i \longrightarrow_w \hat{t'_i}$. By the induction hypothesis $\|u'\| = \|u''\|$ and thus $\|s\| = \|s'\|$. In the second case, $F_\infty(s') = u'' = \hat{u}\{\overline{x_n} \to \overline{t'_n}\}$ where $u[\overline{x_n}] \longrightarrow_w \hat{u}$. Again, the induction hypothesis can be used to show $\|u'\| = \|u''\|$ and the theorem follows.

We now prove part (2); $\|u[\overline{t_n}]\|$ must be greater than 0. Again, let $u' = F_\infty(s)$ and $u'' = F_\infty(s')$. If $u' = u[t_1, \ldots, \hat{t_i}, \ldots, t_n]$ and $\|t'_i\| \neq 0$, then by Lemma 7 $u'' = u[t'_1, \ldots, \hat{t'_i}, \ldots, t'_n]$ where $t'_i \longrightarrow_w \hat{t'_i}$ unless $\|t'_i\| = 0$ and the lemma follows by the induction hypothesis.

If $\|t'_i\| = 0$, consider terms u' and s'. If $\|\hat{t_i}\| > 0$ or $\|t_j\| > \|t'_j\|$ for some $j \neq i$, then the induction hypothesis can be used to show $\|u'\| > \|s'\|$ and therefore $\|s\| = \|u'\| + 1 > \|s'\|$. Otherwise, $\|t_j\| = \|t'_j\|$ for all $j \neq i$ and $\|\hat{t_i}\| = 0 = \|t'_i\|$. Part 1 of this lemma can be used to show that $\|u'\| = \|s'\|$ and thus $\|s\| = \|u'\| + 1 > \|s'\|$. If $u' = \hat{u}\{\overline{x_n} \to \overline{t_n}\}$, then $u'' = \hat{u}\{\overline{x_n} \to \overline{t'_n}\}$ and the lemma follows by the induction hypothesis.

Theorem 5 (Compatibility with Contexts). *For ground terms s and t, such that $head(s)$, $head(t) \notin \{\mathsf{I}, \mathsf{K}, \mathsf{B}, \mathsf{C}, \mathsf{S}\}$, and $s >_{\mathsf{ski}} t$, then $u[s] >_{\mathsf{ski}} u[t]$ for all ground contexts $u[]$.*

Proof. Let $s' = [\![s]\!]$, $t' = [\![t]\!]$ and $u' = [\![u]\!]$. By Lemma 9 Part 2, we have that if $\|s'\| > \|t'\|$, then $\|u'[s']\| > \|u'[t']\|$. Thus, if $s >_{\mathsf{ski}} t$ was derived by R1, $u[s] >_{\mathsf{ski}} u[t]$ follows by R1. Otherwise, $s >_{\mathsf{ski}} t$ is derived by R2 and $\|s'\| = \|t'\|$. By Lemma 9 Part 1, $\|u'[s']\| = \|u'[t']\|$ follows. Thus, $u'[s']$ is compared with $u'[t']$ by R2 and $u[s] >_{\mathsf{ski}} u[t]$ by the compatibility with contexts of $>_{\mathsf{hb}}$.

Lemma 10. $\|s\| > \|t\| \implies \|u\langle\!\langle s\rangle\!\rangle\| > \|u\langle\!\langle t\rangle\!\rangle\|$ *and* $\|s\| = \|t\| \implies \|u\langle\!\langle s\rangle\!\rangle\| = \|u\langle\!\langle t\rangle\!\rangle\|$.

Proof. Proceed by induction on the size of the context u. If u is the empty context, both parts of the theorem hold trivially.

The inductive case is proved for the first implication of the lemma first. If u is not the empty context, $u\langle\!\langle s\rangle\!\rangle$ is of the form $u'\langle\!\langle \zeta\, t_1 \ldots t_{i-1}, s, t_{i+1} \ldots t_n\rangle\!\rangle$. By the definition of a stable subterm ζ cannot be a fully applied combinator and thus by Lemma 5 we have that $\|\zeta\, t_1 \ldots t_{i-1}, s, t_{i+1} \ldots t_n\| = \sum_{j=1 \wedge j \neq i}^{n} \|t_j\| + \|s\| > \sum_{j=1 \wedge j \neq i}^{n} \|t_j\| + \|t\| = \|\zeta\, t_1 \ldots t_{i-1}, t, t_{i+1} \ldots t_n\|$. If ζ is not a combinator, then $\|u'\langle\!\langle \zeta\, t_1 \ldots t_{i-1}, s, t_{i+1} \ldots\, t_n\rangle\!\rangle\| > \|u'\langle\!\langle \zeta\;\, t_1 \ldots t_{i-1}, t,\, t_{i+1} \ldots t_n\rangle\!\rangle\|$ follows from

Lemma 9 Part 2. Otherwise, ζ is a partially applied combinator and u' is a smaller stable context than u. The induction hypothesis can be used to conclude that $\|u'\langle\langle \zeta\, t_1 \ldots t_{i-1}, s, t_{i+1} \ldots t_n\rangle\rangle\| > \|u'\langle\langle \zeta\, t_1 \ldots t_{i-1}, t, t_{i+1} \ldots t_n\rangle\rangle\|$ and thus that $\|u\langle\langle s\rangle\rangle\| > \|u\langle\langle t\rangle\rangle\|$. The proof of the inductive case for the second implication of the lemma is almost identical.

Theorem 6 (Compatibility with Stable Contexts). *For all stable ground contexts $u\langle\langle\rangle\rangle$ and ground terms s and t, if $s >_{\mathsf{ski}} t$ then $u\langle\langle s\rangle\rangle >_{\mathsf{ski}} u\langle\langle t\rangle\rangle$.*

Proof. If $\|s\| > \|t\|$ then by Lemma 10, $\|u\langle\langle s\rangle\rangle\| > \|u\langle\langle t\rangle\rangle\|$ holds and then by an application of R1 we have $u\langle\langle s\rangle\rangle >_{\mathsf{ski}} u\langle\langle t\rangle\rangle$. Otherwise, if $\|s\| = \|t\|$, then by Lemma 10 we have that $\|u\langle\langle s\rangle\rangle\| = \|u\langle\langle t\rangle\rangle\|$. Thus $u\langle\langle s\rangle\rangle$ and $u\langle\langle t\rangle\rangle$ are compared using $>_{\mathsf{hb}}$. By the compatibility with contexts of $>_{\mathsf{hb}}$, $[\![u\langle\langle s\rangle\rangle]\!] >_{\mathsf{hb}} [\![u\langle\langle t\rangle\rangle]\!]$ holds and then by ofan application of R2 $u\langle\langle s\rangle\rangle >_{\mathsf{ski}} u\langle\langle t\rangle\rangle$ is true.

We next prove stability under substitution. In order to prove this, it needs to be shown that for untyped terms s and t and all substitutions σ:

1. *var_cond*(s, t) implies *var_cond*$(s\sigma, t\sigma)$.
2. *var_cond*(s, t) and $\|s\| \geq \|t\|$ imply $\|s\sigma\| \geq \|t\sigma\|$

The first is proved in Lemma 15. A slightly generalised version of (2) is proved in Lemma 14. Lemmas 11–13 are helper lemmas used in the proof of the above two properties.

Lemma 11. *For a single hole context $u\langle\langle\rangle\rangle$ such that the hole does not occur below a fully applied combinator and any term t, $\|u\langle\langle t\rangle\rangle\| = \|u\langle\langle\rangle\rangle\| + \|t\|$.*

Proof. Proof to be found in report.

Lemma 12. *Let $\overline{t_n}$ and $\overline{s_n}$ be terms such that for $n_1 \ldots n_n \in \mathbb{N}$ and for $1 \leq i \leq n$, $\|t_i\| \geq \|s_i\| + n_i$. Further, let $t = t_1\, t_2 \ldots t_n$ and $s = s_1\, s_2 \ldots s_n$. Assume that $semisafe(t, s)$ holds. Then $\|t\| \geq \|s\| + \sum_{i=1}^{n} n_i$.*

Proof. Proof to be found in report.

Lemma 13. *Let t and s be non-ground terms such that $\|t\| \geq \|s\| + m$ for some $m \in \mathbb{N}$ and $safe(t, s)$. Then, for any substitution σ, $\|t\sigma\| \geq \|s\sigma\| + m$ and $safe(t\sigma, s\sigma)$.*

Proof. Proof to be found in report.

Lemma 14. *For terms t and s such that var_cond(t, s) holds and $\|t\| \geq \|s\| + n$ for some $n \in \mathbb{N}$, for all substitutions σ, $\|t\sigma\| \geq \|s\sigma\| + n$.*

Proof. If s and t are ground, the theorem is trivial. If s is ground, then $\|t\sigma\| \geq \|t\| \geq \|s\| + n$. If s is not ground, then *var_cond*(t, s) implies that t is not ground. Therefore, assume that neither is ground. If *head*(s) (and therefore *head*(t) by the variable condition) are fully applied combinators or variables, then *var_cond*(t, s)

implies $safe(t, s)$ and Lemma 13 can be invoked to prove the lemma. Therefore, assume that both have non-variable, non-fully applied combinator heads.

Let $t = u\langle\langle \overline{t_m} \rangle\rangle$ and $s = u'\langle\langle \overline{s_m} \rangle\rangle$ where $\overline{s_m}$ are all the non-ground, top-level, first-order subterms of the form $x\ \overline{args}$ or $\mathcal{C}_{\mathsf{any}}\ \overline{args}$ in s. By the variable condition, we have that there exists a total injective function respecting the given conditions from the s_i to non-ground, top-level, first-order subterms of t of the form $x\ \overline{args}$ or $\mathcal{C}_{\mathsf{any}}\ \overline{args}$. Let $\overline{t_m}$ be the terms related to $\overline{s_m}$ by this function. Without loss of generality, assume that this function relates s_1 to t_1, s_2 to t_2 and so on. For $1 \leq i \leq m$, $\|t_i\| = \|s_i\| + m_i$ for $m_i \in \mathbb{N}$. This follows from the fact that since t_i and s_i are both non-ground and $safe(t_i, s_i)$, we have $semisafe(t_i, s_i)$ and can therefore invoke Lemma 12.

Let $m' = \|u\langle\langle\rangle\rangle\| - \|u'\langle\langle\rangle\rangle\|$. Note that m' could be negative. By Lemma 11, $\|t\| = \|u\langle\langle\rangle\rangle\| + \sum_{i=1}^{m} \|t_i\|$ and $\|s\| = \|u'\langle\langle\rangle\rangle\| + \sum_{i=1}^{m} \|s_i\|$. Thus, $\|t\| = \|s\| + m' + \sum_{i=1}^{m} m_i$. Therefore, $m' + \sum_{i=1}^{m} m_i \geq n$. Lemma 13 can be used to show that for all i, $\|t_i\sigma\| \geq \|s_i\sigma\| + m_i$. Because $u'\langle\langle\rangle\rangle$ is ground, it follows $\|u\sigma\langle\langle\rangle\rangle\| - \|u'\sigma\langle\langle\rangle\rangle\| \geq m'$. To conclude the proof:

$$\begin{aligned}
\|t\sigma\| &= \|u\sigma\langle\langle \overline{t_m\sigma} \rangle\rangle\| \qquad\qquad = \|u\sigma\langle\langle\rangle\rangle\| + \textstyle\sum_{i=1}^{m} \|t_i\sigma\| \\
&\geq \|u'\sigma\langle\langle\rangle\rangle\| + \textstyle\sum_{i=1}^{m} \|s_i\sigma\| + m' + \sum_{i=1}^{m} m_i \\
&\geq \|u'\sigma\langle\langle\rangle\rangle\| + \textstyle\sum_{i=1}^{m} \|s_i\sigma\| + n \quad = \quad \|s\sigma\| + n
\end{aligned}$$

Lemma 15. *For terms t and s such that $var\_cond(t, s)$ holds and for all substitutions σ, $var\_cond(t\sigma, s\sigma)$.*

Proof. Let $t = u\langle\langle \overline{t_m} \rangle\rangle$ and $s = u'\langle\langle \overline{s_m} \rangle\rangle$ where $\overline{s_m}$ are all the non-ground, top-level, first-order subterms of the form $x\ \overline{args}$ or $\mathcal{C}_{\mathsf{any}}\ \overline{args}$ in s. By the variable condition, we have that there exists a total injective function respecting the given conditions from the s_i to non-ground, top-level, first-order subterms of t of the form $x\ \overline{args}$ or $\mathcal{C}_{\mathsf{any}}\ \overline{args}$. Let $\overline{t_m}$ be the terms related to $\overline{s_m}$ by this function. Without loss of generality, assume that this function relates s_1 to t_1, s_2 to t_2 and so on. By the definition of the variable condition, we have that u' must be ground. This implies that any non-ground subterms of $s\sigma$ must be subterms of some $s_i\sigma$ for $1 \leq i \leq m$.

Assume that for some i and $p \in pos(s_i\sigma)$, $s_i\sigma|_p$ is a non-ground, top-level, first-order subterm of the form $x\ \overline{args}$ or $\mathcal{C}_{\mathsf{any}}\ \overline{args}$. We show that $t_i\sigma|_p$ is a non-ground, top-level, first-order subterm of $t\sigma$ and $safe(t_i|_p, s_i|_p)$. This implies the existence of a total, injective function from the multiset of non-ground, top-level first-order subterms in $s\sigma$ to the like multiset of $t\sigma$ in turn proving $var\_cond(t\sigma, s\sigma)$.

From Lemma 13, it can be shown that for $1 \leq i \leq m$, $safe(t_i\sigma, s_i\sigma)$. By the subterm property of safety, this implies that $safe(t_i\sigma|_p, s_i\sigma|_p)$.

To show that $t_i\sigma|_p$ must be a non-ground, top-level, first-order subterm in $t\sigma$, it can be assumed that this not the case. This easily leads to a contradiction with $safe(t_i\sigma, s_i\sigma)$.

Lemma 16. *Let t be a polymorphic term and σ be a substitution. We define a new substitution ρ such that the domain of ρ is $dom(\sigma)$. Define $y\rho = [\![y\sigma]\!]$. For all terms t, $[\![t\sigma]\!] = [\![t]\!]\rho$.*

Proof. Via a straightforward induction on t.

Theorem 7 (Stability under Substitution). *If $s >_{\mathsf{ski}} t$ then $s\sigma >_{\mathsf{ski}} t\sigma$ for all substitutions σ that respect the ghd mapping.*

Proof. Let $s' = [\![s]\!]$ and $t' = [\![t]\!]$. Let ρ be defined as per Lemma 16. First, we show that if R1 was used to derive $s >_{\mathsf{ski}} t$ and thus $\|s'\| > \|t'\|$ then $\|s'\rho\| > \|t'\rho\|$ and thus $s\sigma >_{\mathsf{ski}} t\sigma$ because $[\![s\sigma]\!] = s'\rho$ and $[\![t\sigma]\!] = t'\rho$.

From Lemma 15 and $var\_cond(s', t')$, $var\_cond(s'\rho, t'\rho)$ holds. Furthermore, if $\|s'\| > \|t'\|$, then by Lemma 14 $\|s'\rho\| > \|t'\rho\|$ and $s\sigma >_{\mathsf{ski}} t\sigma$ by an application of R1.

On the other hand, if $\|s'\| = \|t'\|$, then R2 was used to derive $s >_{\mathsf{ski}} t$. By Lemma 14 $\|s'\rho\| \geq \|t'\rho\|$. If $\|s'\rho\| > \|t'\rho\|$, then this is the same as the former case. Otherwise $\|s'\rho\| = \|t'\rho\|$ and $s'\rho$ and $t\rho$ are compared using R2. From the stability under substitution of $>_{\mathsf{hb}}$, $s'\rho >_{\mathsf{hb}} t'\rho$ follows and $s\sigma >_{\mathsf{ski}} t\sigma$ can be concluded.

Theorem 8 (Well-foundedness). *There exists no infinite descending chain of comparisons $s_1 >_{\mathsf{ski}} s_2 >_{\mathsf{ski}} s_3 \cdots$.*

Proof. Assume that such a chain exists. For each $s_i >_{\mathsf{ski}} s_{i+1}$ derived by R1, we have that $\|s_i\| > \|s_{i+1}\|$. For each $s_i >_{\mathsf{ski}} s_{i+1}$ derived by R2, we have that $\|s_i\| = \|s_{i+1}\|$. Therefore the number of times $s_i >_{\mathsf{ski}} s_{i+1}$ by R1 in the infinite chain must be finite and there must exist some m such that for all $n > m$, $s_n >_{\mathsf{ski}} s_{n+1}$ by R2. Therefore, there exists an infinite sequence of $>_{\mathsf{hb}}$ comparisons $[\![s_m]\!] >_{\mathsf{hb}} [\![s_{m+1}]\!] >_{\mathsf{hb}} [\![s_{m+2}]\!] \cdots$. This contradicts the well-foundedness of $>_{\mathsf{hb}}$.

Theorem 9 (Coincidence with First-Order KBO). *Let $>_{\mathsf{fo}}$ be the first-order KBO as described by Becker et al. in [2]. Assume that $>_{\mathsf{ski}}$ and $>_{\mathsf{fo}}$ are parameterised by the same precedence $\succ$ and that $>_{\mathsf{fo}}$ always compares tuples using the lexicographic extension operator. Then $>_{\mathsf{ski}}$ and $>_{\mathsf{fo}}$ always agree on first-order terms.*

Proof. Let $t' = [\![t]\!]$ and $s' = [\![s]\!]$. Since s and t are first-order, $\|s'\| = 0$ and $\|t'\| = 0$. Thus, s' and t' will always be compared by $>_{\mathsf{hb}}$. Since $>_{\mathsf{hb}}$ coincides with $>_{\mathsf{fo}}$ on first-order terms, so does $>_{\mathsf{ski}}$.

5 Examples

To give a flavour of how the ordering behaves, we provide a number of examples.

Example 1. Consider the terms (ignoring type arguments) $t = \mathsf{S}\,(\mathsf{K}\,\mathsf{a})\,\mathsf{b}\,\mathsf{c}$ and $s = \mathsf{f}\,\mathsf{c}\,\mathsf{e}$. From the definition of the translation $[\![]\!]$, we have that $[\![t]\!] = \mathsf{S}\,(\mathsf{K}\,\mathsf{a})\,\mathsf{b}\,\mathsf{c}$ and $[\![s]\!] = \mathsf{f}\,\mathsf{c}\,\mathsf{e}$. Since $\|\mathsf{S}\,(\mathsf{K}\,\mathsf{a})\,\mathsf{b}\,\mathsf{c}\| = 2$ and $\|\mathsf{f}\,\mathsf{c}\,\mathsf{e}\| = 0$, we have that $t >_{\mathsf{ski}} s$.

Example 2. Consider the terms $t = \mathsf{f}\,(\mathsf{g}\,\mathsf{b})\,\mathsf{e}\,\mathsf{d}$ and $s = \mathsf{I}\,\mathsf{a}$. Here $s >_{\mathsf{ski}} t$ despite the fact that s is syntactically smaller than t because s has a maximum reduction of 1 as opposed to 0 of t.

Example 3. Consider terms $t = \mathsf{f}\,(\mathsf{I}\,\mathsf{d})\,(\mathsf{S}\,x\,\mathsf{a}\,\mathsf{b})$ and $s = \mathsf{g}\,(\mathsf{S}\,x\,(\mathsf{h}\,\mathsf{d})\,\mathsf{b})$. The two terms are comparable as the variable condition relates subterm $\mathsf{S}\,x\,(\mathsf{h}\,\mathsf{d})\,\mathsf{b}$ in s to subterm $\mathsf{S}\,x\,\mathsf{a}\,\mathsf{b}$ in t. The unsafe combinator S and variable x are in the same position in each subterm. As $\|t\| > \|s\|$, $t >_{\mathsf{ski}} s$.

Example 4. Consider terms $t = \mathsf{f}\,(\mathsf{I}\,\mathsf{d})\,(\mathsf{S}\,x\,\mathsf{a}\,y)$ and $s = \mathsf{g}\,(\mathsf{S}\,x\,(\mathsf{h}\,y)\,\mathsf{b})$. This is very similar to the previous example, but in this case the terms are incomparable. Let s' be a name for the subterm $(\mathsf{S}\,x\,(\mathsf{h}\,y)\,\mathsf{b})$ in s and t' a name for the subterm $(\mathsf{S}\,x\,\mathsf{a}\,y)$. The variable y occurs in different positions in s' and t'. Therefore, s' cannot be related to t by the variable condition and the two terms are incomparable.

Example 5. Consider terms $t = \mathsf{f}\,(x\,(\mathsf{g}\,(\mathsf{K}\,\mathsf{I}\,\mathsf{a}\,\mathsf{b})))$ and $s = \mathsf{h}\,(\mathsf{I}\,\mathsf{a})\,(x\,\mathsf{c})$. The variable condition holds between t and s by relating $(x\,(\mathsf{g}\,(\mathsf{K}\,\mathsf{I}\,\mathsf{a}\,\mathsf{b})))$ to $(x\,\mathsf{c})$. The combinator I in s is not unsafe and therefore does not need to be related to a combinator in t.

Since $\|t\| = 2 > \|s\| = 1$, $t >_{\mathsf{ski}} s$. Intuitively, this is safe because a substitution for x in t can duplicate $(\mathsf{g}\,(\mathsf{K}\,\mathsf{I}\,\mathsf{a}\,\mathsf{b}))$ whose maximum reduction length is 2 whilst a substitution for x in s can only duplicate c whose maximum reduction length is 0.

6 Conclusion and Discussion

We have presented an ordering that orients all ground instances of S, C, B, K and I axioms left-to-right. The ordering enjoys many other useful properties such as stability under substitution, compatibility with stable contexts, ground totality and transitivity. In as yet unpublished work, we have used this ordering to parameterise a complete superposition calculus for HOL [5]. Lack of full compatibility with context has not been an obstacle. In the standard first-order proof of the completeness of superposition, compatibility with contexts is used in model construction to rule out the need for superposition inferences beneath variables [19]. Thus, by utilising $>_{\mathsf{ski}}$, some superposition is required beneath variables. However, because terms with functional heads are compatible with all contexts, such inference are quite restricted.

The $>_{\mathsf{ski}}$ ordering presented here is able to compare non-ground terms that cannot be compared by any ordering used to parameterise Bentkamp et al.'s lambda superposition calculus [3]. They define terms to be β-equivalence classes. Non-ground terms are compared using a quasiorder, $\succsim$, such that $t \succsim s$ iff for all grounding substitutions θ, $t\theta \succeq s\theta$. Consider terms $t = x\,\mathsf{a}\,\mathsf{b}$ and $s = x\,\mathsf{b}\,\mathsf{a}$ and grounding substitutions $\theta_1 = x \rightarrow \lambda x\,y\,.\mathsf{f}\,y\,x$ and $\theta_2 = x \rightarrow \lambda x\,y\,.\mathsf{f}\,x\,y$. By ground totality of $\succ$ it must be the case that either $\mathsf{f}\,\mathsf{a}\,\mathsf{b} \succ \mathsf{f}\,\mathsf{b}\,\mathsf{a}$ or $\mathsf{f}\,\mathsf{b}\,\mathsf{a} \succ \mathsf{f}\,\mathsf{a}\,\mathsf{b}$. Without loss of generality assume the first. Then, neither $t \succsim s$ nor $s \succsim t$ since $t\theta_1 = \mathsf{f}\,\mathsf{b}\,\mathsf{a} \prec \mathsf{f}\,\mathsf{a}\,\mathsf{b} = s\theta_1$ and $t\theta_2 = \mathsf{f}\,\mathsf{a}\,\mathsf{b} \succ \mathsf{f}\,\mathsf{b}\,\mathsf{a} = s\theta_2$.

The $>_{\mathsf{ski}}$ ordering allows weak reduction (or β-reduction) to be treated as part of the superposition calculus. This allows terms t and t' such that $t \longrightarrow_w^+ t'$ (or $t \longrightarrow_\beta^+ t'$) to be considered separate terms resulting in terms such as t and s given above being comparable. Since $\|t\| = \|s\|$, t and s are compared using $>_{\mathsf{hb}}$ with stability under substitution ensured by the stability under substitution of $>_{\mathsf{hb}}$.

Many of the definitions that have been provided here are conservative and can be tightened to allow the comparison of a far larger class of non-ground terms without losing stability under substitution. We provide an example of how the definition of stable subterm could be refined in our report [6]. In further work, we hope to thoroughly explore such refinements.

Acknowledgements. Thanks to Jasmin Blanchette, Alexander Bentkamp and Petar Vukmirović for many discussions on aspects of this research. We would also like to thank reviewers of this paper, whose comments have done much to shape this paper. The first author thanks the family of James Elson for funding his research.

References

1. Barendregt, H.P.: The Lambda Calculus: Its Syntax and Semantics, 2nd edn. Elsevier Science Publishers B.V., Amsterdam (1984)
2. Becker, H., Blanchette, J.C., Waldmann, U., Wand, D.: A transfinite Knuth–Bendix order for lambda-free higher-order terms. In: de Moura, L. (ed.) CADE 2017. LNCS (LNAI), vol. 10395, pp. 432–453. Springer, Cham (2017). https://doi.org/10.1007/978-3-319-63046-5_27
3. Bentkamp, A., Blanchette, J., Tourret, S., Vukmirović, P., Waldmann, U.: Superposition with lambdas. In: Fontaine, P. (ed.) CADE 2019. LNCS (LNAI), vol. 11716, pp. 55–73. Springer, Cham (2019). https://doi.org/10.1007/978-3-030-29436-6_4
4. Benzmüller, C., Sultana, N., Paulson, L.C., Theiß, F.: The higher-order prover LEO-II. J. Autom. Reasoning **55**(4), 389–404 (2015). https://doi.org/10.1007/s10817-015-9348-y
5. Bhayat, A., Reger, G.: A combinator-based superposition calculus for higher-order logic. In: The 10th International Joint Conference on Automated Reasoning (IJCAR) (2020)
6. Bhayat, A., Reger, G.: A Knuth-Bendix-like ordering for orienting combinator equations (technical report). Technical report, University of Mancester (2020). https://easychair.org/publications/preprint_open/rXSk
7. Blanchette, J.C., Waldmann, U., Wand, D.: A lambda-free higher-order recursive path order. In: Esparza, J., Murawski, A.S. (eds.) FoSSaCS 2017. LNCS, vol. 10203, pp. 461–479. Springer, Heidelberg (2017). https://doi.org/10.1007/978-3-662-54458-7_27
8. Blanqui, F., Jouannaud, J.-P., Rubio, A.: The computability path ordering: the end of a quest. In: Kaminski, M., Martini, S. (eds.) CSL 2008. LNCS, vol. 5213, pp. 1–14. Springer, Heidelberg (2008). https://doi.org/10.1007/978-3-540-87531-4_1
9. Bofill, M., Godoy, G., Nieuwenhuis, R., Rubio, A.: Paramodulation with non-monotonic orderings. In: Proceedings - Symposium on Logic in Computer Science, August 1999

10. Brown, C.E.: Satallax: an automatic higher-order prover. In: Gramlich, B., Miller, D., Sattler, U. (eds.) IJCAR 2012. LNCS (LNAI), vol. 7364, pp. 111–117. Springer, Heidelberg (2012). https://doi.org/10.1007/978-3-642-31365-3_11
11. Czajka, Ł., Kaliszyk, C.: Hammer for Coq: automation for dependent type theory. J. Autom. Reasoning **61**(1), 423–453 (2018)
12. Graf, P.: Substitution tree indexing. In: Hsiang, J. (ed.) RTA 1995. LNCS, vol. 914, pp. 117–131. Springer, Heidelberg (1995). https://doi.org/10.1007/3-540-59200-8_52
13. Hindley, J.R., Seldin, J.P.: Lambda-Calculus and Combinators: An Introduction, 2nd edn. Cambridge University Press, New York (2008)
14. Jouannaud, J.P., Rubio, A.: Polymorphic higher-order recursive path orderings. J. ACM **54**(1) (2007). https://doi.org/10.1145/1206035.1206037
15. Kerber, M.: How to prove higher order theorems in first order logic. In: IJCAI, pp. 137–142, January 1991
16. Kop, C., van Raamsdonk, F.: A higher-order iterative path ordering. In: Cervesato, I., Veith, H., Voronkov, A. (eds.) LPAR 2008. LNCS (LNAI), vol. 5330, pp. 697–711. Springer, Heidelberg (2008). https://doi.org/10.1007/978-3-540-89439-1_48
17. Lindblad, F.: https://github.com/frelindb/agsyHOL. Accessed 25 Sept 2019
18. Meng, J., Paulson, L.C.: Translating higher-order clauses to first-order clauses. J. Autom. Reasoning **40**(1), 35–60 (2008). https://doi.org/10.1007/s10817-007-9085-y
19. Nieuwenhuis, R., Rubio, A.: Paramodulation-based theorem proving. In: Handbook of Automated Reasoning, vol. 1, pp. 371–443. Elsevier Press and MIT press, August 2001. https://doi.org/10.1016/B978-044450813-3/50009-6
20. Sekar, R., Ramakrishnan, I., Voronkov, A.: Term indexing, chap. 26. In: Robinson, A., Voronkov, A. (eds.) Handbook of Automated Reasoning, vol. II, pp. 1853–1964. Elsevier Science (2001)
21. Steen, A.: Extensional paramodulation for higher-order logic and its effective implementation Leo-III. Ph.D. thesis, Freie Universität Berlin (2018)
22. Sultana, N., Blanchette, J.C., Paulson, L.C.: Leo-II and Satallax on the Sledgehammer test bench. J. Appl. Logic **11**(1), 91–102 (2013). https://doi.org/10.1016/j.jal.2012.12.002
23. van Raamsdonk, F., Severi, P., Sørensen, M., Xi, H.: Perpetual reductions in lambda calculus. Inf. Comput. **149**(2), 173–225 (1999). https://doi.org/10.1006/inco.1998.2750

A Combinator-Based Superposition Calculus for Higher-Order Logic

Ahmed Bhayat(✉) and Giles Reger(✉)

University of Manchester, Manchester, UK
ahmed_bhayat@hotmail.com, giles.reger@manchester.ac.uk

Abstract. We present a refutationally complete superposition calculus for a version of higher-order logic based on the combinatory calculus. We also introduce a novel method of dealing with extensionality. The calculus was implemented in the Vampire theorem prover and we test its performance against other leading higher-order provers. The results suggest that the method is competitive.

1 Introduction

First-order superposition provers are often used to reason about problems in extensional higher-order logic (HOL) [19,26]. Commonly, this is achieved by translating the higher-order problem to first-order logic (FOL) using combinators. Such a strategy is sub-optimal as translations generally sacrifice completeness and at times even soundness. In this paper, we provide a modification of first-order superposition that is sound and complete for a combinatory version of HOL. Moreover, it is *graceful* in the sense of that it coincides with standard superposition on purely first-order problems.

The work is complementary to the clausal λ-superposition calculus of Bentkamp et al. [4]. Our approach appears to offer two clear differences. Firstly, as our calculus is based on the combinatory logic and first-order unification, it is far closer to standard first-order superposition. Therefore, it should be easier to implement in state-of-the-art first-order provers. Secondly, the $>_{\mathsf{ski}}$ ordering that we propose to parameterise our calculus with can compare more terms than can be compared by the ordering presented in [4]. On the other hand, we suspect that for problems requiring complex unifiers, our approach will not be competitive with clausal λ-superposition.

Developing a complete and efficient superposition calculus for a combinatory version of HOL poses some difficulties. When working with a monomorphic logic it is impossible to select a finite set of typed combinator axioms that can guarantee completeness for a particular problem [12]. Secondly, using existing orderings, combinator axioms can superpose among themselves, leading to a huge number of consequences of the axioms. If the problem is first-order, these consequences can never interact with non-combinator clauses and are therefore useless.

We deal with both issues in the current work. To circumvent the first issue, we base our calculus on a polymorphic rather than monomorphic first-order logic.

N. Peltier and V. Sofronie-Stokkermans (Eds.): IJCAR 2020, LNAI 12166, pp. 278–296, 2020.
https://doi.org/10.1007/978-3-030-51074-9_16

The second issue can be dealt with by an ordering that orients combinator axioms left-to-right. Consider the **S**-combinator axiom $\mathsf{S}\,x\,y\,z \approx x\,z\,(y\,z)$. Assume that there exists a simplification ordering $\succ$ such that $\mathsf{S}\,x\,y\,z \succ x\,z\,(y\,z)$. Then, since superposition is only carried out on the larger side of literals and not at variables, there can be no inferences between the **S**-axiom and any other combinator axiom. Indeed, in this case the axioms can be removed from the clause set altogether and replaced by an inference rule (Sect. 7).

No ground-total simplification ordering is known that is capable of orienting all axioms for a complete set of combinators.[1] The authors suspect that no such simplification ordering exists. Consider a KBO-like ordering. Since the variable x appears twice on the right-hand side of the **S**-axiom and only once on the left-hand side, the ordering would not be able to orient it. The same is the case for any other combinator which duplicates its arguments.

In other related work [10], we have developed an ordering that enjoys most of the properties of a simplification ordering, but lacks full compatibility with contexts. In particular, the ordering is not compatible with what we call *unstable* contexts. We propose using such an ordering to parameterise the superposition calculus. In the standard proof of the completeness of superposition, compatibility with contexts is used to rule out the need for superposition at or beneath variables. As the ordering doesn't enjoy full compatibility with contexts, limited superposition at and below variables needs to be carried out. This is dealt with by the addition of an extra inference rule to the standard rules of superposition, which we call SubVarSup (Sect. 3).

By turning combinator axioms into rewrite rules, the calculus represents a folding of higher-order unification into the superposition calculus itself. Whilst not as goal-directed as a dedicated higher-order unification algorithm, it is still far more goal-directed than using **SK**-style combinators in superposition provers along with standard orderings. Consider the conjecture $\exists z. \forall xy.\, z\,x\,y \approx \mathsf{f}\,y\,x$. Bentkamp et al. ran an experiment and found that the E prover [25] running on this conjecture supplemented with the **S**- and **K**-combinator axioms had to perform 3756 inferences in order to find a refutation [4]. Our calculus reduces this number to 427 inferences. With the addition of rewrite rules for **C**-, **B**- and **I**-combinators, the required inferences reduces to 18.

We consider likely that for problems requiring 'simple' unifiers, folding unification into superposition will be competitive with higher-order unification whilst providing the advantages that data structures and algorithms developed for first-order superposition can be re-used unchanged. The results of the empirical evaluation of our method can be found in Sect. 8.

2 The Logic

The logic we use is polymorphic applicative first-order logic otherwise known as λ-free (clausal) higher-order logic.

[1] A complete set of combinators is a set of combinators whose members can be composed to form a term extensionally equivalent to any given λ-term.

Syntax. Let $\mathcal{V}_{\mathsf{ty}}$ be a set of type variables and Σ_{ty} be a set of type constructors with fixed arities. It is assumed that a binary type constructor $\rightarrow$ is present in Σ_{ty} which is written infix. The set of types is defined:

Polymorphic Types $\tau ::= \kappa(\overline{\tau_n}) \mid \alpha \mid \tau \rightarrow \tau$ where $\alpha \in \mathcal{V}_{\mathsf{ty}}$ and $\kappa \in \Sigma_{\mathsf{ty}}$

The notation $\overline{t_n}$ is used to denote a tuple or list of types or terms depending on the context. A type declaration is of the form $\Pi\,\overline{\alpha_n}\,.\,\sigma$ where σ is a type and all type variables in σ appear in $\overline{\alpha}$. Let Σ be a set of typed function symbols and $\mathcal{V}$ a set of variables with associated types. It is assumed that Σ contains the following function symbols, known as *basic combinators*:

$$\mathsf{S} : \Pi\alpha\tau\gamma.\,(\alpha \rightarrow \tau \rightarrow \gamma) \rightarrow (\alpha \rightarrow \tau) \rightarrow \alpha \rightarrow \gamma \qquad \mathsf{I} : \Pi\alpha.\,\alpha \rightarrow \alpha$$
$$\mathsf{C} : \Pi\alpha\tau\gamma.\,(\alpha \rightarrow \tau \rightarrow \gamma) \rightarrow \tau \rightarrow \alpha \rightarrow \gamma \qquad \mathsf{K} : \Pi\alpha\gamma.\,\alpha \rightarrow \gamma \rightarrow \alpha$$
$$\mathsf{B} : \Pi\alpha\tau\gamma.\,(\alpha \rightarrow \gamma) \rightarrow (\tau \rightarrow \alpha) \rightarrow \tau \rightarrow \gamma$$

The intended semantics of the combinators is captured by the following *combinator axioms*:

$$\begin{array}{lll} \mathsf{S}\,x\,y\,z = x\,z\,(y\,z) & & \mathsf{I}\,x = x \\ \mathsf{C}\,x\,y\,z = x\,z\,y & & \mathsf{K}\,x\,y = x \\ \mathsf{B}\,x\,y\,z = x\,(y\,z) & & \end{array}$$

The set of terms over Σ and $\mathcal{V}$ is defined below. In what follows, type subscripts are generally omitted.

Terms $\mathcal{T} ::= x \mid \mathsf{f}\langle\overline{\tau_n}\rangle \mid t_{\tau' \rightarrow \tau}\, t'_{\tau'}$ where $\mathsf{f} : \Pi\,\overline{\alpha_n}\,.\,\sigma \in \Sigma, x \in \mathcal{V}$ and $t, t' \in \mathcal{T}$

The type of the term $\mathsf{f}\langle\overline{\tau_n}\rangle$ is $\sigma\{\overline{\alpha_n} \rightarrow \overline{\tau_n}\}$. Terms of the form $t_1\,t_2$ are called applications. Non-application terms are called heads. A term can uniquely be decomposed into a head and n arguments. Let $t = \zeta\ \overline{t'_n}$. Then $head(t) = \zeta$ where ζ could be a variable or constant applied to possibly zero type arguments. The symbol $\mathcal{C}_{\mathsf{any}}$ denotes an arbitrary combinator, whilst $\mathcal{C}_3$ denotes a member of $\{\mathsf{S}, \mathsf{C}, \mathsf{B}\}$. The S-, C- or B-combinators are *fully applied* if they have 3 or more arguments. The K-combinator is fully applied if it has 2 or more arguments and the I is fully applied if it has any arguments. The symbols $\mathcal{C}_{\mathsf{any}}$ and $\mathcal{C}_3$ are only used if the symbols they represent are fully applied. Thus, in $\mathcal{C}_3\ \overline{t_n}$, $n \geq 3$ is assumed. The symbols $x, y, z \ldots$ are reserved for variables, $\mathsf{c}, \mathsf{d}, \mathsf{f} \ldots$ for non-combinator constants and ζ, ξ range over arbitrary function symbols and variables and, by an abuse of notation, at times even terms. A head symbol that is not a combinator applied to type arguments or a variable is called *first-order*.

Positions over Terms: For a term t, if $t \in \mathcal{V}$ or $t = \mathsf{f}\langle\overline{\tau}\rangle$, then $pos(t) = \{\epsilon\}$ (type arguments have no position). If $t = t_1\,t_2$ then $pos(t) = \{\epsilon\} \cup \{i.p \mid 1 \leq i \leq$

2, $p \in pos(t_i)\}$. Subterms at positions of the form $p.1$ are called *prefix* subterms. We define *first-order* subterms inductively as follows. For any term t, t is a first-order subterm of itself. If $t = \zeta\,\overline{t_n}$, where ζ is not a fully applied combinator, then the first-order subterms of each t_i are also first-order subterms of t. The notation $s\langle u\rangle$ is to be read as u is a first-order subterm of s. Note that this definition is subtly different to that in [10] since subterms underneath a fully applied combinator are not considered to be first-order.

Stable Subterms: Let $\mathsf{LPP}(t,p)$ be a partial function that takes a term t, a position p and returns the longest proper prefix p' of p such that $head(t|_{p'})$ is not a partially applied combinator if such a position exists. For a position $p \in pos(t)$, p is a stable position in t if p is not a prefix position and either $\mathsf{LPP}(t,p)$ is not defined or $head(t|_{\mathsf{LPP}(t,p)})$ is not a variable or combinator. A *stable subterm* is a subterm occurring at a stable position. For example, the subterm a is not stable in $\mathsf{f}\,(\mathbf{S}\,\mathsf{a}\,\mathsf{b}\,\mathsf{c})$, $\mathbf{S}\,(\mathbf{S}\,\mathsf{a})\,\mathsf{b}\,\mathsf{c}$ (in both cases, $head(t|_{\mathsf{LPP}(t,p)}) = \mathbf{S}$) and $\mathsf{a}\,\mathsf{c}$ (a is in a prefix position), but is in $\mathsf{g}\,\mathsf{a}\,\mathsf{b}$ and $\mathsf{f}\,(\mathbf{S}\,\mathsf{a})\,\mathsf{b}$. A subterm that is not stable is known as an *unstable* subterm.

The notation $t[u]$ denotes an arbitrary subterm u of t. The notation $t[u_1,\ldots,u_n]_n$, at times given as $t[\,\overline{u}\,]_n$ denotes that the the term t contains n *non-overlapping* subterms u_1 to u_n. By $u[]_n$, we refer to a context with n non-overlapping holes.

Weak Reduction: A term t *weak-reduces* to a term t' in one step (denoted $t \longrightarrow_w t'$) if $t = u[s]_p$ and there exists a combinator axiom $l = r$ and substitution σ such that $l\sigma = s$ and $t' = u[r\sigma]_p$. The term $l\sigma$ in t is called a *weak redex* or just redex. By $\longrightarrow_w^*$, the reflexive transitive closure of $\longrightarrow_w$ is denoted. Weak-reduction is terminating and confluent as proved in [15]. By $(t)\downarrow^w$, we denote the term formed from t by contracting its leftmost redex.

Literals and Clauses: An equation $s \approx t$ is an unordered pair of terms and a literal is an equation or the negation of an equation represented $s \not\approx t$. Let $ax = l \approx r$ be a combinator axiom and $\overline{x_n}$ be a tuple of variables not appearing in ax. Then ax and $l\,\overline{x_n} \approx r\,\overline{x_n}$ for all n are known as *extended combinator axioms*. For example, $\mathbf{I}\,x_1\,x_2 \approx x_1\,x_2$ is an extended combinator axiom. A clause is a multiset of literals represented $L_1 \vee \cdots \vee L_n$ where each L_i is a literal.

Semantics. We follow Bentkamp et al. [6] closely in specifying the semantics. An interpretation is a triple $(\mathcal{U},\mathcal{E},\mathcal{J})$ where $\mathcal{U}$ is a ground-type indexed family of non-empty sets called *universes* and $\mathcal{E}$ is a family of functions $\mathcal{E}_{\tau,\upsilon} : \mathcal{U}_{\tau\to\upsilon} \to (\mathcal{U}_\tau \to \mathcal{U}_\upsilon)$. A *type valuation* ξ is a substitution that maps type variables to ground types and whose domain is the set of all type variables. A type valuation ξ is extended to a *valuation* by setting $\xi(x_\tau)$ to be a member of $\mathcal{U}_{(\tau\xi)}$. An interpretation function $\mathcal{J}$ maps a function symbol $\mathsf{f} : \Pi\,\overline{\alpha_n}\,.\,\sigma$ and a tuple of ground types $\overline{\tau_n}$ to a member of $\mathcal{U}_{(\sigma\{\alpha_i \to \tau_i\})}$. An interpretation is *extensional* if $\mathcal{E}_{\tau,\upsilon}$ is injective for all τ,υ and is *standard* if $\mathcal{E}_{\tau,\upsilon}$ is bijective for all τ,υ.

For an interpretation $\mathcal{I} = (\mathcal{U},\mathcal{E},\mathcal{J})$ and a valuation ξ, a term is denoted as follows: $[\![x]\!]_\mathcal{I}^\xi = \xi(x)$, $[\![\mathsf{f}\langle\,\overline{\tau}\,\rangle]\!]_\mathcal{I}^\xi = \mathcal{J}(\mathsf{f}, [\![\,\overline{\tau}\,]\!]^\xi)$ and $[\![st]\!]_\mathcal{I}^\xi = \mathcal{E}([\![s]\!]_\mathcal{I}^\xi)([\![t]\!]_\mathcal{I}^\xi)$. An equation $s \approx t$ is true in an interpretation $\mathcal{I}$ with valuation function ξ if $[\![s]\!]_{\mathcal{I}}^{\xi}$ and $[\![t]\!]_{\mathcal{I}}^{\xi}$ are the same object and is false otherwise. A disequation $s \not\approx t$ is true if $s \approx t$ is false. A clause is true if one of its literals is true and a clause set is true if every clause in the set is true. An interpretation $\mathcal{I}$ models a clause set N, written $\mathcal{I} \models N$, if N is true in $\mathcal{I}$ for all valuation functions ξ.

As Bentkamp et al. point out in [4] there is a subtlety relating to higher-order models and choice. If, as is the case here, attention is not restricted to models that satisfy the axiom of choice, naive skolemisation is unsound. One solution would be to implement skolemisation with mandatory arguments as explained in [21]. However, the introduction of mandatory arguments considerably complicates both the calculus and the implementation. Therefore, we resort to the same 'trick' as Bentkamp et al., namely, claiming completeness for our calculus with respect to models as described above. This holds since we assume problems to be clausified. Soundness is claimed for the implementation with respect to models that satisfy the axiom of choice and completeness can be claimed if the axiom of choice is added to the clause set.

3 The Calculus

The calculus is modeled after Bentkamp et al.'s intensional non-purifying calculus [6]. The extensionality axiom can be added if extensionality is required. The main difference between our calculus and that of [6] is that superposition inferences are not allowed beneath fully applied combinators and an extra inference rule is added to deal with superposition beneath variables. We name the calculus *clausal combinatory-superposition.*

Term Ordering. We also demand that clausal combinatory-superposition is parameterised by a partial ordering $\succ$ that is well-founded, total on ground terms, stable under substitutions and has the subterm property and which orients all instances of combinator axioms left-to-right. It is an open problem whether a simplification ordering enjoying this last property exists, but it appears unlikely. However, for completeness, compatibility with stable contexts suffices. The $>_{\mathsf{ski}}$ ordering introduced in [10] orients all instances of combinator axioms left-to-right and is compatible with stable contexts. It is *not* compatible with arbitrary contexts. For terms t_1 and t_2 such that $t_1 >_{\mathsf{ski}} t_2$, it is not necessarily the case that $t_1\, u >_{\mathsf{ski}} t_2\, u$ or that $\mathbf{S}\, t_1\, a\, b >_{\mathsf{ski}} \mathbf{S}\, t_2\, a\, b$. We show that by not superposing underneath fully applied combinators and carrying out some restricted superposition beneath variables, this lack of compatibility with arbitrary contexts can be circumvented and does not lead to a loss of completeness. In a number of places in the completeness proof, we assume the following conditions on the ordering (satisfied by the $>_{\mathsf{ski}}$ ordering). It may be possible to relax the conditions at the expense of an increased number of inferences.

P1 For terms t, t' such that $t \longrightarrow_w t'$, then $t \succ t'$
P2 For terms t, t' such that $t \succ t'$ and $head(t')$ is first-order, $u[t] \succ u[t']$

The ordering $\succ$ is extended to literals and clauses using the multiset extension as explained in [22].

Inference Rules. Clausal combinatory-superposition is further parameterised by a selection function that maps a clause to a subset of its negative literals. Due to the requirements of the completeness proof, if a term $t = x\ \overline{s_{n>0}}$ is a maximal term in a clause C, then a literal containing x as a first-order subterm may not be selected. A literal L is *σ-eligible* in a clause C if it is selected or there are no selected literals in C and $L\sigma$ is maximal in $C\sigma$. If σ is the identity substitution it is left implicit. In the latter case, it is *strictly eligible* if it is strictly maximal. A variable x has a *bad* occurrence in a clause C if it occurs in C at an unstable position. Occurrences of x in C at stable positions are *good*.

Conventions: Often a clause is written with a single distinguished literal such as $C' \vee t \approx t'$. In this case:

1. The distinguished literal is always σ-eligible for some σ.
2. The name of the clause is assumed to be the name of the remainder without the dash.
3. If the clause is involved in an inference, the distinguished literal is the literal that takes part.

Positive and negative superposition:

$$\frac{D' \vee t \approx t' \qquad C' \vee [\neg] s\langle u\rangle \approx s'}{(C' \vee D' \vee [\neg] s\langle t'\rangle \approx s')\sigma}\ \textsc{Sup}$$

with the following side conditions:

1. The variable condition (below) holds
2. C is not an extended combinator axiom;
3. $\sigma = mgu(t, u)$;
4. $t\sigma \not\preceq t'\sigma$;
5. $s\langle u\rangle\sigma \not\preceq s'\sigma$;
6. $C\sigma \not\preceq D\sigma$ or D is an extended combinator axiom;
7. $t \approx t'$ is strictly σ-eligible in D;
8. $[\neg]\ s\langle u\rangle \approx s'$ is σ-eligible in C, and strictly σ-eligible if it is positive.

Definition 1. *Let $l = \mathbf{C}_{\mathsf{any}}\ \overline{x_n}$ and $l \approx r$ be an extended combinator axiom. A term $v\ \overline{u_m}$ is* compatible *with $l \approx r$ if $\mathbf{C}_{\mathsf{any}} = \mathsf{I}$ and $m = n$ or if $\mathbf{C}_{\mathsf{any}} = \mathsf{K}$ and $m \geq n - 1$ or if $\mathbf{C}_{\mathsf{any}} \in \{\mathsf{B}, \mathsf{C}, \mathsf{S}\}$ and $m \geq n - 2$.*

Variable Condition: $u \notin \mathcal{V}$. If $u = x\ \overline{s_n}$ and D is an extended combinator axiom, then D and u must be compatible.

Because the term ordering $\succ$ is not compatible with unstable contexts, there are instances when superposition beneath variables must be carried out. The SubVarSup rule deals with this.

$$\frac{D' \vee t \approx t' \qquad C' \vee [\neg]s\langle y\,\overline{u_n}\,\rangle \approx s'}{(C' \vee D' \vee [\neg]s\langle zt'\,\overline{u_n}\,\rangle \approx s')\sigma}\ \textsc{SubVarSup}$$

with the following side conditions in addition to conditions 4 – 8 of Sup:

1. y has another occurrence bad in C;
2. z is a fresh variable;
3. $\sigma = \{y \rightarrow z\,t\}$;
4. t' has a variable or combinator head;
5. $n \leq 1$;
6. D is not an extended combinator axiom.

The EqRes and EqFact inferences:

$$\frac{C' \vee u \not\approx u'}{C'\sigma}\ \textsc{EqRes} \qquad\qquad \frac{C' \vee u' \approx v' \vee u \approx v}{(C' \vee v \not\approx v' \vee u \approx v')\sigma}\ \textsc{EqFact}$$

For both inferences $\sigma = mgu(u, u')$. For EqRes, $u \not\approx u'$ is σ-eligible in the premise. For EqFact, $u'\sigma \not\preceq v'\sigma$, $u\sigma \not\preceq v\sigma$, and $u \approx v$ is σ-eligible in the premise.

In essence, the ArgCong inference allows superposition to take place at prefix positions by 'growing' equalities to the necessary size.

$$\frac{C' \vee s \approx s'}{\begin{array}{c} C'\sigma \vee (s\sigma)x \approx (s'\sigma)x \\ C'\sigma \vee (s\sigma)\,\overline{x_2} \approx (s'\sigma)\,\overline{x_2} \\ C'\sigma \vee (s\sigma)\,\overline{x_3} \approx (s'\sigma)\,\overline{x_3} \\ \vdots \end{array}}\ \textsc{ArgCong}$$

The literal $s \approx s'$ must be σ-eligible in C. Let s and s' be of type $\alpha_1 \rightarrow \cdots \rightarrow \alpha_m \rightarrow \beta$. If β is not a type variable, then σ is the identity substitution and the inference has m conclusions. Otherwise, if β is a type variable, the inference has an infinite number of conclusions. In conclusions where $n > m$, σ is the substitution that maps β to type $\tau_1 \rightarrow \cdots \rightarrow \tau_{n-m} \rightarrow \beta'$ where β' and each τ_i are fresh type variables. In each conclusion, the x_is are variables fresh for C. Note that an ArgCong inference on a combinator axiom results in an extended combinator axiom.

3.1 Extensionality

Clausal combinatory-superposition can be either intensional or extensional. If a conjecture is proved by the intensional version of the calculus, it means that the conjecture holds in all models of the axioms. On the other hand, if a conjecture is proved by the extensional version, it means that the conjecture holds in all extensional models (as defined above). Practically, some domains naturally lend themselves to intensional reasoning whilst other to extensional. For example, when reasoning about programs, we may expect to treat different programs as different entities even if they always produce the same output when provided the same input. For the calculus to be extensional, we provide two possibilities. The first is to add a polymorphic extensionality axiom. Let diff be a polymorphic symbol of type $\Pi\tau_1, \tau_2. (\tau_1 \to \tau_2) \to (\tau_1 \to \tau_2) \to \tau_1$. Then the extensionality axiom can be given as:

$$x\,(\mathsf{diff}\langle\tau_1,\tau_2\rangle\; x\; y) \;\not\approx\; y\,(\mathsf{diff}\langle\tau_1,\tau_2\rangle\; x\; y) \;\vee\; x \approx y$$

However, adding the extensionality axiom to a clause set can be explosive and is not graceful. By any common ordering, the negative literal will be the larger literal and therefore the literal involved in inferences. As it is not of functional type it can unify with terms of atomic type including first-order terms.

In order to circumvent this issue, we developed another method of dealing with extensionality. Unification is replaced by unification with abstraction. During the unification procedure, no attempt is made to unify pairs consisting of terms of functional or variable type. Instead, if the remaining unification pairs can be solved successfully, such pairs are added to the resulting clause as negative constraint literals. This process works in conjunction with the negative extensionality rule presented below.

$$\frac{C' \vee s \not\approx s'}{(C' \vee s\,(\mathsf{sk}\langle\,\overline{\alpha}\,\rangle\;\overline{x}\,) \not\approx s'\,(\mathsf{sk}\langle\,\overline{\alpha}\,\rangle\;\overline{x}\,))\sigma}\;\textsc{NegExt}$$

where $s \not\approx s'$ is σ-eligible in the premise, $\overline{\alpha}$ and $\overline{x}$ are the free type and term variable of the literal $s \not\approx s'$ and σ is the most general type unifier that ensures the well-typedness of the conclusion.

We motivate this second approach to extensionality with an example. Consider the clause set:

$$\mathsf{g}\,x \approx \mathsf{f}\,x \qquad \mathsf{h}\,\mathsf{g} \not\approx \mathsf{h}\,\mathsf{f}$$

equality resolution with abstraction on the second clause produces the clause $\mathsf{g} \not\approx \mathsf{f}$. A NegExt inference on this clause results in $\mathsf{g}\,\mathsf{sk} \not\approx \mathsf{f}\,\mathsf{sk}$ which can superpose with $\mathsf{g}\,x \approx \mathsf{f}\,x$ to produce $\bot$.

The unification with abstraction procedure used here is very similar to that introduced in [24]. Pseudocode for the algorithm can be found in Algorithm 1. The inference rules other than ArgCong and SubVarSup must be modified to utilise unification with abstraction rather than standard unification. We show the updated superposition rule. The remaining rules can be modified along similar lines.

$$\frac{C_1 \vee t \approx t' \qquad C_2 \vee [\neg]s\langle u\rangle \approx s'}{(C_1 \vee C_2 \vee D \vee [\neg]s\langle t'\rangle \approx s')\sigma}\;\textsc{Sup-wA}$$

where D is the possibly empty set of negative literals returned by unification. Sup-wA shares all the side conditions of Sup given above. This method of dealing with extensionality is not complete as shown in Appendix A of [9].

Algorithm 1. Unification algorithm with constraints

function $\mathsf{mgu}_{\mathsf{Abs}}(l, r)$
  let $\mathcal{P}$ be a set of unification pairs; $\mathcal{P} := \{\langle l, r\rangle\}$, $\mathcal{D}$ be a set of disequalities; $\mathcal{D} := \emptyset$
  let θ be a substitution; $\theta := \{\}$
  loop
    if $\mathcal{P}$ is empty **then return** (θ, D), where D is the disjunction of literals in $\mathcal{D}$
    Select a pair $\langle s, t\rangle$ in $\mathcal{P}$ and remove it from $\mathcal{P}$
    if s coincides with t **then** do nothing
    else if s is a variable and s does not occur in t **then** $\theta := \theta \circ \{s \mapsto t\}$; $\mathcal{P} := \mathcal{P}\{s \mapsto t\}$
    else if s is a variable and s occurs in t **then** fail
    else if t is a variable **then** $\mathcal{P} := \mathcal{P} \cup \{\langle t, s\rangle\}$
    else if s and t have functional or variable type **then** $\mathcal{D} := \mathcal{D} \cup \{s \not\approx t\}$
    else if s and t have different head symbols **then** fail
    else if $s = \mathsf{f}\, s_1 \ldots s_n$ and $t = \mathsf{f}\, t_1 \ldots t_n$ for some f **then**
      $\mathcal{P} := \mathcal{P} \cup \{\langle s_1, t_1\rangle, \ldots, \langle s_n, t_n\rangle\}$

4 Examples

We provide some examples of how the calculus works. Some of the examples utilised come from Bentkamp et al.'s paper [4] in order to allow a comparison of the two methods. In all examples, it is assumed that the clause set has been enriched with the combinator axioms.

Example 1. Consider the unsatisfiable clause:

$$x\,\mathsf{a}\,\mathsf{b} \not\approx x\,\mathsf{b}\,\mathsf{a}$$

Superposing onto the left-hand side with the extended $\mathbf{K}$ axiom $\mathbf{K}\,x_1\,x_2\,x_3 \approx x_1\,x_3$ results in the clause $x_1\,\mathsf{b} \not\approx x_1\,\mathsf{a}$. Superposing onto the left-hand side of this clause, this time with the standard $\mathbf{K}$ axiom adds the clause $x \not\approx x$ from which $\perp$ is derived by an EqRes inference.

Example 2. Consider the unsatisfiable clause set where $\mathsf{f}\,\mathsf{a} \succ \mathsf{c}$:

$$\mathsf{f}\,\mathsf{a} \approx \mathsf{c} \qquad \mathsf{h}\,(y\,\mathsf{b})(y\,\mathsf{a}) \not\approx \mathsf{h}\,(\mathsf{g}(\mathsf{f}\,\mathsf{b}))(\mathsf{g}\,\mathsf{c})$$

A Sup inference between the $\mathbf{B}$ axiom $\mathbf{B}\,x_1\,x_2\,x_3 \approx x_1\,(x_2\,x_3)$ and the subterm $y\,\mathsf{b}$ of the second clause adds the clause $\mathsf{h}(x_1(x_2\,\mathsf{b}))(x_1(x_2\,\mathsf{a})) \not\approx \mathsf{h}(\mathsf{g}(\mathsf{f}\,\mathsf{b}))(\mathsf{g}\,\mathsf{c})$ to the set. By superposing onto the subterm $x_2\,\mathsf{a}$ of this clause with the equation $\mathsf{f}\,\mathsf{a} \approx \mathsf{c}$, we derive the clause $\mathsf{h}(x_1(\mathsf{f}\,\mathsf{b}))(x_1\,\mathsf{c}) \not\approx \mathsf{h}(\mathsf{g}(\mathsf{f}\,\mathsf{b}))(\mathsf{g}\,\mathsf{c})$ from which $\perp$ can be derived by an EqRes inference.

Example 3. Consider the unsatisfiable clause set where $\mathsf{f}\,\mathsf{a} \succ \mathsf{c}$. This example is the combinatory equivalent of Bentkamp et al.'s Example 6.

$$\mathsf{f}\,\mathsf{a} \approx \mathsf{c} \qquad \mathsf{h}\,(y\,(\mathbf{B}\,\mathsf{g}\,\mathsf{f})\,\mathsf{a})\,y \not\approx \mathsf{h}\,(\mathsf{g}\,\mathsf{c})\,\mathbf{I}$$

A SUP inference between the extended I axiom $\mathsf{I}\, x_1\, x_2 \approx x_1\, x_2$ and the subterm $y\,(\mathsf{B}\,\mathsf{g}\,\mathsf{f})\,\mathsf{a}$ of the second clause adds the clause $\mathsf{h}\,(\mathsf{B}\,\mathsf{g}\,\mathsf{f}\,\mathsf{a})\,\mathsf{I} \not\approx \mathsf{h}\,(\mathsf{g}\,\mathsf{c})\,\mathsf{I}$ to the set. Superposing onto the subterm $\mathsf{B}\,\mathsf{g}\,\mathsf{f}\,\mathsf{a}$ of this clause with the B axiom results in the clause $\mathsf{h}\,(\mathsf{g}\,(\mathsf{f}\,\mathsf{a}))\,\mathsf{I} \not\approx \mathsf{h}\,(\mathsf{g}\,\mathsf{c})\,\mathsf{I}$. Superposition onto the subterm $\mathsf{f}\,\mathsf{a}$ with the first clause of the original set gives $\mathsf{h}\,(\mathsf{g}\,\mathsf{c}))\,\mathsf{I} \not\approx \mathsf{h}\,(\mathsf{g}\,\mathsf{c})\,\mathsf{I}$ from which $\bot$ can be derived via EQRES.

Note that in Examples 2 and 3, no use is made of SUBVARSUP even though the analogous FLUIDSUP rule in required in Bentkamp et al.'s calculus. We have been unable to develop an example that requires the SUBVARSUP rule even though it is required for the completeness result in Sect. 6.

5 Redundancy Criterion

In Sect. 6, we prove that the calculus is refutationally complete. The proof follows that of Bachmair and Ganzinger's original proof of the completeness of superposition [2], but is presented in the style of Bentkamp et al. [6] and Waldmann [31]. As is normal with such proofs, it utilises the concept of *redundancy* to reduce the number of clauses that must be considered in the induction step during the model construction process.

We define a weaker logic by an encoding $\lfloor\,\rfloor$ of ground terms into non-applicative first-order terms with $\lceil\,\rceil$ as its inverse. The encoding works by indexing each symbol with its type arguments and argument number. For example, $\lfloor \mathsf{f} \rfloor = \mathsf{f}_0$, $\lfloor \mathsf{f}\langle\overline{\tau}\rangle \mathsf{a} \rfloor = \mathsf{f}_1^{\overline{\tau}}(\mathsf{a}_0)$. Terms with fully applied combinators as their head symbols are translated to constants such that syntactically identical terms are translated to the same constant. For example, $\lfloor \mathsf{S}\, t_1\, t_2\, t_3 \rfloor = \mathsf{s}_0$. The weaker logic is known as the *floor logic* whilst the original logic is called the *ceiling logic.* The encoding can be extended to literals and clauses in the obvious manner as detailed in [5]. The function $\lceil\,\rceil$ is used to compare floor terms. More precisely, for floor logic terms t and t', $t \succ t'$ if $\lceil t \rceil \succ \lceil t' \rceil$. It is straightforward to show that the order $\succ$ on floor terms is compatible with all contexts, well-founded, total on ground terms and has the subterm property.

The encoding serves a dual purpose. Firstly, as redundancy is defined with respect to the floor logic, it prevents the conclusion of all ARGCONG from being redundant. Secondly, subterms in the floor logic correspond to first-order subterms in the ceiling logic. This is of critical importance in the completeness proof.

An inference is the ground instance of an inference I if it is equal to I after the application of some grounding substitution θ to the premise(s) and conclusion of I and the result is still an inference.

A ground ceiling clause C is redundant with respect to a set of ground ceiling clauses N if $\lfloor C \rfloor$ is entailed by clauses in $\lfloor N \rfloor$ smaller than itself and the floor of ground instances of extended combinator axioms in $\lfloor N \rfloor$. An arbitrary ceiling clause C is redundant to a set of ceiling clauses N if all its ground instances are redundant with respect to $\mathcal{G}_\Sigma(N)$, the set of all ground instances of clauses in N. *Red*(N) is the set of all clauses redundant with respect to N.

For ground inferences other than ArgCong, an inference with right premise C and conclusion E is redundant with respect to a set of clauses N if $\lfloor E \rfloor$ is entailed by clauses in $\lfloor \mathcal{G}_\Sigma(N) \rfloor$ smaller than $\lfloor C \rfloor$. A non-ground inference is redundant if all its ground instances are redundant.

An ArgCong inference from a combinator axiom is redundant with respect to a set of clauses N if its conclusion is in N. For any other ArgCong inference, it is redundant with respect N if its premise is redundant with respect to N, or its conclusion is in N or redundant with respect to N. A set N is saturated up to redundancy if every inference with premises in N is redundant with respect to N.

6 Refutational Completeness

The proof of refutational completeness of clausal combinatory-superposition is based on the completeness proof the λ-free HOL calculi presented in [6]. We first summarise their proof and then indicate the major places where our proof differs. A detailed version of the proof can be found in our technical report [9].

Let N be a set of clauses saturated to redundancy by one of the λ-free HOL calculi and not containing $\bot$. Then, Bentkmap et al. show that N must have a model. This is done in stages, first building a model R_∞ of $\lfloor \mathcal{G}_\Sigma(N) \rfloor$ and then lifting this to a model of $\mathcal{G}_\Sigma(N)$ and N. Most of the heavy work is in showing R_∞ to be a model of $\lfloor \mathcal{G}_\Sigma(N) \rfloor$. In the standard first-order proof, superposition is ruled out at or beneath variables. Consider a clause C containing a variable x. Since the orderings that parameterise standard first-order superposition are compatible with contexts, we are guaranteed for terms t and t' such that $t \succ t'$ and grounding substitution θ that $C\theta[x \rightarrow t] \succ C\theta[x \rightarrow t']$. Then, the induction hypotheses is used to show that $C\theta[x \rightarrow t']$ is true in the candidate interpretation rendering superposition into x unnecessary. A similar argument works for superposition below variables.

This argument does not work for the λ-free calculi since in their case the ordering is not compatible with arguments. Consider the clause $C = \mathsf{f}\,x \approx \mathsf{g}\,x \vee x\,u \approx v$. For terms t and t' such that $t \succ t'$, it cannot be guaranteed that $C\theta[x \rightarrow t] \succ C\theta[x \rightarrow t']$ since $t'\,u \succ t\,u$ is possible. Therefore, some superposition has to take place at variables. Even in those cases where it can be ruled out, the proof is more complex than the standard first-order proof. Superposition underneath variables can be ruled out entirely.

Returning to our calculus, we face a number of additional difficulties. Since the ordering that we use is not compatible with arguments, but also not compatible with stronger concept of unstable subterms, we cannot rule out superposition beneath a variable. Consider, for example, the clause $C = \mathsf{f}\,x \approx \mathsf{g}\,x \vee y\,x\,u \approx v$ and the grounding substitution $\theta = \{y \rightarrow \mathbf{S}\,\mathsf{a}, x \rightarrow \mathbf{K}\,t\}$. Let θ' be the same as θ, but with x mapped to $\mathbf{K}\,t'$. Even if $t \succ t'$, we do not necessarily have $C\theta \succ C\theta'$ since t occurs at an unstable position. Thus, some superposition below variables must be carried out.

This introduces a new difficulty, namely, showing that superposition below a variable is the ground instance of a SubVarSup rule. In some cases it may not be.

Consider the clause $C = x\,u \approx v$ and the grounding substitution $\theta = \{x \rightarrow \mathsf{f}\,t\,\mathsf{a}\}$. A superposition inference from $C\theta$ with conclusion C' that replaces t with t' is not the ground instance of a SubVarSup from C. The only conclusion of a SubVarSup from C is $z\,t'\,u \approx v$. There is no instantiation of the z variable that can make the term $z\,t'$ equal to $\mathsf{f}\,t'\,\mathsf{a}$. However, mapping z to $\mathsf{C}\,\mathsf{f}\,\mathsf{a}$ results in a term that is equal to $\mathsf{f}\,t'\,\mathsf{a}$ *modulo the combinator axioms*. Let $C'' = z\,t'\,u \approx v\{z \rightarrow \mathsf{C}\,\mathsf{f}\,\mathsf{a}\}$. Since it is the set N that is saturated up to redundancy, we have that all ground instances of the SubVarSup inference are redundant with respect $\mathcal{G}_\Sigma(N)$. For this to imply that C' is redundant with respect to $\mathcal{G}_\Sigma(N)$ requires C'' be rewritable to C' using equations true in $R_{\lfloor C\theta \rfloor}$ (the partial interpretation built from clauses smaller than $C\theta$). This requires that all ground instances of combinator and extended combinator axioms be true in $R_{\lfloor C\theta \rfloor}$ for all clauses C.

This leads to probably the most novel aspect of our proof. We build our candidate interpretations differently to the standard proof by first adding rules derived from combinator axioms. Let R_{ECA} be the set of rewrite rules formed by turning the floor of all ground instances of combinator axioms into left right rewrite rules. Then for all clauses $C \in \lfloor \mathcal{G}_\Sigma(N) \rfloor$, R_C is defined to be $R_{ECA} \cup (\bigcup_{D \prec C} E_D)$. In the detailed proof, we show that R_C and R_∞ are still terminating and confluent.

In lifting R_∞ to be a model of $\mathcal{G}_\Sigma(N)$, we face a further difficulty. This is related to showing that for ceiling terms t, t', u and u', if $[\![\lfloor t \rfloor]\!]^{\xi}_{R_\infty} = [\![\lfloor t' \rfloor]\!]^{\xi}_{R_\infty}$ and $[\![\lfloor u \rfloor]\!]^{\xi}_{R_\infty} = [\![\lfloor u' \rfloor]\!]^{\xi}_{R_\infty}$, then $[\![\lfloor t\,u \rfloor]\!]^{\xi}_{R_\infty} = [\![\lfloor t'\,u' \rfloor]\!]^{\xi}_{R_\infty}$. The difficulty arises because t, t' or both may be of the form $\mathbf{C}_3\,t_1\,t_2$. In such a case rewrites that can be carried out from subterms of $\lfloor t \rfloor$ cannot be carried out from $\lfloor t\,u \rfloor$ because $t\,u$ has a fully applied combinator as its head and therefore is translated to a constant in the floor logic. The fact that the ground instances of all combinator axioms are true in R_∞ comes to the rescue. With these difficulties circumvented, we can prove refutational completeness.

Theorem 1. *For a set of clauses N saturated to redundancy by the above calculus, N has a model iff it does not contain $\bot$. Moreover, if N contains the extensionality axiom, the model is extensional.*

7 Removing Combinator Axioms

Next, we show that it is possible to replace the combinator axioms with a dedicated inference rule. We name the inference Narrow. Unlike the other inference rules, it works at prefix positions. We define *nearly first-order* positions inductively. For any term t, either $t = \zeta\,\overline{t_n}$ where ζ is not a fully applied combinator or $t = \mathbf{C}_{\mathsf{any}}\,\overline{t_n}$. In the first case, the nearly first-order subterms of t are $\zeta\,\overline{t_i}$ for $0 \leq i \leq n$ and all the nearly first-order subterms of the t_i. In the second case, the nearly first-order subterms are $\mathbf{C}_{\mathsf{any}}\,\overline{t_i}$ for $0 \leq i \leq n$. The notation $s\langle\!\langle u \rangle\!\rangle$ is to be read as u is a nearly first-order subterm of s. The Narrow inference:

$$\frac{C' \vee [\neg]s\langle\!\langle u \rangle\!\rangle \approx s'}{(C' \vee [\neg]s\langle\!\langle r \rangle\!\rangle \approx s')\sigma}\ \textsc{Narrow}$$

with the following side conditions:

1. $u \notin \mathcal{V}$
2. Let $l \approx r$ be a combinator axiom. $\sigma = mgu(l, u)$;
3. $s\langle\!\langle u \rangle\!\rangle\sigma \not\preceq s'\sigma$;
4. $[\neg]\, s\langle\!\langle u \rangle\!\rangle \approx s'$ is σ-eligible in C, and strictly σ-eligible if it is positive.

We show that any inference that can be carried out using an extended combinator axiom can be simulated with NARROW proving completeness. It is obvious that an EQRES or EQFACT inference cannot have an extended combinator axiom as its premise. By the SUBVARSUP side conditions, an extended combinator axiom cannot be either of its premises. Thus we only need to show that SUP inferences with extended combinator axioms can be simulated. Note that an extended axiom can only be the left premise of a SUP inference. Consider the following inference:

$$\frac{l \approx r \qquad C' \vee [\neg]s\langle u \rangle|_p \approx s'}{(C' \vee [\neg]s\langle r \rangle \approx s')\sigma}\ \text{SUP}$$

Let $l = \mathbf{S}\,\overline{x_{n>3}}$. By the variable condition, we have that $u = \zeta\,\overline{t_m}$ where $n \geq m \geq n-2$. If $u = y\,\overline{t_{n-2}}$, then $\sigma = \{y \rightarrow \mathbf{S}\,x_1\,x_2, x_3 \rightarrow t_1, \ldots, x_n \rightarrow t_{n-2}\}$. In this case $r\sigma = (x_1\,x_3\,(x_2\,x_3)\,x_4 \ldots x_n)\sigma = x_1\,t_1\,(x_2\,t_1)\,t_2 \ldots t_{n-2}$ and the conclusion of the inference is $(C' \vee [\neg]s\langle x_1\,t_1\,(x_2\,t_1)\,t_2 \ldots t_{n-2}\rangle \approx s')\{y \rightarrow \mathbf{S}\,x_1\,x_2\}$. Now consider the following NARROW inference from C at the nearly first-order subterm $y\,t_1$:

$$\frac{C' \vee [\neg]s\langle\langle\!\langle y\,t_1 \rangle\!\rangle t_2 \ldots t_n\rangle|_p \approx s'}{(C' \vee [\neg]s\langle x_1\,t_1\,(x_2\,t_1)\,t_2 \ldots t_{n-2}\rangle \approx s')\{y \rightarrow \mathbf{S}\,x_1\,x_2\}}\ \text{NARROW}$$

As can be seen, the conclusion of the SUP inference is equivalent to that of the NARROW inference up to variable naming. The same can be shown to be the case where $u = y\,\overline{t_{n-1}}$ or $u = y\,\overline{t_n}$ or $u = \mathbf{S}\,\overline{t_n}$. Likewise, the same can be shown to hold when the $l \approx r$ is an extended $\mathbf{B}, \mathbf{C}, \mathbf{K}$ or $\mathbf{I}$ axiom.

8 Implementation and Evaluation

Clausal combinatory-superposition has been implemented in the Vampire theorem prover [11,17]. The prover was first extended to support polymorphism. This turned out to be simpler than expected with types being turned into terms and type equality checking changing to a unifiability (or matching) check. Applicative terms are supported by the use of a polymorphic function app of type $\Pi\alpha, \beta.\, (\alpha \rightarrow \beta) \rightarrow \alpha \rightarrow \beta$.

As the SUP, EQRES and EQFACT inferences are identical to their first-order counterparts, these required no updating. The NARROW, SUBVARSUP and ARGCONG inferences had to be added to the implementation. Further, though the NEGEXT inference is not required for completeness, empirical results suggest that it is so useful, that it is permanently on in the implementation.

The ArgCong inference implemented in Vampire does not match the rule given in the calculus. The rule provided can have an infinite number of conclusions. In Vampire, we have implemented a version of ArgCong that appends a single fresh variable to each side of the selected literal rather than a tuple and therefore only has a single conclusion. This version matches what was originally in the calculus. Shortly before the submission of this paper, it was discovered that this leads to a subtle issue in the completeness proof and the inference was changed to its current version. We expect to be able to revert to the previous version and fix the proof. As matters stand, Vampire contains a potential source of incompleteness.

A greater challenge was posed by the implementation of the $>_{\mathsf{ski}}$ ordering in the prover. The ordering is based on the length of the longest weak-reduction from a term. In order to increase the efficiency of calculating this quantity, we implemented caching and lazy evaluation. For example, when inserting a term of the form $\mathsf{f}\,t_1\,t_2$ into the term-sharing data structure, a check is made to see if the maximum reduction lengths of t_1 and t_2 have already been calculated. If they have, then the maximum reduction length of the term being inserted is set to the sum of the maximum reduction lengths of t_1 and t_2. If not, it is left unassigned and only calculated at the time it is required.

During the experimentation phase, it was realised that many redundant clauses were being produced due to narrowing. For example, consider the clause $x\,\mathsf{a}\,\mathsf{b} \approx \mathsf{d} \vee \mathsf{f}\,x \approx \mathsf{a}$. Narrowing the first literal with $\mathbf{C}$-axiom results in $x'\,\mathsf{b}\,\mathsf{a} \approx \mathsf{d} \vee \mathsf{f}\,(\mathbf{C}\,x') \approx \mathsf{a}$. A second narrow with the same axiom results in $x''\,\mathsf{a}\,\mathsf{b} \approx \mathsf{d} \vee \mathsf{f}\,(\mathbf{C}\,(\mathbf{C}\,x'')) \approx \mathsf{a}$ which is extensionally equivalent to first clause and therefore redundant. However, it will not be removed by subsumption since it is only equivalent extensionally. To deal with this problem, we implemented some rewrite rules that replace combinator terms with smaller extensionally equivalent terms.[2] For example, any term of the form $\mathbf{C}\,(\mathbf{C}\,t)$ is rewritten to t. There is no guarantee that these rewrites remove all such redundant clauses, but in practice, they appear to help.

To implement unification with abstraction, we reused the method introduced in our previous work relating to the use of substitution trees as filters [8]. In our current context, this involves replacing all subterms of functional or variable sort with special symbols that unify with any term prior to inserting a term into the substitution tree index.

To evaluate our implementation, we ran a number of versions of our prover across two problem sets and compared their performance against that of some of the leading higher-order provers. The first problem set we tested on was the set of all 592 monomorphic, higher-order problems from the TPTP problem library [29] that do not contain first-class boolean subterms. We restricted our attention to monomorphic problems since some of the provers we used in our evaluation do not support polymorphism. The second benchmark set was produced by the Isabelle theorem prover's Sledgehammer system. It contains 1253 benchmarks kindly made available to us by the Matryoshka team and is called SH-λ fol-

[2] Thanks to Petar Vukmirović for suggesting and discussing this idea.

Table 1. Problems proved theorem or unsat

	TPTP TH0 problems		Sh-λ problems	
	Solved	Uniques	Solved	Uniques
Satallax 3.4	473	0	628	5
Leo-III 1.4	**482**	**6**	661	13
Vampire-THF 4.4	472	1	**717**	**14**
Vampire-csup-ninj	470	0 (1)	687	1 (2)
Vampire-csup-ax	469	0 (0)	680	0 (3)
Vampire-csup-abs	472	0 (0)	685	0 (0)
Vampire-csup-prag	475	1 (3)	628	0 (1)
Zipperposition 1.5	476	0	609	6

lowing their naming convention. All tests were run with a CPU time limit of 300. Experiments were performed on StarExec [28] nodes equipped with four 2.40 GHz Intel Xeon CPUs. Our experimental results are publicly available[3].

To compare out current implementation against, we chose the Leo-III, 1.4, Satallax 3.4, Zipperposition 1.5 and Vampire-THF 4.4 provers. These provers achieved the top four spots in the 2019 CASC system competition. Vampire THF 4.4 was developed by the authors, but uses different principles being based on combinatory unification. We compare the performance of these provers against four variants of our current implementation. First, Vampire-csup-ax which implements clausal combinatory-superposition as described above and uses the extensionality axiom. Second, Vampire-csup-abs which deals with extensionality via unification with abstraction. Third, Vampire-csup-ninj which incorporates an inference to synthesise left-inverses for injective functions in a manner similar to Leo-III [26, Section 4.2.5] and finally Vampire-csup-prag which introduces various heuristics to try and control the search space, though at the expense of completeness. For example, it implements a heuristic that restricts the number of narrow steps. It also switches off the SubVarSup rule which is never used in a proof produced by the other variants of Vampire-csup. All four versions are run on top of a first-order portfolio of strategies. These strategies control options such as the saturation algorithm used, which simplification inferences are switched on and so forth. The results of the experiments can be found summarised in Table 1. In brackets, the number of uniques between Vampire-csup versions is provided.

The closeness of the results on the TPTP benchmarks is striking. Out of the 592 benchmarks, 95 are known not to be theorems, leaving 497 problems that could possibly be proved. All the provers are remarkably close to this number and each other. Leo-III which is slightly ahead of the other provers, only manages

[3] https://github.com/vprover/vampire_publications/tree/master/experimental_data/IJCAR-2020-COMB-SUP.

this through function synthesis which is not implemented in any of the other provers.

It is disappointing that Vampire-csup performs worse than its predecessor Vampire-THF 4.4 on Sledgehammer problems. We hypothesise that this is related to the explosion in clauses created as a result of narrowing. Vampire-csup-prag is supposed to control such an explosion, but actually performs worst of all. This is likely due to the fact that it runs a number of lengthy strategies aimed particularly at solving higher-order problems requiring complex unifiers. Interestingly, the pragmatic version solved a difficulty rating 1.00 TPTP problem, namely, NUM829^5.p.

9 Conclusion and Related Work

The combinatory superposition calculus presented here is amongst a small group of complete proof calculi for higher-order logic. This group includes the RUE resolution calculus of Benzmüller which has been implemented in the LEO-II theorem prover [7]. The Satallax theorem prover implements a complete higher-order tableaux calculus [13]. More recently, Bentkamp et al. have developed a complete superposition calculus for clausal HOL [4]. As superposition is one of the most successful calculi in first-order theorem proving [22], their work answered a significant open question, namely, whether superposition could be extended to higher-order logic.

Our work is closely related to theirs, and in some senses, the SUBVARSUP rule of clausal combinatory-superposition mirrors the FLUIDSUP rule of clausal λ-superposition. However, there are some crucial differences. Arguably, the side conditions on SUBVARSUP are tighter than those on FLUIDSUP and some problems such as the one in Example 3 can be solved by clausal combinatory-superposition without the use of SUBVARSUP whilst requiring the use of FLUIDSUP in clausal λ-superposition. Clausal λ-superposition is based on higher-order unification and λ-terms. Our calculus is based on (applicative) first-order terms and first-order unification and implementations can therefore reuse the well-studied data structures and algorithms of first-order theorem proving. On the downside, narrowing terms with combinator axioms is still explosive and results in redundant clauses. It is also never likely to be competitive with higher-order unification in finding complex unifiers. This is particularly the case with recent improvements in higher-order unification being reported [30].

Many other calculi for higher-order theorem proving have been developed, most of them incomplete. Amongst the early calculi to be devised are Andrew's mating calculus [1] and Miller's expansion tree method [20] both linked to tableaux proving. More recent additions include an ordered (incomplete) paramodulation calculus as implemented in the Leo-III prover [27] and a higher-order sequent calculus implemented in the AgsyHOL prover [18]. In previous work, the current authors have extended first-order superposition to use a combinatory unification algorithm [8]. Finally there is ongoing work to extend SMT solving to higher-order logic [3].

There have also been many attempts to prove theorems in HOL by translating to FOL. One of the pioneers in suggesting this approach was Kerber [16]. Since his early work, it has become commonplace to combine a dedicated higher-order theorem prover with a first-order prover used to discharge first-order proof obligations. This is the approach taken by many interactive provers and their associated hammers such as Sledgehammer [23] and CoqHammer [14]. It is also the approach adopted by leading automated higher-order provers Leo-III and Satallax.

In this paper we have presented a complete calculus for a polymorphic, boolean-free, intensional, combinatory formulation of higher-order logic. For the calculus to be extensional, an extensionality axiom can be added maintaining completeness, but losing gracefulness. Alternatively, unification can be turned into unification with abstraction maintaining gracefulness, but losing a completeness guarantee. Experimental results show an implementation of clausal combinatory-superposition to be competitive with leading higher-order provers.

It remains to tune the implementation and calculus. We plan to further investigate the use of heuristics in taming the explosion of clauses that result from narrowing. the heuristics may lead to incompleteness. It would also be of interest to investigate the use of heuristics or even machine learning to guide the prover in selecting specific combinator axioms to narrow a particular clause with. One of the advantages of our calculus is that it does not consider terms modulo β- or weak-reduction. Therefore, theoretically, a larger class of terms should be comparable by the non-ground order than is possible with a calculus that deals with β- or weak-equivalence classes. It remains to implement a stricter version of the $>_{\mathsf{ski}}$ ordering and evaluate its usefulness.

As a next step, we plan to add support for booleans and choice to the calculus. An appealing option for booleans is to extend the unification with abstraction approach currently used for functional extensionality. No attempt would be made to solve unification pairs consisting of boolean terms. Rather, these would be added as negated bi-implications to the result which would then be re-clausified.

Finally, we feel that our calculus complements existing higher-order calculi and presents a particularly attractive option for extending existing first-order superposition provers to dealing with HOL.

Acknowledgements. Thanks to Jasmin Blanchette, Alexander Bentkamp and Petar Vukmirović for many discussions on aspects of this research. We would also like to thank Andrei Voronkov, Martin Riener and Michael Rawson. We are grateful to Visa Nummelin for pointing out the incompleteness of unification with abstraction and providing the counterexample. Thanks is also due to the maintainers of StarExec and the TPTP problem library both of which were invaluable to this research. The first author thanks the family of James Elson for funding his research.

References

1. Andrews, P.B.: On connections and higher-order logic. J. Autom. Reasoning **5**(3), 257–291 (1989)

2. Bachmair, L., Ganzinger, H.: Rewrite-based equational theorem proving with selection and simplification. J. Logic Comput. **4**(3), 217–247 (1994)
3. Barbosa, H., Reynolds, A., El Ouraoui, D., Tinelli, C., Barrett, C.: Extending SMT solvers to higher-order logic. In: Fontaine, P. (ed.) CADE 2019. LNCS (LNAI), vol. 11716, pp. 35–54. Springer, Cham (2019). https://doi.org/10.1007/978-3-030-29436-6_3
4. Bentkamp, A., Blanchette, J., Tourret, S., Vukmirović, P., Waldmann, U.: Superposition with lambdas. In: Fontaine, P. (ed.) CADE 2019. LNCS (LNAI), vol. 11716, pp. 55–73. Springer, Cham (2019). https://doi.org/10.1007/978-3-030-29436-6_4
5. Bentkamp, A., Blanchette, J., Tourret, S., Vukmirović, P., Waldmann, U.: Superposition with lambdas (technical report). Technical report (2019). http://matryoshka.gforge.inria.fr/pubs/lamsup_report.pdf
6. Bentkamp, A., Blanchette, J.C., Cruanes, S., Waldmann, U.: Superposition for lambda-free higher-order logic. In: Galmiche, D., Schulz, S., Sebastiani, R. (eds.) IJCAR 2018. LNCS (LNAI), vol. 10900, pp. 28–46. Springer, Cham (2018). https://doi.org/10.1007/978-3-319-94205-6_3
7. Benzmüller, C., Sultana, N., Paulson, L.C., Theib, F.: The higher-order prover Leo-II. J. Autom. Reasoning **55**(4), 389–404 (2015)
8. Bhayat, A., Reger, G.: Restricted combinatory unification. In: Fontaine, P. (ed.) CADE 2019. LNCS (LNAI), vol. 11716, pp. 74–93. Springer, Cham (2019). https://doi.org/10.1007/978-3-030-29436-6_5
9. Bhayat, A., Reger, G.: A combinator-based superposition calculus for higher-order logic (technical report). Technical report, University of Mancester (2020). https://easychair.org/publications/preprint_open/66hZ
10. Bhayat, A., Reger, G.: A Knuth-Bendix-like ordering for orienting combinator equations. In: The 10th International Joint Conference on Automated Reasoning (IJCAR) (2020)
11. Bhayat, A., Reger, G.: A polymorphic vampire (short paper). In: The 10th International Joint Conference on Automated Reasoning (IJCAR) (2020)
12. Bobot, F., Paskevich, A.: Expressing polymorphic types in a many-sorted language. In: Tinelli, C., Sofronie-Stokkermans, V. (eds.) FroCoS 2011. LNCS (LNAI), vol. 6989, pp. 87–102. Springer, Heidelberg (2011). https://doi.org/10.1007/978-3-642-24364-6_7
13. Brown, C.E.: Satallax: an automatic higher-order prover. In: Gramlich, B., Miller, D., Sattler, U. (eds.) IJCAR 2012. LNCS (LNAI), vol. 7364, pp. 111–117. Springer, Heidelberg (2012). https://doi.org/10.1007/978-3-642-31365-3_11
14. Czajka, Ł., Kaliszyk, C.: Hammer for Coq: automation for dependent type theory. J. Autom. Reasoning **61**(1), 423–453 (2018)
15. Hindley, J.R. Seldin, J.P.: Lambda-Calculus and Combinators: An Introduction, 2nd edn. Cambridge University Press, New York (2008)
16. Kerber, M.: How to prove higher order theorems in first order logic, pp. 137–142, January 1991
17. Kovács, L., Voronkov, A.: First-order theorem proving and Vampire. In: Sharygina, N., Veith, H. (eds.) CAV 2013. LNCS, vol. 8044, pp. 1–35. Springer, Heidelberg (2013). https://doi.org/10.1007/978-3-642-39799-8_1
18. Lindblad, F.: A focused sequent calculus for higher-order logic. In: Demri, S., Kapur, D., Weidenbach, C. (eds.) IJCAR 2014. LNCS (LNAI), vol. 8562, pp. 61–75. Springer, Cham (2014). https://doi.org/10.1007/978-3-319-08587-6_5
19. Meng, J., Paulson, L.C.: Translating higher-order clauses to first-order clauses. J. Autom. Reasoning **40**(1), 35–60 (2008)

20. Miller, D.A.: Proofs in higher-order logic. Ph.D. thesis, University of Pennsylvania (1983)
21. Miller, D.A.: A compact representation of proofs. Stud. Logica **46**(4), 347–370 (1987)
22. Nieuwenhuis, R., Rubio, A.: Paramodulation-based theorem proving, chap. 7. In: Robinson, A., Voronkov, A. (eds.) Handbook of Automated Reasoning, vol. I, pp. 371–443. Elsevier Science (2001)
23. Paulsson, L.C., Blanchette, J.C.: Three years of experience with Sledgehammer, a practical link between automatic and interactive theorem provers. In: IWIL-2010, vol. 1 (2010)
24. Reger, G., Suda, M., Voronkov, A.: Unification with abstraction and theory instantiation in saturation-based reasoning. In: Beyer, D., Huisman, M. (eds.) TACAS 2018. LNCS, vol. 10805, pp. 3–22. Springer, Cham (2018). https://doi.org/10.1007/978-3-319-89960-2_1
25. Schulz, S.: E – a Brainiac theorem prover. AI Commun. **15**(2, 3), 111–126 (2002)
26. Steen, A.: Extensional paramodulation for higher-order logic and its effective implementation Leo-III. Ph.D. thesis, Freie Universität Berlin (2018)
27. Steen, A., Benzmüller, C.: The higher-order prover Leo-III. In: Galmiche, D., Schulz, S., Sebastiani, R. (eds.) IJCAR 2018. LNCS (LNAI), vol. 10900, pp. 108–116. Springer, Cham (2018). https://doi.org/10.1007/978-3-319-94205-6_8
28. Stump, A., Sutcliffe, G., Tinelli, C.: StarExec, a cross community logic solving service (2012). https://www.starexec.org
29. Sutcliffe, G.: The TPTP problem library and associated infrastructure, from CNF to TH0, TPTP v6.4.0. J. Autom. Reasoning **59**(4), 483–502 (2017)
30. Vukmirović, P., Bentkamp, A., Nummelin, V.: Efficient full higher-order unification (2019, unpublished). http://matryoshka.gforge.inria.fr/pubs/hounif_paper.pdf
31. Waldmann, U.: Automated reasoning II. Lecture notes, Max-Planck-Institut für Informatik (2016). http://resources.mpi-inf.mpg.de/departments/rg1/teaching/autrea2-ss16/script-current.pdf

Subsumption Demodulation in First-Order Theorem Proving

Bernhard Gleiss[1], Laura Kovács[1,2], and Jakob Rath[1(✉)]

[1] TU Wien, Vienna, Austria
jakob.rath@tuwien.ac.at
[2] Chalmers University of Technology, Gothenburg, Sweden

Abstract. Motivated by applications of first-order theorem proving to software analysis, we introduce a new inference rule, called subsumption demodulation, to improve support for reasoning with conditional equalities in superposition-based theorem proving. We show that subsumption demodulation is a simplification rule that does not require radical changes to the underlying superposition calculus. We implemented subsumption demodulation in the theorem prover VAMPIRE, by extending VAMPIRE with a new clause index and adapting its multi-literal matching component. Our experiments, using the TPTP and SMT-LIB repositories, show that subsumption demodulation in VAMPIRE can solve many new problems that could so far not be solved by state-of-the-art reasoners.

1 Introduction

For the efficiency of organizing proof search during saturation-based first-order theorem proving, simplification rules are of critical importance. Simplification rules are inference rules that do not add new formulas to the search space, but simplify formulas by deleting (redundant) clauses from the search space. As such, simplification rules reduce the size of the search space and are crucial in making automated reasoning efficient.

When reasoning about properties of first-order logic with equality, one of the most common simplification rules is demodulation [10] for rewriting (and hence simplifying) formulas using unit equalities $l \simeq r$, where l, r are terms and $\simeq$ denotes equality. As a special case of superposition, demodulation is implemented in first-order provers such as E [14], SPASS [21] and VAMPIRE [10]. Recent applications of superposition-based reasoning, for example to program analysis and verification [5], demand however new and efficient extensions of demodulation to reason about and simplify upon conditional equalities $C \rightarrow l \simeq r$, where C is a first-order formula. Such conditional equalities may, for example, encode software properties expressed in a guarded command language, with C denoting a guard (such as a loop condition) and $l \simeq r$ encoding equational properties over program variables. We illustrate the need of considering generalized versions of demodulation in the following example.

N. Peltier and V. Sofronie-Stokkermans (Eds.): IJCAR 2020, LNAI 12166, pp. 297–315, 2020.
https://doi.org/10.1007/978-3-030-51074-9_17

Example 1. Consider the following formulas expressed in the first-order theory of integer linear arithmetic:

$$\begin{aligned} &f(i) \simeq g(i) \\ &0 \leq i < n \rightarrow P(f(i)) \end{aligned} \tag{1}$$

Here, i is an implicitly universally quantified logical variable of integer sort, and n is an integer-valued constant. First-order reasoners will first clausify formulas (1), deriving:

$$\begin{aligned} &f(i) \simeq g(i) \\ &0 \not\leq i \vee i \not< n \vee P(f(i)) \end{aligned} \tag{2}$$

By applying demodulation over (2), the formula $0 \not\leq i \vee i \not< n \vee P(f(i))$ is rewritten[1] using the unit equality $f(i) \simeq g(i)$, yielding the clause $0 \not\leq i \vee i \not< n \vee P(g(i))$. That is, $0 \leq i < n \rightarrow P(g(i))$ is derived from (1) by one application of demodulation.

Let us now consider a slightly modified version of (1), as below:

$$\begin{aligned} &0 \leq i < n \rightarrow f(i) \simeq g(i) \\ &0 \leq i < n \rightarrow P(f(i)) \end{aligned} \tag{3}$$

whose clausal representation is given by:

$$\begin{aligned} &0 \not\leq i \vee i \not< n \vee f(i) \simeq g(i) \\ &0 \not\leq i \vee i \not< n \vee P(f(i)) \end{aligned} \tag{4}$$

It is again obvious that from (3) one can derive the formula $0 \leq i < n \rightarrow P(g(i))$, or equivalently the clause:

$$0 \not\leq i \vee i \not< n \vee P(g(i)) \tag{5}$$

Yet, *one cannot anymore apply demodulation-based simplification over* (4) *to derive such a clause*, as (4) contains no unit equality. □

In this paper we propose a generalized version of demodulation, called *subsumption demodulation*, allowing to rewrite terms and simplify formulas using rewriting based on conditional equalities, such as in (3). To do so, we extend demodulation with subsumption, that is with deciding whether (an instance of a) clause C is a submultiset of a clause D. In particular, the non-equality literals of the conditional equality (i.e., the condition) need to subsume the unchanged literals of the simplified clause. This way, subsumption demodulation can be applied to non-unit clauses and is not restricted to have at least one premise clause that is a unit equality. We show that subsumption demodulation is a simplification rule of the superposition framework (Sect. 4), allowing for example to derive the clause (5) from (4) in one inference step. By properly adjusting clause indexing and multi-literal matching in first-order theorem provers, we provide an efficient implementation of subsumption demodulation in Vampire (Sect. 5) and

[1] Assuming that g is simpler/smaller than f.

evaluate our work against state-of-the-art reasoners, including E [14], SPASS [21], CVC4 [3] and Z3 [7] (Sect. 6).

Related Work. While several approaches generalize demodulation in superposition-based theorem proving, we argue that subsumption demodulation improves existing methods either in terms of applicability and/or efficiency. The AVATAR architecture of first-order provers [19] splits general clauses into components with disjoint sets of variables, potentially enabling demodulation inferences whenever some of these components become unit equalities. Example 1 demonstrates that subsumption demodulation applies in situations where AVATAR does not: in each clause of (4), all literals share the variable i and hence none of the clauses from (4) can be split using AVATAR. That is, AVATAR would not generate unit equalities from (4), and therefore cannot apply demodulation over (4) to derive (5).

The local rewriting approach of [20] requires rewriting equality literals to be maximal[2] in clauses. However, following [10], for efficiency reasons we consider equality literals to be "smaller" than non-equality literals. In particular, the equality literals of clauses (4) are "smaller" than the non-equality literals, preventing thus the application of local rewriting in Example 1.

To the extent of our knowledge, the ordering restrictions on non-unit rewriting [20] do not ensure redundancy, and thus the rule is not a simplification inference rule. Subsumption demodulation includes all necessary conditions and we prove it to be a simplification rule. Furthermore, we show how the ordering restrictions can be simplified which enables an efficient implementation, and then explain how such an implementation can be realized.

We further note that the contextual rewriting rule of [1] is more general than our rule of subsumption demodulation, and has been first implemented in the SATURATE system [12]. Yet, efficiently automating contextual rewriting is extremely challenging, while subsumption demodulation requires no radical changes in the existing machinery of superposition provers (see Sect. 5).

To the best of our knowledge, except SPASS [21] and SATURATE, no other state-of-the-art superposition provers implement variants of conditional rewriting. Subterm contextual rewriting [22] is a refined notion of contextual rewriting and is implemented in SPASS. A major difference of subterm contextual rewriting when compared to subsumption demodulation is that in subsumption demodulation the discovery of the substitution is driven by the side conditions whereas in subterm contextual rewriting the side conditions are evaluated by checking the validity of certain implications by means of a reduction calculus. This reduction calculus recursively applies another restriction of contextual rewriting called recursive contextual ground rewriting, among other standard reduction rules. While subterm contextual rewriting is more general, we believe that the benefit of subsumption demodulation comes with its relatively easy and efficient integration within existing superposition reasoners, as evidenced also in Sect. 6.

[2] w.r.t. clause ordering.

Local contextual rewriting [9] is another refinement of contextual rewriting implemented in Spass. In our experiments it performed similarly to subterm contextual rewriting.

Finally, we note that SMT-based reasoners also implement various methods to efficiently handle conditional equalities [6,13]. Yet, the setting is very different as they rely on the DPLL(T) framework [8] rather than implementing superposition.

Contributions. Summarizing, this paper brings the following contributions.

- To improve reasoning in the presence of conditional equalities, we introduce the new inference rule *subsumption demodulation*, which generalizes demodulation to non-unit equalities by combining demodulation and subsumption (Sect. 4).
- Subsumption demodulation does not require radical changes to the underlying superposition calculus. We implemented subsumption demodulation in the first-order theorem prover Vampire, by extending Vampire with a new clause index and adapting its multi-literal matching component (Sect. 5).
- We compared our work against state-of-the-art reasoners, using the TPTP and SMT-LIB benchmark repositories. Our experiments show that subsumption demodulation in Vampire can solve 11 first-order problems that could so far not be solved by any other state-of-the-art provers, including Vampire, E, Spass, CVC4 and Z3 (Sect. 6).

2 Preliminaries

For simplicity, in what follows we consider standard first-order logic with equality, where equality is denoted by $\simeq$. We support all standard boolean connectives and quantifiers in the language. Throughout the paper, we denote terms by l, r, s, t, variables by x, y, constants by c, d, function symbols by f, g and predicate symbols by P, Q, R, all possibly with indices. Further, we denote literals by L and clauses by C, D, again possibly with indices. We write $s \not\simeq t$ to denote the formula $\neg s \simeq t$. A literal $s \simeq t$ is called an *equality literal.* We consider clauses as multisets of literals and denote by $\subseteq_M$ the subset relation among multisets. A clause that only consists of one equality literal is called a *unit equality.*

An expression E is a term, literal, or clause. We write $E[s]$ to mean an expression E with a particular occurrence of a term s. A *substitution*, denoted by σ, is any finite mapping of the form $\{x_1 \mapsto t_1, \ldots, x_n \mapsto t_n\}$, where $n > 0$. Applying a substitution σ to an expression E yields another expression, denoted by $E\sigma$, by simultaneously replacing each x_i by t_i in E. We say that $E\sigma$ is an instance of E. A *unifier* of two expressions E_1 and E_2 is a substitution σ such that $E_1\sigma = E_2\sigma$. If two expressions have a unifier, they also have a *most general unifier (mgu)*. A *match* of expression E_1 to expression E_2 is a substitution σ such that $E_1\sigma = E_2$. Note that any match is a unifier (assuming the sets of variables in E_1 and E_2 are disjoint), but not vice-versa, as illustrated below.

Example 2. Let E_1 and E_2 be the clauses $Q(x, y) \vee R(x, y)$ and $Q(c, d) \vee R(c, z)$, respectively. The only possible match of $Q(x, y)$ to $Q(c, d)$ is $\sigma_1 = \{x \mapsto c, y \mapsto d\}$. On the other hand, the only possible match of $R(x, y)$ to $R(c, z)$ is $\sigma_2 = \{x \mapsto c, y \mapsto z\}$. As σ_1 and σ_2 are not the same, there is no match of E_1 to E_2. Note however that E_1 and E_2 can be unified; for example, using $\sigma_3 = \{x \mapsto c, y \mapsto d, z \mapsto d\}$.

Superposition Inference System. We assume basic knowledge in first-order theorem proving and superposition reasoning [2,11]. We adopt the notations and the inference system of superposition from [10]. We recall that first-order provers perform inferences on clauses using inference rules, where an *inference* is usually written as:

$$\frac{C_1 \quad \ldots \quad C_n}{C}$$

with $n \geq 0$. The clauses $C_1, \ldots, C_n$ are called the premises and C is the conclusion of the inference above. An inference is *sound* if its conclusion is a logical consequence of its premises. An inference rule is a set of (concrete) inferences and an inference system is a set of inference rules. An inference system is *sound* if all its inference rules are sound.

Modern first-order theorem provers implement the *superposition inference system* for first-order logic with equality. This inference system is parametrized by a *simplification ordering* over terms and a *literal selection function* over clauses. In what follows, we denote by $\succ$ a simplification ordering over terms, that is $\succ$ is a well-founded partial ordering satisfying the following three conditions:

- *stability under substitutions*: if $s \succ t$, then $s\theta \succ t\theta$;
- *monotonicity*: if $s \succ t$, then $l[s] \succ l[t]$;
- *subterm property*: $s \succ t$ whenever t is a proper subterm of s.

The simplification ordering $\succ$ on terms can be extended to a simplification ordering on literals and clauses, using a multiset extension of orderings. For simplicity, the extension of $\succ$ to literals and clauses will also be denoted by $\succ$. Whenever $E_1 \succ E_2$, we say that E_1 is bigger than E_2 and E_2 is smaller than E_1 w.r.t. $\succ$. We say that an equality literal $s \simeq t$ is *oriented*, if $s \succ t$ or $t \succ s$. The literal extension of $\succ$ asserts that negative literals are always bigger than their positive counterparts. Moreover, if $L_1 \succ L_2$, where L_1 and L_2 are positive, then $\neg L_1 \succ L_1 \succ \neg L_2 \succ L_2$. Finally, equality literals are set to be smaller than any literal using a predicate different than $\simeq$.

A *selection function* selects at least one literal in every non-empty clause. In what follows, selected literals in clauses will be underlined: when writing $\underline{L} \vee C$, we mean that (at least) L is selected in $L \vee C$. In what follows, we assume that selection functions are *well-behaved* w.r.t. $\succ$: either a negative literal is selected or all maximal literals w.r.t. $\succ$ are selected.

In the sequel, we fix a simplification ordering $\succ$ and a well-behaved selection function and consider the superposition inference system, denoted by Sup, parametrized by these two ingredients. The inference system SUP for first-order logic with equality consists of the inference rules of Fig. 1, and it is both sound and refutationally complete. That is, if a set S of clauses is unsatisfiable, then the empty clause (that is, the always false formula) is derivable from S in SUP.

– Resolution and Factoring

$$\frac{\underline{L} \vee C_1 \qquad \underline{\neg L'} \vee C_2}{(C_1 \vee C_2)\sigma} \qquad\qquad \frac{\underline{L} \vee \underline{L'} \vee C}{(L \vee C)\sigma}$$

where L is not an equality literal and $\sigma = mgu(L, L')$

– Superposition

$$\frac{\underline{s \simeq t} \vee C_1 \qquad \underline{L[s']} \vee C_2}{(C_1 \vee L[t] \vee C_2)\theta}$$

$$\frac{\underline{s \simeq t} \vee C_1 \qquad \underline{l[s'] \simeq l'} \vee C_2}{(C_1 \vee l[t] \simeq l' \vee C_2)\theta} \qquad \frac{\underline{s \simeq t} \vee C_1 \qquad \underline{l[s'] \not\simeq l'} \vee C_2}{(C_1 \vee l[t] \not\simeq l' \vee C_2)\theta}$$

where s' not a variable, L is not an equality, $\theta = mgu(s, s')$, $t\theta \not\succeq s\theta$ and $l'\theta \not\succeq l[s']\theta$

– Equality Resolution and Equality Factoring

$$\frac{\underline{s \not\simeq s'} \vee C}{C\theta} \qquad\qquad \frac{s \simeq t \vee \underline{s' \simeq t'} \vee C}{(s \simeq t \vee t \not\simeq t' \vee C)\theta}$$

where $\theta = mgu(s, s')$, $t\theta \not\succeq s\theta$ and $t'\theta \not\succeq t\theta$

Fig. 1. The superposition calculus SUP.

3 Superposition-Based Proof Search

We now overview the main ingredients in organizing proof search within first-order provers, using the superposition calculus. For details, we refer to [2,10,11].

Superposition-based provers use *saturation algorithms*: applying all possible inferences of SUP in a certain order to the clauses in the search space until (i) no more inferences can be applied or (ii) the empty clause has been derived. A simple implementation of a saturation algorithm would however be very inefficient as applications of all possible inferences will quickly blow up the search space.

Saturation algorithms can however be made efficient by exploiting a powerful concept of *redundancy*: deleting so-called redundant clauses from the search space by preserving completeness of SUP. A clause C in a set S of clauses (i.e., in the search space) is *redundant* in S, if there exist clauses $C_1, \ldots, C_n$ in S, such that $C \succ C_i$ and $C_1, \ldots, C_n \models C$. That is, a clause C is redundant in S if it is a logical consequence of clauses that are smaller than C w.r.t. $\succ$. It is known that redundant clauses can be removed from the search space without affecting completeness of superposition-based proof search. For this reason, saturation-based theorem provers, such as E, SPASS and VAMPIRE, not only generate new clauses but also delete redundant clauses during proof search by using both *generating* and *simplifying* inferences.

Simplification Rules. A *simplifying inference* is an inference in which one premise C_i becomes redundant after the addition of the conclusion C to the search space, and hence C_i can be deleted. In what follows, we will denote deleted clauses by drawing a line through them and refer to simplifying inferences as *simplification rules.* The premise C_i that becomes redundant is called the *main premise*, whereas other premises are called *side premises* of the simplification rule. Intuitively, a simplification rule simplifies its main premise to its conclusion by using additional knowledge from its side premises. Inferences that are not simplifying are called *generating*, as they generate and add a new clause C to the search space.

In saturation-based proof search, we distinguish between *forward* and *backward* simplifications. During forward simplification, a newly derived clause is simplified using previously derived clauses as side clauses. Conversely, during backward simplification a newly derived clause is used as a side clause to simplify previously derived clauses.

Demodulation. One example of a simplification rule is *demodulation*, or also called *rewriting by unit equalities.* Demodulation is the following inference rule:

$$\frac{l \simeq r \qquad \cancel{L[t] \vee C}}{L[r\sigma] \vee C}$$

where $l\sigma = t$, $l\sigma \succ r\sigma$ and $L[t] \vee C \succ (l \simeq r)\sigma$, for some substitution σ.

It is easy to see that demodulation is a simplification rule. Moreover, demodulation is a special case of a superposition inference where one premise of the inference is deleted. However, unlike a superposition inference, demodulation is not restricted to selected literals.

Example 3. Consider the clauses $C_1 = f(f(x)) \simeq f(x)$ and $C_2 = P(f(f(c))) \vee Q(d)$. Let σ be the substitution $\sigma = \{x \mapsto c\}$. By the subterm property of $\succ$, we have $f(f(c)) \succ f(c)$. Further, as equality literals are smaller than non-equality literals, we have $P(f(f(c))) \vee Q(d) \succ f(f(c)) \simeq f(c)$. We thus apply demodulation and C_2 is simplified into the clause $C_3 = P(f(c)) \vee Q(d)$:

$$\frac{f(f(x)) \simeq f(x) \qquad \cancel{P(f(f(c))) \vee Q(d)}}{P(f(c)) \vee Q(d)}$$

□

Deletion Rules. Even when simplification rules are in use, deleting more/other redundant clauses is still useful to keep the search space small. For this reason, in addition to simplifying and generating rules, theorem provers also use *deletion rules*: a *deletion rule* checks whether clauses in the search space are redundant due to the presence of other clauses in the search space, and removes redundant clauses from the search space.

Given clauses C and D, we say C subsumes D if there is some substitution σ such that $C\sigma$ is a submultiset of D, that is $C\sigma \subseteq_M D$. *Subsumption* is the deletion rule that removes subsumed clauses from the search space.

Example 4. Let $C = P(x) \vee Q(f(x))$ and $D = P(f(c)) \vee P(g(c)) \vee Q(f(c)) \vee Q(f(g(c))) \vee R(y)$ be clauses in the search space. Using $\sigma = \{x \mapsto g(c)\}$, it is easy to see that C subsumes D, and hence D is deleted from the search space. □

4 Subsumption Demodulation

In this section we introduce a new simplification rule, called subsumption demodulation, by extending demodulation to a simplification rule over conditional equalities. We do so by combining demodulation with subsumption checks to find simplifying applications of rewriting by non-unit (and hence conditional) equalities.

4.1 Subsumption Demodulation for Conditional Rewriting

Our rule of subsumption demodulation is defined below.

Definition 1 (Subsumption Demodulation). Subsumption demodulation *is the inference rule:*

$$\frac{l \simeq r \vee C \qquad L[t] \vee D}{L[r\sigma] \vee D} \tag{6}$$

where:

1. $l\sigma = t$,
2. $C\sigma \subseteq_M D$,
3. $l\sigma \succ r\sigma$, *and*
4. $L[t] \vee D \succ (l \simeq r)\sigma \vee C\sigma$.

We call the equality $l \simeq r$ *in the left premise of* (6) *the* rewriting equality *of subsumption demodulation.*

Intuitively, the side conditions 1 and 2 of Definition 1 ensure the soundness of the rule: it is easy to see that if $l \simeq r \vee C$ and $L[t] \vee D$ are true, then $L[r\sigma] \vee D$ also holds. We thus conclude:

Theorem 1 (Soundness). *Subsumption demodulation is sound.*

On the other hand, side conditions 3 and 4 of Definition 1 are vital to ensure that subsumption demodulation is a simplification rule (details follow in Sect. 4.2).

Detecting possible applications of subsumption demodulation involves (i) selecting one equality of the side clause as rewriting equality and (ii) matching each of the remaining literals, denoted C in (6), to some literal in the main clause. Step (i) is similar to finding unit equalities in demodulation, whereas step (ii) reduces to showing that C subsumes parts of the main premise. Informally speaking, subsumption demodulation combines demodulation and subsumption, as discussed in Sect. 5. Note that in step (ii), matching allows any instantiation of C to $C\sigma$ via substitution σ; yet, we we do *not* unify the side and main premises of subsumption demodulation, as illustrated later in Example 7. Furthermore, we need to find a term t in the unmatched part $D \setminus C\sigma$ of the main premise, such that t can be rewritten according to the rewriting equality into $r\sigma$.

As the ordering $\succ$ is partial, the conditions of Definition 1 must be checked a posteriori, that is after subsumption demodulation has been applied with a fixed substitution. Note however that if $l \succ r$ in the rewriting equality, then $l\sigma \succ r\sigma$ for any substitution, so checking the ordering a priori helps, as illustrated in the following example.

Example 5. Let us consider the following two clauses:

$$C_1 = f(g(x)) \simeq g(x) \vee Q(x) \vee R(y)$$
$$C_2 = P(f(g(c))) \vee Q(c) \vee Q(d) \vee R(f(g(d)))$$

By the subterm property of $\succ$, we conclude that $f(g(x)) \succ g(x)$. Hence, the rewriting equality, as well as any instance of it, is oriented.

Let σ be the substitution $\sigma = \{x \mapsto c, y \mapsto f(g(d))\}$. Due to the previous paragraph, we know $f(g(c)) \succ g(c)$ As equality literals are smaller than non-equality ones, we also conclude $P(f(g(c))) \succ f(g(c)) \simeq g(c)$. Thus, we have $P(f(g(c))) \vee Q(c) \vee Q(d) \vee R(f(g(d))) \succ f(g(c)) \simeq g(c) \vee Q(c) \vee R(f(g(d)))$ and we can apply subsumption demodulation to C_1 and C_2, deriving clause $C_3 = P(g(c)) \vee Q(c) \vee Q(d) \vee R(f(g(d)))$.

We note that demodulation cannot derive C_3 from C_1 and C_2, as there is no unit equality. □

Example 5 highlights limitations of demodulation when compared to subsumption demodulation. We next illustrate different possible applications of subsumption demodulation using a fixed side premise and different main premises.

Example 6. Consider the clause $C_1 = f(g(x)) \simeq g(y) \vee Q(x) \vee R(y)$. Only the first literal $f(g(x)) \simeq g(y)$ is a positive equality and as such eligible as rewriting equality. Note that $f(g(x))$ and $g(y)$ are incomparable w.r.t. $\succ$ due to occurrences of different variables, and hence whether $f(g(x))\sigma \succ g(y)\sigma$ depends on the chosen substitution σ.

(1) Consider the clause $C_2 = P(f(g(c))) \vee Q(c) \vee R(c)$ as the main premise. With the substitution $\sigma_1 = \{x \mapsto c, y \mapsto c\}$, we have $f(g(x))\sigma_1 \succ g(x)\sigma_1$ as $f(g(c)) \succ g(c)$ due to the subterm property of $\succ$, enabling a possible application of subsumption demodulation over C_1 and C_2.

(2) Consider now $C_3 = P(g(f(g(c)))) \vee Q(c) \vee R(f(g(c)))$ as the main premise and the substitution $\sigma_2 = \{x \mapsto c, y \mapsto f(g(c))\}$. We have $g(y)\sigma_2 \succ f(g(x))\sigma_2$, as $g(f(g(c)) \succ f(g(c))$. The instance of the rewriting equality is oriented differently in this case than in the previous one, enabling a possible application of subsumption demodulation over C_1 and C_3.
(3) On the other hand, using the clause $C_4 = P(f(g(c))) \vee Q(c) \vee R(z)$ as the main premise, the only substitution we can use is $\sigma_3 = \{x \mapsto c, y \mapsto z\}$. The corresponding instance of the rewriting equality is then $f(g(c)) \simeq g(z)$, which cannot be oriented in general. Hence, subsumption demodulation cannot be applied in this case, even though we can find the matching term $f(g(c))$ in C_4. □

As mentioned before, the substitution σ appearing in subsumption demodulation can only be used to instantiate the side premise, but not for unifying side and main premises, as we would not obtain a simplification rule.

Example 7. Consider the clauses:

$$C_1 = f(c) \simeq c \vee Q(d)$$
$$C_2 = P(f(c)) \vee Q(x)$$

As we cannot match $Q(d)$ to $Q(x)$ (although we could match $Q(x)$ to $Q(d)$), subsumption demodulation is not applicable with premises C_1 and C_2. □

4.2 Simplification Using Subsumption Demodulation

Note that in the special case where C is the empty clause in (6), subsumption demodulation reduces to demodulation and hence it is a simplification rule. We next show that this is the case in general:

Theorem 2 (Simplification Rule). *Subsumption demodulation is a simplification rule and we have:*

$$\frac{l \simeq r \vee C \qquad \cancel{L[t] \vee D}}{L[r\sigma] \vee D}$$

where:

1. $l\sigma = t$,
2. $C\sigma \subseteq_M D$,
3. $l\sigma \succ r\sigma$, and
4. $L[t] \vee D \succ (l \simeq r)\sigma \vee C\sigma$.

Proof. Because of the second condition of the definition of subsumption demodulation, $L[t] \vee D$ is clearly a logical consequence of $L[r\sigma] \vee D$ and $l \simeq r \vee C$. Moreover, from the fourth condition, we trivially have $L[t] \vee D \succ (l \simeq r)\sigma \vee C\sigma$. It thus remains to show that $L[r\sigma] \vee D$ is smaller than $L[t] \vee D$ w.r.t. $\succ$. As

$t = l\sigma \succ r\sigma$, the monotonicity property of $\succ$ asserts that $L[t] \succ L[r\sigma]$, and hence $L[t] \vee D \succ L[r\sigma] \vee D$. This concludes that $L[t] \vee D$ is redundant w.r.t. the conclusion and left-most premise of subsumption demodulation. □

Example 8. By revisiting Example 5, Theorem 2 asserts that clause C_2 is simplified into C_3, and subsumption demodulation deletes C_2 from the search space. □

4.3 Refining Redundancy

The fourth condition defining subsumption demodulation in Definition 1 is required to ensure that the main premise of subsumption demodulation becomes redundant. However, comparing clauses w.r.t. the ordering $\succ$ is computationally expensive; yet, not necessary for subsumption demodulation. Following the notation of Definition 1, let D' such that $D = C\sigma \vee D'$. By properties of multiset orderings, the condition $L[t] \vee D \succ (l \simeq r)\sigma \vee C\sigma$ is equivalent to $L[t] \vee D' \succ (l \simeq r)\sigma$, as the literals in $C\sigma$ occur on both sides of $\succ$. This means, to ensure the redundancy of the main premise of subsumption demodulation, we only need to ensure that there is a literal from $L[t] \vee D$ such that this literal is bigger that the rewriting equality.

Theorem 3 (Refining Redundancy). *The following conditions are equivalent:*

(R1) $L[t] \vee D \succ (l \simeq r)\sigma \vee C\sigma$
(R2) $L[t] \vee D' \succ (l \simeq r)\sigma$

As mentioned in Sect. 4.1, application of subsumption demodulation involves checking that an ordering condition between premises holds (side condition 4 in Definition 1). Theorem 3 asserts that we only need to find a literal in $L[t] \vee D'$ that is bigger than the rewriting equality in order to ensure that the ordering condition is fulfilled. In the next section we show that by re-using and properly changing the underlying machinery of first-order provers for demodulation and subsumption, subsumption demodulation can efficiently be implemented in superposition-based proof search.

5 Subsumption Demodulation in Vampire

We implemented subsumption demodulation in the first-order theorem prover Vampire. Our implementation consists of about 5000 lines of C++ code and is available at:

https://github.com/vprover/vampire/tree/subsumption-demodulation

As for any simplification rule, we implemented the forward and backward versions of subsumption demodulation separately. Our new VAMPIRE options controlling subsumption demodulation are `fsd` and `bsd`, both with possible values `on` and `off`, to respectively enable forward and backward subsumption demodulation.

As discussed in Sect. 4, subsumption demodulation uses reasoning based on a combination of demodulation and subsumption. Algorithm 1 details our implementation for *forward subsumption demodulation.* In a nutshell, given a clause D as main premise, (forward) subsumption demodulation in VAMPIRE consists of the following main steps:

1. *Retrieve candidate clauses* C as side premises of subsumption demodulation (line 1 of Algorithm 1). To this end, we design a new clause index with imperfect filtering, by modifying the subsumption index in VAMPIRE, as discussed later in this section.
2. *Prune candidate clauses* by checking the conditions of subsumption demodulation (lines 3–7 of Algorithm 1), in particular selecting a rewriting equality and matching the remaining literals of the side premise to literals of the main premise. After this, prune further by performing a posteriori checks for orienting the rewriting equality E, and checking the redundancy of the given main premise D. To do so, we revised multi-literal matching and redundancy checking in VAMPIRE (see later).
3. *Build simplified clause* by simplifying and deleting the (main) premise D of subsumption demodulation using (forward) simplification (line 8 of Algorithm 1).

Our implementation of *backward subsumption demodulation* requires only a few changes to Algorithm 1: (i) we use the input clause as side premise C of backward subsumption demodulation and (ii) we retrieve candidate clauses D as potential main premises of subsumption demodulation. Additionally, (iii) instead of returning a single simplified clause D', we record a replacement clause for each candidate clause D where a simplification was possible.

Clause Indexing for Subsumption Demodulation. We build upon the indexing approach [15] used for subsumption in VAMPIRE: the subsumption index in VAMPIRE stores and retrieves candidate clauses for subsumption. Each clause is indexed by exactly one of its literals. In principle, any literal of the clause can be chosen. In order to reduce the number of retrieved candidates, the best literal is chosen in the sense that the chosen literal maximizes a certain heuristic (e.g., maximal weight). Since the subsumption index is not a perfect index (i.e., it may retrieve non-subsumed clauses), additional checks on the retrieved clauses are performed.

Using the subsumption index of VAMPIRE as the clause index for forward subsumption demodulation would however omit retrieving clauses (side premises) in which the rewriting equality is chosen as key for the index, omitting this way a possible application of subsumption demodulation. Hence, we need a new clause index in which the best literal can be adjusted to be the rewriting equality. To

Algorithm 1. Forward Subsumption Demodulation – FSD

```
Input  : Clause D, to be used as main premise
Output: Simplified clause D' if (forward) subsumption demodulation is possible
// Retrieve candidate side premises
candidates := FSDIndex.Retrieve(D)
for each C ∈ candidates do
    while m = FindNextMLMatch(C, D) do
        σ' := m.GetSubstitution()
        E := m.GetRewritingEquality()
        // E is of the form l≃r, for some terms l, r
        if exists term t in D \ Cσ' and substitution σ ⊇ σ' s.t. t = lσ then
            if CheckOrderingConditions(D, E, t, σ) then
                D' := BuildSimplifiedClause(D, E, t, σ)
                return D'
            end
        end
    end
end
```

address this issue, we added a new clause index, called the *forward subsumption demodulation index (FSD index)*, to VAMPIRE, as follows: we index potential side premises either by their best literal (according to the heuristic), the second best literal, or both. If the best literal in a clause C is a positive equality (i.e., a candidate rewriting equality) but the second best is not, C is indexed by the second best literal, and vice versa. If both the best and second best literal are positive equalities, C is indexed by both of them. Furthermore, because the FSD index is exclusively used by forward subsumption demodulation, this index only needs to keep track of clauses that contain at least one positive equality.

In the backward case, we can in fact reuse VAMPIRE's index for backward subsumption. Instead we need to query the index by the best literal, the second best literal, or both (as described in the previous paragraph).

Multi-literal Matching. Similarly to the subsumption index, our new subsumption demodulation index is not a perfect index, that is it performs imperfect filtering for retrieving clauses. Therefore, additional post-checks are required on the retrieved clauses. In our work, we devised a multi-literal matching approach to:

- choose the rewriting equality among the literals of the side premise C, and
- check whether the remaining literals of C can be uniformly instantiated to the literals of the main premise D of subsumption demodulation.

There are multiple ways to organize this process. A simple approach is to (i) first pick any equality of a side premise C as the rewriting equality of subsumption demodulation, and then (ii) invoke the existing multi-literal matching machinery of VAMPIRE to match the remaining literals of C with a subset of literals of D. For the latter step (ii), the task is to find a substitution σ such that $C\sigma$ becomes a submultiset of the given clause D. If the choice of the rewriting equality in step (i) turns out to be wrong, we backtrack. In our work, we revised the existing multi-literal matching machinery of VAMPIRE to a new multi-literal matching approach for subsumption demodulation, by using the steps (i)-(ii) and interleaving equality selection with matching.

We note that the substitution σ in step (ii) above is built in two stages: first we get a partial substitution σ' from multi-literal matching and then (possibly) extend σ' to σ by matching term instances of the rewriting equality with terms of $D \setminus C\sigma$.

Example 9. Let D be the clause $P(f(c,d)) \vee Q(c)$. Assume that our (FSD) clause index retrieves the clause $C = f(x,y) \simeq y \vee Q(x)$ from the search space (line 1 of Algorithm 1). We then invoke our multi-literal matcher (line 3 of Algorithm 1), which matches the literal $Q(x)$ of C to the literal $Q(c)$ of D and selects the equality literal $f(x,y) \simeq y$ of C as the rewriting equality for subsumption demodulation over C and D. The matcher returns the choice of rewriting equality and the partial substitution $\sigma' = \{x \mapsto c\}$. We arrive at the final substitution $\sigma = \{x \mapsto c, y \mapsto d\}$ only when we match the instance $f(x,y)\sigma'$, that is $f(c,y)$, of the left-hand side of the rewriting equality to the literal $f(c,d)$ of D. Using σ, subsumption demodulation over C and D will derive $P(d) \vee Q(c)$, after ensuring that D becomes redundant (line 8 of Algorithm 1). □

We further note that multi-literal matching is an NP-complete problem. Our multi-literal matching problems may have more than one solution, with possibly only some (or none) of them leading to successful applications of subsumption demodulation. In our implementation, we examine all solutions retrieved by multi-literal matching. We also experimented with limiting the number of matches examined after multi-literal matching but did not observe relevant improvements. Yet, our implementation in VAMPIRE also supports an additional option allowing the user to specify an upper bound on how many solutions of multi-literal matching should be examined.

Redundancy Checking. To ensure redundancy of the main premise D after the subsumption demodulation inference, we need to check two properties. First, the instance $E\sigma$ of the rewriting equality E must be oriented. This is a simple ordering check. Second, the main premise D must be larger than the side premise C. Thanks to Theorem 3, this latter condition is reduced to finding a literal among the unmatched part of the main premise D that is bigger than the instance $E\sigma$ of the rewriting equality E.

Example 10. In case of Example 9, the rewriting equality E is oriented and hence $E\sigma$ is also oriented. Next, the literal $P(f(c,d))$ is bigger than $E\sigma$, and hence D is redundant w.r.t. C and D'. □

6 Experiments

We evaluated our implementation of subsumption demodulation in VAMPIRE on the problems of the TPTP [17] (version 7.3.0) and SMT-LIB [4] (release 2019-05-06) repositories. All our experiments were carried out on the StarExec cluster [16].

Benchmark Setup. From the 22,686 problems in the TPTP benchmark set, VAMPIRE can parse 18,232 problems.[3] Out of these problems, we only used those problems that involve equalities as subsumption demodulation is only applicable in the presence of (at least one) equality. As such, we used 13,924 TPTP problems in our experiments.

On the other hand, when using the SMT-LIB repository, we chose the benchmarks from categories LIA, UF, UFDT, UFDTLIA, and UFLIA, as these benchmarks involve reasoning with both theories and quantifiers and the background theories are the theories that VAMPIRE supports. These are 22,951 SMT-LIB problems in total, of which 22,833 problems remain after removing those where equality does not occur.

Comparative Experiments with Vampire. As a first experimental study, we compared the performance of subsumption demodulation in VAMPIRE for different values of `fsd` and `bsd`, that is by using forward (FSD) and/or backward (BSD) subsumption demodulation. To this end, we evaluated subsumption demodulation using the CASC and SMTCOMP schedules of VAMPIRE's portfolio mode. In order to test subsumption demodulation with the portfolio mode, we added the options `fsd` and/or `bsd` to *all* strategies of VAMPIRE. While the resulting strategy schedules could potentially be further improved, it allowed us to test FSD/BSD with a variety of strategies.

Table 1. Comparing VAMPIRE with and without subsumption demodulation on TPTP, using VAMPIRE in portfolio mode.

Configuration	Total	Solved	New (SAT + UNSAT)
VAMPIRE	13,924	9,923	–
VAMPIRE, with FSD	13,924	9,757	20 (3 + 17)
VAMPIRE, with BSD	13,924	9,797	14 (2 + 12)
VAMPIRE, with FSD and BSD	13,924	9,734	30 (6 + 24)

Our results are summarized in Tables 1 and 2. The first column of these tables lists the VAMPIRE version and configuration, where VAMPIRE refers to VAMPIRE in its portfolio mode (version 4.4). Lines 2–4 of these tables use our new VAMPIRE, that is our implementation of subsumption demodulation in VAMPIRE. The

[3] The other problems contain features, such as higher-order logic, that have not been implemented in VAMPIRE yet.

Table 2. Comparing VAMPIRE with and without subsumption demodulation on SMT-LIB, using VAMPIRE in portfolio mode.

Configuration	Total	Solved	New (SAT + UNSAT)
VAMPIRE	22,833	13,705	–
VAMPIRE, with FSD	22,833	13,620	55 (1 + 54)
VAMPIRE, with BSD	22,833	13,632	48 (0 + 48)
VAMPIRE, with FSD and BSD	22,833	13,607	76 (0 + 76)

column "Solved" reports, respectively, the total number of TPTP and SMT-LIB problems solved by the considered VAMPIRE configurations. Column "New" lists, respectively, the number of TPTP and SMT-LIB problems solved by the version with subsumption demodulation but not by the portfolio version of VAMPIRE. This column also indicates in parentheses how many of the solved problems were satisfiable/unsatisfiable.

While in total the portfolio mode of VAMPIRE can solve more problems, we note that this comes at no surprise as the portfolio mode of VAMPIRE is highly tuned using the existing VAMPIRE options. In our experiments, we were interested to see whether subsumption demodulation in VAMPIRE can solve problems that cannot be solved by the portfolio mode of VAMPIRE. Such a result would justify the existence of the new rule because the set of problems that VAMPIRE can solve in principle is increased. In future work, the portfolio mode should be tuned by also taking into account subsumption demodulation, which then ideally leads to an overall increase in performance. The columns "New" of Tables 1 and 2 give indeed practical evidence of the impact of subsumption demodulation: there are 30 new TPTP problems and 76 SMT-LIB problems[4] that the portfolio version of VAMPIRE cannot solve, but forward and backward subsumption demodulation in VAMPIRE can.

New Problems Solved Only by Subsumption Demodulation. Building upon our results from Tables 1 and 2, we analysed how many new problems subsumption demodulation in VAMPIRE can solve when compared to other state-of-the-art reasoners. To this end, we evaluated our work against the superposition provers E (version 2.4) and SPASS (version 3.9), as well as the SMT solvers CVC4 (version 1.7) and Z3 (version 4.8.7). We note however, that when using our 30 new problems from Table 1, we could not compare our results against Z3 as Z3 does not natively parse TPTP. On the other hand, when using our 76 new problems from Table 2, we only compared against CVC4 and Z3, as E and SPASS do not support the SMT-LIB syntax.

Table 3 summarizes our findings. First, 11 of our 30 "new" TPTP problems can only be solved using forward and backward subsumption demodulation in VAMPIRE; none of the other systems were able solve these problems.

[4] The list of these new problems is available at https://gist.github.com/JakobR/605a7b7db010 1259052e137ade54b32c.

Table 3. Comparing VAMPIRE with subsumption demodulation against other solvers, using the "new" TPTP and SMT-LIB problems of Tables 1 and 2 and running VAMPIRE in portfolio mode.

Solver/configuration	TPTP problems	SMT-LIB problems
Baseline: VAMPIRE, with FSD and BSD	30	76
E with `--auto-schedule`	14	–
SPASS (default)	4	–
SPASS (local contextual rewriting)	6	–
SPASS (subterm contextual rewriting)	5	–
CVC4 (default)	7	66
Z3 (default)	–	49
Only solved by VAMPIRE, with FSD and BSD	11	0

Second, while all our 76 "new" SMT-LIB problems can also be solved by CVC4 and Z3 together, we note that out of these 76 problems there are 10 problems that CVC4 cannot solve, and similarly 27 problems that Z3 cannot solve.

Comparative Experiments without AVATAR. Finally, we investigated the effect of subsumption demodulation in VAMPIRE without AVATAR [19]. We used the default mode of VAMPIRE (that is, without using a portfolio approach) and turned off the AVATAR setting. While this configuration solves less problems than the portfolio mode of VAMPIRE, so far VAMPIRE is the only superposition-based theorem prover implementing AVATAR. Hence, evaluating subsumption demodulation in VAMPIRE without AVATAR is more relevant to other reasoners. Further, as AVATAR may often split non-unit clauses into unit clauses, it may potentially simulate applications of subsumption demodulation using demodulation. Table 4 shows that this is indeed the case: with both `fsd` and `bsd` enabled, subsumption demodulation in VAMPIRE can prove 190 TPTP problems and 173 SMT-LIB examples that the default VAMPIRE without AVATAR cannot solve. Again, the column "New" denotes the number of problems solved by the respective configuration but not by the default mode of VAMPIRE without AVATAR.

Table 4. Comparing VAMPIRE in default mode and without AVATAR, with and without subsumption demodulation.

	TPTP problems			SMT-LIB problems		
Configuration	Total	Solved	New (SAT + UNSAT)	Total	Solved	New (SAT + UNSAT)
VAMPIRE	13,924	6,601	–	22,833	9,608	–
VAMPIRE (FSD)	13,924	6,539	152 (13 + 139)	22,833	9,597	134 (1 + 133)
VAMPIRE (BSD)	13,924	6,471	112 (12 + 100)	22,833	9,541	87 (0 + 87)
VAMPIRE (FSD + BSD)	13,924	6,510	190 (15 + 175)	22,833	9,581	173 (1 + 172)

7 Conclusion

We introduced the simplifying inference rule subsumption demodulation to improve support for reasoning with conditional equalities in superposition-based first-order theorem proving. Subsumption demodulation revises existing machineries of superposition provers and can therefore be efficiently integrated in superposition reasoning. Still, the rule remains expensive and does not pay off for all problems, leading to a decrease in total number of solved problems by our implementation in Vampire. However, this is justified because subsumption demodulation also solves many new examples that existing provers, including first-order and SMT solvers, cannot handle. Future work includes the design of more sophisticated approaches for selecting rewriting equalities and improving the imperfect filtering of clauses indexes.

Acknowledgements. This work was funded by the ERC Starting Grant 2014 SYMCAR 639270, the ERC Proof of Concept Grant 2018 SYMELS 842066, the Wallenberg Academy Fellowship 2014 TheProSE, the Austrian FWF research project W1255-N23, the OMAA grant 101öu8, and the WWTF ICT15-193 grant.

References

1. Bachmair, L., Ganzinger, H.: Rewrite-based equational theorem proving with selection and simplification. J. Logic Comput. **4**(3), 217–247 (1994)
2. Bachmair, L., Ganzinger, H., McAllester, D.A., Lynch, C.: Resolution theorem proving. In: Handbook of Automated Reasoning, pp. 19–99 (2001)
3. Barrett, C., et al.: CVC4. In: Gopalakrishnan, G., Qadeer, S. (eds.) CAV 2011. LNCS, vol. 6806, pp. 171–177. Springer, Heidelberg (2011). https://doi.org/10.1007/978-3-642-22110-1_14
4. Barrett, C., Fontaine, P., Tinelli, C.: The Satisfiability Modulo Theories Library (SMT-LIB) (2016). www.SMT-LIB.org
5. Barthe, G., Eilers, R., Georgiou, P., Gleiss, B., Kovács, L., Maffei, M.: Verifying relational properties using trace logic. In: Proceedings of FMCAD, pp. 170–178 (2019)
6. Bjørner, N., Gurfinkel, A., McMillan, K.L., Rybalchenko, A.: Horn clause solvers for program verification. Fields Logic Comput. **II**, 24–51 (2015)
7. de Moura, L., Bjørner, N.: Z3: an efficient SMT solver. In: Ramakrishnan, C.R., Rehof, J. (eds.) TACAS 2008. LNCS, vol. 4963, pp. 337–340. Springer, Heidelberg (2008). https://doi.org/10.1007/978-3-540-78800-3_24
8. Ganzinger, H., Hagen, G., Nieuwenhuis, R., Oliveras, A., Tinelli, C.: DPLL(T): fast decision procedures. In: Alur, R., Peled, D.A. (eds.) CAV 2004. LNCS, vol. 3114, pp. 175–188. Springer, Heidelberg (2004). https://doi.org/10.1007/978-3-540-27813-9_14
9. Hillenbrand, T., Piskac, R., Waldmann, U., Weidenbach, C.: From search to computation: redundancy criteria and simplification at work. In: Voronkov, A., Weidenbach, C. (eds.) Programming Logics: Essays in Memory of Harald Ganzinger, vol. 7797, pp. 169–193. Springer, Heidelberg (2013). https://doi.org/10.1007/978-3-642-37651-1_7

10. Kovács, L., Voronkov, A.: First-order theorem proving and VAMPIRE. In: Sharygina, N., Veith, H. (eds.) CAV 2013. LNCS, vol. 8044, pp. 1–35. Springer, Heidelberg (2013). https://doi.org/10.1007/978-3-642-39799-8_1
11. Nieuwenhuis, R., Rubio, A.: Paramodulation-based theorem proving. In: Handbook of Automated Reasoning, pp. 371–443 (2001)
12. Nivela, P., Nieuwenhuis, R.: Saturation of first-order (constrained) clauses with the saturate system. In: Kirchner, C. (ed.) Rewriting Techniques and Applications, vol. 690, pp. 436–440. Springer, Heidelberg (1993). https://doi.org/10.1007/978-3-662-21551-7_33
13. Reynolds, A., Woo, M., Barrett, C., Brumley, D., Liang, T., Tinelli, C.: Scaling up DPLL(T) string solvers using context-dependent simplification. In: Majumdar, R., Kunčak, V. (eds.) CAV 2017. LNCS, vol. 10427, pp. 453–474. Springer, Cham (2017). https://doi.org/10.1007/978-3-319-63390-9_24
14. Schulz, S., Cruanes, S., Vukmirović, P.: Faster, higher, stronger: E 2.3. In: Fontaine, P. (ed.) CADE 2019. LNCS (LNAI), vol. 11716, pp. 495–507. Springer, Cham (2019). https://doi.org/10.1007/978-3-030-29436-6_29
15. Sekar, R., Ramakrishnan, I.V., Voronkov, A.: Term indexing. In: Robinson, J.A., Voronkov, A. (eds.) Handbook of Automated Reasoning, pp. 1853–1964. Elsevier Science Publishers B.V. (2001)
16. Stump, A., Sutcliffe, G., Tinelli, C.: StarExec: a cross-community infrastructure for logic solving. In: Proceedings of IJCAR, pp. 367–373 (2014)
17. Sutcliffe, G.: The TPTP problem library and associated infrastructure. From CNF to TH0, TPTP v6.4.0. J. Autom. Reasoning **59**(4), 483–502 (2017)
18. Tange, O.: GNU Parallel 2018. Ole Tange (2018)
19. Voronkov, A.: AVATAR: the architecture for first-order theorem provers. In: Biere, A., Bloem, R. (eds.) CAV 2014. LNCS, vol. 8559, pp. 696–710. Springer, Cham (2014). https://doi.org/10.1007/978-3-319-08867-9_46
20. Weidenbach, C.: Combining superposition, sorts and splitting. In: Handbook of Automated Reasoning, pp. 1965–2013 (2001)
21. Weidenbach, C., Dimova, D., Fietzke, A., Kumar, R., Suda, M., Wischnewski, P.: SPASS version 3.5. In: Schmidt, R.A. (ed.) CADE 2009. LNCS (LNAI), vol. 5663, pp. 140–145. Springer, Heidelberg (2009). https://doi.org/10.1007/978-3-642-02959-2_10
22. Weidenbach, C., Wischnewski, P.: Contextual rewriting in SPASS. In: Proceedings of PAAR (2008)

A Comprehensive Framework for Saturation Theorem Proving

Uwe Waldmann[1(✉)], Sophie Tourret[1], Simon Robillard[2], and Jasmin Blanchette[1,3,4]

[1] Max-Planck-Institut für Informatik, Saarland Informatics Campus, Saarbrücken, Germany
{uwe,stourret,jblanche}@mpi-inf.mpg.de
[2] IMT Atlantique, Nantes, France
simon.robillard@imt-atlantique.fr
[3] Vrije Universiteit Amsterdam, Amsterdam, The Netherlands
[4] Université de Lorraine, CNRS, Inria, LORIA, Nancy, France

Abstract. We present a framework for formal refutational completeness proofs of abstract provers that implement saturation calculi, such as ordered resolution or superposition. The framework relies on modular extensions of lifted redundancy criteria. It allows us to extend redundancy criteria so that they cover subsumption, and also to model entire prover architectures in such a way that the static refutational completeness of a calculus immediately implies the dynamic refutational completeness of a prover implementing the calculus, for instance within an Otter or DISCOUNT loop. Our framework is mechanized in Isabelle/HOL.

1 Introduction

In their *Handbook* chapter [5, Sect. 4], Bachmair and Ganzinger remark that "unfortunately, comparatively little effort has been devoted to a formal analysis of redundancy and other fundamental concepts of theorem proving strategies, while more emphasis has been placed on investigating the refutational completeness of a variety of modifications of inference rules, such as resolution." As a remedy, they present an abstract framework for saturation up to redundancy. Briefly, theorem proving derivations take the form $N_0 \rhd N_1 \rhd \cdots$, where N_0 is the initial clause set and each step either adds inferred clauses or deletes redundant clauses. Given a suitable notion of fairness, the limit N_* of a fair derivation is saturated up to redundancy. If the calculus is refutationally complete and N_* does not contain the false clause $\bot$, then N_0 has a model.

Bachmair and Ganzinger also define a concrete prover, RP, based on a first-order ordered resolution calculus and the given clause procedure. However, like all realistic resolution provers, RP implements subsumption deletion. This operation is not covered by the standard definition of redundancy, according to which a clause C is redundant w.r.t. a clause set N if all its ground instances $C\theta$

N. Peltier and V. Sofronie-Stokkermans (Eds.): IJCAR 2020, LNAI 12166, pp. 316–334, 2020.
https://doi.org/10.1007/978-3-030-51074-9_18

are entailed by *strictly* smaller ground instances of clauses belonging to N. As a result, RP-derivations are *not* $\rhd$-derivations, and the framework is *not* applicable.

There are two ways to address this problem. In the *Handbook*, Bachmair and Ganzinger start from scratch and prove the dynamic refutational completeness of RP by relating nonground derivations to ground derivations. This proof, though, turns out to be rather nonmodular—it refers simultaneously to properties of the calculus, to properties of the prover, and to the fairness of the derivations. Extending it to other calculi or prover architectures would be costly. As a result, most authors stop after proving static refutational completeness of their calculi.

An alternative approach is to extend the redundancy criterion so that subsumed clauses become redundant. As demonstrated by Bachmair and Ganzinger in 1990 [3], this is possible by redefining redundancy in terms of closures (C, θ) instead of ground instances $C\theta$. We show that this approach can be generalized and modularized: First, any redundancy criterion that is obtained by lifting a ground criterion can be extended to a redundancy criterion that supports subsumption without affecting static refutational completeness (Sect. 3). Second, by applying this property to labeled formulas, it becomes possible to give generic completeness proofs for prover architectures in a straightforward way.

Most saturation provers implement a variant of the given clause procedure. We present an abstract version of the procedure (Sect. 4) that can be refined to obtain an Otter [18] or DISCOUNT [1] loop and prove it refutationally complete. We also present a generalization that decouples scheduling and computation of inferences, to support orphan deletion [16,25] and dovetailing [9].

When these prover architectures are instantiated with a concrete saturation calculus, the dynamic refutational completeness of the combination follows in a modular way from the properties of the prover architecture and the static refutational completeness proof for the calculus. Thus, the framework is applicable to a wide range of calculi, including ordered resolution [5], unfailing completion [2], standard superposition [4], constraint superposition [19], theory superposition [28], hierarchic superposition [7], and clausal λ-superposition [9].

Detailed proofs are included in a technical report [29], together with more explanations, examples, and discussions. When Schlichtkrull, Blanchette, Traytel, and Waldmann [24] mechanized Bachmair and Ganzinger's chapter using the Isabelle/HOL proof assistant [21], they found quite a few mistakes, including one that compromised RP's dynamic refutational completeness. This motivated us to mechanize our framework as well (Sect. 5).

2 Preliminaries

Inferences and Redundancy. Let A be a set. An A-sequence is a finite sequence $(a_i)_{i=0}^{k} = a_0, a_1, \ldots, a_k$ or an infinite sequence $(a_i)_{i=0}^{\infty} = a_0, a_1, \ldots$ with $a_i \in A$ for all i. We write $(a_i)_{i\geq 0}$ or $(a_i)_i$ for both finite and infinite sequences. Nonempty sequences can be split into a head a_0 and a tail $(a_i)_{i\geq 1}$. Given $\rhd \subseteq A \times A$, a $\rhd$-derivation is a nonempty A-sequence such that $a_i \rhd a_{i+1}$ for all i.

A set $\mathbf{F}$ of *formulas* is a set with a nonempty subset $\mathbf{F}_\bot \subseteq \mathbf{F}$. Elements of $\mathbf{F}_\bot$ represent *false*. Typically, $\mathbf{F}_\bot := \{\bot\}$. In Sect. 4, different elements of $\mathbf{F}_\bot$ will represent different situations in which a contradiction has been derived.

A *consequence relation* $\models$ over $\mathbf{F}$ is a relation $\models \subseteq \mathcal{P}(\mathbf{F}) \times \mathcal{P}(\mathbf{F})$ with the following properties for all $N_1, N_2, N_3 \subseteq \mathbf{F}$:

(C1) $\{\bot\} \models N_1$ for every $\bot \in \mathbf{F}_\bot$;
(C2) $N_2 \subseteq N_1$ implies $N_1 \models N_2$;
(C3) if $N_1 \models \{C\}$ for every $C \in N_2$, then $N_1 \models N_2$;
(C4) if $N_1 \models N_2$ and $N_2 \models N_3$, then $N_1 \models N_3$.

Consequence relations are used to discuss soundness (and the addition of formulas) and to discuss refutational completeness (and the deletion of formulas). An example that requires this distinction is constraint superposition [19], where one uses entailment w.r.t. the set of all ground instances, $\models\!\!\approx$, for soundness, but entailment w.r.t. a subset of those instances, $\models$, for refutational completeness. Some calculus-dependent argument is then necessary to show that refutational completeness w.r.t. $\models$ implies refutational completeness w.r.t. $\models\!\!\approx$.

An $\mathbf{F}$-*inference* ι is a tuple $(C_n, \ldots, C_0) \in \mathbf{F}^{n+1}$, $n \geq 0$. The formulas $C_n, \ldots, C_1$ are called *premises* of ι; C_0 is called the *conclusion* of ι, denoted by $concl(\iota)$. An $\mathbf{F}$-*inference system* Inf is a set of $\mathbf{F}$-inferences. If $N \subseteq \mathbf{F}$, we write $Inf(N)$ for the set of all inferences in Inf whose premises are contained in N, and $Inf(N, M) := Inf(N \cup M) \setminus Inf(N \setminus M)$ for the set of all inferences in Inf such that one premise is in M and the other premises are contained in $N \cup M$.

A *redundancy criterion* for an inference system Inf and a consequence relation $\models$ is a pair $Red = (Red_{\mathrm{I}}, Red_{\mathrm{F}})$, where $Red_{\mathrm{I}} : \mathcal{P}(\mathbf{F}) \to \mathcal{P}(Inf)$ and $Red_{\mathrm{F}} : \mathcal{P}(\mathbf{F}) \to \mathcal{P}(\mathbf{F})$ are mappings that satisfy the following conditions for all N, N':

(R1) if $N \models \{\bot\}$ for some $\bot \in \mathbf{F}_\bot$, then $N \setminus Red_{\mathrm{F}}(N) \models \{\bot\}$;
(R2) if $N \subseteq N'$, then $Red_{\mathrm{F}}(N) \subseteq Red_{\mathrm{F}}(N')$ and $Red_{\mathrm{I}}(N) \subseteq Red_{\mathrm{I}}(N')$;
(R3) if $N' \subseteq Red_{\mathrm{F}}(N)$, then $Red_{\mathrm{F}}(N) \subseteq Red_{\mathrm{F}}(N \setminus N')$ and $Red_{\mathrm{I}}(N) \subseteq Red_{\mathrm{I}}(N \setminus N')$;
(R4) if $\iota \in Inf$ and $concl(\iota) \in N$, then $\iota \in Red_{\mathrm{I}}(N)$.

Inferences in $Red_{\mathrm{I}}(N)$ and formulas in $Red_{\mathrm{F}}(N)$ are called *redundant* w.r.t. N.[1] Intuitively, (R1) states that deleting redundant formulas preserves inconsistency. (R2) and (R3) state that formulas or inferences that are redundant w.r.t. a set N remain redundant if arbitrary formulas are added to N or redundant formulas are deleted from N. (R4) ensures that computing an inference makes it redundant.

We define the relation $\rhd_{Red} \subseteq \mathcal{P}(\mathbf{F}) \times \mathcal{P}(\mathbf{F})$ such that $N \rhd_{Red} N'$ if and only if $N \setminus N' \subseteq Red_{\mathrm{F}}(N')$.

[1] One can find several slightly differing definitions for redundancy criteria, fairness, and saturation in the literature [5,7,28]. However, as shown in the technical report [29], the differences are typically insignificant as far as static or dynamic refutational completeness is concerned. Here we mostly follow Waldmann [28].

Refutational Completeness. Let $\models$ be a consequence relation, let *Inf* be an inference system, and let *Red* be a redundancy criterion for $\models$ and *Inf*.

A set $N \subseteq \mathbf{F}$ is called *saturated* w.r.t. *Inf* and *Red* if $\mathit{Inf}(N) \subseteq \mathit{Red}_{\mathrm{I}}(N)$. The pair $(\mathit{Inf}, \mathit{Red})$ is called *statically refutationally complete* w.r.t. $\models$ if for every saturated set $N \subseteq \mathbf{F}$ such that $N \models \{\bot\}$ for some $\bot \in \mathbf{F}_\bot$, there exists a $\bot' \in \mathbf{F}_\bot$ such that $\bot' \in N$.

Let $(N_i)_i$ be a $\mathcal{P}(\mathbf{F})$-sequence. Its *limit* is the set $N_* := \bigcup_i \bigcap_{j \geq i} N_j$. Its *union* is the set $N_\infty := \bigcup_i N_i$. A sequence is called *fair* if $\mathit{Inf}(N_*) \subseteq \bigcup_i \mathit{Red}_{\mathrm{I}}(N_i)$. The pair $(\mathit{Inf}, \mathit{Red})$ is called *dynamically refutationally complete* w.r.t. $\models$ if for every fair $\rhd_{\mathit{Red}}$-derivation $(N_i)_i$ such that $N_0 \models \{\bot\}$ for some $\bot \in \mathbf{F}_\bot$, we have $\bot' \in N_i$ for some i and some $\bot' \in \mathbf{F}_\bot$. Properties (R1)–(R3) allow the passage from a static set of formulas to a dynamic prover:

Lemma 1. *$(\mathit{Inf}, \mathit{Red})$ is dynamically refutationally complete w.r.t. $\models$ if and only if it is statically refutationally complete w.r.t. $\models$.*

Intersections of Redundancy Criteria. In the sequel, it will be useful to define consequence relations and redundancy criteria as intersections of previously defined consequence relations or redundancy criteria.

Let Q be an arbitrary set, and let $(\models^q)_{q \in Q}$ be a Q-indexed family of consequence relations over $\mathbf{F}$. Then $\models^\cap \,:= \bigcap_{q \in Q} \models^q$ qualifies as a consequence relation. Moreover, let *Inf* be an inference system, and let $(\mathit{Red}^q)_{q \in Q}$ be a Q-indexed family of redundancy criteria, where each $\mathit{Red}^q = (\mathit{Red}^q_{\mathrm{I}}, \mathit{Red}^q_{\mathrm{F}})$ is a redundancy criterion for *Inf* and $\models^q$. Let $\mathit{Red}^\cap_{\mathrm{I}}(N) := \bigcap_{q \in Q} \mathit{Red}^q_{\mathrm{I}}(N)$ and $\mathit{Red}^\cap_{\mathrm{F}}(N) := \bigcap_{q \in Q} \mathit{Red}^q_{\mathrm{F}}(N)$. Then $\mathit{Red}^\cap := (\mathit{Red}^\cap_{\mathrm{I}}, \mathit{Red}^\cap_{\mathrm{F}})$ qualifies as a redundancy criterion for $\models^\cap$ and *Inf*.

Lemma 2. *A set $N \subseteq \mathbf{F}$ is saturated w.r.t. Inf and $\mathit{Red}^\cap$ if and only if it is saturated w.r.t. Inf and Red^q for every $q \in Q$.*

Often, the consequence relations $\models^q$ agree for all $q \in Q$. For calculi where they disagree, such as constraint superposition [19], one can typically demonstrate the static refutational completeness of $(\mathit{Inf}, \mathit{Red}^\cap)$ in the following form:

Lemma 3. *If for every set $N \subseteq \mathbf{F}$ that is saturated w.r.t. Inf and $\mathit{Red}^\cap$ and does not contain any $\bot' \in \mathbf{F}_\bot$ there exists some $q \in Q$ such that $N \not\models^q \{\bot\}$ for some $\bot \in \mathbf{F}_\bot$, then $(\mathit{Inf}, \mathit{Red}^\cap)$ is statically refutationally complete w.r.t. $\models^\cap$.*

3 Lifting

A standard approach for establishing the refutational completeness of a calculus is to first concentrate on the ground case and then lift the results to the nonground case. In this section, we show how to perform this lifting abstractly, given a suitable grounding function $\mathcal{G}$. The function maps every formula $C \in \mathbf{F}$ to a set $\mathcal{G}(C)$ of formulas from a set of formulas $\mathbf{G}$. Depending on the logic and the calculus, $\mathcal{G}(C)$

may be, for example, the set of all ground instances of C or a subset thereof. Similarly, $\mathcal{G}$ maps *FInf*-inferences to sets of *GInf*-inferences.

There are calculi where some *FInf*-inferences ι do not have a counterpart in *GInf*, such as the PosExt inferences of higher-order superposition calculi [9]. In these cases, we set $\mathcal{G}(\iota) = \mathit{undef}$.

Standard Lifting. Given two sets of formulas **F** and **G**, an **F**-inference system *FInf*, a **G**-inference system *GInf*, and a redundancy criterion *Red* for *GInf*, let $\mathcal{G}$ be a function that maps every formula in **F** to a subset of **G** and every **F**-inference in *FInf* to *undef* or to a subset of *GInf*. $\mathcal{G}$ is called a *grounding function* if

(G1) for every $\bot \in \mathbf{F}_\bot$, $\emptyset \neq \mathcal{G}(\bot) \subseteq \mathbf{G}_\bot$;
(G2) for every $C \in \mathbf{F}$, if $\bot \in \mathcal{G}(C)$ and $\bot \in \mathbf{G}_\bot$ then $C \in \mathbf{F}_\bot$;
(G3) for every $\iota \in \mathit{FInf}$, if $\mathcal{G}(\iota) \neq \mathit{undef}$, then $\mathcal{G}(\iota) \subseteq \mathit{Red}_\mathrm{I}(\mathcal{G}(\mathit{concl}(\iota)))$.

$\mathcal{G}$ is extended to sets $N \subseteq \mathbf{F}$ by defining $\mathcal{G}(N) := \bigcup_{C \in N} \mathcal{G}(C)$. Analogously, for a set $I \subseteq \mathit{FInf}$, $\mathcal{G}(I) := \bigcup_{\iota \in I,\, \mathcal{G}(\iota) \neq \mathit{undef}} \mathcal{G}(\iota)$.

Example 4. In standard superposition, **F** is the set of all universally quantified first-order clauses over some signature Σ, **G** is the set of all ground first-order clauses over Σ, and $\mathcal{G}$ maps every clause C to the set of its ground instances $C\theta$ and every superposition inference ι to the set of its ground instances $\iota\theta$.

Let $\mathcal{G}$ be a grounding function from **F** and *FInf* to **G** and *GInf*, and let $\models$ be a consequence relation over **G**. We define the relation $\models_\mathcal{G} \subseteq \mathcal{P}(\mathbf{F}) \times \mathcal{P}(\mathbf{F})$ such that $N_1 \models_\mathcal{G} N_2$ if and only if $\mathcal{G}(N_1) \models \mathcal{G}(N_2)$. We call $\models_\mathcal{G}$ the *$\mathcal{G}$-lifting* of $\models$. It qualifies as a consequence relation over **F** and corresponds to Herbrand entailment. If Tarski entailment (i.e., $N_1 \models_\mathrm{T} N_2$ if and only if any model of N_1 is also a model of N_2) is desired, the mismatch can be repaired by showing that the two notions of entailment are equivalent as far as refutations are concerned.

Let $\mathit{Red} = (\mathit{Red}_\mathrm{I}, \mathit{Red}_\mathrm{F})$ be a redundancy criterion for $\models$ and *GInf*. We define functions $\mathit{Red}_\mathrm{I}^\mathcal{G} : \mathcal{P}(\mathbf{F}) \to \mathcal{P}(\mathit{FInf})$ and $\mathit{Red}_\mathrm{F}^\mathcal{G} : \mathcal{P}(\mathbf{F}) \to \mathcal{P}(\mathbf{F})$ by

$$
\begin{aligned}
&\iota \in \mathit{Red}_\mathrm{I}^\mathcal{G}(N) \text{ if and only if}\\
&\quad \mathcal{G}(\iota) \neq \mathit{undef} \text{ and } \mathcal{G}(\iota) \subseteq \mathit{Red}_\mathrm{I}(\mathcal{G}(N))\\
&\quad \text{or } \mathcal{G}(\iota) = \mathit{undef} \text{ and } \mathcal{G}(\mathit{concl}(\iota)) \subseteq \mathcal{G}(N) \cup \mathit{Red}_\mathrm{F}(\mathcal{G}(N));\\
&C \in \mathit{Red}_\mathrm{F}^\mathcal{G}(N) \text{ if and only if}\\
&\quad \mathcal{G}(C) \subseteq \mathit{Red}_\mathrm{F}(\mathcal{G}(N)).
\end{aligned}
$$

We call $\mathit{Red}^\mathcal{G} := (\mathit{Red}_\mathrm{I}^\mathcal{G}, \mathit{Red}_\mathrm{F}^\mathcal{G})$ the *$\mathcal{G}$-lifting* of *Red*. It qualifies as a redundancy criterion for $\models_\mathcal{G}$ and *FInf*. We get the following folklore theorem:

Theorem 5. *If (GInf, Red) is statically refutationally complete w.r.t.* $\models$, *and if we have* $\mathit{GInf}(\mathcal{G}(N)) \subseteq \mathcal{G}(\mathit{FInf}(N)) \cup \mathit{Red}_\mathrm{I}(\mathcal{G}(N))$ *for every* $N \subseteq \mathbf{F}$ *that is saturated w.r.t. FInf and* $\mathit{Red}^\mathcal{G}$, *then* $(\mathit{FInf}, \mathit{Red}^\mathcal{G})$ *is statically refutationally complete w.r.t.* $\models_\mathcal{G}$.

Adding Tiebreaker Orderings. We now strengthen the $\mathcal{G}$-lifting of redundancy criteria to also support subsumption deletion. Let $\sqsupset$ be a well-founded strict partial ordering on **F**. We define $\mathit{Red}_\mathrm{F}^{\mathcal{G},\sqsupset} : \mathcal{P}(\mathbf{F}) \to \mathcal{P}(\mathbf{F})$ as follows:

$C \in \mathit{Red}_{\mathrm{F}}^{\mathcal{G},\sqsupset}(N)$ if and only if
for every $D \in \mathcal{G}(C)$,
$D \in \mathit{Red}_{\mathrm{F}}(\mathcal{G}(N))$ or there exists $C' \in N$ such that $C \sqsupset C'$ and $D \in \mathcal{G}(C')$.

Notice how $\sqsupset$ is used to break ties between C and C', possibly making C redundant. We call $\mathit{Red}^{\mathcal{G},\sqsupset} := (\mathit{Red}_{\mathrm{I}}^{\mathcal{G}}, \mathit{Red}_{\mathrm{F}}^{\mathcal{G},\sqsupset})$ the $(\mathcal{G},\sqsupset)$-*lifting* of *Red*. We get the previously defined $\mathit{Red}^{\mathcal{G}}$ as a special case of $\mathit{Red}^{\mathcal{G},\sqsupset}$ by setting $\sqsupset := \emptyset$.

We obtain our first main result:

Theorem 6. *Let Red be a redundancy criterion for* $\models$ *and GInf, let* $\mathcal{G}$ *be a grounding function from* $\mathbf{F}$ *and FInf to* $\mathbf{G}$ *and GInf, and let* $\sqsupset$ *be a well-founded strict partial ordering on* $\mathbf{F}$. *Then the* $(\mathcal{G},\sqsupset)$*-lifting* $\mathit{Red}^{\mathcal{G},\sqsupset}$ *of Red is a redundancy criterion for* $\models_{\mathcal{G}}$ *and FInf.*

Observe that $\sqsupset$ appears only in the second component of $\mathit{Red}^{\mathcal{G},\sqsupset} = (\mathit{Red}_{\mathrm{I}}^{\mathcal{G}}, \mathit{Red}_{\mathrm{F}}^{\mathcal{G},\sqsupset})$ and that the definitions of a saturated set and of static refutational completeness do not depend on the second component of a redundancy criterion. The following lemmas are immediate consequences of these observations:

Lemma 7. *A set* $N \subseteq \mathbf{F}$ *is saturated w.r.t. FInf and* $\mathit{Red}^{\mathcal{G},\sqsupset}$ *if and only if it is saturated w.r.t. FInf and* $\mathit{Red}^{\mathcal{G},\emptyset}$.

Lemma 8. *(FInf,* $\mathit{Red}^{\mathcal{G},\sqsupset}$*) is statically refutationally complete w.r.t.* $\models_{\mathcal{G}}$ *if and only if (FInf,* $\mathit{Red}^{\mathcal{G},\emptyset}$*) is statically refutationally complete w.r.t.* $\models_{\mathcal{G}}$.

Combining Lemmas 1 and 8, we obtain our second main result:

Theorem 9. *Let Red be a redundancy criterion for* $\models$ *and GInf, let* $\mathcal{G}$ *be a grounding function from* $\mathbf{F}$ *and FInf to* $\mathbf{G}$ *and GInf, and let* $\sqsupset$ *be a well-founded strict partial ordering on* $\mathbf{F}$. *If (FInf,* $\mathit{Red}^{\mathcal{G},\emptyset}$*) is statically refutationally complete w.r.t.* $\models_{\mathcal{G}}$*, then (FInf,* $\mathit{Red}^{\mathcal{G},\sqsupset}$*) is dynamically refutationally complete w.r.t.* $\models_{\mathcal{G}}$.

Example 10. For resolution or superposition in standard first-order logic, we can define the subsumption quasi-ordering $\gtrsim$ on clauses by $C \gtrsim C'$ if and only if $C = C'\sigma$ for some substitution σ. The subsumption ordering $\gtrdot := {\gtrsim} \setminus {\lesssim}$ is well founded. By choosing $\sqsupset := \gtrdot$, we obtain a criterion $\mathit{Red}^{\mathcal{G},\sqsupset}$ that includes standard redundancy and also supports subsumption deletion. Similarly, for proof calculi modulo associativity and commutativity, we can let $C \gtrsim C'$ be true if there exists a substitution σ such that C equals $C'\sigma$ up to the equational theory.

Example 11. Constraint superposition with ordering constraints [19] is an example of a calculus where the subsumption ordering $\gtrdot$ is not well founded: A ground instance of a constrained clause $C\ [\![K]\!]$ is a ground clause $C\theta$ for which $K\theta$ evaluates to true. Define $\gtrsim$ by stating that $C\ [\![K]\!] \gtrsim C'\ [\![K']\!]$ if and only if every ground instance of $C\ [\![K]\!]$ is a ground instance of $C'\ [\![K']\!]$, and define $\gtrdot := {\gtrsim} \setminus {\lesssim}$. If $\succ$ is a simplification ordering, then $\mathsf{P}(x)\ [\![x \prec \mathsf{b}]\!] \gtrdot \mathsf{P}(x)\ [\![x \prec \mathsf{f(b)}]\!] \gtrdot \mathsf{P}(x)\ [\![x \prec \mathsf{f(f(b))}]\!] \gtrdot \cdots$ is an infinite chain.

Example 12. For higher-order calculi such as higher-order resolution [17] and clausal λ-superposition [9], subsumption is also not well founded, as witnessed by the chain $\mathsf{p}\,x\,x \gtrdot \mathsf{p}\,(x\,\mathsf{a})\,(x\,\mathsf{b}_1) \gtrdot \mathsf{p}\,(x\,\mathsf{a}\,\mathsf{a})\,(x\,\mathsf{b}_1\,\mathsf{b}_2) \gtrdot \cdots$.

Even if the subsumption ordering for some logic is not well founded, as in the two examples above, we can always define $\sqsupset$ as the intersection of the subsumption ordering and an appropriate ordering based on formula sizes or weights.

Conversely, the $\sqsupset$ relation can be more general than subsumption. In Sect. 4, we will use it to justify the movement of formulas between sets in the given clause procedure.

Example 13. For some superposition-based decision procedures [6], one would like to define $\sqsupset$ as the reverse subsumption ordering $\lessdot$ on first-order clauses. Even though $\lessdot$ is not well founded in general, it is well founded on $\{C \in \mathbf{F} \mid D \in \mathcal{G}(C)\}$ for every $D \in \mathbf{G}$. As shown in the technical report [29], our framework can be extended to support this case by defining $Red_{\mathrm{F}}^{\mathcal{G},\sqsupset}$ using a $\mathbf{G}$-indexed family $(\sqsupset_D)_{D\in\mathbf{G}}$ of well-founded strict partial orderings instead of a single $\sqsupset$.

Intersections of Liftings. The above results can be extended in a straightforward way to intersections of lifted redundancy criteria. As before, let $\mathbf{F}$ and $\mathbf{G}$ be two sets of formulas, and let *FInf* be an $\mathbf{F}$-inference system. In addition, let Q be a set. For every $q \in Q$, let $\models^q$ be a consequence relation over $\mathbf{G}$, let $GInf^q$ be a $\mathbf{G}$-inference system, let Red^q be a redundancy criterion for $\models^q$ and $GInf^q$, and let $\mathcal{G}^q$ be a grounding function from $\mathbf{F}$ and *FInf* to $\mathbf{G}$ and $GInf^q$. Let $\sqsupset$ be a well-founded strict partial ordering on $\mathbf{F}$.

For each $q \in Q$, we know by Theorem 6 that the $(\mathcal{G}^q, \emptyset)$-lifting $Red^{q,\mathcal{G}^q,\emptyset} = (Red_{\mathrm{I}}^{q,\mathcal{G}^q}, Red_{\mathrm{F}}^{q,\mathcal{G}^q,\emptyset})$ and the $(\mathcal{G}^q, \sqsupset)$-lifting $Red^{q,\mathcal{G}^q,\sqsupset} = (Red_{\mathrm{I}}^{q,\mathcal{G}^q}, Red_{\mathrm{F}}^{q,\mathcal{G}^q,\sqsupset})$ of Red^q are redundancy criteria for $\models_{\mathcal{G}^q}^q$ and *FInf*. Consequently, the intersections

$$Red^{\cap\mathcal{G},\sqsupset} := (Red_{\mathrm{I}}^{\cap\mathcal{G},\sqsupset}, Red_{\mathrm{F}}^{\cap\mathcal{G},\sqsupset}) := \left(\bigcap\nolimits_{q\in Q} Red_{\mathrm{I}}^{q,\mathcal{G}^q}, \bigcap\nolimits_{q\in Q} Red_{\mathrm{F}}^{q,\mathcal{G}^q,\emptyset}\right) \text{ and}$$
$$Red^{\cap\mathcal{G},\sqsupset} := (Red_{\mathrm{I}}^{\cap\mathcal{G},\sqsupset}, Red_{\mathrm{F}}^{\cap\mathcal{G},\sqsupset}) := \left(\bigcap\nolimits_{q\in Q} Red_{\mathrm{I}}^{q,\mathcal{G}^q}, \bigcap\nolimits_{q\in Q} Red_{\mathrm{F}}^{q,\mathcal{G}^q,\sqsupset}\right)$$

are redundancy criteria for $\models_{\mathcal{G}}^{\cap} := \bigcap_{q\in Q} \models_{\mathcal{G}^q}^q$ and *FInf*.

Theorem 14. *If $(GInf^q, Red^q)$ is statically refutationally complete w.r.t. $\models^q$ for every $q \in Q$, and if for every $N \subseteq \mathbf{F}$ that is saturated w.r.t. FInf and $Red^{\cap\mathcal{G}}$ there exists a q such that $GInf^q(\mathcal{G}^q(N)) \subseteq \mathcal{G}^q(FInf(N)) \cup Red_{\mathrm{I}}^q(\mathcal{G}^q(N))$, then $(FInf, Red^{\cap\mathcal{G}})$ is statically refutationally complete w.r.t. $\models_{\mathcal{G}}^{\cap}$.*

Lemma 15. *A set $N \subseteq \mathbf{F}$ is saturated w.r.t. FInf and $Red^{\cap\mathcal{G},\sqsupset}$ if and only if it is saturated w.r.t. FInf and $Red^{\cap\mathcal{G}}$.*

Lemma 16. *$(FInf, Red^{\cap\mathcal{G},\sqsupset})$ is statically refutationally complete w.r.t. $\models_{\mathcal{G}}^{\cap}$ if and only if $(FInf, Red^{\cap\mathcal{G}})$ is statically refutationally complete w.r.t. $\models_{\mathcal{G}}^{\cap}$.*

Theorem 17. *If $(FInf, Red^{\cap\mathcal{G}})$ is statically refutationally complete w.r.t. $\models_{\mathcal{G}}^{\cap}$, then $(FInf, Red^{\cap\mathcal{G},\sqsupset})$ is dynamically refutationally complete w.r.t. $\models_{\mathcal{G}}^{\cap}$.*

Example 18. Intersections of liftings are needed to support selection functions in superposition [4]. The calculus *FInf* is parameterized by a function *fsel* on the set **F** of first-order clauses that selects a subset of the negative literals in each $C \in \mathbf{F}$. There are several ways to extend *fsel* to a selection function *gsel* on the set **G** of ground clauses such that for every $D \in \mathbf{G}$ there exists some $C \in \mathbf{F}$ such that $D = C\theta$ and D and C have corresponding selected literals. For every such *gsel*, $\models^{gsel}$ is first-order entailment, $GInf^{gsel}$ is the set of ground inferences satisfying *gsel*, and Red^{gsel} is the redundancy criterion for $GInf^{gsel}$. The grounding function $\mathcal{G}^{gsel}$ maps $C \in \mathbf{F}$ to $\{C\theta \in \mathbf{G} \mid \theta \text{ is a substitution}\}$ and $\iota \in FInf$ to the set of ground instances of ι in $GInf^{gsel}$ with corresponding literals selected in the premises. In the static refutational completeness proof, only one *gsel* is needed, but this *gsel* is not known during a derivation, so fairness must be guaranteed w.r.t. $Red_{\mathrm{I}}^{gsel,\mathcal{G}^{gsel}}$ for every possible extension *gsel* of *fsel*. Thus, checking $Red_{\mathrm{I}}^{\cap\mathcal{G}}$ amounts to a worst-case analysis, where we must assume that every ground instance $C\theta$ of a premise $C \in \mathbf{F}$ inherits the selection of C.

Example 19. Intersections of liftings are also necessary for constraint superposition calculi [19]. Here the calculus *FInf* operates on the set **F** of first-order clauses with constraints. For a convergent rewrite system R, $\models^{R}$ is first-order entailment up to R on the set **G** of unconstrained ground clauses, $GInf^{R}$ is the set of ground superposition inferences, and Red^{R} is redundancy up to R. The grounding function $\mathcal{G}^{R}$ maps $C\,[\![K]\!] \in \mathbf{F}$ to $\{D \in \mathbf{G} \mid D = C\theta,\ K\theta = \text{true},\ x\theta \text{ is } R\text{-irreducible for all } x\}$[2] and $\iota \in FInf$ to the set of ground instances of ι where the premises and conclusion of $\mathcal{G}^{R}(\iota)$ are the $\mathcal{G}^{R}$-ground instances of the premises and conclusion of ι. In the static refutational completeness proof, only one particular R is needed, but this R is not known during a derivation, so fairness must be guaranteed w.r.t. $Red_{\mathrm{I}}^{R,\mathcal{G}^{R}}$ for every convergent rewrite system R.

Almost every redundancy criterion for a nonground inference system *FInf* that can be found in the literature can be written as $Red^{\mathcal{G},\emptyset}$ for some grounding function $\mathcal{G}$ from **F** and *FInf* to **G** and *GInf*, and some redundancy criterion *Red* for *GInf*, or as an intersection $Red^{\cap\mathcal{G}}$ of such criteria. By Theorem 17, every static refutational completeness result for *FInf* and $Red^{\cap\mathcal{G}}$—which does not permit the deletion of subsumed formulas during a run—yields immediately a dynamic refutational completeness result for *FInf* and $Red^{\cap\mathcal{G},\sqsupset}$—which permits the deletion of subsumed formulas during a run, provided that they are larger w.r.t. $\sqsupset$.

Adding Labels. In practice, the ordering $\sqsupset$ used in $(\mathcal{G},\sqsupset)$-lifting often depends on meta-information about a formula, such as its age or the way in which it has been processed so far during a derivation. To capture this meta-information, we extend formulas and inference systems in a rather trivial way with labels. As before, let **F**

---

[2] For a variable x that occurs only in positive literals $x \approx t$, the condition is slightly more complicated.

and $\mathbf{G}$ be two sets of formulas, let *FInf* be an $\mathbf{F}$-inference system, let *GInf* be a $\mathbf{G}$-inference system, let $\models \subseteq \mathcal{P}(\mathbf{G}) \times \mathcal{P}(\mathbf{G})$ be a consequence relation over $\mathbf{G}$, let *Red* be a redundancy criterion for $\models$ and *GInf*, and let $\mathcal{G}$ be a grounding function from $\mathbf{F}$ and *FInf* to $\mathbf{G}$ and *GInf*.

Let $\mathbf{L}$ be a nonempty set of *labels*. Define $\mathbf{FL} := \mathbf{F} \times \mathbf{L}$ and $\mathbf{FL}_\bot := \mathbf{F}_\bot \times \mathbf{L}$. Notice that there are at least as many false values in $\mathbf{FL}$ as there are labels in $\mathbf{L}$. We use $\mathcal{M}, \mathcal{N}$ to denote labeled formula sets. Given a set $\mathcal{N} \subseteq \mathbf{FL}$, let $\lfloor \mathcal{N} \rfloor := \{C \mid (C, l) \in \mathcal{N}\}$ denote the set of formulas without their labels. We call an $\mathbf{FL}$-inference system *FLInf* a *labeled version* of *FInf* if it has the following properties:

(L1) for every inference $(C_n, \ldots, C_0) \in \mathit{FInf}$ and every tuple $(l_1, \ldots, l_n) \in \mathbf{L}^n$, there exists an $l_0 \in \mathbf{L}$ and an inference $((C_n, l_n), \ldots, (C_0, l_0)) \in \mathit{FLInf}$;

(L2) if $\iota = ((C_n, l_n), \ldots, (C_0, l_0))$ is an inference in *FLInf*, then $(C_n, \ldots, C_0)$ is an inference in *FInf*, denoted by $\lfloor \iota \rfloor$.

Let *FLInf* be a labeled version of *FInf*. Define $\mathcal{G}_\mathbf{L}$ by $\mathcal{G}_\mathbf{L}((C, l)) := \mathcal{G}(C)$ for every $(C, l) \in \mathbf{FL}$ and by $\mathcal{G}_\mathbf{L}(\iota) := \mathcal{G}(\lfloor \iota \rfloor)$ for every $\iota \in \mathit{FLInf}$. It qualifies as a grounding function from $\mathbf{FL}$ and *FLInf* to $\mathbf{G}$ and *GInf*. Let $\models_{\mathcal{G}_\mathbf{L}}$ be the $\mathcal{G}_\mathbf{L}$-lifting of $\models$. Let $\mathit{Red}^{\mathcal{G}_\mathbf{L}, \emptyset}$ be the $(\mathcal{G}_\mathbf{L}, \emptyset)$-lifting of *Red*. The following lemmas are obvious:

Lemma 20. *If a set* $\mathcal{N} \subseteq \mathbf{FL}$ *is saturated w.r.t. FLInf and* $\mathit{Red}^{\mathcal{G}_\mathbf{L}, \emptyset}$, *then* $\lfloor \mathcal{N} \rfloor \subseteq \mathbf{F}$ *is saturated w.r.t. FInf and* $\mathit{Red}^{\mathcal{G}, \emptyset}$.

Lemma 21. *If* $(\mathit{FInf}, \mathit{Red}^{\mathcal{G}, \emptyset})$ *is statically refutationally complete w.r.t.* $\models_\mathcal{G}$, *then* $(\mathit{FLInf}, \mathit{Red}^{\mathcal{G}_\mathbf{L}, \emptyset})$ *is statically refutationally complete w.r.t.* $\models_{\mathcal{G}_\mathbf{L}}$.

The extension to intersections of redundancy criteria is also straightforward. Let $\mathbf{F}$ and $\mathbf{G}$ be two sets of formulas, and let *FInf* be an $\mathbf{F}$-inference system. Let Q be a set. For every $q \in Q$, let $\models^q$ be a consequence relation over $\mathbf{G}$, let GInf^q be a $\mathbf{G}$-inference system, let Red^q be a redundancy criterion for $\models^q$ and GInf^q, and let $\mathcal{G}^q$ be a grounding function from $\mathbf{F}$ and *FInf* to $\mathbf{G}$ and GInf^q. Then for every $q \in Q$, the $(\mathcal{G}^q, \emptyset)$-lifting $\mathit{Red}^{q, \mathcal{G}^q, \emptyset}$ is a redundancy criterion for the $\mathcal{G}^q$-lifting $\models^q_{\mathcal{G}^q}$, and so $\mathit{Red}^{\cap\mathcal{G}}$ is a redundancy criterion for $\models^\cap_\mathcal{G}$ and *FInf*.

Now let $\mathbf{L}$ be a nonempty set of labels, and define $\mathbf{FL}$, $\mathbf{FL}_\bot$, and *FLInf* as above. For every $q \in Q$, define the function $\mathcal{G}^q_\mathbf{L}$ by $\mathcal{G}^q_\mathbf{L}((C, l)) := \mathcal{G}^q(C)$ for every $(C, l) \in \mathbf{FL}$ and by $\mathcal{G}^q_\mathbf{L}(\iota) := \mathcal{G}^q(\lfloor \iota \rfloor)$ for every $\iota \in \mathit{FLInf}$. Then for every $q \in Q$, the $(\mathcal{G}^q_\mathbf{L}, \emptyset)$-lifting $\mathit{Red}^{q, \mathcal{G}^q_\mathbf{L}} = (\mathit{Red}_\mathrm{I}^{q, \mathcal{G}^q_\mathbf{L}}, \mathit{Red}_\mathrm{F}^{q, \mathcal{G}^q_\mathbf{L}, \emptyset})$ of Red^q is a redundancy criterion for the $\mathcal{G}^q_\mathbf{L}$-lifting $\models^q_{\mathcal{G}^q_\mathbf{L}}$ of $\models^q$ and *FLInf*, and so

$$\mathit{Red}^{\cap\mathcal{G}_\mathbf{L}} := (\mathit{Red}_\mathrm{I}^{\cap\mathcal{G}_\mathbf{L}}, \mathit{Red}_\mathrm{F}^{\cap\mathcal{G}_\mathbf{L}}) := \Big(\bigcap\nolimits_{q \in Q} \mathit{Red}_\mathrm{I}^{q, \mathcal{G}^q_\mathbf{L}}, \bigcap\nolimits_{q \in Q} \mathit{Red}_\mathrm{F}^{q, \mathcal{G}^q_\mathbf{L}, \emptyset}\Big)$$

is a redundancy criterion for $\models^\cap_{\mathcal{G}_\mathbf{L}} := \bigcap_{q \in Q} \models^q_{\mathcal{G}^q_\mathbf{L}}$ and *FLInf*.

Lemma 22. *If a set* $\mathcal{N} \subseteq \mathbf{FL}$ *is saturated w.r.t. FLInf and* $\mathit{Red}^{\cap\mathcal{G}_\mathbf{L}}$, *then* $\lfloor \mathcal{N} \rfloor \subseteq \mathbf{F}$ *is saturated w.r.t. FInf and* $\mathit{Red}^{\cap\mathcal{G}}$.

Theorem 23. *If* $(\mathit{FInf}, \mathit{Red}^{\cap\mathcal{G}})$ *is statically refutationally complete w.r.t.* $\models^\cap_\mathcal{G}$, *then* $(\mathit{FLInf}, \mathit{Red}^{\cap\mathcal{G}_\mathbf{L}})$ *is statically refutationally complete w.r.t.* $\models^\cap_{\mathcal{G}_\mathbf{L}}$.

4 Prover Architectures

We now use the above results to prove the refutational completeness of a popular prover architecture: the given clause procedure [18]. The architecture is parameterized by an inference system and a redundancy criterion. A generalization of the architecture decouples scheduling and computation of inferences.

Given Clause Procedure. Let $\mathbf{F}$ and $\mathbf{G}$ be two sets of formulas, and let *FInf* be an $\mathbf{F}$-inference system without premise-free inferences. Let Q be a set. For every $q \in Q$, let $\models^q$ be a consequence relation over $\mathbf{G}$, let GInf^q be a $\mathbf{G}$-inference system, let Red^q be a redundancy criterion for $\models^q$ and GInf^q, and let $\mathcal{G}^q$ be a grounding function from $\mathbf{F}$ and *FInf* to $\mathbf{G}$ and GInf^q. Assume $(\mathit{FInf}, \mathit{Red}^{\cap\mathcal{G}})$ is statically refutationally complete w.r.t. $\models^{\cap}_{\mathcal{G}}$. Furthermore, let $\mathbf{L}$ be a nonempty set of labels, let $\mathbf{FL} := \mathbf{F} \times \mathbf{L}$, and let the $\mathbf{FL}$-inference system *FLInf* be a labeled version of *FInf*. By Theorem 23, $(\mathit{FLInf}, \mathit{Red}^{\cap\mathcal{G}_{\mathbf{L}}})$ is statically refutationally complete w.r.t. $\models^{\cap}_{\mathcal{G}_{\mathbf{L}}}$.

Let $\doteq$ be an equivalence relation on $\mathbf{F}$, let $\succ\!\!\cdot$ be a well-founded strict partial ordering on $\mathbf{F}$ such that $\succ\!\!\cdot$ is compatible with $\doteq$ (i.e., $C \succ\!\!\cdot D$, $C \doteq C'$, $D \doteq D'$ implies $C' \succ\!\!\cdot D'$), such that $C \doteq D$ implies $\mathcal{G}^q(C) = \mathcal{G}^q(D)$ for all $q \in Q$, and such that $C \succ\!\!\cdot D$ implies $\mathcal{G}^q(C) \subseteq \mathcal{G}^q(D)$ for all $q \in Q$. We define $\succeq\!\!\cdot \;:=\; \succ\!\!\cdot \cup \doteq$. In practice, $\doteq$ is typically α-renaming, $\succ\!\!\cdot$ is either the subsumption ordering $\gtrdot$ (provided it is well founded) or some well-founded ordering included in $\gtrdot$, and for every $q \in Q$, $\mathcal{G}^q$ maps every formula $C \in \mathbf{F}$ to the set of ground instances of C, possibly modulo some theory.

Let $\sqsupset$ be a well-founded strict partial ordering on $\mathbf{L}$. We define the ordering $\sqsupset$ on $\mathbf{FL}$ by $(C, l) \sqsupset (C', l')$ if either $C \succ\!\!\cdot C'$ or else $C \doteq C'$ and $l \sqsupset l'$. By Lemma 16, the static refutational completeness of $(\mathit{FLInf}, \mathit{Red}^{\cap\mathcal{G}_{\mathbf{L}}})$ w.r.t. $\models^{\cap}_{\mathcal{G}_{\mathbf{L}}}$ implies the static refutational completeness of $(\mathit{FLInf}, \mathit{Red}^{\cap\mathcal{G}_{\mathbf{L}},\sqsupset})$, which by Lemma 1 implies the dynamic refutational completeness of $(\mathit{FLInf}, \mathit{Red}^{\cap\mathcal{G}_{\mathbf{L}},\sqsupset})$.

This result may look intimidating, so let us unroll it. The $\mathbf{FL}$-inference system *FLInf* is a labeled version of *FInf*, which means that we get an *FLInf*-inference by first omitting the labels of the premises $(C_n, l_n), \ldots, (C_1, l_1)$, then performing an *FInf*-inference $(C_n, \ldots, C_0)$, and finally attaching an arbitrary label l_0 to the conclusion C_0. Since $\mathcal{G}^q_{\mathbf{L}}$ differs from $\mathcal{G}^q$ only by the omission of the labels and the first components of $\mathit{Red}^{\cap\mathcal{G}_{\mathbf{L}},\sqsupset}$ and $\mathit{Red}^{\cap\mathcal{G}_{\mathbf{L}}}$ agree, we get this result:

Lemma 24. *An FLInf-inference ι is redundant w.r.t. $\mathit{Red}^{\cap\mathcal{G}_{\mathbf{L}},\sqsupset}$ and $\mathcal{N}$ if and only if the underlying FInf-inference $\lfloor\iota\rfloor$ is redundant w.r.t. $\mathit{Red}^{\cap\mathcal{G}}$ and $\lfloor\mathcal{N}\rfloor$.*

Lemma 25. *Let $\mathcal{N} \subseteq \mathbf{FL}$, and let (C, l) be a labeled formula. Then $(C, l) \in \mathit{Red}_{\mathrm{F}}^{\cap\mathcal{G}_{\mathbf{L}},\sqsupset}(\mathcal{N})$ if* (i) *$C \in \mathit{Red}_{\mathrm{F}}^{\cap\mathcal{G}}(\lfloor\mathcal{N}\rfloor)$, or* (ii) *$C \succ\!\!\cdot C'$ for some $C' \in \lfloor\mathcal{N}\rfloor$, or* (iii) *$C \succeq\!\!\cdot C'$ for some $(C', l') \in \mathcal{N}$ with $l \sqsupset l'$.*

The given clause procedure that lies at the heart of saturation provers can be presented and studied abstractly. We assume that the set of labels $\mathbf{L}$ contains at least two values, including a distinguished $\sqsupset$-smallest value denoted by active, and that the labeled version *FLInf* of *FInf* never assigns active to a conclusion.

The state of a prover is a set of labeled formulas. The label identifies to which formula set each formula belongs. The active label identifies the active formula set from the familiar given clause procedure. The other, unspecified formula sets are considered passive. Given a set $\mathcal{N}$ and a label l, we define the projection $\mathcal{N}\!\downarrow_l$ as consisting only of the formulas labeled by l.

The given clause prover GC is defined as the following transition system:

PROCESS $\mathcal{N} \uplus \mathcal{M} \Longrightarrow_{\mathsf{GC}} \mathcal{N} \cup \mathcal{M}'$
where $\mathcal{M} \subseteq \mathit{Red}_{\mathrm{F}}^{\cap\mathcal{G}_{\mathbf{L}},\sqsupset}(\mathcal{N} \cup \mathcal{M}')$ and $\mathcal{M}'\!\downarrow_{\mathsf{active}} = \emptyset$
INFER $\mathcal{N} \uplus \{(C, l)\} \Longrightarrow_{\mathsf{GC}} \mathcal{N} \cup \{(C, \mathsf{active})\} \cup \mathcal{M}$
where $l \neq \mathsf{active}$, $\mathcal{M}\!\downarrow_{\mathsf{active}} = \emptyset$, and
$\mathit{FInf}(\lfloor \mathcal{N}\!\downarrow_{\mathsf{active}} \rfloor, \{C\}) \subseteq \mathit{Red}_{\mathrm{I}}^{\cap\mathcal{G}}(\lfloor \mathcal{N} \rfloor \cup \{C\} \cup \lfloor \mathcal{M} \rfloor)$

The PROCESS rule covers most operations performed in a theorem prover. By Lemma 25, this includes deleting $\mathit{Red}_{\mathrm{F}}^{\cap\mathcal{G}}$-redundant formulas with arbitrary labels and adding formulas that make other formulas $\mathit{Red}_{\mathrm{F}}^{\cap\mathcal{G}}$-redundant (i.e., simplifying w.r.t. $\mathit{Red}_{\mathrm{F}}^{\cap\mathcal{G}}$), by (i); deleting formulas that are $\dot{>}$-subsumed by other formulas with arbitrary labels, by (ii); deleting formulas that are $\dot{\geq}$-subsumed by other formulas with smaller labels, by (iii); and replacing the label of a formula by a smaller label different from active, also by (iii).

INFER is the only rule that puts a formula in the active set. It relabels a passive formula C to active and ensures that all inferences between C and the active formulas, including C itself, become redundant. Recall that by Lemma 24, $\mathit{FLInf}(\mathcal{N}\!\downarrow_{\mathsf{active}}, \{(C, \mathsf{active})\}) \subseteq \mathit{Red}_{\mathrm{I}}^{\cap\mathcal{G}_{\mathbf{L}}}(\mathcal{N} \cup \{(C, \mathsf{active})\} \cup \mathcal{M})$ if and only if $\mathit{FInf}(\lfloor \mathcal{N}\!\downarrow_{\mathsf{active}} \rfloor, \{C\}) \subseteq \mathit{Red}_{\mathrm{I}}^{\cap\mathcal{G}}(\lfloor \mathcal{N} \rfloor \cup \{C\} \cup \lfloor \mathcal{M} \rfloor)$. By property (R4), every inference is redundant if its conclusion is contained in the set of formulas, and typically, inferences are in fact made redundant by adding their conclusions to any of the passive sets. Then, $\lfloor \mathcal{M} \rfloor$ equals $\mathit{concl}(\mathit{FInf}(\lfloor \mathcal{N}\!\downarrow_{\mathsf{active}} \rfloor, \{C\}))$.

Since every $\Longrightarrow_{\mathsf{GC}}$-derivation is also a $\rhd_{\mathit{Red}^{\cap\mathcal{G}_{\mathbf{L}},\sqsupset}}$-derivation and ($\mathit{FLInf}$, $\mathit{Red}^{\cap\mathcal{G}_{\mathbf{L}},\sqsupset}$) is dynamically refutationally complete, it now suffices to show fairness to prove the refutational completeness of GC.

Lemma 26. *Let $(\mathcal{N}_i)_i$ be a $\Longrightarrow_{\mathsf{GC}}$-derivation. If $\mathcal{N}_0\!\downarrow_{\mathsf{active}} = \emptyset$ and $\mathcal{N}_*\!\downarrow_l = \emptyset$ for all $l \neq \mathsf{active}$, then $(\mathcal{N}_i)_i$ is a fair $\rhd_{\mathit{Red}^{\cap\mathcal{G}_{\mathbf{L}},\sqsupset}}$-derivation.*

Theorem 27. *Let $(\mathcal{N}_i)_i$ be a $\Longrightarrow_{\mathsf{GC}}$-derivation, where $\mathcal{N}_0\!\downarrow_{\mathsf{active}} = \emptyset$ and $\mathcal{N}_*\!\downarrow_l = \emptyset$ for all $l \neq \mathsf{active}$. If $\lfloor \mathcal{N}_0 \rfloor \models_{\mathcal{G}}^{\cap} \{\bot\}$ for some $\bot \in \mathbf{F}_\bot$, then some $\mathcal{N}_i$ contains $(\bot', l)$ for some $\bot' \in \mathbf{F}_\bot$ and $l \in \mathbf{L}$.*

Example 28. The following Otter loop [18, Sect. 2.3.1] prover OL is an instance of the given clause prover GC. This loop design is inspired by Weidenbach's prover without splitting from his *Handbook* chapter [30, Tables 4–6]. The prover's state is a five-tuple $N \mid X \mid P \mid Y \mid A$ of formula sets. The N, P, and A sets store the new, passive, and active formulas. The X and Y sets are subsingletons (i.e., sets of at most one element) that can store a chosen new or passive formula. Initial states are of the form $N \mid \emptyset \mid \emptyset \mid \emptyset \mid \emptyset$.

CHOOSEN $N \uplus \{C\} \mid \emptyset \mid P \mid \emptyset \mid A \Longrightarrow_{\mathsf{OL}} N \mid \{C\} \mid P \mid \emptyset \mid A$

DELETEFWD $N \mid \{C\} \mid P \mid \emptyset \mid A \Longrightarrow_{\mathsf{OL}} N \mid \emptyset \mid P \mid \emptyset \mid A$
  if $C \in \mathit{Red}_{\mathrm{F}}^{\cap\mathcal{G}}(P \cup A)$ or $C \mathrel{\dot{\succeq}} C'$ for some $C' \in P \cup A$

SIMPLIFYFWD $N \mid \{C\} \mid P \mid \emptyset \mid A \Longrightarrow_{\mathsf{OL}} N \mid \{C'\} \mid P \mid \emptyset \mid A$
  if $C \in \mathit{Red}_{\mathrm{F}}^{\cap\mathcal{G}}(P \cup A \cup \{C'\})$

DELETEBWDP $N \mid \{C\} \mid P \uplus \{C'\} \mid \emptyset \mid A \Longrightarrow_{\mathsf{OL}} N \mid \{C\} \mid P \mid \emptyset \mid A$
  if $C' \in \mathit{Red}_{\mathrm{F}}^{\cap\mathcal{G}}(\{C\})$ or $C' \mathrel{\dot{\succ}} C$

SIMPLIFYBWDP $N \mid \{C\} \mid P \uplus \{C'\} \mid \emptyset \mid A \Longrightarrow_{\mathsf{OL}} N \cup \{C''\} \mid \{C\} \mid P \mid \emptyset \mid A$
  if $C' \in \mathit{Red}_{\mathrm{F}}^{\cap\mathcal{G}}(\{C, C''\})$

DELETEBWDA $N \mid \{C\} \mid P \mid \emptyset \mid A \uplus \{C'\} \Longrightarrow_{\mathsf{OL}} N \mid \{C\} \mid P \mid \emptyset \mid A$
  if $C' \in \mathit{Red}_{\mathrm{F}}^{\cap\mathcal{G}}(\{C\})$ or $C' \mathrel{\dot{\succ}} C$

SIMPLIFYBWDA $N \mid \{C\} \mid P \mid \emptyset \mid A \uplus \{C'\} \Longrightarrow_{\mathsf{OL}} N \cup \{C''\} \mid \{C\} \mid P \mid \emptyset \mid A$
  if $C' \in \mathit{Red}_{\mathrm{F}}^{\cap\mathcal{G}}(\{C, C''\})$

TRANSFER $N \mid \{C\} \mid P \mid \emptyset \mid A \Longrightarrow_{\mathsf{OL}} N \mid \emptyset \mid P \cup \{C\} \mid \emptyset \mid A$

CHOOSEP $\emptyset \mid \emptyset \mid P \uplus \{C\} \mid \emptyset \mid A \Longrightarrow_{\mathsf{OL}} \emptyset \mid \emptyset \mid P \mid \{C\} \mid A$

INFER $\emptyset \mid \emptyset \mid P \mid \{C\} \mid A \Longrightarrow_{\mathsf{OL}} M \mid \emptyset \mid P \mid \emptyset \mid A \cup \{C\}$
  if $\mathit{FInf}(A, \{C\}) \subseteq \mathit{Red}_{\mathrm{I}}^{\cap\mathcal{G}}(A \cup \{C\} \cup M)$

A reasonable strategy for applying the OL rules is presented below. It relies on a well-founded ordering $\succ$ on formulas to ensure that the backward simplification rules actually "simplify" their target, preventing nontermination of the inner loop. It also assumes that $\mathit{FInf}(N, \{C\})$ is finite if N is finite. Briefly, the strategy corresponds to the regular expression $\big((\text{CHOOSEN};\ \text{SIMPLIFYFWD}^*;\ (\text{DELETEFWD} \mid (\text{DELETEBWDP}^*;\ \text{DELETEBWDA}^*;\ \text{SIMPLIFYBWDP}^*;\ \text{SIMPLIFYBWDA}^*;\ \text{TRANSFER})))^*;\ (\text{CHOOSEP};\ \text{INFER})^?\big)^*$, where ; denotes concatenation and $*$ and ? are given an eager semantics. Simplifications are applicable only if the result is $\succ$-smaller than the original formula. Moreover, CHOOSEC always chooses the oldest formula in N, and the choice of C in CHOOSEP must be fair.

The instantiation of GC relies on five labels $l_1 \sqsupset \cdots \sqsupset l_5 = \mathsf{active}$ representing N, X, P, Y, A. Let $(N_i \mid X_i \mid P_i \mid Y_i \mid A_i)_i$ be a derivation following the strategy, where N_0 is finite and $X_0 = P_0 = Y_0 = A_0 = \emptyset$. We can show that $N_* = X_* = P_* = Y_* = \emptyset$. Therefore, by Theorem 27, OL is dynamically refutationally complete.

In most calculi, *Red* is defined in terms of some total and well-founded ordering $\succ_{\mathbf{G}}$ on $\mathbf{G}$. We can then define $\succ$ so that $C \succ C'$ if the smallest element of $\mathcal{G}^q(C)$ is greater than the smallest element of $\mathcal{G}^q(C')$ w.r.t. $\succ_{\mathbf{G}}$, for some arbitrary fixed $q \in Q$. This allows a wide range of simplifications typically implemented in superposition provers. To ensure fairness when applying CHOOSEP, one approach is to use an $\mathbb{N}$-valued weight function that is strictly antimonotone in the age of the formula [22, Sect. 4]. Another option is to alternate between heuristically choosing n formulas and taking the oldest formula [18, Sect. 2.3.1]. To guarantee soundness, we can require that the formulas added by simplification and INFER are $\models_\approx$-entailed by the formulas in the state before the transition. This can be relaxed to consistency-preservation, e.g., for calculi that perform skolemization.

Example 29. Bachmair and Ganzinger's resolution prover RP [5, Sect. 4.3] is another instance of GC. It embodies both a concrete prover architecture and a concrete inference system: ordered resolution with selection ($\mathsf{O}_S^{\succ}$). States are triples $N \mid P \mid O$ of finite clause sets. The instantiation relies on three labels $l_1 \sqsupset l_2 \sqsupset l_3 = \mathsf{active}$. Subsumption can be supported as described in Example 10.

Delayed Inferences. An *orphan* is a passive formula that was generated by an inference for which at least one premise is no longer active. The given clause prover GC presented above is sufficient to describe a prover based on an Otter loop as well as a basic DISCOUNT loop prover, but to describe a DISCOUNT loop prover with orphan deletion, we need to decouple the scheduling of inferences and their computation. The same scheme can be used for inference systems that contain premise-free inferences or that may generate infinitely many conclusions from finitely many premises. Yet another use of the scheme is to save memory: A delayed inference can be stored more compactly than a new formula, as a tuple of premises together with instructions on how to compute the conclusion.

The lazy given clause prover LGC generalizes GC. It is defined as the following transition system on pairs $(T, \mathcal{N})$, where T ("to do") is a set of inferences and $\mathcal{N}$ is a set of labeled formulas. We use the same assumptions as for GC except that we now permit premise-free inferences in *FInf*. Initially, T consists of all premise-free inferences of *FInf*.

PROCESS $(T, \mathcal{N} \uplus \mathcal{M}) \Longrightarrow_{\mathsf{LGC}} (T, \mathcal{N} \cup \mathcal{M}')$
  where $\mathcal{M} \subseteq \mathit{Red}_{\mathrm{F}}^{\cap\mathcal{G}_{\mathrm{L}},\sqsupset}(\mathcal{N} \cup \mathcal{M}')$ and $\mathcal{M}'\!\downarrow_{\mathsf{active}} = \emptyset$

SCHEDULEINFER $(T, \mathcal{N} \uplus \{(C, l)\}) \Longrightarrow_{\mathsf{LGC}} (T \cup T', \mathcal{N} \cup \{(C, \mathsf{active})\})$
  where $l \neq \mathsf{active}$ and $T' = \mathit{FInf}(\lfloor \mathcal{N}\!\downarrow_{\mathsf{active}} \rfloor, \{C\})$

COMPUTEINFER $(T \uplus \{\iota\}, \mathcal{N}) \Longrightarrow_{\mathsf{LGC}} (T, \mathcal{N} \cup \mathcal{M})$
  where $\mathcal{M}\!\downarrow_{\mathsf{active}} = \emptyset$ and $\iota \in \mathit{Red}_{\mathrm{I}}^{\cap\mathcal{G}}(\lfloor \mathcal{N} \cup \mathcal{M} \rfloor)$

DELETEORPHANS $(T \uplus T', \mathcal{N}) \Longrightarrow_{\mathsf{LGC}} (T, \mathcal{N})$
  where $T' \cap \mathit{FInf}(\lfloor \mathcal{N}\!\downarrow_{\mathsf{active}} \rfloor) = \emptyset$

SCHEDULEINFER relabels a passive formula C to active and puts all inferences between C and the active formulas, including C itself, into the set T. COMPUTEINFER removes an inference from T and makes it redundant by adding appropriate labeled formulas to $\mathcal{N}$ (typically the conclusion of the inference). DELETEORPHANS can delete scheduled inferences from T if some of their premises have been deleted from $\mathcal{N}\!\downarrow_{\mathsf{active}}$ in the meantime. Note that the rule cannot delete premise-free inferences, since the side condition is then vacuously false.

Abstractly, the T component of the state is a set of inferences $(C_n, \ldots, C_0)$. In an actual implementation, it can be represented in different ways: as a set of compactly encoded recipes for computing the conclusion C_0 from the premises $(C_n, \ldots, C_1)$ as in Waldmeister [16], or as a set of explicit formulas C_0 with information about their parents $(C_n, \ldots, C_1)$ as in E [25]. In the latter case, some presimplifications may be performed on C_0; this could be modeled more faithfully by defining T as a set of pairs $(\iota, \mathit{simp}(C_0))$.

Lemma 30. *If $(T_i, \mathcal{N}_i)_i$ is a $\Longrightarrow_{\mathsf{LGC}}$-derivation, then $(\mathcal{N}_i)_i$ is a $\rhd_{\mathit{Red}^{\cap\mathcal{G}_{\mathrm{L}}},\sqsupset}$- derivation.*

Lemma 31. *Let $(T_i, \mathcal{N}_i)_i$ be a $\Longrightarrow_{\mathsf{LGC}}$-derivation. If $\mathcal{N}_0{\downarrow_{\mathsf{active}}} = \emptyset$, $\mathcal{N}_*{\downarrow_l} = \emptyset$ for all $l \neq \mathsf{active}$, T_0 is the set of all premise-free inferences of FInf, and $T_* = \emptyset$, then $(\mathcal{N}_i)_i$ is a fair $\rhd_{\mathit{Red}^{\cap\mathcal{G}_{\mathrm{L}}},\sqsupset}$-derivation.*

Theorem 32. *Let $(T_i, \mathcal{N}_i)_i$ be a $\Longrightarrow_{\mathsf{LGC}}$-derivation, where $\mathcal{N}_0{\downarrow_{\mathsf{active}}} = \emptyset$, $\mathcal{N}_*{\downarrow_l} = \emptyset$ for all $l \neq \mathsf{active}$, T_0 is the set of all premise-free inferences of FInf, and $T_* = \emptyset$. If $\lfloor\mathcal{N}_0\rfloor \models^{\cap}_{\mathcal{G}} \{\bot\}$ for some $\bot \in \mathbf{F}_\bot$, then some $\mathcal{N}_i$ contains $(\bot', l)$ for some $\bot' \in \mathbf{F}_\bot$ and $l \in \mathbf{L}$.*

Example 33. The following DISCOUNT loop [1] prover DL is an instance of the lazy given clause prover LGC. This loop design is inspired by the description of E [25]. The prover's state is a four-tuple $T \mid P \mid Y \mid A$, where T is a set of inferences and P, Y, A are sets of formulas. The T, P, and A sets correspond to the scheduled inferences, the passive formulas, and the active formulas. The Y set is a subsingleton that can store a chosen passive formula. Initial states have the form $T \mid P \mid \emptyset \mid \emptyset$, where T is the set of all premise-free inferences of FInf.

ComputeInfer $T \uplus \{\iota\} \mid P \mid \emptyset \mid A \Longrightarrow_{\mathsf{DL}} T \mid P \mid \{C\} \mid A$
  if $\iota \in \mathit{Red}_{\mathrm{I}}^{\cap\mathcal{G}}(A \cup \{C\})$

ChooseP $T \mid P \uplus \{C\} \mid \emptyset \mid A \Longrightarrow_{\mathsf{DL}} T \mid P \mid \{C\} \mid A$

DeleteFwd $T \mid P \mid \{C\} \mid A \Longrightarrow_{\mathsf{DL}} T \mid P \mid \emptyset \mid A$
  if $C \in \mathit{Red}_{\mathrm{F}}^{\cap\mathcal{G}}(A)$ or $C \mathrel{\dot{\succeq}} C'$ for some $C' \in A$

SimplifyFwd $T \mid P \mid \{C\} \mid A \Longrightarrow_{\mathsf{DL}} T \mid P \mid \{C'\} \mid A$
  if $C \in \mathit{Red}_{\mathrm{F}}^{\cap\mathcal{G}}(A \cup \{C'\})$

DeleteBwd $T \mid P \mid \{C\} \mid A \uplus \{C'\} \Longrightarrow_{\mathsf{DL}} T \mid P \mid \{C\} \mid A$
  if $C' \in \mathit{Red}_{\mathrm{F}}^{\cap\mathcal{G}}(\{C\})$ or $C' \mathrel{\dot{\succ}} C$

SimplifyBwd $T \mid P \mid \{C\} \mid A \uplus \{C'\} \Longrightarrow_{\mathsf{DL}} T \mid P \cup \{C''\} \mid \{C\} \mid A$
  if $C' \in \mathit{Red}_{\mathrm{F}}^{\cap\mathcal{G}}(\{C, C''\})$

ScheduleInfer $T \mid P \mid \{C\} \mid A \Longrightarrow_{\mathsf{DL}} T \cup T' \mid P \mid \emptyset \mid A \cup \{C\}$
  if $T' = \mathit{FInf}(A, \{C\})$

DeleteOrphans $T \uplus T' \mid P \mid Y \mid A \Longrightarrow_{\mathsf{DL}} T \mid P \mid Y \mid A$
  if $T' \cap \mathit{FInf}(A) = \emptyset$

A reasonable strategy for applying the DL rules along the lines of that for OL and with the same assumptions follows: ((ComputeInfer | ChooseP);

SimplifyFwd*; (DeleteFwd | (DeleteBwd*; SimplifyBwd*; DeleteOrphans; ScheduleInfer)))*. In ComputeInfer, the first formula from $T \cup P$, organized as a single queue, is chosen. The instantiation of LGC relies on three labels $l_1 \sqsupset \cdots \sqsupset l_3 = \mathsf{active}$ corresponding to the sets P, Y, A.

Example 34. Higher-order unification can give rise to infinitely many incomparable unifiers. As a result, in clausal λ-superposition [9], performing all inferences between two clauses can lead to infinitely many conclusions, which need to be enumerated fairly. The Zipperposition prover [9] performs this enumeration in an extended DISCOUNT loop. Another instance of infinitary inferences is the n-ary Acycl and Uniq rules of superposition with (co)datatypes [14].

Abstractly, a Zipperposition loop prover ZL operates on states $T \mid P \mid Y \mid A$, where T is organized as a finite set of possibly infinite sequences $(\iota_i)_i$ of inferences, and P, Y, A are as in DL. The ChooseP, DeleteFwd, SimplifyFwd, DeleteBwd, and SimplifyBwd rules are as in DL. The other rules follow:

ComputeInfer $T \uplus \{(\iota_i)_i\} \mid P \mid \emptyset \mid A \Longrightarrow_{\mathsf{ZL}} T \cup \{(\iota_i)_{i \geq 1}\} \mid P \cup \{C\} \mid \emptyset \mid A$
  if $\iota_0 \in \mathit{Red}_{\mathrm{I}}^{\cap \mathcal{G}}(A \cup \{C\})$

ScheduleInfer $T \mid P \mid \{C\} \mid A \Longrightarrow_{\mathsf{ZL}} T \cup T' \mid P \mid \emptyset \mid A \cup \{C\}$
  if T' is a finite set of sequences $(\iota_i^j)_i$ of inferences such that the set of all ι_i^j equals $\mathit{FInf}(A, \{C\})$

DeleteOrphan $T \uplus \{(\iota_i)_i\} \mid P \mid Y \mid A \Longrightarrow_{\mathsf{ZL}} T \mid P \mid Y \mid A$
  if $\iota_i \notin \mathit{FInf}(A)$ for all i

ComputeInfer works on the first element of sequences. ScheduleInfer adds new sequences to T. Typically, these sequences store $\mathit{FInf}(A, \{C\})$, which may be countably infinite, in such a way that all inferences in one sequence have identical premises and can be removed together by DeleteOrphan. To produce fair derivations, a prover needs to choose the sequence in ComputeInfer fairly and to choose the formula in ChooseP fairly, thereby achieving dovetailing.

Example 35. The prover architectures described above can be instantiated with saturation calculi that use a redundancy criterion obtained as an intersection of lifted redundancy criteria. Most calculi are defined in such a way that this requirement is obviously satisfied. The outlier is unfailing completion [2].

Although unfailing completion predates the introduction of Bachmair–Ganzinger-style redundancy, it can be incorporated into that framework by defining that formulas (i.e., rewrite rules and equations) and inferences (i.e., orientation and critical pair computation) are redundant if for every rewrite proof using that rewrite rule, equation, or critical peak, there exists a smaller rewrite proof. The requirement that the redundancy criterion must be obtained by lifting (which is necessary to introduce the labeling) can then be trivially fulfilled by "self-lifting"—i.e., by defining $\mathbf{G} := \mathbf{F}$ and $\cdot\!\succ := \emptyset$ and by taking $\mathcal{G}$ as the function that maps every formula or inference to the set of its α-renamings.

5 Isabelle Development

The framework described in the previous sections has been formalized in Isabelle/HOL [20,21], including all the theorems and lemmas and the prover architectures GC and LGC but excluding the examples. The Isabelle theory files are available in the *Archive of Formal Proofs* [26]. The development is also part of the IsaFoL (Isabelle Formalization of Logic) [12] effort, which aims at developing a reusable computer-checked library of results about automated reasoning.

The development relies heavily on Isabelle's locales [8]. These are contexts that fix variables and make assumptions about these. With locales, the definitions and lemmas look similar to how they are stated on paper, but the proofs often become more complicated: Layers of locales may hide definitions, and often these need to be manually unfolded before the desired lemma can be proved.

We chose to represent basic nonempty sets such as $\mathbf{F}$ and $\mathbf{L}$ by types. It relieved us from having to thread through nonemptiness conditions. Moreover, objects are automatically typed, meaning that lemmas could be stated without explicit hypotheses that given objects are formulas, labels, or indices. On the other hand, for sets such as $\mathbf{F}_\bot$ and *FInf* that are subsets of other sets, it was natural to use simply typed sets. Derivations, which are used to describe the dynamic behavior of a calculus, are represented by the same lazy list codatatype [13] and auxiliary definitions that were used in the mechanization of the ordered resolution prover RP (Example 29) by Schlichtkrull et al. [23,24].

The framework's design and its mechanization were carried out largely in parallel. This resulted in more work on the mechanization side, but it also helped shape the theory itself. In particular, an attempt at verifying RP in Isabelle using an earlier version of the framework made it clear that the theory was not general enough yet to support selection functions (Example 18). In ongoing work, we are completing the RP proof and are developing a verified superposition prover.

6 Conclusion

We presented a formal framework for saturation theorem proving inspired by Bachmair and Ganzinger's *Handbook* chapter [5]. Users can conveniently derive a dynamic refutational completeness result for a concrete prover based on a statically refutationally complete calculus. The key was to strengthen the standard redundancy criterion so that all prover operations, including subsumption deletion, can be justified by inference or redundancy. The framework is mechanized in Isabelle/HOL, where it can be instantiated to verify concrete provers.

To employ the framework, the starting point is a statically complete saturation calculus that can be expressed as the lifting $(\mathit{FInf}, \mathit{Red}^{\mathcal{G}})$ or $(\mathit{FInf}, \mathit{Red}^{\cap\mathcal{G}})$ of a ground calculus $(\mathit{GInf}, \mathit{Red})$, where *Red* qualifies as a redundancy criterion and $\mathcal{G}$ qualifies as a grounding function or grounding function family. The framework can be used to derive two main results:

1. After defining a well-founded ordering $\sqsupset$ that captures subsumption, invoke Theorem 17 to show $(\mathit{FInf}, \mathit{Red}^{\cap\mathcal{G},\sqsupset})$ dynamically complete.

2. Based on the previous step, invoke Theorem 27 or 32 to derive the dynamic completeness of a prover architecture building on the given clause procedure, such as the Otter loop, the DISCOUNT loop, or the Zipperposition loop.

The framework can also help establish the static completeness of the nonground calculus. For many calculi (with the notable exceptions of constraint superposition and hierarchic superposition), Theorem 5 or 14 can be used to lift the static completeness of $(\mathit{GInf}, \mathit{Red})$ to $(\mathit{FInf}, \mathit{Red}^{\mathcal{G}})$ or $(\mathit{FInf}, \mathit{Red}^{\cap\mathcal{G}})$.

The main missing piece of the framework is a generic treatment of clause splitting. The only formal treatment of splitting we are aware of, by Fietzke and Weidenbach [15], hard-codes both the underlying calculus and the splitting strategy. Voronkov's AVATAR architecture [27] is more flexible and yields impressive empirical results, but it offers no dynamic completeness guarantees.

Acknowledgment. We thank Alexander Bentkamp for discussions about prover architectures for higher-order logic and for feedback from instantiating the framework, Mathias Fleury and Christian Sternagel for their help with the Isabelle development, and Robert Lewis, Visa Nummelin, Dmitriy Traytel, and the anonymous reviewers for their comments and suggestions. Blanchette's research has received funding from the European Research Council (ERC) under the European Union's Horizon 2020 research and innovation program (grant agreement No. 713999, Matryoshka). He also benefited from the Netherlands Organization for Scientific Research (NWO) Incidental Financial Support scheme and he has received funding from the NWO under the Vidi program (project No. 016.Vidi.189.037, Lean Forward).

References

1. Avenhaus, J., Denzinger, J., Fuchs, M.: DISCOUNT: a system for distributed equational deduction. In: Hsiang, J. (ed.) RTA 1995. LNCS, vol. 914, pp. 397–402. Springer, Heidelberg (1995). https://doi.org/10.1007/3-540-59200-8_72
2. Bachmair, L., Dershowitz, N., Plaisted, D.A.: Completion without failure. In: Aït-Kaci, H., Nivat, M. (eds.) Rewriting Techniques—Resolution of Equations in Algebraic Structures, vol. 2, pp. 1–30. Academic Press (1989)
3. Bachmair, L., Ganzinger, H.: On restrictions of ordered paramodulation with simplification. In: Stickel, M.E. (ed.) CADE 1990. LNCS, vol. 449, pp. 427–441. Springer, Heidelberg (1990). https://doi.org/10.1007/3-540-52885-7_105
4. Bachmair, L., Ganzinger, H.: Rewrite-based equational theorem proving with selection and simplification. J. Log. Comput. **4**(3), 217–247 (1994)
5. Bachmair, L., Ganzinger, H.: Resolution theorem proving. In: Robinson, A., Voronkov, A. (eds.) Handbook of Automated Reasoning, vol. I, pp. 19–99. Elsevier and MIT Press (2001)
6. Bachmair, L., Ganzinger, H., Waldmann, U.: Superposition with simplification as a decision procedure for the monadic class with equality. In: Gottlob, G., Leitsch, A., Mundici, D. (eds.) KGC 1993. LNCS, vol. 713, pp. 83–96. Springer, Heidelberg (1993). https://doi.org/10.1007/BFb0022557
7. Bachmair, L., Ganzinger, H., Waldmann, U.: Refutational theorem proving for hierarchic first-order theories. Appl. Algebra Eng. Commun. Comput. **5**, 193–212 (1994)
8. Ballarin, C.: Locales: a module system for mathematical theories. J. Autom. Reason. **52**(2), 123–153 (2014)

9. Bentkamp, A., Blanchette, J., Tourret, S., Vukmirović, P., Waldmann, U.: Superposition with lambdas. In: Fontaine, P. (ed.) CADE 2019. LNCS (LNAI), vol. 11716, pp. 55–73. Springer, Cham (2019). https://doi.org/10.1007/978-3-030-29436-6_4
10. Bentkamp, A., Blanchette, J.C., Cruanes, S., Waldmann, U.: Superposition for lambda-free higher-order logic. In: Galmiche, D., Schulz, S., Sebastiani, R. (eds.) IJCAR 2018. LNCS (LNAI), vol. 10900, pp. 28–46. Springer, Cham (2018). https://doi.org/10.1007/978-3-319-94205-6_3
11. Bhayat, A., Reger, G.: A combinator-based superposition calculus for higher-order logic. In: Peltier, N., Sofronie-Stokkermans, V. (eds.) IJCAR 2020. LNCS (LNAI). Springer, Heidelberg (2020)
12. Blanchette, J.C.: Formalizing the metatheory of logical calculi and automatic provers in Isabelle/HOL (invited talk). In: Mahboubi, A., Myreen, M.O. (eds.) CPP 2019, pp. 1–13. ACM (2019)
13. Blanchette, J.C., Hölzl, J., Lochbihler, A., Panny, L., Popescu, A., Traytel, D.: Truly modular (Co)datatypes for Isabelle/HOL. In: Klein, G., Gamboa, R. (eds.) ITP 2014. LNCS, vol. 8558, pp. 93–110. Springer, Cham (2014). https://doi.org/10.1007/978-3-319-08970-6_7
14. Blanchette, J.C., Peltier, N., Robillard, S.: Superposition with datatypes and codatatypes. In: Galmiche, D., Schulz, S., Sebastiani, R. (eds.) IJCAR 2018. LNCS (LNAI), vol. 10900, pp. 370–387. Springer, Cham (2018). https://doi.org/10.1007/978-3-319-94205-6_25
15. Fietzke, A., Weidenbach, C.: Labelled splitting. Ann. Math. Artif. Intell. **55**(1–2), 3–34 (2009)
16. Hillenbrand, T., Löchner, B.: The next WALDMEISTER loop. In: Voronkov, A. (ed.) CADE 2002. LNCS (LNAI), vol. 2392, pp. 486–500. Springer, Heidelberg (2002). https://doi.org/10.1007/3-540-45620-1_38
17. Huet, G.P.: A mechanization of type theory. In: Nilsson, N.J. (ed.) IJCAI 1973, pp. 139–146. William Kaufmann (1973)
18. McCune, W., Wos, L.: Otter—the CADE-13 competition incarnations. J. Autom. Reason. **18**(2), 211–220 (1997)
19. Nieuwenhuis, R., Rubio, A.: Theorem proving with ordering and equality constrained clauses. J. Symb. Comput. **19**(4), 321–351 (1995)
20. Nipkow, T., Klein, G.: Concrete Semantics: With Isabelle/HOL. Springer, Heidelberg (2014). https://doi.org/10.1007/978-3-319-10542-0
21. Nipkow, T., Paulson, L.C., Wenzel, M.: Isabelle/HOL: A proof assistant for higher-order logic. LNCS, vol. 2283. Springer, Heidelberg (2002). https://doi.org/10.1007/3-540-45949-9
22. Schlichtkrull, A., Blanchette, J.C., Traytel, D.: A verified prover based on ordered resolution. In: Mahboubi, A., Myreen, M.O. (eds.) CPP 2019, pp. 152–165. ACM (2019)
23. Schlichtkrull, A., Blanchette, J.C., Traytel, D., Waldmann, U.: Formalization of Bachmair and Ganzinger's ordered resolution prover. Archive of Formal Proofs 2018 (2018). https://www.isa-afp.org/entries/Ordered_Resolution_Prover.html
24. Schlichtkrull, A., Blanchette, J.C., Traytel, D., Waldmann, U.: Formalizing Bachmair and Ganzinger's ordered resolution prover. In: Galmiche, D., Schulz, S., Sebastiani, R. (eds.) IJCAR 2018. LNCS, vol. 10900, pp. 89–107. Springer, Heidelberg (2018). https://doi.org/10.1007/978-3-319-94205-6_7
25. Schulz, S.: E—a brainiac theorem prover. AI Commun. **15**(2–3), 111–126 (2002)
26. Tourret, S.: A comprehensive framework for saturation theorem proving. Arch. Formal Proofs 2020 (2020). https://www.isa-afp.org/entries/Saturation_Framework.shtml

27. Voronkov, A.: AVATAR: the architecture for first-order theorem provers. In: Biere, A., Bloem, R. (eds.) CAV 2014. LNCS, vol. 8559, pp. 696–710. Springer, Cham (2014). https://doi.org/10.1007/978-3-319-08867-9_46
28. Waldmann, U.: Cancellative abelian monoids and related structures in refutational theorem proving (part I). J. Symb. Comput. **33**(6), 777–829 (2002)
29. Waldmann, U., Tourret, S., Robillard, S., Blanchette, J.: A comprehensive framework for saturation theorem proving (technical report). Technical report (2020). http://matryoshka.gforge.inria.fr/pubs/saturate_report.pdf
30. Weidenbach, C.: Combining superposition, sorts and splitting. In: Robinson, A., Voronkov, A. (eds.) Handbook of Automated Reasoning, vol. II, pp. 1965–2013. Elsevier and MIT Press (2001)

Proof Procedures

Possible Models Computation and Revision – A Practical Approach

Peter Baumgartner[1,2](✉)

[1] Data61/CSIRO, Canberra, Australia
Peter.Baumgartner@data61.csiro.au
[2] The Australian National University, Canberra, Australia

Abstract. This paper describes a method of computing plausible states of a system as a logical model. The problem of analyzing state-based systems as they evolve over time has been studied widely in the automated reasoning community (and others). This paper proposes a specific approach, one that is tailored to situational awareness applications. The main contribution is a calculus for a novel specification language that is built around disjunctive logic programming under a possible models semantics, stratification in terms of event times, default negation, and a model revision operator for dealing with incomplete or erroneous events – a typical problem in realistic applications. The paper proves the calculus correct wrt. a formal semantics of the specification language and it describes the calculus' implementation via embedding in Scala. This enables immediate access to rich data structures and external systems, which is important in practice.

1 Introduction

This paper is concerned with logic-based modeling and automated reasoning for estimating the current state of a system as it evolves over time. The main motivation is situational awareness [12], which requires the ability to understand and explain a system's state, at any time, and at a level that matters to the user, even if only partial or incorrect information about the external events leading to that state is available. In a supply chain context, for example, one cannot expect that events are reported correctly and in a timely manner. Sensors may fail, transmission channels are laggy, reports exist only in paper form, not every player is willing to share information, etc. Because of that, it is often impossible to know with full certainty the actual state of the system. The paper addresses this problem and proposes to instead derive a set of *plausible candidate states* as an approximation of ground truth. The states may include consequences relevant for situational awareness, e.g., that a shipment will be late. A human operator may then make decisions or provide additional details, this way closing the loop.

The plausible candidate states are represented as models of a logical specification and a given a set of external timestamped events. The proposed modeling

This research is supported by the Science and Industry Endowment Fund.

N. Peltier and V. Sofronie-Stokkermans (Eds.): IJCAR 2020, LNAI 12166, pp. 337–355, 2020.
https://doi.org/10.1007/978-3-030-51074-9_19

paradigm is logic programming, and models are computed in a bottom-up way. It adopts notions of stratification, default negation and a possible model semantics for its (disjunctive) program rules. Stratification is in terms of event time, with increasing time horizons for anytime reasoning; default negation is needed to reason in absence of information such as event reports; disjunctions are needed to derive alternate candidate states. In order to deal with less-than-perfect event data, the modeling language features a novel model revision operator that allows the programmer to formulate conditions under which a model computation with a corrected set of events should be attempted in an otherwise inconsistent state. The following informal overview illustrates these features.

1.1 Main Ideas and Design Rationale

A model, or program, is comprised of a set of rules of the form *head* $\leftarrow$ *body*. The *head* can be a non-empty disjunction of atoms, or a **fail** head. The former rules open the solution (models) space for a fixed set of external events, while **fail** head rules limit it and specify amended event sets for new solution attempts. The *body* is a conjunction of atoms and negated (via "**not**") conjunctions of atoms. Negation is "default negation", i.e., a closed world assumption is in place for evaluating the latter. Rules may contain first-order variables and must be range restricted. This guarantees that only ground heads can be derived from ground facts when a rule is evaluated in a bottom-up way. Our notion of range restriction is somewhat non-standard, though, and permits extra variables inside negation. These variables are implicitly existentially quantified ("**not** $\exists x \ldots$ ").

We need, however, syntactic restrictions that enforce stratification in terms of "time". This entails that rule evaluation does not depend on facts from the future. In fact, this is a reasonable assumption for situational awareness, whose any-time character requires to understand the current situation based on the available information up to now *only*. Technically, every atom must have a dedicated "time" argument, a $\mathbb{N}$-sorted variable, which, together, with earlier-than constraints (via "$<$" or "$\leq$") enforces stratification. The details of that will have to wait until later (Definition 1). An example rule is

$$\mathsf{hungry}(t, x) \vee \mathsf{thirsty}(t, x) \leftarrow \mathsf{get\_up}(t, x), \mathbf{not}(t - 6 \leq s, s \leq t, \mathsf{meal}(s, x)) \ . \quad (1)$$

It could say "if x gets up at time t and didn't have a meal in 6 hours prior then x is hungry or thirsty at t, *or both*". A set of facts, say, $\{\mathsf{get\_up}(8, \mathsf{bob}), \mathsf{meal}(12, \mathsf{bob})\}$ then entails $\mathsf{hungry}(8, \mathsf{bob}) \vee \mathsf{thirsty}(8, \mathsf{bob})$. Notice that in the relevant rule instance the negated body element $\mathbf{not}(8 - 6 \leq s, s < 8, \mathsf{meal}(s, \mathsf{bob}))$ is satisfied by the facts (using the closed-world assumption), as for the only relevant meal-instance $\mathsf{meal}(12, \mathsf{bob})$ the arithmetic constraint is false. The possible model semantics [26], which we adopt, interprets disjunctions (also) inclusively. Each resulting case $\{\mathsf{hungry}(8, \mathsf{bob})\}$, $\{\mathsf{thirsty}(8, \mathsf{bob})\}$ and $\{\mathsf{hungry}(8, \mathsf{bob}), \mathsf{thirsty}(8, \mathsf{bob})\}$ together with the facts yields a possible model.

Stratification in terms of time makes default negation with existentially quantified variables possible; no need to look into the future. We impose a secondary kind of stratification that also makes it more *efficient*. It rests on distinguishing two types of atoms: EDB atoms and IDB atoms (extensional/intensional database, respectively). EDB atoms are for external events, the given facts, and IDB atoms are for derived facts. A disjunctive *head* can contain IDB atoms only. Now, an IDB atom within a negation has to be strictly earlier ($<$) than the head, while an EDB atom within a negation can be non-strictly earlier ($\leq$). This make sure that a truth value for a negated expression cannot change later, in the course of rule evaluation in increasing time order. The rule (1) above is stratified if meal is EDB, otherwise "$s \leq t$" would have to be "$s < t$". A rule like $\mathsf{hungry}(t, x) \leftarrow \mathsf{get\_up}(t, x), \textbf{not}\, \mathsf{hungry}(t, x)$ cannot be stratified.

Rules with **fail** heads enable the programmer to specify when a (partial) model candidate is unsatisfiable *and to say how to potentially fix this situation.* A rule of the form **fail**() ← *body* without arguments to **fail** is a usual integrity constraint, e.g.:

$$\textbf{fail}() \leftarrow \mathsf{hungry}(t, x), \mathsf{eat}(s, x), t - 4 \leq s, s < t \tag{2}$$

The rule (2) rejects that x is both hungry and has eaten within the last 4 hours. Together with rule (1) and the EDB $\{\mathsf{eat}(7, \mathsf{bob}), \mathsf{get\_up}(8, \mathsf{bob})\}$ one obtains the sole possible model $\{\mathsf{eat}(7, \mathsf{bob}), \mathsf{get\_up}(8, \mathsf{bob}), \mathsf{thirsty}(8, \mathsf{bob})\}$.

The second usage is of the form $\textbf{fail}(+a, -b) \leftarrow \textit{body}$, where a and b are EDB atoms with timestamps not in the future. When the rule body is satisfied, the model computation restarts with a added and b removed from the EDB. For example, the rule

$$\textbf{fail}(-\mathsf{eat}(s, x)) \leftarrow \mathsf{get\_up}(t, x), \mathsf{eat}(s, x), t - 1 \leq s, s < t \tag{3}$$

rejects a model candidate where x has eaten within one hour before getting up. The rule correspondingly removes the eat event from the EDB and the model computation restarts. One could further add

$$\textbf{fail}(-\mathsf{get\_up}(t, x)) \leftarrow \mathsf{get\_up}(t, x), \mathsf{eat}(s, x), t - 1 \leq s, s < t \tag{4}$$

to (1)–(3) as an alternative to fix the problem. The principle is that as soon as the earliest **fail** head is derived, the model candidate at that time is given up. Then alternate model computations are started for *all* **fail** heads derivable for that time. Later times are not considered. In the example, thus, both restarts prescribed by rules (3) and (4) are tried.

1.2 Related Work and Novelty

Assigning models to logic programs as their intended meaning has been studied for decades. We only mention the stable models semantics [14,18], its extension for disjunctive programs [11,15], and the possible model semantics [26,27] as the most relevant in the following discussion. Both ascribe meaning to a given program in terms of minimal models, but differ in the way disjunctive rule heads are interpreted (exclusive vs. inclusive, respectively).

Most reasoning tasks around stable models are rather complex, e.g., model existence for propositional disjunctive programs is Σ_2^P-complete [10]. This complexity translates into generate-and-test algorithms even without default negation. For instance, the tableau calculus in [22] for negation-free programs generates in its branches model candidates, whose minimality need to be tested by a subsequent theorem prover call. Another need for generate-and-test algorithms for stable models comes from default negation. In the general case, these algorithms need to guess a stable model candidate and verify its minimality for a certain negation-free program obtained by simplification with this candidate, the Gelfond-Lifschitz transformation.

In contrast, our approach avoids the intricacies of generate-and-test algorithms. This is achieved by using the possible model semantics [26] and a specific concept of stratification for dealing with default negation. The latter fits in the framework of local stratification [25]. A similar concept of stratification by time has been employed for expressing greedy algorithms in Datalog [30]). The usual stratified case, by predicates, but without quantification within negation was already been considered in [26].

As indicated in Sect. 1.1 our language features a **fail** operator for model revision. This feature is the one that possible stands out most among the other mentioned. A rule $\mathbf{fail}(-\mathsf{p}(x)) \leftarrow \mathsf{q}(x), \mathsf{p}(x)$ applied to the facts $\{\mathsf{q}(\mathsf{a}), \mathsf{p}(\mathsf{a})\}$ derives the model $\{\mathsf{q}(\mathsf{a})\}$. This cannot be achieved without belief revision, as given facts have to be satisfied.

Belief revision [1,24] is the process of changing beliefs to take into account a new piece of information. It has also been studied extensively in a logic programming context and in a general way. For instance, Schwind and Inoue [29] consider the problem of revision by a *program* in a rather expressive setting, generalized logic programs equipped with stable model semantics. The perhaps closest approach to ours are the revision specifications of [19,20]. Revision programs generalize logic programming with stable model semantics by an explicit deletion operator. Each revised model is obtained from the initial interpretation by means of insertions and deletions specified by a Gelfond-Lifschitz type reduced program. In that way, our approach is related, but simpler, as it revises only the EDB and does not require a generate-and-test algorithm. On the other hand, the semantics of our revisions takes timestamps into account, so that intended revisions are only those that are derivable "now".

The focus on the paper is not on situational awareness as such. We merely mention that the problem has attracted interest from a logical perspective. In earlier work [2] we proposed bottom-up model computation with a Hyper Tableaux prover [4,23] as a component for data aggregation. In a related context of conformance checking, the authors of [7] propose if-then rules for validating process execution traces by means of a Prolog interpreter. Other approaches for conformance checking include planning [9] and diagnosis of discrete dynamical systems [8,21].

To sum up, the main novelty of our approach lies in the combination of the possible model semantics with specific concepts of stratification and model revision. The combination is designed to enable simple fixpoint algorithms that are sound and complete for a not too complicated declarative semantics. This is the main theoretical contribution.

On the practical side we offer a (publicly available) implementation of our calculus, as a shallow embedding into the Scala programming language. Somewhat related, a shallow embedding into Scala has been used for monitoring event streams over Allen's temporal interval logic [17]. Yet, it is an uncommon implementation technique for automated reasoning systems. The practical advantages are described in Sect. 6.

2 Preliminaries

We assume the reader is familiar with basic notions of first-order logic and answer set programming. See [6] and [13], respectively, for introductory texts.

A first-order logic signature $\Sigma = \Sigma_P \uplus \Sigma_F$ is comprised of predicate symbols Σ_P and function symbols Σ_F of fixed arities. We assume $\mathbb{N} \subseteq \Sigma_F$, i.e., the natural numbers are also constants of the logical language, and that Σ_P contains the *arithmetic predicate symbols* $\Sigma_\mathbb{N} = \{<, \leq, =, \neq\}$. The *ordinary predicate symbols* are $\Sigma_P \setminus \Sigma_\mathbb{N}$. Let $\mathcal{X}$ be a countably infinite set of variables. Instead of introducing a two-sorted signature we assume informally that all terms and formulas over Σ and $\mathcal{X}$ are built in a sorted way.

The letters s and t usually stand for terms, x and y stand for variables, and p and q for ordinary predicate symbols. We speak of *ordinary atoms* and *arithmetic atoms* depending on whether the predicate symbol is ordinary or arithmetic, respectively. For a set A of atoms let $ord(A)$ be the set of all ordinary atoms in A.

Intuitively, $\mathbb{N}$ represents timestamps (points in time), and $<$ and $\leq$ stand for the strict and non-strict earlier-than relationships, respectively. We assume every ordinary predicate symbol has arity ≥ 1 and its, say, first argument ranges over $\mathbb{N}$. For any ordinary atom $a = p(t_1, \ldots, t_n)$ let $time(a) = t_1$ be its timestamp. The function symbols Σ_F may contain arithmetic operations as needed to compute with timestamps, but Σ_F may not contain uninterpreted operators with $\mathbb{N}$ as the result sort.

We assume the ordinary predicate symbols are partitioned as $\Sigma_P \setminus \Sigma_\mathbb{N} = \Sigma_{\text{EDB}} \uplus \Sigma_{\text{IDB}}$. The symbols in Σ_{EDB} are called *extensional database (EDB) predicates*, and the symbols in Σ_{IDB} are the *intensional database (IDB) predicates*. An *EDB* is a finite set of ground Σ_{EDB}-atoms, and an *IDB* is a finite set of ground Σ_{IDB}-atoms. We may think of an EDB as a timestamped sequence of external events, and an IDB as higher-level conclusions derived from that EDB. Below we will exploit this distinction for computing models in a stratified way and for defining a model revision operator.

As usual, a substitution σ is a mapping from the variables to terms. A substitution is identified with its homomorphic extension to terms. Substitution application is written postfix, i.e., we write $t\sigma$ instead of $\sigma(t)$. The *domain of* σ is the set $dom(\sigma) = \{x \in \mathcal{X} \mid x\sigma \neq x\}$ and is always assumed to be finite.

When z is a term, an atom, a sequence, or a set of those, let $var(z)$ denote the set of variables occurring in z. We say that z is *ground* if $var(z) = \emptyset$. A substitution γ is a *grounding substitution for* z iff $z\gamma$ is ground. In this case $z\gamma$ is also called a *ground instance of* z *(via* γ*)*. Let $gnd(z)$ denote the set of all ground instances of z.

3 Stratified Programs

We are now in a position to define our main modeling tool, a variation on if-then rules as popularized in the area of disjunctive logic programming.

A *positive body* is a list $\vec{b} = b_1, \ldots, b_k$ of atoms with $k \geq 0$. If $k = 0$ then $\vec{b}$ is *empty* otherwise it is *non-empty*. (The list represents a conjunction.) A *negative body literal* is an expression of the form $\textsf{not}\, \vec{b}$, where $\vec{b}$ is a non-empty positive body. A *body* is a list $b_1, \ldots, b_k, \textsf{not}\, \vec{b}_{k+1}, \ldots, \textsf{not}\, \vec{b}_n$ comprised of a (possibly empty) positive body and (possibly zero) negative body literals. It is *variable free* if $var(b_1, \ldots, b_k) = \emptyset$. A *head* is one of the following:

(a) an *ordinary head*: a disjunction $h_1 \vee \cdots \vee h_m$ of IDB atoms, for some $m \geq 1$, or

(b) a *fail head*: an expression of the form $\textsf{fail}(\vec{e})$ where $\vec{e} = \pm_1 e_1, \ldots, \pm_k e_k$, for some $k \geq 0$, EDB atoms e_i and $\pm_i \in \{+, -\}$. If $k = 0$ then $\vec{e}$ is the empty sequence ε, and $\textsf{fail}(\epsilon)$ is usually written as $\textsf{fail}()$.

A *rule* consist of a head H and a body and is commonly written as an implication

$$H \leftarrow b_1, \ldots, b_k, \textsf{not}\, \vec{b}_{k+1}, \ldots, \textsf{not}\, \vec{b}_n\ . \tag{5}$$

By an *ordinary rule* (*fail rule*) we mean a rule with an ordinary head (fail head), respectively. A *fail set* is a (possibly empty) set of ground fail heads.

Let r be a rule (5) and $\vec{b} = b_1, \ldots, b_k$ its positive body. We say that r is *variable free* iff $var(H) \cup var(\vec{b}) = \emptyset$. This notion of variable-freeness is justified by the fact that the extra variables $var(\vec{b}_i) \setminus var(\vec{b})$ in the negative body literals $\textsf{not}\, \vec{b}_i$ are implicitly existentially quantified, see Definition 5 below. We say that r' *is a variable free instance of* r *via* σ iff $r' = r\sigma$ is variable free and $dom(\sigma) = var(H) \cup var(\vec{b})$. Notice that σ must not act on the extra variables as these are shielded by quantification.

A *program* is a set of rules. It is *variable free* if all of its rules are. Semantically, every program R stands for the (possibly infinite) variable free program $vfinst(R)$ that is obtained by taking all variable free instances of all rules in R.

The rules are to be evaluated in a bottom-up way. If a current model candidate satisfies a rule body then its head needs evaluation. An ordinary rule extends the current model according to the possible model semantics as explained below and a fail rule rejects the current model. If a fail rule's head is $\textsf{fail}()$ it acts like a traditional rule with an empty head, as an integrity constraint. If the argument list $\vec{e}$ is non-empty the fail rule "fixes" the current EDB by adding ("+") or removing ("−") EDB atoms and starting a new model computation.

In order to admit effective model computation our rules will be stratified. Stratification means range-restrictedness and other restrictions on variables and negation.

Definition 1 (Stratified rule). *Let r be a rule (5) with positive body $\vec{b} = b_1, \ldots, b_k$ and y be a variable. The rule r is* stratified wrt. *y if there is a $b \in \vec{b}$ such that $time(b) = y$ and the following holds:*

(i) $var(\vec{b}) \subseteq var(ord(\vec{b}))$,
(ii) for every ordinary atom $b \in \vec{b}$,
$time(b) = y$ *or* $time(b) = x$ *and* $x \lhd y \in \vec{b}$ *for some variable x and* $\lhd \in \{<, \leq\}$,
(iii) every negative body literal **not** $\vec{b}_{k+1}, \ldots,$ **not** $\vec{b}_n$ *is stratified, and*
(iv) the head H is stratified.

In the above, a negative body literal **not** $\vec{b}_i$ *is stratified if the following holds:*

(i) $var(\vec{b}_i) \setminus var(\vec{b}) \subseteq var(ord(\vec{b}_i))$,
(ii) for every EDB atom $b \in \vec{b}_i$,
$time(b) = y$ *or* $time(b) = x$ *and* $x \lhd y \in \vec{b}_i$ *for some variable x and* $\lhd \in \{<, \leq\}$, *and*
(iii) for every IDB atom $b \in \vec{b}_i$,
$time(b) = x$ *and* $x < y \in \vec{b}_i$ *for some variable x.*

The head H is stratified if the following holds:

(i) $var(H) \subseteq var(ord(\vec{b}))$,
(ii) if H is an ordinary head $h_1 \vee \cdots \vee h_m$ then $time(h_1) = \cdots = time(h_m) = y$.
(iii) if H is a fail head **fail**$(\vec{e})$ *then for all $\pm e \in \vec{e}$, $time(e)$ is an arithmetic expression and $time(e) \lhd y \in \vec{b}$, for some $\lhd \in \{<, \leq\}$.*

A rule is stratified *if it is stratified wrt. some variable y. A program is stratified if each of its rules is stratified.*[1]

Definition 1 expresses conditions on rules in terms of *time-restrictedness* and *range-restrictedness*. The variable y stands for the latest of all timestamps among all timestamps of the ordinary atoms in the rule body. This is made sure by constraints $x \lhd y$ in the various parts of the definition where $\lhd \in \{<, \leq\}$. More precisely, ordinary atoms in the positive body are timestamped "$\leq$"; ordinary heads are timestamped "y" so that no literals timestamped in the past can be inserted into the model (this would defy stratification); and restarts can modify only the past. For the ordinary atoms in negative body literals we distinguish between EDB and IDB atoms. EDB atoms cannot be derived in heads of rules, which affords "$\leq$", whereas IDB atoms must be "$<$".

[1] Usually, stratification is defined as a property of the program as a whole, via its call-graph.

```
(1) In(time, obj, cont) :-
      Load(time, obj, cont)

(2) In(time, obj, cont) :-
      In(time, obj, c),
      In(time, c, cont)

(3) In(next, obj, cont) :-
      In(time, obj, cont),
      Step(next, time),
      not(Unload(next, obj, cont)),
      not(In(time, obj, c),
          Unload(next, c, cont))

(4) fail(+Unload((time + prev)/2,
                 cont, c)) :-
       Unload(time, obj, cont),
       In(time, cont, c),
       Load(t, cont, c), t < time,
       Step(time, prev)

(5) fail() :-
      Unload(time, obj, cont),
      Step(time, prev),
      not(In(prev, obj, cont))

(6) fail() :-
      Unload(time, obj, cont),
      not(Load(t, obj, cont), t < time)

(7) fail(-Unload(time, obj, cont),
         +Unload(time, o, cont)) :-
      Unload(time, obj, cont),
      not(Load(t, obj, cont), t < time),
      t < time,
      Load(t, o, cont),
      SameBatch(t, b),
      if((b contains obj) and (b contains o))

(8) fail(+Load(t, obj, cont)) :-
      Unload(time, obj, cont),
      not(Load(t, obj, cont), t < time),
      t < time,
      Load(t, o, cont),
      SameBatch(t, b),
      if((b contains obj) and (b contains o))

(9) fail(-Load(t, o, cont),
         +Load(t, obj, cont)) :-
      not(Load(t, obj, cont), t < time),
      Load(t, o, cont),
      t < time,
      SameBatch(t, b),
      if((b contains obj) and (b contains o))
```

Fig. 1. Supply chain program. See Example 2 for explanations.

The remaining conditions force range-restrictedness. In the first part, condition (i) says that every variable in a positive body atom appears also in some ordinary positive atom; similarly for condition (i) for heads. Condition (i) for negative body atoms says that every extra variable in a negative body atom appears also in some of its ordinary body atoms. Together these conditions make sure that matching a rule's ordinary atoms against a ground candidate model always removes all variables. This way, all arithmetic expressions can be evaluated and only ground heads can be derived.

Example 1 (Stratified rule). Let $\Sigma_{\mathrm{EDB}} = \{\mathsf{p}\}$ and $\Sigma_{\mathrm{IDB}} = \{\mathsf{d}\}$. The rule $\mathsf{d}(x_3, x_1) \leftarrow x_1 < x_3, \mathsf{p}(x_1), \mathsf{p}(x_3), \textbf{not}\,(x_1 < x_2, x_2 < x_3, \mathsf{p}(x_2))$ is stratified wrt. x_3 ($= \mathit{time}(\mathsf{p}(x_3))$). It collects in d the timestamps between *consecutive* p-events. For example, given the set $I = \{\mathsf{p}(2), \mathsf{p}(4), \mathsf{p}(7), \mathsf{p}(13)\}$, the rule when applied exhaustively derives $\mathsf{d}(4, 2)$, $\mathsf{d}(7, 4)$ and $\mathsf{d}(13, 7)$ but not, e.g, $\mathsf{d}(7, 2)$. The extra variable x_2 in its negative body literal is implicitly existentially quantified. □

Example 2 (Supply chain). The program in Fig. 1 illustrates a possible use of our approach in a supply chain application. It is written in the concrete input syntax of our implementation, the Fusemate system (Sect. 6).

The signature is $\Sigma_{\text{EDB}} = \{\mathsf{Load}, \mathsf{Unload}, \mathsf{SameBatch}\}$ and $\Sigma_{\text{IDB}} = \{\mathsf{In}\}$. A fluent $\mathsf{Load(time, obj, cont)}$ expresses that at the given time an object obj is loaded into a container cont, similarly for Unload. A fluent $\mathsf{In(time, obj, cont)}$ says that at the given time an object obj is inside the container cont.

With this interpretation, the rules (1) and (2) for the In relation should be obvious. Rule (3) is a frame axiom for the In relation. That is, it states when an In-fluent carries over to the next timestamp: an object remains in a container if neither it nor a container containing it is unloaded from the container. The body atom $\mathsf{Step(time, prev)}$ holds true if prev is the most recent timestamp preceding time. The Step relation is "built-in" into Fusemate for convenience.

Rule (4) fixes the problem of a "missing" unloading event by inserting one into the EDB at a speculated time $(\mathsf{time} + \mathsf{prev})/2$. This rule will become clearer in Example 3 below, where we discuss the program in conjunction with a concrete EDB.

Rule (5) says that only items that are in a container can be unloaded in the next step. Rule (6) demands loading prior to unloading. The other rules will also be discussed below. □

4 Semantics

The possible model semantics [26,27] associates to a disjunctive program sets of possible facts that might have been true in the actual world. (This is already a good fit for situational awareness.) We extend it to our stratified programs with fail rules.

Let $\mathrm{Th}(\Sigma_{\mathbb{N}})$ be the set of all ground time atoms that are true in the standard model of natural number arithmetic. For a set A of ground ordinary atoms define $\mathcal{I}(A) = A \cup \mathrm{Th}(\Sigma_{\mathbb{N}})$ which represents the Herbrand Σ-interpretation that assigns true to a ground atom a if and only if $a \in \mathcal{I}(A)$.

Definition 2 (Rule semantics). *A set A of ground ordinary atoms* satisfies *a variable free body $B = b_1, \ldots, b_k, \mathbf{not}\, \vec{b}_{k+1}, \ldots, \mathbf{not}\, \vec{b}_n$, written as $A \models B$, if $\{b_1, \ldots, b_k\} \subseteq \mathcal{I}(A)$ and $(gnd(\vec{b}_{k+1}) \cup \cdots \cup gnd(\vec{b}_n)) \cap \mathcal{I}(A) = \emptyset$.*

The set A satisfies *a variable free ordinary rule $h_1 \vee \cdots \vee h_m \leftarrow B$ if $A \models B$ entails $\{h_1, \ldots, h_m\} \cap \mathcal{I}(A) \neq \emptyset$. It is a* model *of a set R of variable free ordinary rules, written as $A \models R$, iff it satisfies every rule in R.*

The fail set *of A and a set of variable free fail rules R is the set $F = \{\mathbf{fail}(\vec{e}) \mid$ there is a rule $\mathbf{fail}(\vec{e}) \leftarrow B \in R$ such that $A \models B\}$. This is written as $A \models^{\mathbf{fail}}_R F$.*

Satisfaction of a variable free body according to Definition 2 is equivalent to satisfaction of the first-order logic formula $b_1 \wedge \cdots \wedge b_k \wedge \neg(\exists \vec{x}_{k+1}.\ \wedge \vec{b}_{k+1}) \wedge \cdots \wedge \neg(\exists \vec{x}_n.\ \wedge \vec{b}_n)$ in the interpretation $\mathcal{I}(A)$, where $\vec{x}_j = var(\vec{b}_j)$, for all $j = k+1, \ldots, n.$.

Definition 2 is, in fact, somewhat more general than needed for defining the possible models semantics of logic programs. Possible models interpret disjunctive heads inclusively, in all possible ways. This is expressed in the following definition.

Definition 3 (Split program [26,27]**).** *Let R be a variable free program. A* split program *of R is any program obtained from R by replacing every ordinary rule $h_1 \vee \cdots \vee h_m \leftarrow B$ by the normal ordinary rules (called split rules) $h \leftarrow B$, for every $h \in \mathcal{H}$, where $\mathcal{H}$ is some non-empty subset of $\{h_1, \ldots, h_m\}$.*

In [26,27], Sakama et al. define the possible model semantics (also) for disjunctive programs without negation. Our stratified case admits a similar definition.

For any program R let R^+ (R^-) be the set of all ordinary (fail) rules of R, respectively.

Definition 4 (Satisfication of stratified programs). *Let P be a stratified program, F a fail set, E an EBD and I an IDB. We write $(E, I) \models_P^{\mathsf{fail}} F$ if there is a split program S of vfinst(P) such that all of the following hold:*

(i) $E \cup I \models S^+$ ($E \cup I$ is a model of the ordinary rules)
(ii) $E \cup J \models S^+$ for no $J \subsetneq I$ (I is minimal)
(iii) $E \cup I \models_{S^-}^{\mathsf{fail}} F$ (F is the triggered fail heads)

If $F = \emptyset$ then we say that (E, I) satifies *P and writte $(E, I) \models P$.*

As an example (without time), if $E = \{p\}$ and $R = \{q \vee r \leftarrow p,\ q \leftarrow r,\ s \leftarrow s\}$ then $\{p, q, s\} \models R$, but only $(\{p\}, \{q\})$ and $(\{p\}, \{r, q\})$ satisfy R in the sense of Definition 4.

The purpose of a program P is to compute all extensions (E, I) of a given EDB E that satisfy P. For failed such attempts, P also specifies ways to revise E, if any, as early as possible, leading to new tries. The following Definition 5 make this precise.

For a ground fail head $\mathsf{fail}(\vec{e})$ and ground EDB E let $upd(E, \vec{e})$ be the EDB obtained from E by first adding all EDB atoms e such that $\vec{e}$ contains the expression $+e$, and then deleting all all EDB atoms e such that $\vec{e}$ contains $-e$. For any set A of ground ordinary atoms and $t \in \mathbb{N}$ let $A_{\leq t} = \{a \in A \mid time(a) \leq t\}$; analogously for $A_{<t}$.

Definition 5 (Possible models of stratified programs). *Let P be a stratified program and E_{init} an EDB. Let $\mathcal{E}$ be the smallest set of EDBs containing E_{init} and satisfying, for all $E \in \mathcal{E}$, timestamps t in E, IDBs I and fail sets F:*

If $(E_{\leq t}, I) \models_P^{\mathsf{fail}} F$ and there is no $J \subseteq I$ and $G \neq \emptyset$ such that $(E_{<t}, J) \models_P^{\mathsf{fail}} G$ then $\{upd(E, \vec{e}) \mid \mathsf{fail}(\vec{e}) \in F \text{ and } \vec{e} \neq \varepsilon\} \subseteq \mathcal{E}$.

The set $\mathcal{E}$ is called the restart EDBs induced by P and E_{init}. *Any pair (E, I) such that $E \in \mathcal{E}$ and $(E, I) \models P$ is called a* possible model of P and E_{init}, *written as $(E_{\mathsf{init}}, E, I) \models P$. Let $mods_P(E_{\mathsf{init}}) = \{(E, I) \mid (E_{\mathsf{init}}, E, I) \models P\}$ be all such possible models.*

The set $\mathcal{E}$ in Definition 5 contains a restart EDB E apart from E_{init} if and only if E is obtained from some *earliest time* fail set F from another restart EDB in $\mathcal{E}$. This excludes fail sets that may otherwise be additionally derivable at a later time. This was a design decision in support of "anytime" reasoning, for not having to consider future events.

Example 3 (Supply chain, Example 2 *continued).* Consider the following EDB E_{init} consisting of loading and unloading events:

$$\begin{array}{ll}\mathsf{SameBatch}(10, \mathsf{Set}(\mathsf{tomatoes}, \mathsf{apples})) & \mathsf{Load}(40, \mathsf{container}, \mathsf{ship}) \\ \mathsf{Load}(10, \mathsf{tomatoes}, \mathsf{pallet}) & \mathsf{Unload}(60, \mathsf{apples}, \mathsf{pallet}) \\ \mathsf{Load}(20, \mathsf{pallet}, \mathsf{container}) & \end{array}$$

The intuitive meaning of the Load atoms between times 10 and 40 should be obvious. All what is reported at time 60 is that apples are unloaded from the pallet. However, this is suspicious from a (practical) completeness and consistency perspective. First, it can be alleged that some unloading events went under unreported. Before an item (apples) can be unloaded from a pallet that was loaded earlier into a container, the pallet needs to be unloaded from the container first, and that container must have been unloaded from the ship. Such reports are missing. Second, loading of tomatoes does not go together well with the unloading of apples later. This could be a reporting inconsistency or a reporting incompleteness if indeed apples were (also) loaded earlier.

All these plausible explanations are provided by (E_1, I_1), (E_2, I_2), and (E_3, I_3), the three possible models of E_{init} and the program in Fig. 1. For space reasons we list only their EDB components, which are as follows:

E_1	E_2	E_3
Load(10, tomatoes, pallet)	Load(10, tomatoes, pallet)	Load(10, apples, pallet)
	Load(10, apples, pallet)	
Unload(45, container, ship)	Unload(45, container, ship)	Unload(45, container, ship)
Unload(50, pallet, container)	Unload(50, pallet, container)	Unload(50, pallet, container)
Unload(60, tomatoes, pallet)	Unload(60, apples, pallet)	Unload(60, apples, pallet)

In each of these models, the missing unloading events Unload(45, container, ship) and Unload(50, pallet, container) are added by repeated application of rule (4). Generally speaking, rule (4) inserts an Unload of the "containing container" the object to be unloaded from is in. The rules (7) – (9) all fix the "unloading apples vs. loading tomatoes" problem. Rule (7) leads to (E_1, I_1), rule (8) leads to (E_2, I_2), and rule (9) leads to (E_3, I_3). Each of these rules tests whether an object (apples) is swappable with another object (tomatoes) for the purpose of model revision, which is the case if the SameBatch relation says so. Notice that if E_{init} had, say Unload(60, oranges, pallet) instead of Unload(60, apples, pallet) then none of the rules (7) – (9) is applicable and no possible model exists. □

5 Model Computation

This section introduces our calculus for computing possible models of stratified programs. It borrows some terminology from tableau calculi. A *path* p is a triple

(E, I, t) where E is an EDB, I is an IDB and $t \in \mathbb{N}$ is a timestamp. Intuitively, p represents the interpretation $\mathcal{I}((E \cup I)_{\leq t})$. An *initial path* is of the form $(E, \emptyset, 0)$. A *tableau* is a finite set of paths.[2]

Let $B = b_1, \ldots, b_k, \textsf{not}\, \vec{b}_{k+1}, \ldots, \textsf{not}\, \vec{b}_n$ be the body of a variable free stratified rule and A a set of ground ordinary atoms. A substitution σ with $dom(\sigma) = var(b_1, \ldots, b_k)$ is a *body matcher for B on A*, written as $(B, \sigma) \sqsubseteq A$, if the following holds:

(i) $\{b_1\sigma, \ldots, b_k\sigma\} \subseteq A \cup \mathrm{Th}(\Sigma_\mathbb{N})$, and
(ii) for no $i = k+1 \ldots n$ there is a grounding substitution γ for $\vec{b}_i\sigma$ such that $\vec{b}_i\sigma\gamma \subseteq A \cup \mathrm{Th}(\Sigma_\mathbb{N})$.

Note 1 (Computing body matchers). The definition of body matchers only applies to bodies of stratified rules. It is easy to see that a body matcher σ, if any exists, can be found by computing a simultaneous matching substitution σ for the ordinary atoms among $b_1, \ldots, b_k$ to A. Similarly for the substitution γ in condition (ii). Furthermore, stratification guarantees that all arithmetic atoms for testing conditions (i) and (ii) are necessarily ground and hence can be evaluated. □

An inference rule is a schematic expression of the form $p \Rightarrow p_1, \ldots, p_k$ where p and p_j are paths, for all $1 \leq j \leq k$, where $k \geq 0$. It means that the premise p is to be replaced by the conclusions $p_1, \ldots, p_k$. An *inference* is an instance of an inference rule.

In the following, P is a stratified program and σ is a substitution such that $r\sigma$ is a variable free instance of a rule r that is clear from the context.

Ext: $(E, I, t) \;\Rightarrow\; (E, I \cup \mathcal{H}_1, t),\; \ldots,\; (E, I \cup \mathcal{H}_k, t)$
if P contains an ordinary rule $h_1 \vee \cdots \vee h_k \leftarrow B$ such that
$\{\mathcal{H}_1, \ldots, \mathcal{H}_k\} = \{\mathcal{H} \mid (B, \sigma) \sqsubseteq (E \cup I)_{\leq t}$ and $\emptyset \subsetneq \mathcal{H} \subseteq \{h_1\sigma, \ldots, h_k\sigma\}\}$
Restart: $(E, I, t) \;\Rightarrow\; (upd(E, \vec{e}_1), \emptyset, 0),\; \ldots,\; (upd(E, \vec{e}_k), \emptyset, 0)$
if $k \geq 1$ and $\{\vec{e}_1, \ldots, \vec{e}_k\} = \{\vec{e}\sigma \mid \textsf{fail}(\vec{e}) \leftarrow B \in P,\ \vec{e} \neq \varepsilon$ and $(B, \sigma) \sqsubseteq (E \cup I)_{\leq t}\}$.
Fail: $(E, I, t) \;\Rightarrow$
if P contains a rule $\textsf{fail}() \leftarrow B$ and there is a σ such that $(B, \sigma) \sqsubseteq (E \cup I)_{\leq t}$.
Jump: $(E, I, t) \;\Rightarrow\; (E, I, s)$ if s is the least timestamp in E with $t < s$.

The Ext rule extends I to satisfy all split rules for each case $\mathcal{H}$ of some instance of an ordinary rule in P whose body is satisfied by $(E \cup I)_{\leq t}$; Restart replaces the current path with *all* initial paths as per the non-empty fail rules after Ext is exhausted; Fail also terminates the current path but is to be applied

[2] This terminology is inspired by visualizing a set of paths as a tableau in the usual sense. For that, a path (E, I, t) leads to a branch whose nodes are labeled with the atoms $E \cup I$ and the branch as a whole is labeled with t. Moreover, the way the calculus constructs these paths sets indeed corresponds to a typical tableau construction. See, e.g., [3].

only if Restart doesn't; Jump advances the current time bound t. The following formalizes this intuition.

An initial path $(E, \emptyset, 0)$ is *new wrt. a tableau T* iff there is no I and no t such that $(E, I, t) \in T$. Let E_{init} be an *input EDB*. A *derivation D (from E_{init} and P)* is a sequence $(T)_{i \geq 0}$ of tableaus $D = (T_0 = \{(E_{\mathsf{init}}, \emptyset, 0)\}), T_1, T_2, \ldots$ such that, for all $i \geq 0$, there is a *selected path* $p \in T_i$ and $T_{i+1} = (T_i \setminus \{p\}) \cup \{p_1, \ldots, p_k\}$ where:

(a) $p \Rightarrow p_1, \ldots, p_k$ by Ext and $\{p_1, \ldots, p_k\} \not\subseteq T_i$,
(b) $p \Rightarrow q_1, \ldots, q_m$ by Restart and $\{p_1, \ldots, p_k\} = \{p \in \{q_1, \ldots, q_m\} \mid p \text{ is new wrt. } T_i\}$,
(c) $p \Rightarrow$ by Fail and $k = 0$, or
(d) $p \Rightarrow p_1$ by Jump and $k = 1$.

In addition, the inference rules must be prioritized in this order. That is, if T_{i+1} is obtained from T_i by, say, case (c) , then there is no tableau that can be obtained from T_i by case (a) or case (b) with the same selected path p; analogously for the other cases.

The derivation D is *exhausted* if it is finite and no inference rule is applicable to its final tableau T_n, for no $p \in T_n$. In this case the *computed models of D* is the set $\mathcal{M}(D) = \{(E, I) \mid (E, I, t) \in T_n \text{ for some } t \in \mathbb{N}\}$.

Figure 2 is a graphical illustration of a derivation and its computed models.

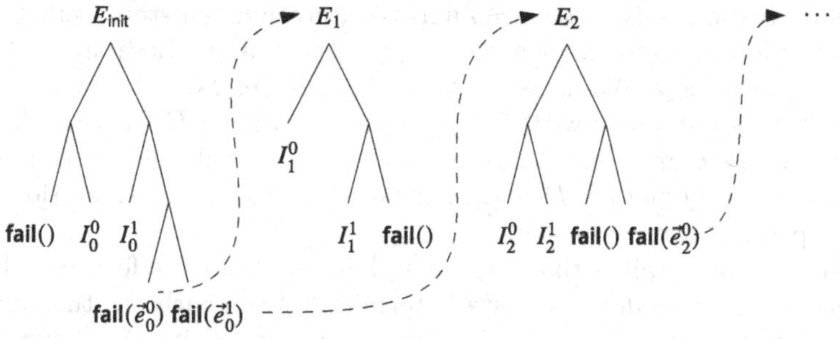

Fig. 2. Illustration of a hypothetical derivation. The root of each sub-tableau is labeled with the EDB in that sub-derivation. The first sub-tableau has two Restart inferences, leading to the second and third sub-tableau, where $E_1 = upd(E_{\mathsf{init}}, \vec{e}_0^0)$, $E_2 = upd(E_{\mathsf{init}}, \vec{e}_0^1)$. The isolated **fail**()s do not cause a Restart, they cause Fail. The computed models are $(E_{\mathsf{init}}, I_0^0)$, $(E_{\mathsf{init}}, I_0^1)$, (E_1, I_1^0), etc.

Theorem 1 (Soundness and completeness). *Assume a signature Σ without k-ary function symbol, for $k > 0$. Let P be a stratified program and E_{init} an EDB. Assume an exhausted derivation D from E_{init} and P. Then $\mathcal{M}(D) = mods_P(E_{\mathsf{init}})$.*

Proof. (Sketch) Let T_n be the final tableau of D. For soundness, assume $\mathcal{M}(D) \neq \emptyset$ and chose any $(E, I) \in \mathcal{M}(D)$ arbitrary. That is, $(E, I, t) \in T_n$, for some t. We have to show $(E, I) \in mods_P(E_{\mathsf{init}})$, equivalently $(E_{\mathsf{init}}, E, I) \models P$.

The EDB E is either E_{init} or derived from E_{init} through, say, $k > 0$ intermediate EDBs by Restarts. By induction on k one can show that, on the semantic side, E is a restart induced by P and E_{init}, i.e., $E \in \mathcal{E}$ in Definition 5. This follows from the definition of derivations. In particular, the earliest-time requirement in Definition 5 is matched by prioritizing Restart over Fail and Jump.

With the EDB E traced down in $\mathcal{E}$, it remains to prove $(E, I) \models P$. With the stratification of P (Definition 1) this is rather straightforward. Range-restrictedness makes sure that only ground heads are derivable. The Ext inference rule achieves on-the-fly splitting and only for those variable-free instances of rules whose body is satisfied, which are the only ones that count (details in [26, 27]). The requirements on EDB/IDB atoms in negative body literals ($\leq$ vs. $<$) in Definition 1 entail that utilizing body matchers in derivations is correct wrt. the rule semantics in Definition 2.

The timed setting requires a layered fixpoint iteration. Stratification makes sure that stepwisely incrementing in derivations the time bound by Jump inferences is all that is needed to comply with the "unstepped" possible model semantics in Definitions 4 and 5. In particular, no ordinary rule can derive in its head a conclusion with a timestamp earlier than the latest timestamp in its body. This makes the derivability relation monotonic wrt. increasing time stamps (usual stratification by predicates is covered in [26]). Moreover, because the given derivation is exhausted, no fixpoint iteration can stop prematurely.

For completeness, assume $mods_P(E_{\mathsf{init}}) \neq \emptyset$ and chose any $(E, I) \in mods_P(E_{\mathsf{init}})$ arbitrary. We have to show $(E, I) \in \mathcal{M}(D)$. The first step is to locate in D the sub-tableau with E at its root, by tracing $E \in \mathcal{E}$ from E_{init}. The next step then is to argue for the completeness of the sub-tableau construction with that fixed E, giving $(E, I) \in T_n$. All that uses similar considerations as the soundness proof above.

One important detail is that Ext is the highest-priority inference rule. This makes sure that no model candidate is terminated too early, so that all possible branching out takes place. As a consequence, *all* possible fail heads for the current time point will be derived. The requirement that there are no proper (non-constant) function symbols make sure that the layered fixpoint computation of derivations terminates and finds (E, I). □

6 Implementation

It is not too difficult to translate the model computation calculus of Sect. 5 into a proof procedure. Tableaux can be represented in a direct way, as a set of paths (E, I, t). In terms of Fig. 2, the proof procedure can implement a one-branch-at-a-time approach for one sub-tableau at a time, for space efficiency, embedded in an adapted given-clause loop algorithm. Only the EDBs, not the full models, need to be remembered to implement the "Progress" condition for derivations (cf. Sect. 5).

A concrete implementation could based on, e.g., hyper resolution with splitting [5] or hyper tableaux [3] calculi. Our implementation however is implemented in an unusual way, by shallow embedding in Scala.[3]

Scala [28] is a modern high-level programming language that combines object-oriented and functional programming styles. It has functions as first-class objects and supports user-definable pre-, post- and infix syntax. With these features, Scala is suitable as a host language for embedding domain-specific languages (DSLs). (See, e.g., [16] for a Scala DSL for runtime verification.) In our case, the logic program rules are nothing but partial functions, instantiating and evaluating a rule body reduces to partial function definedness, and deriving a rule head reduces to executing a partial function on a defined point. An advantage of the DSL approach is that is is easy to interface with external systems, e.g., databases, in particular if they have a Java interface.

Moreover, it is easy to make the full Scala language and its associated data structure libraries available for writing rules. (There is no theoretical problem doing that as long as all Scala expressions are ground and, hence, can be evaluated.) For example, the EDB in Example 3 has the Scala-set forming term Set(tomatoes, apples), and the rules (7) - (9) in Fig. 1 test in their last lines membership in such sets by Scala expressions.

While EDBs are naturally written as Scala source code, logic program rules are usually written in a (much) more convenient syntax and translated into the required format by the Scala macro mechanism. See Listing 1.1 for an example.[4]

Listing 1.1. Sample EDB/IDB declarations and a rule. Some unimportant declarations left away.

```
1 type Time = Int
2 case class Load(time: Time, obj: String, cont: String) extends EDBAtom
3 case class SameBatch(time: Time, objs: Set[String]) extends EDBAtom
4 case class In(time: Time, obj: String, cont: String) extends IDBAtom
5 @rules
6 val rules = List(In(time, obj, cont) :- (In(time, obj, c), In(time, c, cont))
```

In Listing 1.1, line 1 sets the concrete type for time to Int, the Scala integers. A realistic application could use a rich time class like java.time.OffsetDataTime. Lines 2, 3 and 4 define the EDB and IDB signature of the supply chain Example 2 by extension of the Scala classes EDBAtom and IDBAtom. The Load relation says that the object obj was loaded into the container cont at time time. The In relation says that obj was in container cont at time time. The *time* argument must be named time, but all other arguments and their types can be freely chosen. For simplicity we used strings, except for the SameBatch relation, which has a set of strings for its objs parameter.

Line 6 defines a list with one rule that expresses the transitivity of the In relation (rule (2) in Fig. 1). Line 5 is an annotation that tells the compiler to

[3] Our implementation, the Fusemate system, is available at https://bitbucket.csiro.au/users/bau050/repos/fusemate/.

[4] Scala case classes are records in language agnostic terminology.

expand the subsequent definition by a macro named rules. Indeed, without the help of a macro the rule would not compile because of undefined variables in the rule. The macro expansion of the rule in line 6 is the Scala function in Listing 1.2.

Listing 1.2. Macro expansion of the transitivity rule.

```
(I: Set[Lit]) => {
  case List(In(time, obj, c), In(time0, c1, cont)) if
    c == c1 && time == time0 => In(time, obj, cont)
}
```

The anonymous function in Listing 1.2 is passed in its formal parameter I a set of atoms which will always be the "current interpretation" $(E \cup I)_{\leq}t$ where $p = (E, I, t)$ is the current path. The set I is needed for evaluating negative body literals and is not relevant for this example. The function returns a partial function in the form of a case expression. The pattern in the case expression are the *ordinary* atom of the *positive* body literals. These are the ones that need to be matched to the atoms of $(E \cup I)_{\leq}t$ for rule evaluation (see Note 1). The matching is done by applying the partial function to all tuples (of the proper arity) of elements from the current interpretation. If the application succeeds, i.e., if the case pattern match succeeds and the additional if-condition is satisfied, the partial function body (to the right of =>) is executed, which results in an instantiated head. If the head is disjunctive, all non-empty subsets are taken. This gives all split programs (cf. Definition 3) on the fly. The resulting sets are collected in one sweep for each rule and are candidates for extending the current path. **fail**-rules are processed in a similar way.

Notice that the pattern of a case expression needs to be linear, hence the renaming apart of pattern variables and the obvious equalities in the if-condition. Notice also that substitutions are not explicitly represented, they are hidden in the Scala runtime system.

It remains to be explained how arithmetic atoms and negative body literals are macro-expanded. By way of example, consider, say, rule (7) in Fig. 1. Its arithmetic condition $t < time$ is simply conjoined to the **if**-condition of the rule's case expression. An **if**-expression like **if**((b contains obj) and (b contains o)) is a backdoor for adding arbitrary Boolean-valued Scala code to that condition ("contains" belongs to the Scala library and tests set membership). Recall that all variables in arithmetic atoms will always be ground instantiated by matching and hence can be evaluated. This must be be extended to Scala conditions. In the implementation, any such free variable would be detected as a compile time error.

The body literal **not**(load(t, obj, cont), t < time) is expanded into a partial function **case** load(t, obj1, cont1) **if** obj == obj1 && cont == cont1 && t < time => Abort. It is structurally the same as the one in Listing 1.1 except that the binding of the variables obj and cont in the surrounding context need to respected, giving the stated equalities. Now, the **if**-condition of the case expression of the surrounding rule (i.e., rule 7) is conjoined with a Scala expression for testing that the partial function does not return Abort, for no tuple of elements from the set I explained above.

The implementation supports some more features not further discussed here: rules can be defined locally within case classes; literals – possibly negated atoms – can be used anywhere instead of atoms; a "strong fail" head operator terminates a model candidate without restarts, e.g., for classical negation: a and $\neg a$ together is unfixable; and Scala conditions can access the interpretation $(E \cup I)_{\leq t}$ for, e.g., concise data aggregation.

7 Conclusions

This paper presented a novel calculus and implementation for situational awareness applications. The approach is meant to be *practical* in *three* ways: first, realistic situational awareness requires being able to reason with incomplete or erroneous data. Moreover, "anytime" reasoning is needed, meaning that a model can be derived, rejected or repaired at any current time. Our approach supports these needs with a (disjunctive) logic programming framework with timed predicates, stratified negation and a novel model revision operator. Second, thanks to implementation on top of Scala, it is trivial to attach arbitrary Scala code and Java libraries. (It would not be difficult to extend the calculus respectively.) For instance, reading in XML data and making them available as terms (Scala case classes) is easy. Third, we strived for a "cheap" model computation procedure that makes do without additional generate-and-test needs. As such, it is perhaps more adequately seen as study in pushing bottom-up first-order logic model computation technology rather than slimmed down answer-set programming.

As for future work, one interesting idea is to add probabilities to the picture, say, in the way ProbLog extends Prolog. This is obviously useful because, e.g., some explanations (models) or repairs (restarts) are more likely than others. Another idea is to view the model computation as runtime verification. This view suggests that (probabilistic) linear temporal logic could serve as an additional useful high-level specification language component.

Acknowledgements. I thank the reviewers for their constructive comments. Yuzhou Chen discovered an error in an earlier version of the paper.

References

1. Alchourròn, C.E., Gärdenfors, P., Makinson, D.: On the logic of theory change: partial meet contraction and revision functions. J. Symb. Logic **50**, 510–530 (1985)
2. Baader, F., et al.: A novel architecture for situation awareness systems. In: Giese, M., Waaler, A. (eds.) TABLEAUX 2009. LNCS (LNAI), vol. 5607, pp. 77–92. Springer, Heidelberg (2009). https://doi.org/10.1007/978-3-642-02716-1_7
3. Baumgartner, P., Furbach, U., Niemelä, I.: Hyper tableaux. In: Alferes, J.J., Pereira, L.M., Orlowska, E. (eds.) JELIA 1996. LNCS, vol. 1126, pp. 1–17. Springer, Heidelberg (1996). https://doi.org/10.1007/3-540-61630-6_1

4. Baumgartner, P., Furbach, U., Pelzer, B.: Hyper tableaux with equality. In: Pfenning, F. (ed.) CADE 2007. LNCS (LNAI), vol. 4603, pp. 492–507. Springer, Heidelberg (2007). https://doi.org/10.1007/978-3-540-73595-3_36
5. Baumgartner, P., Schmidt, R.: Blocking and other enhancements for bottom-up model generation methods. J. Autom. Reason. **64**, 197–251 (2019). https://doi.org/10.1007/s10817-019-09515-1
6. Bradley, A., Manna, Z.: The Calculus of Computation. Springer, Heidelberg (2007). https://doi.org/10.1007/978-3-540-74113-8
7. Chesani, F., Mello, P., Montali, M., Riguzzi, F., Sebastianis, M., Storari, S.: Checking compliance of execution traces to business rules. In: Ardagna, D., Mecella, M., Yang, J. (eds.) BPM 2008. LNBIP, vol. 17, pp. 134–145. Springer, Heidelberg (2009). https://doi.org/10.1007/978-3-642-00328-8_13
8. Cordier, M., Thiébaux, S.: Event-based diagnosis for evolutive systems. In: Proceedings of 5th International Workshop on Principles of Diagnosis, pp. 64–69 (1994)
9. De Giacomo, G., Maggi, F.M., Marrella, A., Sardina, S.: Computing trace alignment against declarative process models through planning. In: Proceedings of 26th International Conference on Automated Planning and Scheduling (ICAPS) (2016)
10. Eiter, T., Gottlob, G.: Complexity results for disjunctive logic programming and application to nonmonotonic logics. In: Proceedings of the 1993 International Symposium on Logic Programming, ILPS 1993, pp. 266–278. MIT Press, Cambridge (1993)
11. Eiter, T., Gottlob, G., Mannila, H.: Disjunctive datalog. ACM Trans. Database Syst. **22**, 364–418 (2001)
12. Endsley, M.: Toward a theory of situation awareness in dynamic systems. Hum. Factors J.: J. Hum. Factors Ergon. Soc. **37**(1), 32–64 (1995)
13. Gelfond, M.: Answer sets. In: van Harmelen, F., Lifschitz, V., Porter, B. (eds.) Handbook of Knowledge Representation, Foundations of Artificial Intelligence, vol. 3, pp. 285–316. Elsevier, Amsterdam (2008)
14. Gelfond, M., Lifschitz, V.: The stable model semantics for logic programming. In: Kowalski, R., Bowen, K. (eds.) Proceedings of the 5th International Conference on Logic Programming, Seattle, pp. 1070–1080 (1988)
15. Gelfond, M., Lifschitz, V.: Classical negation in logic programs and disjunctive databases. New Gener. Comput. **9**, 365–385 (1991). https://doi.org/10.1007/BF03037169
16. Havelund, K., Joshi, R.: Modeling rover communication using hierarchical state machines with scala. In: Tonetta, S., Schoitsch, E., Bitsch, F. (eds.) SAFECOMP 2017. LNCS, vol. 10489, pp. 447–461. Springer, Cham (2017). https://doi.org/10.1007/978-3-319-66284-8_38
17. Kauffman, S., Havelund, K., Joshi, R.: `nfer` – a notation and system for inferring event stream abstractions. In: Falcone, Y., Sánchez, C. (eds.) RV 2016. LNCS, vol. 10012, pp. 235–250. Springer, Cham (2016). https://doi.org/10.1007/978-3-319-46982-9_15
18. Lifschitz, V.: Action languages, answer sets, and planning. In: Apt, K.R., Marek, V.W., Truszczynski, M., Warren, D.S. (eds.) The Logic Programming Paradigm: A 25-Year Perspective, pp. 357–373. Springer, Heidelberg (1999). https://doi.org/10.1007/978-3-642-60085-2_16
19. Marek, V.W., Truszczyński, M.: Revision specifications by means of programs. In: MacNish, C., Pearce, D., Pereira, L.M. (eds.) JELIA 1994. LNCS, vol. 838, pp. 122–136. Springer, Heidelberg (1994). https://doi.org/10.1007/BFb0021968
20. Marek, V.W., Truszczyński, M.: Revision programming. Theoret. Comput. Sci. **190**(2), 241–277 (1998)

21. McIlraith, S.: Toward a theory of diagnosis, testing and repair. In: Proceedings of 5th International Workshop on Principles of Diagnosis, pp. 185–192 (1994)
22. Niemelä, I.: A tableau calculus for minimal model reasoning. In: Miglioli, P., Moscato, U., Mundici, D., Ornaghi, M. (eds.) TABLEAUX 1996. LNCS, vol. 1071, pp. 278–294. Springer, Heidelberg (1996). https://doi.org/10.1007/3-540-61208-4_18
23. Pelzer, B., Wernhard, C.: System description: E- KRHyper. In: Pfenning, F. (ed.) CADE 2007. LNCS (LNAI), vol. 4603, pp. 508–513. Springer, Heidelberg (2007). https://doi.org/10.1007/978-3-540-73595-3_37
24. Peppas, P.: Chapter 8 belief revision. In: van Harmelen, F., Lifschitz, V., Porter, B. (eds.) Handbook of Knowledge Representation, Foundations of Artificial Intelligence, vol. 3, pp. 317–359. Elsevier, Amsterdam (2008)
25. Przymusinski, T.C.: Chapter 5 - on the declarative semantics of deductive databases and logic programs. In: Minker, J. (ed.) Foundations of Deductive Databases and Logic Programming, pp. 193–216. Morgan Kaufmann, Burlington (1988)
26. Sakama, C.: Possible model semantics for disjunctive databases. In: Kim, W., Nicholas, J.M., Nishio, S. (eds.) Proceedings First International Conference on Deductive and Object-Oriented Databases (DOOD-89), pp. 337–351. Elsevier (1990)
27. Sakama, C., Inoue, K.: An alternative approach to the semantics of disjunctive logic programs and deductive databases. J. Autom. Reason. **13**, 145–172 (1994). https://doi.org/10.1007/BF00881915
28. The Scala Programming Language. https://www.scala-lang.org
29. Schwind, N., Inoue, K.: Characterization of logic program revision as an extension of propositional revision. Theory Pract. Logic Program. **16**(1), 111–138 (2016)
30. Zaniolo, C.: Expressing and supporting efficiently greedy algorithms as locally stratified logic programs. Technical Communications of ICLP 2015 1433 (2015)

SGGS Decision Procedures

Maria Paola Bonacina and Sarah Winkler(✉)

Università degli Studi di Verona, Verona, Italy
{mariapaola.bonacina,sarahmaria.winkler}@univr.it

Abstract. SGGS (Semantically-Guided Goal-Sensitive reasoning) is a conflict-driven first-order theorem-proving method which is refutationally complete and model complete in the limit. These features make it attractive as a basis for decision procedures. In this paper we show that SGGS decides the *stratified fragment* which generalizes EPR, the *PVD fragment*, and a new fragment that we dub *restrained*. The new class has the *small model property*, as the size of SGGS-generated models can be upper-bounded, and is also decided by hyperresolution and ordered resolution. We report on experiments with a termination tool implementing a restrainedness test, and with an SGGS prototype named Koala.

1 Introduction

Many applications of automated reasoning require to combine the decidability of satisfiability with an expressive logic. Since first-order theorem proving is only semidecidable, the quest for decidable fragments of first-order logic is key in advancing the field, and many classes of formulæ were shown decidable. Without claiming completeness (see [15,18,22] for surveys), we mention: the *Bernays-Schönfinkel class*, also known as EPR for *effectively propositional* [3,8,19,37,39]; the *Ackermann class* [2,23]; the *monadic class* with and without equality [2,5,23]; the *positive variable dominated* (PVD) fragment [17]; the *two-variable* fragment (FO^2) [21]; the *guarded* fragment [4,20]; the *modal fragment* [9,31], which is included in the EPR [22], FO^2 [32], and guarded [4] fragments; and the *stratified* fragment [1,25,36], which generalizes EPR. However, many theorem proving problems from the practice fall in none of these classes.

Example 1. Problem HWV036-2 from TPTP 7.3.0 [41] specifies a full-adder in 51 clauses, including for instance:

$$\neg\mathsf{and}_{\mathsf{ok}}(x) \lor \neg\mathsf{1}(\mathsf{in}_1(x)) \lor \neg\mathsf{1}(\mathsf{in}_2(x)) \lor \mathsf{1}(\mathsf{out}_1(x)), \quad \neg\mathsf{lor}(x) \lor \mathsf{or}_{\mathsf{ok}}(x) \lor \mathsf{error}(x),$$
$$\neg\mathsf{halfadd}(x) \lor \mathsf{connection}(\mathsf{in}_1(x), \mathsf{in}_1(\mathsf{or}_1(x))), \quad \neg\mathsf{fulladd}(x) \lor \mathsf{halfadd}(\mathsf{h}_1(x)).$$

This set is satisfiable, which means that termination of a theorem prover is *a priori* not guaranteed. However, it is neither EPR ($\exists^*\forall^*\varphi$ formulæ with φ quantifier- and function-free), nor Ackermann ($\exists^*\forall\exists^*\varphi$ formulæ with φ as above) nor FO^2 (only two variables, no functions), nor monadic (only unary predicates, no functions). One can also check that it is neither guarded nor stratified.

This research was funded in part by grant "Ricerca di base 2017" of the Università degli Studi di Verona. Authors are listed alphabetically.

N. Peltier and V. Sofronie-Stokkermans (Eds.): IJCAR 2020, LNAI 12166, pp. 356–374, 2020.
https://doi.org/10.1007/978-3-030-51074-9_20

As refutational completeness guarantees termination on unsatisfiable inputs, if one can prove termination of an inference system on satisfiable inputs in a certain class, any strategy given by that inference system and a fair search plan is a decision procedure for satisfiability in that class. Here we consider *Semantically-Guided Goal-Sensitive reasoning* (SGGS) [13,14], that is a refutationally complete instance-based theorem-proving method especially suitable for decision procedures: SGGS is *model-based* (it searches for a model by building candidates), *semantically guided* (the search is guided by a fixed initial interpretation), *conflict-driven* (it applies inferences such as resolution only to explain conflicts), *proof confluent* (it never needs to undo inferences), and *model complete in the limit* (if the input is satisfiable, the limit of the derivation represents a model), so that model generation is guaranteed if termination is.

This paper shows that SGGS decides the *stratified fragment* [1], which includes EPR and finds application in verification [1,25,36], and the *PVD fragment.* Then, we discover a *new decidable class* named the *restrained fragment,* and show that SGGS, ordered resolution, and hyperresolution all decide it. Since it is possible to compute *bounds on the size of SGGS-generated models*, this new class enjoys the *small model property.* We give a sufficient condition for membership in the restrained fragment that can be tested *automatically* by termination tools for rewriting. The relevance of this new class is evaluated empirically by applying this test to problems in TPTP. For instance, the axiomatization in Example 1, as well as all the TPTP problems including it, turn out to be restrained. We also summarize the outcomes of experiments with an SGGS prototype, named Koala, built reusing code from Konstantin Korovin's iProver [24,26].

The paper is structured as follows. Since the stratified fragment has sorts, Sect. 2 presents SGGS for a language with sorts. Section 3 shows that SGGS decides the stratified fragment. In Sect. 4 we define the restrained fragment, establish the small model property, and prove that SGGS decides both this class and PVD. Ordered resolution and hyperresolution also decide restrained sets (Sect. 5). The experimental results are reported in Sect. 6, and Sect. 7 concludes the paper.

2 Preliminaries: SGGS for Many-Sorted Logic

Let S be a set of clauses in many-sorted logic with non-empty sorts (there is a ground term for every sort). We use a, b for constants, P, Q for predicates, f, g for functions, w, x, y, z for variables, t, u for terms, L, M for literals, $at(L)$ for L's atom, C, D for clauses, $\mathcal{V}ar(C)$ for the set of variables in C, α, σ for substitutions, I, J for interpretations, and we extend the at notation to sets of literals, clauses, and sets of clauses. C^+ and C^- are the disjunctions of the positive and negative literals in C, respectively; C is *positive* if $C = C^+$ and *negative* if $C = C^-$.

In SGGS, a clause C may have a constraint A, written $A \rhd C$. An atomic *SGGS constraint* is *true*, *false*, $t \equiv u$, and $top(t) = f$, where $\equiv$ is syntactic identity, and $top(t)$ is the top symbol of term t. The negation, conjunction, and disjunction of constraints is a constraint. Constraints in *standard form* are *true*, *false*, and conjunctions of distinct atomic constraints $x \not\equiv y$ and $top(x) \neq f$. Substitutions are *sort-preserving* ($x\sigma$ has the same sort as x) so that instantiation respects sorts. The set $Gr(A \rhd C)$ of *constrained ground instances* (cgi) of $A \rhd C$ is the set of ground instances of C that satisfy A. Literals $A \rhd L$ and $B \rhd M$ *intersect* if $at(Gr(A \rhd L)) \cap at(Gr(B \rhd M)) \neq \emptyset$, and are *disjoint* otherwise.

Example 2. In a signature with sorts $\{s_1, s_2\}$ and symbols $\mathsf{a}\colon s_1$, $\mathsf{b}\colon s_2$, $\mathsf{f}\colon s_1 \rightarrow s_2$, and $\mathsf{P} \subseteq s_2 \times s_2$, the only term of sort s_1 is a, and b and $\mathsf{f}(\mathsf{a})$ are the only terms of sort s_2. Thus, $Gr(\mathsf{P}(x,y)) = \{\mathsf{P}(\mathsf{b},\mathsf{b}), \mathsf{P}(\mathsf{f}(\mathsf{a}),\mathsf{b}), \mathsf{P}(\mathsf{b},\mathsf{f}(\mathsf{a})), \mathsf{P}(\mathsf{f}(\mathsf{a}),\mathsf{f}(\mathsf{a}))\}$. For $\mathsf{P}(\mathsf{f}(x), y)$, with $x\colon s_1$ and $y\colon s_2$, constraint $top(x) \neq \mathsf{a}$ is unsatisfiable, while $top(y) \neq \mathsf{a}$ is valid. Then, $top(x) \neq \mathsf{a} \rhd \mathsf{P}(\mathsf{f}(x), y)$ is equivalent to $false \rhd \mathsf{P}(\mathsf{f}(x), y)$ and has no cgi's, while $top(y) \neq \mathsf{a} \rhd \mathsf{P}(\mathsf{f}(x), y)$ is equivalent to $true \rhd \mathsf{P}(\mathsf{f}(x), y)$, or simply $P(\mathsf{f}(x), y)$, and has all cgi's, namely $P(\mathsf{f}(\mathsf{a}), \mathsf{b})$ and $P(\mathsf{f}(\mathsf{a}), \mathsf{f}(\mathsf{a}))$.

SGGS is *semantically guided* by an *initial interpretation* I: unless $I \models S$, SGGS seeks a model of S, by building candidate partial interpretations different from I, and using I as default to complete them. If the empty clause $\bot$ arises in the process, unsatisfiability is reported. If I is the *all-negative interpretation* I^- that makes all negative literals true, SGGS tries to discover which positive literals need to be true to satisfy S, and dually if I is the *all-positive interpretation* I^+. While I can be any Herbrand interpretation, I^+ and I^- suffice in this paper.

SGGS works with a *trail* of clauses $\Gamma = A_1 \rhd C_1[L_1], \ldots, A_n \rhd C_n[L_n]$, where $C[L]$ means that literal $L \in C$ is *selected* in C. The length of Γ and its prefix of length j are denoted $|\Gamma|$ and $\Gamma|_j$, respectively. An SGGS-trail Γ represents a partial interpretation $I^p(\Gamma)$: if Γ is empty, denoted ε, $I^p(\Gamma) = \emptyset$; otherwise, $I^p(\Gamma) = I^p(\Gamma|_{n-1}) \cup pcgi(A_n \rhd L_n, \Gamma)$, where pcgi abbreviates *proper constrained ground instances*. A pcgi of $A_n \rhd C_n[L_n]$ is a cgi $C[L]$ that is not satisfied by $I^p(\Gamma|_{n-1})$ (i.e., $I^p(\Gamma|_{n-1}) \cap C[L] = \emptyset$) and can be satisfied by adding L as $\neg L \notin I^p(\Gamma|_{n-1})$. For the selected literal, $pcgi(A_n \rhd L_n, \Gamma) = \{L : \ C[L] \in pcgi(A_n \rhd C_n[L_n], \Gamma)\}$. $I^p(\Gamma)$ is completed into an interpretation $I[\Gamma]$ by consulting I for the truth value of any literal undefined in $I^p(\Gamma)$.

A literal L is *uniformly false* in an interpretation J, if all $L' \in Gr(L)$ are false in J. Then, L is said to be *I-false* if it is uniformly false in I, and *I-true* if it is true in I. SGGS requires that if a clause in Γ has I-false literals, one is selected, so as to differentiate $I[\Gamma]$ from I. A clause whose literals are all I-true is an *I-all-true clause*, and only in such a clause an I-true literal is selected.

A *conflict clause* is one whose literals are all uniformly false in $I[\Gamma]$. SGGS ensures that every I-all-true clause $C[L]$ in Γ is either a conflict clause or the *justification* of its selected literal L, meaning that all literals of $C[L]$ except L are uniformly false in $I[\Gamma]$, so that L must be true in $I[\Gamma]$ to satisfy $C[L]$. In the latter case $C[L]$ is in the *disjoint prefix* of Γ, denoted $dp(\Gamma)$, which is the longest prefix such that $pcgi(A \rhd C[L], \Gamma) = Gr(A \rhd C[L])$ for all its clauses $A \rhd C[L]$.

An *SGGS-derivation* is a series of trails $\Gamma_0 \vdash \Gamma_1 \vdash \ldots \Gamma_j \vdash \ldots$, where $\Gamma_0 = \varepsilon$, and $\forall j,\ j > 0$, an SGGS-inference generates Γ_j from Γ_{j-1} and S. If $\bot \notin \Gamma$

and $I[\Gamma] \not\models S$, SGGS has *two ways to make progress.* If $\Gamma = dp(\Gamma)$, the trail is in order, but $I[\Gamma] \not\models C'$ for some $C' \in Gr(C)$ and $C \in S$. Then, SGGS applies *SGGS-extension* to generate from C and Γ a clause $A \rhd E$, such that E is an instance of C and $C' \in Gr(A \rhd E)$. If $\Gamma \neq dp(\Gamma)$, the trail needs repair: either there is a conflict, or there are intersections between selected literals to be removed by *SGGS-splitting.* The *SGGS-extension rules* specialize the *SGGS-extension scheme* ([14, Def. 12]) of which we give here the instance for I based on sign:

Definition 1. *Given input clause set S and trail Γ, if there is a clause $C \in S$ such that for all its I-true literals $L_1, \ldots, L_n$ $(n \geq 0)$ there are clauses $B_1 \rhd D_1[M_1], \ldots, B_n \rhd D_n[M_n]$ in $dp(\Gamma)$, such that literals $M_1, \ldots, M_n$ are I-false, and $\forall j$, $1 \leqslant j \leqslant n$, $L_j\alpha = \neg M_j\alpha$ with simultaneous most general unifier (mgu) α, then* SGGS-extension *adds $A \rhd E = (\bigwedge_{j=1}^{n} B_j\alpha) \rhd C\alpha$ to Γ.*

SGGS-splitting decomposes a clause into instances to isolate and remove intersections between literals, and it is the only rule that introduces constraints. Splitting a ground clause is trivial and never done. Let $A \rhd C[L]$ be a clause where A is satisfiable. Roughly speaking (see [14, Sect. 3.2]), a *partition* of $A \rhd C[L]$ is a set $\{A_i \rhd C_i[L_i]\}_{i=1}^{n}$ such that $Gr(A \rhd C) = \bigcup_{i=1}^{n}\{Gr(A_i \rhd C_i)\}$ and the literals $A_i \rhd L_i$ are pairwise disjoint. Adding predicate symbol $\mathsf{Q} \subseteq s_1 \times s_2$ to Example 2, a partition of $[\mathsf{P}(\mathsf{f}(x), y)] \vee \mathsf{Q}(x, y)$ is $\{[\mathsf{P}(\mathsf{f}(x), \mathsf{b})] \vee \mathsf{Q}(x, \mathsf{b}),\ [\mathsf{P}(\mathsf{f}(x), \mathsf{f}(\mathsf{a}))] \vee \mathsf{Q}(x, \mathsf{f}(\mathsf{a}))\}$. Given trail clauses $A \rhd C[L]$ and $B \rhd D[M]$, a *splitting* of C by D, denoted *split*(C, D), is a partition of $A \rhd C[L]$ such that $at(Gr(A_j \rhd L_j))$ for some j is the intersection of $A \rhd L$ and $B \rhd M$, and all other $A_i \rhd L_i$ are disjoint from $B \rhd M$. *SGGS-splitting* replaces $A \rhd C[L]$ with *split*(C, D). Not all clauses in *split*(C, D) need to be kept for completeness (see [14, Sect. 4.2]).

Clause $A_n \rhd C_n[L_n]$ is *disposable,* if $I^p(\Gamma|_{n-1}) \models A_n \rhd C_n[L_n]$, and *SGGS-deletion* removes *all* disposable clauses from the trail. The following example shows that SGGS halts, if applied to the EPR set used to show that another semantically-guided method, hyperresolution, cannot decide EPR.

Example 3. The set S consists of four clauses ([18, Ex. 4.8] and [15, Ex. 3.17]):

$$\mathsf{P}(x, x, \mathsf{a}) \quad (i), \qquad \mathsf{P}(x, y, w) \vee \mathsf{P}(y, z, w) \vee \neg\mathsf{P}(x, z, w) \quad (ii),$$
$$\neg\mathsf{P}(x, x, \mathsf{b}) \quad (iii), \qquad \mathsf{P}(x, z, w) \vee \neg\mathsf{P}(x, y, w) \vee \neg\mathsf{P}(y, z, w) \quad (iv).$$

SGGS with all-negative initial interpretation I^- yields the following derivation:

$$\begin{aligned}
\Gamma_0 : \varepsilon \ \vdash \Gamma_1 &: [\mathsf{P}(x, x, \mathsf{a})] && \text{extend } (i)\\
\vdash \Gamma_2 &: [\mathsf{P}(x, x, \mathsf{a})],\ \mathsf{P}(x, y, \mathsf{a}) \vee [\mathsf{P}(y, x, \mathsf{a})] \vee \neg\mathsf{P}(x, x, \mathsf{a}) && \text{extend } (ii)\\
\vdash \Gamma_3 &: [\mathsf{P}(x, x, \mathsf{a})],\ \mathsf{P}(x, x, \mathsf{a}) \vee [\mathsf{P}(x, x, \mathsf{a})] \vee \neg\mathsf{P}(x, x, \mathsf{a}),\\
&\quad y \neq x \rhd \mathsf{P}(x, y, \mathsf{a}) \vee [\mathsf{P}(y, x, \mathsf{a})] \vee \neg\mathsf{P}(x, x, \mathsf{a}) && \text{split}\\
\vdash \Gamma_4 &: [\mathsf{P}(x, x, \mathsf{a})],\\
&\quad y \neq x \rhd \mathsf{P}(x, y, \mathsf{a}) \vee [\mathsf{P}(y, x, \mathsf{a})] \vee \neg\mathsf{P}(x, x, \mathsf{a}) && \text{delete}
\end{aligned}$$

Since $I^- \not\models P(x, x, \mathsf{a})$, SGGS-extension puts it on the trail. As $I[\Gamma_1]$ satisfies $P(x, x, \mathsf{a})$, but no other positive literal, $I[\Gamma_1] \not\models (ii)$. Thus, SGGS-extension unifies the third literal of clause (ii) with $[\mathsf{P}(x, x, \mathsf{a})]$ on the trail, producing Γ_2, where an I^--false (i.e., positive) literal is selected in the added clause (choosing the other makes no difference). As the selected literals intersect, the second clause gets split, yielding Γ_3. The second clause in Γ_3 is disposable and SGGS-deletion removes it. Since $I[\Gamma_4] \models S$, the derivation halts reporting satisfiable. In contrast, hyperresolution generates infinitely many clauses of the form $\mathsf{P}(x_1, x_2, \mathsf{a}) \vee \mathsf{P}(x_2, x_3, \mathsf{a}) \vee \cdots \vee \mathsf{P}(x_{n-1}, x_n, \mathsf{a}) \vee \mathsf{P}(x_n, x_1, \mathsf{a})$.

If a clause $A \rhd C[L]$ added by SGGS-extension is in conflict with $I[\Gamma]$ and $C[L]$ contains I-false literals, *SGGS-resolution explains the conflict* by resolving $A \rhd C[L]$ with a justification in $dp(\Gamma)$: the resolvent is also a conflict clause that replaces $A \rhd C[L]$. As SGGS-extension (see [14, Def. 19]) ensures that all I-false literals of a conflict clause can be resolved away, conflict explanation generates either $\bot$ or an I-all-true conflict clause $B \rhd D[M]$. The conflict represented by $B \rhd D[M]$ is *solved* by moving (*SGGS-move*) $B \rhd D[M]$ to the left of the clause in $dp(\Gamma)$ whose selected literal makes M uniformly false in $I[\Gamma]$: the effect is to *flip* M from being uniformly false to being an implied literal.

Fairness of an SGGS-derivation involves several properties: SGGS-deletion and other clause removals are applied eagerly; trivial splitting is avoided; progress is made whenever possible; every SGGS-extension generating a conflict clause is *bundled* with explanation and conflict-solving inferences to eliminate the conflict before new extensions occur; and inferences applying to shorter prefixes of the trail are never neglected in favor of others applying to longer prefixes (see [14, Defs. 32, 37, and 39]). The *limit* of a fair derivation $\Gamma_0 \vdash \Gamma_1 \vdash \ldots \Gamma_j \vdash \Gamma_{j+1} \vdash \ldots$ is the longest trail Γ_∞ such that $\forall i,\ i \leqslant |\Gamma_\infty|$, there is an n_i such that $\forall j$, $j \geqslant n_i$, if $|\Gamma_j| \geq i$ then $\Gamma_j|_i$ is equivalent to $\Gamma_\infty|_i$ (see [14, Def. 50]). In words, all prefixes of the trail stabilize eventually. Both derivation and Γ_∞ may be infinite, but if the derivation halts at stage k, $\Gamma_\infty = \Gamma_k$. The following results employ an *SGGS-suitable* (i.e., total and extending the size ordering, hence well-founded) ordering on ground atoms (see [14, Def. 16]) and a *convergence ordering* $>^c$ on SGGS-trails (see [14, Def. 46]):

- *Completeness*: For all input clause sets S, initial interpretations I, and fair SGGS-derivations, if S is satisfiable, $I[\Gamma_\infty] \models S$, and if S is unsatisfiable, $\bot \in \Gamma_k$ for some k (see [14, Thm. 9 and 11]).
- *Descending chain theorem*: A fair SGGS-derivation forms a descending chain $\Gamma_0 >^c \Gamma_1 >^c \ldots >^c \Gamma_j >^c \Gamma_{j+1} \ldots$ (see [14, Thm. 8]).
- *Finiteness of descending chains of length-bounded trails*: A chain $\Gamma_0 >^c \Gamma_1 >^c \ldots \Gamma_j >^c \Gamma_{j+1} \ldots$ where $\forall j,\ j \geq 0,\ |\Gamma_j| \leqslant n$, for some $n \geq 0$, is finite (see [14, Thm. 6 and Cor. 2]).

The gist of this paper is to find fragments where the length of SGGS-trails is bounded so that termination of fair derivations is guaranteed.

3 SGGS Decides the Stratified Fragment

A way to ensure termination is to restrict an inference engine to produce only terms or atoms from a finite set $\mathcal{B}$, usually called a *basis*. For SGGS, let $\mathcal{B}$ be a finite subset of the Herbrand base $\mathcal{A}$ of the input clause set S.

Definition 2. *An SGGS-trail* $\Gamma = A_1 \rhd C_1[L_1], \ldots, A_n \rhd C_n[L_n]$ is in $\mathcal{B}$ *if for all* i*,* $1 \leqslant i \leqslant n$*,* $at(Gr(A_i \rhd C_i)) \subseteq \mathcal{B}$*.*

An SGGS-derivation *is in* $\mathcal{B}$ if all its trails are.

Lemma 1. *If a fair SGGS-derivation* $\Gamma_0 \vdash \Gamma_1 \vdash \ldots \Gamma_j \vdash \Gamma_{j+1} \vdash \ldots$ *is in a finite basis* $\mathcal{B}$*, then for all* j*,* $j \geqslant 0$*,* $|\Gamma_j| \leqslant |\mathcal{B}|+1$*, and if the derivation halts at stage* k*,* $k \geqslant 0$*, then* $|\Gamma_k| \leqslant |\mathcal{B}|$*.*

Proof. SGGS cannot do worse than generating a ground trail where every atom in $\mathcal{B}$ appears selected with either sign: any trail with non-ground clauses will be shorter, because a non-ground clause covers many (possibly infinitely many) ground instances. By fairness, if the trail contains an intersection given by clauses $C[L]$ and $D[L]$, or $C[L]$ and $D[\neg L]$ with $L \in \mathcal{B}$, the clause on the right is either deleted eagerly by SGGS-deletion, or replaced with a resolvent by SGGS-resolution before SGGS-extension applies. Thus, there can be at most one such intersection, and the first claim follows. The second claim holds, because the intersection is removed by fairness prior to termination. □

By the descending chain theorem and the finiteness of descending chains of length-bounded trails, the following general result follows:

Theorem 1. *A fair SGGS-derivation in a finite basis is finite.*

In order to apply this result, we need to find fragments that admit a finite basis. We begin with the *stratified fragment.* A signature is *stratified*, if there is a well-founded ordering $<_s$ on sorts, and for all functions $f\colon s_1 \times \cdots \times s_n \rightarrow s$, it holds that $s <_s s_i$ for all $1 \leqslant i \leqslant n$ [1,25,36]. Thus, there are *no cycles over sorts* when applying functions. The signature from Example 2 is stratified with ordering $s_1 >_s s_2$. If a sentence over a stratified signature belongs to the $\exists^*\forall^*$ fragment, Skolemization only introduces constants and preserves stratification [25]. If there is only one sort, this fragment reduces to EPR, because stratification over a single sort implies that there are no function symbols. However, also stratified sentences with a prefix other than $\exists^*\forall^*$ can yield stratified clauses [33].

Example 4. Assume a stratified signature with sorts s_1 and s_2 such that $s_1 <_s s_2$, and symbols $\mathsf{f}\colon s_1 \rightarrow s_2$, and $\mathsf{P} \subseteq s_2 \times s_2$. The Skolemization of $\forall x \exists y.\, \mathsf{P}(\mathsf{f}(x), y)$ preserves stratification, as clause $\mathsf{P}(\mathsf{f}(x), \mathsf{g}(x))$ with Skolem symbol $\mathsf{g}\colon s_1 \rightarrow s_2$ is still stratified. On the other hand, the Skolemization of $\forall x \exists y.\, \mathsf{P}(\mathsf{f}(y), x)$ yields $\mathsf{P}(\mathsf{f}(\mathsf{g}(x)), x)$ with Skolem symbol $\mathsf{g}\colon s_2 \rightarrow s_1$, so that stratification is lost.

Given a set S of clauses whose signature is stratified, or stratified clause set for short, the Herbrand universe $\mathcal{H}$ and the Herbrand base $\mathcal{A}$ are *finite*, because stratification prevents building terms of unbounded depth [1]. Therefore, it suffices to pick $\mathcal{A}$ itself as the finite basis for Theorem 1.

Theorem 2. *Any fair SGGS-derivation from a stratified clause set* S *halts, is a refutation if* S *is unsatisfiable, and constructs a model of* S *if* S *is satisfiable.*

However, SGGS-derivations can get exponentially long.

Example 5. Consider the following clause set S_k describing a k-digits binary counter [38, Def. 2.4.10]. Let Q be a predicate symbol of arity k, and for all i, $1 \leqslant i \leqslant k$, let $\overline{0}_i$, $\overline{1}_i$, and $\overline{x}_i$ be i-tuples of 0's, 1's, and distinct variables $x_1, \ldots, x_i$, respectively. S_k consists of the $k+2$ clauses, for $1 \leqslant m \leqslant k$,

$$C_0\colon \mathsf{Q}(\overline{0}_k) \quad C_m\colon \neg\mathsf{Q}(\overline{x}_m, 0, \overline{1}_{k-m-1}) \vee \mathsf{Q}(\overline{x}_m, 1, \overline{0}_{k-m-1}) \quad C_{k+1}\colon \neg\mathsf{Q}(\overline{1}_k)$$

so that it is in EPR. Guided by I^-, SGGS generates a derivation

$$\begin{array}{lr}
\Gamma_0\colon \varepsilon \vdash \Gamma_1\colon [\mathsf{Q}(\overline{0}_k)] & \mathsf{extend}\ (C_0)\\
\vdash \Gamma_2\colon \ldots, \neg\mathsf{Q}(\overline{0}_k) \vee [\mathsf{Q}(\overline{0}_{k-1}, 1)] & \mathsf{extend}\ (C_{k-1})\\
\vdash \Gamma_3\colon \ldots, \neg\mathsf{Q}(\overline{0}_{k-1}, 1) \vee [\mathsf{Q}(\overline{0}_{k-2}, 1, 0)] & \mathsf{extend}\ (C_{k-2})\\
\vdash \Gamma_4\colon \ldots, \neg\mathsf{Q}(\overline{0}_{k-2}, 1, 0) \vee [\mathsf{Q}(\overline{0}_{k-2}, 1, 1)] & \mathsf{extend}\ (C_{k-1})\\
\ldots & \ldots\\
\vdash \Gamma_{2^k-1}\colon \ldots, \neg\mathsf{Q}(\overline{1}_{k-2}, 0, 1) \vee [\mathsf{Q}(\overline{1}_{k-1}, 0)] & \mathsf{extend}\ (C_{k-2})\\
\vdash \Gamma_{2^k}\colon \ldots, \neg\mathsf{Q}(\overline{1}_{k-1}, 0) \vee [\mathsf{Q}(\overline{1}_k)] & \mathsf{extend}\ (C_{k-1})\\
\vdash \Gamma_{2^k+1}\colon \ldots, \neg\mathsf{Q}(\overline{1}_{k-1}, 0) \vee [\mathsf{Q}(\overline{1}_k)],\ [\neg\mathsf{Q}(\overline{1}_k)] & \mathsf{extend}\ (C_{k+1})
\end{array}$$

that simulates binary counting by adding a clause in each of these 2^k+1 steps till a conflict emerges. Then it takes another 2^{k+1} steps to detect unsatisfiability:

$$\begin{array}{lr}
\vdash \Gamma_{2^k+2}\colon \ldots, [\neg\mathsf{Q}(\overline{1}_k)],\ \neg\mathsf{Q}(\overline{1}_{k-1}, 0) \vee [\mathsf{Q}(\overline{1}_k)] & \mathsf{move}\\
\vdash \Gamma_{2^k+3}\colon \ldots, [\neg\mathsf{Q}(\overline{1}_k)],\ [\neg\mathsf{Q}(\overline{1}_{k-1}, 0)] & \mathsf{resolve}\\
\vdash \Gamma_{2^k+2}\colon \ldots, [\neg\mathsf{Q}(\overline{1}_{k-1}, 0)],\ \neg\mathsf{Q}(\overline{1}_{k-2}, 0, 1) \vee [\mathsf{Q}(\overline{1}_{k-1}, 0)],\ [\neg\mathsf{Q}(\overline{1}_k)] & \mathsf{move}\\
\vdash \Gamma_{2^k+3}\colon \ldots, [\neg\mathsf{Q}(\overline{1}_{k-1}, 0)],\ [\neg\mathsf{Q}(\overline{1}_{k-2}, 0, 1)],\ [\neg\mathsf{Q}(\overline{1}_k)] & \mathsf{resolve}\\
\ldots & \ldots\\
\vdash \Gamma_{2^{k+2}}\colon [\neg\mathsf{Q}(\overline{0}_k)],\ [\mathsf{Q}(\overline{0}_k)], \ldots & \mathsf{move}\\
\vdash \Gamma_{2^{k+2}+1}\colon \bot, \ldots & \mathsf{resolve}
\end{array}$$

Similar to positive resolution[1] [38, Thm. 2.4.12] or SCL [19], SGGS behaves exponentially, whereas resolution offers a refutation in $2k+1$ steps [38, Thm. 2.4.11], which shows that in EPR ground resolution (same as positive resolution for S_k) can do exponentially worse than resolution [34]. Encoding S_k in propositional logic requires exponentially many clauses, as each clause C_m, $1 \leqslant m \leqslant k$, has 2^m ground instances that need to be modeled as distinct propositional clauses. This indicates that generating instances is not good for this example.

4 SGGS Decides the Restrained Fragment

We begin with the notion of *ground-preserving* clause, which is convenient for sign-based semantic guidance.

[1] Every resolution step has a positive parent.

Definition 3. *A clause C is* positively ground-preserving *if* $\mathcal{V}ar(C) \subseteq \mathcal{V}ar(C^-)$, *and* negatively ground-preserving *if* $\mathcal{V}ar(C) \subseteq \mathcal{V}ar(C^+)$. *A set of clauses is* positively/negatively ground-preserving *if all its clauses are, and* ground-preserving *if it is positively or negatively ground-preserving.*

For example, $\neg \mathsf{P}(x, y, z) \vee \mathsf{Q}(y) \vee \mathsf{Q}(\mathsf{f}(z))$ and $\neg \mathsf{Q}(x) \vee \neg \mathsf{Q}(y)$ are positively ground-preserving, while the clauses in Example 5 are both positively and negatively ground-preserving. We say that I is *suitable* for a ground-preserving set S if either I is I^- and S is positively ground-preserving, or I is I^+ and S is negatively ground-preserving. If S is positively ground-preserving, its positive clauses are ground, positive hyperresolution only generates ground clauses ([12], Lem. 3), and similarly for the negative variant. We show that this also holds for SGGS.

Lemma 2. *If the input clause set S is ground-preserving and the initial interpretation is suitable for S, any fair SGGS-derivation from S is ground.*

Proof. We consider S positively ground-preserving and I^- (for the dual case one exchanges the signs). The proof is by induction on the length n of the derivation. The base case ($n = 0$) is vacuously true. The induction hypothesis is that the claim holds for a derivation of length n producing trail Γ. Let $\Gamma \vdash \Gamma'$ be the (n+1)-th step. Since Γ is ground, $\Gamma \vdash \Gamma'$ cannot be a splitting step, because any splitting of a ground clause yields the clause itself, and fairness excludes such trivial splittings ([14, Defs. 32, 47, and 49]). If $\Gamma \vdash \Gamma'$ is an SGGS-resolution step, it is a ground resolution step, and also Γ' is ground. If $\Gamma \vdash \Gamma'$ is an SGGS-extension step, it adds an instance $C\alpha$ of a clause $C \in S$, where α is the simultaneous mgu of *all* I^--true (i.e., negative) literals $L_1, \ldots, L_n$ in C with as many I^--false (i.e., positive) selected literals $M_1, \ldots, M_n$ in Γ (see Def. 1). Since Γ is ground by induction hypothesis, the clauses containing $M_1, \ldots, M_n$ are ground and do not have constraints. Thus, $L_1\alpha, \ldots, L_n\alpha$ are also ground. The I^--false literals of $C\alpha$ are ground, because C is positively ground-preserving (i.e., $\mathcal{V}ar(C) \subseteq \mathcal{V}ar(C^-)$), so that all its variables get grounded by α. Hence $C\alpha$ and Γ' are ground. □

The next example illustrates Lemma 2 and gives the intuition for restrainedness.

Example 6. Assume a positively ground-preserving set S which includes:

$$\mathsf{P}(\mathsf{s}^{10}(\mathsf{0}), \mathsf{s}^9(\mathsf{0})) \quad (i), \quad \neg \mathsf{P}(\mathsf{s}(\mathsf{s}(x)), y) \vee \mathsf{P}(x, \mathsf{s}(y)) \quad (ii), \quad \neg \mathsf{P}(\mathsf{s}(\mathsf{0}), \mathsf{0}) \quad (iii),$$

and I^- is the initial interpretation. SGGS starts with an extension that puts the positive clause $\mathsf{P}(10, 9)$ on the trail, where we write n for $\mathsf{s}^n(\mathsf{0})$. Subsequent extensions unify the negative literal in clause (ii) with some positive ground literal on the trail, so that new literals in added clauses are positive:

$$\begin{aligned} \Gamma_0 : \varepsilon \vdash\ & \Gamma_1 : [\mathsf{P}(10, 9)] \\ \vdash\ & \Gamma_2 : [\mathsf{P}(10, 9)],\ \neg \mathsf{P}(10, 9) \vee [\mathsf{P}(8, 10)] \\ \vdash\ & \Gamma_3 : [\mathsf{P}(10, 9)],\ \neg \mathsf{P}(10, 9) \vee [\mathsf{P}(8, 10)],\ \neg \mathsf{P}(8, 10) \vee [\mathsf{P}(6, 11)]. \end{aligned}$$

The positive literals have decreasing number of symbols, matching the fact that $\mathsf{P}(\mathsf{s}(\mathsf{s}(x)), y) \succ \mathsf{P}(x, \mathsf{s}(y))$ in (*ii*) for $\succ$ any lexicographic path ordering (LPO).

This suggests to strengthen ground-preservingness with an ordering (unrelated to the SGGS-suitable ordering of Sect. 2) to get a finite basis.

Definition 4. *A quasi-ordering $\succeq$ on terms and atoms is* restraining, *if (i) it is stable under substitution, (ii) the strict ordering $\succ = \succeq \setminus \preceq$ is well-founded, and (iii) the equivalence $\approx = \succeq \cap \preceq$ has finite equivalence classes.*

Note that Condition (i) implies that $\succ$ and $\approx$ are stable under substitution. From now on, $\succeq$ is a restraining quasi-ordering. Let $\mathcal{A}_S$ be the set of ground atoms occurring in S, and $\mathcal{A}_S^{\preceq}$ the subset of the Herbrand base $\mathcal{A}$ of ground atoms upper-bounded by $\mathcal{A}_S$, so $\mathcal{A}_S^{\preceq} = \{L : L \in \mathcal{A}, \exists M \in \mathcal{A}_S \text{ with } M \succeq L\}$. By Conditions (ii) and (iii) in Definition 4, $\mathcal{A}_S^{\preceq}$ is finite and thus can serve as basis.

Definition 5. *A clause C is* (strictly) positively restrained *if it is positively ground preserving, and for all non-ground literals $L \in C^+$ there is a literal $M \in C^-$ such that $at(M) \succeq at(L)$ ($at(M) \succ at(L)$). A set of clauses is* positively restrained *if all its clauses are.*

Negatively restrained clauses and clause sets are defined similarly, and a set of clauses is *restrained* if it is positively or negatively restrained. The set in Example 6 is strictly positively restrained. We see next the role of the *quasi-ordering*.

Example 7. Problem PLA030-1 in TPTP is neither stratified, nor monadic, nor guarded. Its clause $\mathsf{differ}(x, y) \vee \neg\mathsf{differ}(y, x)$ cannot be shown *strictly* restrained. Let $\succ_{\mathsf{acrpo}}$ be an AC-compatible [40] recursive path ordering with differ as an AC-symbol, meaning associative-commutative. The quasi-ordering $\succeq_{\mathsf{acrpo}}$, built from $\succ_{\mathsf{acrpo}}$ and the AC-equivalence $\approx_{\mathsf{AC}}$ that has finite equivalence classes, satisfies $\mathsf{differ}(x, y) \succeq_{\mathsf{acrpo}} \mathsf{differ}(y, x)$, and shows that PLA030-1 is negatively restrained.

Restrainedness is undecidable in general, but decidable for fixed, suitable orderings. If S is restrained, a fair SGGS-derivation will be in $\mathcal{A}_S^{\preceq}$.

Lemma 3. *If the input clause set S is restrained and the initial interpretation is suitable for S, any fair SGGS-derivation from S is in $\mathcal{A}_S^{\preceq}$.*

Proof. We consider S positively ground-preserving and I^- (for the dual case one exchanges the signs). Since the set is restrained hence ground-preserving, the derivation is ground by Lemma 2 (†). The proof is by induction on the length n of the derivation, and it follows the same pattern as that of Lemma 2. Let $\Gamma \vdash \Gamma'$ be the $(n{+}1)$-th step. By induction hypothesis, Γ is in $\mathcal{A}_S^{\preceq}$. If $\Gamma \vdash \Gamma'$ is an SGGS-resolution step, it is a ground resolution step which does not generate new atoms, and also Γ' is in $\mathcal{A}_S^{\preceq}$. If $\Gamma \vdash \Gamma'$ is an SGGS-extension step, it adds an instance $C\alpha$ of a clause $C \in S$, where α is the simultaneous mgu of all I^--true (i.e., negative) literals $\neg L_1, \ldots, \neg L_n$ in C with as many I^--false (i.e., positive) selected literals $M_1, \ldots, M_n$ in Γ. The literals $M_1, \ldots, M_n$ are ground by (†), and by induction hypothesis they are in $\mathcal{A}_S^{\preceq}$. We have to show $at(C\alpha) \subseteq \mathcal{A}_S^{\preceq}$.

- For the negative literals $\neg L_1\alpha, \ldots, \neg L_n\alpha$ we have $L_i\alpha = M_i\alpha = M_i \in \mathcal{A}_S^{\preceq}$.
- Let L be a literal in C^+. If L is ground, then $L\alpha = L \in \mathcal{A}_S \subseteq \mathcal{A}_S^{\preceq}$. If L is not ground, by positive restrainedness there exists a $\neg L_i$, $1 \leqslant i \leqslant n$, such that $L_i \succeq L$. By stability, $L_i\alpha \succeq L\alpha$. Since for all i, $1 \leqslant i \leqslant n$, $M_i \in \mathcal{A}_S^{\preceq}$ and $M_i = M_i\alpha = L_i\alpha \succeq L\alpha$, we have $L\alpha \in \mathcal{A}_S^{\preceq}$. □

Therefore, as $\mathcal{A}_S^{\preceq}$ is finite, Theorem 1 applies.

Theorem 3. *Any fair SGGS-derivation from a restrained clause set S with a suitable initial interpretation for S halts, is a refutation if S is unsatisfiable, and constructs a model of S if S is satisfiable.*

Restrainedness also makes it possible to derive an upper bound on the cardinality of a single-sorted model, defined as the cardinality of its domain. Let $\mathcal{H}_S^{\preceq}$ be the set $\mathcal{H}_S^{\preceq} = \{t : t \text{ is a strict subterm of } L \text{ for some } L \in \mathcal{A}_S^{\preceq}\}$.

Theorem 4. *A restrained satisfiable clause set S has a model of cardinality at most $|\mathcal{H}_S^{\preceq}| + 1$ that can be extracted from the limit of any fair SGGS-derivation from S with suitable initial interpretation I.*

Proof. By Theorem 3 the derivation halts with some trail Γ, and by Lemmas 2 and 3, Γ contains only ground clauses whose atoms are in $\mathcal{A}_S^{\preceq}$. Since SGGS is model complete, $I[\Gamma] \models S$. Consider the following interpretation J with domain $\mathcal{H}_S^{\preceq} \uplus \{u\}$, where u is a new constant symbol: for every constant symbol c we set $c^J = c$ if $c \in \mathcal{H}_S^{\preceq}$, and $c^J = u$ otherwise. For every n-ary ($n \geqslant 1$) function symbol f, we set $f^J(t_1, \ldots, t_n) = f(t_1^J, \ldots, t_n^J)$ if $f(t_1, \ldots, t_n) \in \mathcal{H}_S^{\preceq}$, and $f^J(t_1, \ldots, t_n) = u$ otherwise. For every predicate symbol P, $(t_1, \ldots, t_n) \in P^J$ if $I[\Gamma] \models P(t_1, \ldots, t_n)$. Note that J is well-defined because if $f(t_1, \ldots, t_n) \in \mathcal{H}_S^{\preceq}$ then $t_1, \ldots, t_n$ are also, hence all terms are interpreted in $\mathcal{H}_S^{\preceq} \uplus \{u\}$. As J agrees with $I[\Gamma]$ on all atoms, $J \models S$, and it has cardinality $|\mathcal{H}_S^{\preceq}| + 1$ by construction. □

Therefore the restrained fragment also enjoys the *small model property*.

Example 8. The satisfiable clause set S (PUZ054-1 in TPTP) extending Example 6:

$$\mathsf{P}(\mathsf{s}^{10}(0), \mathsf{s}^9(0)), \quad \neg\mathsf{P}(\mathsf{s}(\mathsf{s}(x)), y) \vee \mathsf{P}(x, \mathsf{s}(y)), \quad \neg\mathsf{P}(x, \mathsf{s}(\mathsf{s}(y))) \vee \mathsf{P}(x, \mathsf{s}(y)),$$
$$\neg\mathsf{P}(\mathsf{s}(0), 0), \quad \neg\mathsf{P}(\mathsf{s}(x), \mathsf{s}(y)) \vee \mathsf{P}(\mathsf{s}(x), y),$$

is neither in EPR, nor in FO^2, nor in the monadic class. However, it can be shown strictly positively restrained by a Knuth-Bendix ordering (KBO) $\succ$ with empty precedence and weights $w(\mathsf{P}) = 0$ and $w_0 = w(\mathsf{s}) = w(0) = 1$, where w_0 is the weight of variables. The largest atom in $\mathcal{A}_S = \{\mathsf{P}(\mathsf{s}^{10}(0), \mathsf{s}^9(0)),\ \mathsf{P}(\mathsf{s}(0), 0)\}$ has weight $w(\mathsf{P}(\mathsf{s}^{10}(0), \mathsf{s}^9(0))) = 21$. No atom L of the form $\mathsf{P}(\mathsf{s}^n(0), \mathsf{s}^m(0))$ in $\mathcal{A}_S^{\preceq}$ can have a subterm $\mathsf{s}^k(0)$ with $k > 19$, because otherwise $w(L) > w(\mathsf{P}(\mathsf{s}^{10}(0), \mathsf{s}^9(0)))$. Therefore, we have $\mathcal{H}_S^{\preceq} = \{\mathsf{s}^i(0) \,:\, 0 \leqslant i \leqslant 19\}$ and there is a model of cardinality at most 21 by Theorem 4.

We next consider PVD [17]. Let depth(C) be the maximum depth of an atom in clause C, and depth$_x(C)$ the maximum occurrence depth in C of $x \in \mathcal{V}ar(C)$.

Definition 6. *A clause set S is in PVD if every clause $C \in S$ is positively ground-preserving and $\forall x \in Var(C^+)$ it holds that depth$_x(C^+) \leqslant$ depth$_x(C^-)$.*

Let $\mathcal{A}_d$ be the subset of the Herbrand base $\mathcal{A}$ containing all ground atoms whose depth does not exceed d.

Lemma 4. *If the input clause set S is in PVD and has maximal depth d then any fair SGGS-derivation from S using I^- is in $\mathcal{A}_d$.*

Proof. Since S is in PVD and hence ground-preserving, by Lemma 2 the derivation is ground (†). The proof is by induction on the length n of the derivation, and follows that of Lemma 3 except for the extension case in the inductive step. Let $\Gamma \vdash \Gamma'$ be an SGGS-extension step as described in the proof of Lemma 3. The positive literals $M_1, \ldots, M_n$ are ground by (†) and by induction hypothesis in $\mathcal{A}_d$. Since $L_i\alpha = M_i\alpha = M_i$ for all i, $1 \leqslant i \leqslant n$, the atoms $L_i\alpha$ in $C^-\alpha$ are ground and in $\mathcal{A}_d$. Let L be a literal in C^+. Since S is in PVD, every variable in L occurs at a lower or equal depth in some L_i $(1 \leqslant i \leqslant n)$. Therefore, $L\alpha$ is ground and its depth cannot exceed that of $L_i\alpha$, so that $L\alpha \in \mathcal{A}_d$. □

Since $\mathcal{A}_d$ is a finite basis, termination follows from Theorem 1.

Theorem 5. *Any fair SGGS-derivation using I^- from a PVD set S halts, is a refutation if S is unsatisfiable, and constructs a model of S if S is satisfiable.*

5 Ordered Resolution Decides the Restrained Fragment

In this section we work with the positively restrained fragment; the case for the negatively restrained one is symmetric. Let $>$ be a stable and well-founded ordering on literals, such that positive literals are maximal only in positive clauses, so that ordered resolution is positive ordered resolution. In this way, the ordering embeds the suitable sign-based semantic guidance for the positively restrained fragment. The ordering $>$ could be the extension of the restraining ordering $\succ$ (see Definition 5) to literals, but does not have to be. We use the following notations [18]: $Res_>(S)$ denotes the set of ordered resolvents generated from parents in S; $R^0_>(S) = S$, $R^{k+1}_>(S) = R^k_>(S) \cup Res_>(R^k_>(S))$, and $R^*_>(S) = \bigcup_{k \geqslant 0} R^k_>(S)$. We begin with an auxiliary lemma.

Lemma 5. *If S is positively restrained, then for all $C \in R^*_>(S)$, for all $L \in C^+$ either (i) $L \in \mathcal{A}_S^{\preceq}$, or (ii) $at(M) \succeq at(L)$ for some $M \in C^-$.*

Proof. The proof is by induction on the stage k of the construction of $R^*_>(S)$. For $k=0$, the clauses in $R^0_>(S) = S$ satisfy the claim by the definitions of restrainedness, $\mathcal{A}_S$, and $\mathcal{A}_S^{\preceq}$. The induction hypothesis is that all clauses in $R^k_>(S)$ satisfy the claim. For the inductive step, let $C\sigma \vee D\sigma$ be a resolvent in $Res_>(R^k_>(S))$ generated from parents $\neg L \vee C$ and $L' \vee D$, where $\neg L\sigma$ and $L'\sigma$ are $>$-maximal literals, and $L\sigma = L'\sigma$ for mgu σ. The clause $L' \vee D$ must be positive, otherwise $L'\sigma$ cannot be $>$-maximal. Since $L' \vee D \in R^k_>(S)$, by induction hypothesis $at(L' \vee D) \subseteq \mathcal{A}_S^{\preceq}$ (†), which means $L' \vee D$ is ground, $(L' \vee D)\sigma = L' \vee D$, and all atoms in $D\sigma$ are in $\mathcal{A}_S^{\preceq}$. For the positive literals in $C\sigma$, let $M\sigma$ be one of them, so $M \in C^+$. Since $\neg L \vee C$ is in $R^k_>(S)$, by induction hypothesis, either (i) $M \in \mathcal{A}_S^{\preceq}$, or (ii) $M' \succeq M$ for some negative literal $\neg M'$ in $\neg L \vee C$. In case (i), M is ground, $M\sigma = M$, and $M\sigma \in \mathcal{A}_S^{\preceq}$. In case (ii), if $\neg M'$ occurs in C then $\neg M'\sigma \in C\sigma$, and $M'\sigma \succeq M\sigma$ holds by stability, so that the claim holds. Otherwise, $\neg M'$ is the resolved-upon literal $\neg L$ with $L\sigma = L'\sigma$. Thus, $L = M' \succeq M$, which implies $L\sigma \succeq M\sigma$ by stability, and since L' is ground, $L'\sigma = L'$. By (†), $L' \in \mathcal{A}_S^{\preceq}$. Since $L' = L'\sigma = L\sigma \succeq M\sigma$, we have $M\sigma \in \mathcal{A}_S^{\preceq}$ by definition of $\mathcal{A}_S^{\preceq}$. □

Theorem 6. *Any fair ordered resolution derivation using $>$ from a restrained clause set S terminates, and is a refutation if S is unsatisfiable.*

Proof. Let S be positively restrained. We show that $R^*_>(S)$ is finite. The claim then follows from refutational completeness of ordered resolution. We first show the following (†): for every ordered resolvent $C\sigma \vee D\sigma$ in $R^*_>(S)$ from parents $\neg L \vee C$ and $L' \vee D$ and mgu σ, $L' \vee D$ is ground and positive, and $C\sigma \vee D\sigma$ has strictly fewer negative literals than $\neg L \vee C$. Indeed, by the definition of ordered resolution, $\neg L\sigma$ and $L'\sigma$ are $>$-maximal in $(\neg L \vee C)\sigma$ and $(L' \vee D)\sigma$, respectively. The clause $L' \vee D$ must be positive, otherwise $L'\sigma$ cannot be $>$-maximal. By Lemma 5, every positive literal in a positive clause is in $\mathcal{A}_S^{\preceq}$, and therefore it is ground. Thus, in the resolvent $C\sigma \vee D\sigma$ the literals in $D\sigma$ are positive, hence $C\sigma \vee D\sigma$ has fewer negative literals than $\neg L \vee C$.

Now suppose that $R^*_>(S)$ is infinite. Observation (†) reveals that the number of negative literals decreases with every resolution step. Hence, if $R^*_>(S)$ is infinite, it must contain infinitely many positive clauses. By Lemma 5, these positive clauses are ground, and all their atoms are in the finite basis $\mathcal{A}_S^{\preceq}$. As repeated literals in ground clauses disappear by merging, the number of ground clauses that can be built from $\mathcal{A}_S^{\preceq}$ is finite. This contradicts $R^*_>(S)$ being infinite. □

This result extends to other positive resolution strategies.

Corollary 1. *Hyperresolution and $>$-ordered resolution with negative selection decide the positively restrained fragment.*

The next example shows that SGGS can be exponentially more efficient than these saturation-based resolution strategies because it is model-based.

Example 9. Consider the following parametric clause set S_n consisting of $n+1$ clauses, using i+1-ary predicates P_i and constants c_i, for all i, $0 \leqslant i \leqslant n$:

$$\begin{array}{lr}
\mathsf{P}_0(\mathsf{c}_0) \vee \mathsf{P}_0(\mathsf{c}_1) \vee \cdots \vee \mathsf{P}_0(\mathsf{c}_n) & (C_0), \\
\neg\mathsf{P}_0(x_1) \vee \mathsf{P}_1(x_1, \mathsf{c}_0) \vee \mathsf{P}_1(x_1, \mathsf{c}_1) \vee \cdots \vee \mathsf{P}_1(x_1, \mathsf{c}_n) & (C_1), \\
\neg\mathsf{P}_1(x_1, x_2) \vee \mathsf{P}_2(x_1, x_2, \mathsf{c}_0) \vee \cdots \vee \mathsf{P}_2(x_1, x_2, \mathsf{c}_n) & (C_2), \\
\cdots & \cdots \\
\neg\mathsf{P}_{n-1}(x_1, \ldots, x_n) \vee \mathsf{P}_n(x_1, \ldots, x_n, \mathsf{c}_0) \vee \cdots \vee \mathsf{P}_n(x_1, \ldots, x_n, \mathsf{c}_n) & (C_n).
\end{array}$$

The set S_n is positively restrained by an LPO with precedence $\mathsf{P}_0 > \cdots > \mathsf{P}_n > \mathsf{c}_i$ for all i, $0 \leqslant i \leqslant n$. SGGS with I^- simply selects one positive literal per clause and detects satisfiability after $n+1$ SGGS-extension steps, producing for instance the model where $\mathsf{P}_0(\mathsf{c}_0), \mathsf{P}_1(\mathsf{c}_0, \mathsf{c}_0), \ldots, \mathsf{P}_n(\mathsf{c}_0, \ldots, \mathsf{c}_0)$ are true and all other atoms are false. A saturation by any of the above positive resolution strategies produces exponentially many clauses, because for all i, $0 \leqslant i \leqslant n$, all n positive literals in C_i unify with the negative literal in C_{i+1}, generating n^{i+1} positive clauses, so that the clause count is given by $\sum_{k=1}^{n} n^k$.

6 Experiments

We begin by reducing positive restrainedness (the negative case is similar) to termination of rewrite systems, so that termination tools can yield a partial practical test. Since restrainedness employs a quasi-ordering, we consider unsorted rewrite systems $\mathcal{R}$ and $\mathcal{E}$, and *rewriting modulo* $\rightarrow_{\mathcal{R}/\mathcal{E}}$ defined by $\leftrightarrow^*_{\mathcal{E}} \circ \rightarrow_{\mathcal{R}} \circ \leftrightarrow^*_{\mathcal{E}}$.

Definition 7. *Given a clause set S, a pair of rewrite systems $(\mathcal{R}_S, \mathcal{E}_S)$ is* positively restraining *for S if for all clauses $C \in S$ and all non-ground literals $L \in C^+$, there is a rule $at(M) \rightarrow at(L)$ in $\mathcal{R}_S \cup \mathcal{E}_S$ for some literal $M \in C^-$.*

Let $(\mathcal{R}_S, \mathcal{E}_S)$ be positively restraining for S. For instance, for Example 8, $\mathcal{R}_S$ will consist of $\mathsf{P}(\mathsf{s}(\mathsf{s}(x)), y) \rightarrow \mathsf{P}(x, \mathsf{s}(y))$, $\mathsf{P}(x, \mathsf{s}(\mathsf{s}(y))) \rightarrow \mathsf{P}(x, \mathsf{s}(y))$, and $\mathsf{P}(\mathsf{s}(x), \mathsf{s}(y)) \rightarrow \mathsf{P}(\mathsf{s}(x), y)$, while $\mathcal{E}_S = \emptyset$. A common situation is that $\mathcal{E}_S$ has permutative rules, as in Example 7 where $\mathsf{differ}(x, y) \rightarrow \mathsf{differ}(y, x)$ will be in $\mathcal{E}_S$.

Lemma 6. *(1) If $\rightarrow_{\mathcal{R}_S}$ is terminating and $\mathcal{E}_S = \emptyset$, clause set S is strictly positively restrained. (2) If $\rightarrow_{\mathcal{R}_S/\mathcal{E}_S}$ is terminating, $\mathcal{V}ar(t) = \mathcal{V}ar(u)$ for all $t \rightarrow u$ in $\mathcal{E}_S$, and $\leftrightarrow^*_{\mathcal{E}}$ has finite equivalence classes, S is positively restrained.*

Proof. (1) Since $\rightarrow_{\mathcal{R}_S}$ is terminating, for all $t \rightarrow u$ in $\mathcal{R}_S$, $\mathcal{V}ar(u) \subseteq \mathcal{V}ar(t)$, so that S is positively ground-preserving. S is strictly positively restrained by the quasi-ordering $\rightarrow^*_{\mathcal{R}_S}$: indeed, $\rightarrow^*_{\mathcal{R}_S}$ is stable, $\rightarrow^+_{\mathcal{R}_S}$ is well-founded, and the equivalence classes of $\rightarrow^*_{\mathcal{R}_S} \cap {}_{\mathcal{R}_S}^{*}\!\!\leftarrow$, which is identity, are finite. (2) Since $\rightarrow_{\mathcal{R}_S/\mathcal{E}_S}$ is terminating, for all $t \rightarrow u$ in $\mathcal{R}_S$, $\mathcal{V}ar(u) \subseteq \mathcal{V}ar(t)$, and for all $t \rightarrow u$ in $\mathcal{E}_S$, $\mathcal{V}ar(u) = \mathcal{V}ar(t)$ by hypothesis, so that S is positively ground-preserving. S is positively restrained by $\rightarrow^*_{\mathcal{R}_S/\mathcal{E}_S}$: indeed, $\rightarrow^*_{\mathcal{R}_S/\mathcal{E}_S}$ is stable, $\rightarrow^+_{\mathcal{R}_S/\mathcal{E}_S}$ is well-founded, and the equivalence classes of $\leftrightarrow^*_{\mathcal{E}}$ are finite by hypothesis. $\square$

For the experiments, given a clause set S, a script named StoR generates rewrite systems $\mathcal{R}_S$ and $\mathcal{E}_S$ that can be fed to a termination tool. The source of clause sets is TPTP 7.2.0 and the termination tool is $\mathsf{T_TT_2}$ [27]. All problems in the FOF category are transformed into conjunctive normal form, excluding those with equality (whether sets with equality can be restrained is a topic for future work). Besides 1,539 inputs where either StoR or $\mathsf{T_TT_2}$ timed out, out of the remaining 3,462 problems, $\mathsf{T_TT_2}$ found 313 restrained ones. For those still undetermined, we tested whether it is sufficient to flip the sign of all literals with a certain predicate to get a restrained problem, which succeeded in 36 cases, for a total of 349 restrained problems. Of these, 277 are positively restrained, 181 negatively restrained, and 109 are both; 74 are ground, 232 are PVD, 277 are stratified, 252 are EPR, 169 are monadic, 204 are FO^2, 209 are guarded, but 43 problems do not fall in any of these classes, and therefore, to the best of our knowledge, they are proved decidable here for the first time. The average TPTP rating of the 349 problems is 0.1, and that of the 43 problems is 0.015, where 0.1 means that the problem can be solved by most provers. However, the group of 349 includes hard problems such as instances of the binary counter problem in Example 5 (MSC015-1.n), and Rubik's cube problems (e.g., PUZ052-1). For example, MSC015-1.030 is restrained and has rating 1.00, that is, no theorem prover could solve it so far within the timeout allowed in the CASC competition.

Example 10. Problem HWV036-2 (cf. Example 1) is a set of axioms with no ground atoms, so that $\mathcal{H}_S^{\preceq}$ is empty, and the model constructed according to Theorem 4 is trivial. Several other problems combine this set with ground clauses to prove theorems from those axioms. For example, HWV008-2.002 adds 23 ground clauses. As we found a terminating positively restraining rewrite system for HWV008-2.002, this problem, as well as HWV036-2, is strictly restrained.

An SGGS prototype named Koala was built reusing code for basic data structures, term indexing, and type inference from iProver [24,26]. In Koala, the SGGS-trail is represented as a list of constrained clauses, with constraints maintained in *standardized form* (see Sect. 2 and [14, Sect. 7]), and selected literals stored in a *discrimination tree*, since SGGS-extension requires to find selected literals that unify with literals in an input clause (cf. Definition 1). Koala takes sorts into account when checking satisfiability of constraints (e.g., Example 2), and implements a *fair search plan* which ensures that all derivations are fair (see Sect. 2). The *SGGS-suitable ordering* is a KBO with built-in precedence, $w_0 = 1$, and weight 1 for all function and predicate symbols in order to extend the size ordering. Koala also sorts by this ordering the clauses in a splitting, according to the SGGS notion of *preferred clause* in a splitting [14, Def. 22]).

Koala picks either I^- or I^+ as initial interpretation, based on whether the problem is positively or negatively ground-preserving, which overapproximates positively or negatively restrained. Koala implements the above mentioned sign-flipping test to obtain more ground-preserving sets. In order to recognize stratified input problems, one can compute the sort dependency graph and check for acyclicity [25], or let Koala apply type inference. For all experiments with Koala, the time-out was 300 sec of wall clock time. Out of the above 349 restrained problems, Koala shows that 50 are satisfiable and 283 unsatisfiable. Of 351 PVD problems, not including the 232 ones in the set of 349 restrained problems, Koala discovers 232 unsatisfiable and 76 satisfiable instances. Of the 1,246 stratified problems found in the FOF category by the above acyclicity test, Koala solves 643 unsatisfiable and 277 satisfiable instances. The list of the 349 restrained sets with their properties and restraining rewrite systems, as well as detailed results of the experiments with Koala are available.[2]

7 Discussion

SGGS [13,14] is an attractive candidate for decision procedures, because it is conflict-driven and it builds models. In this paper, we showed that if the generated clauses are in a *finite basis*, SGGS is guaranteed to terminate, and we instantiated this result to yield SGGS decision procedures for the *stratified fragment*, *PVD*, and the newly introduced *restrained fragment*, all without equality.

While also Inst-Gen [24] and the model evolution calculus (MEC) [7] decide the stratified fragment [25], this is not the case for the restrained fragment: for instance, if Inst-Gen uses an unfortunate literal selection in Example 8, it does not halt. SGGS avoids this phenomenon thanks to semantic guidance. Since MEC starts with I^+ as candidate model, it may not terminate on satisfiable negatively restrained sets such as Example 7. Indeed, E-Darwin [6] does not halt on this example that Koala solves in a few seconds.

However, it is generally difficult to tame a refutationally complete first-order inference system to yield decision procedures. For instance, SGGS does not halt given the following set of clauses that belongs to several decidable fragments.[3]

Example 11. The following set is in the Ackermann, monadic, and FO^2 classes:

$$\mathsf{P}(0) \quad (i) \qquad \mathsf{P}(x) \vee \mathsf{P}(\mathsf{f}(x)) \quad (ii) \qquad \neg\mathsf{P}(x) \vee \neg\mathsf{P}(\mathsf{f}(x)) \quad (iii),$$

where membership in the Ackermann class stems from the Skolemization of $\exists v \forall x \exists y. P(v) \wedge (P(x) \vee P(y)) \wedge (\neg P(x) \vee \neg P(y))$. It has the finite model property, as witnessed by model $\mathcal{I}$ with domain $\{0,1\}$, $\mathsf{f}^{\mathcal{I}}(x) = 1 - x$, $0^{\mathcal{I}} = 0$, and $\mathsf{P}^{\mathcal{I}} = \{0\}$. With I^-, SGGS performs an infinite series of non-conflicting extensions:

$$\begin{array}{llr}
\varepsilon & \vdash [\mathsf{P}(0)] & \text{extend } (i)\\
 & \vdash [\mathsf{P}(0)],\ \neg\mathsf{P}(0) \vee [\neg\mathsf{P}(\mathsf{f}(0))] & \text{extend } (iii)\\
 & \vdash [\mathsf{P}(0)],\ \neg\mathsf{P}(0) \vee [\neg\mathsf{P}(\mathsf{f}(0))],\ \mathsf{P}(\mathsf{f}(0)) \vee [\mathsf{P}(\mathsf{f}(\mathsf{f}(0)))] & \text{extend } (ii)\\
 & \vdash \ldots &
\end{array}$$

It can be shown that SGGS does not terminate with I^+ either. Inst-Gen may terminate if appropriate candidate models are constructed, but not in general. Ordered resolution terminates if the ordering satisfies $\neg\mathsf{P}(\mathsf{f}(x)) > \neg\mathsf{P}(x)$.

[2] http://profs.scienze.univr.it/winkler/sggsdp/.
[3] Renate A. Schmidt and Marco Voigt suggested this example to the first author.

The stratified fragment was presented as the first of three fragments of increasing expressivity, named St_0, St_1, and St_2, and suitable to capture Alloy specifications [1]. If the generic predicate symbols P of Example 4 is replaced with equality for sort s_2, the second sentence of Example 4 becomes $\forall x \exists y.\, f(y) \simeq x$, which states that x is in the image of f: this formula separates St_0 and St_1, since it is allowed in St_1, but not in St_0. The LISBQ logic for the verification of unbounded data structures [29] can be translated into St_0 with equality [1].

The *restrained fragment* is defined starting from a notion of *ground-preservingness*, which is convenient for positive or negative strategies. Negative ground-preservingness was used for Horn theories with equality [28] and [10, Sect. 5.2]. Positive ground-preservingness, also named *range restrictedness* [15], is implicit in PVD and recent extensions [30]. Unlike these settings, the restrained fragment does not limit literal depth. Positive ground-preservingness was used to show that DPLL($\Gamma+\mathcal{T}$), where Γ is an inference system including hyperresolution, superposition with negative selection, and simplification, decides *essentially finite* (only one monadic function f with finite range) and positively ground-preserving axiomatizations, provided that *speculative axioms* $f^j(x) \simeq f^k(x)$ $(j > k)$ are tried for increasing values of j and k ([12], Thm. 7 and Lem. 3). Simplification applies the speculative axioms to limit the depth of generated terms. Without this feature, it is not surprising that SGGS does not halt.

Example 12. Consider a positively ground-preserving variant of Example 11:

$$\mathsf{P}(0) \quad (i) \qquad \neg\mathsf{P}(x) \vee \mathsf{P}(\mathsf{f}(\mathsf{f}(x))) \quad (ii) \qquad \neg\mathsf{P}(x) \vee \neg\mathsf{P}(\mathsf{f}(x)) \quad (iii)$$

The finite model property implies that it is essentially finite. SGGS terminates with neither I^- nor I^+ as initial interpretation. With I^- it generates:

$$\begin{array}{llr}
\varepsilon & \vdash [\mathsf{P}(0)] & \text{extend } (i)\\
 & \vdash [\mathsf{P}(0)],\ \neg\mathsf{P}(0) \vee [\mathsf{P}(\mathsf{f}^2(0))] & \text{extend } (ii)\\
 & \vdash [\mathsf{P}(0)],\ \neg\mathsf{P}(0) \vee [\mathsf{P}(\mathsf{f}^2(0))],\ \neg\mathsf{P}(\mathsf{f}^2(0)) \vee [\mathsf{P}(\mathsf{f}^4(0))] & \text{extend } (ii)\\
 & \vdash \ldots &
\end{array}$$

Positive hyperresolution generates the series $\{\mathsf{P}(\mathsf{f}^{2k}(0))\}_{k \geq 0}$, which is essentially the same behavior as SGGS. DPLL($\Gamma+\mathcal{T}$) tries $\mathsf{f}(x) \simeq x$, detects a conflict, backtracks, tries $\mathsf{f}^2(x) \simeq x$, and halts reporting satisfiability.

Ensuring termination by restricting new (i.e., non-input) terms to come from a finite basis is common for conflict-driven decision procedures [11,16]. A major direction for future work towards decision procedures for richer languages is the integration of SGGS with CDSAT [11], in order to endow SGGS with equality and CDSAT with quantifiers. Speculative inferences and initial interpretations not based on sign are additional leads. For the restrained fragment, one may consider its relations with other decidable fragments and its relevance to applications beyond TPTP. While positive clauses in a restrained set are ground, one may study the decidability of sets where positive clauses are not necessarily ground, but admit a *restraining* rewrite system (Definition 7) such that *narrowing* halts. Techniques to detect the termination of narrowing are known [35].

Although Koala is only a prototype, the experiments show potential and allow us to identify critical issues for the performance of an SGGS prover. For example, instance generation by SGGS-extension is a bottleneck for problems with many input clauses, and forms of *caching* should be considered to avoid repeating computations. Further development of Koala and more experiments may contribute to discover classes of first-order problems where the conflict-driven style of SGGS reasoning is especially rewarding.

References

1. Abadi, A., Rabinovich, A., Sagiv, M.: Decidable fragments of many-sorted logic. J. Symb. Comput. **45**(2), 153–172 (2010)
2. Ackermann, W.: Solvable Cases of the Decision Problem. North Holland, Amsterdam (1954)
3. Alagi, G., Weidenbach, C.: NRCL - a model building approach to the Bernays-Schönfinkel fragment. In: Lutz, C., Ranise, S. (eds.) FroCoS 2015. LNCS (LNAI), vol. 9322, pp. 69–84. Springer, Cham (2015). https://doi.org/10.1007/978-3-319-24246-0_5
4. Andréka, H., van Benthem, J., Nemeti, I.: Modal logics and bounded fragments of predicate logic. J. Phil. Logic **27**(3), 217–274 (1998)
5. Bachmair, L., Ganzinger, H., Waldmann, U.: Superposition with simplification as a decision procedure for the monadic class with equality. In: Gottlob, G., Leitsch, A., Mundici, D. (eds.) KGC 1993. LNCS, vol. 713, pp. 83–96. Springer, Heidelberg (1993). https://doi.org/10.1007/BFb0022557
6. Baumgartner, P., Fuchs, A., Tinelli, C.: Implementing the model evolution calculus. Int. J. Artif. Intell. Tools **15**(1), 21–52 (2006)
7. Baumgartner, P., Tinelli, C.: The model evolution calculus as a first-order DPLL method. Artif. Intell. **172**(4–5), 591–632 (2008)
8. Bernays, P., Schönfinkel, M.: Zum Entscheidungsproblem der mathematischen Logik. Math. Annalen **99**, 342–372 (1928)
9. Blackburn, P., de Rijke, M., Venema, Y.: Modal Logic. Cambridge University Press, Cambridge (2001)
10. Bonacina, M.P., Dershowitz, N.: Canonical ground Horn theories. In: Voronkov, A., Weidenbach, C. (eds.) Programming Logics. LNCS, vol. 7797, pp. 35–71. Springer, Heidelberg (2013). https://doi.org/10.1007/978-3-642-37651-1_3
11. Bonacina, M.P., Graham-Lengrand, S., Shankar, N.: Conflict-driven satisfiability for theory combination: transition system and completeness. J. Autom. Reason. **64**(3), 579–609 (2020)
12. Bonacina, M.P., Lynch, C.A., de Moura, L.: On deciding satisfiability by theorem proving with speculative inferences. J. Autom. Reason. **47**(2), 161–189 (2011)
13. Bonacina, M.P., Plaisted, D.A.: Semantically-guided goal-sensitive reasoning: model representation. J. Autom. Reason. **56**(2), 113–141 (2016)
14. Bonacina, M.P., Plaisted, D.A.: Semantically-guided goal-sensitive reasoning: inference system and completeness. J. Autom. Reason. **59**(2), 165–218 (2017)
15. Caferra, R., Leitsch, A., Peltier, N.: Automated Model Building. Kluwer, Alphen aan den Rijn (2004)

16. de Moura, L., Jovanović, D.: A model-constructing satisfiability calculus. In: Giacobazzi, R., Berdine, J., Mastroeni, I. (eds.) VMCAI 2013. LNCS, vol. 7737, pp. 1–12. Springer, Heidelberg (2013). https://doi.org/10.1007/978-3-642-35873-9_1
17. Fermüller, C., Leitsch, A.: Model building by resolution. In: Börger, E., Jäger, G., Kleine Büning, H., Martini, S., Richter, M.M. (eds.) CSL 1992. LNCS, vol. 702, pp. 134–148. Springer, Heidelberg (1993). https://doi.org/10.1007/3-540-56992-8_10
18. Fermüller, C., Leitsch, A., Hustadt, U., Tammet, T.: Resolution decision procedures. In: Handbook of Automated Reasoning, pp. 1791–1849. Elsevier and MIT Press (2001)
19. Fiori, A., Weidenbach, C.: SCL clause learning from simple models. In: Fontaine, P. (ed.) CADE 2019. LNCS (LNAI), vol. 11716, pp. 233-249. Springer, Cham (2019). https://doi.org/10.1007/978-3-030-29436-6_14
20. Ganzinger, H., de Nivelle, H.: A superposition decision procedure for the guarded fragment with equality. In: Proceedings of LICS-14. IEEE Computer Society Press (1999)
21. Grädel, E., Kolaitis, P., Vardi, M.: On the decision problem for two-variable first-order logic. Bull. Symb. Log. **3**, 53–69 (1997)
22. Hustadt, U., Schmidt, R.A., Georgieva, L.: A survey of decidable first-order fragments and description logics. J. Relat. Methods Comput. Sci. **1**, 251–276 (2004)
23. Joyner Jr., W.H.: Resolution strategies as decision procedures. J. ACM **23**(3), 398–417 (1976)
24. Korovin, K.: Inst-Gen – a modular approach to instantiation-based automated reasoning. In: Voronkov, A., Weidenbach, C. (eds.) Programming Logics. LNCS, vol. 7797, pp. 239–270. Springer, Heidelberg (2013). https://doi.org/10.1007/978-3-642-37651-1_10
25. Korovin, K.: Non-cyclic sorts for first-order satisfiability. In: Fontaine, P., Ringeissen, C., Schmidt, R.A. (eds.) FroCoS 2013. LNCS (LNAI), vol. 8152, pp. 214–228. Springer, Heidelberg (2013). https://doi.org/10.1007/978-3-642-40885-4_15
26. Korovin, K.: iProver – an instantiation-based theorem prover for first-order logic (v2.8) (2018). http://www.cs.man.ac.uk/korovink/iprover/index.html
27. Korp, M., Sternagel, C., Zankl, H., Middeldorp, A.: Tyrolean Termination Tool 2. In: Treinen, R. (ed.) RTA 2009. LNCS, vol. 5595, pp. 295–304. Springer, Heidelberg (2009). https://doi.org/10.1007/978-3-642-02348-4_21
28. Kounalis, E., Rusinowitch, M.: On word problems in Horn theories. J. Symb. Comput. **11**(1–2), 113–128 (1991)
29. Lahiri, S., Qadeer, S.: Back to the future: revisiting precise program verification using SMT solvers. In: Proceedings of POPL-31, pp. 171–182. ACM (2004)
30. Lamotte-Schubert, M., Weidenbach, C.: BDI: a new decidable clause class. J. Log. Comput. **27**(2), 441–468 (2017)
31. Lewis, C.I., Langford, C.H.: Symbolic Logic. Dover Publications Inc., Mineola (1932)
32. Lutz, C., Sattler, U., Wolter, F.: Modal logic and the two-variable fragment. In: Fribourg, L. (ed.) CSL 2001. LNCS, vol. 2142, pp. 247–261. Springer, Heidelberg (2001). https://doi.org/10.1007/3-540-44802-0_18
33. McMillan, K.: Developing distributed protocols with Ivy. Slides (2019). http://vmcaischool19.tecnico.ulisboa.pt/
34. Navarro, J.A., Voronkov, A.: Proof systems for effectively propositional logic. In: Armando, A., Baumgartner, P., Dowek, G. (eds.) IJCAR 2008. LNCS (LNAI), vol. 5195, pp. 426–440. Springer, Heidelberg (2008). https://doi.org/10.1007/978-3-540-71070-7_36

35. Nishida, N., Vidal, G.: Termination of narrowing via termination of rewriting. Appl. Algebr. Eng. Commun. **21**(3), 177–225 (2010)
36. Padon, O., McMillan, K., Panda, A., Sagiv, M., Shoham, S.: Ivy: Safety verification by interactive generalization. In: SIGPLAN Not. vol. 51, no. 6, pp. 614–630 (2016)
37. Piskac, R., de Moura, L., Bjørner, N.: Deciding effectively propositional logic using DPLL and substitution sets. J. Autom. Reason. **44**(4), 401–424 (2010)
38. Plaisted, D.A., Zhu, Y.: The Efficiency of Theorem Proving Strategies. Friedr. Vieweg, Hamburg (1997)
39. Ramsey, F.: On a problem in formal logic. Proc. Lond. Math. Soc. **30**, 264–286 (1930)
40. Rubio, A.: A fully syntactic AC-RPO. Inform. Comput. **178**(2), 515–533 (2002)
41. Sutcliffe, G.: The TPTP problem library and associated infrastructure from CNF to TH0, TPTP v6.4.0. J. Autom. Reason. **59**(4), 483–502 (2017)

Integrating Induction and Coinduction via Closure Operators and Proof Cycles

Liron Cohen[1] and Reuben N. S. Rowe[2](✉)

[1] Department of Computer Science, Ben-Gurion University, Beersheba, Israel
cliron@cs.bgu.ac.il

[2] Department of Computer Science, Royal Holloway University of London, Egham, UK
reuben.rowe@rhul.ac.uk

Abstract. Coinductive reasoning about infinitary data structures has many applications in computer science. Nonetheless developing natural proof systems (especially ones amenable to automation) for reasoning about coinductive data remains a challenge. This paper presents a minimal, generic formal framework that uniformly captures applicable (i.e. finitary) forms of inductive and coinductive reasoning in an intuitive manner. The logic extends transitive closure logic, a general purpose logic for inductive reasoning based on the transitive closure operator, with a dual 'co-closure' operator that similarly captures applicable coinductive reasoning in a natural, effective manner. We develop a sound and complete non-well-founded proof system for the extended logic, whose cyclic subsystem provides the basis for an effective system for automated inductive and coinductive reasoning. To demonstrate the adequacy of the framework we show that it captures the canonical coinductive data type: streams.

1 Introduction

The principle of induction is used widely in computer science for reasoning about data types such as numbers or lists. The lesser-known principle of coinduction is used for reasoning about coinductive data types, which are data structures containing non-well-founded elements, e.g. infinite streams or trees [7,25,27,32,35,37,44,46,48]. A duality between the two principles is observed when formulating them within an algebraic, or categorical, framework [49]. However, such formulation does not account well for the way these principles are commonly used in deduction, where there is a mismatch in how they are usually applied.

Due to this tension between the abstract theory of coalgebras and its implementation in formal frameworks [41], coinductive reasoning is generally not fully and naturally incorporated into major proof assistants (e.g. Coq [7], Nuprl [20], Agda [8], Idris [9] and Dafny [36]). Even in notable exceptions such as [33,36,38,44] the combination of induction and coinduction is not intuitively accounted for. The standard approach in such formalisations is to define

N. Peltier and V. Sofronie-Stokkermans (Eds.): IJCAR 2020, LNAI 12166, pp. 375–394, 2020.
https://doi.org/10.1007/978-3-030-51074-9_21

inductive data with constructors and coinductive data with destructors, or observations [1]. In this paper we propose an alternative approach to formally integrating induction and coinduction that clearly reveals the duality between the two principles. Our approach has the advantage that the same signature is shared for both inductive and coinductive data, making certain aspects of the relationship between the two principles more apparent. To achieve this, we extend and combine two powerful frameworks: semantically we follow the approach of transitive closure logic, a generic logic for expressing inductive structures [3,14–16,31,39,51]; for deduction, we adopt non-well-founded proof theory [2,5,10–12,17–19,23,24,26,50,55]. This combination captures the intuitive dynamics of inductive and coinductive reasoning, reflecting how these principles are understood and applied in practice.

Transitive closure (RTC) logic minimally extends first-order logic by adding a single, intuitive notion: an operator, RTC, for forming the (reflexive) transitive closures of an arbitrary formula (more precisely, of the binary relation induced by the formula). This operator alone is sufficient for capturing all finitary induction schemes within a single, unified language (unlike other systems that are *a priori* parametrized by a set of inductive definitions [12,40,42,58]). Transitive closures arise as least fixed points of certain composition operators. In this paper we extend RTC logic with the semantically dual notion: an operator, RTC^{op}, for forming *greatest* fixed points of these same composition operators.[1] We call these *transitive co-closures*, and show that they are equally as intuitive. Just as transitive closure captures induction, we show that transitive co-closure facilitates coinductive definitions and reasoning.

Non-well-founded proof theory formalises the infinite-descent style of induction. It enables a separation between local steps of deductive inference and global well-foundedness arguments (i.e. induction), which are encoded in traces of formulas through possibly infinite derivations. A major benefit of these systems is that inductive invariants do not need to be explicit. On the other hand, existing approaches for combining induction and coinduction rely on making (co)invariants explicit within proofs [4,30,59]. In previous work, a non-well-founded proof system for RTC logic was developed [17,18]. In this paper, we show that the meaning of the transitive co-closure operator can be captured proof-theoretically using inference rules having *the exact same structure*, with soundness now requiring infinite *a*scent (i.e. showing productivity) rather than *de*scent. What obtains is a proof system in which induction and coinduction are smoothly integrated, and which very clearly highlights their similarities. Their differences are also thrown into relief, consisting in the way formulas are traced in a proof derivation. Specifically, traces of RTC formulas show that certain infinite paths *cannot* exist (induction is well-founded), while traces of RTC^{op} formulas show that other infinite paths *must* exist (coinduction is productive).

To demonstrate that our system naturally captures patterns of mixed inductive/coinductive reasoning, we formalise one of the most well-known examples of a coinductive data type: streams. In particular, we consider two illustrative

[1] The notation RTC^{op} comes from the categorical notion of the opposite (dual) category.

examples: transitivity of the lexicographic ordering on streams; and transitivity of the substream relation. Both are known to be hard to prove. Our system handles these without recourse to general fixpoint operators or algebraic structures.

The transitive (co-)closure framework is contained in the first-order mu-calculus [43], but offers several advantages. The concept of transitive (co-)closure is intuitively simpler than that of general fixed-point operators, and does not require any syntactic restrictions to ensure monotonicity. Our framework is also related, but complementary to logic programming with coinductive interpretations [52,53] and its coalgebraic semantics [34]. Logic programs, built from Horn clauses, have a fixed intended domain (*viz.* Herbrand universes), and the semantics of mixing inductive and coinductive interpretations is subtle. Our framework, on the other hand, uses a general syntax that can freely mix closures and co-closures, and its semantics considers all first-order models. Furthermore, the notion of proof in our setting is more general than the (semantic) notion of proof in logic programming, in which, for instance, there is no analogous concept of global trace condition.

Outline. Section 2 presents the syntax and semantics of the extended logic, RTcC. Section 3 describes how streams and their properties can be expressed in RTcC. Section 4 presents non-well-founded proof systems for RTcC, showing soundness and completeness. Section 5 then illustrates how the examples of Sect. 3 are formalised in this system. Section 6 concludes with directions for future work.

2 RTcC Logic: Syntax and Semantics

Transitive closure (RTC) logic [3,15] extends the language of first-order logic with a predicate-forming operator, RTC, for denoting the (reflexive) transitive closures of (binary) relations. In this section we extend RTC logic into what we call *transitive (co-)closure* (RTcC) *logic*, by adding a single *transitive co-closure* operator, RTC^{op}. Roughly speaking, whilst the RTC operator denotes the set of all pairs that are related via a *finite* chain (or path), the RTC^{op} operator gives the set of all pairs that are 'related' via a *possibly infinite* chain. In Sect. 3 we show that this allows capturing coinductive definitions and reasoning.

For simplicity of presentation we assume (as is standard practice) a designated equality symbol. Note also that we use the *reflexive* transitive closure; however the reflexive and non-reflexive forms are equivalent in the presence of equality.

Definition 1 (RTcC **Formulas**). *Let s, t and P range over the terms and predicate symbols, respectively, of a first-order signature Σ. The language $\mathcal{L}_{\mathsf{RTcC}}$ (of formulas over Σ) is given by the following grammar:*

$$\varphi, \psi ::= s = t \mid P(t_1, \ldots, t_n) \mid \neg\varphi \mid \forall x \,.\, \varphi \mid \exists x \,.\, \varphi \mid \varphi \wedge \psi \mid \varphi \vee \psi \mid \varphi \rightarrow \psi \mid (RTC_{x,y}\, \varphi)(s,t) \mid (RTC^{\mathsf{op}}_{x,y}\, \varphi)(s,t)$$

where the variables x and y in the formulas $(RTC_{x,y}\, \varphi)(s,t)$ and $(RTC^{\mathsf{op}}_{x,y}\, \varphi)(s,t)$ must be distinct and are bound in the subformula φ, referred to as the body.

The semantics of formulas is an extension of the standard semantics of first-order logic. We write M and ν to denote a first-order structure over a (non-empty) domain D and a valuation of variables in D, respectively. We denote by $\nu[x_1 := d_n, \ldots, x_n := d_n]$ the valuation that maps x_i to d_i for each i and behaves as ν otherwise. We write $\varphi\{^{t_1}/_{x_1}, \ldots, {}^{t_n}/_{x_n}\}$ for the result of simultaneously substituting each t_i for the free occurrences of x_i in φ. We use $(\boldsymbol{d}_i)_{i \leq n}$ to denote a *non-empty* sequence of elements $d_1, \ldots, d_n$; and $(\boldsymbol{d}_i)_{i>0}$ for a (countably) *infinite* sequence of elements $d_1, d_2, \ldots$. We use $\equiv$ to denote syntactic equality.

Definition 2 (Semantics). *Let M be a structure for $\mathcal{L}_{\mathsf{RTcC}}$, and ν a valuation in M. The satisfaction relation $M, \nu \models \varphi$ extends the standard satisfaction relation of classical first-order logic with the following clauses:*

$$M, \nu \models (RTC_{x,y}\, \varphi)(s,t) \Leftrightarrow$$
$$\exists (\boldsymbol{d}_i)_{i \leq n} \,.\, d_1 = \nu(s) \wedge d_n = \nu(t) \wedge \forall i < n \,.\, M, \nu[x := d_i, y := d_{i+1}] \models \varphi$$

$$M, \nu \models (RTC^{\mathsf{op}}_{x,y}\, \varphi)(s,t) \Leftrightarrow$$
$$\exists (\boldsymbol{d}_i)_{i > 0} \,.\, d_1 = \nu(s) \wedge \forall i > 0 \,.\, d_i = \nu(t) \vee M, \nu[x := d_i, y := d_{i+1}] \models \varphi$$

Intuitively, the formula $(RTC_{x,y}\, \varphi)(s,t)$ asserts that there is a (possibly empty) finite φ-path from s to t. The formula $(RTC^{\mathsf{op}}_{x,y}\, \varphi)(s,t)$ asserts that either there is a (possibly empty) finite φ-path from s to t, or an infinite φ-path starting at s.

We can connect these closure operators to the general theory of fixed points, with $(RTC_{x,y}\, \varphi)$ and $(RTC^{\mathsf{op}}_{x,y}\, \varphi)$ denoting, respectively, the least and greatest fixed points of a certain operator on binary relations.

Definition 3 (Composition Operator). *Given a binary relation X, we define an operator Ψ_X on binary relations, which post-composes its input with X, by: $\Psi_X(R) = X \cup (X \circ R) = \{(a,c) \mid (a,c) \in X \vee \exists b \,.\, (a,b) \in X \wedge (b,c) \in R\}$.*

Notice that the set of all binary relations (over some given domain) forms a complete lattice under the subset ordering $\subseteq$. Moreover, composition operators Ψ_X are monotone w.r.t. $\subseteq$. Thus we have the following standard results, from the Knaster–Tarski theorem. For any binary relation X, the least fixed point $\mathsf{lfp}(\Psi_X)$ of Ψ_X is given by $\mathsf{lfp}(\Psi_X) = \bigcap\{R \mid \Psi_X(R) \subseteq R\}$, i.e. the intersection of all its prefixed points. Dually, the greatest fixed point $\mathsf{gfp}(\Psi_X)$ of Ψ_X is given by the union of all its postfixed points, i.e. $\mathsf{gfp}(\Psi_X) = \bigcup\{R \mid R \subseteq \Psi_X(R)\}$. Via the usual notion of formula definability, RTC and RTC^{op} are easily seen to be fixed point operators. For a model M and valuation ν, denote the binary relation defined by a formula φ with respect to x and y by $[\![\varphi]\!]^{M,\nu}_{x,y} = \{(a,b) \mid M, \nu[x := a, y := b] \models \varphi\}$.

Proposition 1. *The following hold.*

(i) $M, \nu \models (RTC_{x,y}\, \varphi)(s,t)$ *iff* $\nu(s) = \nu(t)$ *or* $(\nu(s), \nu(t)) \in \mathsf{lfp}(\Psi_{[\![\varphi]\!]^{M,\nu}_{x,y}})$.
(ii) $M, \nu \models (RTC^{\mathsf{op}}_{x,y}\, \varphi)(s,t)$ *iff* $\nu(s) = \nu(t)$ *or* $(\nu(s), \nu(t)) \in \mathsf{gfp}(\Psi_{[\![\varphi]\!]^{M,\nu}_{x,y}})$.

Note that labelling the co-closure 'transitive' is justified since, for any model M, valuation ν, and formula φ, the relation $\mathbf{gfp}(\Psi_{[\![\varphi]\!]^{M,\nu}_{x,y}})$ is indeed transitive.

The RTC^{op} operator enjoys dualisations of properties governing the transitive closure operator (see, e.g., [16, Proposition 3]) *that are either symmetrical, or involve the first component.* This is because the semantics of the RTC^{op} has an embedded asymmetry between the arguments. Reasoning about closures is based on decomposition into one step and the remaining path. For RTC, this decomposition can be done in both directions, but for RTC^{op} it can only be done in one direction.

Proposition 2. *The following formulas, connecting the two operators, are valid.*

i) $(RTC_{x,y}\,\varphi)(s,t) \rightarrow (RTC^{\mathsf{op}}_{x,y}\,\varphi)(s,t)$
ii) $\neg(RTC_{x,y}\,\neg\varphi)(s,t) \rightarrow (RTC^{\mathsf{op}}_{x,y}\,\varphi)(s,t)$
iii) $\neg(RTC^{\mathsf{op}}_{x,y}\,\neg\varphi)(s,t) \rightarrow (RTC_{x,y}\,\varphi)(s,t)$
iv) $((RTC^{\mathsf{op}}_{x,y}\,\varphi)(s,t) \wedge \exists z.\neg(RTC^{\mathsf{op}}_{x,y}\,\varphi)(s,z)) \rightarrow (RTC_{x,y}\,\varphi)(s,t)$
v) $((RTC^{\mathsf{op}}_{x,y}\,\varphi)(s,t) \wedge \neg(RTC^{\mathsf{op}}_{x,y}\,\varphi \wedge y \neq t)(s,t)) \rightarrow (RTC_{x,y}\,\varphi)(s,t)$

Note that the converse of these properties do not hold in general, thus they do not provide characterisations of one operator in terms of the other. A counter-example for the converses of (ii) and (iii) can be obtained by taking φ to be $x = y$. Then, for any domain D, the formulas $(RTC_{x,y}\,\neg\varphi)$, $(RTC^{\mathsf{op}}_{x,y}\,\varphi)$, and $(RTC^{\mathsf{op}}_{x,y}\,\neg\varphi)$ all denote the full binary relation $D \times D$, while $(RTC_{x,y}\,\varphi)$ denotes the identity relation on D.

3 Streams in RTcC Logic

This section demonstrates the adequacy of RTcC logic for formalising and reasoning about coinductive data types. As claimed by Rutten: "streams are the best known example of a final coalgebra and offer a perfect playground for the use of coinduction, both for definitions and for proofs." [47]. Hence, in this section and Sect. 5 we illustrate that RTcC logic naturally captures the stream data type (see, e.g., [29,48]).

3.1 The Stream Datatype

We formalise streams as infinite lists, using a signature consisting of the standard list constructors: the constant nil and the (infix) binary function symbol '::', traditionally referred to as 'cons'. These are axiomatized by:

$$\mathsf{nil} = e :: \sigma \Rightarrow \quad (1) \qquad e :: \sigma = e' :: \sigma' \Rightarrow e = e' \quad (2) \qquad e :: \sigma = e' :: \sigma' \Rightarrow \sigma = \sigma' \quad (3)$$

Note that for simplicity of presentation we have not specified that the elements of possibly infinite lists should be any particular sort (e.g. numbers). Thus, the theory of streams we formulate here is generic in this respect. To refer specifically to streams over a particular domain, we could use a multisorted signature containing a Base sort, in addition to the sort List^∞ of possibly infinite lists, with nil a constant of type List^∞ and :: a function of type $\mathsf{Base} \times \mathsf{List}^\infty \longrightarrow \mathsf{List}^\infty$. Nonetheless, we do use the following conventions for formalising streams in this section and in Sect. 5. For variables and terms ranging over Base we use $a, b, c, \ldots$ and $e, e', \ldots$, respectively; and for variables and terms ranging over possibly infinite lists we use $x, y, z, \ldots$ and $\sigma, \sigma', \ldots$, respectively.

The (graphs of) the standard head (hd) and tail (tl) functions are definable[2] by $\mathsf{hd}(\sigma) = e \overset{\text{def}}{:=} \exists x.\sigma = e :: x$ and $\mathsf{tl}(\sigma) = \sigma' \overset{\text{def}}{:=} \exists a.\sigma = a :: \sigma'$. Finite and possibly infinite lists can be defined by using the transitive closure and co-closure operators, respectively, as follows.

$$\mathsf{List}(\sigma) \overset{\text{def}}{:=} (RTC_{x,y}\, \mathsf{tl}(x) = y)(\sigma, \mathsf{nil})$$
$$\mathsf{List}^\infty(\sigma) \overset{\text{def}}{:=} (RTC^{\mathsf{op}}_{x,y}\, \mathsf{tl}(x) = y)(\sigma, \mathsf{nil})$$

Roughly speaking, these formulas assert that we can perform some number of successive tail decompositions of the term σ. For the RTC formula, this decomposition must reach the second component, nil, in a finite number of steps. For the RTC^{op} formula, on the other hand, the decomposition is not required to reach nil but, in case it does not, must be able to continue indefinitely.

To define the notion of a necessarily infinite list (i.e. a stream), we specify in the body that, at each step, the decomposition of the stream cannot actually reach nil (abbreviating $\neg(s = t)$ by $s \neq t$). Moreover, since we are using reflexive forms of the operators we must also stipulate that nil itself is not a stream.

$$\mathsf{Stream}(\sigma) \overset{\text{def}}{:=} (RTC^{\mathsf{op}}_{x,y}\, \mathsf{tl}(x) = y \wedge y \neq \mathsf{nil})(\sigma, \mathsf{nil}) \wedge \sigma \neq \mathsf{nil}$$

This technique—of specifying that a single step cannot reach nil and then taking nil to be the terminating case in the RTC^{op} formula—is a general method we will use in order to restrict attention to the infinite portion in the induced semantics of an RTC^{op} formula. To this end, we define the following notation.

$$\overline{\varphi}^{\mathtt{inf}}_{x,y}(\sigma) \overset{\text{def}}{:=} (RTC^{\mathsf{op}}_{x,y}\, (\varphi \wedge y \neq \mathsf{nil}))(\sigma, \mathsf{nil}) \wedge \sigma \neq \mathsf{nil}$$

3.2 Relations and Operations on Streams

We next show that RTcC also naturally captures properties of streams. Using the RTC operator we can (inductively) define the extension relation $\lhd$ on possibly infinite lists as follows:

$$\sigma \lhd \sigma' \overset{\text{def}}{:=} (RTC_{x,y}\, \mathsf{tl}(x) = y)(\sigma, \sigma')$$

This asserts that σ extends σ', i.e. that σ is obtained from σ' by prepending some finite sequence of elements to σ'. Equivalently, σ' is obtained by some finite number of tail decompositions from σ: that is, σ' is a suffix of σ.

[2] Although $\mathsf{hd}(\sigma)$ and $\mathsf{tl}(\sigma)$ could have been defined as terms using Russell's ι operator, we opted for the above definition for simplicity of the proof theory.

We next formalise some standard predicates.

$$\mathsf{Contains}(e,\sigma) \overset{\texttt{def}}{:=} \exists x\,.\,\sigma \lhd x \wedge \mathsf{hd}(x) = e$$
$$\mathsf{Const}(e,\sigma) \overset{\texttt{def}}{:=} \overline{(x = e :: y)}^{\texttt{inf}}_{x,y}(\sigma)$$
$$\mathsf{Const}_{\overrightarrow{\infty}}(\sigma) \overset{\texttt{def}}{:=} \exists x\,.\,\sigma \lhd x \wedge \exists a\,.\,\mathsf{Const}(a,x)$$

$\mathsf{Contains}(e,\cdot)$ defines the possibly infinite lists that contain the element denoted by e; $\mathsf{Const}(e,\cdot)$ defines the constant stream consisting of the element denoted by e; and $\mathsf{Const}_{\overrightarrow{\infty}}$ defines streams that are eventually constant.

We next consider how (functional) relations on streams can be formalised in RTcC, using some illustrative examples. To capture these we need to use ordered pairs. For this, we use the notation $\langle u, v\rangle$ for $u :: (v :: \mathsf{nil})$,[3] then abbreviate $(RTC_{w,w'}\, \exists u,u',v,v'\,.\,w = \langle u,v\rangle \wedge w' = \langle u',v'\rangle \wedge \varphi)$ by $(RTC_{\langle u,v\rangle,\langle u',v'\rangle}\, \varphi)$ (and similarly for RTC^{op} formulas), and also write $\overline{\varphi}^{\texttt{inf}}_{\langle x_1,x_2\rangle,\langle y_1,y_2\rangle}(\langle\sigma,\sigma'\rangle)$ to stand for $(RTC^{\mathsf{op}}_{\langle x_1,x_2\rangle,\langle y_1,y_2\rangle}\,(\varphi \wedge y_1 \neq \mathsf{nil} \wedge y_2 \neq \mathsf{nil}))(\langle\sigma,\sigma'\rangle,\langle\mathsf{nil},\mathsf{nil}\rangle) \wedge \sigma \neq \mathsf{nil} \wedge \sigma' \neq \mathsf{nil}$.

Append and Periodicity. With ordered pairs, we can inductively define (the graph of) the function that appends a possibly infinite list to a finite list.

$$\sigma_1 {}^\frown \sigma_2 = \sigma_3 \overset{\texttt{def}}{:=}$$
$$(RTC_{\langle x_1,x_2\rangle,\langle y_1,y_2\rangle}\,\exists a\,.\,x_1 = a :: y_1 \wedge x_2 = a :: y_2)(\langle\sigma_1,\sigma_3\rangle,\langle\mathsf{nil},\sigma_2\rangle)$$

We remark that the formulas $\sigma \lhd \sigma'$ and $\exists z\,.\,z {}^\frown \sigma' = \sigma$ are equivalent. To define this as a *function* requires also proofs that the defined relation is total and functional. However, this is generally straightforward when the body formula is deterministic, as is the case in all the examples we present here. Other standard operations on streams, such as element-wise operations, are also definable in RTcC as (functional) relations. For example, assuming a unary function $\oplus$, we can coinductively define its elementwise extension to streams $\oplus_\infty$ as follows.

$$\oplus_\infty(\sigma) = \sigma' \overset{\texttt{def}}{:=} \overline{(\exists a\,.\,x_1 = a :: y_1 \wedge x_2 = \oplus(a) :: y_2)}^{\texttt{inf}}_{\langle x_1,x_2\rangle,\langle y_1,y_2\rangle}(\langle\sigma,\sigma'\rangle)$$

As an example of mixing induction and coinduction, we can express a predicate coinductively defining the *periodic* streams using the append function.

$$\mathsf{Periodic}(\sigma) \overset{\texttt{def}}{:=} \exists z\,.\,z \neq \mathsf{nil} \wedge \overline{(z {}^\frown y = x)}^{\texttt{inf}}_{x,y}(\sigma)$$

Lexicographic Ordering. The lexicographic order on streams extends pointwise an order on the underlying elements. Thus, we assume a binary relation symbol $\leq$ with the standard axiomatisation of a (non-strict) partial order.

$$\Rightarrow e \leq e \qquad e \leq e', e' \leq e'' \Rightarrow e \leq e'' \qquad e \leq e', e' \leq e \Rightarrow e = e'$$

[3] Here we use the fact that '::' behaves as a pairing function. In other languages one might need to add a function $\langle\cdot,\cdot\rangle$, and (axiomatically) restrict the semantics to structures that interpret it as a pairing function. Note that incorporating pairs is equivalent to taking $2n$-ary operators RTC_n and RTC^{op}_n for every $n \geq 1$.

The lexicographic ordering relation $\leq_\ell$ is captured as follows, where we use $e < e'$ as an abbreviation for $e \leq e' \land e \neq e'$.

$$\sigma \leq_\ell \sigma' \stackrel{\text{def}}{:=} (RTC^{\mathsf{op}}_{\langle x_1,x_2\rangle,\langle y_1,y_2\rangle}\, \psi_\ell)(\langle \sigma, \sigma'\rangle, \langle \mathsf{nil}, \mathsf{nil}\rangle)$$
$$\text{where } \psi_\ell \equiv \exists a, b, z_1, z_2 \,.\, x_1 = a :: z_1 \land x_2 = b :: z_2 \land$$
$$\mathsf{Stream}(z_1) \land \mathsf{Stream}(z_2) \land (a < b \lor (a = b \land z_1 = y_1 \land z_2 = y_2))$$

The semantics of the RTC^{op} operator require an infinite sequence of pairs such that, until $\langle \mathsf{nil}, \mathsf{nil}\rangle$ is reached, each two consecutive pairs are related by ψ_ℓ. This formula states that if the heads of the lists in the first pair are equal, the next pair of lists in the infinite sequence is their two tails, thus the lexicographic relation must also hold of them. Otherwise, if the head of the first is less than that of the second, nothing is required of the tails, i.e. they may be any streams.

Substreams. We consider one stream to be a substream of another if the latter contains every element of the former in the same order (although it may contain other elements too). Equivalently, the latter is obtained by inserting some (possibly infinite) number of finite sequences of elements in between those of the former. This description makes it clearer that defining this relation involves mixing (or, rather, nesting) induction and coinduction. We formalise the substream relation, $\succcurlyeq$ using the *inductive* extension relation $\lhd$ to capture the inserted finite sequences, wrapping it within a *coinductive* definition using the RTC^{op} operator.

$$\sigma \succcurlyeq \sigma' \stackrel{\text{def}}{:=} \overline{\psi_{\succcurlyeq}}^{\,\mathtt{inf}}_{\langle x_1,x_2\rangle,\langle y_1,y_2\rangle}(\langle \sigma, \sigma'\rangle)$$
$$\text{where } \psi_{\succcurlyeq} \equiv \exists a \,.\, x_1 \lhd a :: y_1 \land x_2 = a :: y_2$$

On examination, one can observe that this relation is transitive. However, proving this is non-trivial and, unsurprisingly, involves applying both induction and coinduction. In Sect. 5, we give a proof of the transitivity of $\succcurlyeq$ in RTcC. This relation was also considered at length in [6, §5.1.3] where it is formalised in terms of *selectors*, which form streams by picking out certain elements from other streams. The treatment in [6] requires some heavy (coalgebraic) metatheory. While our proof in Sect. 5 requires some (fairly obvious) lemmas, the basic structure of the (co)inductive reasoning required is made plain by the cycles in the proof. Furthermore, the RTcC presentation seems to enable a more intuitive understanding of the nature of the coinductive definitions and principles involved.

4 Proof Theory

We now present a non-well-founded proof system for RTcC, which extends (an equivalent of) the non-well-founded proof system considered in [17,18] for transitive closure logic (i.e. the RTC-fragment of RTcC).

$$\frac{}{\Gamma \Rightarrow \Delta, (RTC_{x,y}\,\varphi)(s,s)} \tag{4}$$

$$\frac{\Gamma \Rightarrow \Delta, \varphi\{s/x, r/y\} \quad \Gamma \Rightarrow \Delta, (RTC_{x,y}\,\varphi)(r,t)}{\Gamma \Rightarrow \Delta, (RTC_{x,y}\,\varphi)(s,t)} \tag{5}$$

$$\frac{\Gamma, s = t \Rightarrow \Delta \quad \Gamma, \varphi\{s/x, z/y\}, (RTC_{x,y}\,\varphi)(z,t) \Rightarrow \Delta}{\Gamma, (RTC_{x,y}\,\varphi)(s,t) \Rightarrow \Delta}\,(\dagger) \tag{6}$$

$$\frac{}{\Gamma \Rightarrow \Delta, (RTC^{\mathsf{op}}_{x,y}\,\varphi)(s,s)} \tag{7}$$

$$\frac{\Gamma \Rightarrow \Delta, \varphi\{s/x, r/y\} \quad \Gamma \Rightarrow \Delta, (RTC^{\mathsf{op}}_{x,y}\,\varphi)(r,t)}{\Gamma \Rightarrow \Delta, (RTC^{\mathsf{op}}_{x,y}\,\varphi)(s,t)} \tag{8}$$

$$\frac{\Gamma, s = t \Rightarrow \Delta \qquad \Gamma, \varphi\{s/x, z/y\}, (RTC^{\mathsf{op}}_{x,y}\,\varphi)(z,t) \Rightarrow \Delta}{\Gamma, (RTC^{\mathsf{op}}_{x,y}\,\varphi)(s,t) \Rightarrow \Delta}\,(\ddagger) \tag{9}$$

where: ($\dagger$) $z \notin \mathsf{fv}(\Gamma, \Delta, (RTC_{x,y}\,\varphi)(s,t))$; and ($\ddagger$) $z \notin \mathsf{fv}(\Gamma, \Delta, (RTC^{\mathsf{op}}_{x,y}\,\varphi)(s,t))$.

Fig. 1. Proof rules of $\mathsf{RTcC}^{\infty}_{G}$

4.1 A Non-well-Founded Proof System

In non-well-founded proof systems, e.g. [2,5,10–12,23,24,50], proofs are allowed to be infinite, i.e. non-well-founded trees, but they are subject to the restriction that every infinite path in the proof admits some infinite progress, witnessed by tracing terms or formulas. The infinitary proof system for RTcC logic is defined as an extension of $\mathcal{LK}_=$, the sequent calculus for classical first-order logic with equality and substitution [28,56].[4] Sequents are expressions of the form $\Gamma \Rightarrow \Delta$, for finite sets of formulas Γ and Δ. We abbreviate Γ, Δ and Γ, φ by $\Gamma \cup \Delta$ and $\Gamma \cup \{\varphi\}$, respectively, and write $\mathsf{fv}(\Gamma)$ for the set of free variables of the formulas in Γ. A sequent $\Gamma \Rightarrow \Delta$ is valid if and only if the formula $\bigwedge_{\varphi \in \Gamma} \varphi \to \bigvee_{\psi \in \Delta} \psi$ is.

Definition 4 ($\mathsf{RTcC}^{\infty}_{G}$). *The proof system* $\mathsf{RTcC}^{\infty}_{G}$ *is obtained by adding to* $\mathcal{LK}_=$ *the proof rules given in Fig. 1.*

Rules (6), and (8) are the unfolding rules for the two operators that represent the induction and coinduction principles in the system, respectively. The proof rules for both operators have exactly the same form, and so the reader may wonder what it is, then, that distinguishes the behaviour of the two operators. The difference proceeds from the way the decomposition of the corresponding formulas is traced in the non-well-founded proof system. For induction, *RTC*

[4] Unlike in the original system, here we take $\mathcal{LK}_=$ to include the substitution rule.

formulas on the left-hand side of the sequents are traced through Rule (6); for coinduction, RTC^{op} formulas on the right-hand side of sequents are traced through Rule (8).

Definition 5 (RTcC_G^∞ **Pre-proofs).** *An* RTcC_G^∞ pre-proof *is a rooted, possibly non-well-founded (i.e. infinite) derivation tree constructed using the* RTcC_G^∞ *proof rules. A* path *in a pre-proof is a possibly infinite sequence* $S_0, S_1, \ldots (, S_n)$ *of sequents with* S_0 *the root of the proof, and* S_{i+1} *a premise of* S_i *for each* $i < n$.

We adopt the usual proof-theoretic notions of formula occurrence and sub-occurrence, and of ancestry between formulas [13]. A formula occurrence is called a *proper* formula if it is not a sub-occurrence of any formula.

Definition 6 ((Co-)Traces). *A* trace *(resp.* co-trace*) is a possibly infinite sequence* $\tau_1, \tau_2, \ldots (, \tau_n)$ *of proper* RTC *(resp.* RTC^{op}*) formula occurrences in the left-hand (resp, right-hand) side of sequents in a pre-proof such that* τ_{i+1} *is an immediate ancestor of* τ_i *for each* $i > 0$*. If the trace (resp. co-trace) contains an infinite number of formula occurrences that are principal for instances of Rule (6) (resp. Rule (8)), then we say that it is* infinitely progressing.

As usual in non-well-founded proof theory, we use the notion of (co-)trace to define a *global trace condition*, distinguishing certain 'valid' pre-proofs.

Definition 7 (RTcC_G^∞ **Proofs).** *An* RTcC_G^∞ *proof is a pre-proof in which every infinite path has a tail followed by an infinitely progressing (co-)trace.*

In general, one cannot reason effectively about infinite proofs, as found in RTcC_G^∞. In order to do so our attention has to be restricted to those proof trees which are finitely representable. That is, the *regular* infinite proof trees, containing only finitely many *distinct* subtrees. They can be specified as systems of recursive equations or, alternatively, as cyclic *graphs* [22]. One way of formalising such proof graphs is as standard proof trees containing open nodes (called buds), to each of which is assigned a syntactically equal internal node of the proof (called a companion). The restriction to cyclic proofs provides the basis for an effective system for automated inductive and coinductive reasoning. The system RTcC_G^∞ can naturally be restricted to a cyclic proof system for RTcC logic as follows.

Definition 8 (Cyclic Proofs). *The cyclic proof system* RTcC_G^ω *for* RTcC *logic is the subsystem of* RTcC_G^∞ *comprising of all and only the finite and* regular *infinite proofs (i.e. proofs that can be represented as finite, possibly cyclic, graphs).*[5]

It is decidable whether a cyclic pre-proof satisfies the global trace condition, using a construction involving an inclusion between Büchi automata [10,54]. However since this requires complementing Büchi automata (a PSPACE procedure), RTcC_G^ω is not a proof system in the Cook-Reckhow sense [21]. Notwithstanding, checking the trace condition for cyclic proofs found in practice is not prohibitive [45,57].

[5] Note that in [17,18] RTC_G^ω denoted the *full* infinitary system for the RTC-fragment.

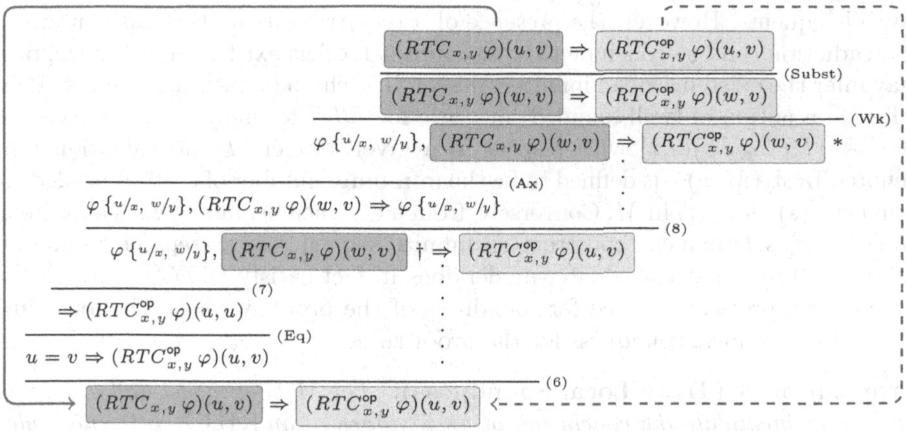

Fig. 2. Proof in $\mathsf{RTcC}^{\omega}_{G}$ of $(RTC_{x,y}\,\varphi)(u,v) \Rightarrow (RTC^{\mathsf{op}}_{x,y}\,\varphi)(u,v)$

Although $\mathsf{RTcC}^{\infty}_{G}$ is complete (cf. Theorem 2 below) $\mathsf{RTcC}^{\omega}_{G}$ is *not*, since arithmetic can be encoded in RTcC logic and the set of $\mathsf{RTcC}^{\omega}_{G}$ proofs is recursively enumerable.[6] Nonetheless, $\mathsf{RTcC}^{\omega}_{G}$ is adequate for RTcC logic in the sense that it suffices for proving the standard properties of the operators, as in, e.g., Proposition 2.

Example 1. Figure 2 demonstrates an $\mathsf{RTcC}^{\omega}_{G}$ proof that the transitive closure is contained within the transitive co-closure. Notice that the proof has a single cycle, and thus a single infinite path. Following this path, there is both a trace (consisting of the highlighted *RTC* formulas, on the left-hand side of sequents) which progresses on traversing Rule (6) (marked †), and a co-trace (consisting of the highlighted RTC^{op} forumlas, on the right-hand side of sequents), which progresses on traversing Rule (8) (marked $*$). Thus, Fig. 2 can be seen both as a proof by induction and a proof by coinduction. It exemplifies how naturally such reasoning can be captured within $\mathsf{RTcC}^{\omega}_{G}$.

A salient feature of non-well-founded proof systems, including this one, is that (co)induction invariants need not be mentioned explicitly, but instead are encoded in the cycles of a proof. This facilitates the automation of such reasoning, as the invariants may be interactively constructed during a proof-search process.

4.2 Soundness

To show soundness, i.e. that all derived sequents are valid, we establish that the infinitely progressing (co-)traces in proofs preclude the existence of counter-models. By local soundness of the proof rules, any given counter-model for a sequent derived by a proof identifies an infinite path in the proof consisting of

[6] The *RTC*-fragment of $\mathsf{RTcC}^{\omega}_{G}$ was shown complete for a Henkin-style semantics [17].

invalid sequents. However, the presence of a (co-)trace along this path entails a contradiction (and so conclude that no counter-models exist). From a trace, one may infer the existence of an infinitely descending chain of natural numbers. This relies on a notion of (well-founded) measure for *RTC* formulas, *viz.* the measure of $\phi \equiv (RTC_{x,y}\,\varphi)(s,t)$ with respect to a given model M and valuation ν—denoted by $\delta_\phi(M,\nu)$—is defined to be the minimum number of φ-steps needed to connect $\nu(s)$ and $\nu(t)$ in M. Conversely, from a co-trace beginning with a formula $(RTC^{\mathsf{op}}_{x,y}\,\varphi)(s,t)$ one can construct an infinite sequence of φ-steps beginning at s, i.e. a witness that the counter-model does in fact satisfy $(RTC^{\mathsf{op}}_{x,y}\,\varphi)(s,t)$.

The key property needed for soundness of the proof system is the following strong form of local soundness for the proof rules.

Proposition 3 (Trace Local Soundness). *Let M be a model and ν a valuation that invalidate the conclusion of an instance of an* RTcC^∞_G *inference rule; then there exists a valuation ν' that invalidates some premise of the inference rule such that the following hold.*

1. *If (τ,τ') is a trace following the path from the conclusion to the invalid premise, then $\delta_{\tau'}(M,\nu') \leq \delta_\tau(M,\nu)$; moreover $\delta_{\tau'}(M,\nu') < \delta_\tau(M,\nu)$ if the rule is an instance of (6) and τ is the principal formula.*
2. *If (τ,τ') is a co-trace following the path from the conclusion to the invalid premise, with $\tau \equiv (RTC^{\mathsf{op}}_{x,y}\,\varphi)(s,t)$ and $\tau' \equiv (RTC^{\mathsf{op}}_{x,y}\,\varphi')(r,t')$, then: (a) $M,\nu[x:=d,y:=d'] \models \varphi$ if and only if $M,\nu'[x:=d,y:=d'] \models \varphi'$, for all elements d and d' in M; and (b) $M,\nu' \models \varphi\,\{{}^s/_x, {}^r/_y\}$ if τ is the principal formula of an instance of (8), and $\nu(s) = \nu'(r)$ otherwise.*

The global soundness of the proof system then follows.

Theorem 1 (Soundness of RTcC^∞_G). *Sequents derivable in* RTcC^∞_G *are valid.*

Proof. Take a proof deriving $\Gamma \Rightarrow \Delta$. Suppose, for contradiction, that there is a model M and valuation ν_1 invalidating $\Gamma \Rightarrow \Delta$. Then by Proposition 3 there exists an infinite path of sequents $(\boldsymbol{S}_i)_{i>0}$ in the proof and an infinite sequence of valuations $(\boldsymbol{\nu}_i)_{i>0}$ such that M and ν_i invalidate S_i for each $i > 0$. Since the proof must satisfy the global trace condition, this infinite path has a tail $(\boldsymbol{S}_i)_{i>k}$ followed by an infinitely progressing (co-)trace $(\boldsymbol{\tau}_i)_{i>0}$.

- If $(\boldsymbol{\tau}_i)_{i>0}$ is a trace, Proposition 3 implies an infinitely descending chain of natural numbers: $\delta_{\tau_1}(M_{k+1},\nu_{k+1}) \leq \delta_{\tau_2}(M_{k+2},\nu_{k+2}) \leq \dots$
- If $(\boldsymbol{\tau}_i)_{i>0}$ is a co-trace, with $\tau_1 \equiv (RTC^{\mathsf{op}}_{x,y}\,\varphi)(s,t)$ and $M,\nu_{k+1} \not\models \tau_1$, then Proposition 3 entails that there is an infinite sequence of terms $t_0,t_1,t_2,\dots$ with $s \equiv t_0$ such that $M,\nu_{k+1}[x{:=}\nu_{k+1}(t_j), y{:=}\nu_{k+1}(t_{j+1})] \models \varphi$ for each $j \geq 0$. That is, it follows from Definition 2 that $M,\nu_{k+1} \models (RTC^{\mathsf{op}}_{x,y}\,\varphi)(s,t)$.

In both cases we have a contradiction, so conclude that $\Gamma \Rightarrow \Delta$ is valid. □

Since every RTcC^ω_G proof is also an RTcC^∞_G proof, soundness of RTcC^ω_G is an immediate corollary.

Corollary 1. *A sequent $\Gamma \Rightarrow \Delta$ is valid if there is an* RTcC^ω_G *proof deriving it.*

4.3 Completeness

The completeness proof for RTcC_G^∞ is obtained by extending the completeness proof of the *RTC*-fragment of RTcC_G^∞ found in [17,18], which, in turn, follows a standard technique used in e.g. [12]. We next outline the core of the proof, full details can be found in the appendix.

Roughly speaking, for a given sequent $\Gamma \Rightarrow \Delta$ one constructs a 'search tree' which corresponds to an exhaustive search strategy for a cut-free proof for the sequent. Search trees are, by construction, recursive and cut-free. In case the search tree is not an RTcC_G^∞ proof (and there are no open nodes) it must contain some *untraceable* infinite branch, i.e. one that does not satisfy the global trace condition. We then collect the formulas occurring along such an untraceable branch to construct a (possibly infinite) 'sequent', $\Gamma_\omega \Rightarrow \Delta_\omega$ (called a *limit sequent*), and construct the Herbrand model M_ω of open terms quotiented by the equalities it contains. That is, taking $\sim$ to be the smallest congruence on terms such that $s \sim t$ whenever $s = t \in \Gamma_\omega$, the elements of M_ω are $\sim$-equivalence classes and every k-ary relation symbol q is interpreted as $\{([t_1], \ldots, [t_k]) \mid q(t_1, \ldots, t_k) \in \Gamma_\omega\}$ (here $[t]$ denotes the $\sim$-equivalence class containing t). This model, together with the valuation ν_ω defined by $\nu_\omega(x) = [x]$ for all variables x, can be shown to invalidate the sequent $\Gamma \Rightarrow \Delta$. The completeness result therefore follows.

Theorem 2 (Completeness). *All valid sequents are derivable in* RTcC_G^∞.

Proof. Given any sequent S, if some search tree for S is not an RTcC_G^∞ proof then it has an untraceable branch, and the model M_ω and valuation ν_ω constructed from the corresponding limit sequent invalidate S. Thus if S is valid, then the search tree is a recursive RTcC_G^∞ proof deriving S. □

We obtain admissibility of cut for the full infinitary system as the search tree, by construction, is cut-free. Since the construction of the search tree does not necessarily produce RTcC_G^ω pre-proofs, we do not obtain a regular completeness result using this technique.

Corollary 2 (Cut admissibility). *Cut is admissible in* RTcC_G^∞.

5 Proving Properties of Streams

We now demonstrate how (co)inductive reasoning about streams and their properties is formalised in the cyclic fragment of the proof system presented above. For the sake of clarity, in the derivations below we elide detailed applications of the proof rules (including the axioms for list constructors), instead indicating the principal rules involved at each step. We also elide (using '...') formulas in sequents that are not relevant to the local reasoning at that point.

Transitivity of Lexicographic Ordering. Fig. 3 outlines the main structure of an RTcC_G^ω proof deriving the sequent $x \leq_\ell y, y \leq_\ell z \Rightarrow x \leq_\ell z$, where x, y,

$$\cfrac{\cfrac{(\exists R)/(Ax)}{\vdots}}{a = c, \ldots \Rightarrow \psi_\ell \left\{ {}^x/_{x_1}, {}^z/_{x_2}, {}^{x'}/_{y_1}, {}^{z'}/_{y_2} \right\}} \quad \cfrac{\cfrac{x \leq_\ell y, y \leq_\ell z \Rightarrow \boxed{x \leq_\ell z}}{x' \leq_\ell y', y' \leq_\ell z', \ldots \Rightarrow \boxed{x' \leq_\ell z'}} \text{(Wk)/(Subst)} \ast}{a = c, \ldots, x' \leq_\ell y', y' \leq_\ell z' \Rightarrow \boxed{x \leq_\ell z}} \text{(8)}$$

$$\cfrac{\cfrac{\cfrac{(\exists R)/(Ax)}{\vdots}}{a < c, \ldots \Rightarrow \psi_\ell \left\{ {}^x/_{x_1}, {}^z/_{x_2}, {}^{\mathsf{nil}}/_{y_1}, {}^{\mathsf{nil}}/_{y_2} \right\}} \quad \cfrac{\ldots \Rightarrow \mathsf{nil} \leq_\ell \mathsf{nil}}{} \text{(7)}}{\cfrac{a < c, \ldots \Rightarrow x \leq_\ell z \qquad\qquad \times 3}{}} \text{(8)}$$

$$\cfrac{x = a :: x', y = b :: y', z = c :: z', a < b \lor (a = b \land x' = x'' \land y' = y_1''),}{\cfrac{b < c \lor (b = c \land y' = y_2'' \land z' = z''), x'' \leq_\ell y_1'', y_2'' \leq_\ell z'', \ldots \Rightarrow \boxed{x \leq_\ell z}}{\mathcal{U}_\ell(x, y, x', y'), \mathcal{U}_\ell(y, z, y'', z') \Rightarrow \boxed{x \leq_\ell z}}} \begin{array}{l}(\lor L)\\[4pt](2),(3)\end{array}$$

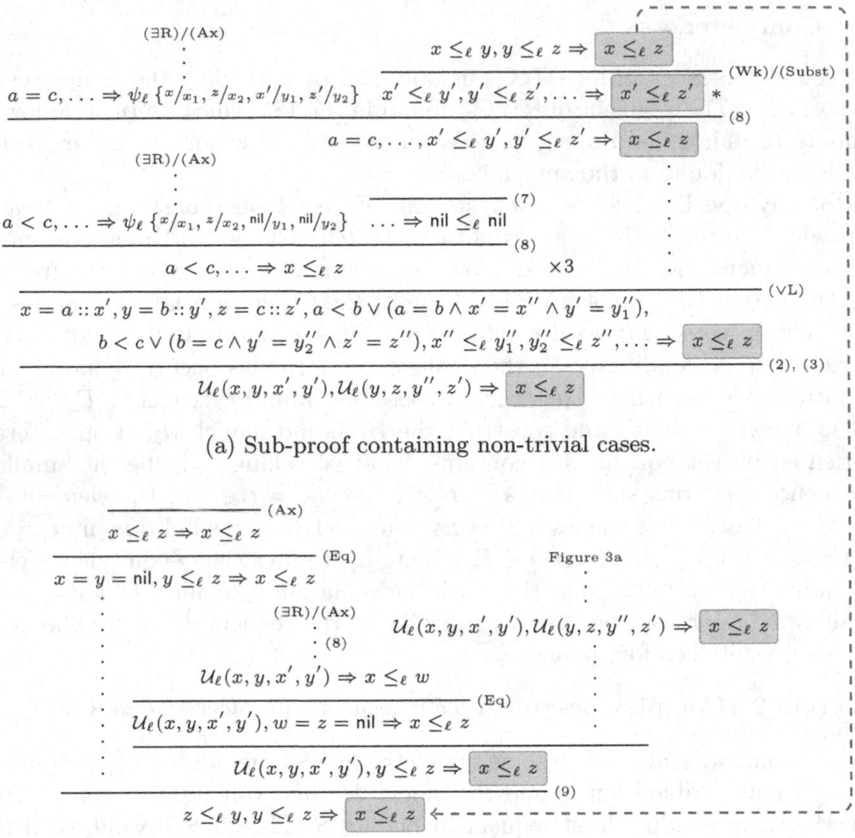

(a) Sub-proof containing non-trivial cases.

$$\cfrac{\cfrac{\cfrac{x \leq_\ell z \Rightarrow x \leq_\ell z}{x = y = \mathsf{nil}, y \leq_\ell z \Rightarrow x \leq_\ell z} \text{(Eq)}}{\vdots} \text{(Ax)}}{\cfrac{\cfrac{\cfrac{\mathcal{U}_\ell(x, y, x', y') \Rightarrow x \leq_\ell w}{\mathcal{U}_\ell(x, y, x', y'), w = z = \mathsf{nil} \Rightarrow x \leq_\ell z} \text{(Eq)}}{\mathcal{U}_\ell(x, y, x', y'), y \leq_\ell z \Rightarrow \boxed{x \leq_\ell z}}}{z \leq_\ell y, y \leq_\ell z \Rightarrow \boxed{x \leq_\ell z}} \leftarrow} \text{(9)}}$$

$$\text{Figure 3a}$$
$$\vdots$$
$$\mathcal{U}_\ell(x, y, x', y'), \mathcal{U}_\ell(y, z, y'', z') \Rightarrow \boxed{x \leq_\ell z}$$
$$\vdots$$

(b) Root of the proof, with trivial cases.

Fig. 3. High-level structure of an RTcC_G^ω proof of transitivity of $\leq_\ell$.

and z are distinct variables. All other variables in Fig. 3 are freshly introduced. $\mathcal{U}_\ell(\sigma_1, \sigma_2, \sigma_1', \sigma_2')$ abbreviates the set $\{\psi_\ell \{{}^{\sigma_1}/_{x_1}, {}^{\sigma_2}/_{x_2}, {}^{\sigma_1'}/_{y_1}, {}^{\sigma_2'}/_{y_2}\}, \sigma_1' \leq_\ell \sigma_2'\}$ (i.e. the result of unfolding the step case of the formula $\sigma_1 \leq_\ell \sigma_2$ using σ_1' and σ_2' as the intermediate terms).

The proof begins by unfolding the definitions of $x \leq_\ell y$ and $y \leq_\ell z$, shown in Fig. 3b. The interesting part is the sub-proof shown in Fig. 3a, when each of the lists is not nil. Here, we perform case splits on the relationship between the head elements a, b, and c. For the case $a = c$, i.e. the heads are equal, when unfolding the formula $x \leq_\ell z$ on the right-hand side, we instantiate the second components of the RTC^{op} formula to be the tails of the streams, x' and z'. In the left-hand premise we must show $\psi_\ell \{{}^x/_{x_1}, {}^z/_{x_2}, {}^{x'}/_{y_1}, {}^{z'}/_{y_2}\}$, which can be done by matching with formulas already present in the sequent. The right-hand premise must derive $x' \leq_\ell z'$, i.e. the tails are lexicographically related. This is where we apply the coinduction principle, by renaming the variables and forming a cycle in

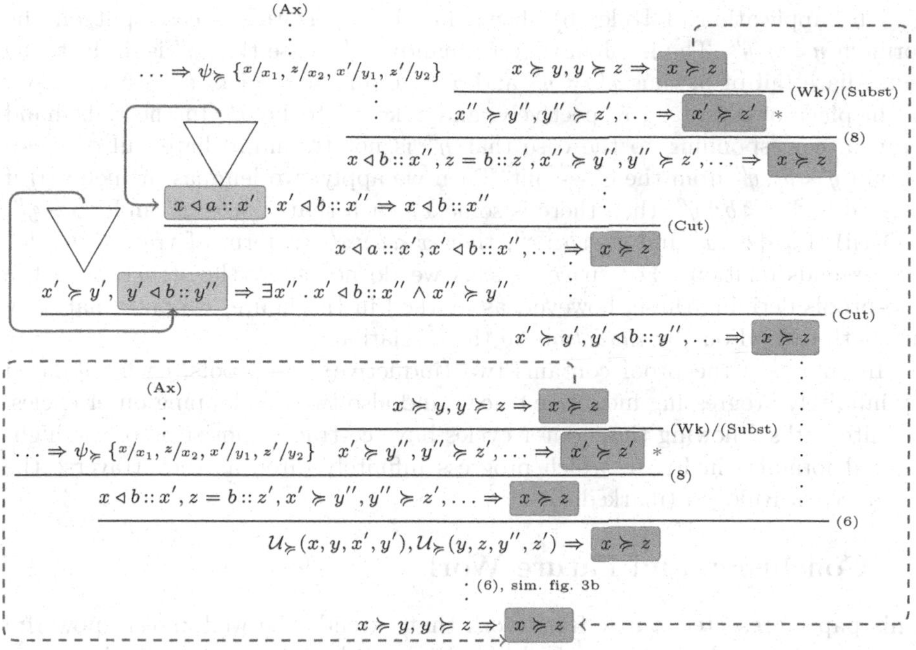

Fig. 4. High-level structure of an RTcC_G^ω proof of transitivity of $\succcurlyeq$.

the proof back to the root. This does indeed produce a proof, since we can form a co-trace by following the formulas $x \leq_\ell z, \ldots, x' \leq_\ell z'$ on the right-hand side of sequents along this cycle. This co-trace progresses as it traverses the instance of Rule (8) each time around the cycle (marked $*$).

Transitivity of the Substream Relation. Fig. 4 outlines the structure of an RTcC_G^ω proof of the sequent $x \succcurlyeq y, y \succcurlyeq z \Rightarrow x \succcurlyeq z$, for distinct variables x, y, and z. As above, other variables are freshly introduced, and we use $\mathcal{U}_{\succcurlyeq}(\sigma_1, \sigma_2, \sigma_1', \sigma_2')$ to denote the set $\{\psi_{\succcurlyeq}\{\sigma_1/x_1, \sigma_2/x_2, \sigma_1'/y_1, \sigma_2'/y_2\}, \sigma_1' \succcurlyeq \sigma_2'\}$ (i.e. the result of unfolding the step-case of the formula $\sigma_1 \succcurlyeq \sigma_2$ using σ_1' and σ_2' as the intermediate terms).

The reflexive cases are handled similarly to the previous example. Again, the work is in proving the step cases. After unfolding both $x \succcurlyeq y$ and $y \succcurlyeq z$, we obtain $x' \succcurlyeq y'$ and $y'' \succcurlyeq z'$, as part of $\mathcal{U}_{\succcurlyeq}(x, y, x', y')$ and $\mathcal{U}_{\succcurlyeq}(y, z, y'', z')$, respectively. We also have (for fresh variables a and b) that: (i) $x \triangleleft a :: x'$; (ii) $y = a :: y'$ (y' is the immediate tail of y); (iii) $y \triangleleft b :: y''$ (y'' is some tail of y); and (iv) $z = b :: z'$ (z' is the immediate tail of z). Ultimately, we are looking to obtain $x \triangleleft b :: x''$ and $x'' \succcurlyeq y''$ (for some tail x''), so that we can unfold the formula $x \succcurlyeq z$ on the right-hand side to obtain $x'' \succcurlyeq z'$ and thus be able to form a (coinductive) cycle.

The application of Rule (6) shown in Fig. 4 performs a case-split on the formula $y \lhd b :: y''$. The left-hand branch handles the case that y'' is, in fact, the immediate tail of y; thus $y' = y''$ and $a = b$, and so we can substitute b and y'' in place of a and y', respectively, and take x'' to be x'. In the right-hand branch, corresponding to the case that y'' is not the immediate tail of y, we obtain $y' \lhd b :: y''$ from the case-split. Then we apply two lemmas; namely: (i) if $x' \succcurlyeq y'$ and $y' \lhd b :: y''$, then there is some x'' such that $x' \lhd b :: x''$ and $x'' \succcurlyeq y''$; and (ii) if $x \lhd a :: x'$ and $x' \lhd b :: x''$, then $x \lhd b :: x''$ (a form of transitivity for the extends relation). For space reasons we do not show the structure of the sub-proofs deriving these, however, as marked in the figure, we note that they are both carried out by *induction* on the $\lhd$ relation.

In summary the proof contains two (inductive) sub-proofs, each validated by infinitely progressing inductive traces, and also two overlapping outer cycles. Infinite paths following these outer cycles have co-traces consisting of the highlighted formulas in Fig. 4, which progress infinitely often as they traverse the instances of Rule (8) (marked $*$).

6 Conclusion and Future Work

This paper presented a new framework that extends the well-known, powerful transitive closure logic with a dual transitive co-closure operator. An infinitary proof system for the logic was developed and shown to be sound and complete. Its cyclic subsystem was shown to be powerful enough for reasoning over streams, and in particular automating combinations of inductive and coinductive arguments.

Much remains to be done to fully develop the new logic and its proof theory, and to study its implications. Although we have shown that our framework captures many interesting properties of the canonical coinductive data type, streams, a primary task for future research is to formally characterise its ability to capture finitary coinductive definitions in general. In particular, it seems plausible that RTcC is a good candidate setting in which to look for characterisations that complement and bridge existing results for coinductive data in automata theory and coalgebra. That is, it may potentially mirror (and also perhaps even replace) the role that monadic second order logic plays for (ω-)regular languages.

Another important research task is to further develop the structural proof theory of the systems $\mathsf{RTcC}^{\infty}_{G}$ and $\mathsf{RTcC}^{\omega}_{G}$ in order to describe the natural process and dynamics of inductive and coinductive reasoning. This includes properties such as cut elimination, admissibility of rules, regular forms for proofs, focussing, and proof search strategies. For example, syntactic cut elimination for non-well-founded systems has been studied extensively in the context of linear logic [5, 26]. The basic approach would seem to work for RTcC, however, one expects that cut-elimination will *not* preserve regularity.

Through the proofs-as-programs paradigm (a.k.a. the Curry-Howard correspondence) our proof-theoretic synthesis of induction and coinduction has a number of applications that invite further investigation. Namely, our framework provides a general setting for verifying program correctness against specifications of coinductive (safety) and inductive (liveness) properties. Implementing proof-search procedures can lead to automation, as well as correct-by-construction synthesis of programs operating on (co)inductive data. Finally, grounding proof assistants in our framework will provide a robust, proof-theoretic basis for mechanistic coinductive reasoning.

Acknowledgements. We are grateful to Alexandra Silva for valuable coinductive reasoning examples, and Juriaan Rot for helpful comments and pointers. We also extend thanks to the anonymous reviewers for their questions and comments.

References

1. Abel, A., Pientka, B.: Well-founded recursion with copatterns and sized types. J. Funct. Program. **26**, e2 (2016). https://doi.org/10.1017/S0956796816000022
2. Afshari, B., Leigh, G.E.: Cut-free completeness for modal mu-calculus. In: Proceedings of the 32nd Annual ACM/IEEE Symposium on Logic in Computer Science (LICS 2017), Reykjavik, Iceland, 20–23 June 2017, pp. 1–12 (2017). https://doi.org/10.1109/LICS.2017.8005088
3. Avron, A.: Transitive closure and the mechanization of mathematics. In: Kamareddine, F.D. (ed.) Thirty Five Years of Automating Mathematics, Applied Logic Series. APLS 2013, vol. 28, pp. 149–171. Springer, Netherlands (2003). https://doi.org/10.1007/978-94-017-0253-9_7
4. Baelde, D.: Least and greatest fixed points in linear logic. ACM Trans. Comput. Log. **13**(1), 2:1–2:44 (2012). https://doi.org/10.1145/2071368.2071370
5. Baelde, D., Doumane, A., Saurin, A.: Infinitary proof theory: the multiplicative additive case. In: Proceedings of the 25th EACSL Annual Conference on Computer Science Logic (CSL 2016), 29 August–1 September 2016, Marseille, France, pp. 42:1–42:17 (2016). https://doi.org/10.4230/LIPIcs.CSL.2016.42
6. Basold, H.: Mixed Inductive-Coinductive Reasoning Types, Programs and Logic. Ph.D. thesis, Radboud University (2018). https://hdl.handle.net/2066/190323
7. Bertot, Y., Casteran, P.: Interactive Theorem Proving and Program Development. Springer, Heidelberg (2004). https://doi.org/10.1007/978-3-662-07964-5
8. Bove, A., Dybjer, P., Norell, U.: A brief overview of Agda – a functional language with dependent types. In: Berghofer, S., Nipkow, T., Urban, C., Wenzel, M. (eds.) TPHOLs 2009. LNCS, vol. 5674, pp. 73–78. Springer, Heidelberg (2009). https://doi.org/10.1007/978-3-642-03359-9_6
9. Brady, E.: Idris, a general-purpose dependently typed programming language: design and implementation. J. Funct. Program. **23**, 552–593 (2013). https://doi.org/10.1017/S095679681300018X
10. Brotherston, J.: Formalised inductive reasoning in the logic of bunched implications. In: Nielson, H.R., Filé, G. (eds.) SAS 2007. LNCS, vol. 4634, pp. 87–103. Springer, Heidelberg (2007). https://doi.org/10.1007/978-3-540-74061-2_6
11. Brotherston, J., Bornat, R., Calcagno, C.: Cyclic proofs of program termination in separation logic. In: Proceedings of the 35th ACM SIGPLAN-SIGACT Symposium on Principles of Programming Languages (POPL 2008), pp. 101–112 (2008). https://doi.org/10.1145/1328438.1328453
12. Brotherston, J., Simpson, A.: Sequent calculi for induction and infinite descent. J. Log. Comput. **21**(6), 1177–1216 (2010). https://doi.org/10.1093/logcom/exq052

13. Buss, S.R.: Handbook of proof theory. In: Studies in Logic and the Foundations of Mathematics. Elsevier Science (1998)
14. Cohen, L.: Completeness for ancestral logic via a computationally-meaningful semantics. In: Schmidt, R.A., Nalon, C. (eds.) TABLEAUX 2017. LNCS (LNAI), vol. 10501, pp. 247–260. Springer, Cham (2017). https://doi.org/10.1007/978-3-319-66902-1_15
15. Cohen, L., Avron, A.: Ancestral logic: a proof theoretical study. In: Kohlenbach, U., Barceló, P., de Queiroz, R. (eds.) WoLLIC 2014. LNCS, vol. 8652, pp. 137–151. Springer, Heidelberg (2014). https://doi.org/10.1007/978-3-662-44145-9_10
16. Cohen, L., Avron, A.: The middle ground-ancestral logic. Synthese, 1–23 (2015). https://doi.org/10.1007/s11229-015-0784-3
17. Cohen, L., Rowe, R.N.S.: Uniform inductive reasoning in transitive closure logic via infinite descent. In: Proceedings of the 27th EACSL Annual Conference on Computer Science Logic (CSL 2018), 4–7 September 2018, Birmingham, UK, pp. 16:1–16:17 (2018). https://doi.org/10.4230/LIPIcs.CSL.2018.16
18. Cohen, L., Rowe, R.N.S.: Non-well-founded proof theory of transitive closure logic. Trans. Comput. Logic (2020, to appear). https://arxiv.org/pdf/1802.00756.pdf
19. Cohen, L., Rowe, R.N.S., Zohar, Y.: Towards automated reasoning in Herbrand structures. J. Log. Comput. **29**(5), 693–721 (2019). https://doi.org/10.1093/logcom/exz011
20. Constable, R.L., et al.: Implementing Mathematics with the Nuprl Proof Development System. Prentice-Hall Inc, Upper Saddle River (1986)
21. Cook, S.A., Reckhow, R.A.: The relative efficiency of propositional proof systems. J. Symbolic Log. **44**(1), 36–50 (1979). https://doi.org/10.2307/2273702
22. Courcelle, B.: Fundamental properties of infinite trees. Theor. Comput. Sci. **25**, 95–169 (1983). https://doi.org/10.1016/0304-3975(83)90059-2
23. Das, A., Pous, D.: Non-Wellfounded Proof Theory for (Kleene+Action)(Algebras+Lattices). In: Proceedings of the 27th EACSL Annual Conference on Computer Science Logic (CSL 2018), pp. 19:1–19:18 (2018). https://doi.org/10.4230/LIPIcs.CSL.2018.19
24. Doumane, A.: Constructive completeness for the linear-time μ-calculus. In: Proceedings of the 32nd Annual ACM/IEEE Symposium on Logic in Computer Science (LICS 2017), pp. 1–12 (2017). https://doi.org/10.1109/LICS.2017.8005075
25. Endrullis, J., Hansen, H., Hendriks, D., Polonsky, A., Silva, A.: A coinductive framework for infinitary rewriting and equational reasoning. In: 26th International Conference on Rewriting Techniques and Applications (RTA 2015), vol. 36, pp. 143–159 (2015). https://doi.org/10.4230/LIPIcs.RTA.2015.143
26. Fortier, J., Santocanale, L.: Cuts for circular proofs: semantics and cut-elimination. In: Rocca, S.R.D. (ed.) Computer Science Logic 2013 (CSL 2013). Leibniz International Proceedings in Informatics (LIPIcs), vol. 23, pp. 248–262. Dagstuhl, Germany (2013). https://doi.org/10.4230/LIPIcs.CSL.2013.248
27. Gapeyev, V., Levin, M.Y., Pierce, B.C.: Recursive subtyping revealed. J. Funct. Program. **12**(6), 511–548 (2002). https://doi.org/10.1017/S0956796802004318
28. Gentzen, G.: Untersuchungen über das Logische Schließen. I. Mathematische Zeitschrift **39**(1), 176–210 (1935). https://doi.org/10.1007/BF01201353
29. Hansen, H.H., Kupke, C., Rutten, J.: Stream differential equations: specification formats and solution methods. In: Logical Methods in Computer Science, vol. 13(1), February 2017. https://doi.org/10.23638/LMCS-13(1:3)2017
30. Heath, Q., Miller, D.: A proof theory for model checking. J. Autom. Reasoning **63**(4), 857–885 (2019). https://doi.org/10.1007/s10817-018-9475-3

31. Immerman, N.: Languages that capture complexity classes. SIAM J. Comput. **16**(4), 760–778 (1987). https://doi.org/10.1137/0216051
32. Jacobs, B., Rutten, J.: A tutorial on (co) algebras and (co) induction. Bull. Eur. Assoc. Theor. Comput. Sci. **62**, 222–259 (1997)
33. Jeannin, J.B., Kozen, D., Silva, A.: CoCaml: functional programming with regular coinductive types. Fundamenta Informaticae **150**, 347–377 (2017). https://doi.org/10.3233/FI-2017-1473
34. Komendantskaya, E., Power, J.: Coalgebraic semantics for derivations in logic programming. In: Corradini, A., Klin, B., Cîrstea, C. (eds.) CALCO 2011. LNCS, vol. 6859, pp. 268–282. Springer, Heidelberg (2011). https://doi.org/10.1007/978-3-642-22944-2_19
35. Kozen, D., Silva, A.: Practical coinduction. Math. Struct. Comput. Sci. **27**(7), 1132–1152 (2017). https://doi.org/10.1017/S0960129515000493
36. Leino, R., Moskal, M.: Co-induction simply: automatic co-inductive proofs in a program verifier. Technical report MSR-TR-2013-49, Microsoft Research, July 2013. https://www.microsoft.com/en-us/research/publication/co-induction-simply-automatic-co-inductive-proofs-in-a-program-verifier/
37. Leroy, X., Grall, H.: Coinductive big-step operational semantics. Inf. Comput. **207**(2), 284–304 (2009). https://doi.org/10.1016/j.ic.2007.12.004
38. Lucanu, D., Roşu, G.: CIRC: a circular coinductive prover. In: Mossakowski, T., Montanari, U., Haveraaen, M. (eds.) CALCO 2007. LNCS, vol. 4624, pp. 372–378. Springer, Heidelberg (2007). https://doi.org/10.1007/978-3-540-73859-6_25
39. Martin, R.M.: A homogeneous system for formal logic. J. Symbolic Log. **8**(1), 1–23 (1943). https://doi.org/10.2307/2267976
40. Martin-Löf, P.: Hauptsatz for the intuitionistic theory of iterated inductive definitions. In: Fenstad, J.E. (ed.) Proceedings of the Second Scandinavian Logic Symposium, Studies in Logic and the Foundations of Mathematics, vol. 63, pp. 179–216. Elsevier (1971). https://doi.org/10.1016/S0049-237X(08)70847-4
41. McBride, C.: Let's see how things unfold: reconciling the infinite with the intensional (extended abstract). In: Kurz, A., Lenisa, M., Tarlecki, A. (eds.) CALCO 2009. LNCS, vol. 5728, pp. 113–126. Springer, Heidelberg (2009). https://doi.org/10.1007/978-3-642-03741-2_9
42. McDowell, R., Miller, D.: Cut-elimination for a logic with definitions and induction. Theor. Comput. Sci. **232**(1–2), 91–119 (2000). https://doi.org/10.1016/S0304-3975(99)00171-1
43. Park, D.M.R.: Finiteness is mu-ineffable. Theor. Comput. Sci. **3**(2), 173–181 (1976). https://doi.org/10.1016/0304-3975(76)90022-0
44. Roşu, G., Lucanu, D.: Circular coinduction: a proof theoretical foundation. In: Kurz, A., Lenisa, M., Tarlecki, A. (eds.) CALCO 2009. LNCS, vol. 5728, pp. 127–144. Springer, Heidelberg (2009). https://doi.org/10.1007/978-3-642-03741-2_10
45. Rowe, R.N.S., Brotherston, J.: Automatic cyclic termination proofs for recursive procedures in separation logic. In: Proceedings of the 6th ACM SIGPLAN Conference on Certified Programs and Proofs (CPP 2017), Paris, France, 16–17 January 2017, pp. 53–65 (2017). https://doi.org/10.1145/3018610.3018623
46. Rutten, J.: Universal coalgebra: a theory of systems. Theor. Comput. Sci. **249**(1), 3–80 (2000)
47. Rutten, J.: On Streams and Coinduction (2002). https://homepages.cwi.nl/~janr/papers/files-of-papers/CRM.pdf
48. Rutten, J.: The Method of Coalgebra: Exercises in Coinduction. CWI, Amsterdam (2019)

49. Sangiorgi, D., Rutten, J.: Advanced Topics in Bisimulation and Coinduction, 1st edn. Cambridge University Press, Cambridge (2011)
50. Santocanale, L.: A calculus of circular proofs and its categorical semantics. In: Nielsen, M., Engberg, U. (eds.) FoSSaCS 2002. LNCS, vol. 2303, pp. 357–371. Springer, Heidelberg (2002). https://doi.org/10.1007/3-540-45931-6_25
51. Shapiro, S.: Foundations Without Foundationalism : A Case for Second-order Logic. Clarendon Press, Oxford Logic Guides (1991)
52. Simon, L., Bansal, A., Mallya, A., Gupta, G.: Co-logic programming: extending logic programming with coinduction. In: Arge, L., Cachin, C., Jurdziński, T., Tarlecki, A. (eds.) ICALP 2007. LNCS, vol. 4596, pp. 472–483. Springer, Heidelberg (2007). https://doi.org/10.1007/978-3-540-73420-8_42
53. Simon, L., Mallya, A., Bansal, A., Gupta, G.: Coinductive logic programming. In: Etalle, S., Truszczyński, M. (eds.) ICLP 2006. LNCS, vol. 4079, pp. 330–345. Springer, Heidelberg (2006). https://doi.org/10.1007/11799573_25
54. Simpson, A.: Cyclic arithmetic is equivalent to peano arithmetic. In: Esparza, J., Murawski, A.S. (eds.) FoSSaCS 2017. LNCS, vol. 10203, pp. 283–300. Springer, Heidelberg (2017). https://doi.org/10.1007/978-3-662-54458-7_17
55. Sprenger, C., Dam, M.: On the structure of inductive reasoning: circular and tree-shaped proofs in the μcalculus. In: Gordon, A.D. (ed.) FoSSaCS 2003. LNCS, vol. 2620, pp. 425–440. Springer, Heidelberg (2003). https://doi.org/10.1007/3-540-36576-1_27
56. Takeuti, G.: Proof Theory. Dover Books on Mathematics. Dover Publications, Incorporated, New York (2013)
57. Tellez, G., Brotherston, J.: Automatically verifying temporal properties of pointer programs with cyclic proof. In: de Moura, L. (ed.) CADE 2017. LNCS (LNAI), vol. 10395, pp. 491–508. Springer, Cham (2017). https://doi.org/10.1007/978-3-319-63046-5_30
58. Tiu, A.: A Logical Framework For Reasoning About Logical Specifications. Ph.D. thesis, Penn. State University (2004)
59. Tiu, A., Momigliano, A.: Cut elimination for a logic with induction and co-induction. J. Appl. Log. **10**(4), 330–367 (2012). https://doi.org/10.1016/j.jal.2012.07.007

Logic-Independent Proof Search in Logical Frameworks (Short Paper)

Michael Kohlhase[1], Florian Rabe[1], Claudio Sacerdoti Coen[2], and Jan Frederik Schaefer[1(✉)]

[1] Computer Science, FAU Erlangen-Nürnberg, Erlangen, Germany
jan.frederik.schaefer@fau.de
[2] Department of Computer Science and Engineering, Università di Bologna, Bologna, Italy

Abstract. Logical frameworks like LF allow to specify the syntax and (natural deduction) inference rules for syntax/proof-checking a wide variety of logical systems. A crucial feature that is missing for prototyping logics is a way to specify basic proof automation. We try to alleviate this problem by generating λProlog (ELPI) inference predicates from logic specifications and controlling them by logic-independent helper predicates that encapsulate the prover characteristics. We show the feasibility of the approach with three experiments: We directly automate ND calculi, we generate tableau theorem provers and model generators.

1 Introduction and Related Work

Logical frameworks like LF [HHP93] and λProlog [Mil] enable prototyping and analyzing logical systems, using high-level declarative logic definitions based on higher-order abstract syntax. Building theorem provers automatically from declarative logic definitions has been a long-standing research goal. But currently, logical framework-induced fully logic-independent proof support is generally limited to proof checking and simple search.

Competitive proof support, on the other hand, is either highly optimized for very specific logics, most importantly propositional logics or untyped first-order logic. Generic approaches have so far been successful for logics without binding (and thus quantifiers) such as the tableaux prover generation in MetTeL2 [TSK]. Logical frameworks shine when applied to logics with binding, for which specifying syntax and calculus is substantially more difficult. However, while the Isabelle framework was designed as such a generic prover [Pau93], it is nowadays primarily used for one specific logic (Isabelle/HOL). On the other hand, there has been an explosion of logical systems, often domain-specific, experimental, or

The authors gratefully acknowledge project support by German Research Council (DFG) grants KO 2428/13-1 and RA-1872/3-1 OAF as well as EU Horizon 2020 grant ERI 676541 OpenDreamKit.

N. Peltier and V. Sofronie-Stokkermans (Eds.): IJCAR 2020, LNAI 12166, pp. 395–401, 2020.
https://doi.org/10.1007/978-3-030-51074-9_22

otherwise restricted to small user communities that cannot sustain the development of a practical theorem prover. To gain theorem proving support for such logics, proof obligations can be shipped to existing provers via one of the TPTP languages, or the logics may be defined as DSLs inside existing provers as is commonly done using Coq [Coq15], Isabelle [Pau94], or Leo [Ben+08]. If applicable, these approaches are very successful. But they are also limited by the language and proof strategy of the host system, which can preclude fully exploring the design space for logics and prover.

We investigate this question by combining the advantages of two logical frameworks: To define logics, we use the implementation of LF in MMT [Rab17, Rab18]. MMT is optimized for specifying and prototyping logics, providing in particular type reconstruction, module system, and graphical user interface. Then we generate ELPI theorem provers from these logic definitions. ELPI [SCT15] is an extension of λProlog with constraint programming via user-defined rules, macros, type abbreviations, optional polymorphic typing and more. ELPI is optimized for fast execution of logical algorithms such as type inference, unification, or proof search, and it allows prototyping such systems much more rapidly than traditional imperative or functional languages. Both MMT and ELPI were designed to be flexible and easy to integrate with other systems. Our approach is logic-independent and applicable to any logic defined in LF. Concretely, we evaluate our systems by generating provers for the highly modular suite of logic definitions in the LATIN atlas [Cod+11], which includes e.g. first- and higher-order and modal logics and various dependent type theories. These logic definitions can be found at [LATIN] and the generated ELPI provers in [GEP].

We follow the approach proposed by Miller et al. in the ProofCert project [CMR13] but generalize it to non-focused logics. The key idea is to translate each rule R of the deduction system to an ELPI clause for the provability predicate, whose premises correspond to the premises of R. The provability predicate has an additional argument that represents a proof certificate and each clause has a new premise that is a predicate, called its *helper predicate*, that relates the proof certificates of the premises to the one of the conclusion. Following [CMR13], the definitions of the certificates and the helper predicates are initially left open, and by providing different instances we can implement different theorem provers. In the simplest case, the additional premise acts as a guard that determines if and when the theorem prover should use a rule. It can also suggest which formulas to use during proof search when the rule is not analytic. This allows implementing strategies such as iterative deepening or backchaining. Alternatively, the helper predicates can be used to track information in order to return information such as the found proof. These can be combined modularly with minimal additional work, e.g., to return the proof term found by a backchaining prover or to run a second prover on a branch where the first one failed.

[CMR13] and later works by the authors use focusing as a preliminary requirement. While focusing was the source of inspiration of the whole technique and brings great benefits by reduction of search space, we show here that focusing is not really a requirement and that we can achieve comparable reductions in search space by designing certificates that impose the two phases of proof search in focused calculi a posteriori. By not using focusing, our work makes the approach much more accessible as focusing is largely unknown in many communities (e.g., in linguistics). Moreover, [CMR13] was motivated by applications to one specific logic: classical/intuitionist focused first-order logic. While it is clear in theory that the same methodology can be applied to other logics, in practice we need to build tools to test and benchmark the approach in those logics. Here the possibility of generating many different provers for LF-defined logics in a completely uniform way is an important novelty of our work.

This paper is a short version of [Koh+20] which contains additional details.

2 Natural Deduction Provers

Logic Definitions in MMT/LF. While our approach is motivated by and applicable to very complex logics, including e.g. dependent type theories, it is easier to present our ideas by using a very simple running example. Concretely, we will use the language features of conjunction and untyped universal quantification. Their formalized syntax is shown below relative to base theories for propositions and terms.

$$\begin{aligned}&\texttt{Props} = \{\texttt{prop} : \texttt{type}\}\quad \texttt{Terms} = \{\texttt{term} : \texttt{type}\}\\&\texttt{Conj} = \{\textbf{include}\,\texttt{Props},\ \texttt{and} : \texttt{prop} \to \texttt{prop} \to \texttt{prop}\}\\&\texttt{Univ} = \{\textbf{include}\,\texttt{Props},\ \textbf{include}\,\texttt{Terms},\ \texttt{univ} : (\texttt{term} \to \texttt{prop}) \to \texttt{prop}\}\end{aligned}$$

Below we extend these with the respective natural deduction rules relative to a theory ND that introduces a judgment to map a proposition $F : \texttt{prop}$ to the type $\texttt{ded}\, F$ of proofs of F:

$$\begin{aligned}&\texttt{ND} = \{\textbf{include}\,\texttt{Props},\ \textbf{judg}\,\texttt{ded} : \texttt{prop} \to \texttt{type}\}\\&\texttt{ConjND} = \{\textbf{include}\,\{\texttt{ND},\texttt{Conj}\},\ \texttt{andEl} : \Pi_{A,B:\texttt{prop}}\texttt{ded}\,(A \wedge B) \to \texttt{ded}\,A,\ \ldots\}\end{aligned}$$

For brevity, we only give some of the rules and use the usual notations for the constants and and univ. Note that **judg** tags ded as a judgment: while this is irrelevant for LF type checking, it allows our theorem provers to distinguish the data from the judgment level. Concretely, type declarations are divided into *data* types (such as prop and term) and *judgments* (such as ded). And term declarations are divided, by looking at their return type, into *data constructors* (such as and and univ) and *rules* (such as andEl and univE).

Generating ELPI Provers. Our LF-based formalizations of logics define the well-formed proofs, but implementations of LF usually do not offer proof search control that would allow for automation. Therefore, we systematically translate every LF theory into an ELPI file. ELPI is similarly expressive as LF so that a naive approach could simply translate the andEl rule to the ELPI statement

$$\texttt{ded}\,A\ \ \texttt{:-}\ \ \texttt{ded}\,(\texttt{and}\,A\,B)$$

Note how the Π-bound variables of the LF rule (which correspond to implicit arguments that LF implementations reconstruct) simply become free variables for ELPI's Prolog engine to instantiate. However, this would not yield a useful theorem prover at all — instead, the depth-first search behavior of Prolog would easily lead to divergence. Therefore, to control proof search, we introduce additional arguments as follows:

- An n-ary judgment like `ded` becomes a $(1+n)$-ary predicate in ELPI. The new argument, called a *proof certificate*, can record information about the choice of rules and arguments to be used during proof search.
- A rule r with n premises (i.e., with n arguments remaining after discarding the implicit ones) becomes an ELPI statement with $1+n$ premises.

The additional premise is a predicate named r_{help}, i.e., we introduce one helper predicate for each rule; it receives all certificates and formulas of the rule as input. A typical use for r_{help} is to disable or enable r according to the certificate provided in the input and, if enabled, extract from this certificate those for the recursive calls. An alternative use is to synthesize a certificate in output from the output certificates of the recursive calls, which allows recording a successful proof search attempt. This is possible because λProlog is based on relations, and it is not fixed a priori which arguments are to be used as input and which as output. The two uses can even be combined by providing half-instantiated certificates in input that become fully instantiated in output.

For our running example, the following ELPI rule is generated along with a number of helper predicates:

$$\texttt{ded}\, X_2\, F \;\; \texttt{:-} \;\; \texttt{help/andEl}\, X_2\, F\, G\, X_1,\;\; \texttt{ded}\, X_1\, (\texttt{and}\, F\, G).$$

Iterative Deepening. Iterative deepening is a very simple mechanism to control the proof search and avoid divergence. Here the certificate simply contains an integer indicating the remaining search depth. A top-level loop (not shown here) just repeats proof search with increasing initial depth. Due to its simplicity, we can easily generate the necessary helper predicate automatically:

$$\texttt{help/andEl}\, (\texttt{idcert}\, X_3)\, F\, G\, (\texttt{idcert}\, X_2) \;\; \texttt{:-} \;\; X_3 > 0,\;\; X_2\; \texttt{is}\; X_3 - 1.$$

Backchaining. Here, the idea is to be more cautious when trying to use non-analytic elimination rules like `andEl`, whose premises contain a sub-formula not present in the conclusion. To avoid wrongly guessing these, Miller [CMR13] employs a focused logic where forward and backward search steps are alternated. We reproduce a similar behavior for our simpler unfocused logic by programming the helper to trigger forward reasoning search and by automatically generating forward reasoning clauses for some of our rules:

$$\begin{array}{l}\texttt{help/andEl}\, (\texttt{bccert}\, X_3)\, F\, G\, (\texttt{bccert}\, (\texttt{bc/fwdLocked}\, X_2)) \\ \quad \texttt{:-}\;\; \texttt{bc/val}\, X_3\, X_4,\;\; X_4 > 0,\;\; X_2\; \texttt{is}\; X_4 - 1,\;\; \texttt{bc/fwdable}\, (\texttt{and}\, F\, G). \\ \texttt{bc/fwdable} \;\; \texttt{:-}\;\; \texttt{ded/hyp}\, \_\, T,\;\; \texttt{bc/aux}\, T\, A.\end{array}$$

Here we use two predicates that are defined once and for all, i.e., logic-independently: `bc/fwdable` (`and` F G) asks for a forward reasoning proof of (`and` F G); and `ded/hyp` _ T recovers an available hypothesis T.

Finally, bc/aux T A proves A from T using forward reasoning steps. Its definition picks up on annotations in LF that mark forward rules, and if andEl is annotated accordingly, we automatically generate the forward-reasoning clause below, which says that X_5 is provable from (and F G) if it is provable from F:

$$\texttt{bc/aux}\ (\texttt{and}\ F\ G)\ X_5\ \ \texttt{:-}\ \ \texttt{bc/aux}\ F\ X_5.$$

3 Tableau Provers

Logic Definitions in MMT/LF. The formalizations of the tableau rules for our running example are given below. The general idea is to represent each branch of a tableau as an LF context; the unary judgments $1\,A$ and $0\,A$ represent the presence of the signed formula A on the branch, and the judgment $\bot$ represents its closability. Thus, the type $0\,A \to \bot$ represents that A can be proved. For example, the rule and0 below states: if $0\,(A \wedge B)$ is on a branch, then that branch can be closed if the two branches extending it with $0\,A$ resp. $0\,B$ can.

$$\begin{aligned}
&\texttt{Tab} = \{\textbf{include}\,\texttt{Props},\ \textbf{judg}\,1 : \texttt{prop} \to \texttt{type},\ \textbf{judg}\,0 : \texttt{prop} \to \texttt{type},\\
&\quad \textbf{judg}\,\bot : \texttt{type},\ \texttt{close} : \Pi_{A:\texttt{prop}} 1\,A \to 0\,A \to \bot\}\\
&\texttt{ConjTab} = \{\textbf{include}\,\texttt{Tab},\ \textbf{include}\,\texttt{Conj},\\
&\quad \texttt{and0} : \Pi_{A,B:\texttt{prop}}\, 0\,(A \wedge B) \to (0\,A \to \bot) \to (0\,B \to \bot) \to \bot,\ \ldots\}\\
&\texttt{UnivTab} = \{\ldots,\ \texttt{forall1} : \Pi_P \Pi_{X:\texttt{term}}\, 1(\forall P) \to (1(P X) \to \bot) \to \bot\}
\end{aligned}$$

Generating ELPI Provers. We use the same principle to generate ELPI statements, i.e., every LF-judgment receives an additional argument and every LF-rule an additional premise.

To generate a **tableau prover**, we use the additional arguments to track the current branch. This allows recording how often a rule has been applied in order to prioritize rule applications. For first-order logic, this is only needed to allow applying the relevant quantifier rules more than once.

For theorem proving, branches that are abandoned when the depth-limit is reached represent failed proof attempts. But we can just as well use the prover as a **model generator**: here we modify the helper predicates in such a way that abandoning an unclosed branch returns that branch. Thus, the overall run returns the list of open branches, i.e., the potential models.

Note that the ND theorem prover from Sect. 2 is strong enough to prove the tableau rules admissible for the logics we experimented with. If this holds up, it makes prototyping proof support for logic experiments much more convenient.

Application to Natural Language Understanding. Part of the motivation for the work reported here was to add an inference component — a tableau machine for natural language pragmatics — to our Grammatical/Logical Framework [KS19], which combines syntactic processing via the LF-based GF [Ran04] with MMT to obtain an NL interpretation pipeline that is parametric in the target logic. A variant of the model generator discussed above—where we extend the helper predicate to choose a currently preferred branch when a resource bound in saturation is reached—yields an NL understanding framework that combines anaphor resolution, world-knowledge, and belief revision as in [KK03] with support for changing and experimenting with the target logic. A demo of the envisioned system can be found at [GD].

4 Conclusion and Future Work

We have revisited the question of generating theorem provers from declarative logic definitions in logical frameworks. We believe that, after studying such frameworks for the last few decades, the community has now understood them well enough and implemented them maturely enough to have a serious chance at succeeding. The resulting provers will never be competitive with existing state-of-the-art provers optimized for a particular logic, but the expressivity and flexibility of these frameworks allows building practically relevant proof support for logics that would otherwise have no support at all.

Our infrastructure already scales well to large collections of logics and multiple prover strategies, and we have already used it successfully to rapidly prototype a theorem prover in a concrete natural language understanding application. In the future, we will develop stronger proof strategies, in particular better support for equational reasoning and advanced type systems. We will also integrate the ELPI-based theorem provers as a backend for MMT/LF in order to provide both automated and interactive proof support directly in the graphical user interface. A key question will be how the customization of the theorem prover can be integrated with the logic definitions (as we already did by annotating forward rules) without losing the declarative flavor of LF.

References

[Ben+08] Benzmüller, C., Paulson, L.C., Theiss, F., Fietzke, A.: LEO-II - A Cooperative Automatic Theorem Prover for Classical Higher-Order Logic (System Description). In: Armando, A., Baumgartner, P., Dowek, G. (eds.) IJCAR 2008. LNCS (LNAI), vol. 5195, pp. 162–170. Springer, Heidelberg (2008). https://doi.org/10.1007/978-3-540-71070-7_14

[CMR13] Chihani, Z., Miller, D., Renaud, F.: Checking foundational proof certificates for first-order logic. In: Blanchette, J., Urban, J. (eds.) Proof Exchange for Theorem Proving. EasyChair, pp. 58–66 (2013)

[Cod+11] Codescu, M., Horozal, F., Kohlhase, M., Mossakowski, T., Rabe, F.: Project Abstract: Logic Atlas and Integrator (LATIN). In: Davenport, J.H., Farmer, W.M., Urban, J., Rabe, F. (eds.) CICM 2011. LNCS (LNAI), vol. 6824, pp. 289–291. Springer, Heidelberg (2011). https://doi.org/10.1007/978-3-642-22673-1_24

[Coq15] Coq Development Team: The Coq proof assistant: reference manual. Technical report INRIA (2015)

[GD] GLIF Demo. https://gl.kwarc.info/COMMA/glif-demo-ijcar-2020. Accessed 27 Jan 2020

[GEP] Generated ELPI Provers. https://gl.mathhub.info/MMT/LATIN2/tree/devel/elpi. Accessed 26 Jan 2020

[HHP93] Harper, R., Honsell, F., Plotkin, G.: A framework for defining logics. J. Assoc. Comput. Mach. **40**(1), 143–184 (1993)

[KK03] Kohlhase, M., Koller, A.: Resource-adaptive model generation as a performance model. Log. J. IGPL **11**(4), 435–456 (2003). http://jigpal.oxfordjournals.org/cgi/content/abstract/11/4/435

[Koh+20] Kohlhase, M., et al.: Logic-independent proof search in logical frameworks (extended report). Extended Report of Conference Submission (2020). https://kwarc.info/kohlhase/submit/mmtelpi.pdf

[KS19] Kohlhase, M., Schaefer, J.F.: GF + MMT = GLF - from language to semantics through LF. In: Miller, D., Scagnetto, I., (eds.) Proceedings of the Fourteenth Workshop on Logical Frameworks and Meta-Languages: Theory and Practice (LFMTP 2019). Electronic Proceedings in Theoretical Computer Science (EPTCS), vol. 307, pp. 24–39 (2019). https://doi.org/10.4204/EPTCS.307.4

[LATIN] LATIN2 - Logic Atlas Version 2. https://gl.mathhub.info/MMT/LATIN2. Accessed 02 June 2017

[Mil] Miller, D: λProlog. http://www.lix.polytechnique.fr/Labo/Dale.Miller/lProlog/

[Pau93] Paulson, L.C.: Isabelle: The Next 700 Theorem Provers. In: arXiv CoRR cs.LO/9301106 (1993). https://arxiv.org/abs/cs/9301106

[Pau94] Paulson, L.C. (ed.): Isabelle: A Generic Theorem Prover. LNCS, vol. 828. Springer, Heidelberg (1994). https://doi.org/10.1007/BFb0030541

[Rab17] Rabe, F.: How to identify, translate, and combine logics? J. Log. Comput. **27**(6), 1753–1798 (2017)

[Rab18] Rabe, F.: A modular type reconstruction algorithm. ACM Trans. Comput. Log. **19**(4), 1–43 (2018)

[Ran04] Ranta, A.: Grammatical framework, a type-theoretical grammar formalism. J. Funct. Programm. **14**(2), 145–189 (2004)

[SCT15] Coen, C.S., Tassi, E.: The ELPI system (2015). https://github.com/LPCIC/elpi

[TSK] Tishkovsky, D., Schmidt, R.A., Khodadadi, M.: MetTeL2: towards a tableau prover generation platform. In: Proceedings of the Third Workshop on Practical As- pects of Automated Reasoning (PAAR-2012), p. 149 (2012)

Layered Clause Selection for Theory Reasoning (Short Paper)

Bernhard Gleiss[1(✉)] and Martin Suda[2]

[1] TU Wien, Vienna, Austria
bgleiss@forsyte.at
[2] Czech Technical University in Prague, Prague, Czech Republic

Abstract. Explicit theory axioms are added by a saturation-based theorem prover as one of the techniques for supporting theory reasoning. While simple and effective, adding theory axioms can also pollute the search space with many irrelevant consequences. As a result, the prover often gets lost in parts of the search space where the chance to find a proof is low. In this paper, we describe a new strategy for controlling the amount of reasoning with explicit theory axioms. The strategy refines a recently proposed two-layer-queue clause selection and combines it with a heuristic measure of the amount of theory reasoning in the derivation of a clause. We implemented the new strategy in the automatic theorem prover VAMPIRE and present an evaluation showing that our work dramatically improves the state-of-the-art clause-selection strategy in the presence of theory axioms.

1 Introduction

Thanks to recent advances, saturation-based theorem provers are increasingly used to reason about problems requiring quantified theory-reasoning [4,6]. One of the standard techniques to enable such reasoning is to automatically add first-order axiomatisations of theories detected in the input [14,18]. For example, (incomplete) axiomatisations of integer and real arithmetic or McCarthy's axioms of the theory of arrays [15] are routinely used. While this simple technique is often effective, we observed (see also [21]) two problems inherent to the solution: First, explicit axioms blow up the search space in the sense that a huge amount of consequences can additionally be generated. This happens since theory axioms are often repeatedly combined with certain clauses or among themselves, effectively creating cyclic patterns in the derivation. Most of these consequences would immediately be classified as practically useless by humans. Second, many of the resulting consequences have small weight. This has the unfortunate effect that the age-weight clause selection heuristic [16], predominantly used by saturation-based theorem provers for guiding the exploration of the search-space, often selects these theory-focused consequences. This way the prover is getting lost in parts of the search space where the chance of finding a proof is low.

N. Peltier and V. Sofronie-Stokkermans (Eds.): IJCAR 2020, LNAI 12166, pp. 402–409, 2020.
https://doi.org/10.1007/978-3-030-51074-9_23

In this paper, we propose to limit the exploration of theory-focused consequences by extending clause selection to take into account the amount of theory reasoning in the derivation of a clause. Our solution consists of two parts. First, we propose an efficiently computable feature of clauses, which we call *th-distance*, that measures the amount of theory reasoning in the derivation of a clause (Sect. 3). Second, we turn to the general problem of incorporating a feature to a clause selection strategy. There has been an ongoing interest in this problem [24,25,28]. We take inspiration from the layered clause selection approach presented in [28] and introduce the refined notion of *multi-split queues*, which present a principled solution to the incorporation problem (Sect. 2). We finally obtain a clause selection strategy for theory reasoning by instantiating multi-split queues with the feature *th-distance*. We implemented the resulting clause selection in the state-of-the-art saturation-based theorem prover VAMPIRE [14], and evaluate its benefits on a relevant subset of the SMT-LIB benchmark (Sect. 4).

Related Work. There are different approaches to adding support for theory reasoning to saturation-based theorem provers, either by extending the prover's inference system with dedicated inference rules [2,10,12,13] or using even more fundamental design changes [1,7,20,22]. While such solutions can result in very efficient reasoning procedures, their development is incredibly challenging and their implementation is a huge effort. As a result, only a few theories are covered by such approaches, in contrast to our technique, which applies to arbitrary theories. In particular, our technique can be used by non-experts on custom theory-domains coming from applications for which no dedicated solution exists. Our work has similar motivation to [21], where the authors use the set-of-support strategy [30] to limit the amount of reasoning performed with pure theory consequences. However, unlike our technique, they do not impose any limit on clauses whose derivation contains at least one non-theory-axiom.

Contributions. The summarized contributions of this paper are:

- A new approach for building clause selection strategies from clause features, based on *multi-split queues*.
- A new clause selection strategy for theory reasoning based on the instantiation of multi-split queues with the *th-distance*-feature measuring the amount of theory reasoning in the derivation of a clause. Our solution applies to arbitrary theories and does not require fundamental changes to the implementation of clause selection.
- An implementation of the introduced clause selection strategy in the state-of-the-art theorem prover VAMPIRE.
- An experimental evaluation confirming the effectiveness of the technique, by improving on the existing heuristics by up to 37 % on a relevant set of benchmarks.

2 Layered Clause Selection

We assume the reader to be familiar with the saturation-based theorem proving technology (see, e.g. [3,17]) and, in particular, with clause selection, the procedure for deciding, at each iteration of a saturation algorithm, which of the currently passive clauses to next select for activation, i.e. for participation in inferences with the previously activated clauses. To agree on terminology, we start this section by recalling clause selection by age and weight. We then move on to explaining layered clause selection.

The two most important features of a clause for clause selection are 1) its *age*, typically implemented using an ever-increasing "date of birth" timestamp, and 2) *weight*, which refers to the number of symbols occurring in the clause. A theorem prover prefers to select clauses that are old, which implicitly corresponds to a breadth-first search strategy, and clauses that are light, which is a form of best-first search (clauses with few symbols are cheaper to process, tend to be stronger simplifiers, and are intuitively closer to the ultimate target, the empty clause). In practice, the best performance is achieved by combining these two criteria [16,25]. This is achieved by storing the passive clauses in two *queues*, one sorted by age and the other by weight, and setting a ratio to specify how the selection alternates between picking from these two queues.

Layered Selection. In the system description of GKC [28], Tammet describes an idea of using two layers of queues to organise clause selection. The first layer relies on the just-described combination of selection by age and weight. In the second layer, clauses are split into disjoint groups using a certain property (e.g., "being derived from the *goal* or not" could define two groups), each group is represented by two sub-queues of the first layer, and the decision from which group to select the next clause is dictated by a new second-layer ratio. Although Tammet does not expand much on the insights behind using the layered approach, he reports it highly beneficial for the performance of GKC. In our understanding, the additional layer (in principle, there could be more than two) provides a clean way of incorporating into clause selection a new notion of what a preferred clause should be, without a priori disturbing the already established and tuned primary approach, such as selection by age and weight.[1]

Our preliminary experiments with the idea (instantiated with the derived-from-the-goal property) found it useful, but not as powerful as other goal-directed heuristics in VAMPIRE. In particular, finding a universally good ratio between the "good" clauses and the "bad" ones seemed hard. What we propose here instead (and what also led in our experiment to a greater performance gain) is to instead organise the clauses into groups with "good" ones and "all". Here the second group contains all the passive clauses and essentially represents a fallback to the original single-layer strategy. The advantage of this new take on layered selection is that a bad clause is only selected if 1) it is time to try a bad

[1] A known alternative [25] is to adapt the formula for computing weight to include a term for penalising bad clauses and still rely on selection by age and this new refined notion of weight. (See also the *non-goal weight coefficient* in [27].).

clause according to the second-layer ratio and 2) the best bad clause is also the current overall best according to the age-weight perspective. This makes picking a good second-layer ratio much easier. In particular, one can "smoothly" move (by changing the second-layer ratio) from a high preference for the "all" second-layer queue towards selecting more "good" clauses without necessarily having to select any "bad" ones.

Multi-split Queues. We propose multi-split queues to realize layered selection with second layer groups defined by a real-valued clause feature.

Definition 1. *Let μ be a real-valued clause evaluation feature such that preferable clauses have low value of $\mu(C)$. Let the* cutoffs $c_1, \ldots, c_k$ *be monotonically increasing real numbers with $c_k = \infty$, and let the ratio $r_1 : \ldots : r_k$ be a list of positive integer values. These together determine a layered selection scheme with k groups $\mathcal{C}_i = \{C | \mu(C) \leq c_i\}$ for $i = 1, \ldots, k$, such that we select from the i-th group with a frequency $r_i / (\Sigma_{j=1}^{k} r_j)$.*

It is easy to see that multi-split queues generalise the binary "good" vs "all" arrangement, since, thanks to monotonicity of the cutoffs, we have $\mathcal{C}_i \subseteq \mathcal{C}_{i+1}$. Moreover, since $c_k = \infty$, $\mathcal{C}_k$ will contain all the passive clauses.

3 Theory Part

In this section, we instantiate the idea of multi-split queues from Sect. 2 with a concrete clause evaluation feature, which measures the amount of theory reasoning in the derivation of a clause. We assume that the initial clauses given to the saturation algorithm, which we simply refer to as *axioms*, consists of non-theory axioms obtained by classifying the input problem and theory axioms added to facilitate theory reasoning.

We start by defining the fraction of theory reasoning in the derivation of a general clause. This relies on counting the number of theory axioms, resp. the number of all axioms, in the derivation-tree using running sums.

Definition 2. *For a theory axiom C, define both $thAx(C)$ and $allAx(C)$ as 1. For a non-theory axiom C, define $thAx(C)$ as 0 and $allAx(C)$ as 1. For a derived clause C with parent clauses $C_1, \ldots, C_n$, define $thAx(C)$ as $\sum_i thAx(C_i)$ and $allAx(C)$ as $\sum_i allAx(C_i)$. Finally, we set $frac(C) := thAx(C)/allAx(C)$.*

Assume now that for a given problem we expect (based on domain knowledge and experience) the fraction of theory reasoning in the final refutation *frac*($\bot$) to be at most $1/d$, for a positive integer d. Our clause evaluation feature *th-distance* measures how much *frac*(C) exceeds the expected "maximally allowed" fraction $1/d$. More precisely, *th-distance* counts the number of non-theory axioms which the derivation of C would additionally need to contain to achieve a ratio $1/d$.

Definition 3. *The th-distance: Clauses $\rightarrow \mathbb{N}$ is defined as*

$$th-distance(C) := max(thAx(C) \cdot d - allAx(C), 0).$$

Our heuristic is based on the idea that a clause with small *th-distance* is more likely to contribute to the refutation than a clause with high *th-distance*. We therefore want to ensure that clause selection focuses on selecting clauses C with a low value *th-distance*(C). We realize this with the multi-split queues (see Sect. 2), instantiating the clause evaluation feature μ by *th-distance*, resulting in a second layer clause selection strategy with parameters d, $c_1, \ldots, c_k$ and $r_1 : \ldots : r_k$.

4 Experiments

We implemented the heuristic described in Sect. 3 in VAMPIRE (version 4.4). Our newly added implementation consists of about 900 lines of C++ code and is compatible with both the LRS saturation algorithm [23] and AVATAR [29].

For evaluation, we used the following subset of the most recent version (as of January 2020) of SMTLIB [5]: We took all the problems from the sub-logics that contain quantification and theories, such as LIA, LRA, NRA, ALIA, UFDT, ... except for those requiring bit-vector (BV) or floating-point (FP) reasoning, currently not supported by VAMPIRE. Subsequently, we excluded problems known to be satisfiable and those that were provable using VAMPIRE's default strategy in 10 s either without adding theory axioms or while performing clause selection by age only. This way, we obtained 20 795 problems.[2]

Table 1. Comparing clause selection strategies on VAMPIRE's default configuration.

Strategy	d-value	Cutoffs	Ratio	Refuted	Δbase	Δbase%
`default`	–	–	–	886	0	0.0
`layered2`	10	$23, \infty$	33:8	1112	226	25.5
`layered3`	7	$0, 30, \infty$	16:8:1	1170	284	32.1
`layered4`	8	$16, 41, 59, \infty$	84:9:2:2	1176	290	32.7

As a first experiment, we compared the number of problems solved in 10 s by the default strategy[3] and its various extensions by multi-split queues defined in Sect. 3.[4] The d-value, cutoffs and ratio values for the heuristic were selected by educated guessing and randomised hill-climbing. Table 1 lists results of the best obtained configurations. It can be seen that already with two second layer queues a substantial improvement of 25.5% over the default is achieved. Moreover, while it is increasingly more difficult to choose good values for the many parameters defining a configuration with multiple queues, their use further significantly improves the number of problems solved.

[2] A list of the selected problems along with other information needed to reproduce our experiments can be found at https://git.io/JvqhP.

[3] The default strategy uses AVATAR [29], the LRS saturation algorithm [23] and an age-weight ratio of 1:1.

[4] The experiment was run on our local server with Intel Xeon 2.3 GHz processors.

Table 2. Comparing clause selection strategies on VAMPIRE's portfolio configuration.

Strategy	*d*-value	Cutoffs	Ratio	Refuted	Uniques
`SMTCOMP2019`	–	–	–	5479	194
`SMTCOMP2019+layered4`	8	$16, 41, 59, \infty$	84:9:2:2	5629	344

In a second experiment,[5] we ran VAMPIRE's strategy schedule for SMTCOMP 2019 [11] on our problems and also the same schedule additionally imposing the most successful second-layer clause selection scheme `layered4` from the first experiment. The time limit was 500 s per problem. Table 2 shows the results.

We can see that the version with second-layer queues improved over the standard schedule by 150 solved problems. This is a very significant result, suggesting the achieved control of theory reasoning is incredibly helpful. Moreover, one should keep in mind that strategies in a schedule are carefully selected to complement each other and even locally good changes in the strategies often destroy this complementarity (cf., e.g., [19,21]). In our case, however, we achieve an improvement despite this looming negative effect. Finally, it is very likely that a new schedule, constructed while taking our new technique into account, will be able to additionally cover some of the 194 problems currently only solved by the unaltered schedule.

5 Conclusion

We introduced a new clause selection heuristic for reasoning in the presence of explicit theory axioms. The heuristic is based on the combination of multi-split queues and a new clause-feature measuring the amount of theory reasoning in the derivation of a clause. Our experiments show that the new heuristic significantly improves the existing state-of-the-art clause selection strategy. As future work, we want to extend layered clause selection with new clause-features and combine it with the machine-learning-based approach in the style of ENIGMA [8].

References

1. Althaus, E., Kruglov, E., Weidenbach, C.: Superposition modulo linear arithmetic SUP(LA). In: Ghilardi, S., Sebastiani, R. (eds.) FroCoS 2009. LNCS (LNAI), vol. 5749, pp. 84–99. Springer, Heidelberg (2009). https://doi.org/10.1007/978-3-642-04222-5_5
2. Bachmair, L., Ganzinger, H.: Ordered chaining calculi for first-order theories of transitive relations. J. ACM **45**(6), 1007–1049 (1998)
3. Bachmair, L., Ganzinger, H., McAllester, D.A., Lynch, C.: Resolution theorem proving. In: Robinson, J.A., Voronkov, A. (eds.) Handbook of Automated Reasoning, vol. 2, pp. 19–99. Elsevier and MIT Press, Cambridge (2001)

[5] The second experiment was run on the StarExec cluster [26] with 2.4 GHz processors.

4. Backes, J., et al.: Reachability analysis for AWS-based networks. In: Dillig, I., Tasiran, S. (eds.) CAV 2019. LNCS, vol. 11562, pp. 231–241. Springer, Cham (2019). https://doi.org/10.1007/978-3-030-25543-5_14
5. Barrett, C., Fontaine, P., Tinelli, C.: The Satisfiability Modulo Theories Library (SMT-LIB) (2016). www.SMT-LIB.org
6. Barthe, G., Eilers, R., Georgiou, P., Gleiss, B., Kovcs, L., Maffei, M.: Verifying relational properties using trace logic. In: 2019 Formal Methods in Computer Aided Design (FMCAD), pp. 170–178, October 2019
7. Baumgartner, P., Waldmann, U.: Hierarchic superposition with weak abstraction. In: Bonacina, M.P. (ed.) CADE 2013. LNCS (LNAI), vol. 7898, pp. 39–57. Springer, Heidelberg (2013). https://doi.org/10.1007/978-3-642-38574-2_3
8. Chvalovský, K., Jakubuv, J., Suda, M., Urban, J.: ENIGMA-NG: efficient neural and gradient-boosted inference guidance for E. In: Fontaine [9], pp. 197–215
9. Fontaine, P. (ed.): CADE 2019. LNCS (LNAI), vol. 11716. Springer, Cham (2019). https://doi.org/10.1007/978-3-030-29436-6
10. Gupta, A., Kovács, L., Kragl, B., Voronkov, A.: Extensional crisis and proving identity. In: Cassez, F., Raskin, J.-F. (eds.) ATVA 2014. LNCS, vol. 8837, pp. 185–200. Springer, Cham (2014). https://doi.org/10.1007/978-3-319-11936-6_14
11. Hadarean, L., Hyvarinen, A., Niemetz, A., Reger, G.: 14th International Satisfiability Modulo Theories Competition (SMT-COMP 2019) (2019). https://smt-comp.github.io/2019/
12. Kotelnikov, E., Kovács, L., Reger, G., Voronkov, A.: The vampire and the FOOL. In: Avigad, J., Chlipala, A. (eds.) Proceedings of the 5th ACM SIGPLAN Conference on Certified Programs and Proofs, Saint Petersburg, FL, USA, 20–22 January 2016, pp. 37–48. ACM (2016)
13. Kovács, L., Robillard, S., Voronkov, A.: Coming to terms with quantified reasoning. In: Proceedings of the 44th ACM SIGPLAN Symposium on Principles of Programming Languages. POPL 2017, New York, NY, USA, pp. 260–270. Association for Computing Machinery (2017)
14. Kovács, L., Voronkov, A.: First-order theorem proving and VAMPIRE. In: Sharygina, N., Veith, H. (eds.) CAV 2013. LNCS, vol. 8044, pp. 1–35. Springer, Heidelberg (2013). https://doi.org/10.1007/978-3-642-39799-8_1
15. Mccarthy, J.: Towards a mathematical science of computation. In: IFIP Congress, pp. 21–28. North-Holland (1962)
16. McCune, W.: Otter 3.0 reference manual and guide. Technical report ANL-94/6, Argonne National Laboratory (1994)
17. Overbeek, R.A.: A new class of automated theorem-proving algorithms. J. ACM **21**(2), 191–200 (1974)
18. Prevosto, V., Waldmann, U.: SPASS+T. In: Sutcliffe, G., Schmidt, R., Schulz, S. (eds.) Proceedings of the FLoC 2006 Workshop on Empirically Successful Computerized Reasoning, 3rd International Joint Conference on Automated Reasoning, number 192 in CEUR Workshop Proceedings, pp. 19–33 (2006)
19. Rawson, M., Reger, G.: Old or heavy? Decaying gracefully with age/weight shapes. In: Fontaine [9], pp. 462–476
20. Reger, G., Bjorner, N., Suda, M., Voronkov, A.: AVATAR modulo theories. In: Benzmüller, C., Sutcliffe, G., Rojas, R. (eds.) GCAI 2016. 2nd Global Conference on Artificial Intelligence, EPiC Series in Computing, vol. 41, pp. 39–52. EasyChair (2016)

21. Reger, G., Suda, M.: Set of support for theory reasoning. In: Eiter, T., Sands, D., Sutcliffe, G., Voronkov, A. (eds.) IWIL@LPAR 2017 Workshop and LPAR-21 Short Presentations, Maun, Botswana, 7–12 May 2017, vol. 1. Kalpa Publications in Computing. EasyChair (2017)
22. Reger, G., Suda, M., Voronkov, A.: Unification with abstraction and theory instantiation in saturation-based reasoning. In: Beyer, D., Huisman, M. (eds.) TACAS 2018. LNCS, vol. 10805, pp. 3–22. Springer, Cham (2018). https://doi.org/10.1007/978-3-319-89960-2_1
23. Riazanov, A., Voronkov, A.: Limited resource strategy in resolution theorem proving. J. Symb. Comput. **36**(1–2), 101–115 (2003)
24. Schulz, S., Cruanes, S., Vukmirovic, P.: Faster, higher, stronger: E 2.3. In Fontaine [9], pp. 495–507
25. Schulz, S., Möhrmann, M.: Performance of clause selection heuristics for saturation-based theorem proving. In: Olivetti, N., Tiwari, A. (eds.) IJCAR 2016. LNCS (LNAI), vol. 9706, pp. 330–345. Springer, Cham (2016). https://doi.org/10.1007/978-3-319-40229-1_23
26. Stump, A., Sutcliffe, G., Tinelli, C.: StarExec, a cross community logic solving service (2012). https://www.starexec.org
27. Suda, M.: Aiming for the goal with SInE. In: Kovacs, L., Voronkov, A. (eds.) Vampire 2018 and Vampire 2019. The 5th and 6th Vampire Workshops. EPiC Series in Computing, vol. 71, pp. 38–44. EasyChair (2020)
28. Tammet, T.: GKC: a reasoning system for large knowledge bases. In: Fontaine [9], pp. 538–549
29. Voronkov, A.: AVATAR: the architecture for first-order theorem provers. In: Biere, A., Bloem, R. (eds.) CAV 2014. LNCS, vol. 8559, pp. 696–710. Springer, Cham (2014). https://doi.org/10.1007/978-3-319-08867-9_46
30. Wos, L., Robinson, G.A., Carson, D.F.: Efficiency and completeness of the set of support strategy in theorem proving. J. ACM **12**(4), 536–541 (1965)

Non Classical Logics

Description Logics with Concrete Domains and General Concept Inclusions Revisited

Franz Baader(✉) and Jakub Rydval

Institute of Theoretical Computer Science, TU Dresden, Dresden, Germany
{franz.baader,jakub.rydval}@tu-dresden.de

Abstract. Concrete domains have been introduced in the area of Description Logic to enable reference to concrete objects (such as numbers) and predefined predicates on these objects (such as numerical comparisons) when defining concepts. Unfortunately, in the presence of general concept inclusions (GCIs), which are supported by all modern DL systems, adding concrete domains may easily lead to undecidability. One contribution of this paper is to strengthen the existing undecidability results further by showing that concrete domains even weaker than the ones considered in the previous proofs may cause undecidability. To regain decidability in the presence of GCIs, quite strong restrictions, in sum called ω-admissibility, need to be imposed on the concrete domain. On the one hand, we generalize the notion of ω-admissibility from concrete domains with only binary predicates to concrete domains with predicates of arbitrary arity. On the other hand, we relate ω-admissibility to well-known notions from model theory. In particular, we show that finitely bounded, homogeneous structures yield ω-admissible concrete domains. This allows us to show ω-admissibility of concrete domains using existing results from model theory.

Keywords: Description logic · Concrete domains · GCIs · ω-admissibility · Homogeneity · Finite boundedness · Decidability · Constraint satisfaction

1 Introduction

Description Logics (DLs) [3,7] are a well-investigated family of logic-based knowledge representation languages, which are frequently used to formalize ontologies for application domains such as the Semantic Web [27] or biology and medicine [26]. To define the important notions of such an application domain as formal concepts, DLs state necessary and sufficient conditions for an individual to belong to a concept. These conditions can be Boolean combinations of atomic properties required for the individual (expressed by concept names) or

Supported by DFG GRK 1763 (QuantLA) and TRR 248 (cpec, grant 389792660).

N. Peltier and V. Sofronie-Stokkermans (Eds.): IJCAR 2020, LNAI 12166, pp. 413–431, 2020.
https://doi.org/10.1007/978-3-030-51074-9_24

properties that refer to relationships with other individuals and their properties (expressed as role restrictions). For example, the concept of a father that has only daughters can be formalized by the concept description

$$C := \neg \mathit{Female} \sqcap \exists \mathit{child}.\mathit{Human} \sqcap \forall \mathit{child}.\mathit{Female},$$

which uses the concept names *Female* and *Human* and the role name *child* as well as the concept constructors negation ($\neg$), conjunction ($\sqcap$), existential restriction ($\exists r.D$), and value restriction ($\forall r.D$). The GCIs

$$\mathit{Human} \sqsubseteq \forall \mathit{child}.\mathit{Human} \quad \text{and} \quad \exists \mathit{child}.\mathit{Human} \sqsubseteq \mathit{Human}$$

say that humans have only human children, and they are the only ones that can have human children.

DL systems provide their users with reasoning services that allow them to derive implicit knowledge from the explicitly represented one. In our example, the above GCIs imply that elements of our concept C also belong to the concept $D := \mathit{Human} \sqcap \forall \mathit{child}.\mathit{Human}$, i.e., C is subsumed by D w.r.t. these GCIs. A specific DL is determined by which kind of concept constructors are available. A major goal of DL research was and still is to find a good compromise between expressiveness and the complexity of reasoning, i.e., to locate DLs that are expressive enough for interesting applications, but still have inference problems (like subsumption) that are decidable and preferably of a low complexity. For the DL $\mathcal{ALC}$, in which all the concept descriptions used in the above example can be expressed, the subsumption problem w.r.t. GCIs is ExpTime-complete [7].

Classical DLs like $\mathcal{ALC}$ cannot refer to concrete objects and predefined relations over these objects when defining concepts. For example, a constraint stating that parents are strictly older than their children cannot be expressed in $\mathcal{ALC}$. To overcome this deficit, a scheme for integrating certain well-behaved concrete domains, called admissible, into $\mathcal{ALC}$ was introduced in [4], and it was shown that this integration leaves the relevant inference problems (such as subsumption) decidable. Basically, admissibility requires that the set of predicates of the concrete domain is closed under negation and that the constraint satisfaction problem (CSP) for the concrete domain is decidable. However, in this setting, GCIs were not considered since they were not a standard feature of DLs then,[1] though a combination of concrete domains and GCIs would be useful in many applications. For example, using the syntax employed in [33] and also in the present paper, the above constraint regarding the age of parents and their children could be expressed by the GCI $\mathit{Human} \sqcap \exists \mathit{child}\, \mathit{age}, \mathit{age}.[>] \sqsubseteq \bot$, which says that there cannot be a human whose age is smaller than the age of one of his or her children. Here $\bot$ is the bottom concept, which is always interpreted as the empty set, *age* is a concrete feature that maps from the abstract domain populating concepts into the concrete domain of natural numbers, and $>$ is the usual greater predicate on the natural numbers.

[1] Actually, they were introduced (with a different name) at about the same time as concrete domains [2,38].

A first indication that concrete domains might be harmful for decidability was given in [6], where it was shown that adding transitive closure of roles to $\mathcal{ALC}(\mathbf{R})$, the extension of $\mathcal{ALC}$ by an admissible concrete domain $\mathbf{R}$ based on real arithmetics, renders the subsumption problem undecidable. The proof of this result uses a reduction from the Post Correspondence Problem (PCP). It was shown in [31] that this proof can be adapted to the case where transitive closure of roles is replaced by GCIs, and it actually works for considerably weaker concrete domain, such as the rational numbers $\mathbb{Q}$ or the natural numbers $\mathbb{N}$ with a unary predicate $=_0$ for equality with zero, a binary equality predicate $=$, and a unary predicate $+_1$ for incrementation. In [7] it is shown, by a reduction from the halting problem of two-register machines, that undecidability even holds without binary equality. In the present paper, we will improve on this result by showing that, even if $=_0$ is removed as well, undecidability still holds, and that the same is true if we replace $+_1$ with $+$.

To regain decidability, one can either impose syntactic restriction on how the DL can interact with the concrete domain [22,36]. The main idea is here to disallow paths (such as *child age* in our example), which has the effect that concrete domain predicates cannot compare properties (such as the age) of different individuals. The other option is to impose stronger restrictions than admissibility on the concrete domain. After first positive results for specific concrete domains (e.g., a concrete domain over the rational numbers with order and equality [30,32]), the notion of ω-admissible concrete domains was introduced in [33], and it was shown (by designing a tableau-based decision procedure) that integrating such a concrete domain into $\mathcal{ALC}$ leaves reasoning decidable also in the presence of GCIs. In the present paper, we generalize the notion of ω-admissibility and the decidability result from concrete domains with only binary predicates as in [33] to concrete domains with predicates of arbitrary arity. But the main contribution of this paper is to show that there is a close relationship between ω-admissibility and well-known notions from model theory. In particular, we show that finitely bounded, homogeneous structures yield ω-admissible concrete domains. This allows us to locate new ω-admissible concrete domains using existing results from model theory. Due to space constraints, we cannot prove all our results in detail here. Complete proofs and more examples of ω-admissible concrete domains can be found in [8].

2 Preliminaries

We write $[n]$ for the set $\{1, \ldots, n\}$. Given a set A, the *diagonal* on A is defined as the binary relation $\Delta_A := \{(a, a) \mid a \in A\}$. The *kernel* of a mapping $f \colon A \to B$, denoted by $\ker f$, is the equivalence relation $\{(a, a') \in A \times A \mid f(a) = f(a')\}$.

From a mathematical point of view, concrete domains are relational structures. A *(relational) signature* τ is a set of *predicate symbols*, each with an associated natural number called its *arity*. A *relational τ-structure* $\mathbf{A}$ consists of a set A (the *domain*) together with relations $R^{\mathbf{A}} \subseteq A^k$ for each k-ary symbol $R \in \tau$. We often describe structures by listing their domain and relations, e.g., we write $\mathbf{Q} = (\mathbb{Q}; <)$ for the relational structure whose domain is the set of rational numbers $\mathbb{Q}$, and which has the usual smaller relation $<$ on $\mathbb{Q}$ as its only relation.[2] An *expansion* of the τ-structure $\mathbf{A}$ is a σ-structure $\mathbf{B}$ with $A = B$, $\tau \subseteq \sigma$, and $R^{\mathbf{B}} = R^{\mathbf{A}}$ for each relation symbol $R \in \tau$. Conversely, we call $\mathbf{A}$ a reduct of $\mathbf{B}$.

One possibility to obtain an expansion of a τ-structure is to use formulas of *first-order logic* (FO) over the signature τ to define new predicates, where a formula with k free variables defines a k-ary predicate in the obvious way. We assume that equality $=$ as well as the symbol `false` for falsity is always available. Thus, atomic formulas are of the form `false`, $x_i = x_j$, and $R(x_1, \ldots, x_k)$ for some k-ary $R \in \tau$ and variables $x_1, \ldots, x_k$. The FO *theory* of a structure is the set of all FO sentences that are true in the structure. In addition to full FO, we also use standard fragments of FO such as the *existential positive* ($\exists^+$), the *quantifier-free* (qf), and the *primitive positive* (pp) *fragment*. The existential positive fragment consists of formulas built using conjunction, disjunction, and existential quantification only. The quantifier-free fragment consists of Boolean combinations of atomic formulas, and the primitive positive fragment of existentially quantified conjunctions of atomic formulas. Let Σ be a set of FO formulas and $\mathbf{D}$ a structure. We say that a relation over D has a Σ *definition* in $\mathbf{D}$ if it is of the form $\{t \in D^k \mid \mathbf{D} \models \phi(t)\}$ for some $\phi \in \Sigma$. We refer to this relation by $\phi^{\mathbf{D}}$. For example, the formula $y < x \vee x = y$ is an existential positive formula and, interpreted in the structure $\mathbf{Q}$, it clearly defines the binary relation $\geq$ on $\mathbb{Q}$. This shows that $\geq$ is $\exists^+$ definable in $\mathbf{Q}$. An example of a pp formula is the formula $\exists y.\, x = y$, which defines the unary relation interpreted as the whole domain $\mathbb{Q}$.

A *homomorphism* $h\colon \mathbf{A} \to \mathbf{B}$ for τ-structures $\mathbf{A}, \mathbf{B}$ is a mapping $h\colon A \to B$ that *preserves* each relation of $\mathbf{A}$, i.e., $(a_1, \ldots, a_k) \in R^{\mathbf{A}}$ for some k-ary relation symbol $R \in \tau$ implies $(h(a_1), \ldots, h(a_k)) \in R^{\mathbf{B}}$. We write $\mathbf{A} \to \mathbf{B}$ if $\mathbf{A}$ homomorphically maps to $\mathbf{B}$ and $\mathbf{A} \not\to \mathbf{B}$ otherwise. We say that $\mathbf{A}$ and $\mathbf{B}$ are *homomorphically equivalent* if $\mathbf{A} \to \mathbf{B}$ and $\mathbf{B} \to \mathbf{A}$. An *endomorphism* is a homomorphism from a structure to itself. By an *embedding* we mean an injective homomorphism that additionally satisfies the only if direction in the definition of a homomorphism, i.e., it also preserves the complements of relations. We write $\mathbf{A} \hookrightarrow \mathbf{B}$ if $\mathbf{A}$ embeds into $\mathbf{B}$. A *substructure* of $\mathbf{A}$ is a structure $\mathbf{B}$ over $B \subseteq A$ such that the natural inclusion map $i\colon A \to B$ is an embedding. We call $\mathbf{A}$ an *extension* of $\mathbf{B}$. An *isomorphism* is a surjective embedding. We say that two structures $\mathbf{A}$ and $\mathbf{B}$ are *isomorphic* and write $\mathbf{A} \cong \mathbf{B}$ if there exists an isomorphism from $\mathbf{A}$ to $\mathbf{B}$. An *automorphism* is an isomorphism from a structure into itself.

The definition of admissibility of a concrete domain in [4] requires that the constraint satisfaction problem for this structure is decidable. Let $\mathbf{D}$ be a structure with a finite relational signature τ. The *constraint satisfaction problem* of $\mathbf{D}$, short CSP($\mathbf{D}$), is the following decision problem:

[2] By a slight abuse of notation, we use $<$ instead of $<^{\mathbf{Q}}$ to denote also the interpretation of the predicate symbol $<$ in $\mathbf{Q}$.

INPUT: A finite τ-structure $\mathbf{A}$.
QUESTION: Does $\mathbf{A}$ homomorphically map to $\mathbf{D}$?

Formally, CSP($\mathbf{D}$) is the class of all finite τ-structures that homomorphically map to $\mathbf{D}$. We call $\mathbf{D}$ the *template* of CSP($\mathbf{D}$). A *solution* for an instance $\mathbf{A}$ of the CSP is a homomorphism $h\colon \mathbf{A} \to \mathbf{D}$.

It is easy to see that deciding whether a CSP instance admits a solution amounts to evaluating a pp sentences in the template and vice versa [10]. For example, verifying whether the structure $\mathbf{A} = (\{a_1, a_2, a_3\}; <^{\mathbf{A}})$ with $<^{\mathbf{A}} := \{(a_1, a_2), (a_2, a_3), (a_3, a_1)\}$ homomorphically maps into $\mathbf{Q}$ is the same as checking whether the pp sentence $\exists x_1.\exists x_2.\exists x_3.(x_1 < x_2 \wedge x_2 < x_3 \wedge x_3 < x_1)$ is true in $\mathbf{Q}$.

The CSP for $\mathbf{Q}$ is in P since a structure $\mathbf{A} = (A, <^{\mathbf{A}})$ can homomorphically be mapped into $\mathbb{Q}$ iff it does not contain a $<$-cycle, i.e., there are is no $n \geq 1$ and elements $a_0, \dots, a_{n-1} \in A$ such that $a_0 <^{\mathbf{A}} \dots <^{\mathbf{A}} a_{n-1} <^{\mathbf{A}} a_0$. Testing whether such a cycle exists can be done in logarithmic space since it requires solving the reachability problem in a directed graph (digraph). In the example above, we obviously have a cycle, and thus this instance of CSP($\mathbf{Q}$) has no solution.

The definition of admissibility in [4] actually also requires that the predicates are closed under negation and that there is a predicate for the whole domain. We have already seen that the negation $\geq$ of $<$ is $\exists^+$ definable in $\mathbf{Q}$ and that the predicate for the whole domain is pp definable. The negation of this predicate has the pp definition $x < x$. The following lemma implies that the expansion of $\mathbf{Q}$ by these predicates still has a decidable CSP.[3]

Lemma 1 ([10]). *Let* $\mathbf{C}, \mathbf{D}$ *be structures over the same domain with finite signatures. If the relations of* $\mathbf{C}$ *have a pp definition in* $\mathbf{D}$*, then CSP*($\mathbf{C}$) $\leq_{\text{PTIME}}$ *CSP*($\mathbf{D}$)*; if they have an* $\exists^+$ *definition in* $\mathbf{D}$*, then CSP*($\mathbf{C}$) $\leq_{\text{NPTIME}}$ *CSP*($\mathbf{D}$).

3 DLs with Concrete Domains

As in [4] and [33], we extend the well-known DL $\mathcal{ALC}$ with concrete domains. We assume that the reader is familiar with the syntax and semantics of $\mathcal{ALC}$ (as, e.g., defined in [7]), and thus only show how both need to be extended to accommodate a concrete domain $\mathbf{D}$. In the general definition, we allow reference to Σ definable predicates for a fragment Σ of FO rather than just the elements of τ. For technical reasons, we must, however, restrict the arities of definable predicates by a fixed upper bound d. Given a τ-structure $\mathbf{D}$ with finite relational signature τ, a set Σ of FO formulas over the signature τ, and a bound $d \geq 1$ on the arity of the Σ-definable predicates, we obtain a DL $\mathcal{ALC}^d_\Sigma(\mathbf{D})$, which extends $\mathcal{ALC}$ as follows.

From a syntactic point of view, we assume that the set of role names $\mathsf{N_R}$ contains a set of functional roles $\mathsf{N_{fR}} \subseteq \mathsf{N_R}$, and that in addition we have a set of

[3] The lemma actually only yields an NP decision procedure for this CSP, but it is easy to see that the above polynomial-time cycle-checking algorithm can be adapted such that it also works for the expanded structure.

feature names N_F, which provide the connection between the abstract and the concrete domain. A *path* is of the form $r\,f$ or f where $r \in N_R$ and $f \in N_F$. In our example in the introduction, *age* is a feature name and *child age* is a path. The DL $\mathcal{ALC}^d_\Sigma(\mathbf{D})$ extends $\mathcal{ALC}$ with two new concept constructors:

$$\exists p_1, \ldots, p_k. [\phi(x_1, \ldots, x_k)] \quad \text{and} \quad \forall p_1, \ldots, p_k. [\phi(x_1, \ldots, x_k)],$$

where $k \leq d$, $p_1, \ldots, p_k$ are paths, and $\phi(x_1, \ldots, x_k)$ is a formula in Σ with k free variables, defining a k-ary predicate on D. As usual, a TBox is defined to be a finite set of GCIs $C \sqsubseteq D$, where C, D are $\mathcal{ALC}^d_\Sigma(\mathbf{D})$ concept descriptions.

Regarding the semantics, we consider interpretations $\mathcal{I} = (\Delta^\mathcal{I}, \cdot^\mathcal{I})$, consisting of a non-empty set $\Delta^\mathcal{I}$ and an interpretation function $\cdot^\mathcal{I}$, which interprets concept names A as subsets $A^\mathcal{I}$ of $\Delta^\mathcal{I}$ and role names r as binary relations $r^\mathcal{I}$ on $\Delta^\mathcal{I}$, with the restriction that $r^\mathcal{I}$ is functional for $r \in N_{fR}$, i.e., $(d, e) \in r^\mathcal{I}$ and $(d, e') \in r^\mathcal{I}$ imply $e = e'$. In addition, features $f \in N_F$ are interpreted as functional binary relations $f^\mathcal{I} \subseteq \Delta^\mathcal{I} \times D$. We extend the interpretation function to paths of the form $p = r\,f$ by setting

$$p^\mathcal{I} = \{(d, d') \mid \text{there is } d'' \in \Delta^\mathcal{I} \text{ such that } (d, d'') \in r^\mathcal{I} \text{ and } (d'', d') \in f^\mathcal{I}\}.$$

For the concept constructors of $\mathcal{ALC}$, the extension of the interpretation function to complex concepts is defined in the usual way. The new concrete domain constructors are interpreted as follows:

$$(\exists p_1, \ldots, p_k. [\phi(x_1, \ldots, x_k)])^\mathcal{I} = \{d \in \Delta^\mathcal{I} \mid \text{there exist } d_1, \ldots, d_k \in D \text{ such that } (d, d_i) \in p_i^\mathcal{I} \text{ for all } i \in [k] \text{ and } \mathbf{D} \models \phi(d_1, \ldots, d_k)\},$$

$$(\forall p_1, \ldots, p_k. [\phi(x_1, \ldots, x_k)])^\mathcal{I} = \{d \in \Delta^\mathcal{I} \mid \text{for all } d_1, \ldots, d_k \in D \text{ we have that } (d, d_i) \in p_i^\mathcal{I} \text{ for all } i \in [k] \text{ implies } \mathbf{D} \models \phi(d_1, \ldots, d_k)\}.$$

As usual, an interpretation $\mathcal{I}$ is a model a TBox $\mathcal{T}$ if it satisfies all the GCIs in $\mathcal{T}$, where $\mathcal{I}$ satisfies the GCI $C \sqsubseteq D$ if $C^\mathcal{I} \subseteq D^\mathcal{I}$ holds. The $\mathcal{ALC}^d_\Sigma(\mathbf{D})$ concept description C is satisfiable w.r.t. $\mathcal{T}$ if there is a model $\mathcal{I}$ of $\mathcal{T}$ such that $C^\mathcal{I} \neq \emptyset$. Since all Boolean operators are available in $\mathcal{ALC}^d_\Sigma(\mathbf{D})$, the subsumption problem mentioned in the introduction and the satisfiability problem are inter-reducible in polynomial time [7].

As a convention, we write $\mathcal{ALC}(\mathbf{D})$ instead of $\mathcal{ALC}^d_\Sigma(\mathbf{D})$ if d is the maximal arity of the predicates in τ and Σ consists of all atomic τ-formulas not using the equality predicate.

3.1 Undecidable DLs with Concrete Domains

We show by a reduction from the halting problem of two-register machines that concept satisfiability in $\mathcal{ALC}(\mathbf{D})$ is undecidable if $\mathbf{D}$ is a structure with domain $\mathbb{Q}$, $\mathbb{Z}$, or $\mathbb{N}$ whose only predicate is the binary predicate $+_1$, which is interpreted as incrementation (i.e., it consists of the tuples $(m, m+1)$ for numbers m in the respective domain).

Our proof is an adaptation of the undecidability proof in [7] to the case where no zero test $=_0$ is available.[4] A *two-register machine* (2RM) is a pair (Q, P)

[4] A similar trick for zero test elimination is used in the proof of Proposition 1 in [16].

with *states* $Q = \{q_0, \ldots, q_\ell\}$ and a sequence of *instructions* $P = I_0, \ldots, I_{\ell-1}$. By definition, q_0 is the initial state and q_ℓ the halting state. In state q_i ($i < \ell$) the instruction I_i must be applied. Instructions come in two varieties. An *incrementation instruction* is of the form $I = +(p, q)$ where $p \in \{1, 2\}$ is the register number and q is a state. This instruction increments (the content of) register p and then goes to state q. A *decrementation instruction* is of the form $I = -(p, q, q')$ where $p \in \{1, 2\}$ and q, q' are states. This instruction decrements register p and goes to state q if the content of register p is not zero; otherwise, it leaves register p as it is and goes to state q'. It is well-known that the problem of deciding whether a given 2RM halts on input $(0, 0)$ is undecidable [35].

Proposition 1. *If* $\mathbf{D}$ *is* $(\mathbb{Q}, +_1)$, $(\mathbb{Z}, +_1)$, *or* $(\mathbb{N}, +_1)$, *then concept satisfiability in* $\mathcal{ALC}(\mathbf{D})$ *w.r.t. TBoxes is undecidable.*

Proof. Let (Q, P) be an arbitrary 2RM. We define a concept C and a TBox $\mathcal{T}$ in such a way that every model of $\mathcal{T}$ in which C is non-empty represents the computation of (Q, P) on the input $(0, 0)$. For every state q_i we introduce a concept name Q_i. We also introduce two concept names Z_1, Z_2 to indicate a positive zero test for the first and second register, respectively. In addition, we introduce a functional role $g \in \mathsf{N_{fR}}$ representing the transitions between configurations of the 2RM. For $p \in \{1, 2\}$, we have features $r_p \in \mathsf{N_F}$ representing the content of register p. However, since our concrete domain does not have the predicate $=_0$, we cannot enforce that, in our representation of the initial configuration, r_1 and r_2 have value zero. What we can ensure, though, is that their value is the same number, which we can store in a concrete feature $z \in \mathsf{N_F}$. The idea is now that register p of the machine actually contain the value of r_p offset with the value of z. We also need auxiliary concrete features $s_1, s_2, s \in \mathsf{N_F}$, which respectively refer to the successor values of r_1, r_2, z. They are needed to express equality using $+_1$.

The following GCI ensures that the elements of C represent the initial configuration together with appropriate values for the auxiliary features:

$$C \sqsubseteq Q_0 \sqcap \exists r_1, s_1. [+_1] \sqcap \exists r_2, s_2. [+_1] \sqcap \exists z, s_1. [+_1] \sqcap \exists z, s_2. [+_1] \sqcap \exists z, s. [+_1].$$

Next, the GCI $\top \sqsubseteq \exists gz, s. [+_1] \sqcap \exists gz, gs. [+_1]$ guarantees that the value z of an individual carries over to its g-successor. We denote the second value in $\{1, 2\}$ beside p by $\widehat{p}$, i.e., $\widehat{p} = 3 - p$. To enforce that the incrementation instructions are executed correctly, for every instruction $I_i = +(p, q_j)$, we include in $\mathcal{T}$ the GCI

$$Q_i \sqsubseteq \exists r_p, gr_p. [+_1] \sqcap \exists gr_{\widehat{p}}, s_{\widehat{p}}. [+_1] \sqcap \exists s_p, gs_p. [+_1] \sqcap \exists r_{\widehat{p}}, gs_{\widehat{p}}. [+_1] \sqcap \exists g. Q_j$$

The GCIs $Z_p \sqsubseteq \exists z, s_p. [+_1]$, $\exists z, s_p. [+_1] \sqsubseteq Z_p$ ensure that Z_p represents a positive zero test for register p. Note that, for individuals for which values for s, z, s_p, r_p are defined, the negation of Z_p is semantically equivalent to a negative zero test for register p. To enforce that decrementation is executed correctly, for every instruction $I_i = -(p, q_j, q_k)$, we include in $\mathcal{T}$ the GCIs

$$Q_i \sqcap Z_p \sqsubseteq \exists gr_p, s_p. [+_1] \sqcap \exists gr_{\widehat{p}}, s_{\widehat{p}}. [+_1] \sqcap \exists r_p, gs_p. [+_1] \sqcap \exists r_{\widehat{p}}, gs_{\widehat{p}}. [+_1] \sqcap \exists g. Q_k,$$
$$Q_i \sqcap \neg Z_p \sqsubseteq \exists gr_p, r_p. [+_1] \sqcap \exists gr_{\widehat{p}}, s_{\widehat{p}}. [+_1] \sqcap \exists gs_p, s_p. [+_1] \sqcap \exists r_{\widehat{p}}, gs_{\widehat{p}}. [+_1] \sqcap \exists g. Q_j.$$

Finally, we include the GCI $Q_\ell \sqsubseteq \bot$, which states that the halting state is never reached. It is now easy to see that the computation of (Q, P) on $(0, 0)$ does not reach the halting state iff C is satisfiable w.r.t. $\mathcal{T}$. □

Note that, even though our proof of Proposition 1 uses a functional role g to represent the transitions between configurations of a given two-register machine, the reduction also works if g is assumed to be an arbitrary role. One simply must use additional universal quantification to ensure that all the g-successors of an individual behave the same (i.e., for every existential quantification in the current proof we add the corresponding universal quantification).

It turns out that undecidability also holds if we use the ternary predicate $+$ rather than the binary predicate $+_1$. Intuitively, with $+$ we can easily test for 0 since m is 0 iff $m + m = m$. Instead of incrementation by 1, we can then use addition of a fixed non-zero number (see [8] for a detailed proof).

Proposition 2. *If* $\mathbf{D}$ *is* $(\mathbb{Q}, +)$, $(\mathbb{Z}, +)$, *or* $(\mathbb{N}, +)$, *then concept satisfiability in* $\mathcal{ALC}(\mathbf{D})$ *w.r.t. TBoxes is undecidable.*

3.2 ω-admissible Concrete Domains

To regain decidability in the presence of GCIs and concrete domains, the notion of ω-admissible concrete domains was introduced in [33]. We generalize this notion and the decidability result from concrete domains with only binary predicates as in [33] to concrete domains with predicates of arbitrary arity.

We say that a countable structure $\mathbf{D}$ has *homomorphism compactness* if, for every countable structure $\mathbf{B}$, it holds that $\mathbf{B} \to \mathbf{D}$ iff $\mathbf{A} \to \mathbf{D}$ for every finite structure $\mathbf{A}$ with $\mathbf{A} \hookrightarrow \mathbf{B}$. A relational τ-structure $\mathbf{D}$ satisfies

(JE) if, for some $d \geq 2$, $\bigcup\{R^{\mathbf{D}} \mid R \in \tau,\ R^{\mathbf{D}} \subseteq D^k\} = D^k$ if $k \leq d$ and $\emptyset$ else;
(PD) if $R^{\mathbf{D}} \cap R'^{\mathbf{D}} = \emptyset$ for all pairwise distinct $R, R' \in \tau$;
(JD) if $\bigcup\{R^{\mathbf{D}} \mid R \in \tau,\ R^{\mathbf{D}} \subseteq \Delta_D\} = \Delta_D$.

Here JE stands for "jointly exhaustive," PD for "pairwise disjoint," and JD for "jointly diagonal." Note that JD was not considered in [33]. We include it here since it makes the comparison with known notions from model theory easier. In addition, all the ω-admissible concrete domains considered in [33] satisfy JD.

A relational τ-structure $\mathbf{D}$ is a *patchwork* if it is JDJEPD, and for all finite JEPD τ-structures $\mathbf{A}, \mathbf{B}_1, \mathbf{B}_2$ with $e_1 \colon \mathbf{A} \hookrightarrow \mathbf{B}_1$, $e_2 \colon \mathbf{A} \hookrightarrow \mathbf{B}_2$, $\mathbf{B}_1 \to \mathbf{D}$ and $\mathbf{B}_2 \to \mathbf{D}$, there exist $f_1 \colon \mathbf{B}_1 \to \mathbf{D}$ and $f_2 \colon \mathbf{B}_2 \to \mathbf{D}$ with $f_1 \circ e_1 = f_2 \circ e_2$.

Definition 1. *The relational structure* $\mathbf{D}$ *is* ω-admissible *if CSP*($\mathbf{D}$) *is decidable,* $\mathbf{D}$ *has homomorphism compactness, and* $\mathbf{D}$ *is a patchwork.*

The idea is now that one can use disjunctions of atomic formulas of the same arity within concrete domain restrictions. We refer to the set of all FO τ-formulas of the form $R_1(x_1, \ldots, x_k) \vee \cdots \vee R_m(x_1, \ldots, x_k)$ for $R_1, \ldots, R_m$ k-ary predicates in τ by $\vee^+$. The following theorem is proved in [8] by extending the tableau-based decision procedure given in [33] to our more general definition of ω-admissibility. Note that the proof of correctness of this procedure does not depend on JD.

Theorem 1. *Let* **D** *be an* ω*-admissible* τ*-structure with at most d-ary relations for some* $d \geq 2$*. Then concept satisfiability in* $\mathcal{ALC}^{d}_{\vee+}(\mathbf{D})$ *w.r.t. TBoxes is decidable.*

The main motivation for the definition of ω-admissible concrete domains in [33] was that they can capture qualitative calculi of time and space. In particular, it was shown in [33] that Allen's interval logic [1] as well as the region connection calculus RCC8 [37] can be represented as ω-admissible concrete domains. To the best of our knowledge, no other ω-admissible concrete domains have been exhibited in the literature since then.

4 A Model-Theoretic Approach Towards ω-admissibility

In this section, we introduce several model-theoretic properties of relational structures and show their connection to ω-admissibility. This allows us to formulate sufficient conditions for ω-admissibility using well-know notions from model theory, and thus to employ existing model-theoretic results to find new ω-admissible concrete domains.

ω-categoricity. We start with introducing ω-categoricity since it gives us homomorphism compactness "for free." A structure is *ω-categorical* if its first-order theory has exactly one countable model up to isomorphism. For example, it is well-known that $\mathbf{Q}$ is, up to isomorphism, the only countable dense linear order without lower or upper bound. This result, which clearly implies that $\mathbf{Q}$ is ω-categorical, is due to Cantor.

For every structure **A**, the set of all its automorphisms forms a permutation group, which we denote by Aut(**A**) (see Theorem 1.2.1 in [25]). Every relation with an FO definition in **A** is easily seen to be preserved by Aut(**A**). For ω-categorical structures, the other direction holds as well.

Theorem 2 (Engeler, Ryll-Nardzewski, Svenonius [25]**).** *For a countably infinite structure* **A** *with a countable signature, the following are equivalent:*

1. **A** *is* ω*-categorical.*
2. *For every* $k \geq 1$*, only finitely many k-ary relations are FO definable in* **A**.
3. *Every relation over A preserved by Aut(***A***) is FO definable in* **A**.

The following corollary to this theorem establishes the first link between model theory and ω-admissibility.

Corollary 1 (Lemma 3.1.5 in [10]**).** *Every* ω*-categorical structure has homomorphism compactness.*

In order to obtain JDJEPD, we replace the original relations of a given ω-categorical τ-structure **A** with appropriate first-order definable ones, using the results of Theorem 2. The *orbit* of a tuple $(a_1, \ldots, a_k) \in A^k$ under the natural action of Aut(**A**) on A^k is the set $\{(g(a_1), \ldots, g(a_n)) \mid g \in \mathrm{Aut}(\mathbf{A})\}$. By Theorem 2, the set of all at most k-ary relations FO definable in **A** is finite for every $k \in \mathbb{N}$. Since every such set is closed under intersections, it contains finitely many minimal non-empty relations. Since every relation over A that is preserved by all automorphisms of $\mathbf{A}$ is FO definable in $\mathbf{A}$, these minimal elements are precisely the orbits of tuples over A under the natural action of $\text{Aut}(\mathbf{A})$.

Definition 2. *For a given arity bound $d \geq 2$, the* d-reduct *of the ω-categorical τ-structure $\mathbf{A}$, denoted by $\mathbf{A}^{\leqslant d}$, is the relational structure over A whose relations are all orbits of at most d-ary tuples over A under $Aut(\mathbf{A})$. We denote the signature of $\mathbf{A}^{\leqslant d}$ by $\tau^{\leqslant d}$.*

It is easy to see that $\mathbf{A}^{\leqslant d}$ is JDJEPD, and that every at most d-ary relation over A FO definable in $\mathbf{A}$ can be obtained as a disjunction of atomic formulas built using the symbols in $\tau^{\leqslant d}$. As an example, consider the ω-categorical structure $\mathbf{Q}$. The orbits of k-tuples of elements of $\mathbb{Q}$ can be defined by quantifier-free formulas that are conjunctions of atoms of the form $x_i = x_j$ or $x_i < x_j$. For example, the orbit of the tuple $(2, 3, 2, 5)$ consists of all tuples $(q_1, q_2, q_3, q_4) \in \mathbb{Q}^4$ that satisfy the formula $x_1 < x_2 \wedge x_1 = x_3 \wedge x_2 < x_4$ if x_i is replaced by q_i for $i = 1, \ldots, 4$. The FO definable k-ary relations in $\mathbf{Q}$ are obtained as unions of these orbits, where the defining formula is then the disjunction of the formulas defining the respective orbits. Since these formulas are quantifier-free, this also shows that $\mathbf{Q}$ admits quantifier elimination. Recall that a τ-structure admits *quantifier elimination* if for every FO τ-formula there exists a quantifier-free (qf) τ-formula that defines the same relation over this structure.

Homogeneity. To obtain the patchwork property, we restrict the attention to homogeneous structures. A structure $\mathbf{A}$ is *homogeneous* if every isomorphism between finite substructures of $\mathbf{A}$ extends to an automorphism of $\mathbf{A}$.

Theorem 3 ([25]). *A countable relational structure with a finite signature is homogeneous iff it is ω-categorical and admits quantifier elimination.*

Since $\mathbf{Q}$ is ω-categorical and admits quantifier elimination, it is thus homogeneous. This can, however, also easily be shown directly without using the theorem. In fact, given finite substructures $\mathbf{B}$ and $\mathbf{C}$ of $\mathbf{Q}$ and an isomorphism between them, we know that B consists of finitely many elements $p_1, \ldots, p_n$ and C of the same number of elements $q_1, \ldots, q_n$ such that $p_1 < \ldots < p_n$, $q_1 < \ldots < q_n$, and the isomorphism maps p_i to q_i (for $i = 1, \ldots, n$). It is now easy to see that $<$ is also a dense linear order without lower or upper bound on the sets $\{p \mid p < p_1\}$ and $\{q \mid q < q_1\}$, and thus there is an order isomorphism between these sets. The same is true for the pairs of sets $\{p \mid p_i < p < p_{i+1}\}$ and $\{q \mid q_i < q < q_{i+1}\}$, and for the pair $\{p \mid p_n < p\}$ and $\{q \mid q < q_n\}$. Using the isomorphisms between these pairs, we can clearly put together an isomorphisms from $\mathbf{Q}$ to $\mathbf{Q}$ that extends the original isomorphism from $\mathbf{B}$ to $\mathbf{C}$.

Countable homogeneous structures can be obtained as Fraïssé limits of amalgamation classes. A class $\mathcal{K}$ of relational τ-structures has the *amalgamation property* (AP) if, for every $\mathbf{A}, \mathbf{B}_1, \mathbf{B}_2 \in \mathcal{K}$ with $e_1 \colon \mathbf{A} \hookrightarrow \mathbf{B}_1$ and $e_2 \colon \mathbf{A} \hookrightarrow \mathbf{B}_2$ there exists $\mathbf{C} \in \mathcal{K}$ with $f_1 \colon \mathbf{B}_1 \hookrightarrow \mathbf{C}$ and $f_2 \colon \mathbf{B}_2 \hookrightarrow \mathbf{C}$ such that $f_1 \circ e_1 = f_2 \circ e_2$. A class $\mathcal{K}$ of finite relational structures with a countable signature τ is called an *amalgamation class* if it has AP, is closed under applying isomorphisms and taking substructures, and contains only countably many structures up to isomorphism. We denote by Age$(\mathbf{A})$ the class of all finite structures that embed into the structure $\mathbf{A}$.

Theorem 4 (Fraïssé [25]). *Let $\mathcal{K}$ be an amalgamation class of τ-structures. Then there exists a homogeneous countable τ-structure $\mathbf{A}$ with Age$(\mathbf{A}) = \mathcal{K}$. The structure $\mathbf{A}$ is unique up to isomorphism and referred to as the* Fraïssé limit *of $\mathcal{K}$. Conversely, Age$(\mathbf{A})$ for a countable homogeneous structure $\mathbf{A}$ with a countable signature is an amalgamation class.*

For our running example $\mathbf{Q} = (\mathbb{Q}, <)$, we have that Age$(\mathbf{Q})$ consists of all finite linear orders, and thus by Fraïssé's theorem this class of structures is an amalgamation class. In addition, $\mathbf{Q}$ is the Fraïssé limit of this class. Proposition 3 below shows that there is a close connection between AP and the patchwork property. Its proof uses the following lemma, whose proof can be found in [8].

Lemma 2. *Let $\mathbf{A}, \mathbf{B}$ be two JEPD τ-structures, such that $\mathbf{B}$ is JD, and $f \colon \mathbf{A} \to \mathbf{B}$ a homomorphism. Then f preserves the complements of all relations of $\mathbf{A}$ and* $\ker f = \bigcup\{R^{\mathbf{A}} \mid R \in \tau$ *and* $R^{\mathbf{B}} \subseteq \triangle_B\}$.

Proposition 3. *Let $\mathbf{D}$ be a JDJEPD τ-structure. Then $\mathbf{D}$ is a patchwork iff Age$(\mathbf{D})$ has AP.*

Proof. For simplification purposes, every statement indexed by i is suppose to hold for both $i \in \{1, 2\}$. First, suppose that Age$(\mathbf{D})$ has AP. Let $\mathbf{A}, \mathbf{B}_1, \mathbf{B}_2$ be finite JEPD τ-structures with $e_i \colon \mathbf{A} \hookrightarrow \mathbf{B}_i$ and $h_i \colon \mathbf{B}_i \to \mathbf{D}$. We must show that there exist $f_i \colon \mathbf{B}_i \to \mathbf{D}$ with $f_1 \circ e_1 = f_2 \circ e_2$. Let $\widehat{\mathbf{A}}_1$ and $\widehat{\mathbf{A}}_2$ be the substructures of $\mathbf{D}$ on $(h_1 \circ e_1)(A)$ and $(h_2 \circ e_2)(A)$, respectively. Clearly both $\widehat{\mathbf{A}}_1$ and $\widehat{\mathbf{A}}_2$ are JDJEPD, because they are substructures of $\mathbf{D}$. Due to Lemma 2, we have $\widehat{\mathbf{A}}_1 \cong \widehat{\mathbf{A}}_2$, because both $h_1 \circ e_1$ and $h_2 \circ e_2$ preserve the complements of all relations of $\mathbf{A}$ and $\ker h_1 \circ e_1 = \bigcup\{R^{\mathbf{A}} \mid R \in \tau \text{ and } R^{\mathbf{D}} \subseteq \triangle_D\} = \ker h_2 \circ e_2$.

However, what we want is an isomorphism that commutes with $h_1 \circ e_1$ and $h_2 \circ e_2$. Consider the map $g \colon \widehat{A}_1 \to \widehat{A}_2$ given by $g\big((h_1 \circ e_1)(a)\big) := (h_2 \circ e_2)(a)$. It is well defined, because $\ker h_1 \circ e_1 = \ker h_2 \circ e_2$. Now, for every $R \in \tau$ and $((h_1 \circ e_1)(a_1), \ldots, (h_1 \circ e_1)(a_k)) \in R^{\widehat{\mathbf{A}}_1}$, we have $(a_1, \ldots, a_k) \in R^{\mathbf{A}}$, because $h_1 \circ e_1$ preserves the complements of all relations of $\mathbf{A}$ due to Lemma 2. But this implies $((h_2 \circ e_2)(a_1), \ldots, (h_2 \circ e_2)(a_k)) \in R^{\widehat{\mathbf{A}}_2}$, because $h_2 \circ e_2$ is a homomorphism. By Lemma 2, g preserves the complements of all relations of $\widehat{\mathbf{A}}_1$ and

$$\ker g = \bigcup\{R^{\widehat{\mathbf{A}}_1} \mid R \in \tau \text{ and } R^{\widehat{\mathbf{A}}_2} \subseteq \triangle_{\widehat{A}_2}\} = \triangle_{\widehat{A}_1}.$$

Hence g is an isomorphism that additionally satisfies $g \circ h_1 \circ e_1 = h_2 \circ e_2$. Let $\widehat{\mathbf{B}}_1$ and $\widehat{\mathbf{B}}_2$ be the substructures of $\mathbf{D}$ on $h_1(B_1)$ and $h_2(B_2)$, respectively. Now consider the inclusions $\widehat{e}_i \colon \widehat{\mathbf{A}}_i \hookrightarrow \widehat{\mathbf{B}}_i$. Since Age$(\mathbf{D})$ has AP, there exists

$\mathbf{C} \in \mathrm{Age}(\mathbf{D})$ together with $\widehat{f}_i\colon \widehat{\mathbf{B}}_i \hookrightarrow \mathbf{C}$ and $e\colon \mathbf{C} \hookrightarrow \mathbf{D}$ such that $\widehat{f}_1 \circ \widehat{e}_1 = \widehat{f}_2 \circ \widehat{e}_2 \circ g$. We define the homomorphisms $f_i\colon \mathbf{B}_i \to \mathbf{D}$ by $f_i := e \circ \widehat{f}_i \circ h_i$. Then, for every $a \in A$,

$$\begin{aligned} f_1 \circ e_1(a) &= e \circ \widehat{f}_1 \circ h_1 \circ e_1(a) = e \circ \widehat{f}_1 \circ \widehat{e}_1 \circ h_1 \circ e_1(a) \\ &= e \circ \widehat{f}_2 \circ \widehat{e}_2 \circ g \circ h_1 \circ e_1(a) = e \circ \widehat{f}_2 \circ \widehat{e}_2 \circ h_2 \circ e_2(a) \\ &= e \circ \widehat{f}_2 \circ h_2 \circ e_2(a) = f_2 \circ e_2(a). \end{aligned}$$

Hence $\mathbf{D}$ is a patchwork. For the other direction, suppose that $\mathbf{D}$ is a patchwork. Let $\mathbf{A}, \mathbf{B}_1, \mathbf{B}_2$ be finite τ-structures with $e_i\colon \mathbf{A} \hookrightarrow \mathbf{B}_i$ and $h_i\colon \mathbf{B}_i \hookrightarrow \mathbf{D}$. Since $\mathbf{B}_1$ and $\mathbf{B}_2$ are isomorphic to substructures of $\mathbf{D}$, they are clearly JEPD. Thus, as $\mathbf{D}$ is a patchwork, there exist homomorphisms $f_i\colon \mathbf{B}_i \to \mathbf{D}$ with $f_1 \circ e_1 = f_2 \circ e_2$. By Lemma 2, the f_i preserve the complements of all relations of $\mathbf{B}_i$, and

$$\ker f_i = \bigcup \{R^{\mathbf{B}_i} \mid R \in \tau \text{ and } R^{\mathbf{D}} \subseteq \Delta_D\} = \ker h_i = \Delta_{B_i}.$$

This means that f_i are embeddings. We obtain AP for Age ($\mathbf{D}$) by choosing $\mathbf{C}$ to be the substructure of $\mathbf{D}$ on $f_2(B_1) \cup f_1(B_2)$. □

Recall that, to obtain JDJEPD, we actually need to take the d-reduct of a given ω-categorical structure, rather than the structure itself. Fortunately, homogeneity transfers from $\mathbf{D}$ to $\mathbf{D}^{\leqslant d}$ (see [8] for the proof).

Lemma 3. *Let $\mathbf{D}$ be a countable homogeneous structure with a finite relational signature τ. Then $\mathbf{D}^{\leqslant d}$ is homogeneous for every d that exceeds or is equal to the maximal arity of the symbols from τ.*

Finite Boundedness. The only property of ω-admissible structures we have not yet considered in this section is the decidability of the CSP. One possibility to achieve this is to consider finitely bounded structures. For a class $\mathcal{N}$ of τ-structures, we denote by $\mathrm{Forb}_e(\mathcal{N})$ the class of all finite τ structures not embedding any member of $\mathcal{N}$. We say that a structure $\mathbf{A}$ is *finitely bounded* if its signature is finite and Age ($\mathbf{A}$) $= \mathrm{Forb}_e(\mathcal{N})$ for a finite $\mathcal{N}$ [14]. Note that $\mathbf{A}$ is finitely bounded iff there exists a universal FO sentence $\Phi(\mathbf{A})$ s.t. $\mathbf{B} \in$ Age ($\mathbf{A}$) iff $\mathbf{B} \models \Phi(\mathbf{A})$ [8].

The structure $\mathbf{Q}$ is finitely bounded. To show this, we can use the set $\mathcal{N}$ consisting of the four structures depicted in Fig. 1: the self loop, the 2-cycle, the 3-cycle, and two isolated vertices. We must show that Age ($\mathbf{Q}$) $= \mathrm{Forb}_e(\mathcal{N})$. Clearly, none of the structures in $\mathcal{N}$ embeds into a linear order, which shows Age ($\mathbf{Q}$) $\subseteq \mathrm{Forb}_e(\mathcal{N})$. Conversely, assume that $\mathbf{A}$ is an element of $\mathrm{Forb}_e(\mathcal{N})$. We must show that $<^{\mathbf{A}}$ is a linear order. Since $\mathcal{N}$ contains the self loop, we have $(a, a) \notin <^{\mathbf{A}}$ for all $a \in A$, which shows that $<^{\mathbf{A}}$ is irreflexive. For distinct elements $a, b \in A$, we must have $a <^{\mathbf{A}} b$ or $b <^{\mathbf{A}} a$ since otherwise the structure consisting of two isolated vertices could be embedded into $\mathbf{A}$. This shows that any two distinct elements are comparable w.r.t. $<^{\mathbf{A}}$. To show that $<^{\mathbf{A}}$ is transitive,

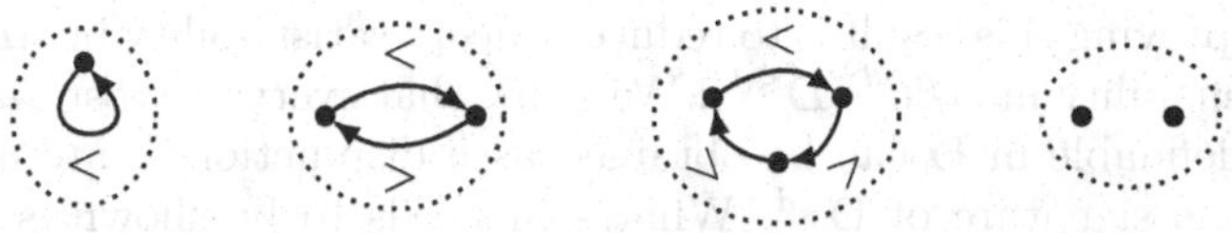

Fig. 1. A set of four forbidden substructures for $\mathbf{Q} = (\mathbb{Q}; <)$.

assume that $a <^{\mathbf{A}} b$ and $b <^{\mathbf{A}} c$ holds. Since the 2-cycle does not embed into $\mathbf{A}$, a and c must be distinct, and are thus comparable. We cannot have $c <^{\mathbf{A}} a$ since then we could embed the 3-cycle into $\mathbf{A}$. Consequently, we must have $a <^{\mathbf{A}} c$, which proves transitivity. This show that $\mathbf{A}$ is a linear order. As formula $\Phi(\mathbf{Q})$ we can take the conjunction of the usual axioms defining linear orders.

Finitely bounded structures are interesting since their CSP and their first-order theory are decidable. The first result can, e.g., be found in [13] (Theorem 4) and the second result is stated in [28,29] (see [8] for a detailed proof).

Proposition 4. *Let* $\mathbf{D}$ *be finitely bounded homogeneous structure with* $|D| > 1$. *Then CSP*($\mathbf{D}$) *is decidable in* NP *and the FO theory of* $\mathbf{D}$ *is* PSPACE-*complete.*

The following proposition, whose proof can be found in [8], implies that Proposition 4 applies not only to a given finitely bounded homogeneous structure $\mathbf{D}$, but also to its d-reduct $\mathbf{D}^{\leqslant d}$.

Proposition 5. *Let* $\mathbf{A}$ *be a finitely bounded homogeneous structure and* $\mathbf{B}$ *a structure with the same domain and finitely many relations that are FO definable in* $\mathbf{A}$. *Then* $\mathbf{B}$ *is a reduct of a finitely bounded homogeneous structure.*

We are now ready to formulate our first sufficient condition for ω-admissibility.

Theorem 5. *Let* $\mathbf{D}$ *be a finitely bounded homogeneous relational structure with at most* d*-ary relations for some* $d \geq 2$. *Then* $\mathbf{D}^{\leqslant d}$ *is* ω*-admissible.*

Proof. It follows directly from its definition that $\mathbf{D}^{\leqslant d}$ is JEPD. Since $d \geq 2$, it is clearly also JD. By Lemma 3, $\mathbf{D}^{\leqslant d}$ is homogeneous and ω-categorical. Thus $\mathbf{D}$ has homomorphism compactness by Corollary 1. By Theorem 4, Age $(\mathbf{D}^{\leqslant d})$ has AP. Thus $\mathbf{D}^{\leqslant d}$ is a patchwork by Proposition 3. By Proposition 5, Lemma 1, and Proposition 4, CSP($\mathbf{D}^{\leqslant d}$) is in NP. Hence $\mathbf{D}^{\leqslant d}$ is ω-admissible. □

This theorem, together with Theorem 1, immediately yields decidability for $\mathcal{ALC}^{d}_{\vee+}(\mathbf{D}^{\leqslant d})$. The following corollary shows that we can even allow for arbitrary FO definable relations with arity bounded by d in the concrete domain.

Corollary 2. *Let* $\mathbf{D}$ *be a finitely bounded homogeneous relational structure with at most* d*-ary relations for some* $d \geq 2$. *Then concept satisfiability in* $\mathcal{ALC}^{d}_{FO}(\mathbf{D})$ *w.r.t. TBoxes is decidable.*

The idea for proving this result is to reduce concept satisfiability in $\mathcal{ALC}^d_{\mathrm{FO}}(\mathbf{D})$ to concept satisfiability in $\mathcal{ALC}^d_{\vee+}(\mathbf{D}^{\leqslant d})$. We know that every at most d-ary relation over D FO definable in $\mathbf{D}$ can be obtained as a disjunction of atomic formulas built using the signature of $\mathbf{D}^{\leqslant d}$. What still needs to be shown is that, given a first-order formula in the signature of $\mathbf{D}$ with at most d free variables, this disjunction can effectively be computed (see [8] for how this can be proved).

Cores. Finally, we consider the situation where we have a homogeneous relational structure $\mathbf{D}$ with finitely many at most d-ary relations that is not finitely bounded, but which we know (by some other means) to have a decidable CSP. In this situation, we can show decidability for $\mathcal{ALC}^d_{\exists+}(\mathbf{D})$ under one additional assumption. A structure $\mathbf{D}$ is called a *core* if every endomorphism from $\mathbf{D}$ to itself is an embedding. It was shown in [9] that, if $\mathbf{D}$ is a homogeneous core, then the orbits of tuples over D under $\mathrm{Aut}(\mathbf{D})$ are pp definable in $\mathbf{D}$. As an easy consequence of this result, we obtain our second sufficient condition for ω-admissibility (see [8] for the proof).

Theorem 6. *Let $\mathbf{D}$ be a homogeneous relational structure with finitely many at most d-ary relations for some $d \geq 2$ that is a core and has a decidable CSP. Then $\mathbf{D}^{\leqslant d}$ is ω-admissible.*

By showing that concept satisfiability in $\mathcal{ALC}^d_{\exists+}(\mathbf{D})$ can effectively be reduced to concept satisfiability in $\mathcal{ALC}^d_{\vee+}(\mathbf{D}^{\leqslant d})$, we obtain the following decidability result (see [8] for the proof).

Corollary 3. *Let $\mathbf{D}$ be a homogeneous relational structure with finitely many at most d-ary relations for some $d \geq 2$ that is a core and has a decidable CSP. Then concept satisfiability in $\mathcal{ALC}^d_{\exists+}(\mathbf{D})$ w.r.t. TBoxes is decidable.*

5 Application and Discussion

In this section, we discuss how the results of Sect. 4 can be used to obtain specific ω-admissible concrete domains. But let us first start with a caveat.

Finiteness of Signature Matters. In Corollary 2 and Corollary 3, the signature of the structure $\mathbf{D}$ is required to be finite. This restriction is needed to obtain decidability. For instance, the expansion of the structure $(\mathbb{Z}; +_1)$ from Sect. 3.1 by all relations $+_k = \{(m, n) \in \mathbb{Z}^2 \mid m + k = n\}$ for $k \in \mathbb{Z}$ is homogeneous, and satisfiability of finite conjunctions of constraints is decidable in this structure. However, we have seen in Proposition 1 that reasoning with $(\mathbb{Z}; +_1)$ as a concrete domain w.r.t. TBoxes is undecidable.

(Un)decidability of the Conditions. If one intends to use Theorem 5 to obtain an ω-admissible concrete domain, one could start with selecting a finite set $\mathcal{N}$ of bounds, i.e., forbidden τ-substructures, for a finite signature τ. The question is then whether $\mathcal{N}$ really induces a finitely bounded structure, i.e., whether

there is a τ-structure $\mathbf{D}$ such that Age $(\mathbf{D}) = \text{Forb}_e(\mathcal{N})$. This question is in general undecidable. In fact, it is shown in [17] that the *joint embedding property (JEP)* is undecidable for classes of structures that are definable by finitely many bounds. In addition, it is known that a class of structures definable by finitely many bounds has JEP iff this class is the age of some countably infinite structure (see [25], Theorem 6.1.1). However, if one restricts the attention to binary signatures, then it is decidable whether a class of the form $\text{Forb}_e(\mathcal{N})$ has AP [12]. If this is the case, then the Fraïssé limit $\mathbf{D}$ of $\text{Forb}_e(\mathcal{N})$ is a finitely bounded homogeneous structure satisfying Age $(\mathbf{D}) = \text{Forb}_e(\mathcal{N})$ by Theorem 4.

Reproducing Known Results. The examples for ω-admissible concrete domains given in [33] were RCC8 and Allen's interval algebra, for which the patchwork property is proved "by hand" in [33]. Given Theorem 5, we obtain these results as a consequence of known results from model theory. It was shown in [15] that RCC8 has a representation by a homogeneous structure $\mathbf{R}$ with a finite relational signature (Theorem 2 in [15]). Since Age $(\mathbf{R})$ has a finite universal axiomatization (Definition 3 in [15]), $\mathbf{R}$ is finitely bounded. For Allen's interval algebra, it was shown in [24] that it has a representation by a homogeneous structure $\mathbf{A}$ with a finite relational signature. Since Age $(\mathbf{A})$ has a finite universal axiomatization, $\mathbf{A}$ is finitely bounded. Our running example $\mathbf{Q} = (\mathbb{Q}, <)$ also satisfies the preconditions of Theorem 5, and thus Corollary 2 yields decidability of $\mathcal{ALC}^d_{\text{FO}}(\mathbf{Q})$ with TBoxes. For $\mathbf{Q}$ extended just with $>, \leq, \geq, =, \neq$, decidability was proved in [30], using an automata-based procedure. Our results show that there is also a tableau-based decision procedure for this logic.

Expansions, Disjoint Unions, and Products. When modelling concepts in a DL with concrete domain $\mathbf{D}$, it is often useful to be able to refer to specific elements d of the domain, i.e., to have unary predicate symbols $=_d$ that are interpreted as $\{d\}$. We can show that the class of reducts of finitely bounded homogeneous structures is closed under expansion by finitely many such relations [8].

It would also be useful to be able to refer to predicates of different concrete domains (say RCC8 *and* Allen) when defining concepts. In [5], it was shown that admissible concrete domains are closed under disjoint union. We can prove the corresponding result for finitely bounded homogeneous structures [8]. Using disjoint union to refer to several concrete domain works well if the paths employed in concrete domain constructors contain only functional roles, which is the case considered in [5]. However, if we allow for non-functional roles in paths, then using disjoint union is not appropriate. In general, if R, R' are two binary relations over D and $r \in \mathsf{N_R} \setminus \mathsf{N_{fR}}$, then the situation where an individual x has an r-successor y with features related through both R and R' cannot be described using the disjoint union of $(D; R)$ and $(D; R')$ as a concrete domain (see [8] for details).

To overcome this problem, we propose to use the so-called full product [10]. Let $\mathbf{A}_1, \ldots, \mathbf{A}_k$ be finitely many structures with disjoint relational signatures $\tau_1, \ldots, \tau_k$. The *full product* of $\mathbf{A}_1, \ldots, \mathbf{A}_k$, denoted by $\mathbf{A}_1 \boxtimes \cdots \boxtimes \mathbf{A}_k$, has as its domain the Cartesian product $A := A_1 \times \cdots \times A_k$ and as its signature the union of the signatures τ_i. For $a \in A_1 \times \cdots \times A_k$ and $i \in [k]$, we denote the ith component of the tuple a by $a[i]$. The relations are defined as follows:

$$R^{\mathbf{A}_1 \times \cdots \times \mathbf{A}_k} := \{(a_1, \ldots, a_n) \in A^n \mid (a_1[i], \ldots, a_n[i]) \in R^{\mathbf{A}_i}\}$$

for every $i \in [k]$ and every n-ary $R \in \tau_i$. We show in [8] that the full product preserves homogeneity and finite boundedness, and thus the prerequisites for Theorem 5 and Corollary 2 are preserved under building the full product.

Proposition 6. *Let $\mathbf{A}_1, \ldots, \mathbf{A}_k$ be finitely bounded homogeneous structures with disjoint relational signatures $\tau_1, \ldots, \tau_k$ such that, for $i \in [k]$, τ_i contains the symbol $=_i$, which is defined in $\mathbf{A}_i$ as $\triangle_{A_i}$. Then $\mathbf{A}_1 \boxtimes \cdots \boxtimes \mathbf{A}_k$ is a finitely bounded homogeneous structure.*

Together with Proposition 4 this also yields a general complexity result for combinations of constraints over several finitely bounded homogeneous templates. Such combinations were previously considered in the literature in special cases; for example, for RCC8 and Allen [21].

Henson Digraphs. A directed graph is a *tournament* if every two distinct vertices in it are connected by exactly one directed edge. In [23], Henson proved that there are uncountably many homogeneous directed graphs by showing that, for any set $\mathcal{N}$ of finite tournaments (plus the loop and the 2-cycle) such that no member of $\mathcal{N}$ is embeddable into any other member of $\mathcal{N}$, $\mathrm{Forb}_e(\mathcal{N})$ is an amalgamation class whose Fraïssé limit is a homogeneous directed graph. Furthermore, the Fraïssé limits for two distinct sets of such tournaments are distinct as well. In the literature, such directed graphs are often called *Henson digraphs* [34]. If $\mathbf{G}$ is a Henson digraph, then $\mathrm{Age}(\mathbf{G}) = \mathrm{CSP}(\mathbf{G})$.[5] Clearly, only countably many Henson digraphs can have a decidable CSP. Beside the finitely bounded ones (see Proposition 4), there is an interesting example constructed using the infinite set of non-isomorphic tournaments from Henson's original proof of uncountability. Consider the tournaments $\mathbf{T}_1, \mathbf{T}_2, \ldots$ with domains $[2], [3], \ldots$ such that the edge relation of $\mathbf{T}_n$ consists of the edges (i, j) for every $j = i + 1$ with $0 \leq i \leq n$, $(0, n+1)$, and (j, i) for every $j > i + 1$ with $(i, j) \neq (0, n+1)$. It was shown in [11] that the CSP of the Henson digraph corresponding to $\mathcal{N} := \{\mathbf{T}_1, \mathbf{T}_2, \ldots\}$ is coNP-complete. This digraph is a homogeneous core, and its CSP is decidable. Thus, it satisfies the requirements of Corollary 3. However, it is clearly not finitely bounded, and thus does not satisfy the requirements of Corollary 2. Conversely, it is known that the random graph is finitely bounded and homogeneous [25], but it is not a core [9]. This shows that the class of structures covered by Corollary 3 is incomparable with the one covered by Corollary 2.

6 Conclusion

We have shown that ω-admissibility, which was introduced in the DL community to obtain decidable extensions of DLs by concrete domains, is closely related to

[5] One direction is obvious, the other holds because homomorphisms between directed graphs cannot contract any edges.

well-known notions from model theory. Given the fact that a large number of homogeneous structures are known from the literature [34] and that homogeneous and finitely bounded structures play an important rôle in the CSP community, we believe that our work will turn out to be useful for locating new ω-admissible concrete domains.

This is not the first model-theoretic description of a sufficient condition for decidability of reasoning in DLs with concrete domains in the presence of TBoxes. The *existence of homomorphism is definable (EHD)* property was used in [19] to obtain decidability results for DLs with concrete domains. However, the way the concrete domain is integrated into the DL in [19] is different from the classical one employed by us and used in all other papers on DLs with concrete domains. In [19], constraints are always placed along a linear path stemming from a single individual, which is rather similar to the use of constraints in temporal logics [18,20]. In contrast, in the classical setting of DLs with concrete domains, one can compare feature values of siblings of an individual.

References

1. Allen, J.F.: Maintaining knowledge about temporal intervals. Commun. ACM **26**(11), 832–843 (1983)
2. Baader, F., Bürckert, H.J., Hollunder, B., Nutt, W., Siekmann, J.H.: Concept logics. In: Lloyd, J.W. (ed.) Computational Logics. ESPRIT Basic Research Series, pp. 177–201. Springer, Heidelberg (1990). https://doi.org/10.1007/978-3-642-76274-1_10
3. Baader, F., Calvanese, D., McGuinness, D., Nardi, D., Patel-Schneider, P.F. (eds.): The Description Logic Handbook: Theory, Implementation, and Applications. Cambridge University Press, Cambridge (2003)
4. Baader, F., Hanschke, P.: A schema for integrating concrete domains into concept languages. In: Proceedings of the 12th International Joint Conference on Artificial Intelligence (IJCAI 1991), pp. 452–457 (1991). Long version available as [5]
5. Baader, F., Hanschke, P.: A scheme for integrating concrete domains into concept languages. Technical report RR-91-10, Deutsches Forschungszentrum für Künstliche Intelligenz (DFKI) (1991). https://lat.inf.tu-dresden.de/research/reports/1991/DFKI-RR-91-10.pdf
6. Baader, F., Hanschke, P.: Extensions of concept languages for a mechanical engineering application. In: Jürgen Ohlbach, H. (ed.) GWAI 1992. LNCS, vol. 671, pp. 132–143. Springer, Heidelberg (1993). https://doi.org/10.1007/BFb0018999
7. Baader, F., Horrocks, I., Lutz, C., Sattler, U.: An Introduction to Description Logic. Cambridge University Press, Cambridge (2017)
8. Baader, F., Rydval, J.: Using model-theory to find ω-admissible concrete domains. LTCS-Report 20-01, Chair of Automata Theory, Institute of Theoretical Computer Science, Technische Universität Dresden, Dresden, Germany (2020). https://tu-dresden.de/inf/lat/reports#BaRy-LTCS-20-01
9. Bodirsky, M.: The core of a countably categorical structure. In: Diekert, V., Durand, B. (eds.) STACS 2005. LNCS, vol. 3404, pp. 110–120. Springer, Heidelberg (2005). https://doi.org/10.1007/978-3-540-31856-9_9
10. Bodirsky, M.: Complexity classification in infinite-domain constraint satisfaction. Habilitation thesis, Université Diderot - Paris 7 (2012). https://arxiv.org/pdf/1201.0856.pdf

11. Bodirsky, M., Grohe, M.: Non-dichotomies in constraint satisfaction complexity. In: Aceto, L., Damgård, I., Goldberg, L.A., Halldórsson, M.M., Ingólfsdóttir, A., Walukiewicz, I. (eds.) ICALP 2008. LNCS, vol. 5126, pp. 184–196. Springer, Heidelberg (2008). https://doi.org/10.1007/978-3-540-70583-3_16
12. Bodirsky, M., Knäuer, S., Starke, F.: AMSNP: a tame fragment of existential second-order logic. In: Proceedings of the 16th Conference on Computability in Europe - Beyond the Horizon of Computability (CiE 2020). Lecture Notes in Computer Science. Springer (2020). https://arxiv.org/abs/2001.08190.pdf
13. Bodirsky, M., Mottet, A.: Reducts of finitely bounded homogeneous structures, and lifting tractability from finite-domain constraint satisfaction. In: Proceedings of the 31st Annual ACM/IEEE Symposium on Logic in Computer Science (LICS 2016), pp. 623–632. ACM/IEEE (2016)
14. Bodirsky, M., Nešetřil, J.: Constraint satisfaction with countable homogeneous templates. J. Logic Comput. **16**(3), 359–373 (2006)
15. Bodirsky, M., Wölfl, S.: RCC8 is polynomial on networks of bounded treewidth. In: Proceedings of the 22nd International Joint Conference on Artificial Intelligence (IJCAI 2011). IJCAI/AAAI (2011)
16. Bojańczyk, M., Segoufin, L., Toruńczyk, S.: Verification of database-driven systems via amalgamation. In: Proceedings of the 32nd ACM SIGMOD-SIGACT-SIGART Symposium on Principles of Database Systems (PODS 2013), pp. 63–74. ACM (2013)
17. Braunfeld, S.: The undecidability of joint embedding and joint homomorphism for hereditary graph classes. Discrete Math. Theoret. Comput. Sci. **21**(2) (2019). https://arxiv.org/pdf/1903.11932.pdf
18. Carapelle, C., Feng, S., Kartzow, A., Lohrey, M.: Satisfiability of ECTL* with local tree constraints. Theory Comput. Syst. **61**(2), 689–720 (2017). https://doi.org/10.1007/s00224-016-9724-y
19. Carapelle, C., Turhan, A.Y.: Description logics reasoning wrt general TBoxes is decidable for concrete domains with the EHD-property. In: Proceedings of the 22nd European Conference on Artificial Intelligence (ECAI 2016), pp. 1440–1448. IOS Press (2016)
20. Demri, S., D'Souza, D.: An automata-theoretic approach to constraint LTL. Inf. Comput. **205**(3), 380–415 (2007)
21. Gerevini, A., Nebel, B.: Qualitative spatio-temporal reasoning with RCC-8 and Allen's interval calculus: Computational complexity. In: Proceedings of the 15th European Conference on Artificial Intelligence (ECAI 2002), pp. 312–316. IOS Press (2002)
22. Haarslev, V., Möller, R., Wessel, M.: The description logic $\mathcal{ALCNH}_{R+}$ extended with concrete domains: a practically motivated approach. In: Goré, R., Leitsch, A., Nipkow, T. (eds.) IJCAR 2001. LNCS, vol. 2083, pp. 29–44. Springer, Heidelberg (2001). https://doi.org/10.1007/3-540-45744-5_4
23. Henson, C.W.: Countable homogeneous relational structures and $\aleph_0$-categorical theories. J. Symb. Logic **37**(3), 494–500 (1972)
24. Hirsch, R.: Relation algebras of intervals. Artif. Intell. **83**(2), 267–295 (1996)
25. Hodges, W.: A Shorter Model Theory. Cambridge University Press, Cambridge (1997)
26. Hoehndorf, R., Schofield, P.N., Gkoutos, G.V.: The role of ontologies in biological and biomedical research: a functional perspective. Brief. Bioinform. **16**(6), 1069–1080 (2015)
27. Horrocks, I., Patel-Schneider, P.F., van Harmelen, F.: From SHIQ and RDF to OWL: the making of a web ontology language. J. Web Semant. **1**(1), 7–26 (2003)

28. Klin, B., Lasota, S., Ochremiak, J., Torunczyk, S.: Homomorphism problems for first-order definable structures. In: 36th IARCS Annual Conference on Foundations of Software Technology and Theoretical Computer Science (FSTTCS 2016). Schloss Dagstuhl-Leibniz-Zentrum für Informatik (2016)
29. Kopczyński, E., Toruńczyk, S.: LOIS: an application of SMT solvers. In: King, T., Piskac, R. (eds.) Proceedings of the 14th International Workshop on Satisfiability Modulo Theories (SMT@IJCAR 2016). CEUR Workshop Proceedings, vol. 1617, pp. 51–60. CEUR-WS.org (2016)
30. Lutz, C.: Interval-based temporal reasoning with general TBoxes. In: Nebel, B. (ed.) Proceedings of the 17th International Joint Conference on Artificial Intelligence (IJCAI 2001), pp. 89–94. Morgan Kaufmann, San Mateo (2001)
31. Lutz, C.: NExpTime-complete description logics with concrete domains. In: Goré, R., Leitsch, A., Nipkow, T. (eds.) IJCAR 2001. LNCS, vol. 2083, pp. 45–60. Springer, Heidelberg (2001). https://doi.org/10.1007/3-540-45744-5_5
32. Lutz, C.: Adding numbers to the $\mathcal{SHIQ}$ description logic–first results. In: Proceedings of the 8th International Conference on Principles of Knowledge Representation and Reasoning (KR 2002), pp. 191–202. Morgan Kaufmann, Los Altos (2002)
33. Lutz, C., Milicic, M.: A tableau algorithm for description logics with concrete domains and general TBoxes. J. Autom. Reason. **38**(1–3), 227–259 (2007). https://doi.org/10.1007/s10817-006-9049-7
34. Macpherson, D.: A survey of homogeneous structures. Discrete Math. **311**(15), 1599–1634 (2011)
35. Minsky, M.L.: Computation: Finite and Infinite Machines. Prentice-Hall, Englewood Cliffs (1967)
36. Pan, J.Z., Horrocks, I.: Reasoning in the $\mathcal{SHOQ}(D_n)$ description logic. In: Horrocks, I., Tessaris, S. (eds.) Proceedings of the 2002 Description Logic Workshop (DL 2002). CEUR Workshop Proceedings, vol. 53. CEUR-WS.org (2002)
37. Randell, D.A., Cui, Z., Cohn, A.G.: A spatial logic based on regions and connection. In: Proceedings of the 3rd International Conference on the Principles of Knowledge Representation and Reasoning (KR 1992), pp. 165–176. Morgan Kaufmann, Los Altos (1992)
38. Schild, K.: A correspondence theory for terminological logics: preliminary report. In: Mylopoulos, J., Reiter, R. (eds.) Proceedings of the 12th International Joint Conference on Artificial Intelligence (IJCAI 1991), pp. 466–471. Morgan Kaufmann (1991)

A Formally Verified, Optimized Monitor for Metric First-Order Dynamic Logic

David Basin, Thibault Dardinier, Lukas Heimes, Srđan Krstić(✉), Martin Raszyk(✉), Joshua Schneider(✉), and Dmitriy Traytel(✉)

Institute of Information Security, Department of Computer Science,
ETH Zürich, Zurich, Switzerland
{srdan.krstic,martin.raszyk,joshua.schneider,traytel}@inf.ethz.ch

Abstract. Runtime monitors for rich specification languages are sophisticated algorithms, especially when they are heavily optimized. To gain trust in them and safely explore the space of possible optimizations, it is important to verify the monitors themselves. We describe the development and correctness proof in Isabelle/HOL of a monitor for metric first-order dynamic logic. This monitor significantly extends previous work on formally verified monitors by supporting aggregations, regular expressions (the dynamic part), and optimizations including multi-way joins adopted from databases and a new sliding window algorithm.

1 Introduction

As the complexity of IT systems increases, so does the complexity and importance of their verification. Research in runtime verification (RV) has developed well-established formal techniques that can often be applied more easily than traditional formal methods such as model checking. RV is based on dynamic analysis, trading off completeness for efficiency. It is mechanized using *monitors*, which are algorithms that search sequences of events, either offline from log files or online, for patterns indicating faults.

Monitors must be trusted when they are used as verifiers. This trust can be justified by checking the monitors themselves for correctness [16,17,31,36,41,42,44,45,49]. Recently, a simplified version of the algorithm used in the MonPoly tool [8,9] has been formalized and proved correct in Isabelle/HOL [45] (Sect. 2). MonPoly and its formal counterpart, called VeriMon, are both monitors for metric first-order temporal logic (MFOTL). However, VeriMon only supports a restricted fragment of this logic and lacks many optimizations that are necessary for an acceptable and competitive performance.

We present a formally verified monitor, VeriMon+, that substantially extends and improves VeriMon. VeriMon+ closes all expressiveness gaps between MonPoly and VeriMon. It supports aggregation operators like sum and average [7] similar to those found in database query languages, arbitrary negations of closed formulas, the unbounded ○ (Next) operator, and constraints involving terms (e.g., $P(x) \wedge y = x + 2$). Due to space limitations, our focus (Sect. 3) will be

N. Peltier and V. Sofronie-Stokkermans (Eds.): IJCAR 2020, LNAI 12166, pp. 432–453, 2020.
https://doi.org/10.1007/978-3-030-51074-9_25

primarily on aggregations, our largest addition. Moreover, VeriMon+ exceeds MFOTL in expressiveness by featuring a significantly richer specification language, metric first-order *dynamic* logic (MFODL). To our knowledge, it is the first monitor for MFODL with past and bounded future operators (Sect. 4). This logic combines MFOTL with regular expressions, similar to linear dynamic logic [22] but enriched with metric constraints, aggregations, and first-order quantification.

We have also implemented and proved correct several new optimizations. First, to speed up the evaluation of conjunctions, we integrated an efficient algorithm for multi-way joins [38,39], which we generalized to include anti-joins (Sect. 5). Second, we developed a specialized sliding window algorithm to evaluate the Since and Until operators more efficiently (Sect. 6). VeriMon+ is executable via the generation of OCaml code from Isabelle. To this end, we augmented the code generation setup for IEEE floating point numbers in OCaml [50] with a linear ordering, which is needed for efficient set and mapping data structures.

The result of our efforts is both a verified monitor and a tool for evaluating unverified monitors. Since MFODL is extremely expressive, this gives us very wide scope. For example, we discovered previously unknown bugs in MonPoly via differential testing (Sect. 7), extending a previous case study [45]. As this experience suggests, and we firmly believe, formal verification is the most reliable way to obtain correct, optimized monitors.

In sum, our main contribution is a verified monitor for MFODL with aggregations, a highly expressive specification language that combines regular expressions and first-order temporal logic. Our monitor includes optimizations that are novel in the context of first-order monitoring. Our formalization is publicly available [20,21].

Related Work. We refer to a recent book [4] for an introduction to runtime verification. The main families of specification languages in this domain are extensions of LTL [13,26,46], automata [3], stream expressions [19], and rule systems [23]. We combine two expressive temporal logics and their corresponding monitoring algorithms. MFOTL, implemented in MonPoly [7–9], supports first-order quantification over parametrized events, but it cannot express all regular patterns. Metric dynamic logic (MDL), implemented in Aerial [12], supports regular expressions, but it is not first-order. VeriMon+ is based on VeriMon [45], which only supports a fragment of MFOTL and is inefficient (Sect. 7). We refer to [45, Section 1] for an overview of related monitor formalizations. Relational database systems have been formalized by Malecha et al. [33] and by Benzaken et al. [15]. These works use binary joins only, which are not worst-case optimal.

Another efficient first-order monitor, DejaVu [25], supports past-only first-order temporal logic. It uses binary decision diagrams (BDDs) and does not restrict the use of negation, unlike MonPoly, which uses finite tables. DejaVu's performance is incomparable to MonPoly's and it is unclear whether multi-way joins can improve conjunctions of BDDs. Aerial and VeriMon+ evaluate regular expressions using derivatives [2,18], which also have been used for timed regular

```
datatype data = Int int | Flt double | Str string     type_synonym ts = nat
type_synonym db = (string × data list) set            typedef trace = {s :: (db × ts) stream. trace s}
datatype trm = V nat | C data | trm + trm | ...        typedef I = {(a :: nat, b :: enat). a ≤ b}
datatype frm = string(trm list) | trm ≈ trm | trm ≺ trm | trm ≼ trm
  | ¬ frm | ∃ frm | frm ∨ frm | frm ∧ frm | ●_I frm | ○_I frm | frm S_I frm | frm U_I frm
fun etrm :: data list ⇒ trm ⇒ data where
  etrm v (V x) = v ! x | etrm v (C x) = x | etrm v (t1 + t2) = etrm v t1 + etrm v t2 | ...
fun sat :: trace ⇒ data list ⇒ nat ⇒ frm ⇒ bool where
  sat σ v i (r(ts)) = ((r, map (etrm v) ts) ∈ Γ σ i)   | sat σ v i (t1 ≈ t2) = (etrm v t1 = etrm v t2)
  | sat σ v i (t1 ≺ t2) = (etrm v t1 < etrm v t2)       | sat σ v i (t1 ≼ t2) = (etrm v t1 ≤ etrm v t2)
  | sat σ v i (¬φ) = (¬sat σ v i φ)                     | sat σ v i (∃φ) = (∃z. sat σ (z#v) i φ)
  | sat σ v i (α∨β) = (sat σ v i α ∨ sat σ v i β)       | sat σ v i (α∧β) = (sat σ v i α ∧ sat σ v i β)
  | sat σ v i (●_I φ) = (case i of 0 ⇒ False | j+1 ⇒ T σ i − T σ j ∈_I I ∧ sat σ v j φ)
  | sat σ v i (○_I φ) = (T σ (i+1) − T σ i ∈_I I ∧ sat σ v (i+1) φ)
  | sat σ v i (α S_I β) = (∃j ≤ i. T σ i − T σ j ∈_I I ∧ sat σ v j β ∧ (∀k ∈ {j <.. i}. sat σ v k α))
  | sat σ v i (α U_I β) = (∃j ≥ i. T σ j − T σ i ∈_I I ∧ sat σ v j β ∧ (∀k ∈ {i ..< j}. sat σ v k α))
```

Fig. 1. Syntax and semantics of MFOTL as presented in [45], with additions in gray

expressions [47]. Quantified regular expressions [1,34] extend regular expressions with data and aggregations. They can be evaluated efficiently, but can neither express metric constraints nor future modalities directly.

2 A Verified Monitor for Metric First-Order Temporal Logic

VeriMon [45] is a formally verified monitor for a large fragment of MFOTL [8]. The monitor takes an MFOTL formula, which may be open, and incrementally processes an infinite stream of time-stamped events. It outputs for every stream position the set of variable assignments that satisfy the formula. Thus, the monitor can be used to extract data from the stream. Typically, one is interested in the violations of a property specified as an MFOTL formula, which can be obtained by monitoring the negated formula.

We give an overview of MFOTL and VeriMon. We also cover some of the smaller additions in our new monitor, VeriMon+, highlighted in gray. For readability, we liberally use abbreviations and symbolic notation, departing mildly from Isabelle's syntax.

Figure 1 shows MFOTL's syntax and semantics. Events have a name (*string*) and a list of parameters of type *data*. In VeriMon+, *data* is a disjoint union of integers, double-precision floats, and strings. Multiple events are grouped together into a database (*db*) if they are considered to occur simultaneously. We call an infinite stream of databases, augmented with their corresponding time-stamps, an event stream or *trace*. Time-stamps (*ts*) are modeled as natural numbers (*nat*). We write $\mathsf{T}\ \sigma\ i$ to denote the time-stamp of the ith database $\Gamma\ \sigma\ i$ of the event stream σ. The predicate trace expresses that the time-stamps are monotone, i.e., $\mathsf{T}\,\sigma\, i \leq \mathsf{T}\,\sigma\,(i{+}1)$ for all $i \geq 0$, and always eventually strictly increasing, i.e., $\forall t.\ \exists i.\ t < \mathsf{T}\,\sigma\, i$. Consecutive time-points i can have the same time-stamp.

Terms and formulas are represented by the datatypes *trm* and *frm*, respectively. Our formalization uses de Bruijn indices for free and bound variables (constructor V). In examples, we prefer the standard named syntax (and omit V). The type $\mathcal{I}$ models nonempty, possibly unbounded intervals over *nat*. We write $n \in_{\mathcal{I}} I$ for n's membership in I, and $[a, b]$ for the unique interval satisfying $n \in_{\mathcal{I}} [a, b]$ iff $a \leq n \leq b$. The right bound b is of type *enat*, i.e., either a natural number or infinity ∞ for an unbounded interval.

The functions etrm and sat (Fig. 1) define MFOTL's semantics. Both take a variable assignment v, which is a list of type *data list* whose ith element $v\ !\ i$ is the value assigned to the variable with index i. The function etrm evaluates terms under a given assignment. The expression $\mathsf{sat}\ \sigma\ v\ i\ \varphi$ is true iff the formula φ is satisfied by v at time-point i in the trace σ. VeriMon+ adds arithmetic operators and type conversions to terms, as well as the predicates $\prec$ and $\preceq$. Their semantics on *data* is lifted from the corresponding operations on integers, floats, and strings, whenever they are meaningful. The ordering $\leq$ on *data* is total: strings are compared lexicographically and $\mathsf{Int}\ i < \mathsf{Flt}\ f < \mathsf{Str}\ s$.

VeriMon computes sets of satisfactions (i.e., satisfying assignments) by recursion over the formula's structure. It represents these sets as finite tables, to which it applies standard relational operations such as the natural join ($\bowtie$) and union. Tables are sets of tuples, which are lists of optional *data* values; missing values are denoted by $\bot$. This representation allows us to use tuples with the same length across subformulas with different free variables. The predicate wf_tuple defines the well-formed tuples for a given length n and a set of variables V. We also refer to V as the columns of a tuple (or table).

definition wf_tuple :: *nat* $\Rightarrow$ *nat set* $\Rightarrow$ *tuple* $\Rightarrow$ *bool* **where**
  $\mathsf{wf\_tuple}\ n\ V\ v = (\mathsf{length}\ v = n \wedge (\forall x < n.\ v\,!\,x = \bot \longleftrightarrow x \notin V))$

The set of satisfactions may be infinite. VeriMon supports only a fragment of MFOTL for which all computed tables are finite. The predicate safe (omitted) defines the monitorable fragment [45]. It accepts only certain combinations of operators and constrains the free variables of subformulas. Also, the intervals of all U operators must be bounded.

VeriMon's interface consists of two functions init :: *frm* $\Rightarrow$ *mstate* and step :: *db* $\times$ *ts* $\Rightarrow$ *mstate* $\Rightarrow$ (*nat* $\times$ *table*) *list* $\times$ *mstate*. The former initializes the monitor's state, and the latter updates it with a new time-stamped database to report any new satisfactions. We require that satisfactions are reported for every time-point and in order. Note that a formula containing a future operator such as U cannot necessarily be evaluated at time-point i after observing the ith database. Therefore, the output for several time-points may become available at once, so step returns a list of pairs of time-points and tables.

We describe the evaluation of $\alpha\,\mathsf{S}_{[a,\,b]}\,\beta$ in more detail. This formula is equivalent to the disjunction of $\alpha\,\mathsf{S}_{[c,\,c]}\,\beta$ for all c such that $a \leq c \leq b$. Suppose that the most recent time-point is i with time-stamp τ. The monitor's state for $\alpha\,\mathsf{S}_{[a,\,b]}\,\beta$

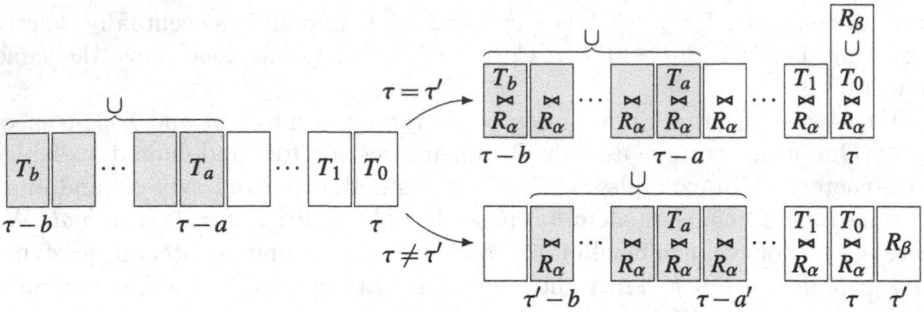

Fig. 2. Simplified state of a Since operator and its update

consists of a list of tables T_c with the satisfactions of $\alpha\, \mathsf{S}_{[c,\,c]}\, \beta$ at time-point i, along with the corresponding time-stamps $\tau - c$. VeriMon also stores the satisfactions T_c (and time-stamps) for $0 \leq c < a$, which are not yet in the interval. Figure 2 (left) depicts a state, where we assume for simplicity that we store a table for every time-stamp between $\tau - b$ and τ. (In reality, time-stamps not in the trace do not have a corresponding entry in this list.) The state is updated for every new time-point with time-stamp τ', for which we already know the satisfactions R_α and R_β of the subformulas α and β. In Fig. 2, we distinguish whether τ' equals τ (otherwise $\tau' > \tau$ by monotonicity). The update consists of three steps: (1) remove tables that fall out of the interval; (2) evaluate the conjunction of each remaining table with R_α using a relational join; and (3) add the new tuples from R_β, either by inserting them into the most recent table T_0 or by adding a new table, depending on whether τ' equals τ. Finally, we take the union of all tables within the interval to obtain the satisfactions of $\alpha\, \mathsf{S}_{[a,b]}\, \beta$.

We summarize VeriMon's correctness, which we also prove for VeriMon+. It relates the monitor's implementation to its specification $\mathsf{verdicts} :: \mathit{frm} \Rightarrow (\mathit{db} \times \mathit{ts})\ \mathit{list} \Rightarrow (\mathit{nat} \times \mathit{tuple})\ \mathit{set}$, which defines the expected output on a stream prefix. The first result shows that verdicts characterizes an MFOTL monitor, where $\mathsf{prefix}\ \pi\ \sigma$ means that π is a prefix of σ, and $\mathsf{map\ the}\ v$ converts v to an assignment by mapping $\bot$ to an unspecified value.

Lemma 1 (**[45], Lemma 2**). *Suppose that* safe φ *is true. Then,* verdicts φ *is sound and eventually complete, i.e., for all prefixes* π *of trace* σ*, time-points* i*, and tuples* v*,*

(a) $(i, v) \in \mathsf{verdicts}\ \varphi\ \pi \longrightarrow \mathsf{sat}\ \sigma\ (\mathsf{map\ the}\ v)\ i\ \varphi$*, and*
(b) $i < \mathsf{length}\ \pi \wedge \mathsf{wf\_tuple}\ (\mathsf{nfv}\ \varphi)\,(\mathsf{fv}\ \varphi)\ v \wedge (\forall \sigma'.\ \mathsf{prefix}\ \pi\ \sigma' \longrightarrow \mathsf{sat}\ \sigma'\ (\mathsf{map\ the}\ v)\ i\ \varphi) \longrightarrow (\exists \pi'.\ \mathsf{prefix}\ \pi'\ \sigma \wedge (i, v) \in \mathsf{verdicts}\ \varphi\ \pi')$.

Above, $\mathsf{nfv}\ \varphi$ is the smallest number larger than all free variables of φ, written $\mathsf{fv}\ \varphi$. The next result establishes the implementation's correctness using the state invariant $\mathsf{wf\_mstate} :: \mathit{frm} \Rightarrow (\mathit{db} \times \mathit{ts})\ \mathit{list} \Rightarrow \mathit{mstate} \Rightarrow \mathit{bool}$ (omitted). Let set convert lists into sets, $\mathsf{last\_ts}\ \pi$ be the last time-stamp in π, and $\pi_1 @ \pi_2$ be the concatenation of π_1 and π_2.

Theorem 1 **([45], Theorem 1).** *The initialization* init *establishes the invariant and the update* step *preserves the invariant and its output can be described in terms of* verdicts*:*

(a) *If* safe φ, *then* wf_mstate φ [] (init φ).
(b) *Let* step (db, τ) $mst = (A, mst')$. *If* wf_mstate φ π mst *and* last_ts $\pi \leq \tau$, *then* $(\bigcup(i, V) \in$ set $A.\ \{(i, v) \mid v \in V\})$ = verdicts φ $(\pi\ @\ [(db, \tau)])$ − verdicts φ π *and* wf_mstate φ $(\pi\ @\ [(db, \tau)])$ mst'.

3 Aggregations

Basin et al. [7] extended MFOTL with a generic aggregation operator. This operator was inspired by the group-by clause and aggregation functions of SQL. It first partitions the satisfying assignments of its subformula into groups, and then computes a summary value, such as count, sum, or average, for each group. We formalized the aggregation operator's semantics, added an evaluation algorithm to VeriMon+, and proved its correctness.

Consider the formula $s \leftarrow$ Sum $x; x.\ P(g, x)$. The aggregation operator $s \leftarrow$ Sum $x; x$ has four parameters: a result variable (s), the aggregation type (Sum), an aggregation term (first x), and a list of variables that are bound by the operator and thus excluded from grouping (second x). When evaluated, the above formula yields a set of tuples (s, g). There is one such tuple for every value of g with at least one P event that has g's value as its first parameter. The values of g partition the satisfactions of $P(g, x)$ into groups. For every group, the sum over the values of x in that group is assigned to the variable s.

We added the constructor $nat \leftarrow agg\_op\ trm; nat.\ frm$ to frm. Consider the instance $y \leftarrow \Omega\ t; b.\ \varphi$. The operator binds b variables simultaneously in the formula φ and in the term t, over which we aggregate. In examples, we list the bound variables explicitly instead of writing the number b. The remaining free variables (possibly none) of φ are used for grouping. The variable y receives the result of the aggregation operation $\Omega = (\omega, d)$, where ω is one of Cnt (count), Min, Max, Sum, Avg (average), or Med (median). The default value d, which we usually omit, determines the result for empty groups (e.g., 0 for Cnt). The formula's free variables are those of φ excluding the b bound variables, plus y.

Figure 3 shows the semantics of the aggregation operator $y \leftarrow \Omega\ t; b.\ \varphi$. The assignment v determines both a group and a candidate value $v\ !\ y$ for the aggregation's result on that group. The sat function checks whether the value is correct. First, it computes the set M, which encodes a multiset in the form of pairs (x, c), where c is x's multiplicity. This multiset contains the values of the term t under all assignments $z\ @\ v$ that satisfy φ, where z is an assignment to the bound variables. The expression card$^\infty$ Z stands for the cardinality of Z when it is finite, and ∞ otherwise. Then, sat compares $v\ !\ y$ to the result of the aggregation operation Ω on M, which is given by agg_op Ω M (omitted).

We extended the safe predicate with sufficient conditions that describe when the aggregation formula $y \leftarrow \Omega\ t; b.\ \varphi$ has finitely many satisfactions. We require that φ satisfies safe, that the variable y is not free in φ excluding the b bound variables, and that all bound variables and the variables in t occur free in φ. We

$\textbf{fun}\ \mathsf{sat} :: \mathit{trace} \Rightarrow \mathit{data\ list} \Rightarrow \mathit{nat} \Rightarrow \mathit{frm} \Rightarrow \mathit{bool}\ \textbf{where}\ \ldots$
$|\ \mathsf{sat}\ \sigma\ v\ i\ (y \leftarrow \Omega\ t; b.\ \varphi) = (\underline{\mathrm{let}}\ M = \{(x, \mathsf{card}^{\infty}\ Z) \mid x\ Z.$
$Z = \{z.\ \mathsf{length}\ z = b \wedge \mathsf{sat}\ \sigma\ (z\ @\ v)\ i\ \varphi \wedge \mathsf{etrm}\ (z\ @\ v)\ t = x\} \wedge Z \neq \{\}\}$
$\underline{\mathrm{in}}\ (M = \{\} \longrightarrow \mathsf{fv}\ \varphi \subseteq \{0\,..{<}\,b\}) \wedge v\,!\,y = \mathsf{agg\_op}\ \Omega\ M)$

$\textbf{fun}\ \mathsf{eval\_agg} :: \mathit{nat} \Rightarrow \mathit{bool} \Rightarrow \mathit{nat} \Rightarrow \mathit{agg\_op} \Rightarrow \mathit{nat} \Rightarrow \mathit{trm} \Rightarrow \mathit{table} \Rightarrow \mathit{table}\ \textbf{where}$
$\mathsf{eval\_agg}\ n\ g_0\ y\ \Omega\ b\ t\ R = (\underline{\mathrm{if}}\ g_0 \wedge R = \{\}\ \underline{\mathrm{then}}\ \mathsf{singleton\_table}\ n\ y\ (\mathsf{agg\_op}\ \Omega\ \{\})\ \underline{\mathrm{else}}$
$(\lambda k.\ \underline{\mathrm{let}}\ G = \{v \in R \mid \mathsf{drop}\ b\ v = k\};\ M = (\lambda x.\ (x, \mathsf{card}^{\infty}\ \{v \in G \mid \mathsf{meval\_trm}\ t\ v = x\}))\ `$
$(\mathsf{meval\_trm}\ t)\ `\ G\ \underline{\mathrm{in}}\ k[y := \mathsf{Some}\ (\mathsf{agg\_op}\ \Omega\ M)])\ `\ (\mathsf{drop}\ b)\ `\ R)$

Fig. 3. Semantics and evaluation of the aggregation operator

adopted the convention [7] that an aggregation formula is not satisfied when M is empty, unless all free variables of φ are bound by the operator. Otherwise, there would be infinitely many groups (and hence, satisfactions) with the aggregate value agg_op Ω {}, assuming that φ is safe.

Figure 3 also defines eval_agg, which evaluates the aggregation operator. It takes a table R with φ's satisfactions, and returns a table with the aggregation operator's satisfactions. The first argument n controls the length of the tuples in the tables (Sect. 2). The argument g_0 specifies whether all free variables of φ are bound by the operator. The remaining arguments y, Ω, b, and t are those of the operator. We write f ‘ X for the image of X under f.

In eval_agg, we first check whether $g_0 \wedge R = \{\}$ is true to handle the special case mentioned above. (The expression singleton_table n y a is a table with a single tuple of length n that assigns a to variable y.) Otherwise, we compute the aggregate value separately for each group k. The set of groups is obtained by discarding the first b values of each tuple in R. To every group k, we apply the lambda-term to augment the tuple with the aggregate value. The set G contains all tuples in the group. Note that these tuples extend k with assignments to the b bound variables. Then, we compute the image of G under the term t, which is evaluated by meval_trm :: *trm* ⇒ *tuple* ⇒ *data* (omitted). Finally, we obtain the multiset M by counting how many tuples in G map to each value in the image.

4 Regular Expressions

VeriMon+ extends VeriMon's language by generalizing MFOTL's temporal operators to regular expressions. The resulting metric first-order dynamic logic (MFODL) can be seen [24, §3.16] as the "supremum" (in the sense of combining features) of metric dynamic logic (MDL) [12] and MFOTL [8]. Peycheva's master's thesis [40] develops a monitor for past-only MFODL. We give the first formal definition of MFODL with past and future operators. We also define a fragment whose formulas can be evaluated using finite relations (Sect. 4.1). This fragment guides our evaluation algorithm's design (Sect. 4.2).

Figure 4 (left) defines the syntax and semantics of our variant of regular expressions. The type *re* is parametrized by a type variable *'a*, which is used

$$
\begin{array}{l}
\textbf{datatype}\ \textit{'a re} = \star^{\textit{nat}} \mid \textit{'a}? \mid \textit{'a re} + \textit{'a re} \mid \textit{'a re} \cdot \textit{'a re} \mid (\textit{'a re})^* \\
\textbf{fun}\ \mathsf{match} :: (\textit{nat} \Rightarrow \textit{'a} \Rightarrow \textit{bool}) \Rightarrow \textit{'a re} \Rightarrow \textit{nat} \otimes \textit{nat}\ \textbf{where} \\
\quad \mathsf{match}\ \textit{test}\ (\star^k) = \{(i,j) \mid j = i+k\} \\
\mid \mathsf{match}\ \textit{test}\ (x?) = \{(i,i) \mid \textit{test}\ i\ x\} \\
\mid \mathsf{match}\ \textit{test}\ (r+s) = \mathsf{match}\ \textit{test}\ r \cup \mathsf{match}\ \textit{test}\ s \\
\mid \mathsf{match}\ \textit{test}\ (r \cdot s) = \mathsf{match}\ \textit{test}\ r \bullet \mathsf{match}\ \textit{test}\ s \\
\mid \mathsf{match}\ \textit{test}\ (r^*) = (\mathsf{match}\ \textit{test}\ r)^* \\
\textbf{datatype}\ \textit{frm} = \ldots \mid \blacktriangleleft_{\mathcal{I}}\ (\textit{frm re}) \mid \triangleright_{\mathcal{I}}\ (\textit{frm re}) \\
\textbf{fun}\ \mathsf{sat} :: \textit{trace} \Rightarrow \textit{data list} \Rightarrow \textit{nat} \Rightarrow \textit{frm} \Rightarrow \textit{bool}\ \textbf{where}\ \ldots \\
\mid \mathsf{sat}\ \sigma\ v\ i\ (\blacktriangleleft_I\ r) = (\exists j \le i.\ \mathsf{T}\ \sigma\ i - \mathsf{T}\ \sigma\ j \in_{\mathcal{I}} I \wedge (j,i) \in \mathsf{match}\ (\mathsf{sat}\ \sigma\ v)\ r) \\
\mid \mathsf{sat}\ \sigma\ v\ i\ (\triangleright_I\ r) = (\exists j \ge i.\ \mathsf{T}\ \sigma\ j - \mathsf{T}\ \sigma\ i \in_{\mathcal{I}} I \wedge (i,j) \in \mathsf{match}\ (\mathsf{sat}\ \sigma\ v)\ r)
\end{array}
$$

$$
\begin{array}{lcl}
\bullet_I\ \alpha & \rightsquigarrow & \blacktriangleleft_I\ (\alpha? \cdot \star) \\
\alpha\ \mathsf{S}_I\ \beta & \rightsquigarrow & \blacktriangleleft_I\ (\beta? \cdot (\star \cdot \alpha?)^*) \\
\bigcirc_I\ \alpha & \rightsquigarrow & \triangleright_I\ (\star \cdot \alpha?) \\
\alpha\ \mathsf{U}_I\ \beta & \rightsquigarrow & \triangleright_I\ ((\alpha? \cdot \star)^* \cdot \beta?)
\end{array}
$$

Fig. 4. Syntax and semantics of MFODL (left) and conversion of MFOTL into MFODL (right)

in the _? constructor. The semantics is given by match and assigns to each expression a binary relation ($\otimes$) on natural numbers. Intuitively, a pair $(i,\ j)$ is in the relation assigned to r when r matches the portion of a trace from i to j. The trace notion is abstracted away in match via the argument *test*, which indicates whether a parameter of type $'a$ may advance past a given point.

In more detail, the wildcard operator $\star^k$ matches all pairs $(i,\ j)$, where $j = i+k$; we write $\star$ for the useful special case $\star^1$. The test $x?$ only matches pairs of the form $(i,\ i)$ that pass $\textit{test}\ i\ x$. The semantics of alternation (+) as union ($\cup$), concatenation ($\cdot$) as relation composition ($\bullet$), and Kleene star (_*) as reflexive-transitive closure (_*) is standard.

Figure 4 (left) also shows *frm*'s extension with two constructors that use regular expressions. The regular expression's parameter nests a recursive occurrence of *frm*, i.e., our regular expressions' leaves are formulas, which in turn may further nest regular expressions, and so on. MDL's syntax is often presented as a mutually recursive datatype [12]. Our nested formulation is beneficial because it lets us formalize regular expressions independently, for use in different applications (e.g., monitors for MDL and MFODL).

In terms of their semantics, the two new operators naturally generalize the S_I and U_I operators. The past match operator $\blacktriangleleft_I\ r$ is satisfied at i if there is an earlier time-point j subject to the same temporal constraint I as in the satisfaction of S_I and moreover the regular expression r matches from j to i. For the future match operator $\triangleright_I\ r$, the situation is symmetric with the existentially quantified j being a future time-point. In both cases, the *test* parameter of match is recursively instantiated with the satisfaction predicate sat.

We can embed MFOTL into MFODL by expressing the temporal operators using semantically equivalent formulas built from regular expressions (Fig. 4, right). Thus, we could in principle remove the operators $\bullet$, S, $\bigcirc$, and U from *frm* and use regular expressions instead. We prefer to keep these operators in *frm* as this allows us to optimize their evaluation in a way that is not available for the more general match operators (Sect. 6).

$\textbf{datatype}\ \mathit{context} = \mathsf{past} \mid \mathsf{futu} \qquad \textbf{datatype}\ \mathit{mode} = \mathsf{strict} \mid \mathsf{lax}$

$\textbf{fun}\ \mathsf{safe} :: \mathit{context} \Rightarrow \mathit{mode} \Rightarrow \mathit{frm}\ \mathit{re} \Rightarrow \mathit{bool}\ \textbf{where}$

$\quad \mathsf{safe}\ \_\ \_\ (\star^k) = \mathsf{True}$

$\mid \mathsf{safe}\ \_\ m\ ((\neg\varphi)?) = (m = \mathsf{lax} \wedge \mathsf{safe}\ \varphi)$

$\mid \mathsf{safe}\ \_\ \_\ (\varphi?) = \mathsf{safe}\ \varphi$

$\mid \mathsf{safe}\ c\ m\ (r+s) = ((m = \mathsf{lax} \vee \mathsf{fv}\ r = \mathsf{fv}\ s) \wedge \mathsf{safe}\ c\ m\ r \wedge \mathsf{safe}\ c\ m\ s)$

$\mid \mathsf{safe}\ \mathsf{futu}\ m\ (r \cdot s) = ((m = \mathsf{lax} \vee \mathsf{fv}\ r \subseteq \mathsf{fv}\ s) \wedge \mathsf{safe}\ \mathsf{futu}\ \mathsf{lax}\ r \wedge \mathsf{safe}\ \mathsf{futu}\ m\ s)$

$\mid \mathsf{safe}\ \mathsf{past}\ m\ (r \cdot s) = ((m = \mathsf{lax} \vee \mathsf{fv}\ s \subseteq \mathsf{fv}\ r) \wedge \mathsf{safe}\ \mathsf{past}\ m\ r \wedge \mathsf{safe}\ \mathsf{past}\ \mathsf{lax}\ s)$

$\mid \mathsf{safe}\ c\ m\ (r^*) = (m = \mathsf{lax} \wedge \mathsf{safe}\ c\ m\ r)$

$\textbf{fun}\ \mathsf{safe} :: \mathit{frm} \Rightarrow \mathit{bool}\ \textbf{where}\ \ldots$

$\mid \mathsf{safe}\ (\blacktriangleleft_I\ r) = \mathsf{safe}\ \mathsf{past}\ \mathsf{strict}\ r \qquad \mid \mathsf{safe}\ (\triangleright_{[a,b]}\ r) = \mathsf{safe}\ \mathsf{futu}\ \mathsf{strict}\ r \wedge b < \infty$

Fig. 5. Safety conditions for MFODL

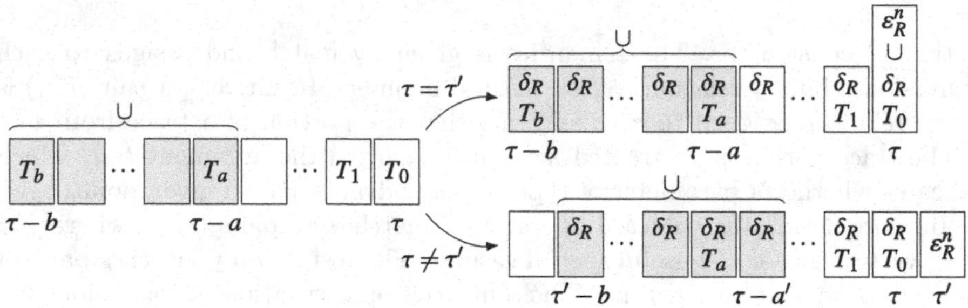

Fig. 6. Simplified state of a past match operator and its update

We conclude MFODL's introduction with an example. Many systems for user authentication follow a policy like: "A user should not be able to authenticate after entering the wrong password three times in a row within the last 10 minutes." We write $✗(u)$ for the event "User u entered the wrong password" and $✓(u)$ for "User u has successfully authenticated." Additionally, we abbreviate $\varphi? \cdot \star$ by φ. (This abbreviation is only used when φ appears in a regular expression position, e.g., as an argument of $\cdot$). Then the formula

$$✓(u) \wedge \blacktriangleleft_{[0,600]} \left(✗(u) \cdot (\neg ✓(u))^* \cdot ✗(u) \cdot (\neg ✓(u))^* \cdot ✗(u) \cdot (\neg ✓(u))^*\right)$$

expresses this policy's violations: its satisfying assignments are precisely the users that successfully authenticate after entering wrong credentials for three times in the last 600 seconds, without intermediate successful authentications. We can express this property in MFOTL using three nested S operators, one for each of the $✗(u)$ subformulas. Yet, it is unclear which intervals to put as arguments to S beyond the fact that they should sum up to 600. The rather impractical solution exploits that there are only finitely many ways to split the intervals due to their bounds being natural numbers and constructs the disjunction of all possible splits (180 901 in this case). MFODL remediates this infeasible construction.

4.1 Finitely Evaluable Regular Expressions

Following MonPoly's design [8], VeriMon+ represents all sets of satisfying assignments with finite tables. The databases occurring in the trace are all finite, yet their combination may not be. Therefore, MonPoly and VeriMon+ work with syntactic restrictions that ensure that all sets that arise are finite. For example, negation must occur under a conjunction $\alpha \wedge \neg\beta$, where the free variables of β, written $\mathsf{fv}\ \beta$, are contained in those of α. We say that $\neg\beta$ is guarded by α and compute $\alpha \wedge \neg\beta$ as the anti-join ($\triangleright$) of the corresponding tables. For disjunctions $\alpha \vee \beta$, we must have $\mathsf{fv}\ \alpha = \mathsf{fv}\ \beta$. Similar restrictions also apply to temporal operators: to evaluate $\alpha\, \mathsf{S}_I\, \beta$ and $\alpha\, \mathsf{U}_I\, \beta$ we require $\mathsf{fv}\ \alpha \subseteq \mathsf{fv}\ \beta$.

We derive a new sufficient criterion for match operators to have finitely many satisfying assignments. To develop some intuition, we first consider several examples that result in infinite tables. The first example is any expression with a Kleene star as the topmost operator. The formula $\varphi = \blacktriangleleft_{[0,b]} (r^*)$ is satisfied at all points i for all assignments v (regardless of r's free variables) since $0 \in_{\mathcal{I}} I$ and any (i, i) matches r^*. Thus, when we evaluate φ at i, we can choose i as the witness for the existential quantifier in the definition of sat. It follows that Kleene stars must be guarded by a finite table.

The union of two finite tables is finite only if the tables have the same columns (assuming an infinite domain *data*). This explains the requirement for the subformulas of $\vee$ to have the same variables, but a similar requirement is needed for the $+$ of regular expressions. Perhaps more surprisingly, concatenation can also hide a union: consider $\varphi = \blacktriangleleft_{[0,b]} (r \cdot s^*)$ and assume that s matches (j, i) for some $j < i$. By the semantics of concatenation, we can split the satisfactions of φ into those that use s^*'s matching pair (i, i) (i.e., the satisfying assignments of $\blacktriangleleft_I r$ at i) and those that do not. To combine these assignments it seems necessary to take the union of the satisfaction of φ and $\blacktriangleleft_I r$, which in turn requires these formulas to have the same free variables, or equivalently $\mathsf{fv}\ s \subseteq \mathsf{fv}\ r$ (overloading notation to apply fv to regular expression). The future match operator behaves symmetrically, requiring the side condition $\mathsf{fv}\ r \subseteq \mathsf{fv}\ s$ for $\triangleright_{[0,b]} (r^* \cdot s)$.

MonPoly also allows the left subformula of S and U to be negated: $(\neg\alpha)\ \mathsf{S}_I\ \beta$ and $(\neg\alpha)\ \mathsf{U}_I\ \beta$. Hence, we should support the MFODL variants $\blacktriangleleft_I (\beta? \cdot (\star \cdot (\neg\alpha)?)^*)$ and $\triangleright_I (((\neg\alpha)? \cdot \star)^* \cdot \beta?)$, but also generalize these patterns to flexibly support negated tests.

Our solution to these issues comprises the predicates shown in Fig. 5. The safe predicate on regular expressions is parametrized by two flags: *context* distinguishing whether the expression occurs under a past or a future match operator and *mode* determining whether the tests may be negated and other safety conditions relaxed. The most interesting cases are those for concatenation. There, in addition to the fv side conditions, only one argument is checked recursively in the same mode as the overall expression. The other argument is checked using the lax mode, in which side conditions are skipped, except for the requirement that (possibly negated) formulas under the test operators are safe. The context parameter dictates which argument keeps, and which changes, the mode.

$$\begin{array}{l}\textbf{fun}\ \varepsilon_{\mathsf{lax}} :: \mathit{table} \Rightarrow (\mathit{frm} \Rightarrow \mathit{table})\\ \quad \Rightarrow \mathit{frm\ re} \Rightarrow \mathit{table}\ \textbf{where}\\ \quad \varepsilon_{\mathsf{lax}}\ X\ R\ (\star^k) = (\underline{\mathsf{if}}\ k = 0\ \underline{\mathsf{then}}\ X\ \underline{\mathsf{else}}\ \{\})\\ \mid \varepsilon_{\mathsf{lax}}\ X\ R\ ((\neg\varphi)?) = X \triangleright R\ \varphi\\ \mid \varepsilon_{\mathsf{lax}}\ X\ R\ (\varphi?) = X \bowtie R\ \varphi\\ \mid \varepsilon_{\mathsf{lax}}\ X\ R\ (r+s) = \varepsilon_{\mathsf{lax}}\ X\ R\ r \cup \varepsilon_{\mathsf{lax}}\ X\ R\ s\\ \mid \varepsilon_{\mathsf{lax}}\ X\ R\ (r \cdot s) = \varepsilon_{\mathsf{lax}}\ X\ R\ r \bowtie \varepsilon_{\mathsf{lax}}\ X\ R\ s\\ \mid \varepsilon_{\mathsf{lax}}\ X\ R\ (r^*) = X\end{array}$$

$$\begin{array}{l}\textbf{fun}\ \varepsilon_{\mathsf{strict}} :: \mathit{nat} \Rightarrow (\mathit{frm} \Rightarrow \mathit{table})\\ \quad \Rightarrow \mathit{frm\ re} \Rightarrow \mathit{table}\ \textbf{where}\\ \quad \varepsilon_{\mathsf{strict}}\ n\ R\ (\star^k) =\\ \qquad (\underline{\mathsf{if}}\ k = 0\ \underline{\mathsf{then}}\ \{\overbrace{(\bot, \ldots, \bot)}^{n}\}\ \underline{\mathsf{else}}\ \{\})\\ \mid \varepsilon_{\mathsf{strict}}\ n\ R\ (\varphi?) = R\ \varphi\\ \mid \varepsilon_{\mathsf{strict}}\ n\ R\ (r+s) = \varepsilon_{\mathsf{strict}}\ n\ R\ r \cup \varepsilon_{\mathsf{strict}}\ n\ R\ s\\ \mid \varepsilon_{\mathsf{strict}}\ n\ R\ (r \cdot s) = \varepsilon_{\mathsf{lax}}\ (\varepsilon_{\mathsf{strict}}\ n\ R\ r)\ R\ s\end{array}$$

$$\begin{array}{l}\textbf{fun}\ \delta :: (\mathit{frm\ re} \Rightarrow \mathit{frm\ re}) \Rightarrow (\mathit{frm} \Rightarrow \mathit{table}) \Rightarrow (\mathit{frm\ re} \Rightarrow \mathit{table}) \Rightarrow \mathit{frm\ re} \Rightarrow \mathit{table}\ \textbf{where}\\ \quad \delta\ \kappa\ R\ T\ (\star^k) = (\underline{\mathsf{if}}\ k = 0\ \underline{\mathsf{then}}\ \{\}\ \underline{\mathsf{else}}\ T\ (\kappa\ (\star^{k-1})))\\ \mid \delta\ \kappa\ R\ T\ (\varphi?) = \{\}\\ \mid \delta\ \kappa\ R\ T\ (r+s) = \delta\ \kappa\ R\ T\ r \cup \delta\ \kappa\ R\ T\ s\\ \mid \delta\ \kappa\ R\ T\ (r \cdot s) = \delta\ (\lambda t.\ \kappa\ (r \cdot t))\ R\ T\ s \cup \varepsilon_{\mathsf{lax}}\ (\delta\ \kappa\ R\ T\ r)\ R\ s\\ \mid \delta\ \kappa\ R\ T\ (r^*) = \delta\ (\lambda t.\ \kappa\ (r^* \cdot t))\ R\ T\ r\end{array}$$

Fig. 7. The core evaluation functions for MFODL

4.2 Evaluation Algorithm

The evaluation algorithm's structure for the past match operator $◀_I\ r$ (Fig. 6) closely resembles the evaluation of $\alpha\ \mathsf{S}_I\ \beta$ (Fig. 2). What is different is the data that is stored for each time-stamp and the way we update it. For S, each stored table T_c corresponds to the satisfactions of $\alpha\ \mathsf{S}_{[c,\,c]}\ \beta$. For $◀_I\ r$, each T_c is a mapping from a regular expression s to the table denoting the satisfactions of $◀_{[c,\,c]}\ s$. (We represent mappings here by plain functions for readability.) Clearly, this mapping's domain must be finite. We restrict it to the finite set $\Delta(r)$ of right partial derivatives [2,12] of the overall regular expression r, which correspond to the states of a non-deterministic automaton that matches r from right to left.

Partial derivatives allow us to extend satisfactions of $◀_{[c,\,c]}\ s$ for $s \in \Delta(r)$ at time-point i to satisfactions of $◀_{[c+(\tau_{i+1}-\tau_i),\,c+(\tau_{i+1}-\tau_i)]}\ s$ for $s \in \Delta(r)$ at time-point $i+1$. The Since operator's counterpart of this extension is the join with R_α, the new satisfactions of α, which is performed for all T_cs for every update. Here, the extension function δ_R inputs a function R assigning the new satisfactions for all tests occurring in r (possibly with a negation stripped) and updates the mapping T_c. It is defined as $\delta_R\ T = (\lambda s.\ \delta\ \mathsf{id}\ R\ T\ s)$ where δ is defined recursively on the structure of regular expressions as shown in Fig. 7. The first parameter of δ uses continuation passing style. It builds up a regular expression context that we use when evaluating the leaves. It is thus guaranteed that if we apply δ to any regular expression $s \in \Delta(r)$, all calls to T will apply T to some $s' \in \Delta(r)$.

The function δ uses the recursive function $\varepsilon_{\mathsf{lax}}$ in its definition. This function computes the assignments that give rise to matches of the form (i, i) under the assumption that a guard (in form of the table X) is given. For δ, the recursive call acts as $\varepsilon_{\mathsf{lax}}$'s guard.

The function $\varepsilon_{\mathsf{strict}}$ is used to update the state with satisfying assignments at the newly added time-point (Fig. 6). It is only specified for expressions satisfying

$$\textbf{type\_synonym}\ \mathit{atable} = \mathit{nat\ set} \times \mathit{table} \qquad \textbf{type\_synonym}\ \mathit{query} = \mathit{atable\ set}$$
$$\textbf{fun} \downarrow :: \mathit{atable} \Rightarrow \mathit{nat\ set} \Rightarrow \mathit{atable}\ \textbf{where}\ (U, A) \downarrow V = (U \cap V, \{v \downarrow V \mid v \in A\})$$
$$\textbf{fun} \downarrow :: \mathit{query} \Rightarrow \mathit{nat\ set} \Rightarrow \mathit{query}\ \textbf{where}\ Q \downarrow V = \{(U, A) \downarrow V \mid (U, A) \in Q\}$$
$$\textbf{fun}\ \mathsf{extend} :: \mathit{nat\ set} \Rightarrow \mathit{query} \Rightarrow \mathit{nat\ set} \times \mathit{tuple} \Rightarrow \mathit{query}\ \textbf{where}$$
$$\quad \mathsf{extend}\ V\ Q\ (T, t) = \{(U, \{v \in A \mid \forall i \in T \cap U.\ t\,!\,i = v\,!\,i\}) \downarrow V \mid (U, A) \in Q\}$$

$$\textbf{fun}\ \mathsf{generic\_join} :: \mathit{nat\ set} \Rightarrow \mathit{query} \Rightarrow \mathit{query} \Rightarrow \mathit{table}\ \textbf{where}$$
$$\quad \mathsf{generic\_join}\ V\ Q_{pos}\ Q_{neg} = \underline{\text{if}}\ |V| \le 1\ \underline{\text{then}}\ (\textstyle\bigcap(\_, X) \in Q_{pos}.\ X) - (\textstyle\bigcup(\_, Y) \in Q_{neg}.\ Y)\ \underline{\text{else}}$$
$$\quad\quad \underline{\text{let}}\ (I, J) = \mathsf{getIJ}\ V\ Q_{pos}\ Q_{neg};$$
$$\quad\quad\quad Q^I_{pos} = \{(V, X) \in Q_{pos} \mid V \cap I \neq \{\}\} \downarrow I; \quad Q^I_{neg} = \{(U, X) \in Q_{neg} \mid U \subseteq I\};$$
$$\quad\quad\quad A^I = \mathsf{generic\_join}\ I\ Q^I_{pos}\ Q^I_{neg};$$
$$\quad\quad\quad Q^J_{pos} = \{(V, X) \in Q_{pos} \mid V \cap J \neq \{\}\}; \quad Q^J_{neg} = Q_{neg} - Q^I_{neg};$$
$$\quad\quad\quad R = \{(t, \mathsf{generic\_join}\ J\ (\mathsf{extend}\ J\ Q^J_{pos}\ (I, t))\ (\mathsf{extend}\ J\ Q^J_{neg}\ (I, t)) \mid t \in A^I\}$$
$$\quad\quad \underline{\text{in}}\ (\textstyle\bigcup(t, A) \in R.\ \{\mathsf{the}\ (\mathsf{join1}\ (v, t)) \mid v \in A\})$$

Fig. 8. Multi-way join algorithm

safe past strict and uses $\varepsilon_{\mathsf{lax}}$ for subexpressions that only satisfy safe past lax. The recursive structure of $\varepsilon_{\mathsf{strict}}$ and $\varepsilon_{\mathsf{lax}}$ follows the one of safe past. We write $\varepsilon^n_R = (\lambda r.\ \varepsilon_{\mathsf{strict}}\ n\ R\ r)$ and use $\cup$ to denote the pointwise union of mappings in Fig. 6. The Since operator's counterpart of this update is the addition of the satisfactions for the subformula β (Fig. 2).

The above description just sketches our evaluation algorithm and our formalization provides full details. Our proofs establish the monitor's overall correctness, which amounts to the same statement as Theorem 1 but now covers the syntax and semantics extended with the match operators (and aggregations). In particular, the formalization also includes the future match operators for which the evaluation uses similar ideas (partial derivatives), but in a symmetric fashion following the definition of safe future.

5 Multi-way Join

The natural join $\bowtie$ is a central operation in first-order monitors. Not only is it used to evaluate conjunctions; temporal operators also crucially rely on it. Despite this operation's importance, both MonPoly and VeriMon naively compute $A \bowtie B$ as nested unions: $\bigcup v \in A.\ \bigcup w \in B.\ \lceil \mathsf{join1}\ (v, w) \rceil$, where $\mathsf{join1}$ joins two tuples v and w if possible, and $\lceil \_ \rceil$ converts the optional result into a set. In this section, we describe a recent development from database theory that we formalize and extend to optimize the computation of joins.

Ngo et al. [37] and Veldhuizen [48] have developed worst-case optimal multi-way join algorithms that compute the natural join of multiple tables. Here, optimality means that the algorithm never constructs an intermediate result that is larger than the maximum size of all input tables and the overall output. This strictly improves over any evaluation plan using binary joins: There are tables A, B, and C such that the size of $A \bowtie B \bowtie C$ is linear in $|A| = |B| = |C|$, but any plan constructs a quadratic intermediate result from the binary join it evaluates first [39, Fig. 2]. The key idea of the multi-way join is to build the result table column-wise, adding one or more columns at a time, while taking all tables that refer to the currently added columns into account. All intermediate results are restrictions of the overall result to the processed columns, and thus not larger than the overall result.

Figure 8 shows our formalization of the multi-way join algorithm following Ngo et al.'s unified presentation [39] but generalizing it to support anti-joins $\triangleright$; these additions are highlighted in gray. A *query* is a set of *atable*s, i.e., tables annotated with the columns (represented by *nat*) they have. The main function, generic_join, takes as input a set of columns V and two queries Q_{pos} and Q_{neg}. It computes the multi-way join of Q_{pos} while subtracting the tuples of tables in Q_{neg}. For example, generic_join $\{a, b, c, d\}$ $\{(\{a, b\}, A), (\{b, c\}, B), (\{c, d\}, C)\}$ $\{(\{d\}, D), (\{a, c\}, E)\}$ computes $A \bowtie B \bowtie C \triangleright D \triangleright E$.

The algorithm proceeds by recursion on V. The base case in which V is empty or a singleton set is evaluated directly using intersections and unions. We first describe the recursive structure of the original algorithm [39], obtained by ignoring the highlighted anti-join additions in Fig. 8. The algorithm is parametrized by the getIJ function, which partitions V into two nonempty sets I and J that each determine the number of columns and the order in which they are added. Ngo et al. [39] show how different multi-way join algorithms [37,48] can be obtained by using specific instances of getIJ. We use a heuristic to pick first the column i that maximizes the number of tuples in Q_{pos} it affects (by setting $I = \{i\}$). The partitioning only affects performance, not correctness.

Once I and J are fixed, the algorithm constructs a reduced query Q^I_{pos} by focusing on tables that have a column in I. Furthermore, it restricts their columns to I via the overloaded notation $\_ \downarrow I$, which denotes the restriction of tuples (by setting the optional *data* values for columns outside I to $\bot$ [45]), annotated tables, and queries (Fig. 8).

Next, Q^I_{pos} is evaluated recursively, yielding table A^I with columns I. We now consider tables that have a column in J. This yields a second reduced query Q^J_{pos}, which is, however, not restricted to J. Keeping the columns I in Q^J_{pos} allows us to focus on tuples in Q^J_{pos} that match some $t \in A^I$, i.e., coincide with t for all values in columns I. The function extend performs this matching. For each tuple $t \in A^I$, it creates the query extend J Q^J_{pos} (I, t) consisting of tables from Q^J_{pos} restricted to t-matching tuples (in database terminology this is a semi-join) further restricted to columns J. These queries are again solved recursively, each resulting in a table A_t with columns J. The final step consists of merging the tuples t with A_t. Since t and A have disjoint columns I and J, the function call join1 (v, t) will return some result (which we extract via the) for all $v \in A$.

We extend the algorithm to support anti-joins by introducing a second query Q_{neg}, which we think of as being negated. It is not possible to split Q_{neg}'s tables column-wise. Instead, our generalization processes tables with columns U from Q_{neg} once the positive query has accumulated a superset of U as its columns. This is an improvement over the naive strategy of computing Q_{pos} first and only then removing tuples from it.

The correctness of generic_join relies on several side conditions, e.g., no input table may have zero columns and V must be the union of the columns in the positive query. A wrapper function mwjoin takes care of these corner cases, e.g., by computing V from Q_{pos} and Q_{neg}. We omit mwjoin's straightforward definition, but show its correctness property (which only differs from generic_join's correctness by having fewer assumptions):

$$\begin{aligned}
&Q_{pos} \neq \{\} \wedge (\forall (V, A) \in Q_{pos} \cup Q_{neg}.\ (\forall v \in A.\ \mathsf{wf_tuple}\ n\ V\ v) \wedge (\forall x \in V.\ x < n)) \longrightarrow \\
&z \in \mathsf{mwjoin}\ Q_{pos}\ Q_{neg} \longleftrightarrow \mathsf{wf_tuple}\ n\ (\textstyle\bigcup (V,\ _) \in Q_{pos}.\ V)\ z\ \wedge \\
&\qquad (\forall (V, A) \in Q_{pos}.\ z \downarrow V \in A) \wedge (\forall (U, B) \in Q_{neg}.\ z \downarrow U \notin B)
\end{aligned}$$

In words: whenever Q_{pos} is nonempty and all tables in Q_{pos} and Q_{neg} fit their declared columns, a tuple z belongs to the output of mwjoin iff it has the correct columns and matches all positive tables from Q_{pos} and does not match any negative ones from Q_{neg}.

The multi-way join algorithm is integrated in VeriMon+ by adding a new constructor Ands :: *frm list* ⇒ *frm* to the formula datatype. At least one of the subformulas of Ands must be non-negated, and the columns of the negative subformulas must be a subset of the positive ones. Since MonPoly's parser, which we reuse in VeriMon+, generates formulas with binary conjunctions, we have defined a semantics-preserving preprocessing function convert_multiway (omitted), which rewrites nested binary conjunctions into Ands.

6 Sliding Window Algorithm

To evaluate the temporal operators S and U, VeriMon computes the union of tables that are associated with time-stamps within the operator's interval. These sets of time-stamps often overlap between consecutive monitor steps. The *sliding window algorithm* (SWA) [10] is an efficient algorithm for combining the elements of overlapping sequences with an associative operator. It improves over the naive approach that recomputes the combination (here, the union) from scratch for every sequence. MonPoly uses SWA for the special cases $\blacklozenge_I \beta = \mathsf{TT}\ \mathsf{S}_I\ \beta$ and $\lozenge_I \beta = \mathsf{TT}\ \mathsf{U}_I\ \beta$, where $\mathsf{TT} = \exists x.\ x \approx x$. However, SWA was not designed for the evaluation of arbitrary S and U operators. For these, the tables in the sequence must be joined with the left subformula's results in every monitor step. In a separate work [27,28], we formally verified SWA's functional correctness (but not its optimality) and extended it with a join operation to support arbitrary S and U operators.

SWA is overly general: it supports any associative operator, not just the union of tables. We conjecture that the generic SWA algorithm is not optimal in the special case needed for S and U. To optimize the evaluation of the S and U operators in VeriMon+, we abstracted the individual steps of their evaluation in one locale for each of them (Sect. 6.1). We then instantiated the locales with specialized sliding window algorithms (Sect. 6.2). Due to space limitations, we only describe the optimization for the Since operator here.

```
locale msaux = fixes valid_msaux :: args ⇒ ts ⇒ 'msaux ⇒ mslist ⇒ bool
  and init_msaux :: args ⇒ 'msaux
  and add_new_ts :: args ⇒ ts ⇒ 'msaux ⇒ 'msaux
  and join_msaux :: args ⇒ table ⇒ 'msaux ⇒ 'msaux
  and add_new_table :: args ⇒ table ⇒ 'msaux ⇒ 'msaux
  and result_msaux :: args ⇒ 'msaux ⇒ table
```

Fig. 9. The locale for evaluating the Since operator (assumptions omitted)

6.1 Integration into the Monitor

Recall the evaluation of Since in VeriMon (Sect. 2). First, VeriMon updates the operator's state with a new time-stamp τ' and the satisfactions R_α and R_β for the subformulas α and β. Second, it evaluates the state to obtain the satisfactions for $\alpha \, \mathsf{S}_I \, \beta$.

Let *mslist* denote the type of the S operator's state in VeriMon. In VeriMon+, we define a locale msaux that abstracts the update and evaluation of an optimized state *'msaux* and relates the optimized state to VeriMon's original *mslist* state (Fig. 9). We provide additional constant arguments for evaluating the S operator in a record *args*. It consists of the S operator's interval, the arguments to wf_tuple characterizing the satisfactions of the two subformulas, and a Boolean value denoting whether the left subformula occurs negated. The predicate valid_msaux relates an optimized state to VeriMon's state with respect to the given *args* and a current time-stamp. The function init_msaux returns an initial optimized state. The next three functions add_new_ts, join_msaux, and add_new_table correspond to the three steps in which VeriMon's state is updated (Sect. 2), except that now they act on the optimized state. Finally, result_msaux evaluates the optimized state to obtain the satisfactions of the S operator at the current time-point. The omitted locale assumptions state that all operations preserve valid_msaux and that result_msaux returns the union computed on any VeriMon state related by valid_msaux.

6.2 The Specialized Algorithm

VeriMon's state for the S operator consists of a list of tables T_c with the satisfactions of formulas $\alpha \, \mathsf{S}_{[c,c]} \, \beta$, along with the corresponding time-stamps. VeriMon stores the satisfactions T_c (and time-stamps) for all c that do not exceed the interval's upper bound.

In our optimized state, we partition the list of tables T_c into a list data_prev for time-stamps that are not yet in the interval and a list data_in for time-stamps that are already in the interval. The state also contains a mapping tuple_in that assigns to each tuple occurring in some table T_c in the interval the latest time-stamp in the interval for which this tuple occurs in the respective table. Finally, the state contains a mapping tuple_since that assigns to each tuple occurring in some table T_c in the entire state the *earliest* time-stamp for which this tuple occurs in the respective table. (For efficiency, we delete tuples from tuple_since

ts	*db*	step	data_prev	data_in	tuple_in	tuple_since
		init_msaux	$[]$	$[]$	$\{\}$	$\{\}$
1	$\{Q(a), Q(b), Q(c)\}$	add_new_table	$[(1, \{a, b, c\})]$	$[]$	$\{\}$	$\{a \mapsto 1, b \mapsto 1, c \mapsto 1\}$
2	$\{P(b), P(c)\}$	join_msaux	$[(1, \{a, b, c\})]$	$[]$	$\{\}$	$\{b \mapsto 1, c \mapsto 1\}$
		add_new_table	$[(1, \{a, b, c\}), (2, \{\})]$	$[]$	$\{\}$	$\{b \mapsto 1, c \mapsto 1\}$
3	$\{P(b), P(c), Q(a), Q(b)\}$	add_new_ts	$[(2, \{\})]$	$[(1, \{a, b, c\})]$	$\{b \mapsto 1, c \mapsto 1\}$	$\{b \mapsto 1, c \mapsto 1\}$
		add_new_table	$[(2, \{\}), (3, \{a, b\})]$	$[(1, \{a, b, c\})]$	$\{b \mapsto 1, c \mapsto 1\}$	$\{a \mapsto 3, b \mapsto 1, c \mapsto 1\}$
7	$\{P(a)\}$	add_new_ts	$[]$	$[(3, \{a, b\})]$	$\{a \mapsto 3, b \mapsto 3\}$	$\{a \mapsto 3, b \mapsto 1, c \mapsto 1\}$
		join_msaux	$[]$	$[(3, \{a, b\})]$	$\{a \mapsto 3\}$	$\{a \mapsto 3\}$
		add_new_table	$[(7, \{\})]$	$[(3, \{a, b\})]$	$\{a \mapsto 3\}$	$\{a \mapsto 3\}$

Fig. 10. An example of updating the optimized state for the formula $P(x)\ \mathsf{S}_{[2,4]}\ Q(x)$

lazily, i.e., only at defined garbage collection points, such that the mapping may even contain tuples from some T_c that already has fallen out of the interval.)

The state is initialized via init_msaux to consist of empty lists and empty mappings. The function add_new_ts drops tables from data_in that fall out of the interval based on a newly received time-stamp. It also removes those tuples from tuple_in whose latest occurrence (which is stored in this mapping) has fallen out of the interval. Then it moves tables that newly enter the interval from data_prev to data_in, and updates the tuples from these moved tables in tuple_in to the most recent time-stamp τ for which they now occur in the interval, but only if tuple_since maps the tuple to a time-stamp that is at most τ.

The function join_msaux only modifies the mappings tuple_since and tuple_in by removing tuples that are not matched by any tuple in the given table R_α. The function add_new_table appends the new table R_β to data_prev (or directly data_in, if $0 \in_{\mathcal{I}} I$), adds the tuples from R_β that were not in tuple_since to that mapping, and, if $0 \in_{\mathcal{I}} I$, updates the tuples from R_β in the mapping tuple_in to the current time-stamp. Finally, result_msaux returns the keys of the mapping tuple_in, in particular without computing any unions. In other words, tuple_in contains precisely the tuples that are in the interval and have not been removed by joins. Crucially, and unlike in VeriMon's state, the join operation does not change the tables T_c in our optimized state. This functionality is implemented more efficiently by filtering the two mappings tuple_since and tuple_in.

Example. Figure 10 shows how the optimized state for the formula $P(x)\ \mathsf{S}_{[2,4]}\ Q(x)$ is updated. In total, four time-points are processed. The first two columns show the time-stamp and database for each time-point. The other columns show the state after applying the step named in the third column. Each step corresponds to a function in the msaux locale. The satisfactions $\{\}, \{\}, \{b, c\}, \{a\}$ returned by result_msaux can be read off from the mapping tuple_in after each time-point's last step. We omit steps that do not change the state.

The first row shows the initial state. For the first time-point, the steps add_new_ts with time-stamp 1 and join_msaux with the table $\{\}$ (as there are no P events) do not change the initial state. Then, add_new_table appends the table $\{a, b, c\}$ with the parameters of the Q events to data_prev (as $0 \notin_{\mathcal{I}} [2,4]$) and adds its elements to tuple_since.

For the second time-point, VeriMon+ applies add_new_ts with time-stamp 2. Again this step has no effect: data_prev's first entry is not moved to data_in as the difference $2-1$ to the current time-stamp is not in $[2,4]$. Next, join_msaux with the table $\{b, c\}$ (from the P events) removes a from tuple_since, but not from data_prev. Finally, add_new_table appends the table $\{\}$ (as there are no Q events) to data_prev.

For the third time-point, add_new_ts moves data_prev's first entry to data_in because the time-stamp difference $3-1$ is in $[2,4]$. The values b, c of that entry are added to tuple_in because tuple_since maps them to a time-stamp ≤ 1. Note that a is not added, as it is not contained in tuple_since. The join_msaux step with the table $\{b, c\}$ does not change the state. The add_new_table step appends $\{a, b\}$ to data_prev. Now, a is added to tuple_since, whereas b is already contained in tuple_since and its value is not updated.

When the fourth time-point is processed, the first two observed time-stamps fall out of the interval and add_new_ts discards their entries from data_prev and data_in, and their values from tuple_in but not from tuple_since. As before, the last table $\{a, b\}$ in data_prev is moved to data_in and its elements are added to tuple_in. As time has progressed by more than the upper bound of the interval $[2,4]$, join_msaux triggers garbage collection, which removes the key c from tuple_since. The join operation further removes b from tuple_in and tuple_since. Finally, add_new_table appends $\{\}$ to data_prev.

7 Evaluation

We perform two kinds of experiments. First, we carry out differential testing [35] of VeriMon+ against three (unverified) state-of-the-art monitors: MonPoly [9], Aerial [11], and Hydra [43]. Second, we compare VeriMon+'s performance to these monitors on representative formulas. VeriMon+ reuses MonPoly's log and formulas parsers and user interface. The verified monitor's code extracted from Isabelle is integrated with these unverified components in about 170 lines of unverified OCaml code. Our implementation and our experiments are available [5]. Of the above monitors, only VeriMon+ supports full MFODL. MonPoly supports a monitorable fragment of MFOTL with bounded future operators and aggregations. Aerial and Hydra support the propositional fragment of MFODL.

Differential Testing. To validate the results produced by unverified monitors, we generate random stream prefixes and formulas, invoke the monitors, and compare their results to VeriMon+'s. For this purpose, we developed a random stream and formula generator. It takes as parameters the formula size (in terms of number of operators) and the number of free variables that occur in the formula. The generator can be configured to generate formulas within the fragments of MFODL supported by the different monitors we evaluate.

Our tests uncovered several classes of inputs where MonPoly's output deviated from VeriMon+'s. Here, we show one example and refer to our extended report [6] for a comprehensive overview. Namely, formulas of the form $m \leftarrow \Omega\ x; x.\ \blacklozenge_I \alpha$, where $\mathsf{fv}\ \alpha = \{x, y\}$, $\Omega \in \{\mathsf{Min}, \mathsf{Max}\}$, and $0 \notin_{\mathcal{I}} I$, were evaluated in MonPoly using a specialized algorithm, which incorrectly updated the satisfactions of α when they fell out of the interval I.

Aerial's and Hydra's output mostly coincided with VeriMon+'s. However, we noticed that Hydra's output is not as eager as it could be at the end of the stream prefix. For example, $\triangleright_{[1,1]} (\star \cdot (\mathsf{TT} \vee \triangleright_{[1,1]} \star)?)$ is satisfied at time-point 0 of the prefix $(\{\}, 0), (\{\}, 1)$ due to the existence of time-point 1, where TT can be evaluated. The subformula $\triangleright_{[1,1]} \star$ cannot be evaluated at time-point 1. This prevents Hydra from outputting this verdict at 0.

Performance Evaluation. To assess VeriMon+'s performance, we selected four formulas, shown in Fig. 11, which exercise the optimizations (multi-way join and sliding window) and the language features (aggregations and regular expressions) we have introduced. The formula $\mathsf{Star}(N)$ is derived from the *star* conjunctive query, commonly used as a benchmark for joins [14]. We use it to evaluate our multi-way join (for $N = 10$) and sliding window (for $N = 30$) implementations. The formula Top is a commonly used aggregation query, which computes the most frequently occurring value of the event P's second parameter. Finally, Alt checks if events P and Q alternate over the last 10 time units.

We generate random stream prefixes with a time span of 60 time units containing events P, Q, and R, each with two integer parameters sampled uniformly at random from the set $\{1, 2, \ldots, 10^9\}$. Our stream generator is parametrized by the event rate (i.e., by the number of events with the same time-stamp). Since VeriMon+ reuses MonPoly's formula and log parsing infrastructure, there is an additional (conceptually unnecessary) overhead caused by converting the data structures to match the appropriate interfaces. In cases where the monitoring task is easy, this becomes the bottleneck and MonPoly performs better than VeriMon+. To make the monitoring task difficult for $\mathsf{Star}(10)$, we sample the value of the first parameter of each event (the common variable x) using the Zipf distribution. Thus, some parameter values occur frequently. This results in large intermediate tables, which are problematic for binary joins.

Figure 11 shows that VeriMon+ outperforms MonPoly on the $\mathsf{Star}(N)$ formulas. The results confirm the feasibility of monitoring aggregations and regular expressions with VeriMon+. Specialized algorithms remain more performant on problems in their domain.

Formula		Star(10)			Star(30)			Top		Alt		
Monitor		MonPoly	VeriMon+	VeriMon	MonPoly	VeriMon+	VeriMon	MonPoly	VeriMon+	Aerial	Hydra	VeriMon+
Event rate	50	0.0/6.2	0.1/9.3	0.2/9.5	0.1/6.4	0.2/12.0	3.0/12.4	0.2/8.9	6.0/10.4	0.3/5.8	0.2/3.2	2.1/8.6
	100	0.1/7.0	0.2/12.0	0.9/16.5	0.1/7.0	0.3/13.4	10.7/24.2	0.3/10.0	29.9/12.6	0.4/5.8	0.2/3.2	3.4/8.6
	200	0.5/9.1	0.3/12.1	6.2/47.4	0.4/9.2	0.7/18.7	50.1/48.5	0.9/9.9	to	0.6/6.0	0.2/3.1	7.2/8.8
	500	6.0/9.8	1.3/16.3	so	2.5/12.9	1.9/32.5	so	2.9/13.5	to	1.1/6.3	0.2/3.2	18.0/8.7
	1000	38.0/12.8	2.5/22.2	so	11.6/17.9	5.2/58.0	so	10.9/22.4	to	1.7/6.4	0.3/3.1	34.1/8.8
	2000	to	5.9/36.5	so	to	11.9/106.7	so	22.0/34.2	to	3.1/6.3	0.4/3.2	to
	4000	to	15.0/65.0	so	to	22.8/206.0	so	50.8/62.1	to	5.0/6.5	0.6/3.3	to

$$\mathsf{Star}(N) \equiv (\blacklozenge_{[0,N]} P(x,y) \land Q(x,z)) \land \lozenge_{[0,N]} R(x,w) \qquad \mathsf{Alt} \equiv \blacktriangleleft_{[10,10]} (P? \cdot \star \cdot Q? \cdot \star)^*$$

$$\mathsf{Top} \equiv (m \leftarrow \mathsf{Max}\; s; v.\ (s \leftarrow \mathsf{Cnt}\; id; id. \blacklozenge P(id,v))) \land (m \leftarrow \mathsf{Cnt}\; id; id. \blacklozenge P(id,v))$$

Fig. 11. Time (s)/memory (MB) usage of the monitors (to = timeout of 60s, so = stack overflow)

8 Conclusion

We have presented a verified monitor, competitive with the state-of-the-art, for the expressive specification language metric first-order dynamic logic. Our formalization comprises roughly 15 000 lines of Isabelle code, distributed over the four features we presented: regular expressions (2 000), terms and aggregations (750), multi-way join (3 300), and the sliding window algorithm (3 000). Isabelle extracts a 7 500 line OCaml program from our formalization. This code includes efficient libraries representing sets and mappings via red–black trees introduced transparently into the formalization via the Containers framework [32]. We also use and extend a formalization of IEEE floating point numbers [50].

We have made additional contributions from the algorithmic perspective. Our monitor is the first monitoring algorithm for MFODL with aggregations. Moreover, our specialized sliding window algorithm improves over the existing generic algorithm [10]. Our usage of multi-way joins in the context of first-order monitoring is also novel, as is our extension of the multi-way join algorithm to handle anti-joins. It would be interesting to investigate the optimality of this extension and further consider a multi-way-like evaluation of an arbitrary Boolean combination of finite tables.

Our focus was on extending the verified monitor's specification language and improving its algorithms. As next steps, we plan to further improve performance by refining our algorithms to imperative data structures following Lammich's methodology [29,30].

Acknowledgment. We thank the anonymous IJCAR reviewers for their helpful comments. This research is supported by the US Air Force grant "Monitoring at Any Cost" (FA9550-17-1-0306) and by the Swiss National Science Foundation grant "Big Data Monitoring" (167162). The authors are listed in alphabetical order.

References

1. Alur, R., Fisman, D., Raghothaman, M.: Regular programming for quantitative properties of data streams. In: Thiemann, P. (ed.) ESOP 2016. LNCS, vol. 9632, pp. 15–40. Springer, Heidelberg (2016). https://doi.org/10.1007/978-3-662-49498-1_2

2. Antimirov, V.M.: Partial derivatives of regular expressions and finite automaton constructions. Theoret. Comput. Sci. **155**(2), 291–319 (1996). https://doi.org/10.1016/0304-3975(95)00182-4
3. Barringer, H., Falcone, Y., Havelund, K., Reger, G., Rydeheard, D.: Quantified event automata: towards expressive and efficient runtime monitors. In: Giannakopoulou, D., Méry, D. (eds.) FM 2012. LNCS, vol. 7436, pp. 68–84. Springer, Heidelberg (2012). https://doi.org/10.1007/978-3-642-32759-9_9
4. Bartocci, E., Falcone, Y. (eds.): Lectures on Runtime Verification - Introductory and Advanced Topics. LNCS, vol. 10457. Springer, Cham (2018). https://doi.org/10.1007/978-3-319-75632-5
5. Basin, D., et al.: VeriMon+: implementation and case study associated with this paper (2020). https://bitbucket.org/jshs/monpoly/downloads/verimonplus.zip
6. Basin, D., et al.: A formally verified, optimized monitor for metric first-order dynamic logic (extended report) (2020). https://people.inf.ethz.ch/trayteld/papers/ijcar20-verimonplus/verimonplus_report.pdf
7. Basin, D., Klaedtke, F., Marinovic, S., Zălinescu, E.: Monitoring of temporal first-order properties with aggregations. Form Methods Syst. Des. **46**(3), 262–285 (2015). https://doi.org/10.1007/s10703-015-0222-7
8. Basin, D., Klaedtke, F., Müller, S., Zălinescu, E.: Monitoring metric first-order temporal properties. J. ACM **62**(2), 15:1–15:45 (2015). https://doi.org/10.1145/2699444
9. Basin, D., Klaedtke, F., Zălinescu, E.: The MonPoly monitoring tool. In: Reger, G., Havelund, K. (eds.) RV-CuBES 2017. Kalpa Publications in Computing, vol. 3, pp. 19–28. EasyChair (2017)
10. Basin, D., Klaedtke, F., Zălinescu, E.: Greedily computing associative aggregations on sliding windows. Inf. Process. Lett. **115**(2), 186–192 (2015). https://doi.org/10.1016/j.ipl.2014.09.009
11. Basin, D., Krstić, S., Traytel, D.: AERIAL: almost event-rate independent algorithms for monitoring metric regular properties. In: Reger, G., Havelund, K. (eds.) RV-CuBES 2017. Kalpa Publications in Computing, vol. 3, pp. 29–36. EasyChair (2017)
12. Basin, D., Bhatt, B.N., Krstić, S., Traytel, D.: Almost event-rate independent monitoring. Form. Methods Syst. Des. **54**(3), 449–478 (2019). https://doi.org/10.1007/s10703-018-00328-3
13. Bauer, A., Küster, J.-C., Vegliach, G.: From propositional to first-order monitoring. In: Legay, A., Bensalem, S. (eds.) RV 2013. LNCS, vol. 8174, pp. 59–75. Springer, Heidelberg (2013). https://doi.org/10.1007/978-3-642-40787-1_4
14. Beame, P., Koutris, P., Suciu, D.: Communication steps for parallel query processing. J. ACM **64**(6), 40:1–40:58 (2017). https://doi.org/10.1145/3125644
15. Benzaken, V., Contejean, É., Keller, C., Martins, E.: A Coq formalisation of SQL's execution engines. In: Avigad, J., Mahboubi, A. (eds.) ITP 2018. LNCS, vol. 10895, pp. 88–107. Springer, Cham (2018). https://doi.org/10.1007/978-3-319-94821-8_6
16. Blech, J.O., Falcone, Y., Becker, K.: Towards certified runtime verification. In: Aoki, T., Taguchi, K. (eds.) ICFEM 2012. LNCS, vol. 7635, pp. 494–509. Springer, Heidelberg (2012). https://doi.org/10.1007/978-3-642-34281-3_34
17. Bohrer, B., Tan, Y.K., Mitsch, S., Myreen, M.O., Platzer, A.: VeriPhy: verified controller executables from verified cyber-physical system models. In: Foster, J.S., Grossman, D. (eds.) PLDI 2018, pp. 617–630. ACM (2018). https://doi.org/10.1145/3192366.3192406
18. Brzozowski, J.A.: Derivatives of regular expressions. J. ACM **11**(4), 481–494 (1964). https://doi.org/10.1145/321239.321249

19. D'Angelo, B., et al.: LOLA: runtime monitoring of synchronous systems. In: TIME 2005, pp. 166–174. IEEE Computer Society (2005). https://doi.org/10.1109/TIME.2005.26
20. Dardinier, T.: Formalization of multiway-join algorithms. Archive of Formal Proofs (2019). https://isa-afp.org/entries/Generic_Join.html
21. Dardinier, T., Heimes, L., Raszyk, M., Schneider, J., Traytel, D.: Formalization of an optimized monitoring algorithm for metric first-order dynamic logic with aggregations. Archive of Formal Proofs (2020). https://isa-afp.org/entries/MFODL_Monitor_Optimized.html
22. De Giacomo, G., Vardi, M.Y.: Linear temporal logic and linear dynamic logic on finite traces. In: Rossi, F. (ed.) IJCAI 2013, pp. 854–860. IJCAI/AAAI (2013)
23. Havelund, K.: Rule-based runtime verification revisited. STTT **17**(2), 143–170 (2015). https://doi.org/10.1007/s10009-014-0309-2
24. Havelund, K., Leucker, M., Reger, G., Stolz, V.: A shared challenge in behavioural specification (Dagstuhl Seminar 17462). Dagstuhl Rep. **7**(11), 59–85 (2017). https://doi.org/10.4230/DagRep.7.11.59
25. Havelund, K., Peled, D.: Efficient runtime verification of first-order temporal properties. In: Gallardo, M.M., Merino, P. (eds.) SPIN 2018. LNCS, vol. 10869, pp. 26–47. Springer, Cham (2018). https://doi.org/10.1007/978-3-319-94111-0_2
26. Havelund, K., Roşu, G.: Synthesizing monitors for safety properties. In: Katoen, J.-P., Stevens, P. (eds.) TACAS 2002. LNCS, vol. 2280, pp. 342–356. Springer, Heidelberg (2002). https://doi.org/10.1007/3-540-46002-0_24
27. Heimes, L.: Extending and optimizing a verified monitor for metric first-order temporal logic. Bachelor's thesis, Department of Computer Science, ETH Zürich (2019)
28. Heimes, L., Schneider, J., Traytel, D.: Formalization of an algorithm for greedily computing associative aggregations on sliding windows. Archive of Formal Proofs (2020). https://isa-afp.org/entries/Sliding_Window_Algorithm.html
29. Lammich, P.: Generating verified LLVM from Isabelle/HOL. In: Harrison, J., O'Leary, J., Tolmach, A. (eds.) ITP 2019. LIPIcs, vol. 141, pp. 22:1–22:19. Schloss Dagstuhl - Leibniz-Zentrum für Informatik (2019). https://doi.org/10.4230/LIPIcs.ITP.2019.22
30. Lammich, P.: Refinement to imperative HOL. J. Autom. Reasoning **62**(4), 481–503 (2019). https://doi.org/10.1007/s10817-017-9437-1
31. Laurent, J., Goodloe, A., Pike, L.: Assuring the guardians. In: Bartocci, E., Majumdar, R. (eds.) RV 2015. LNCS, vol. 9333, pp. 87–101. Springer, Cham (2015). https://doi.org/10.1007/978-3-319-23820-3_6
32. Lochbihler, A.: Light-weight containers for Isabelle: efficient, extensible, nestable. In: Blazy, S., Paulin-Mohring, C., Pichardie, D. (eds.) ITP 2013. LNCS, vol. 7998, pp. 116–132. Springer, Heidelberg (2013). https://doi.org/10.1007/978-3-642-39634-2_11
33. Malecha, J.G., Morrisett, G., Shinnar, A., Wisnesky, R.: Toward a verified relational database management system. In: Hermenegildo, M.V., Palsberg, J. (eds.) POPL 2010, pp. 237–248. ACM (2010. https://doi.org/10.1145/1706299.1706329
34. Mamouras, K., Raghothaman, M., Alur, R., Ives, Z.G., Khanna, S.: StreamQRE: modular specification and efficient evaluation of quantitative queries over streaming data. In: Cohen, A., Vechev, M.T. (eds.) PLDI 2017, pp. 693–708. ACM (2017). https://doi.org/10.1145/3062341.3062369
35. McKeeman, W.M.: Differential testing for software. Digit. Tech. J. **10**(1), 100–107 (1998)

36. Mitsch, S., Platzer, A.: ModelPlex: verified runtime validation of verified cyber-physical system models. Form. Methods Syst. Des. **49**(1–2), 33–74 (2016). https://doi.org/10.1007/s10703-016-0241-z
37. Ngo, H.Q., Porat, E., Ré, C., Rudra, A.: Worst-case optimal join algorithms: [extended abstract]. In: Benedikt, M., Krötzsch, M., Lenzerini, M. (eds.) PODS 2012, pp. 37–48. ACM (2012). https://doi.org/10.1145/2213556.2213565
38. Ngo, H.Q., Porat, E., Ré, C., Rudra, A.: Worst-case optimal join algorithms. J. ACM **65**(3), 16:1–16:40 (2018). https://doi.org/10.1145/3180143
39. Ngo, H.Q., Ré, C., Rudra, A.: Skew strikes back: new developments in the theory of join algorithms. SIGMOD Rec. **42**(4), 5–16 (2013). https://doi.org/10.1145/2590989.2590991
40. Peycheva, G.: Real-time verification of datacenter security policies via online log analysis. Master's thesis, ETH Zürich (2018)
41. Pike, L., Niller, S., Wegmann, N.: Runtime verification for ultra-critical systems. In: Khurshid, S., Sen, K. (eds.) RV 2011. LNCS, vol. 7186, pp. 310–324. Springer, Heidelberg (2012). https://doi.org/10.1007/978-3-642-29860-8_23
42. Pike, L., Wegmann, N., Niller, S., Goodloe, A.: Experience report: a do-it-yourself high-assurance compiler. In: Thiemann, P., Findler, R.B. (eds.) ICFP 2012, pp. 335–340. ACM (2012). https://doi.org/10.1145/2364527.2364553
43. Raszyk, M., Basin, D., Krstić, S., Traytel, D.: Multi-head monitoring of metric temporal logic. In: Chen, Y.-F., Cheng, C.-H., Esparza, J. (eds.) ATVA 2019. LNCS, vol. 11781, pp. 151–170. Springer, Cham (2019). https://doi.org/10.1007/978-3-030-31784-3_9
44. Rizaldi, A., et al.: Formalising and monitoring traffic rules for autonomous vehicles in Isabelle/HOL. In: Polikarpova, N., Schneider, S. (eds.) IFM 2017. LNCS, vol. 10510, pp. 50–66. Springer, Cham (2017). https://doi.org/10.1007/978-3-319-66845-1_4
45. Schneider, J., Basin, D., Krstić, S., Traytel, D.: A formally verified monitor for metric first-order temporal logic. In: Finkbeiner, B., Mariani, L. (eds.) RV 2019. LNCS, vol. 11757, pp. 310–328. Springer, Cham (2019). https://doi.org/10.1007/978-3-030-32079-9_18
46. Thati, P., Roşu, G.: Monitoring algorithms for metric temporal logic specifications. Electron. Notes Theoret. Comput. Sci. **113**, 145–162 (2005). https://doi.org/10.1016/j.entcs.2004.01.029
47. Ulus, D.: MONTRE: a tool for monitoring timed regular expressions. In: Majumdar, R., Kunčak, V. (eds.) CAV 2017. LNCS, vol. 10426, pp. 329–335. Springer, Cham (2017). https://doi.org/10.1007/978-3-319-63387-9_16
48. Veldhuizen, T.L.: Triejoin: a simple, worst-case optimal join algorithm. In: Schweikardt, N., Christophides, V., Leroy, V. (eds.) ICDT 2014, pp. 96–106. OpenProceedings.org (2014). https://doi.org/10.5441/002/icdt.2014.13
49. Völlinger, K.: Verifying the output of a distributed algorithm using certification. In: Lahiri, S., Reger, G. (eds.) RV 2017. LNCS, vol. 10548, pp. 424–430. Springer, Cham (2017). https://doi.org/10.1007/978-3-319-67531-2_29
50. Yu, L.: A formal model of IEEE floating point arithmetic. Archive of Formal Proofs (2013). https://isa-afp.org/entries/IEEE_Floating_Point.html

Constructive Hybrid Games

Rose Bohrer[1(✉)] and André Platzer[1,2]

[1] Computer Science Department, Carnegie Mellon University, Pittsburgh, USA
aplatzer@cs.cmu.edu
[2] Fakultät für Informatik, Technische Universität München, Munich, Germany

Abstract. Hybrid games combine discrete, continuous, and adversarial dynamics. Differential game logic (dGL) enables proving (classical) existence of winning strategies. We introduce *constructive differential game logic* (CdGL) for hybrid games, where proofs that a player can win the game correspond to *computable* winning strategies. This constitutes the logical foundation for synthesis of correct control and monitoring code for safety-critical cyber-physical systems. Our contributions include novel semantics as well as soundness and consistency.

Keywords: Game logic · Constructive logic · Hybrid games · Dependent types

1 Introduction

Differential Game Logic (dGL) provides a calculus for proving the (classical) existence of winning strategies for hybrid games [42], whose mixed discrete, continuous, and adversarial dynamics are compelling models for cyber-physical systems (CPSs). Classical existence does *not* necessarily imply that the resulting winning strategies are computable, however. To overcome this challenge, this paper introduces *Constructive Differential Game Logic* (CdGL) with a Curry-Howard correspondence: constructive proofs for constructive hybrid games correspond to programs implementing their winning strategies. We develop a new type-theoretic semantics which elucidates this correspondence and an operational semantics which describes the execution of strategies. Besides its theoretical appeal, this Curry-Howard interpretation provides the foundation for proof-driven synthesis methods, which excel at synthesizing expressive classes of games for which synthesis and correctness require interactive proof. Hybrid games are a compelling domain for proof-based synthesis both because many CPS applications are safety-critical or even life-critical, such as transportation systems, energy systems, and medical devices and because the combination of discrete, continuous, and adversarial dynamics makes verification and synthesis undecidable in both theory and practice. Our example model and proof, while short, lay the groundwork for future case studies.

This research was sponsored by the AFOSR under grant number FA9550-16-1-0288 and the Alexander von Humboldt Foundation. The first author was also funded by an NDSEG Fellowship.

N. Peltier and V. Sofronie-Stokkermans (Eds.): IJCAR 2020, LNAI 12166, pp. 454–473, 2020.
https://doi.org/10.1007/978-3-030-51074-9_26

Challenges and Contributions. In addition to dGL [42], we build directly on Constructive Game Logic (CGL) [9] for discrete games. Compared to CGL, we target a domain with readily-available practical applications (hybrid games), and introduce new type-theoretic and operational semantics which complement the realizability semantics of CGL while making Curry-Howard particularly clear and providing a simple notion of strategy execution. We overcome the following challenges in the process:

- Our semantics must carefully capture the meaning of constructive hybrid game strategies, including strategies for differential equations (ODEs).
- Soundness must be justified *constructively.* We adapt previous arguments to use constructive analysis [6,12] by appealing to constructive formalizations of ODEs [17,34]. This adaptation to our new semantics makes it possible to simplify statements of some standard lemmas.
- We study 1D driving control as an example, which demonstrates the strengths of both games and constructivity. Games and constructivity both introduce uncertainties: A player is uncertain how their opponent will play, while constructive real-number comparisons are never sure of exact equality. These uncertainties demand more nuanced proof invariants, but these nuances improve our fidelity to real systems.

These contributions are of likely interest to several communities. Other constructive program logics could reuse our semantic approach. Our example uses reach-avoid proofs for hybrid games, a powerful, under-explored [48] approach.

2 Related Work

We discuss related works on games, constructive logic, and hybrid systems.

Games in Logic. Propositional GL was introduced by Parikh [39]. GL is a *program logic* in the spirit of Hoare calculi [26] or especially dynamic logics (DL) [47]: modalities capture the effect of game execution. GLs are unique in their clear delegation of strategy to the *proof* language rather than the *model* language, allowing succinct game specifications with sophisticated winning strategies. Succinct specifications are important: specifications are *trusted* because proving the *wrong theorem* would not ensure correctness. Relatives without this separation include SL [14], ATL [2], CATL [27], SDGL [23], structured strategies [49], DEL [3,5,56], evidence logic [4], and Angelic Hoare Logic [35].

Constructive Modal Logics. We are interested in the semantics of games, thus we review constructive modal semantics generally. This should not be confused with game semantics [1], which give a semantics to programs *in terms of* games. The main semantic approaches for constructive modal logics are intuitionistic Kripke semantics [58] and realizability semantics [32,38]. CGL [9] used a realizability semantics which operate on a state, reminiscent of state in Kripke semantics, whereas we interpret CdGL formulas into type theory.

Modal Curry-Howard is relatively little-studied, and each author has their own emphasis. Explicit proof terms are considered for CGL [9] and a small fragment thereof [30]. Others [13,18,59] focus on intuitionistic semantics for their logics, fragments of CGL. Our semantics should be of interest for these fragments. We omit proof terms for space. CdGL proof terms would extend CGL proof terms [9] with a constructive version of existing classical ODE proof terms [8]. Propositional modal logic [37] has been interpreted as a type system.

Hybrid Systems Synthesis. Hybrid games synthesis is one motivation of this work. Synthesis of hybrid *systems* (1-player games) is an active area. The unique strength of proof-based synthesis is expressiveness: it can synthesize every provable system. CdGL proofs support first-order regular games with first-order (e.g., semi-algebraic) initial and goal regions. While synthesis and proof are both undecidable, interactive proof for undecidable logics is well-understood. The ModelPlex [36] synthesizer for CdGL's classical systems predecessor dL [44] recently added [11] proof-based synthesis to improve expressiveness. CdGL aims to provide a computational foundation for a more systematic proof-based synthesizer in the more general context of games.

Fully automatic synthesis, in contrast, restricts itself to small fragments in order to sidestep undecidability. Studied classes include rectangular hybrid games [25], switching systems [52], linear systems with polyhedral sets [31,52], and discrete abstractions [20,21]. A well-known [55] *systems* synthesis approach translates specifications *into* finite-alternation games. Arbitrary first-order games are our *source* rather than *target* language. Their approach is only known to terminate for simpler classes [50,51].

3 Constructive Hybrid Games

Hybrid games in CdGL are 2-player, zero-sum, and perfect-information, where continuous subgames are ordinary differential equations (ODEs) whose duration is chosen by a player. Hybrid games should not be confused with *differential games* which compete continuously [29,43]. The players considered in this paper are Angel and Demon where the player currently controlling choices is always called Angel, while the player waiting to play is always called Demon. For any game α and formula ϕ, the modal formula $\langle\alpha\rangle\phi$ says Angel can play α to ensure postcondition ϕ, while $[\alpha]\phi$ says Demon can play α to ensure postcondition ϕ. These generalize safety and liveness modalities from DL. Dual games α^d, unique to GLs, take turns by switching the Angel and Demon roles in game α. The Curry-Howard interpretation of a proof of a CdGL modality $\langle\alpha\rangle\phi$ or $[\alpha]\phi$ is a program which performs each player's winning strategy. Games can have several winning strategies, each corresponding to a different proof and a different program.

3.1 Syntax of CdGL

We introduce the language of CdGL with three classes of expressions e: terms f, g, games α, β, and formulas ϕ, ψ. We characterize terms semantically for the sake of generality: a shallow embedding of CdGL inside a proof assistant might use the host language for terms. For games and formulas, we find it more convenient to explicitly and syntactically define a closed language.

A (scalar) semantic term is a function from states to reals, which are understood constructively *à la* Bishop [6,12]. We use Bishop-style real analysis because it preserves many classical intuitions (e.g., uncountability) about $\mathbb{R}$ while ensuring computability. Type-2 [57] computability requires that all *functions on real numbers* are computable to arbitrary precision if represented as streams of bits, yet computability *does not* require that variables range over *only* computable reals. It is a theorem [57] that all such computable functions are continuous, but not always Lipschitz-continuous nor differentiable.

We introduce commonly used term constructs, which are not exhaustive because the language of terms is open. The simplest terms are *game variables* $x, y \in \mathcal{V}$ where $\mathcal{V}$ is the (at most countable) set of variable identifiers. The game variables, which are mutable, contain the state of the game, which is globally scoped. For every base game variable x there is a primed counterpart x' whose purpose within an ODE is to track the time derivative of x. Real-valued terms f, g are simply type-2 computable functions, usually from states to reals. It is occasionally useful for f to return a tuple of reals, which are computable when every component is computable. Since terms are functions, operators are combinators: $f + g$ is a function which sums the results of f and g.

Definition 1 (Terms). *A term f, g is any computable function over the game state. The following constructs appear in this paper:*

$$f, g \ ::= \ \cdots \mid c \mid x \mid f + g \mid f \cdot g \mid f/g \mid \min(f, g) \mid \max(f, g) \mid (f)'$$

where $c \in \mathbb{R}$ is a real literal, x a game variable, $f + g$ a sum, $f \cdot g$ a product, and f/g is real division of f by g. Divisors g are assumed to be nonzero. Minimum and maximum of terms f and g are written $\min(f, g)$ *and* $\max(f, g)$*. Any differentiable term f has a definable (Sect. 4.2) spatial differential term $(f)'$, which agrees with the time derivative within an ODE.*

CdGL is constructive, so Angel strategies make choices computably. Until his turn, Demon just observes Angel's choices, and does not care whether Angel made them computably. We discuss game-playing informally here, then formally in Sect. 4. In red are the ODE and dual games, which respectively distinguish hybrid games from discrete games and games from systems.

Definition 2 (Games). *The set of* games *α, β is defined recursively as such:*

$$\alpha, \beta \ ::= \ ?\phi \mid x := f \mid x := * \mid x' = f \,\&\, \psi \mid \alpha \cup \beta \mid \alpha; \beta \mid \alpha^* \mid \alpha^d$$

The *test game* $?\phi$, is a no-op if Angel proves ϕ, else Demon wins by default since Angel "broke the rules". A deterministic assignment $x := f$ updates game variable x to the value of term f. Nondeterministic assignments $x := *$ ask Angel to compute the new value of $x : \mathbb{R}$, i.e., Angel's strategy for $x := *$ is a term whose value is assigned to x. The ODE game $x' = f \,\&\, \psi$ evolves ODE $x' = f$ for duration $d \geq 0$ chosen by Angel such that Angel proves the domain constraint formula

ψ is true throughout. We require that term f is effectively-locally-Lipschitz on domain ψ, meaning that at every state satisfying ψ, a neighborhood and coefficient L can be constructed such that L is a Lipschitz constant of f in the neighborhood. Effective local Lipschitz continuity guarantees unique solutions exist by constructive Picard-Lindelöf [34]. ODEs are explicit-form, so no primed variable y' for $y \in \mathcal{V}$ is mentioned in f or ψ. Systems of ODEs are supported, we present single equations for readability. In the choice game $\alpha \cup \beta$, Angel chooses whether to play game α or game β. In the sequential composition game $\alpha; \beta$, game α is played first, then β from the resulting state. In the repetition game α^*, Angel chooses after each repetition of α whether to continue playing, but must not repeat α infinitely. The exact number of repetitions is not known in advance, because it may depend on Demon's reactions. In the dual game α^d, Angel takes the Demon role and vice-versa while playing α. Demon strategies "wait" until a dual game α^d is encountered, then play an Angelic strategy for α. We parenthesize games with braces $\{\alpha\}$ when necessary.

Definition 3 (CdGL Formulas). *The CdGL formulas ϕ (also ψ) are:*

$$\phi ::= \langle\alpha\rangle\phi \mid [\alpha]\phi \mid f \sim g$$

Above, $f \sim g$ is a comparison formula for $\sim \in \{\leq, <, =, \neq, >, \geq\}$. The defining formulas of CdGL (and GL) are the modalities $\langle\alpha\rangle\phi$ and $[\alpha]\phi$. These mean that Angel or Demon respectively have a *constructive* strategy to play hybrid game α and prove postcondition ϕ. We do not develop modalities for existence of classical strategies because those cannot be synthesized to executable code.

Standard connectives are defined from games and comparisons. Verum ($\mathtt{tt}$) is defined $1 > 0$ and falsum ($\mathtt{ff}$) is $0 > 1$. Conjunction $\phi \wedge \psi$ is defined $\langle ?\phi\rangle\psi$, disjunction $\phi \vee \psi$ is defined $\langle ?\phi \cup ?\psi\rangle\mathtt{tt}$, and implication $\phi \rightarrow \psi$ is defined $[?\phi]\psi$. Real quantifiers $\forall x\, \phi$ and $\exists x\, \phi$ are defined $[x := *]\phi$ and $\langle x := *\rangle\phi$, respectively. As usual, equivalence $\phi \leftrightarrow \psi$ reduces to $(\phi \rightarrow \psi) \wedge (\psi \rightarrow \phi)$, negation $\neg\phi$ is defined as $\phi \rightarrow \mathtt{ff}$, and inequality is defined by $f \neq g \equiv \neg(f = g)$. Semantics and proof rules are needed only for core constructs, but we use derived constructs when they improve readability. Keep these definitions in mind, because the semantics and rules for some game connectives mirror first-order connectives.

For convenience, we also write derived operators where Demon is given control of a single choice before returning control to Angel. The *Demonic choice* $\alpha \cap \beta$, defined $\{\alpha^d \cup \beta^d\}^d$, says Demon chooses which branch to take, but Angel controls the subgames. *Demonic repetition* $\alpha^\times$ is defined likewise by $\{\{\alpha^d\}^*\}^d$.

We write ϕ_x^y (likewise for α and f) for the *renaming* of variable x for y and vice versa in formula ϕ, and write ϕ_x^f for the result of *substitution* of term f for game variable x in ϕ, if the substitution is admissible (Definition 12 on page 14).

3.2 Example Game

We give an example game and theorem statements, proven in [10]. Automotive systems are a major class of CPS. As a simple indicative example we consider time-triggered 1-dimensional driving with adversarial timing. For maximum time T between control cycles, we let Demon choose any duration in $[0, T]$. When we need to prohibit pathological "Zeno" behaviors while keeping constraints realistic, we can further restrict $t \in [T/2, T]$.

We write x for the current position of the car, v for its velocity, a for the acceleration, $A > 0$ for the maximum positive acceleration, and $B > 0$ for the maximum braking rate. We assume $x = v = 0$ initially to simplify arithmetic. In time-triggered control, the controller runs at least once every $T > 0$ time units. Time and physics are continuous, T gives an upper bound on how often the controller runs. Local clock t marks the current time within the current timestep, then resets at each step. The control game (ctrl) says Angel can pick any acceleration a that is physically achievable ($-B \le a \le A$). The clock t is then reinitialized to 0. The plant game (plant) says Demon can evolve physics for duration $t \in [0, T]$ such that $v \ge 0$ throughout, then returns control to Angel.

Typical theorems in DLs and GLs are *safety* and *liveness*: are unsafe states always avoided and are desirable goals eventually reached? Safety and liveness of the 1D *system* has been proved previously: safe driving (safety) never goes past goal g, while live driving eventually reaches g (liveness).

$$\mathsf{pre} \equiv T > 0 \land A > 0 \land B > 0 \land v = 0 \land x = 0 \quad \mathsf{post} \equiv (g = x \land v = 0)$$
$$\mathsf{ctrl} \equiv a := *;\ ?-B \le a \le A;\ t := 0$$
$$\mathsf{plant} \equiv \{t' = 1, x' = v, v' = a \,\&\, t \le T \land v \ge 0\}^d$$
$$\mathsf{safety} \equiv \mathsf{pre} \to \langle(\mathsf{ctrl}; \mathsf{plant})^\times\rangle\, x \le g$$
$$\mathsf{liveness} \equiv \mathsf{pre} \to \langle(\mathsf{ctrl}; \mathsf{plant}; \{?t \ge T/2\}^d)^*\rangle\, x \ge g$$

Liveness theorem liveness requires a lower time bound ($\{?t \ge T/2\}^d$) to rule out Zeno strategies where Demon "cheats" by exponentially decreasing durations to essentially freeze the progress of time. The limit $t \ge T/2$ is chosen for simplicity. Safety theorem safety omits this constraint because even Zeno behaviors are safe.

Safety and liveness theorems, if designed carelessly, have trivial solutions including but not limited to Zeno behaviors. It is safe to remain at $x = 0$ and is live to maintain $a = A$, but not vice-versa. In contrast to DLs, GLs easily express the requirement that *the same* strategy is both safe and live: we must remain safe *while* reaching the goal. We use this *reach-avoid* specification because it is immune to trivial solutions. We give a new reach-avoid result for 1D driving.

Example 4 (Reach-avoid). The following is provable in dGL and CdGL:

$$\mathsf{reachAvoid} \equiv \mathsf{pre} \to \langle\{\mathsf{ctrl}; \mathsf{plant}; ?x \le g; \{?t \ge T/2\}^d\}^*\rangle \mathsf{post}$$

Angel *reaches* $g = x \land v = 0$ while safely *avoiding* states where $x \le g$ does not hold. Angel is safe at *every* iteration for *every* time $t \in [0, T]$, thus safe *throughout* the game. The (dual) test $?t \ge T/2$ appears second, allowing Demon to win if Angel violates safety during $t < T/2$.

1D driving is well-studied for classical systems, but the constructive reach-avoid proof [10] is subtle. The proof constructs an envelope of safe upper and live lower bounds on velocity as a function of position (Fig. 1). The blue point indicates where Angel must begin to brake to ensure time-triggered safety. It is surprising that Angel can achieve postcondition $g = x \land v = 0$, given that trichotomy $(f < g \lor f = g \lor f > g)$ is constructively invalid. The key [10] is that comparison *terms* $\min(f, g)$ and $\max(f, g)$ *are* exact in Type 2 computability where bits of min and max may be computed lazily. Our exact result encourages us that constructivity is not overly burdensome in practice. When decidable comparisons $(f < g + \delta \lor f > g)$ *are* needed, the alternative is a weaker guarantee $g - \varepsilon \leq x \leq g$ for parameter $\varepsilon > 0$. This relaxation is often enough to make the theorem provable, and reflects the fact that real agents only expect to reach their goal within finite precision.

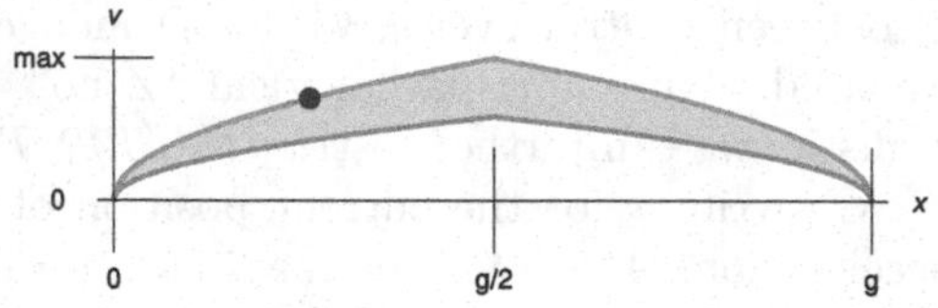

Fig. 1. Safe driving envelope (Color figure online)

4 Type-Theoretic Semantics

In this section, we define the semantics of hybrid games and game formulas in type theory. We start with assumptions on the underlying type theory.

4.1 Type Theory Assumptions

We assume a Calculus of Inductive and Coinductive Constructions (CIC)-like type theory [15,16,54] with polymorphism and dependency. We write M for terms and $\Delta \vdash M : \tau$ to say M has type τ in CIC context Δ. We assume first-class (indexed [19]) inductive and coinductive types. We write τ for type families and κ for kinds: type families inhabited by other type families. Inductive type families are written $\mu t : \kappa.\, \tau$, which denotes the *smallest* solution $\mathtt{ty}$ of kind κ to the fixed-point equation $\mathtt{ty} = \tau_t^{\mathtt{ty}}$. Coinductive type families are written $\rho t : \kappa.\, \tau$, which denotes the *largest* solution $\mathtt{ty}$ of kind κ to the fixed-point equation $\mathtt{ty} = \tau_t^{\mathtt{ty}}$. Type-expression τ must be monotone in t so smallest and largest solutions exist by Knaster-Tarski [24, Thm. 1.12]. Proof assistants like Coq reject definitions where monotonicity requires nontrivial proof; we did not mechanize our proofs because they use such definitions.

We use one predicative universe which we write $\mathbb{T}$ and Coq writes `Type 0`. Predicativity is an important assumption because our semantic definition is a large elimination, a feature known to interact dangerously with impredicativity. We write $\Pi x : \tau_1.\, \tau_2$ for a dependent function type with argument named x of type τ_1 and where return type τ_2 may mention x. We write $\Sigma x : \tau_1.\, \tau_2$ for a dependent pair type with left component named x of type τ_1 and right component of type τ_2, possibly mentioning x. These specialize to the simple function

$\tau_1 \Rightarrow \tau_2$ and product types $\tau_1 * \tau_2$ respectively when x is not mentioned in τ_2. Lambdas $(\lambda x : \tau.\, M)$ inhabit dependent function types. Pairs (M, N) inhabit dependent pair types. Application is $M\ N$. Let-binding unpacks pairs, whose left and right projection are $\pi_L M$ and $\pi_R M$. We write $\tau_1 + \tau_2$ for a disjoint union inhabited by $\ell \cdot M$ and $r \cdot M$, and write $\mathsf{case}\ A\ \mathsf{of}\ p \Rightarrow B \mid q \Rightarrow C$ for its case analysis.

We assume a real number type $\mathbb{R}$ and a Euclidean state type $\mathfrak{S}$. The positive real numbers are written $\mathbb{R}_{>0}$, nonnegative reals $\mathbb{R}_{\geq 0}$. We assume scalar and vector sums, products, inverses, and units. States s, t support operations $s\ x$ and $\mathsf{set}\ s\ x\ v$ which respectively retrieve the value of variable x in $s : \mathfrak{S}$ or update it to v. The usual axioms of setters and getters [22] are satisfied. We write $\mathfrak{s}$ for the distinguished variable of type $\mathfrak{S}$ representing the current state. We will find it useful to consider the semantics of an expression both at current state $\mathfrak{s}$ and at states s, t defined in terms of $\mathfrak{s}$ (e.g., $\mathsf{set}\ \mathfrak{s}\ x\ 5$).

4.2 Semantics of CdGL

Terms f, g are type-theoretic functions of type $\mathfrak{S} \Rightarrow \mathbb{R}$. We will need differential terms $(f)'$, a definable term construct when f is differentiable. Not every term f need be differentiable, so we give a *virtual* definition, defining when $(f)'$ is equal to some term g. If $(f)'$ does not exist, then $(f)' = g$ is not provable. We define the (total) differential as the Euclidean dot product $(\cdot)$ of the gradient (variable name: ∇) with s', which is the vector of values $s\,x'$ assigned to primed variables x'. To show that ∇ is the gradient, we define the gradient as a limit, which we express in (ε, δ) style. In this definition, f and g are scalar-valued, and the minus symbol is used for both scalar and vector difference.

$$\begin{aligned}((f)'\ s = g\ s) \equiv\ & \Sigma\nabla : \mathbb{R}^{|s'|}.\, (g\ s = \nabla \cdot s') * \Pi\varepsilon : \mathbb{R}_{>0}.\, \Sigma\delta : \mathbb{R}_{>0}.\, \Pi r : \mathfrak{S}.\\ & (\|r - s\| < \delta) \Rightarrow |f\ r - f\ s - \nabla \cdot (r - s)| \leq \varepsilon \|r - s\|\end{aligned}$$

For practical proofs, a library of standard rules for automatic, syntactic differentiation of common arithmetic operations [7] could be proven.

The interpretation $\ulcorner\phi\urcorner : \mathfrak{S} \Rightarrow \mathbb{T}$ of formula ϕ is a predicate over states. A predicate of kind $\mathfrak{S} \Rightarrow \mathbb{T}$ is also understood as a *region*, e.g., $\ulcorner\phi\urcorner$ is the region containing states where ϕ is provable. A CdGL context Γ is interpreted over a uniform state term $s : \mathfrak{S}$ where $\mathfrak{s} : \mathfrak{S} \vdash s : \mathfrak{S}$, i.e., s usually mentions $\mathfrak{s}$. We define $\ulcorner\Gamma\urcorner(s)$ to be the CIC context containing $\mathfrak{s} : \mathfrak{S}$ and $\ulcorner\phi\urcorner\ s$ for each $\phi \in \Gamma$. The sequent $(\Gamma \vdash \phi)$ is *valid* if there exists M where $\ulcorner\Gamma\urcorner(\mathfrak{s}) \vdash M : (\ulcorner\phi\urcorner\ \mathfrak{s})$. Formula ϕ is *valid* iff sequent $(\cdot \vdash \phi)$ is valid. That is, a valid formula is provable in every state with a common proof term M. The witness may inspect the state, but must do so constructively. Formula semantics employ the Angelic and Demonic semantics of games, which determine how to win a game α whose postcondition is ϕ. We write $\langle\langle\alpha\rangle\rangle : (\mathfrak{S} \Rightarrow \mathbb{T}) \Rightarrow (\mathfrak{S} \Rightarrow \mathbb{T})$ for the Angelic semantics of α and $[[\alpha]] : (\mathfrak{S} \Rightarrow \mathbb{T}) \Rightarrow (\mathfrak{S} \Rightarrow \mathbb{T})$ for its Demonic semantics.

Definition 5 (Formula semantics). *Angel and Demon strategies for a hybrid game α with goal region P are inhabitants of $\langle\langle\alpha\rangle\rangle\ P$ and $[[\alpha]]\ P$, respectively.*

$$\ulcorner [\alpha]\phi \urcorner\ s = [[\alpha]]\ \ulcorner \phi \urcorner\ s \quad \ulcorner \langle\alpha\rangle\phi \urcorner\ s = \langle\langle\alpha\rangle\rangle\ \ulcorner \phi \urcorner\ s \quad \ulcorner f \sim g \urcorner\ s = ((f\ s) \sim (g\ s))$$

Modality $\langle\alpha\rangle\phi$ is provable in s when $\langle\langle\alpha\rangle\rangle\ \ulcorner\phi\urcorner\ s$ is inhabited so Angel has an α strategy from s to reach region $\ulcorner\phi\urcorner$ on which ϕ is provable. Modality $[\alpha]\phi$ is provable in s when $[[\alpha]]\ \ulcorner\phi\urcorner\ s$ is inhabited so Demon has an α strategy from s to reach region $\ulcorner\phi\urcorner$ on which ϕ is provable. For $\sim\ \in \{\leq, <, =, \neq, >, \geq\}$, the values of f and g are compared at state s in $f \sim g$. The game and formula semantics are simultaneously inductive. In each case, the connectives which define $\langle\langle\alpha\rangle\rangle$ and $[[\alpha]]$ are duals, because $[\alpha]\phi$ and $\langle\alpha\rangle\phi$ are dual. Below, P refers to the goal region of the game and s to the initial state.

Definition 6 (Angel semantics). *We define $\langle\langle\alpha\rangle\rangle : (\mathfrak{S} \Rightarrow \mathbb{T}) \Rightarrow (\mathfrak{S} \Rightarrow \mathbb{T})$ inductively (by a large elimination) on α:*

$$\begin{aligned}
\langle\langle ?\psi\rangle\rangle\ P\ s &= \ulcorner\psi\urcorner\ s * P\ s\\
\langle\langle x := f\rangle\rangle\ P\ s &= P\ (\mathsf{set}\ s\ x\ (f\ s))\\
\langle\langle x := *\rangle\rangle\ P\ s &= \Sigma v : \mathbb{R}.\ P\ (\mathsf{set}\ s\ x\ v)\\
\langle\langle \alpha \cup \beta\rangle\rangle\ P\ s &= \langle\langle\alpha\rangle\rangle\ P\ s + \langle\langle\beta\rangle\rangle\ P\ s\\
\langle\langle \alpha ; \beta\rangle\rangle\ P\ s &= \langle\langle\alpha\rangle\rangle\ (\langle\langle\beta\rangle\rangle\ P)\ s
\end{aligned}$$

$$\begin{aligned}
\langle\langle \alpha^d\rangle\rangle\ P\ s &= [[\alpha]]\ P\ s\\
\langle\langle x' = f \,\&\, \psi\rangle\rangle\ P\ s &= \Sigma d : \mathbb{R}_{\geq 0}.\ \Sigma sol : [0,d] \Rightarrow \mathbb{R}.\\
&\quad (sol, s, d \vDash x' = f)\\
&\quad * (\Pi t : [0,d].\ \ulcorner\psi\urcorner\ (\mathsf{set}\ s\ x\ (sol\ t)))\\
&\quad * P\ (\mathsf{set}\ s\ (x, x')\\
&\qquad (sol\ d, f\ (\mathsf{set}\ s\ x\ (sol\ d))))
\end{aligned}$$

$$\langle\langle \alpha^*\rangle\rangle\ P\ s = \big(\mu\tau' : (\mathfrak{S} \Rightarrow \mathbb{T}).\ \lambda t : \mathfrak{S}.\ (P\ t \Rightarrow \tau'\ t) + (\langle\langle\alpha\rangle\rangle\ \tau'\ t \Rightarrow \tau'\ t)\big)\ s$$

Angel wins $\langle ?\psi\rangle P$ by proving both ψ and P at s. Angel wins the deterministic assignment $x := f$ by performing the assignment, then proving P. Angel wins nondeterministic assignment $x := *$ by constructively choosing a value v to assign, then proving P. Angel wins $\alpha \cup \beta$ by choosing between playing α or β, then winning that game. Angel wins $\alpha; \beta$ if she wins α with the postcondition of winning β. Angel wins α^d if she wins α in the Demon role. Angel wins ODE game $x' = f \,\&\, \psi$ by choosing some solution sol of some duration d which satisfies the ODE and domain constraint throughout and the postcondition ϕ at time d. While top-level postconditions rarely mention x', intermediate invariant steps do, thus x and x' both are updated in the postcondition. The construct $(sol, s, d \vDash x' = f)$, saying sol solves $x' = f$ from state s for time d, is defined:

$$(sol, s, d \vDash x' = f) \equiv \big((s\ x = sol\ 0) * \Pi r : [0,d].\ ((sol)'\ r = f\ (\mathsf{set}\ s\ x\ (sol\ r)))\big)$$

Note that variable sol stands for a function of the host theory, all of which are computable and therefore continuous. When $(sol, s, d \vDash x' = f)$ holds, sol is also continuously differentiable. Constructive Picard-Lindelöf [34] constructs a solution for every effectively-locally-Lipschitz ODEs, which need not have a closed form. The proof calculus we introduce in Sect. 5 includes both solution-based proof rules, which are useful for ODEs with simple closed forms, and invariant-based rules, which enable proof even when closed forms do not exist.

Angel strategies for α^* are inductively defined: either choose to stop the loop and prove P now, else play a round of α before repeating inductively. By Knaster-Tarski [24, Thm. 1.12], this least fixed point exists because the interpretation of a game is monotone in its postcondition (Lemma 7).

Lemma 7 (Monotonicity). *Let* $P, Q : \mathfrak{S} \Rightarrow \mathbb{T}$. *If* $\mathfrak{s} : \mathfrak{S}, P\ s \vdash M : Q\ s$ *then there exists a term* N *such that* $\mathfrak{s} : \mathfrak{S}, [[\alpha]]\ P\ s \vdash N : [[\alpha]]\ Q\ s$

Definition 8 (Demon semantics). *We define* $[[\alpha]] : (\mathfrak{S} \Rightarrow \mathbb{T}) \Rightarrow (\mathfrak{S} \Rightarrow \mathbb{T})$ *inductively (by a large elimination) on* α*:*

$$[[?\psi]]\ P\ s = \ulcorner\psi\urcorner\ s \Rightarrow P\ s$$
$$[[x := f]]\ P\ s = P\ (\mathsf{set}\ s\ x\ (f\ s))$$
$$[[x := *]]\ P\ s = \Pi v : \mathbb{R}.\, P\ (\mathsf{set}\ s\ x\ v)$$
$$[[\alpha \cup \beta]]\ P\ s = [[\alpha]]\ P\ s * [[\beta]]\ P\ s$$
$$[[\alpha; \beta]]\ P\ s = [[\alpha]]\ ([[\beta]]\ P)\ s$$
$$[[\alpha^d]]\ P\ s = \langle\langle\alpha\rangle\rangle\ P\ s$$
$$\begin{aligned}[[x' = f \,\&\, \psi]]\ P\ s = {} & \Pi d : \mathbb{R}_{\geq 0}.\, \Pi sol : [0, d] \Rightarrow \mathbb{R}.\\ & (sol, s, d \vDash x' = f)\\ & \Rightarrow (\Pi t : [0, d].\, \ulcorner\psi\urcorner\ (\mathsf{set}\ s\ x\ (sol\ t)))\\ & \Rightarrow P\ (\mathsf{set}\ s\ (x, x')\\ & \qquad (sol\ d, f\ (\mathsf{set}\ s\ x\ (sol\ d))))\end{aligned}$$
$$[[\alpha^*]]\ P\ s = \big(\rho\tau' : (\mathfrak{S} \Rightarrow \mathbb{T}).\, \lambda t : \mathfrak{S}.\, (\tau'\ t \Rightarrow [[\alpha]]\ \tau'\ t) * (\tau'\ t \Rightarrow P\ t)\big)\ s$$

Demon wins $[?\psi]P$ by proving P under assumption ψ, which Angel must provide (Sect. 7). Demon's deterministic assignment is identical to Angel's. Demon wins $x := *$ by proving ψ for *every* choice of x. Demon wins $\alpha \cup \beta$ with a pair of winning strategies. Demon wins $\alpha; \beta$ by winning α with a postcondition of winning β. Demon wins α^d if he can win α after switching roles with Angel. Demon wins $x' = f \,\&\, \psi$ if for an arbitrary duration and arbitrary solution which satisfy the domain constraint, he can prove the postcondition. Demon wins $[\alpha^*]P$ if he can prove P no matter how many times Angel makes him play α. Demon repetition strategies are coinductive using some invariant τ'. When Angel decides to stop the loop, Demon responds by proving P from τ'. Whenever Angel chooses to continue, Demon proves that τ' is preserved. Greatest fixed points exist by Knaster-Tarski [24, Thm. 1.12] using Lemma 7.

It is worth comparing the Angelic and Demonic semantics of $x := *$. An Angel strategy says how to compute x. A Demon strategy simply accepts $x \in \mathbb{R}$ as its input, even uncomputable numbers. This is because Angel strategies supply a computable real while Demon acts with computable outputs given real *inputs*. In general, each strategy is constructive but permits its opponent to play classically. In the cyber-physical setting, the opponent is indeed rarely a computer.

5 Proof Calculus

To enable direct syntactic proof, we give a natural deduction-style system for CdGL. We write $\Gamma = \psi_1, \ldots, \psi_n$ for a context of formulas and $\Gamma \vdash \phi$ for the natural-deduction sequent with conclusion ϕ and context Γ. We begin with rules shared by CGL [9] and CdGL, then give the ODE rules. We write Γ_x^y for the renaming of game variable x to y and vice versa in context Γ. Likewise Γ_x^f is the substitution of term f for game variable x. To avoid repetition, we write $\llparenthesis\alpha\rrparenthesis\phi$ to indicate that the same rule applies for $\langle\alpha\rangle\phi$ and $[\alpha]\phi$. These rules write $[\![\alpha\rangle\!\rangle\phi$ for the dual of $\llparenthesis\alpha\rrparenthesis\phi$. We write FV($e$), BV($\alpha$), and MBV($\alpha$) for the free variables of expression e, bound variables of game α, and must-bound variables of game α respectively, i.e., variables which *might* influence the meaning of an expression, might be modified during game execution, or are written during *every* execution.

$$([\cup]\mathrm{I})\ \frac{\Gamma\vdash[\alpha]\phi\quad \Gamma\vdash[\beta]\phi}{\Gamma\vdash[\alpha\cup\beta]\phi}\qquad ([\cup]\mathrm{E1})\ \frac{\Gamma\vdash[\alpha\cup\beta]\phi}{\Gamma\vdash[\alpha]\phi}\qquad ([\cup]\mathrm{E2})\ \frac{\Gamma\vdash[\alpha\cup\beta]\phi}{\Gamma\vdash[\beta]\phi}$$

$$(\langle\cup\rangle\mathrm{I1})\ \frac{\Gamma\vdash\langle\alpha\rangle\phi}{\Gamma\vdash\langle\alpha\cup\beta\rangle\phi}\qquad (\langle\cup\rangle\mathrm{I2})\ \frac{\Gamma\vdash\langle\beta\rangle\phi}{\Gamma\vdash\langle\alpha\cup\beta\rangle\phi}\qquad (\mathrm{hyp})\ \frac{}{\Gamma,\phi\vdash\phi}$$

$$(\langle?\rangle\mathrm{I})\ \frac{\Gamma\vdash\phi\quad \Gamma\vdash\psi}{\Gamma\vdash\langle?\phi\rangle\psi}\qquad (\langle?\rangle\mathrm{E1})\ \frac{\Gamma\vdash\langle?\phi\rangle\psi}{\Gamma\vdash\phi}\qquad (\langle?\rangle\mathrm{E2})\ \frac{\Gamma\vdash\langle?\phi\rangle\psi}{\Gamma\vdash\psi}$$

$$([?]\mathrm{I})\ \frac{\Gamma,\phi\vdash\psi}{\Gamma\vdash[?\phi]\psi}\qquad ([?]\mathrm{E})\ \frac{\Gamma\vdash[?\phi]\psi\quad \Gamma\vdash\phi}{\Gamma\vdash\psi}$$

$$(\langle\cup\rangle\mathrm{E})\ \frac{\Gamma\vdash\langle\alpha\cup\beta\rangle\phi\quad \Gamma,\langle\alpha\rangle\phi\vdash\psi\quad \Gamma,\langle\beta\rangle\phi\vdash\psi}{\Gamma\vdash\psi}$$

Fig. 2. CdGL proof calculus: propositional game rules

Figure 2 gives the propositional game rules. Rule [?]E is modus ponens and [?]I is implication introduction because $[?\phi]\psi$ is implication. Angelic choices are disjunctions introduced by $\langle\cup\rangle$I1 and $\langle\cup\rangle$I2 and case-analyzed by $\langle\cup\rangle$E. Angelic tests and Demonic choices are conjunctions introduced by $\langle?\rangle$I and $[\cup]$I, eliminated by $\langle?\rangle$E1, $\langle?\rangle$E2, $[\cup]$E1, and $[\cup]$E2. Rule hyp applies an assumption.

$$([{:}*]\mathrm{I})\ \frac{\Gamma_x^y\vdash\phi}{\Gamma\vdash[x:=*]\phi}\qquad ([{:}*]\mathrm{E})\ \frac{\Gamma\vdash[x:=*]\phi}{\Gamma\vdash\phi_x^f}$$

$$(\langle{:}*\rangle\mathrm{I})\ \frac{\Gamma\vdash\langle x:=f\rangle\phi}{\Gamma\vdash\langle x:=*\rangle\phi}\qquad (\langle{:}*\rangle\mathrm{E})\ \frac{\Gamma\vdash\langle x:=*\rangle\phi\quad \Gamma\vdash\forall x\,(\phi\to\psi)}{\Gamma\vdash\psi}\ (x\notin\mathrm{FV}(\psi))$$

$$(\llparenthesis;\rrparenthesis\mathrm{I})\ \frac{\Gamma\vdash\llparenthesis\alpha\rrparenthesis\llparenthesis\beta\rrparenthesis\phi}{\Gamma\vdash\llparenthesis\alpha;\beta\rrparenthesis\phi}\qquad (\mathrm{M})\ \frac{\Gamma\vdash\llparenthesis\alpha\rrparenthesis\phi\quad \Gamma_{\mathrm{BV}(\alpha)}^{\boldsymbol{y}},\phi\vdash\psi}{\Gamma\vdash\llparenthesis\alpha\rrparenthesis\psi}$$

$$(\llparenthesis{:=}\rrparenthesis\mathrm{I})\ \frac{\Gamma_x^y,x=f_x^y\vdash\phi}{\Gamma\vdash\llparenthesis x:=f\rrparenthesis\phi}\qquad (\llparenthesis^d\rrparenthesis\mathrm{I})\ \frac{\Gamma\vdash[\![\alpha\rangle\!\rangle\phi}{\Gamma\vdash\llparenthesis\alpha^d\rrparenthesis\phi}$$

Fig. 3. CdGL proof calculus: first-order games (y fresh, f computable, ϕ_x^f admissible)

Figure 3 covers assignment, choice, sequencing, duals, and monotonicity. Angelic games have injectors ($\langle*\rangle$S, $\langle*\rangle$G) and case analysis ($\langle*\rangle$E). Duality $\llparenthesis^d\rrparenthesis$I switches players by switching modalities. Sequential games ($\llparenthesis;\rrparenthesis$I) are decomposed as nested modalities.

Monotonicity (M) is Lemma 7 in rule form. The second premiss writes $\Gamma \frac{y}{\mathrm{BV}(\alpha)}$ to indicate that the bound variables of α must be freshly renamed in Γ for soundness. Rule M is used for generalization because all GLs are subnormal, lacking axiom K (modal modus ponens) and necessitation. Common uses include concise right-to-left symbolic execution proofs and, in combination with $\langle\!\![;]\!\!\rangle$I, Hoare-style sequential composition reasoning.

Nondeterministic assignments quantify over real-valued game variables. Assignments $\langle\!\![:=]\!\!\rangle$I remember the initial value of x in fresh variable y ($\Gamma\frac{y}{x}$) for sake of completeness, then provide an assumption that x has been assigned to f. Skolemization $[:*]$I bound-renames x to y in Γ, written $\Gamma\frac{y}{x}$. Specialization $[:*]$E instantiates x to a term f by substituting ϕ_x^f. Existentials are introduced by giving a witness f in $\langle:*\rangle$I. Herbrandization $\langle:*\rangle$E unpacks existentials, soundness requires x is not free in ψ.

$$(\langle*\rangle\mathrm{E})\ \frac{\Gamma\vdash\langle\alpha^*\rangle\phi\quad \Gamma,\phi\vdash\psi\quad \Gamma,\langle\alpha\rangle\langle\alpha^*\rangle\phi\vdash\psi}{\Gamma\vdash\psi}\qquad ([*]\mathrm{E})\ \frac{\Gamma\vdash[\alpha^*]\phi}{\Gamma\vdash\phi\wedge[\alpha][\alpha^*]\phi}$$

$$(\langle*\rangle\mathrm{S})\ \frac{\Gamma\vdash\phi}{\Gamma\vdash\langle\alpha^*\rangle\phi}\qquad (\langle*\rangle\mathrm{G})\ \frac{\Gamma\vdash\langle\alpha\rangle\langle\alpha^*\rangle\phi}{\Gamma\vdash\langle\alpha^*\rangle\phi}\qquad (\mathrm{FP})\ \frac{\Gamma\vdash\langle\alpha^*\rangle\phi\quad \phi\vdash\psi\quad \langle\alpha\rangle\psi\vdash\psi}{\Gamma\vdash\psi}$$

$$([*]\mathrm{R})\ \frac{\Gamma\vdash\phi\wedge[\alpha][\alpha^*]\phi}{\Gamma\vdash[\alpha^*]\phi}\ (\mathrm{loop})\ \frac{\Gamma\vdash J\quad J\vdash[\alpha]J\quad J\vdash\phi}{\Gamma\vdash[\alpha^*]\phi}$$

$$(\langle*\rangle\mathrm{I})\ \frac{\Gamma\vdash\varphi\quad \varphi,\mathbf{0}\succcurlyeq\mathcal{M}\vdash\phi\quad \varphi,(\mathcal{M}\succ\mathbf{0}\wedge\mathcal{M}_0=\mathcal{M})\vdash\langle\alpha\rangle(\varphi\wedge\mathcal{M}_0\succ\mathcal{M})}{\Gamma\vdash\langle\alpha^*\rangle\phi}$$

Fig. 4. CdGL proof calculus: loops ($\mathcal{M}_0$ fresh)

Figure 4 provides rules for repetitions. In rule $\langle*\rangle$I, $\mathcal{M}$ indicates an arbitrary termination metric where $\succ$ and $\succcurlyeq$ denote strict and nonstrict comparison in an arbitrary (effectively) well-founded [28] partial order. Metavariable $\mathbf{0}$ represents a terminal value at which iteration stops; we will choose $\mathbf{0}=0$ in our example, but $\mathbf{0}$ need not be 0 in general. $\mathcal{M}_0$ is a fresh variable which remembers $\mathcal{M}$. Angel plays α^* by repeating an α strategy which always decreases the termination metric. Angel maintains a formula φ throughout, and stops once $0\succcurlyeq\mathcal{M}$. The postcondition need only follow from termination condition $0\succcurlyeq\mathcal{M}$ and convergence formula φ. Simple real comparisons $x\geq y$ are not well-founded, but inflated comparisons like $x\geq y+1$ are. Well-founded metrics ensure convergence in finitely (but often unboundedly) many iterations. In the simplest case, $\mathcal{M}$ is a real-valued term. Generalizing $\mathcal{M}$ to tuples enables, e.g., lexicographic termination metrics. For example, the metric in the proof of Example 4 is the distance to the goal, which must decrease by some minimum amount each iteration.

Repetition games can be folded and unfolded ($[*]$E, $[*]$R). Rule FP says $\langle\alpha^*\rangle\phi$ is a least pre-fixed-point. It works backwards: first show ψ holds after α^*, then preserve ψ when each iteration is unwound. Rule loop is the repetition invariant rule. Demonic repetition is eliminated by [*]E.

Like any first-order program logic, CdGL proofs contain first-order reasoning at the leaves. Decidability of constructive real arithmetic is an open problem [33], so first-order facts are proven manually in practice. Our semantics embed CdGL into type theory; we defer first-order arithmetic proving to the host theory. Even effectively-well-founded $\succcurlyeq$ need not have decidable guards $(\mathbf{0} \succ \mathcal{M} \lor \mathcal{M} \succcurlyeq \mathbf{0})$ since exact comparisons are not computable [6]. We may not be able to distinguish $\mathcal{M} = \mathbf{0}$ from very small positive values of $\mathcal{M}$, leading to one unnecessary loop iteration, after which $\mathcal{M}$ is certainly $\mathbf{0}$ and the loop terminates. Comparison up to $\varepsilon > 0$ is decidable [12] $(f > g \lor (f < g + \varepsilon))$.

$$\text{(DC)}\ \frac{\Gamma \vdash [x'=f \,\&\, \psi]R \quad \Gamma \vdash [x'=f \,\&\, \psi \land R]\phi}{\Gamma \vdash [x'=f \,\&\, \psi]\phi} \qquad \text{(DI)}\ \frac{\Gamma \vdash \phi \quad \Gamma \vdash \forall x\, (\psi \to [x' := f](\phi)')}{\Gamma \vdash [x'=f \,\&\, \psi]\phi}$$

$$\text{(DG)}\ \frac{\Gamma \vdash \exists y\, [x'=f, y' = a(x)y + b(x) \,\&\, \psi]\phi}{\Gamma \vdash [x'=f \,\&\, \psi]\phi} \qquad \text{(DW)}\ \frac{\Gamma \vdash \forall x \forall x'\, (\psi \to \phi)}{\Gamma \vdash [x'=f \,\&\, \psi]\phi}$$

$$\text{(DV)}\ \frac{\begin{array}{c}\psi, h \geq g \vdash \phi \quad \Gamma \vdash d > 0 \land \varepsilon > 0 \land h - g \geq -d\varepsilon \\ \Gamma \vdash \langle t := 0; \{t'=1, x'=f \,\&\, \psi\}\rangle t \geq d \quad \Gamma \vdash [x'=f]((h)' - (g)') \geq \varepsilon\end{array}}{\Gamma \vdash \langle x'=f \,\&\, \psi\rangle\phi}$$

$$\text{(bsolve)}\ \frac{\Gamma \vdash \forall t : \mathbb{R}_{\geq 0}\ ((\forall r : [0,t]\ [t := r; x := sln]\psi) \to [x := sln; x' := f]\phi)}{\Gamma \vdash [x'=f \,\&\, \psi]\phi}$$

$$\text{(dsolve)}\ \frac{\Gamma \vdash \exists t : \mathbb{R}_{\geq 0}\ ((\forall r : [0,t]\ \langle t := r; x := sln\rangle\psi) \land \langle x := sln; x' := f\rangle\phi)}{\Gamma \vdash \langle x'=f \,\&\, \psi\rangle\phi}$$

Fig. 5. CdGL proof calculus: ODEs. In bsolve and dsolve, sln solves $x' = f$ globally, t and r fresh, $x' \notin \mathrm{FV}(\phi)$

Figure 5 gives the ODE rules, which are a constructive version of those from dGL [42]. For nilpotent ODEs such as the plant of Example 4, reasoning via solutions is possible. Since CdGL supports nonlinear ODEs which often do not have closed-form solutions, we provide invariant-based rules, which are complete [46] for invariants of polynomial ODEs. *Differential induction* DI [41] says ϕ is an invariant of an ODE if it holds initially and if its *differential formula* [41] $(\phi)'$ holds throughout, for example $(f \geq g)' \equiv ((f)' \geq (g)')$. Soundness of DI requires differentiability, and $(\phi)'$ is not provable when ϕ mentions nondifferentiable terms. *Differential cut* DC proves R invariant, then adds it to the domain constraint. *Differential weakening* DW says that if ϕ follows from the domain constraint, it holds throughout the ODE. *Differential ghosts* DG permit us to augment an ODE system with a fresh dimension y, which enables [46] proofs of otherwise unprovable properties. We restrict the right-hand side of y to be linear in y and (uniformly) continuous in x because soundness requires that ghosting y' does not change the duration of an ODE. A linear right-hand side is guaranteed to be Lipschitz on the whole existence interval of equation $x' = f$, thus ensuring an unchanged duration by (constructive) Picard-Lindelöf [34]. *Differential variants* [41,53] DV is an Angelic counterpart to DI. The schema parameters d and

ε must not mention x, x', t, t'. To show that f eventually exceeds g, first choose a duration d and a sufficiently high minimum rate ε at which $h - g$ will change. Prove that $h - g$ decreases at rate at least ε and that the ODE has a solution of duration d satisfying constraint ψ. Thus at time d, both $h \geq g$ and its provable consequents hold. Rules bsolve and dsolve assume as a side condition that sln is the unique solution of $x' = f$ on domain ψ. They are convenient for ODEs with simple solutions, while invariant reasoning supports complicated ODEs.

6 Theory: Soundness

Following constructive counterparts of classical soundness proofs for dGL, we prove that the CdGL proof calculus is sound: provable formulas are true in the CIC semantics. For the sake of space, we give statements and some outlines here, reporting all proofs and lemmas elsewhere [10]. Similar lemmas have been used to prove soundness of dGL [45], but our new semantics lead to simpler statements for Lemmas 10 and 11. The coincidence property for terms is not proved but assumed, since we inherit a semantic treatment of terms from the host theory. Let $s\frac{y}{x}$ be s with the values of x and y swapped. Let s^f_x be $\mathsf{set}\ s\ x\ (f\ s)$. Defined CIC term $s \overset{V}{=} t \leftrightarrow *_{x \in V}(s\ x = t\ x)$ says s and t agree on all $x \in V$.

Lemma 9 (Uniform renaming). *Let $M\frac{y}{x}$ rename x and y in proof term M.*

- *If* $\ulcorner\Gamma\urcorner(s) \vdash M : (\ulcorner\phi\urcorner\ s)$ *then* $\ulcorner\Gamma\frac{y}{x}\urcorner(s\frac{y}{x}) \vdash M\frac{y}{x} : (\ulcorner\phi\frac{y}{x}\urcorner\ s\frac{y}{x})$.

Lemma 10 (Coincidence). *Assume $s \overset{V}{=} t$ where $V \supseteq FV(\Gamma) \cup FV(\phi)$.*

- *If* $\ulcorner\Gamma\urcorner(s) \vdash M : (\ulcorner\phi\urcorner\ s)$ *then exists N such that* $\ulcorner\Gamma\urcorner(t) \vdash N : (\ulcorner\phi\urcorner\ t)$.

Lemma 11 (Bound effect). *Let $P : \mathfrak{S} \Rightarrow \mathbb{T}$ and let $V \subseteq BV(\alpha)^{\mathsf{C}}$, the complement of bound variables of α.*

- *There exists M such that* $\ulcorner\Gamma\urcorner(s) \vdash M : (\langle\langle\alpha\rangle\rangle\ P\ s)$ *iff there exists N such that* $\ulcorner\Gamma\urcorner(s) \vdash N : (\langle\langle\alpha\rangle\rangle\ (\lambda t.\ P\ t * s \overset{V}{=} t)\ s)$.
- *There exists M such that* $\ulcorner\Gamma\urcorner(s) \vdash M : ([[\alpha]]\ P\ s)$ *iff there exists N such that* $\ulcorner\Gamma\urcorner(s) \vdash N : ([[\alpha]]\ (\lambda t.\ P\ t * s \overset{V}{=} t)\ s)$.

Definition 12 (Term substitution admissibility [40, Def. 6]). *For a formula ϕ, (likewise for context Γ, term f, and game α) we say ϕ^f_x is* admissible *if x never appears free in ϕ under a binder of $\{x\} \cup FV(f)$.*

Lemma 13 (Term substitution). *Let M^f_x substitute f for x in proof term M. Let Γ^f_x and ϕ^f_x be admissible.*

- *If* $\ulcorner\Gamma\urcorner(s^f_x) \vdash M : (\ulcorner\phi\urcorner\ s^f_x)$ *then* $\ulcorner\Gamma^f_x\urcorner(s) \vdash M^f_x : (\ulcorner\phi^f_x\urcorner s)$.

The converse implication also holds, though its witness is not necessarily M.

Soundness of the proof calculus follows from the lemmas, and soundness of the ODE rules employing several known results from constructive analysis.

Theorem 14 (Soundness). *If $\Gamma \vdash M : \phi$ holds, then sequent $(\Gamma \vdash \phi)$ is valid. As a special case, if $\cdot \vdash M : \phi$ holds, then formula ϕ is valid.*

Proof Sketch. By induction on the derivation. The assignment case holds by Lemma 13 and Lemma 9. Lemma 10 and Lemma 11 are applied when maintaining truth of a formula across changing state. The equality and inequality cases of DI and DV employ the constructive mean-value theorem [10, Thm. 21], which has been formalized, e.g., in Coq [17]. Rules DW, bsolve, and dsolve follow from the semantics of ODEs. Rule DC uses the fact that prefixes of solutions are solutions. Rule DG uses constructive Picard-Lindelöf [34], which constitutes an algorithm for arbitrarily approximating the solution of any Lipschitz ODE, with a convergence rate depending on its Lipschitz constant. □

We have shown that every provable formula is true in the type-theoretic semantics. Because the soundness proof is constructive, it amounts to an extraction algorithm from CdGL into type theory: for each CdGL proof, there exists a program in type theory which inhabits the corresponding type of the semantics.

7 Theory: Extraction and Execution

Another perspective on constructivity is that provable properties must have witnesses. We show Existential and Disjunction properties providing witnesses for existentials and disjunctions. For modal formulas $\langle\alpha\rangle\phi$ and $[\alpha]\phi$ we show proofs can be *used as* winning strategies: a big-step operational semantics play allows playing strategies against each other to extract a proof that their goals hold in some final state t. Our presentation is more concise than defining the language, semantics, and properties of strategies, while providing key insights.

Lemma 15. (Existential Property). *Let $s : \mathfrak{S}$. If $\ulcorner\Gamma\urcorner(s) \vdash M : (\ulcorner\exists x\,\phi\urcorner\ s)$ then there exist terms $f : \mathbb{R}$ and N such that $\ulcorner\Gamma\urcorner(s) \vdash N : (\ulcorner\phi_x^f\urcorner\ s)$.*

Lemma 16. (Disjunction Property). *If $\ulcorner\Gamma\urcorner(s) \vdash M : (\ulcorner\phi \lor \psi\urcorner\ s)$ then there exists a proof term N such that $\ulcorner\Gamma\urcorner(s) \vdash N : (\ulcorner\phi\urcorner\ s)$ or $\ulcorner\Gamma\urcorner(s) \vdash N : (\ulcorner\psi\urcorner\ s)$.*

The proofs follow their counterparts in type theory. The Disjunction Property considers truth at a *specific state.* Validity of $\phi \lor \psi$ does *not* imply validity of either ϕ or ψ. For example, $x < 1 \lor x > 0$ is valid, but its disjuncts are not.

Function play below gives a big-step semantics: Angel and Demon strategies as and ds for respective goals ϕ and ψ in game α suffice to construct a final state t satisfying both. By parametricity, t was found by playing α, because play cannot inspect P and Q, thus can only prove them via as and ds.

$$\begin{aligned}\mathsf{play} :\ & \Pi\alpha : \mathsf{Game}.\,\Pi P, Q : (\mathfrak{S} \Rightarrow \mathbb{T}).\,\Pi s : \mathfrak{S}.\\ & (\langle\!\langle\alpha\rangle\!\rangle\ P\ s \Rightarrow [[\alpha]]\ Q\ s \Rightarrow \Sigma t : \mathfrak{S}.\,P\ t * Q\ t)\end{aligned}$$

Applications of play are written $\mathsf{play}_\alpha\ s\ \mathsf{as}\ \mathsf{ds}$ (P and Q implicit). Game consistency (Corollary 17) is by play and consistency of type theory. Note that α^d is played by swapping the Angel and Demon strategies in α.

$$\begin{aligned}
\mathsf{play}_{x:=f}\ s\ \mathsf{as}\ \mathsf{ds} &= (\mathsf{let}\, t\ =\ \mathsf{set}\ s\ x\ (f\ s)\,\mathsf{in}\,(t,(\mathsf{as}\ t,\mathsf{ds}\ t)))\\
\mathsf{play}_{x:=*}\ s\ \mathsf{as}\ \mathsf{ds} &= \mathsf{let}\, t\ =\ \mathsf{set}\ s\ x\ \pi_L\mathsf{as}\,\mathsf{in}\,(t,(\pi_R\mathsf{as},\mathsf{ds}\ \pi_L\mathsf{as}))\\
\mathsf{play}_{x'=f\,\&\,\psi}\ s\ \mathsf{as}\ \mathsf{ds} &= \mathsf{let}\,(d,\mathrm{sol},\mathrm{solves},c,p)\ =\ \mathsf{as}\ s\,\mathsf{in}\\
&\qquad(\mathsf{set}\ s\ x\ (\mathrm{sol}\ d),(p,\mathsf{ds}\ d\ \mathrm{sol}\ \mathrm{solves}\ c))\\
\mathsf{play}_{?\phi}\ s\ \mathsf{as}\ \mathsf{ds} &= (s,(\pi_R\mathsf{as},\mathsf{ds}\ (\pi_L\mathsf{as})))\\
\mathsf{play}_{\alpha\cup\beta}\ s\ \mathsf{as}\ \mathsf{ds} &= \mathsf{case}\ (\mathsf{as}\ s)\ \mathsf{of}\\
&\qquad \mathsf{as}' \Rightarrow \mathsf{play}_\alpha\ s\ \mathsf{as}'\ (\pi_L\mathsf{ds})\\
&\qquad \mid \mathsf{as}' \Rightarrow \mathsf{play}_\beta\ s\ \mathsf{as}'\ (\pi_R\mathsf{ds})\\
\mathsf{play}_{\alpha;\beta}\ s\ \mathsf{as}\ \mathsf{ds} &= (\mathsf{let}\,(t,(\mathsf{as}',\mathsf{ds}'))\ =\ \mathsf{play}_\alpha\ s\ \mathsf{as}\ \mathsf{ds}\,\mathsf{in}\ \mathsf{play}_\beta\ t\ \mathsf{as}'\ \mathsf{ds}')\\
\mathsf{play}_{\alpha^*}\ s\ \mathsf{as}\ \mathsf{ds} &= \mathsf{case}\ (\mathsf{as}\ s)\ \mathsf{of}\\
&\qquad \mathsf{as}' \Rightarrow (s,(\mathsf{as}',\pi_L\mathsf{ds}))\\
&\qquad \mid \mathsf{as}' \Rightarrow \mathsf{let}\,(t,(\mathsf{as}'',\mathsf{ds}''))\ =\ \mathsf{play}_\alpha\ s\ \mathsf{as}'\ (\pi_R\mathsf{ds})\ \mathsf{in}\\
&\qquad\qquad \mathsf{play}_{\alpha^*}\ t\ \mathsf{as}''\ \mathsf{ds}''\\
\mathsf{play}_{\alpha^d}\ s\ \mathsf{as}\ \mathsf{ds} &= \mathsf{play}_\alpha\ s\ \mathsf{ds}\ \mathsf{as}
\end{aligned}$$

Corollary 17. (Consistency). *It is never the case that both* $\ulcorner\langle\alpha\rangle\phi\urcorner\ s$ *and* $\ulcorner[\alpha]\neg\phi\urcorner\ s$ *are inhabited.*

Proof. Suppose $\mathsf{as} : \ulcorner\langle\alpha\rangle\phi\urcorner\ s$ and $\mathsf{ds} : \ulcorner[\alpha]\neg\phi\urcorner\ s$, then $\pi_R(\mathsf{play}_\alpha\ s\ \mathsf{as}\ \mathsf{ds}) : \bot$, contradicting consistency of type theory. □

The play semantics show how strategies can be executed. Consistency is a theorem which ought to hold in any GL and thus helps validate our semantics.

8 Conclusion and Future Work

We extended Constructive Game Logic CGL to CdGL for constructive *hybrid* games. We contributed new semantics. We presented a natural deduction proof calculus for CdGL and used it to prove reach-avoid correctness of 1D driving with adversarial timing. We showed soundness and constructivity results.

The next step is to implement a proof checker, game interpreter, and synthesis tool for CdGL. Function play is the high-level interpreter algorithm, while synthesis would commit to one Angel strategy and allow black-box Demon implementations for an external environment. Angel strategies are positive and are synthesized by extracting witnesses from each introduction rule. Demonic invariants and test conditions describe allowed observable behaviors. Demon strategies are negative and characterized by observable behaviors, so it suffices to monitor their compliance with invariants and test conditions extracted from the proof.

Acknowledgements. We thank Jon Sterling for suggestions regarding our choice of type theory and for references to the literature. We thank the anonymous reviewers for their helpful feedback.

References

1. Abramsky, S., Jagadeesan, R., Malacaria, P.: Full abstraction for PCF. Inf. Comput. **163**(2), 409–470 (2000). https://doi.org/10.1006/inco.2000.2930
2. Alur, R., Henzinger, T.A., Kupferman, O.: Alternating-time temporal logic. J. ACM **49**(5), 672–713 (2002). https://doi.org/10.1145/585265.585270
3. Benthem, J.: Logic of strategies: what and how? In: van Benthem, J., Ghosh, S., Verbrugge, R. (eds.) Models of Strategic Reasoning. LNCS, vol. 8972, pp. 321–332. Springer, Heidelberg (2015). https://doi.org/10.1007/978-3-662-48540-8_10
4. van Benthem, J., Pacuit, E.: Dynamic logics of evidence-based beliefs. Stud. Logica. **99**(1–3), 61–92 (2011). https://doi.org/10.1007/s11225-011-9347-x
5. van Benthem, J., Pacuit, E., Roy, O.: Toward a theory of play: a logical perspective on games and interaction. Games (2011). https://doi.org/10.3390/g2010052
6. Bishop, E.: Foundations of Constructive Analysis. McGraw-Hill, New York (1967)
7. Bohrer, R., Fernández, M., Platzer, A.: dL_ι: definite descriptions in differential dynamic logic. In: Fontaine, P. (ed.) CADE 2019. LNCS (LNAI), vol. 11716, pp. 94–110. Springer, Cham (2019). https://doi.org/10.1007/978-3-030-29436-6_6
8. Bohrer, R., Platzer, A.: Toward structured proofs for dynamic logics. CoRR abs/1908.05535 (2019), http://arxiv.org/abs/1908.05535
9. Bohrer, R., Platzer, A.: Constructive game logic. ESOP 2020. LNCS, vol. 12075, pp. 84–111. Springer, Cham (2020). https://doi.org/10.1007/978-3-030-44914-8_4
10. Bohrer, R., Platzer, A.: Constructive hybrid games. CoRR abs/2002.02536 (2020), https://arxiv.org/abs/2002.02536
11. Bohrer, R., Tan, Y.K., Mitsch, S., Myreen, M.O., Platzer, A.: VeriPhy: verified controller executables from verified cyber-physical system models. In: Grossman, D. (ed.) PLDI, pp. 617–630. ACM (2018). https://doi.org/10.1145/3192366.3192406
12. Bridges, D.S., Vita, L.S.: Techniques of Constructive Analysis. Springer, New York (2007). https://doi.org/10.1007/978-0-387-38147-3
13. Celani, S.A.: A fragment of intuitionistic dynamic logic. Fundam. Inform. **46**(3), 187–197 (2001). http://content.iospress.com/articles/fundamenta-informaticae/fi46-3-01
14. Chatterjee, K., Henzinger, T.A., Piterman, N.: Strategy logic. In: Caires, L., Vasconcelos, V.T. (eds.) CONCUR 2007. LNCS, vol. 4703, pp. 59–73. Springer, Heidelberg (2007). https://doi.org/10.1007/978-3-540-74407-8_5
15. Coquand, T., Huet, G.P.: The calculus of constructions. Inf. Comput. **76**(2/3), 95–120 (1988). https://doi.org/10.1016/0890-5401(88)90005-3
16. Coquand, T., Paulin, C.: Inductively defined types. In: Martin-Löf, P., Mints, G. (eds.) COLOG 1988. LNCS, vol. 417, pp. 50–66. Springer, Heidelberg (1990). https://doi.org/10.1007/3-540-52335-9_47
17. Cruz-Filipe, L., Geuvers, H., Wiedijk, F.: C-CoRN, the constructive Coq repository at Nijmegen. In: Asperti, A., Bancerek, G., Trybulec, A. (eds.) MKM 2004. LNCS, vol. 3119, pp. 88–103. Springer, Heidelberg (2004). https://doi.org/10.1007/978-3-540-27818-4_7
18. Degen, J., Werner, J.: Towards intuitionistic dynamic logic. Log. Log. Philos. **15**(4), 305–324 (2006). https://doi.org/10.12775/LLP.2006.018

19. Dybjer, P.: Inductive families. Formal Asp. Comput. **6**(4), 440–465 (1994). https://doi.org/10.1007/BF01211308
20. Filippidis, I., Dathathri, S., Livingston, S.C., Ozay, N., Murray, R.M.: Control design for hybrid systems with TuLiP: the temporal logic planning toolbox. In: Conference on Control Applications, pp. 1030–1041. IEEE (2016). https://doi.org/10.1109/CCA.2016.7587949
21. Finucane, C., Jing, G., Kress-Gazit, H.: LTLMoP: experimenting with language, temporal logic and robot control. In: IROS, pp. 1988–1993. IEEE (2010). https://doi.org/10.1109/IROS.2010.5650371
22. Foster, J.N.: Bidirectional programming languages. Technical report MS-CIS-10-08, Department of Computer & Information Science, University of Pennsylvania, Philadelphia, PA, March 2010
23. Ghosh, S.: Strategies made explicit in dynamic game logic. In: Workshop on Logic and Intelligent Interaction at ESSLLI, pp. 74–81 (2008)
24. Harel, D., Kozen, D., Tiuryn, J.: Dynamic Logic. MIT Press, Cambridge (2000)
25. Henzinger, T.A., Horowitz, B., Majumdar, R.: Rectangular hybrid games. In: Baeten, J.C.M., Mauw, S. (eds.) CONCUR 1999. LNCS, vol. 1664, pp. 320–335. Springer, Heidelberg (1999). https://doi.org/10.1007/3-540-48320-9_23
26. Hoare, C.A.R.: An axiomatic basis for computer programming. Commun. ACM **12**(10), 576–580 (1969). https://doi.org/10.1145/363235.363259
27. van der Hoek, W., Jamroga, W., Wooldridge, M.J.: A logic for strategic reasoning. In: Dignum, F., Dignum, V., Koenig, S., Kraus, S., Singh, M.P., Wooldridge, M.J. (eds.) AAMAS. ACM (2005). https://doi.org/10.1145/1082473.1082497
28. Hofmann, M., van Oosten, J., Streicher, T.: Well-foundedness in realizability. Arch. Math. Log. **45**(7), 795–805 (2006). https://doi.org/10.1007/s00153-006-0003-5
29. Isaacs, R.: Differential Games: A Mathematical Theory with Applications to Warfare and Pursuit, Control and Optimization. Series in Applied Mathematics (SIAM), Wiley, New York (1965)
30. Kamide, N.: Strong normalization of program-indexed lambda calculus. Bull. Sect. Log. Univ. Łódź **39**(1–2), 65–78 (2010)
31. Kloetzer, M., Belta, C.: A fully automated framework for control of linear systems from temporal logic specifications. IEEE Trans. Automat. Control **53**(1), 287–297 (2008). https://doi.org/10.1109/TAC.2007.914952
32. Lipton, J.: Constructive Kripke semantics and realizability. In: Moschovakis, Y. (ed.) Logic From Computer Science, pp. 319–357. Springer, New York (1992). https://doi.org/10.1007/978-1-4612-2822-6_13
33. Lombardi, H., Mahboubi, A.: Théories géométriques pour l'algèbre des nombres réels. Contemp. Math. **697**, 239–264 (2017)
34. Makarov, E., Spitters, B.: The Picard algorithm for ordinary differential equations in Coq. In: Blazy, S., Paulin-Mohring, C., Pichardie, D. (eds.) ITP 2013. LNCS, vol. 7998, pp. 463–468. Springer, Heidelberg (2013). https://doi.org/10.1007/978-3-642-39634-2_34
35. Mamouras, K.: Synthesis of strategies using the Hoare logic of angelic and demonic nondeterminism. Log. Methods Comput. Sci. **12**(3), 1–41 (2016). https://doi.org/10.2168/LMCS-12(3:6)2016
36. Mitsch, S., Platzer, A.: ModelPlex: verified runtime validation of verified cyber-physical system models. Form. Methods Syst. Des. **49**(1), 33–74 (2016). https://doi.org/10.1007/s10703-016-0241-z

37. Murphy VII, T., Crary, K., Harper, R., Pfenning, F.: A symmetric modal lambda calculus for distributed computing. In: LICS. IEEE (2004), https://doi.org/10.1109/LICS.2004.1319623
38. van Oosten, J.: Realizability: a historical essay. Math. Structures Comput. Sci. **12**(3), 239–263 (2002). https://doi.org/10.1017/S0960129502003626
39. Parikh, R.: Propositional game logic. In: FOCS, pp. 195–200. IEEE (1983). https://doi.org/10.1109/SFCS.1983.47
40. Platzer, A.: Differential dynamic logic for hybrid systems. J. Autom. Reas. **41**(2), 143–189 (2008). https://doi.org/10.1007/s10817-008-9103-8
41. Platzer, A.: Differential-algebraic dynamic logic for differential-algebraic programs. J. Log. Comput. **20**(1), 309–352 (2010). https://doi.org/10.1093/logcom/exn070
42. Platzer, A.: Differential game logic. ACM Trans. Comput. Log. **17**(1), 1:1-1:51 (2015). https://doi.org/10.1145/2817824
43. Platzer, A.: Differential hybrid games. ACM Trans. Comput. Log. **18**(3), 19:1-19:44 (2017). https://doi.org/10.1145/3091123
44. Platzer, A.: Logical Foundations of Cyber-Physical Systems. Springer, Cham (2018). https://doi.org/10.1007/978-3-319-63588-0
45. Platzer, A.: Uniform substitution for differential game logic. In: Galmiche, D., Schulz, S., Sebastiani, R. (eds.) IJCAR 2018. LNCS (LNAI), vol. 10900, pp. 211–227. Springer, Cham (2018). https://doi.org/10.1007/978-3-319-94205-6_15
46. Platzer, A., Tan, Y.K.: Differential equation invariance axiomatization. J. ACM **67**, 1 (2020). https://doi.org/10.1145/3380825
47. Pratt, V.R.: Semantical considerations on Floyd-Hoare logic. In: FOCS, pp. 109–121. IEEE (1976). https://doi.org/10.1109/SFCS.1976.27
48. Quesel, J.-D., Platzer, A.: Playing hybrid games with KeYmaera. In: Gramlich, B., Miller, D., Sattler, U. (eds.) IJCAR 2012. LNCS (LNAI), vol. 7364, pp. 439–453. Springer, Heidelberg (2012). https://doi.org/10.1007/978-3-642-31365-3_34
49. Ramanujam, R., Simon, S.E.: Dynamic logic on games with structured strategies. In: Brewka, G., Lang, J. (eds.) Knowledge Representation, pp. 49–58. AAAI Press (2008). http://www.aaai.org/Library/KR/2008/kr08-006.php
50. Shakernia, O., Pappas, G.J., Sastry, S.: Semi-decidable synthesis for triangular hybrid systems. In: Di Benedetto, M.D., Sangiovanni-Vincentelli, A. (eds.) HSCC 2001. LNCS, vol. 2034, pp. 487–500. Springer, Heidelberg (2001). https://doi.org/10.1007/3-540-45351-2_39
51. Shakernia, O., Pappas, G.J., Sastry, S.: Decidable controller synthesis for classes of linear systems. In: Lynch, N., Krogh, B.H. (eds.) HSCC 2000. LNCS, vol. 1790, pp. 407–420. Springer, Heidelberg (2000). https://doi.org/10.1007/3-540-46430-1_34
52. Taly, A., Tiwari, A.: Switching logic synthesis for reachability. In: Carloni, L.P., Tripakis, S. (eds.) EMSOFT, pp. 19–28. ACM (2010). https://doi.org/10.1145/1879021.1879025
53. Tan, Y.K., Platzer, A.: An axiomatic approach to liveness for differential equations. In: ter Beek, M.H., McIver, A., Oliveira, J.N. (eds.) FM 2019. LNCS, vol. 11800, pp. 371–388. Springer, Cham (2019). https://doi.org/10.1007/978-3-030-30942-8_23
54. The Coq development team: The Coq proof assistant reference manual (2019). https://coq.inria.fr/
55. Tomlin, C.J., Lygeros, J., Sastry, S.S.: A game theoretic approach to controller design for hybrid systems. Proc. IEEE **88**(7), 949–970 (2000)
56. Van Benthem, J.: Games in dynamic-epistemic logic. Bull. Econ. Res. **53**(4), 219–248 (2001)

57. Weihrauch, K.: Computable Analysis - An Introduction. Texts in Theoretical Computer Science, Springer, Heidelberg (2000). https://doi.org/10.1007/978-3-642-56999-9
58. Wijesekera, D.: Constructive modal logics I. Ann. Pure Appl. Log. **50**(3), 271–301 (1990). https://doi.org/10.1016/0168-0072(90)90059-B
59. Wijesekera, D., Nerode, A.: Tableaux for constructive concurrent dynamic logic. Ann. Pure Appl. Log. **135**(1–3), 1–72 (2005). https://doi.org/10.1016/j.apal.2004.12.001

Formalizing a Seligman-Style Tableau System for Hybrid Logic (Short Paper)

Asta Halkjær From[1(✉)], Patrick Blackburn[2], and Jørgen Villadsen[1]

[1] Technical University of Denmark, Kongens Lyngby, Denmark
ahfrom@dtu.dk
[2] Roskilde University, Roskilde, Denmark

Abstract. Hybrid logic is modal logic enriched with names for worlds. We formalize soundness and completeness proofs for a Seligman-style tableau system for hybrid logic in the proof assistant Isabelle/HOL. The formalization shows how to lift certain rule restrictions, thereby simplifying the original un-formalized proof. Moreover, the completeness proof we formalize is synthetic which suggests we can extend this work to prove a wider range of results about hybrid logic.

Keywords: Isabelle/HOL · Hybrid logic · Soundness · Completeness

1 Introduction

Hybrid logic extends ordinary modal logic with nominals, a special sort of propositional symbol true at exactly one world. Nominals, and the satisfaction operators they give rise to, make hybrid logic well-suited for different applications ranging from temporal logic [4] to epistemic logics for social networks [22]. The description logics underlying the Web Ontology Language and applications in biomedical informatics [16] can also be seen as forms of hybrid logic [2].

ST is a sound and complete tableau system for hybrid logic. It is known to terminate when five restrictions are imposed on the rules, and one key rule is split into three cases [5]. Two completeness proofs exist for ST, a synthetic one that does not cover the rule restrictions [17] and a complex translation-based proof that does [5]. In this paper we modify ST and three of its restrictions slightly, and use the proof assistant Isabelle/HOL to show that we can lift these restrictions by (a) formally proving the admissibility of their unrestricted versions, and (b) formalizing a synthetic completeness proof for the modified calculus.

Isabelle is a generic proof assistant and Isabelle/HOL is its instance based on higher-order logic [20]. Proof assistants like Isabelle provide tools to express mathematical statements and proofs in a formal language that can be mechanically verified; all proofs presented here have been checked in this manner. The full formalization, 4396 lines, is available in the Archive of Formal Proofs which keeps refereed submissions up to date with the current Isabelle version [13]. The formalization was developed for the first author's MSc thesis [15]. We chose Isabelle/HOL because it is the proof assistant we know best.

N. Peltier and V. Sofronie-Stokkermans (Eds.): IJCAR 2020, LNAI 12166, pp. 474–481, 2020.
https://doi.org/10.1007/978-3-030-51074-9_27

2 Syntax and Semantics

The well-formed formulas of basic hybrid logic are defined as follow. We use the letter x for propositional symbols and i, a and b for nominals.

$$\phi, \psi ::= x \mid i \mid \neg\phi \mid \phi \vee \psi \mid \Diamond\phi \mid @_i\phi$$

The language is interpreted on Kripke models $\mathfrak{M}$, consisting of a frame (W, R) and a valuation of propositional symbols V. Here W is a non-empty set of worlds and R is a binary accessibility relation between them. To interpret the nominals we use an assignment g mapping nominals to elements of W; if $g(i) = w$ then we say that nominal i denotes w. Formula satisfiability is defined as follows:

$$\begin{array}{lll}
\mathfrak{M}, g, w \models x & \text{iff} & w \in V(x) \\
\mathfrak{M}, g, w \models i & \text{iff} & g(i) = w \\
\mathfrak{M}, g, w \models \neg\phi & \text{iff} & \mathfrak{M}, g, w \not\models \phi \\
\mathfrak{M}, g, w \models \phi \vee \psi & \text{iff} & \mathfrak{M}, g, w \models \phi \text{ or } \mathfrak{M}, g, w \models \psi \\
\mathfrak{M}, g, w \models \Diamond\phi & \text{iff} & \text{for some } w', wRw' \text{ and } \mathfrak{M}, g, w' \models \phi \\
\mathfrak{M}, g, w \models @_i\phi & \text{iff} & \mathfrak{M}, g, g(i) \models \phi
\end{array}$$

An expression of the form $@_i\phi$ is called a satisfaction statement, and such statements are true iff ϕ is true at the world denoted by nominal i. Note two important special cases: $@_i a$ says that the nominals i and a denote the same world, and $@_i\Diamond b$ says that the world denoted by i has access to the world denoted by b.

3 A Seligman-Style Tableau System

Many proof systems for hybrid logic exist; see Blackburn et al. [5] for discussion. These typically work by manipulating only formulas prefixed by satisfaction operators, which gives a global flavour to proofs, however the tableau system we formalize here manipulates *arbitrary* formulas. It is an adaptation of system ST, due to Blackburn et al. [5], which was inspired by Jeremy Seligman's local natural deduction and sequent calculus systems for hybrid logic [23,24].

The tableau rules are based on a subdivision of tableau branches into blocks. Each pair of blocks is separated by a horizontal line and the first formula on each block is a nominal dubbed the opening nominal. The intuition is that the formulas on a block are true in the world denoted by its opening nominal. We assume that the initial block, like the rest, is always named (this is our first modification of the original ST system). This assumption simplifies the formalization, as we can now model all blocks as lists of formulas paired with an opening nominal, and a branch as a list of blocks. If Θ is a branch and ϕ occurs on an i-block in Θ then we say that ϕ occurs at i in Θ. We occasionally refer to the opening nominal of a block as its name or type.

The rules are given in Fig. 1: the first three are propositional, the three below are for working with the blocks, and the four to the right apply to the hybrid logic connectives. The input to the rule is given above the vertical line and the output below it. Above every input formula, we write the opening nominal of the block it occurs on. Similarly, the opening nominal of the output block is the first thing below the line. If the opening nominals match, then the output block may be the same as an input block. In the formalization we model the rules as an inductively defined set of branches that have closing extensions.

$$
\frac{\begin{array}{c} a \\ \phi \vee \psi \end{array}}{\begin{array}{cc} \phi & \psi \end{array}} \qquad
\frac{\begin{array}{c} a \\ \neg(\phi \vee \psi) \end{array}}{\begin{array}{c} a \\ \neg\phi \\ \neg\psi \end{array}} \qquad
\frac{\begin{array}{c} a \\ \neg\neg\phi \end{array}}{\begin{array}{c} a \\ \phi \end{array}} \qquad
\frac{\begin{array}{c} a \\ \Diamond\phi \end{array}}{\begin{array}{c} a \\ \Diamond i \\ @_i\phi \end{array}} \qquad
\frac{\begin{array}{cc} a & a \\ \neg\Diamond\phi & \Diamond i \end{array}}{\begin{array}{c} a \\ \neg@_i\phi \end{array}}
$$

$$(\vee) \qquad\qquad (\neg\vee) \qquad\qquad (\neg\neg) \qquad\qquad (\Diamond)^1 \qquad\qquad (\neg\Diamond)$$

$$
\frac{\;}{\begin{array}{c} \\ i \end{array}} \qquad
\frac{\begin{array}{ccc} b & b & a \\ i & \phi & i \end{array}}{\begin{array}{c} a \\ \phi \end{array}} \qquad
\frac{\begin{array}{cc} i & i \\ \phi & \neg\phi \end{array}}{\times} \qquad
\frac{\begin{array}{c} b \\ @_a\phi \end{array}}{\begin{array}{c} a \\ \phi \end{array}} \qquad
\frac{\begin{array}{c} b \\ \neg@_a\phi \end{array}}{\begin{array}{c} a \\ \neg\phi \end{array}}
$$

$$\mathsf{GoTo}^2 \qquad\qquad \mathsf{Nom} \qquad\qquad \mathrm{Closing} \qquad\qquad (@) \qquad\qquad (\neg@)$$

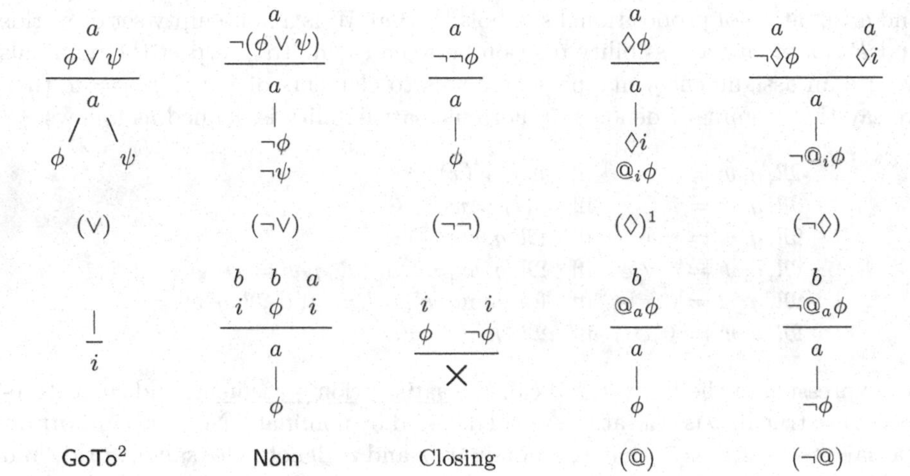

[1] i is fresh, ϕ is not a nominal.
[2] i is not fresh.

Fig. 1. Tableau rules.

Consider the $(\neg\neg)$ rule: if $\neg\neg\phi$ occurs on an a-block and the current block is an a-block, then ϕ is a legal extension of the branch. For the Nom rule, nominal i occurs at both a and b, so they must denote the same world and copying ϕ from a b-block to the current a-block is legal. Here we also differ from the original ST: we do not require the shared nominal i to occur on the current block as this would be a problem for our Strengthening Lemma in Sect. 4. The GoTo rule allows us to change perspective from one world to another by starting a new block with an opening nominal that already occurs somewhere on the branch.

A branch closes if the same formula occurs on the same type of block both positively and negatively, and a tableau closes if all its branches do. If a closed tableau can be obtained starting from the branch Θ, then we write $\vdash \Theta$. If Θ is a branch and the current block has opening nominal a, then we write the extension of Θ by ϕ as $\phi -_a \Theta$ to resemble the extensions in Fig. 1.

The original ST has five restrictions, called R1-R5 [5]. Restriction R3 is unnecessary in our system since it applies to an omitted rule that names the initial block. Restriction R4 forbids applying GoTo twice in a row and formalizing it is left for future work. Here are our adaptations of the three remaining restrictions:

R1 The output of a rule must include a formula new to the current block type.
R2 The ($\Diamond$) rule can only be applied to input $\Diamond\phi$ on an a-block if it is not already witnessed on a.
R5 (@) and ($\neg$@) can only be applied to premises i and $@_i\phi$ ($\neg @_i\phi$) when the current block is an i-block.

The formula ϕ is new to a in Θ if ϕ does not occur at a in Θ. A formula $\Diamond\phi$ is witnessed at a in Θ if for some witnessing nominal i, both $\Diamond i$ and $@_i\phi$ occur at a in Θ. The original R2 restriction states that the ($\Diamond$) rule cannot be applied twice to the same formula occurrence, but formalizing this would require keeping track of previous rule applications. We already keep track of the branch so we prefer the R2 presented here. Our version of the @-rules already satisfy the R5 restriction.

4 Main Results

Theorem 1 (Soundness). *If $\vdash \Theta$ where Θ consists of just $\neg\phi$ on an i-block and i does not occur in ϕ, then ϕ is valid.*

Proof. Similar to the original soundness proof by Blackburn et al. [5]. □

The following lemma allows us to derive rules unrestricted by R1:

Lemma 1 (Strengthening). *If an extension is not new then it is redundant. That is, if $\vdash \phi -_a \Theta$ and ϕ occurs at a in Θ then $\vdash \Theta$.*

Proof. The existing ϕ can be used as rule input in place of the extension. □

To lift R2 we use the following substitution lemma where $\phi\sigma$ and $\Theta\sigma$ are obtained from ϕ and Θ, respectively, by replacing every nominal i with $\sigma(i)$.

Lemma 2 (Substitution). *Let σ be a substitution function whose domain and codomain coincide. If $\vdash \Theta$ then $\vdash \Theta\sigma$.*

Proof. By induction on the derivation of $\vdash \Theta$ for arbitrary σ. In the ($\Diamond$) case we assume that $\Diamond\phi$ occurs at a in Θ and need to derive $\vdash \Theta\sigma$ from $\vdash (@_i\phi -_a \Diamond i -_a \Theta)\sigma'$ where i is some nominal fresh to Θ and we get to pick σ'.

By assumption, $\Diamond\phi$ is unwitnessed at a in Θ but since the substitution can collapse formulas, $\Diamond(\phi\sigma)$ may be witnessed in $\Theta\sigma$ by some witnessing nominal j. In this case, where restriction R2 prevents us from applying the ($\Diamond$) rule, we let $\sigma' = \sigma(i := j)$ in the induction hypothesis. Since i occurs only in the extension the rest of the branch is unaffected by this choice, $\Theta(\sigma(i := j)) = \Theta\sigma$, but now the extension occurs elsewhere at a and the Nom rule justifies it. □

Lemma 3 (Unrestricted ($\Diamond$)). *If $\Diamond\phi$ occurs at a in Θ, i is fresh and ϕ is not a nominal then we can derive $\vdash \Theta$ from a witnessing extension $\vdash @_i\phi -_a \Diamond i -_a \Theta$.*

Proof. If $\Diamond\phi$ is already witnessed at a in Θ then use Lemma 2 to make i coincide with the existing witnessing nominal and justify the extension by Nom. □

If Θ consists of blocks $B_1, B_2, \ldots, B_n$, let $\text{Blocks}(\Theta) = \{B_1, B_2, \ldots, B_n\}$. The substitution lemma allows us to prove the following:

Lemma 4 (Branch structure). *Given infinitely many nominals, we can add, contract and rearrange blocks: If $\vdash \Theta$ and $\text{Blocks}(\Theta) \subseteq \text{Blocks}(\Theta')$ then $\vdash \Theta'$.*

Proof. By induction on the derivation of $\vdash \Theta$ for arbitrary Θ'. □

Lemma 5 (Unrestricted (@) (and ($\neg$@))). *If $\vdash \phi -_a \Theta$, $@_i\phi$ occurs at b in Θ and i occurs at a then $\vdash \Theta$.*

Proof. Figure 2 shows the derivation where each new branch to the right is known by Lemma 4 to still have a closing extension. □

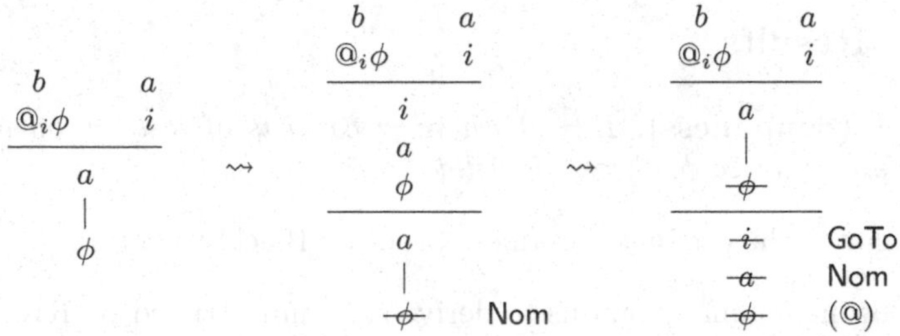

Fig. 2. Deriving the unrestricted (@) rule.

Theorem 2 (Completeness). *If ϕ is valid then $\vdash \Theta$ where Θ consists of a single block with ϕ on it.*

Proof. Essentially a modification of the proof for ST by Jørgensen et al. [17], since our system is similar, and we have proved we can lift our restrictions. □

We remark that the completeness proof is an example of what are known as synthetic approaches to completeness [11,25], which involve reasoning about maximal consistent sets and their properties. However the completeness proof for ST distinguishes itself by using maximal sets of entire blocks rather than plain formulas. One component of the proof is a definition of when such a set of blocks is a Hintikka set and thus satisfiable [17]. In the formalization [13] we precisely formulate this definition in the formal language of Isabelle/HOL and in doing so we discovered a shortcoming in the definition given by Jørgensen et al. Essentially, their requirement on propositional symbols fails to take so-called equivalence of nominals into account, making their model valuation incompatible with their model existence result.

5 Related Work

Linker formalizes in Isabelle/HOL a semantic embedding of a spatio-temporal multi-modal logic designed for reasoning about motorway traffic which includes a hybrid logic-inspired at-operator [18]. Linker and Hilscher give a sound labelled natural deduction proof system for a version of the logic without the hybrid extension [19]. Doczkal and Smolka formalize hybrid logic with nominals but no special operators in constructive type theory using the proof assistant Coq. They do not define a proof system but give algorithmic proofs of small model theorems and computational decidability of satisfiability, validity, and equivalence of formulas [10]. The present work appears to be the first proof system for hybrid logic with a formalized completeness proof.

Formalizations of completeness proofs in Isabelle exist for, among others, a tableau system and a one-sided sequent calculus for first-order logic [14], a natural deduction system for first-order logic [3], a Hilbert system for epistemic logic [12], and the first-order resolution calculus [21]. Blanchette et al. give abstract proofs of soundness and completeness that can be instantiated for a range of Gentzen and tableau systems for various flavors of first-order logic [7]. Moreover, Blanchette gives an overview of the formalized metatheory of various logical calculi and automatic provers in Isabelle [6].

6 Future Work

We are currently working on restricting the GoTo and Nom rules to ensure termination; previous (un-formalized) work has shown via translation to and from a different system that completeness can be preserved and that the resulting system is terminating [5]. We would like to show termination directly via a decreasing length argument in the style of Bolander and Blackburn's work on an internalized labelled tableau system [8]. Given a sound, complete and terminating system we want to verify an algorithm based on it and use it as a decision procedure for basic hybrid logic. Moreover, as the completeness proof that we formalized is based on reasoning about maximal consistent sets and their properties, it should be possible to extend it to other key results for hybrid logic which have been proved by similar forms of reasoning, notably interpolation results [1].

7 Conclusion

We have presented a tableau system for basic hybrid logic whose soundness and completeness has been formalized in Isabelle/HOL. Moreover, we have shown how to lift certain restrictions on the rules so that an existing completeness proof could be formalized and applied. The fact that the completeness proof we formalized is a synthetic proof suggests that it can be extended to a number of other key results for hybrid logic that can be found in the literature.

References

1. Areces, C., Blackburn, P., Marx, M.: Hybrid logics: characterization, interpolation and complexity. J. Symb. Logic **66**(3), 977–1010 (2001)
2. Areces, C.E.: Logic engineering: the case of description and hybrid logics. Ph.D. thesis, Institute for Logic, Language and Computation, Amsterdam, The Netherlands (2000)
3. Berghofer, S.: First-Order Logic According to Fitting. Archive of Formal Proofs, August 2007. http://isa-afp.org/entries/FOL-Fitting.html. Formal proof development
4. Blackburn, P.: Representation, reasoning, and relational structures: a hybrid logic manifesto. Logic J. IGPL **8**(3), 339–365 (2000). https://doi.org/10.1093/jigpal/8.3.339
5. Blackburn, P., Bolander, T., Braüner, T., Jørgensen, K.F.: Completeness and termination for a Seligman-style tableau system. J. Logic Comput. **27**(1), 81–107 (2017). https://doi.org/10.1093/logcom/exv052
6. Blanchette, J.C.: Formalizing the metatheory of logical calculi and automatic provers in Isabelle/HOL (invited talk). In: Mahboubi, A., Myreen, M.O. (eds.) Proceedings of the 8th ACM SIGPLAN International Conference on Certified Programs and Proofs. CPP 2019. pp. 1–13. ACM (2019)
7. Blanchette, J.C., Popescu, A., Traytel, D.: Soundness and completeness proofs by coinductive methods. J. Autom. Reasoning **58**(1), 149–179 (2016). https://doi.org/10.1007/s10817-016-9391-3
8. Bolander, T., Blackburn, P.: Termination for hybrid tableaus. J. Logic Comput. **17**(3), 517–554 (2007). https://doi.org/10.1093/logcom/exm014
9. Braüner, T.: Hybrid Logic and its Proof-Theory. Springer, Dordrecht (2010)
10. Doczkal, C., Smolka, G.: Constructive formalization of hybrid logic with eventualities. In: Jouannaud, J.-P., Shao, Z. (eds.) CPP 2011. LNCS, vol. 7086, pp. 5–20. Springer, Heidelberg (2011). https://doi.org/10.1007/978-3-642-25379-9_3
11. Fitting, M.: Proof Methods for Modal and Intuitionistic Logics, vol. 169. Springer, Heidelberg (1983). https://doi.org/10.1007/978-94-017-2794-5
12. From, A.H.: Epistemic Logic. Archive of Formal Proofs, October 2018. http://isa-afp.org/entries/Epistemic_Logic.html. Formal proof development
13. From, A.H.: Formalizing a Seligman-Style Tableau System for Hybrid Logic. Archive of Formal Proofs, December 2019. http://isa-afp.org/entries/Hybrid_Logic.html. Formal proof development
14. From, A.H.: A Sequent Calculus for First-Order Logic. Archive of Formal Proofs, July 2019. http://isa-afp.org/entries/FOL_Seq_Calc1.html. Formal proof development
15. From, A.H.: Hybrid logic. Master's thesis, Technical University of Denmark, January 2020
16. Horrocks, I., Glimm, B., Sattler, U.: Hybrid logics and ontology languages. Electron. Notes Theoret. Comput. Sci. **174**(6), 3–14 (2007). https://doi.org/10.1016/j.entcs.2006.11.022
17. Jørgensen, K.F., Blackburn, P., Bolander, T., Braüner, T.: Synthetic completeness proofs for Seligman-style tableau systems. In: Proceedings of the 11th Conference on Advances in Modal Logic, held in Budapest, Hungary, 30 August–2 September 2016, vol. 11, pp. 302–321 (2016)
18. Linker, S.: Hybrid Multi-Lane Spatial Logic. Archive of Formal Proofs, November 2017. http://isa-afp.org/entries/Hybrid_Multi_Lane_Spatial_Logic.html. Formal proof development

19. Linker, S., Hilscher, M.: Proof theory of a multi-lane spatial logic. Log. Methods Comput. Sci. **11**(3) (2015). https://doi.org/10.2168/LMCS-11(3:4)2015
20. Nipkow, T., Wenzel, M., Paulson, L.C. (eds.): Isabelle/HOL - A Proof Assistant for Higher-Order Logic. LNCS, vol. 2283. Springer, Heidelberg (2002). https://doi.org/10.1007/3-540-45949-9
21. Schlichtkrull, A.: The Resolution Calculus for First-Order Logic. Archive of Formal Proofs, June 2016. http://isa-afp.org/entries/Resolution_FOL.html. Formal proof development
22. Seligman, J., Liu, F., Girard, P.: Facebook and the epistemic logic of friendship. In: Schipper, B.C. (ed.) Proceedings of the 14th Conference on Theoretical Aspects of Rationality and Knowledge (TARK 2013), Chennai, India, pp. 229–238 (2013)
23. Seligman, J.: The logic of correct description. In: de Rijke, M. (ed.) Advances in Intensional Logic. Applied Logic Series, vol. 7, pp. 107–135. Springer, Dordrecht (1997). https://doi.org/10.1007/978-94-015-8879-9_5
24. Seligman, J.: Internalization: the case of hybrid logics. J. Logic Comput. **11**(5), 671–689 (2001). https://doi.org/10.1093/logcom/11.5.671
25. Smullyan, R.M.: First-Order Logic. Dover Publications, New York (1995)

NP Reasoning in the Monotone μ-Calculus

Daniel Hausmann(✉) and Lutz Schröder(✉)

Friedrich-Alexander-Universität Erlangen-Nürnberg, Erlangen, Germany
{daniel.hausmann,lutz.schroeder}@fau.de

Abstract. Satisfiability checking for monotone modal logic is known to be (only) NP-complete. We show that this remains true when the logic is extended with alternation-free fixpoint operators as well as the universal modality; the resulting logic – the *alternation-free monotone μ-calculus with the universal modality* – contains both concurrent propositional dynamic logic (CPDL) and the alternation-free fragment of game logic as fragments. We obtain our result from a characterization of satisfiability by means of Büchi games with polynomially many Eloise nodes.

1 Introduction

Monotone modal logic differs from normal modal logics (such as K [4], equivalent to the standard description logic $\mathcal{ALC}$ [1]) by giving up distribution of conjunction over the box modality, but retaining monotonicity of the modalities. Its semantics is based on (monotone) neighbourhood models instead of Kripke models. Monotone modalities have been variously used as epistemic operators that restrict the combination of knowledge by epistemic agents [27]; as next-step modalities in the evolution of concurrent systems, e.g. in *concurrent propositional dynamic logic* (CPDL) [24]; and as game modalities in systems where one transition step is determined by moves of two players, e.g. in Parikh's *game logic* [7,12,20,23]. The monotonicity condition suffices to enable formation of fixpoints; one thus obtains the *monotone μ-calculus* [8], which contains both CPDL and game logic as fragments (indeed, the recent proof of completeness of game logic [7] is based on embedding game logic into the monotone μ-calculus).

While many modal logics (including $K/\mathcal{ALC}$) have PSPACE-complete satisfiability problems in the absence of fixpoints, it is known that satisfiability in monotone modal logic is only NP-complete [27] (the lowest possible complexity given that the logic has the full set of Boolean connectives). In the present paper, we show that the low complexity is preserved under two extensions that usually cause the complexity to rise from PSPACE-complete to EXPTIME-complete: Adding the *universal modality* (equivalently global axioms or, in description logic parlance, a general TBox) and alternation-free *fixpoints*; that is, we show that satisfiability checking in the *alternation-free fragment of the monotone μ-calculus with the universal modality* [8] is only NP-complete. This logic subsumes both CPDL and the alternation-free fragment of game logic [23]. Thus, our results

N. Peltier and V. Sofronie-Stokkermans (Eds.): IJCAR 2020, LNAI 12166, pp. 482–499, 2020.
https://doi.org/10.1007/978-3-030-51074-9_28

imply that satisfiability checking in these logics is only NP-complete (the best previously known upper bound being ExpTime in both cases [20,23,24]); for comparison, standard propositional dynamic logic (PDL) and in fact already the extension of K with the universal modality are ExpTime-hard. (Our results thus seemingly contradict previous results on ExpTime-completeness of CPDL. However, these results rely on embedding standard PDL into CPDL, which requires changing the semantics of CPDL to interpret atomic programs as sequential programs, i.e. as relations rather than neighbourhood systems [24].) Our results are based on a variation of the game-theoretic approach to μ-calculi [19]. Specifically, we reduce satisfiability checking to the computation of winning regions in a satisfiability game that has exponentially many Abelard-nodes but only polynomially many Eloise-nodes, so that history-free winning strategies for Eloise have polynomial size. From this approach we also derive a polynomial model property.

Organization. We recall basics on fixpoints and games in Sect. 2, and the syntax and semantics of the monotone μ-calculus in Sect. 3. We discuss a key technical tool, *formula tracking*, in Sect. 4. We adapt the standard tableaux system to the monotone μ-calculus in Sect. 5. In Sect. 6, we establish our main results using a game characterization of satisfiability.

2 Notation and Preliminaries

We fix basic concepts and notation on fixpoints and games.

Fixpoints. Let U be a set; we write $\mathcal{P}(U)$ for the powerset of U. Let $f : \mathcal{P}(U) \to \mathcal{P}(U)$ be a *monotone* function, i.e. $f(A) \subseteq f(B)$ whenever $A \subseteq B \subseteq U$. By the Knaster-Tarski fixpoint theorem, the *greatest* (GFP) and *least* (LFP) *fixpoints* of f are given by

$$\mathsf{LFP}\, f = \bigcap\{V \subseteq U \mid f(V) \subseteq V\} \qquad \mathsf{GFP}\, f = \bigcup\{V \subseteq U \mid V \subseteq f(V)\},$$

and are thus also the least prefixpoint ($f(V) \subseteq V$) and greatest postfixpoint ($V \subseteq f(V)$) of f, respectively. For $V \subseteq U$ and $n \in \mathbb{N}$, we define $f^n(V)$ as expected by $f^0(V) = V$ and $f^{n+1}(V) = f(f^n(V))$. If U is finite, then $\mathsf{LFP}\, f = f^{|U|}(\emptyset)$ and $\mathsf{GFP}\, f = f^{|U|}(U)$ by Kleene's fixpoint theorem.

Infinite Words and Games. We denote the sets of finite and infinite sequences of elements of a set U by U^* and U^ω, respectively. We often view sequences $\tau = u_1, u_2, \ldots$ in U^* (or U^ω) as partial (total) functions $\tau : \mathbb{N} \rightharpoonup U$, writing $\tau(i) = u_i$. We write

$$\mathsf{Inf}(\tau) = \{u \in U \mid \forall i \in \mathbb{N}.\, \exists j > i.\, \tau(j) = u\}$$

for the set of elements of U that occur infinitely often in τ.

A *Büchi game* $G = (V, E, v_0, F)$ consists of a set V of *nodes*, partitioned into the sets $V_\exists$ of Eloise-nodes and $V_\forall$ of Abelard-nodes, a set $E \subseteq V \times V$ of *moves*, an *initial node* v_0, and a set $F \subseteq V$ of *accepting* nodes. We write

$E(v) = \{v' \mid (v, v') \in E\}$. For simplicity, assume that $v_0 \in V_\exists$ and that the game is *alternating*, i.e. $E(v) \subseteq V_\forall$ for all $v \in V_\exists$, and $E(v) \subseteq V_\exists$ for all $v \in V_\forall$ (our games will have this shape). A *play* of G is a sequence $\tau = v_0, v_1, \ldots$ of nodes such that $(v_i, v_{i+1}) \in E$ for all $i \geq 0$ and τ is either infinite or ends in a node without outgoing moves. Eloise *wins* a play τ if and only if τ is finite and ends in an Abelard-node or τ is infinite and $\mathsf{Inf}(\tau) \cap F \neq \emptyset$, that is, τ infinitely often visits an accepting node. A *history-free* Eloise-*strategy* is a partial function $s : V_\exists \rightharpoonup V$ such that $s(v_0)$ is defined, and whenever $s(v)$ is defined, then $(v, s(v)) \in E$ and $s(v')$ is defined for all $v' \in E(s(v))$. A play $v_0, v_1, \ldots$ is an *s-play* if $v_{i+1} = s(v_i)$ whenever $v_i \in V_\exists$. We say that s is a *winning strategy* if Eloise wins every s-play, and that Eloise *wins* G if there is a winning strategy for Eloise. Büchi games are history-free determined, i.e. in every Büchi game, one of the players has a history-free winning strategy [17].

3 The Monotone μ-Calculus

We proceed to recall the syntax and semantics of the monotone μ-calculus.

Syntax. We fix countably infinite sets P, A and V of *atoms*, *atomic programs* and *(fixpoint) variables*, respectively; we assume that P is closed under duals (i.e. atomic negation), i.e. $p \in \mathsf{P}$ implies $\overline{p} \in \mathsf{P}$, where $\overline{\overline{p}} = p$. Formulae of the *monotone μ-calculus* (in negation normal form) are then defined by the grammar

$$\psi, \phi := \bot \mid \top \mid p \mid \psi \wedge \phi \mid \psi \vee \phi \mid \langle a \rangle \psi \mid [a]\psi \mid X \mid \eta X.\psi$$

where $p \in \mathsf{P}$, $a \in \mathsf{A}$, $X \in \mathsf{V}$; throughout, we use $\eta \in \{\mu, \nu\}$ to denote extremal fixpoints. As usual, μ and ν are understood as taking least and greatest fixpoints, respectively, and bind their variables, giving rise to the standard notion of *free variable* in a formula ψ. We write $\mathsf{FV}(\psi)$ for the set of free variables in ψ, and say that ψ is *closed* if $\mathsf{FV}(\psi) = \emptyset$. Negation $\neg$ is not included but can be defined by taking negation normal forms as usual, with $\neg p = \overline{p}$. We refer to formulae of the shape $[a]\phi$ or $\langle a \rangle \phi$ as *(a-)modal literals.* As indicated in the introduction, the *modalities* $[a]$, $\langle a \rangle$ have been equipped with various readings, recalled in more detail in Example 7.

Given a closed formula ψ, the *closure* $\mathsf{cl}(\psi)$ of ψ is defined to be the least set of formulae that contains ψ and satisfies the following closure properties:

$$\begin{aligned} &\text{if } \psi_1 \wedge \psi_2 \in \mathsf{cl}(\psi) \text{ or } \psi_1 \vee \psi_2 \in \mathsf{cl}(\psi), \text{ then } \{\psi_1, \psi_2\} \subseteq \mathsf{cl}(\psi),\\ &\text{if } \langle a \rangle \psi_1 \in \mathsf{cl}(\psi) \text{ or } [a]\psi_1 \in \mathsf{cl}(\psi), \text{ then } \psi_1 \in \mathsf{cl}(\psi),\\ &\text{if } \eta X.\psi_1 \in \mathsf{cl}(\psi), \text{ then } \psi_1[\eta X.\psi_1/X] \in \mathsf{cl}(\psi), \end{aligned}$$

where $\psi_1[\eta X.\psi_1/X]$ denotes the formula that is obtained from ψ_1 by replacing every free occurrence of X in ψ_1 with $\eta X.\psi_1$. Note that all formulae in $\mathsf{cl}(\psi)$ are closed. We define the *size* $|\psi|$ of ψ as $|\psi| = |\mathsf{cl}(\psi)|$. A formula ψ is *guarded* if whenever $\eta X.\,\phi \in \mathsf{cl}(\psi)$, then all free occurrences of X in ϕ are under the scope of at least one modal operator. *We generally restrict to guarded formulae*;

see however Remark 1. A closed formula ψ is *clean* if all fixpoint variables in ψ are bound by exactly one fixpoint operator. Then $\theta(X)$ denotes *the* subformula $\eta X.\phi$ that binds X in ψ, and X is a least (greatest) fixpoint variable if $\eta = \mu$ ($\eta = \nu$). We define a partial order $\geq_\mu$ on the least fixpoint variables in ρ_1 and ρ_0 by $X \geq_\mu Y$ iff $\theta(Y)$ is a subformula of $\theta(X)$ and $\theta(Y)$ is not in the scope of a greatest fixpoint operator within $\theta(X)$ (i.e. there is no greatest fixpoint operator between μX and μY). The *index* $\mathsf{idx}(X)$ of such a fixpoint variable X is

$$\mathsf{idx}(X) = |\{Y \in \mathsf{V} \mid Y \geq_\mu X\}|,$$

For a subformula ϕ of ψ, we write $\mathsf{idx}(\phi) = \max\{\mathsf{idx}(X) \mid X \in \mathsf{FV}(\phi)\}$. We denote by $\theta^*(\phi_0)$ the closed formula that is obtained from a subformula ϕ_0 of ψ by repeatedly replacing free variables X with $\theta(X)$. Formally, we define $\theta^*(\phi_0)$ as $\phi_{|\mathsf{FV}(\phi_0)|}$, where ϕ_{i+1} is defined inductively from ϕ_i. If ϕ_i is closed, then put $\phi_{i+1} = \phi_i$. Otherwise, pick the variable $X_i \in \mathsf{FV}(\phi_i)$ with the greatest index and put $\phi_{i+1} = \phi_i[\theta(X_i)/X_i]$. Then $\mathsf{idx}(\phi_{i+1}) < \mathsf{idx}(\phi_i)$, so $\theta^*(\phi_0)$ really is closed; moreover, one can show that $\theta^*(\phi_0) \in \mathsf{cl}(\psi)$. A clean formula is *alternation-free* if none of its subformulae contains both a free least and a free greatest fixpoint variable. Finally, ψ is *irredundant* if $X \in \mathsf{FV}(\phi)$ whenever $\eta X.\phi \in \mathsf{cl}(\psi)$.

Remark 1. We have defined the size of formulae as the cardinality of their closure, implying a very compact representation [3]. Our upper complexity bounds thus become *stronger*, i.e. they hold even for this small measure of input size. Moreover, the restriction to guarded formulae is then without loss of generality, since one has a *guardedness transformation* that transforms formulae into equivalent guarded ones, with only polynomial blowup of the closure [3].

Semantics. The monotone μ-calculus is interpreted over *neighbourhood models* (or *epistemic structures* [27]) $F = (W, N, I)$ where $N : \mathsf{A} \times W \to 2^{(2^W)}$ assigns to each atomic program a and each state w a set $N(a, w) \subseteq 2^W$ of *a-neighbourhoods* of w, and $I : \mathsf{P} \to 2^W$ interprets propositional atoms such that $I(p) = W \setminus I(\overline{p})$ for $p \in \mathsf{P}$ (by 2, we denote the set $\{\bot, \top\}$ of Boolean truth values, and 2^W is the set of maps $W \to 2$, which is in bijection with the powerset $\mathcal{P}(W)$). Given such an F, each formula ψ is assigned an *extension* $[\![\psi]\!]_\sigma \subseteq W$ that additionally depends on a *valuation* $\sigma : V \to 2^W$, and is inductively defined by

$$\begin{aligned}
[\![p]\!]_\sigma &= I(p) & [\![X]\!]_\sigma &= \sigma(X)\\
[\![\psi \wedge \phi]\!]_\sigma &= [\![\psi]\!]_\sigma \cap [\![\phi]\!]_\sigma & [\![\psi \vee \phi]\!]_\sigma &= [\![\psi]\!]_\sigma \cup [\![\phi]\!]_\sigma\\
[\![\langle a\rangle\psi]\!]_\sigma &= \{w \in W \mid \exists S \in N(a,w).\, S \subseteq [\![\psi]\!]_\sigma\} & [\![\mu X.\psi]\!]_\sigma &= \mathsf{LFP}[\![\psi]\!]_\sigma^X\\
[\![[a]\psi]\!]_\sigma &= \{w \in W \mid \forall S \in N(a,w).\, S \cap [\![\psi]\!]_\sigma \neq \emptyset\} & [\![\nu X.\psi]\!]_\sigma &= \mathsf{GFP}[\![\psi]\!]_\sigma^X
\end{aligned}$$

where, for $U \subseteq W$ and fixpoint variables $X, Y \in \mathsf{V}$, we put $[\![\psi]\!]_\sigma^X(U) = [\![\psi]\!]_{\sigma[X \mapsto U]}$, $(\sigma[X \mapsto U])(X) = U$ and $(\sigma[X \mapsto U])(Y) = \sigma(Y)$ if $X \neq Y$. We omit the dependence on F in the notation $[\![\phi]\!]_\sigma$, and when necessary clarify the underlying neighbourhood model by phrases such as 'in F'. If ψ is closed, then its extension does not depend on the valuation, so we just write $[\![\psi]\!]$. A closed formula ψ is *satisfiable* if there is a neighbourhood model F such that $[\![\psi]\!] \neq \emptyset$ in F; in this case, we also say that ψ is *satisfiable over* F. Given a set Ψ of closed formulae, we write $[\![\Psi]\!] = \bigcap_{\psi \in \Psi} [\![\psi]\!]$. An *(infinite) path* through a neighbourhood model (W, N, I) is a sequence $x_0, x_1, \ldots$ of states $x_i \in W$ such that for all $i \geq 0$, there are $a \in \mathsf{A}$ and $S \in N(a, x_i)$ such that $x_{i+1} \in S$.

The soundness direction of our game characterization will rely on the following immediate property of the semantics, which may be seen as soundness of a modal tableau rule [5].

Lemma 2 *[27, Proposition 3.8]. If $[a]\phi \wedge \langle a \rangle \psi$ is satisfiable over a neighbourhood model F, then $\phi \wedge \psi$ is also satisfiable over F.*

Remark 3. The dual box and diamond operators $[a]$ and $\langle a \rangle$ are completely symmetric, and indeed the notation is not uniform in the literature. Our use of $[a]$ and $\langle a \rangle$ is generally in agreement with work on game logic [20] and CPDL [24]; in work on monotone modal logics and the monotone μ-calculus, the roles of box and diamond are often interchanged [8,26,27].

Remark 4. The semantics may equivalently be presented in terms of *monotone* neighbourhood models, where the set of a-neighbourhoods of a state is required to be upwards closed under subset inclusion [8,12,20,23]. In this semantics, the interpretation of $\langle a \rangle \phi$ simplifies to just requiring that the extension of ϕ is an a-neighbourhood of the current state. We opt for the variant where upwards closure is instead incorporated into the interpretation of the modalities, so as to avoid having to distinguish between monotone neighbourhood models and their representation as upwards closures of (plain) neighbourhood models, e.g. in small model theorems.

We further extend the expressiveness of the logic (see also Remark 9) by adding *global assumptions* or equivalently the *universal modality*:

Definition 5 (Global assumptions). Given a closed formula ϕ, a *ϕ-model* is a neighbourhood model $F = (W, N, I)$ in which $[\![\phi]\!] = W$. A formula ψ is *ϕ-satisfiable* if ψ is satisfiable over some ϕ-model; in this context, we refer to ϕ as the *global assumption*, and to the problem of deciding whether ψ is ϕ-satisfiable as *satisfiability checking under global assumptions.*

We also define an extension of the monotone μ-calculus, the *monotone μ-calculus with the universal modality*, by adding two alternatives

$$\cdots \mid [\forall]\phi \mid [\exists]\phi$$

to the grammar, in both alternatives restricting ϕ to be closed. The definition of the semantics over a neighbourhood model (W, N, I) and valuation σ is correspondingly extended by $[\![[\forall]\phi]\!]_\sigma = W$ if $[\![\phi]\!]_\sigma = W$, and $[\![[\forall]\phi]\!]_\sigma = \emptyset$ otherwise; dually, $[\![[\exists]\phi]\!]_\sigma = W$ if $[\![\phi]\!]_\sigma \neq \emptyset$, and $[\![[\exists]\phi]\!]_\sigma = \emptyset$ otherwise. That is, $[\forall]\phi$ says that ϕ holds in all states of the model, and $[\exists]\phi$ that ϕ holds in some state.

Remark 6. In description logic, global assumptions are typically called (general) *TBoxes* or *terminologies* [1]. For many next-step modal logics (i.e. modal logics without fixpoint operators), satisfiability checking becomes harder under global assumptions. A typical case is the standard modal logic K (corresponding to the description logic $\mathcal{ALC}$), in which (plain) satisfiability checking is PSPACE-complete [15] while satisfiability checking under global assumptions is EXPTIME-complete [6,9]. Our results show that such an increase in complexity does not happen for monotone modalities.

For purposes of satisfiability checking, the universal modality and global assumptions are mutually reducible in a standard manner, where the non-trivial direction (from the universal modality to global assumptions) is by guessing beforehand which subformulae $[\forall]\phi$, $[\exists]\phi$ hold (see also [11]).

Example 7. 1. In *epistemic logic*, neighbourhood models have been termed *epistemic structures* [27]. In this context, the $a \in \mathsf{A}$ are thought of as *agents*, the a-neighbourhoods of a state w are the facts known to agent a in w, and correspondingly the reading of $[a]\phi$ is 'a knows ϕ'. The use of (monotone) neighbourhood models and the ensuing failure of normality imply that agents can still weaken facts that they know but are not in general able to combine them, i.e. knowing ϕ and knowing ψ does not entail knowing $\phi \wedge \psi$ [27].

2. In *concurrent propositional dynamic logic (CPDL)*, a-neighbourhoods of a state are understood as sets of states that can be reached concurrently, while the choice between several a-neighbourhoods of a state models sequential non-determinism. CPDL indexes modalities over composite *programs* α, formed using tests $?\phi$ and the standard operations of propositional dynamic logic (PDL) (union $\cup$, sequential composition ';', and Kleene star $(-)^*$) and additionally intersection $\cap$. It forms a sublogic of Parikh's game logic, recalled next, and thus in particular translates into the monotone μ-calculus.

   As indicated in the introduction, CPDL satisfiability checking has been shown to be EXPTIME-complete [24], seemingly contradicting our results (Corollary 25). Note however that the interpretation of atomic programs in CPDL, originally defined in terms of neighbourhood systems [24, p. 453], is, for purposes of the EXPTIME-hardness proof, explicitly changed to relations [24, pp. 458–459]; EXPTIME-hardness then immediately follows since PDL becomes a sublogic of CPDL [24, p. 461]. Our NP bound applies to the original semantics.

3. *Game logic* [20,23] extends CPDL by a further operator on programs, *dualization* $(-)^d$, and reinterprets programs as games between two players *Angel* and *Demon*; in this view, dualization just corresponds to swapping the roles of the players. In comparison to CPDL, the main effect of dualization is that one obtains an additional *demonic iteration operator* $(-)^\times$, distinguished from standard iteration $(-)^*$ by letting Demon choose whether or not to continue the iteration. A game logic formula is *alternation-free* if it does not contain nested occurrences (unless separated by a test) of $(-)^\times$ within $(-)^*$ or vice versa [23].

Enqvist et al. [7] give a translation of game logic into the monotone μ-calculus that is quite similar to Pratt's [25] translation of PDL into the standard μ-calculus. The translation $(-)^\sharp$ is defined by commutation with all Boolean connectives and by

$$(\langle\gamma\rangle\phi)^\sharp = \tau_\gamma(\phi^\sharp),$$

in mutual recursion with a function τ_γ that translates the effect of applying $\langle\gamma\rangle$ into the monotone μ-calculus. (Boxes $[\gamma]$ can be replaced with $\langle\gamma^d\rangle$). We refrain from repeating the full definition of τ_γ by recursion over γ; some key clauses are

$$\tau_{\gamma\cap\delta}(\psi)=\tau_\gamma(\psi)\wedge\tau_\delta(\psi) \quad \tau_{\gamma^*}(\psi)=\mu X.\,(\psi\vee\tau_\gamma(X)) \quad \tau_{\gamma^\times}(\psi) = \nu Y.\,(\psi\wedge\tau_\gamma(Y))$$

where in the last two clauses, X and Y are chosen as fresh variables (for readability, we gloss over a more precise treatment of this point given in [7]). The first clause (and a similar one for $\cup$) appear at first sight to cause exponential blowup but recall that we measure the size of formulae by the cardinality of their closure; in this measure, there is in fact no blowup. The translated formula $\psi^\sharp$ need not be guarded as the clauses for * and $^\times$ can introduce unguarded fixpoint variables; as mentioned in Remark 1, we can however apply the guardedness transformation, with only quadratic blowup of the closure [3].

Under this translation, the alternation-free fragment of game logic ends up in the (guarded) alternation-free fragment of the monotone μ-calculus.

For later use, we note

Lemma 8. *The monotone μ-calculus with the universal modality has the finite model property.*

Proof (sketch). We reduce to global assumptions as per Remark 6, and proceed by straightforward adaptation of the translations of monotone modal logic [14] and game logic [23] into the *relational* μ-calculus, thus inheriting the finite model property [2]. This translation is based on turning neighbourhoods into additional states, connected to their elements via a fresh relation e. Then, e.g., the monotone modality $[a]$ (in our notation, cf. Remark 3) is translated into $[a]\langle e\rangle$ (relational modalities). Moreover, we translate a global assumption ϕ into a formula saying that all reachable states satisfy ϕ, expressed in the μ-calculus in a standard fashion. $\square$

Remark 9. In the relational μ-calculus, we can encode a modality $\boxtimes$ 'in all reachable states', generalizating the AG operator from CTL. As already indicated in the proof of Lemma 8, this modality allows for a straightforward reduction of satisfiability under global assumptions to plain satisfiablity in the relational μ-calculus: A formula ψ is satisfiable under the global assumption ϕ iff $\psi\wedge\boxtimes\phi$ is satisfiable, where 'if' is shown by restricting the model to reachable states. To motivate separate consideration of the universal modality, we briefly argue why an analogous reduction does not work in the monotone μ-calculus.

It is not immediately clear what reachability would mean on neighbourhood models. We can however equivalently rephrase the definition of $\boxtimes$ in the relational case to let $\boxtimes\phi$ mean 'the present state is contained in a submodel in which every state satisfies ϕ', where as usual a submodel of a relational model C with state set W is a model C' with state set $W' \subseteq W$ such that the graph of the inclusion $W' \hookrightarrow W$ is a bisimulation from C' to C. We thus refer to $\boxtimes$ as the *submodel modality.* This notion transfers to neighbourhood models using a standard notion of bisimulation: A *monotone bisimulation* [21,22] between neighbourhood models (W_1, N_1, I_1), (W_2, N_2, I_2) is a relation $S \subseteq W_1 \times W_2$ such that whenever $(x, y) \in S$, then

- for all $a \in \mathsf{A}$ and $A \in N_1(a, x)$, there is $B \in N_2(a, y)$ such that for all $v \in B$, there is $u \in A$ such that $(u, v) \in S$,
- for all $a \in \mathsf{A}$ and $B \in N_2(a, y)$, there is $A \in N_1(a, x)$ such that for all $u \in A$, there is a $v \in B$ such that $(u, v) \in S$,
- for all $p \in \mathsf{P}$, $x \in I_1(p)$ if and only if $y \in I_2(p)$.

We then define a *submodel* of a neighbourhood model $F = (W, N, I)$ to be a neigbourhood model $F' = (W', N', I')$ such that $W' \subseteq W$ and the graph of the inclusion $W' \hookrightarrow W$ is a monotone bisimulation between F' and F. If $(x, y) \in S$ for some monotone bisimulation S, then x and y satisfy the same formulae in the monotone μ-calculus [8]. It follows that the submodel modality $\boxtimes$ on neighbourhood models, defined verbatim as in the relational case, allows for the same reduction of satisfiability under global assumptions as the relational submodel modality.

However, the submodel modality fails to be expressible in the monotone μ-calculus, as seen by the following example. Let $F_1 = (W_1, N_1, I_1)$, $F_2 = (W_2, N_2, I_2)$ be the neighbourhood models given by $W_1 = \{x_1, u_1, v_{11}, v_{12}\}$, $W_2 = \{x_2, u_2, v_2\}$, $N_1(a, x_1) = \{\{u_1, v_{11}\}, \{v_{11}, v_{12}\}\}$, $N_2(a, x_2) = \{\{v_2\}\}$, $N_1(b, y) = N_2(b, z) = \emptyset$ for $(b, y) \neq (a, x_1)$, $(b, z) \neq (a, x_2)$, $I_1(p) = \{x_1, v_{11}, v_{12}\}$, $I_2(p) = \{x_2, v_2\}$, and $I_1(q) = I_2(q) = \emptyset$ for $q \neq p$. Then $S = \{(x_1, x_2), (u_1, u_2), (v_{11}, v_2), (v_{12}, v_2)\}$ is a monotone bisimulation, so x_1 and x_2 satisfy the same formulae in the monotone μ-calculus. However, x_2 satisfies $\boxtimes p$ because x_2 is contained in a submodel with set $\{x_2, v_2\}$ of states, while x_1 does not satisfy $\boxtimes p$.

4 Formula Tracking

A basic problem in the construction of models in fixpoint logics is to avoid infinite unfolding of least fixpoints, also known as infinite deferral. To this end, we use a *tracking function* to follow formulae along paths in prospective models. For unrestricted μ-calculi, infinite unfolding of least fixpoints is typically detected by means of a *parity condition* on tracked formulae. However, since we restrict to alternation-free formulae, we can instead use a *Büchi condition* to detect (and then reject) such infinite unfoldings by sequential *focussing* on sets of formulae in the spirit of focus games [16].

We fix *closed, clean, irredundant formulae* ρ_1 *and* ρ_0, aiming to check ρ_0-satisfiability of ρ_1; we also require both ρ_1 and ρ_0 to be alternation-free. We put $\mathsf{cl} = \mathsf{cl}(\rho_1) \cup \mathsf{cl}(\rho_0)$ and $n = |\mathsf{cl}|$.

Next we formalize our notion of deferred formulae that originate from least fixpoints; these are the formulae for which infinite unfolding has to be avoided.

Definition 10 (Deferrals). A formula $\phi \in \mathsf{cl}(\psi)$ is a *deferral* if there is a subformula χ of ψ such that $\mathsf{FV}(\chi)$ contains a least fixpoint variable and $\phi = \theta^*(\chi)$. We put $\mathsf{dfr} = \{\phi \in \mathsf{cl} \mid \phi \text{ is a deferral}\}$.

Since we assume formulas to be alternation-free, no formula $\nu X.\,\phi$ is a deferral.

Example 11. For $\psi = \mu X.\,[b]\top \vee \langle a\rangle X$, the formula $[b]\top \vee \langle a\rangle\psi = ([b]\top \vee \langle a\rangle X)[\psi/X] \in \mathsf{cl}(\psi)$ is a deferral since X is a least fixpoint variable and occurs free in $[b]\top \vee \langle a\rangle X$; the formula $[b]\top$ on the other hand is not a deferral since it cannot be obtained from a subformula of ψ by replacing least fixpoint variables with their binding fixpoint formulae.

We proceed to define a tracking function that nondeterministically tracks deferrals along paths in neighbourhood models.

Definition 12 (Tracking function). We define an alphabet $\Sigma = \Sigma_p \cup \Sigma_m$ for traversing the closure, separating propositional and modal traversal, by

$$\begin{aligned}\Sigma_p &= \{(\phi_0 \wedge \phi_1, 0), (\phi_0 \vee \phi_1, b), (\eta X.\phi_0, 0) \mid \\ &\qquad\qquad \phi_0, \phi_1 \in \mathsf{cl}, \eta \in \{\mu, \nu\}, X \in \mathsf{V}, b \in \{0,1\}\}\\ \Sigma_m &= \{(\langle a\rangle\phi_0, [a]\phi_1) \mid \phi_0, \phi_1 \in \mathsf{cl}, a \in \mathsf{A}\}.\end{aligned}$$

The *tracking function* $\delta : \mathsf{dfr} \times \Sigma \to \mathcal{P}(\mathsf{dfr})$ is defined by $\delta(\mathsf{foc}, (\chi, b)) = \{\mathsf{foc}\}$ for $\mathsf{foc} \in \mathsf{dfr}$ and $(\chi, b) \in \Sigma_p$ with $\chi \neq \mathsf{foc}$, and by

$$\delta(\mathsf{foc}, (\mathsf{foc}, b)) = \begin{cases} \{\phi_0, \phi_1\} \cap \mathsf{dfr} & \mathsf{foc} = \phi_0 \wedge \phi_1 \\ \{\phi_b\} \cap \mathsf{dfr} & \mathsf{foc} = \phi_0 \vee \phi_1 \\ \{\phi_0[\theta(X)/X]\} & \mathsf{foc} = \mu X.\,\phi_0 \end{cases}$$

$$\delta(\mathsf{foc}, (\langle a\rangle\phi_0, [a]\phi_1)) = \begin{cases} \{\phi_0\} & \mathsf{foc} = \langle a\rangle\phi_0 \\ \{\phi_1\} & \mathsf{foc} = [a]\phi_1, \end{cases}$$

noting that $\mu X.\,\phi_0 \in \mathsf{dfr}$ implies $\phi_0[\theta(X)/X] \in \mathsf{dfr}$. We extend δ to sets $\mathsf{Foc} \subseteq \mathsf{dfr}$ by putting $\delta(\mathsf{Foc}, l) = \bigcup_{\mathsf{foc} \in \mathsf{Foc}} \delta(\mathsf{foc}, l)$ for $l \in \Sigma$. We also extend δ to words in the obvious way; e.g. we have $\delta(\mathsf{foc}, w) = [a]\mathsf{foc}$ for

$$\mathsf{foc} = \mu X.\,(p \vee (\langle b\rangle p \wedge [a]X)) \quad w = (\mathsf{foc}, 0), (p \vee (\langle b\rangle p \wedge [a]\mathsf{foc}), 1), (\langle b\rangle p \wedge [a]\mathsf{foc}, 0).$$

Remark 13. Formula tracking can be modularized in an elegant way by using tracking automata to accept exactly the paths that contain infinite deferral of some least fixpoint formula. While tracking automata for the full μ-calculus are in general nondeterministic parity automata [10], the automata for alternation-free formulae are nondeterministic co-Büchi automata. Indeed, our tracking function δ can be seen as the transition function of a nondeterministic co-Büchi automaton with set dfr of states and alphabet Σ in which all states are accepting. Since models have to be constructed from paths that do *not* contain infinite deferral of some least fixpoint formula, the tracking automata have to be complemented in an intermediate step (which is complicated by nondeterminism) when checking satisfiability. In our setting, the complemented automata are deterministic Büchi automata obtained using a variant of the Miyano-Hayashi construction [18] similarly as in our previous work on ExpTime satisfiability checking in alternation-free coalgebraic μ-calculi [13]; for brevity, this determinization procedure remains implicit in our satisfiability games (see Definition 20 below).

To establish our game characterization of satisfiability, we combine the tracking function δ with a function for propositional transformation of formula sets guided by letters from Σ_p, embodied in the function γ defined next.

Definition 14 (Propositional transformation). We define $\gamma : \mathcal{P}(\mathsf{cl}) \times \Sigma_p \to \mathcal{P}(\mathsf{cl})$ on $\Gamma \subseteq \mathsf{cl}$, $(\chi, b) \in \Sigma_p$ by $\gamma(\Gamma, (\chi, b)) = \Gamma$ if $\chi \notin \Gamma$, and

$$\gamma(\Gamma, (\chi, b)) = (\Gamma \setminus \chi) \cup \begin{cases} \{\phi_b\} & \chi = \phi_0 \vee \phi_1 \\ \{\phi_0, \phi_1\} & \chi = \phi_0 \wedge \phi_1 \\ \{\phi_0[\eta X.\, \phi_0/X]\} & \chi = \eta X.\, \phi_0 \end{cases}$$

if $\chi \in \Gamma$. We extend γ to words over Σ_p in the obvious way.

Example 15. Take $\phi = \chi \vee \nu X.\, (\psi_1 \wedge \psi_2)$ and $w = (\chi \vee \nu X.\, (\psi_1 \wedge \psi_2), 1), (\nu X. (\psi_1 \wedge \psi_2), 0), ((\psi_1 \wedge \psi_2)[\theta(X)/X], 1)$. The letter $(\chi \vee \nu X.\, (\psi_1 \wedge \psi_2), 1)$ picks the right disjunct and $(\nu X.\, (\psi_1 \wedge \psi_2), 0)$ passes through the fixpoint operator νX to reach $\psi_1 \wedge \psi_2$; the letter $(\psi_1 \wedge \psi_2, 1)$ picks both conjuncts. Thus,

$$\gamma(\{\phi\}, w) = \{\psi_1[\theta(X)/X], \psi_2[\theta(X)/X]\}.$$

5 Tableaux

As a stepping stone between neighbourhood models and satisfiability games, we now introduce *tableaux* which are built using a variant of the standard *tableau rules* [10] (each consisting of one *premise* and a possibly empty set of *conclusions*), where the *modal rule* ($\langle a \rangle$) reflects Lemma 2, taking into account the global assumption ρ_0:

$$(\bot)\ \frac{\Gamma, \bot}{} \qquad (\lightning)\ \frac{\Gamma, p, \overline{p}}{} \qquad (\wedge)\ \frac{\Gamma, \phi_1 \wedge \phi_2}{\Gamma, \phi_1, \phi_2}$$

$$(\vee)\ \frac{\Gamma, \phi_1 \vee \phi_2}{\Gamma, \phi_1 \qquad \Gamma, \phi_2} \qquad (\langle a \rangle)\ \frac{\Gamma, \langle a \rangle \phi_1, [a]\phi_2}{\phi_1, \phi_2, \rho_0} \qquad (\eta)\ \frac{\Gamma, \eta X.\, \phi}{\Gamma, \phi[\eta X.\, \phi/X]}$$

(for $a \in \mathsf{A}$, $p \in \mathsf{P}$); we usually write rule instances with premise Γ and conclusion $\Theta = \Gamma_1, \ldots, \Gamma_n$ ($n \leq 2$) inline as (Γ/Θ). Looking back at Sect. 4, we see that letters in Σ designate rule applications, e.g. the letter $(\langle a \rangle \phi_1, [a]\phi_2) \in \Sigma_m$ indicates application of $(\langle a \rangle)$ to formulae $\langle a \rangle \phi_1$, $[a]\phi_2$.

Definition 16 (Tableaux). Let states denote the set of *(formal) states*, i.e. sets $\Gamma \subseteq \mathsf{cl}$ such that $\bot \notin \Gamma$, $\{p, \overline{p}\} \nsubseteq \Gamma$ for all $p \in \mathsf{P}$, and such Γ does not contain formulae that contain top-level occurrences of the operators $\wedge$, $\vee$, or ηX.

A *pre-tableau* is a directed graph (W, L), consisting of a finite set W of nodes labelled with subsets of cl by a labeling function $l : W \to \mathcal{P}(\mathsf{cl})$, and of a relation $L \subseteq W \times W$ such that for all nodes $v \in W$ with label $l(v) = \Gamma \in \mathsf{states}$ and all applications $(\Gamma/\Gamma_1, \ldots, \Gamma_n)$ of a tableau rule to Γ, there is an edge $(v, w) \in L$ such that $l(w) = \Gamma_i$ for some $1 \leq i \leq n$. For nodes with label $l(v) = \Gamma \notin \mathsf{states}$, we require that there is exactly one node $w \in W$ such that $(v, w) \in L$; then we demand that there is an application $(\Gamma/\Gamma_1, \ldots, \Gamma_n)$ of a non-modal rule to Γ such that $l(w) = \Gamma_i$ for some $1 \leq i \leq n$. Finite or infinite words over Σ then encode sequences of rule applications (and choices of conclusions for disjunctions). That is, given a starting node v, they encode *branches* with root v, i.e. (finite or infinite) paths through (W, L) that start at v.

Deferrals are tracked along rule applications by means of the function δ. E.g. for a deferral $\phi_0 \vee \phi_1$, the letter $l = (\phi_0 \vee \phi_1, 0)$ identifies application of $(\vee)$ to $\{\phi_0 \vee \phi_1\}$, and the choice of the left disjunct; then $\phi_0 \vee \phi_1$ is tracked from a node with label $\Gamma \cup \{\phi_0 \vee \phi_1\}$ to a successor node with label $\Gamma \cup \{\phi_0\}$ if $\phi_0 \in \delta(\phi_0 \vee \phi_1, l)$, that is, if ϕ_0 is a deferral. A *trace* (of ϕ_0) along a branch with root v (whose label contains ϕ_0) encoded by a word w is a (finite or infinite) sequence $t = \phi_0, \phi_1 \ldots$ of formulae such that $\phi_{i+1} \in \delta(\phi_i, w(i))$. A *tableau* is a finite pre-tableau in which all traces are finite, and a tableau is a *tableau for* ρ_1 if some node label contains ρ_1.

Given a tableau (W, L) and a node $v \in W$, let $\mathsf{tab}(v)$ (for *tableau timeout*) denote the least number m such that for all formulae ϕ in $l(v)$ and all branches of (W, L) that are rooted at v, all traces of ϕ along the branch have length at most m; such an m always exists by the definition of tableaux.

To link models and tableaux, we next define an inductive measure on unfolding of least fixpoint formulae in models.

Definition 17 (Extension under timeouts). Let k be the greatest index of any least fixpoint variable in ψ, and let (W, N, I) be a finite neighbourhood model. Then a *timeout* is a vector $\overline{m} = (m_1, \ldots, m_k)$ of natural numbers $m_i \leq |W|$. For $1 \leq i \leq k$ such that $m_i > 0$, we put $\overline{m}@i = (m_1, \ldots, m_{i-1}, m_i - 1, |W|, \ldots, |W|)$. Then $\overline{m} >_l \overline{m}@i$, where $>_l$ denotes lexicographic ordering. For $\phi \in \mathsf{cl}$, we inductively define the *extension* $[\![\phi]\!]_{\overline{m}}$ *under timeout* $\overline{m}$ by $[\![\phi]\!]_{\overline{m}} = [\![\phi]\!]$

for $\phi \notin \mathsf{dfr}$ and, for $\phi \in \mathsf{dfr}$, by

$$\begin{aligned}
[\![\psi_0 \wedge \psi_1]\!]_{\overline{m}} &= [\![\psi_0]\!]_{\overline{m}} \cap [\![\psi_1]\!]_{\overline{m}} \\
[\![\psi_0 \vee \psi_1]\!]_{\overline{m}} &= [\![\psi_0]\!]_{\overline{m}} \cup [\![\psi_1]\!]_{\overline{m}} \\
[\![\langle a \rangle \psi]\!]_{\overline{m}} &= \{w \in W \mid \exists S \in N(a, w).\, S \subseteq [\![\psi]\!]_{\overline{m}}\} \\
[\![[a]\psi]\!]_{\overline{m}} &= \{w \in W \mid \forall S \in N(a, w).\, S \cap [\![\psi]\!]_{\overline{m}} \neq \emptyset\} \\
[\![\mu X.\psi]\!]_{\overline{m}} &= \begin{cases} \emptyset & m_{\mathsf{idx}(X)} = 0 \\ [\![\psi[\mu X.\psi / X]]\!]_{\overline{m}@\mathsf{idx}(X)} & m_{\mathsf{idx}(X)} > 0 \end{cases}
\end{aligned}$$

using the lexicographic ordering on $(\overline{m}, \phi)$ as the termination measure; crucially, unfolding least fixpoints reduces the timeout. We extend this definition to sets of formulae by $[\![\Psi]\!]_{\overline{m}} = \bigcap_{\phi \in \Psi} [\![\phi]\!]_{\overline{m}}$ for $\Psi \subseteq \mathrm{cl}$.

Lemma 18. *In finite neighbourhood models, we have that for all $\psi \in \mathsf{cl}$ there is some timeout $\overline{m}$ such that*

$$[\![\psi]\!] \subseteq [\![\psi]\!]_{\overline{m}}.$$

Proof. Let W be the set of states. Since $[\![\psi]\!]_{\overline{m}}$ is defined like $[\![\psi]\!]$ in all cases but one, we only need to consider the inductive case where $\psi = \mu X.\psi_0 \in \mathsf{cl}$. We show that for all *subformulae* ϕ of ψ_0, for all timeout vectors $\overline{m} = (m_1, \ldots, m_k)$ such that $m_i = |W|$ for all i such that ϕ does not contain a free variable with index at least i, we have

$$[\![\phi]\!]_{\sigma(\overline{m})} \subseteq [\![\theta^*(\phi)]\!]_{\overline{m}}, \tag{1}$$

where $\sigma(\overline{m})$ maps each $Y \in \mathsf{FV}(\phi)$ to $([\![\phi_1]\!]^Y_{\sigma(\overline{m})})^{m_{\mathsf{idx}(Y)}}(\emptyset)$ where $\theta(Y) = \mu Y.\, \phi_1$ (this is a recursive definition of $\sigma(\overline{m})$ since the value of $\sigma(\overline{m})$ for Y is overwritten in $([\![\phi_1]\!]^Y_{\sigma(\overline{m})})^{m_{\mathsf{idx}(Y)}}(\emptyset)$ so that this set depends only on values m_i such that $i < \mathsf{idx}(Y)$). This shows that the claimed property $[\![\mu X.\, \psi_0]\!] \subseteq [\![\mu X.\, \psi_0]\!]_{\overline{m}}$ holds for $\overline{m} = (|W|, \ldots, |W|)$: By Kleene's fixpoint theorem,

$$\begin{aligned}
[\![\mu X.\psi_0]\!] = ([\![\psi_0]\!]^X)^{|W|}(\emptyset) &= [\![\psi_0]\!]_{[X \mapsto ([\![\psi_0]\!]^X)^{|W|-1}(\emptyset)]} = [\![\psi_0]\!]_{\sigma(\overline{m}@\mathsf{idx}(X))} \\
&\subseteq [\![\theta^*(\psi_0)]\!]_{\overline{m}@\mathsf{idx}(X)} = [\![\mu X.\psi_0]\!]_{\overline{m}},
\end{aligned}$$

using (1) in the second-to-last step.

The proof of (1) is by induction on ϕ; we do only the non-trivial cases. If $\theta^*(\phi) \notin \mathsf{dfr}$, then ϕ is closed (since ψ_0 does not contain a free greatest fixpoint variable and since ϕ is not in the scope of a greatest fixpoint operator within ψ_0, that is, no free greatest fixpoint variable is introduced during the induction). Hence we have $[\![\phi]\!]_{\sigma(\overline{m})} = [\![\phi]\!]$ and $[\![\theta^*(\phi)]\!]_{\overline{m}} = [\![\theta^*(\phi)]\!] = [\![\phi]\!]$ so that we are done. If $\phi = Y$, then $[\![Y]\!]_{\sigma(\overline{m})} = ([\![\phi_1]\!]^Y_{\sigma(\overline{m})})^{m_{\mathsf{idx}(Y)}}(\emptyset)$ where $\theta(Y) = \mu Y.\, \phi_1$. If $m_{\mathsf{idx}(Y)} = 0$, then $([\![\phi_1]\!]^Y_{\sigma(\overline{m})})^{m_{\mathsf{idx}(Y)}}(\emptyset) = \emptyset$ so there is nothing to show. If

$m_{\mathsf{idx}(Y)} > 0$, then

$$\begin{aligned}(\llbracket\phi_1\rrbracket^Y_{\sigma(\overline{m})})^{m_{\mathsf{idx}(Y)}}(\emptyset) &= \llbracket\phi_1\rrbracket^Y_{\sigma(\overline{m})}((\llbracket\phi_1\rrbracket^Y_{\sigma(\overline{m})})^{m_{\mathsf{idx}(Y)}-1}(\emptyset))\\ &= \llbracket\phi_1\rrbracket_{(\sigma(\overline{m}))[Y\mapsto(\llbracket\phi_1\rrbracket^Y_\sigma)^{m_{\mathsf{idx}(Y)}-1}(\emptyset)]}\\ &\subseteq \llbracket\theta^*(\phi_1)\rrbracket_{\overline{m}@\mathsf{idx}(Y)}\\ &= \llbracket\theta^*(\phi_1[\mu Y.\,\phi_1/Y])\rrbracket_{\overline{m}@\mathsf{idx}(Y)}\\ &= \llbracket\theta^*(\mu Y.\,\phi_1)\rrbracket_{\overline{m}} = \llbracket\theta^*(Y)\rrbracket_{\overline{m}},\end{aligned}$$

where the inclusion is by the inductive hypothesis. If $\phi = \mu Y.\,\phi_1$, then

$$\begin{aligned}\llbracket\mu Y.\,\phi_1\rrbracket_{\sigma(\overline{m})} = (\llbracket\phi_1\rrbracket^Y_{\sigma(\overline{m})})^{|W|}(\emptyset) &= \llbracket\phi_1\rrbracket^Y_{\sigma(\overline{m})}((\llbracket\phi_1\rrbracket^Y_{\sigma(\overline{m})})^{|W|-1}(\emptyset))\\ &= \llbracket\phi_1\rrbracket_{(\sigma(\overline{m}))[Y\mapsto(\llbracket\phi_1\rrbracket^Y)^{|W|-1}(\emptyset)]}\\ &\subseteq \llbracket\theta^*(\phi_1)\rrbracket_{\overline{m}@\mathsf{idx}(Y)}\\ &= \llbracket\theta^*(\phi_1[\theta(Y)/Y])\rrbracket_{\overline{m}@\mathsf{idx}(Y)}\\ &= \llbracket\theta^*(\mu Y.\,\phi_1)\rrbracket_{\overline{m}},\end{aligned}$$

where the first equality is by Kleene's fixpoint theorem and the inclusion is by the inductive hypothesis since $m_{\mathsf{idx}(Y)} = |W|$ (hence $m@\mathsf{idx}(Y)_{\mathsf{idx}(Y)} = |W| - 1$) by assumption as $\mu Y.\,\phi_1$ does not contain a free variable with index at least $\mathsf{idx}(Y)$. □

To check satisfiability it suffices to decide whether a tableau exists:

Theorem 19. *There is a tableau for ρ_1 if and only if ρ_1 is ρ_0-satisfiable.*

Proof. Let ρ_1 be ρ_0-satisfiable. Then there is a finite neighbourhood model (W, N, I) such that $W \subseteq \llbracket\rho_0\rrbracket$ and $W \cap \llbracket\rho_1\rrbracket \neq \emptyset$ by Lemma 8. We define a tableau over the set $V = \{(x, \Psi, w) \in W \times \mathcal{P}(\mathsf{cl}) \times \Sigma_p^* \mid x \in \llbracket\Psi\rrbracket, \mathsf{u}(\Psi) \leq 3n - |w|\}$, where $\mathsf{u}(\Psi)$ denotes the sum of the numbers of unguarded operators in formulae from Ψ. Let $(x, \Psi, w) \in V$. For $\Psi \notin \mathsf{states}$, $\mathsf{u}(\Psi) \leq 3n - |w|$, we pick some $\phi \in \Psi$, distinguishing cases. If $\phi = \phi_0 \wedge \phi_1$ or $\phi = \eta X.\,\phi_0$, then we put $b = 0$. If $\phi = \phi_0 \vee \phi_1$, then let $\overline{m}$ be the least timeout such that $x \in \llbracket\phi\rrbracket_{\overline{m}}$ (such $\overline{m}$ exists by Lemma 18). Then there is some $b' \in \{0, 1\}$ such that $x \in \llbracket\phi_{b'}\rrbracket_{\overline{m}}$ and we put $b = b'$. In any case, we put $l = (\phi, b)$ and add an edge from (x, Ψ, w) to $(x, \gamma(\Psi, l), (w, l))$ to L, having $\mathsf{u}(\gamma(\Psi, l)) = \mathsf{u}(\Psi) - 1$ and hence $\mathsf{u}(\gamma(\Psi, l)) \leq 3n - |w, l|$. Since we are interested in the nodes that can be reached from nodes (x, Ψ, ϵ) with $|\Psi| \leq 3$ and since each formula in Ψ contains at most n unguarded operators, we indeed only have to construct nodes (x, Ψ', w) with $\mathsf{u}(\Psi) \leq 3n - |w|$ before reaching state labeled nodes. For $\Psi \in \mathsf{states}$ and $\{\langle a\rangle\phi_0, [a]\phi_1\} \subseteq \Psi$, let $\overline{m}$ be the least timeout such that $x \in \llbracket\{\langle a\rangle\phi_0, [a]\phi_1\}\rrbracket_{\overline{m}}$ (again, such $\overline{m}$ exists by Lemma 18). Then there is $S \in N(a, x)$ such that $S \subseteq \llbracket\phi_0\rrbracket_{\overline{m}} \cap \llbracket\rho_0\rrbracket$ and $S \cap \llbracket\phi_1\rrbracket_{\overline{m}} \neq \emptyset$. Pick $y \in \llbracket\phi_0\rrbracket_{\overline{m}} \cap \llbracket\phi_1\rrbracket_{\overline{m}} \cap \llbracket\rho_0\rrbracket$ and add an edge from (x, Ψ, w) to $(y, \{\phi_0, \phi_1, \rho_0\}, \epsilon)$ to L, having $\mathsf{u}(\{\phi_0, \phi_1, \rho_0\}) \leq 3n - |\epsilon|$. Define the label $l(x, \Psi, w)$ of nodes (x, Ψ, w) to be just Ψ. Then the structure (V, L) indeed is a tableau for ρ_1: Since there is $z \in W \cap \llbracket\rho_1\rrbracket$, we have $(z, \{\rho_1\}, \epsilon) \in V$, that is, ρ_1 is contained in the label of some node in (V, L). The requirements of tableaux for matching rule applications are satisfied by construction of L. It remains to show that each trace in (V, L) is finite. So let $\tau = \phi_0, \phi_1, \ldots$ be a trace of some formula ϕ_0 along some branch (encoded by a word w) that is rooted at some node $v = (x, \Psi, w') \in V$ such that $\phi_0 \in \Psi$. Let $\overline{m}$ be the least timeout such that $x \in [\![\phi_0]\!]_{\overline{m}}$ (again, such $\overline{m}$ exists by Lemma 18). For each i, we have $\phi_{i+1} \in \delta(\phi_i, w(i))$ and there are $(x_i, \Psi_i, w_i) \in V$ and $\overline{m}_i$ such that $\phi_i \in \Psi_i$ and $x_i \in [\![\phi_i]\!]_{\overline{m}_i}$. We have chosen disjuncts and modal successors in a minimal fashion, that is, in such a way that we always have $\overline{m}_{i+1} \leq_l \overline{m}_i$. Since traces can be infinite only if they contain infinitely many unfolding steps for some least fixpoint, τ is finite or some least fixpoint is unfolded in it infinitely often. In the former case, we are done. In the latter case, each unfolding of a least fixpoint by Definition 17 reduces some digit of the timeout so that we have an infinite decreasing chain $\overline{m}_{i_1} >_l \overline{m}_{i_2} >_l \cdots$ with $i_{j+1} > i_j$ for all j, which is a contradiction to $<_l$ being a well-order.

For the converse direction, let (V, L) be a tableau for ρ_1 labeled by some function $l : V \to \mathcal{P}(\mathsf{cl})$. We construct a model (W, N, I) over the set

$$W = \{x \in V \mid l(x) \in \mathsf{states}\}.$$

For $p \in \mathsf{P}$, put $I(p) = \{x \in W \mid p \in l(x)\}$; since (V, L) is a tableau, the rules $(\bot)$ and $(\lightning)$ do not match the label of any node, so we have $I(p) = W \setminus I(\overline{p})$, as required. Let $x \in W$ and $a \in \mathsf{A}$. If $l(x)$ contains no a-box literal, then we put $N(a, x) = \{\emptyset\}$. If $l(x)$ contains some a-box literal but no a-diamond literal, then we put $N(a, x) = \emptyset$. Otherwise, let the a-modalities in $l(x)$ be exactly $\langle a \rangle \chi_1, \ldots, \langle a \rangle \chi_o, [a]\psi_1, \ldots, [a]\psi_m$. We put $N(a, x) = \{\{y_{1,1}, \ldots, y_{1,m}\}, \ldots, \{y_{o,1}, \ldots, y_{o,m}\}\}$, where, for $1 \leq i \leq o$, $1 \leq j \leq m$, the state $y_{i,j}$ is picked minimally with respect to tab among all nodes z such that $(x, z) \in L$ and $\{\chi_i, \psi_j, \rho_0\} = l(z)$; such $y_{i,j}$ with minimal tableau timeout always exists since (V, L) is a tableau. It remains to show that (W, N, I) is a ρ_0-model for ρ_1. We put

$$\widehat{[\![\phi]\!]} = \{x \in W \mid l(x) \vdash_{\mathsf{PL}} \phi\}$$

for $\phi \in \mathsf{cl}$, where $\vdash_{\mathsf{PL}}$ denotes *propositional entailment* (modal literals $[a]\phi$, $\langle a \rangle \phi$ are regarded as propositional atoms and $\eta X.\, \phi$ and $\phi[\eta X.\, \phi / X]$ entail each other). Since we have $W \subseteq \widehat{[\![\rho_0]\!]}$ and $W \cap \widehat{[\![\rho_1]\!]} \neq \emptyset$ by definition of (W, N, I) and tableaux, it suffices to show that we have $\widehat{[\![\phi]\!]} \subseteq [\![\phi]\!]$ for all $\phi \in \mathsf{cl}$. The proof of this is by induction on ϕ, using coinduction in the case for greatest fixpoint formulae and a further induction on tableau timeouts in the case for least fixpoint formulae. □

6 Satisfiability Games

We now define a game characterizing ρ_0-satisfiability of ρ_1. Player Eloise tries to establish the existence of a tableau for ρ_1 using only polynomially many supporting points for her reasoning. To this end, sequences of propositional reasoning steps are contracted into single Eloise-moves. Crucially, the limited branching of monotone modalities (and the ensuing limited need for tracking of deferrals) has a restricting effect on the nondeterminism that the game needs to take care of.

Definition 20 (Satisfiability games). We put $U = \{\Psi \subseteq \mathsf{cl} \mid 1 \leq |\Psi| \leq 2\}$ (note $|U| \leq n^2$) and $Q = \{\mathsf{Foc} \subseteq \mathsf{dfr} \mid |\mathsf{Foc}| \leq 2\}$. The *$\rho_0$-satisfiability game for ρ_1* is the Büchi game $G = (V, E, v_0, F)$ with set $V = V_\exists \cup V_\forall$ of nodes (with the union made disjoint by markers omitted in the notation) where $V_\exists = \{(\Psi, \mathsf{Foc}) \in U \times Q \mid \mathsf{Foc} \subseteq \Psi\}$ and $V_\forall = \mathsf{states} \times Q$, with initial node $v_0 = (\{\rho_1\}, \emptyset) \in V_\exists$, and with set $F = \{(\Psi, \mathsf{Foc}) \in V_\exists \mid \mathsf{Foc} = \emptyset\}$ of accepting nodes. The set E of moves is defined by

$$
\begin{aligned}
E(\Psi, \mathsf{Foc}) &= \{(\gamma(\Psi \cup \{\rho_0\}, w), \delta(\mathsf{Foc}, w)) \in \mathsf{states} \times Q \mid w \in (\Sigma_p)^*, |w| \leq 3n\} \\
E(\Gamma, \mathsf{Foc}) &= \{\, (\{\phi_0, \phi_1\}, \mathsf{Foc}') \in U \times Q \mid \{\langle a\rangle\phi_0, [a]\phi_1\} \subseteq \Gamma, \\
&\qquad \text{if } \mathsf{Foc} \neq \emptyset, \text{ then } \mathsf{Foc}' = \delta(\mathsf{Foc}, (\langle a\rangle\phi_0, [a]\phi_1)), \\
&\qquad \text{if } \mathsf{Foc} = \emptyset, \text{ then } \mathsf{Foc}' = \{\phi_0, \phi_1\} \cap \mathsf{dfr}\,\}
\end{aligned}
$$

for $(\Psi, \mathsf{Foc}) \in V_\exists$, $(\Gamma, \mathsf{Foc}) \in V_\forall$.

Thus, Eloise steers the propositional evolution of formula sets into formal states, keeping track of the focussed formulae, while Abelard picks an application of Lemma 2, and resets the focus set after it is *finished*, i.e. becomes $\emptyset$; Eloise wins plays in which the focus set is finished infinitely often.

Remark 21. It is crucial that while the game has exponentially many Abelard-nodes, there are only polynomially many Eloise-nodes. In fact, all Eloise-nodes (Ψ, Foc) have $\mathsf{Foc} \subseteq \Psi$, so the game has at most $4|U| \leq 4n^2$ Eloise-nodes.

Next we prove the correctness of our satisfiability games.

Theorem 22. *There is a tableau for ρ_1 if and only if* Eloise *wins G.*

Proof. Let s be a winning strategy for Eloise in G with which she wins every node in her winning region $\mathsf{win}_\exists$. For $v = (\Psi, \mathsf{Foc}) \in \mathsf{win}_\exists$, we let $w_{s(v)}$ denote a fixed propositional word such that $s(\Psi, \mathsf{Foc}) = (\gamma(\Psi, w_{s(v)}), \delta(\mathsf{Foc}, w_{s(v)}))$, that is, a witness word for the move of Eloise that s prescribes at v. We construct a tableau (W, L) over the set

$$
\begin{aligned}
W = \{(\gamma(\Psi \cup \{\rho_0\}, w'), \delta(\mathsf{Foc}, w')) \in \mathcal{P}(\mathsf{cl}) \times \mathcal{P}(\mathsf{dfr}) \mid \\
(\Psi, \mathsf{Foc}) \in \mathsf{win}_\exists, w' \text{ is a prefix of } w_{s(\Psi,\mathsf{Foc})}\}.
\end{aligned}
$$

We define the label $l(\Phi, \mathsf{Foc}')$ of nodes from $(\Phi, \mathsf{Foc}') \in W$ to be just Φ. Let $(\Phi, \mathsf{Foc}') = (\gamma(\Psi \cup \{\rho_0\}, w'), \delta(\mathsf{Foc}, w')) \in W$. If w' is a proper prefix of $w_{s(\Psi,\mathsf{Foc})}$, then we have $\Phi \notin \mathsf{states}$; let $l \in \Sigma_p$ be *the* letter such that (w', l) is a prefix of $w_{s(\Psi,\mathsf{Foc})}$ and add the pair $((\Phi, \mathsf{Foc}'), (\gamma(\Phi, l), \delta(\mathsf{Foc}', l)))$ to L. If $w' = w_{s(\Psi,\mathsf{Foc})}$, then we have $\Phi \in \mathsf{states}$. For $\{\langle a\rangle\phi, [a]\chi\} \subseteq \Phi$, we distinguish cases. If $\mathsf{Foc}' = \emptyset$, then put $\mathsf{Foc}'' = \{\phi, \chi\} \cap \mathsf{dfr}$; otherwise, put $\mathsf{Foc}'' = \delta(\mathsf{Foc}', (\langle a\rangle\phi, [a]\chi))$. Then add the pair $((\Phi, \mathsf{Foc}'), (\{\phi, \chi, \rho_0\}, \mathsf{Foc}''))$ to L, having $(\{\phi, \chi\}, \mathsf{Foc}'') \in \mathsf{win}_\exists$.

It remains to show that all traces in (W, L) are finite. So let $\tau = \phi_0, \phi_1, \ldots$ be a trace along some branch (encoded by a word w) that is rooted at some node $(\Phi, \mathsf{Foc}') \in W$ such that $\phi_0 \in \Phi$. By construction, this branch gives rise to an s-play $(\Psi_0, \mathsf{Foc}_0), (\Gamma_0, \mathsf{Foc}'_0), (\Psi_1, \mathsf{Foc}_1), (\Gamma_1, \mathsf{Foc}'_1), \ldots$ that starts at $(\Psi_0, \mathsf{Foc}_0) = (\Psi, \mathsf{Foc})$. Let i be the least position such that $\mathsf{Foc}'_i = \emptyset$ (i exists because s is a winning strategy). Since the ϕ_j are tracked along rule applications, we have $\phi_i \in \Gamma_i$; hence $\phi_{i+1} \in \mathsf{Foc}_{i+1}$. Let i' be the least position greater than i such that $\mathsf{Foc}'_{i'} = \emptyset$ (again, i' exists because s is a winning strategy). Between $(\Psi_{i+1}, \mathsf{Foc}_{i+1})$ and $(\Gamma_{i'}, \mathsf{Foc}'_{i'})$, all formulae from Foc_{i+1} (including ϕ_{i+1}) are transformed to a non-deferral by the formula manipulations encoded in w. In particular, the trace τ ends between $\mathsf{node}(\Psi_i, \mathsf{Foc}_i)$ and $\mathsf{node}(\Psi_{i'}, \mathsf{Foc}_{i'})$, and hence is finite.

For the converse direction, let (W, L) be tableau for ρ_1, labeled with $l : W \to \mathcal{P}(\mathsf{cl})$. We extract a strategy s for Eloise in G. A game node (Ψ, Foc) is *realized* if there is $v \in W$ such that $\Psi \subseteq l(v)$; then we say that v *realizes* the game node. For all realized game nodes (Ψ, Foc), we pick a realizing tableau node $v(\Psi, \mathsf{Foc})$ that is minimal with respect to tab among the tableau nodes that realize (Ψ, Foc). Then we construct a propositional word $w = l_0, l_1, \ldots$ as follows, starting with $\Psi_0 = \Psi \cup \{\rho_0\}$ and $v_0 = v(\Psi, \mathsf{Foc})$. For $i \geq 0$, pick some non-modal letter $l_i = (\phi, b)$ such that $\phi \in \Psi_i$, $b \in \{0, 1\}$ and such that $l(v_{i+1}) = \gamma(\Psi_i, l_i)$ where $v_{i+1} \in W$ is *the* node such that $(v_i, v_{i+1}) \in L$. Such a letter l_i exists since (W, L) is a tableau. By guardedness of fixpoint variables, this process will eventually terminate with a word $w = l_0, l_1, \ldots, l_m$ such that $m \leq 3n$, since Ψ_0 contains at most three formulae and each formula contains at most n unguarded operators. Put $s(\Psi, \mathsf{Foc}) = (\gamma(\Psi \cup \{\rho_0\}, w), \delta(\mathsf{Foc}, w))$, having $\gamma(\Psi \cup \{\rho_0\}, w) = l(v_m)$. It remains to show that s is a winning strategy. So let $\tau = (\Psi_0, \mathsf{Foc}_0), (\Gamma_0, \mathsf{Foc}'_0), (\Psi_1, \mathsf{Foc}_1), (\Gamma_1, \mathsf{Foc}'_1), \ldots$ be an s-play, where $\Psi_0 = \{\rho_1\}$ and $\mathsf{Foc}_0 = \emptyset$. It suffices to show that for all i such that $\mathsf{Foc}_i \neq \emptyset$, there is $j \geq i$ such that $\mathsf{Foc}_j = \emptyset$. So let $\mathsf{Foc}_i \neq \emptyset$ and let w_i be the word that is constructed in the play from (Ψ_i, Foc_i) on. Since (W, L) is a tableau and since s has been constructed using realizing nodes with minimal tableau timeouts, all traces of formulae from Foc_i along the branch that is encoded by w_i are finite. Let j be the least number such that all such traces have ended after $2j$ further moves from (Ψ_i, Foc_i). Then we have $\mathsf{Foc}'_{i+j} = \emptyset$, as required. □

Corollary 23. *Every satisfiable formula of size n in the alternation-free monotone μ-calculus with the universal modality has a model of size at most $4n^2$.*

Corollary 24. *The satisfiability checking problem for the alternation-free monotone μ-calculus with the universal modality is in* NP *(hence* NP*-complete).*

Proof. Guess a winning strategy s for Eloise in G and verify that s is a winning strategy. Verification can be done in polynomial time since the structure obtained from G by imposing s is of polynomial size and since the admissibility of single moves can be checked in polynomial time. □

By the translations recalled in Example 7, we obtain moreover

Corollary 25. *Satisfiability-checking in concurrent propositional dynamic logic CPDL and in the alternation-free fragment of game logic is in* NP *(hence* NP*-complete).*

7 Conclusion

We have shown that satisfiability checking in the alternation-free fragment of the monotone μ-calculus with the universal modality is only NP-complete, even when formula size is measured as the cardinality of the closure. Via straightforward translations (which have only quadratic blow-up under the mentioned measure of formula size), it follows that both concurrent propositional dynamic logic (CPDL) and the alternation-free fragment of game logic are also only NP-complete under their original semantics, i.e. with atomic programs interpreted as neighbourhood structures (they become ExpTime-complete when atomic programs are interpreted as relations). We leave as an open problem whether the upper bound NP extends to the full monotone μ-calculus, for which the best known upper bound thus remains ExpTime, by results on the coalgebraic μ-calculus [5], or alternatively by the translation into the relational μ-calculus that we give in the proof of the finite model property (Lemma 8).

References

1. Baader, F., Calvanese, D., McGuinness, D.L., Nardi, D., Patel-Schneider, P.F. (eds.): The Description Logic Handbook. Cambridge University Press, Cambridge (2003)
2. Bradfield, J., Stirling, C.: Modal μ-calculi. In: Handbook of Modal Logic, pp. 721–756. Elsevier (2006)
3. Bruse, F., Friedmann, O., Lange, M.: On guarded transformation in the modal μ-calculus. Log. J. IGPL **23**(2), 194–216 (2015)
4. Chellas, B.F.: Modal Logic. Cambridge University Press, Cambridge (1980)
5. Cîrstea, C., Kupke, C., Pattinson, D.: EXPTIME tableaux for the coalgebraic μ-calculus. Log. Methods Comput. Sci. **7**(3:03), 1–33 (2011)
6. Donini, F.: Complexity of reasoning. In: Baader, F., et al. [1], pp. 96–136
7. Enqvist, S., Hansen, H.H., Kupke, C., Marti, J., Venema, Y.: Completeness for game logic. In: Logic in Computer Science (LICS 2019), pp. 1–13. IEEE (2019)
8. Enqvist, S., Seifan, F., Venema, Y.: Monadic second-order logic and bisimulation invariance for coalgebras. In: Logic in Computer Science (LICS 2015), pp. 353–365. IEEE (2015)
9. Fischer, M., Ladner, R.: Propositional dynamic logic of regular programs. J. Comput. Syst. Sci. **18**, 194–211 (1979)
10. Friedmann, O., Lange, M.: Deciding the unguarded modal μ-calculus. J. Appl. Non-Classical Log. **23**, 353–371 (2013)
11. Goranko, V., Passy, S.: Using the universal modality: gains and questions. J. Log. Comput. **2**, 5–30 (1992)
12. Hansen, H., Kupke, C.: A coalgebraic perspective on monotone modal logic. In: Adámek, J., Milius, S. (eds.) Coalgebraic Methods in Computer Science (CMCS 2004) of ENTCS, vol. 106, pp. 121–143. Elsevier (2004)

13. Hausmann, D., Schröder, L., Egger, C.: Global caching for the alternation-free coalgebraic μ-calculus. In: Concurrency Theory (CONCUR 2016) of LIPIcs, vol. 59, pp. 34:1–34:15. Schloss Dagstuhl - Leibniz-Zentrum für Informatik (2016)
14. Kracht, M., Wolter, F.: Normal monomodal logics can simulate all others. J. Symb. Log. **64**(1), 99–138 (1999)
15. Ladner, R.: The computational complexity of provability in systems of modal propositional logic. SIAM J. Comput. **6**, 467–480 (1977)
16. Lange, M., Stirling, C.: Focus games for satisfiability and completeness of temporal logic. In: Logic in Computer Science (LICS 2001), pp. 357–365. IEEE Computer Society (2001)
17. Mazala, R.: Infinite games. In: Grädel, E., Thomas, W., Wilke, T. (eds.) Automata Logics, and Infinite Games. LNCS, vol. 2500, pp. 23–38. Springer, Heidelberg (2002). https://doi.org/10.1007/3-540-36387-4_2
18. Miyano, S., Hayashi, T.: Alternating finite automata on ω-words. Theory Comput. Sci. **32**, 321–330 (1984)
19. Niwinski, D., Walukiewicz, I.: Games for the μ-calculus. Theory Comput. Sci. **163**, 99–116 (1996)
20. Parikh, R.: The logic of games and its applications. Ann. Discrete Math. **24**, 111–140 (1985)
21. Pauly, M.: Bisimulation for general non-normal modal logic. Unpublished manuscript (1999)
22. Pauly, M.: Logic for social software. Ph.D. thesis, Universiteit van Amsterdam (2001)
23. Pauly, M., Parikh, R.: Game logic - an overview. Stud. Log. **75**(2), 165–182 (2003)
24. Peleg, D.: Concurrent dynamic logic. J. ACM **34**, 450–479 (1987)
25. Pratt, V.: A decidable mu-calculus: preliminary report. In: Foundations of Computer Science (FOCS 1981), pp. 421–427. IEEE Computer Society (1981)
26. van Benthem, J., Bezhanishvili, N., Enqvist, S.: A propositional dynamic logic for instantial neighborhood semantics. Stud. Log. **107**(4), 719–751 (2019)
27. Vardi, M.: On the complexity of epistemic reasoning. In: Logic in Computer Science (LICS 1989), pp. 243–252. IEEE Computer Society (1989)

Soft Subexponentials and Multiplexing

Max Kanovich[1,2], Stepan Kuznetsov[3,2(✉)], Vivek Nigam[5,4], and Andre Scedrov[6,2]

[1] University College London, London, UK
m.kanovich@ucl.ac.uk
[2] National Research University Higher School of Economics, Moscow, Russia
[3] Steklov Mathematical Institute of RAS, Moscow, Russia
sk@mi.ras.ru
[4] Federal University of Paraíba, João Pessoa, Brazil
nigam@fortiss.org
[5] fortiss GmbH, Munich, Germany
[6] University of Pennsylvania, Philadelphia, USA
scedrov@math.upenn.edu

Abstract. Linear logic and its refinements have been used as a specification language for a number of deductive systems. This has been accomplished by carefully studying the structural restrictions of linear logic modalities. Examples of such refinements are subexponentials, light linear logic, and soft linear logic. We bring together these refinements of linear logic in a non-commutative setting. We introduce a non-commutative substructural system with subexponential modalities controlled by a minimalistic set of rules. Namely, we disallow the contraction and weakening rules for the exponential modality and introduce two primitive subexponentials. One of the subexponentials allows the multiplexing rule in the style of soft linear logic and light linear logic. The second subexponential provides the exchange rule. For this system, we construct a sequent calculus, establish cut elimination, and also provide a complete focused proof system. We illustrate the expressive power of this system by simulating Turing computations and categorial grammar parsing for compound sentences. Using the former, we prove undecidability results. The new system employs Lambek's non-emptiness restriction, which is incompatible with the standard (sub)exponential setting. Lambek's restriction is crucial for applications in linguistics: without this restriction, categorial grammars incorrectly mark some ungrammatical phrases as being correct.

1 Introduction

For the specification of deductive systems, linear logic [4,5], and a number of refinements of linear logic have been proposed, such as commutative [23,25] and non-commutative [11,12] subexponentials, light linear logic [7], soft linear logic [16], and easy linear logic [10]. The key difference between these refinements is their treatment of the linear logic exponentials, !, ?. These refinements

N. Peltier and V. Sofronie-Stokkermans (Eds.): IJCAR 2020, LNAI 12166, pp. 500–517, 2020.
https://doi.org/10.1007/978-3-030-51074-9_29

allow, *e.g.*, a finer control on the structural rules, *i.e.*, weakening, contraction and exchange rules, and how exponentials affect the sequent antecedent. For example [12], we proposed a logical framework with commutative and non-commutative subexponentials, applying it for applications in type-logical grammar. In particular, we demonstrated that this logical framework can be used to "type" correctly sentences that were not able before with previous logical frameworks, such as Lambek calculus.

However, as we have shown recently, our logical framework in [12] is incompatible with the important Lambek's non-emptiness property [15]. This property, which requires all antecedents to be non-empty, is motivated by linguistic applications of Lambek-like calculi. Namely, it prevents the system from recognizing ("typing") incorrectly formed sentences as grammatically correct. We discuss these linguistic issues in detail in Sect. 3. The lack of Lambek's restriction means that the logical framework proposed in our previous work is too expressive, typing incorrectly sentences.

To address this problem, we propose a new non-commutative proof system, called SLLM, that admits Lambek's non-emptiness condition and at the same time is expressive enough to type correctly sentences in our previous work. This system is also still capable of modelling computational processes, as we show in Sect. 5.1 on the example of Turing computations.

In particular, SLLM takes inspiration from the following refinements of linear logic: subexponentials, by allowing two types of subexponentials, ! and ∇; soft linear logic, which contributes a version of the multiplexing rule, $!_L$, shown below to the left; and light linear logic, which contributes the two right subexponentials rules, $!_R, \nabla_R$, shown below to the right.

$$\frac{\Gamma, F, \ldots, F, \Delta \to G}{\Gamma, !F, \Delta \to G}\; !_L \qquad \frac{F \to G}{!F \to !G}\; !_R \qquad \frac{F \to G}{\nabla F \to \nabla G}\; \nabla_R$$

In our version of the system, the premise of the rule $!_L$ does not allow zero instances of F. Hence, ! is a relevant subexponential as discussed in [12].

This rule is used to type sentences correctly, while the rules $!_R$ and ∇_R are used to maintain Lambek's condition. SLLM contains, therefore, soft subexponentials and multiplexing.

Our main contributions are summarized below:

- **Admissibility of Cut Rule:** We introduce the proof system SLLM in Sect. 2. We also prove that it has basic properties, namely admissibility of Cut Rule and the substitution property. The challenge is to ensure a reasonable balance between the expressive power of systems and complexity of their implementation, and in particular, to circumvent the difficulties caused by linear logic contraction and weakening rules.
- **Lambek's Non-Emptiness Condition:** We demonstrate in Sect. 3 that SLLM (and thus also SLLMF) admits Lambek's non-emptiness condition. This means that SLLM cannot be used to "type" incorrect sentences.
  We demonstrate this by means of some examples.

- **Focused Proof System:** We introduce in Sect. 4 a focused proof system (SLLMF) proving that it is sound and complete with respect to SLLM. The focused proof system differs from the focused proof system in our previous work [12] by allowing a subexponential that can contract, but not weaken nor be exchanged. Such subexponentials were not allowed in the proof system introduced in [12]. A key insight is to keep track on when a formula necessarily has to be introduced in a branch and when not.
- **Complexity:** We investigate in Sect. 5 the complexity of SLLM. We first demonstrate that the provability problem for SLLM is undecidable in general and identify a fragment for which it is decidable.

Finally, in Sect. 6, we conclude by pointing to related and future work.

Table 1. Lambek Calculus: A non-commutative version of ILL

$$\frac{}{A \to A}\ I$$

$$\frac{\Phi \to A \quad \Sigma_1, B, \Sigma_2 \to C}{\Sigma_1, \Phi, A \setminus B, \Sigma_2 \to C}\ \setminus_L \qquad \frac{A, \Sigma \to B}{\Sigma \to A \setminus B}\ \setminus_R\ (\Sigma \text{ is not empty})$$

$$\frac{\Phi \to A \quad \Sigma_1, B, \Sigma_2 \to C}{\Sigma_1, B \,/\, A, \Phi, \Sigma_2 \to C}\ /_L \qquad \frac{\Sigma, A \to B}{\Sigma \to B \,/\, A}\ /_R\ (\Sigma \text{ is not empty})$$

$$\frac{\Sigma_1, A, B, \Sigma_2 \to C}{\Sigma_1, A \cdot B, \Sigma_2 \to C}\ \cdot_L \qquad \frac{\Sigma_1 \to A \quad \Sigma_2 \to B}{\Sigma_1, \Sigma_2 \to A \cdot B}\ \cdot_R$$

Table 2. SLLM: Lambek calculus with multiplexing.

$$\frac{}{A \to A}\ I$$

$$\frac{\Phi \to A \quad \Sigma_1, B, \Sigma_2 \to C}{\Sigma_1, \Phi, A \setminus B, \Sigma_2 \to C}\ \setminus_L \qquad \frac{A, \Sigma \to B}{\Sigma \to A \setminus B}\ \setminus_R\ (\Sigma \text{ is not empty})$$

$$\frac{\Phi \to A \quad \Sigma_1, B, \Sigma_2 \to C}{\Sigma_1, B \,/\, A, \Phi, \Sigma_2 \to C}\ /_L \qquad \frac{\Sigma, A \to B}{\Sigma \to B \,/\, A}\ /_R\ (\Sigma \text{ is not empty})$$

$$\frac{\Sigma_1, A, B, \Sigma_2 \to C}{\Sigma_1, A \cdot B, \Sigma_2 \to C}\ \cdot_L \qquad \frac{\Sigma_1 \to A \quad \Sigma_2 \to B}{\Sigma_1, \Sigma_2 \to A \cdot B}\ \cdot_R$$

$$\frac{\Gamma_1, \overbrace{A, A, \ldots, A}^{k \text{ times}}, \Gamma_2 \to C}{\Gamma_1, !A, \Gamma_2 \to C}\ !_L\ \underline{(k \geq 1)} \qquad \frac{A \to C}{!A \to !C}\ !_R$$

$$\frac{\Gamma_1, A, \Gamma_2 \to C}{\Gamma_1, \nabla A, \Gamma_2 \to C}\ \nabla_L \qquad \frac{A \to C}{\nabla A \to \nabla C}\ \nabla_R$$

$$\frac{\Gamma_1, \Gamma_2, \nabla A, \Gamma_3 \to C}{\Gamma_1, \nabla A, \Gamma_2, \Gamma_3 \to C} \quad \text{and} \quad \frac{\Gamma_1, \nabla A, \Gamma_2, \Gamma_3 \to C}{\Gamma_1, \Gamma_2, \nabla A, \Gamma_3 \to C}\ \nabla_E$$

2 The Non-commutative System SLLM with Multiplexing

As the non-commutative source, we take the Lambek calculus [17], Table 1, the well-known fundamental system in linguistic foundations.

The proof system introduced here, SLLM, extends the Lambek calculus by adding two new connectives (subexponentials) ! and ∇ and their rules in Table 2.

Drawing inspiration from commutative logics such as *linear logic* [6], *light linear logic* [7], *soft linear logic* [16], and *easy linear logic* [10], here we introduce our primitive non-commutative modalities $!A$ and ∇A controlled by *a minimalistic set of rules*.

Multiplexing Rule (local):

$$\frac{\Gamma_1, \overbrace{A, A, \ldots, A}^{k \text{ times}}, \Gamma_2 \to C}{\Gamma_1, !A, \Gamma_2 \to C}\ !_L\ \underline{(k \geq 1)} \tag{1}$$

Informally, $!A$ stands for: "*any* *positive number* of copies of A *at the same position*".

Remark 1. In contrast to soft linear logic and light linear logic, where Weakening is one of the necessary ingredients, here we exclude the Weakening case: $(k = 0)$.

Unlike Contraction rule that can be recursively reused, and $!A$ keeps the subexponential (it is introduced by Dereliction), our Multiplexing can only be used *once* with all copies provided immediately at the same time in one go, and the subexponentials get removed. Thus, if one wishes to reuse Multiplexing further in the proof, nested subexponentials would be needed (like $!!A$ for two levels of Multiplexing).

The second subexponential, ∇A, provides the exchange rule.
Exchange Rule (non-local):

$$\frac{\Gamma_1, \Gamma_2, \nabla A, \Gamma_3 \to C}{\Gamma_1, \nabla A, \Gamma_2, \Gamma_3 \to C} \quad \text{and} \quad \frac{\Gamma_1, \nabla A, \Gamma_2, \Gamma_3 \to C}{\Gamma_1, \Gamma_2, \nabla A, \Gamma_3 \to C}\ \nabla_E \tag{2}$$

and Dereliction Rule (local):

$$\frac{\Gamma_1, A, \Gamma_2 \to C}{\Gamma_1, \nabla A, \Gamma_2 \to C}\ \nabla_L \tag{3}$$

"∇A can be thought of as storing a missing candidate A in a fixed local storage, with the ability to deliver A *to the right place* at the appropriate time."

Remark 2. Notwithstanding that the traditional Promotion rule $\frac{\Gamma \to C}{!\Gamma \to !C}$ is accepted in linear logic as well as in soft linear logic, we confine ourselves to the restricted light linear logic promotion $\frac{A \to C}{!A \to !C}$, in order to guarantee cut admissibility for the non-commutative SLLM (cf. [11]). E.g., with $\frac{\Gamma \to C}{!\Gamma \to !C}$ the sequent

$$!B,\ !(B\backslash C) \to (C \cdot C) \tag{4}$$

is derivable by cut

$$\dfrac{\dfrac{B, B \setminus C \to C}{!B, !(B \setminus C) \to !C} \quad !C \to C \cdot C}{!B, !(B \setminus C) \to C \cdot C}\ Cut$$

but, to finalize a cut-free proof for (4), we need commutativity:

$$\dfrac{\dfrac{\dfrac{B \to B \quad B, C, B \setminus C \to C \cdot C \ \text{[ok with } C, B, B \setminus C \to C \cdot C]}{B, B, B \setminus C, B \setminus C \to C \cdot C}\ !_L}{B, B, !(B \setminus C) \to C \cdot C}\ !_L}{!B, !(B \setminus C) \to C \cdot C}$$

The following theorem states that SLLM enjoys cut admissibility and the substitution property.

Theorem 1.

(a) *The calculus* SLLM *enjoys admissibility of the Cut Rule:*

$$\dfrac{\Pi \to A \quad \Gamma_1, A, \Gamma_2 \to C}{\Gamma_1, \Pi, \Gamma_2 \to C}\ Cut \tag{5}$$

(b) *Given an atomic p, let the sequent $\Gamma(p) \to C(p)$ be derivable in the calculus. Then for any formula B, $\Gamma(B) \to C(B)$ is also derivable in the calculus. Here by $\Gamma(B)$ and $C(B)$ we denote the result of replacing all occurrences of p by B in $\Gamma(p)$ and $C(p)$, resp.*

Proof.

(a) By reductions. E.g.,

$$\dfrac{\dfrac{B \to A}{!B \to !A}\ !_R \quad \dfrac{\Gamma_1, \overbrace{A, A, \ldots, A}^{k \text{ times}}, \Gamma_2 \to C}{\Gamma_1, !A, \Gamma_2 \to C}\ !_L}{\Gamma_1, !B, \Gamma_2 \to C}\ Cut$$

is reduced to

$$\dfrac{\dfrac{B \to A \quad \Gamma_1, A, A, \ldots, A, \Gamma_2 \to C}{\Gamma_1, B, B, \ldots, B, \Gamma_2 \to C}\ Cut\ (k \text{ times})}{\Gamma_1, !B, \Gamma_2 \to C}\ !_L$$

(b) By induction.

3 Linguistic Motivations

In this Section we illustrate how (and why) our modalities provide parsing complex and compound sentences in a natural language.

We start with standard examples, which go back to Lambek [17]; for in-depth discussion of linguistic matters we refer to standard textbooks [3,19,21].

The sentence "*Bob sent the letter yesterday*" is grammatical, and the following "type" specification is provable in Lambek calculus, Table 1.

$$N, (N \setminus S) / N, N, V \setminus V \to S$$

Here the 'syntactical type' N stands for nouns "the letter" and "Bob", and $((N\backslash S)/N)$, i.e., (V/N), for the transitive verb "sent", and $(V\backslash V)$ for the verb modifier "yesterday", where $V = (N\backslash S)$. The whole sentence is of type S.

Lambek's non-emptiness restriction is important for correctness of Lambek's approach to modelling natural language syntax. Without this restriction Lambek grammars *overgenerate,* that is, recognize ungrammatical phrases as if they were correct. The standard example [19, § 2.5] is as follows: *"very interesting book"* is a correct noun phrase, established by the following derivation:

$$(N / N) / (N / N), N / N, N \to N.$$

The sequent above is derivable in the Lambek calculus. Without Lambek's restriction, however, one can also derive

$$(N / N) / (N / N), N \to N$$

(since $\to N / N$ is derivable with an empty antecedent). This effect is unwanted, since the corresponding phrase *"very book"* is ungrammatical. Thus, Lambek's non-emptiness restriction is a highly desired property for linguistic applications.

Fortunately, SLLM enjoys Lambek's non-emptiness property. That is:

Theorem 2. *The calculus* SLLM *provides Lambek's non-emptiness restriction: If a sequent $\Gamma \to C$ is derivable in the calculus then the list of formulas Γ is not empty.*

Proof. The crucial point is that, in the absence of Weakening, $!A$ never happens to produce the empty list. □

Theorems 1 and 2 show how our new system SLLM resolves the issues discussed in [15] for the case of the Lambek calculus extended with a full-power exponential in linear logic style. Namely, as we show in that article, no reasonable extension of the Lambek calculus with the exponential modality can have the three properties simultaneously:

- cut elimination;
- substitution;
- Lambek's restriction.

Moreover, as also shown in [15], for the one-variable fragment the same happens to the *relevant* subexponential, which allows contraction and permutation, but not weakening. Our new system overcomes these issues by refining the rules for subexponentials.

Now let us show how one can use subexponentials of SLLM to model more complicated sentences. Our analysis shares much with that of Morrill and Valentín [22]. Unlike ours, the systems in [22] do not enjoy Lambek's restriction.

(1) The noun phrase: "*the letter that Bob sent yesterday,*" is grammatical, so that its "type" specification (6) should be provable in Lambek calculus or alike:

$$N, ((N \setminus N) / S'), N, (V / N), (V \setminus V) \to N \qquad (6)$$

Here $((N \setminus N) / S'$ stands for a subordinating connective "that".
As a type for the whole dependent clause,

"*Bob sent _ yesterday,*"

we take some S', not a full S, because the direct object, "the letter", is missing.
Our solution refines the approach of [2,20]. We mark the missing item, the direct object "the letter" of type N, by a specific formula ∇N stored *at a fixed local position* and by means of Exchange Rule (2) and Dereliction Rule (3) deliver the missing N to the right place with providing

$$N, (V / N), N, (V \setminus V) \to S,$$

which is the type specification for the sentence "*Bob sent the letter yesterday*" completed with the direct object, "the letter".
By taking $S' = (\nabla N \setminus S)$, we can prove (6):

"*Bob sent the letter yesterday*"

$$\dfrac{\dfrac{\dfrac{N, V / N, N, V \setminus V \to S}{\nabla N, N, V / N, V \setminus V \to S}\ \nabla_L, \nabla_E}{N, V / N, V \setminus V \to S'} \qquad N, N \setminus N \to N}{N, (N \setminus N) / S', N, V / N, V \setminus V \to N}$$

Remark 3. If we allowed Weakening, we would prove the ungrammatical "*the letter that Bob sent the letter yesterday,*"

(2) The noun phrase: "*the letter that Bob sent without reading*" is grammatical despite two missing items: the direct object to "sent" and the direct object to "reading", resp. Hence, the corresponding "type" specification (7) should be provable in Lambek calculus or alike:

$$N, ((N \setminus N) / S''), \Delta_1, \Delta_2 \to N \qquad (7)$$

Here N stands for "the letter", and $((N \setminus N) / S'')$ for "that", and some Δ_1 for "Bob sent", and Δ_2 for "without reading".
As a type for the whole dependent clause,

"*Bob sent __ without reading __*"

we have to take some S'', not a full S, to respect the fact that this time two items are missing: the direct objects to "sent" and "reading", resp.
To justify "*the letter that Bob sent _ without reading _*" with its (7), in addition to the ∇-rules, we invoke Multiplexing Rule (1).
The correlation between the *full* S and S'' *with its multiple holes* is given by:

$$S'' = ((!\nabla N) \setminus S) \tag{8}$$

As compared with the previous case of S' representing one missing item ∇N, the S'' here is dealing with $!\nabla N$ which provides two copies of ∇N to represent two missing items.
Then the proof for (7) is as:

✓ "*Bob sent the letter without reading it*"

$$\dfrac{\dfrac{\dfrac{\dfrac{\Delta_1, N, \Delta_2, N \to S}{\nabla N, \nabla N, \Delta_1, \Delta_2 \to S}\ \nabla_L, \nabla_E}{!\nabla N, \Delta_1, \Delta_2 \to S}\ !_L}{\Delta_1, \Delta_2 \to S''} \qquad N, N \setminus N \to N}{N, (N \setminus N) / S'', \Delta_1, \Delta_2 \to N}$$

4 Focused Proof System

This section introduces a sound and complete focused proof system SLLMF for SLLM. Focusing [1] is a discipline on proofs that reduces proof search space. We take an intermediate step, by first introducing the proof system SLLM# that handles the non-determinism caused by the multiplexing rule.

4.1 Handling Local Contraction

For bottom-up proof-search, the multiplexing rule

$$\dfrac{\Gamma, F, \ldots, F, \Delta \to G}{\Gamma, !F, \Delta \to G}$$

has a great deal of *don't know non-determinism* as one has to decide how many copies of F appears in its premise. This decision affects provability as each formula has to be necessarily be used in the proof, *i.e.*, they cannot be weakened.

To address this problem, we take a lazy approach by using two new connectives $\sharp^*$ and $\sharp^+$. The formula $\sharp^* F$ denotes zero or more local copies of F, and $\sharp^+ F$ one of more copies of F. We construct the proof system SLLM# from SLLM as follows. It contains all rules of SLLM, except the rule $!_L$, which is replaced by the following rules

$$\frac{\Gamma, \sharp^+ F, \Delta}{\Gamma, !F, \Delta \to G} \quad \frac{\Gamma, F, \sharp^* F, \Delta \to G}{\Gamma, \sharp^+ F, \Delta \to G} \quad \frac{\Gamma, F, \Delta \to G}{\Gamma, \sharp^* F, \Delta \to G} \quad \frac{\Gamma, \Delta \to G}{\Gamma, \sharp^* F, \Delta \to G}$$

Notice that there is no need for explicit contraction and only $\sharp^*$ allows for weakening. We accommodate contraction into the introduction rules, namely, by modifying the rules where there is context splitting, such as in the rules $\backslash_L$. In particular, one should decide in which branch a formula C is necessarily be used. This is accomplished by using adequately $\sharp^+ F$ and $\sharp^* F$. For example, some rules for $\backslash_L$ are shown below:

$$\frac{\sharp^* C, \Gamma_2 \to F \quad \Gamma_1, \sharp^+ C, G, \Gamma_3 \to H}{\Gamma_1, \sharp^+ C, \Gamma_2, F \backslash G, \Gamma_3 \to H}$$

$$\frac{\sharp^+ C, \Gamma_2 \to F \quad \Gamma_1, \sharp^* C, G, \Gamma_3 \to H}{\Gamma_1, \sharp^+ C, \Gamma_2, F \backslash G, \Gamma_3 \to H}$$

$$\frac{\sharp^* C, \Gamma_2 \to F \quad \Gamma_1, \sharp^* C, G, \Gamma_3 \to H}{\Gamma_1, \sharp^* C, \Gamma_2, F \backslash G, \Gamma_3 \to H}$$

In the first rule, the C is necessarily used in the right premise, while in the left premise one can chose to use C or not. SLLM# contains similar symmetric rules where C is necessarily moved to the left premise. Also it contains the corresponding rules for $/_L$.

SLLM# also include the following more refined right-introduction rules for ! and ∇. where Γ_1^*, Γ_2^* are lists containing only formulas of the form $\sharp^* H$.

$$\frac{F \to G}{\Gamma_1^*, !F, \Gamma_2^* \to !G} \quad \frac{F \to G}{\Gamma_1^*, \nabla F, \Gamma_2^* \to \nabla G}$$

Notice how the decision that all formulas in Γ_1, Γ_2 represent zero copies is made in the rules above.

Theorem 3. *Let Γ, G be a list of formulas not containing $\sharp^*$ nor $\sharp^+$. A sequent $\Gamma \to G$ is provable in the* SLLM *if and only if it is provable in* SLLM#.

Completeness follows from straightforward proof by induction on the size of proofs. One needs to slightly generalize the inductive hypothesis. Soundness follows from the fact that contractions are local and can be permuted below every other rule.

4.2 Focused Proof System

First proposed by Andreoli [1] for Linear Logic, focused proof systems reduce proof search space by distinguishing rules which have don't know non-determinism, classified as *positive*, from rules which have don't care non-determinism, classified as *negative*. We classify the rules $\cdot_R, \backslash_L, /_L$ as positive and the rules $\cdot_L, \backslash_R, /_R$ as negative. Non-atomic formulas of the form $F \cdot G$ are classified as positive, $\nabla F, \sharp^* F, \sharp^+ F$ and $!F$ are classified as modal formulas, and formulas of the form $F \backslash G$ and F / G as negative.

The focused proof system manipulates the following types of sequents, where Γ_1 and Γ_2 are possibly empty lists of non-positive formulas, Θ is *multiset* of formulas, and N_N is a non-negative formula. Intitively, Θ will contain all formulas of the form ∇F. As they allow exchange rule, their collection can be treated as a multiset. Γ_1, Γ_2 contain formulas that cannot be introduced by negative rules.

- **Negative Phase:** $\Theta : [\Gamma_1], \Uparrow\Delta, [\Gamma_2] \rightarrow [N_N]$ and $\Theta : [\Gamma_1], \Uparrow\Delta, [\Gamma_2] \rightarrow \Uparrow F$. Intuitively, the formulas in Δ and F are eagerly introduced whenever they negative rules are applicable, as one does not lose completeness in doing so.
- **Border:** $\Theta : [\Gamma_1] \rightarrow [N_N]$. These are sequents for which no negative rule can be applied. At this moment, one has to decide on a formula starting a positive phase.
- **Positive Phase:** $\Theta : [\Gamma_1], \Downarrow F, [\Gamma_2] \rightarrow [N_N]$ and $\Theta : [\Gamma_1] \rightarrow \Downarrow F$, where only the formula F is focused on.

The focused proof system SLLMF is composed by the rules in Figs. 1 and 2. Intuitively, reaction rules R_{L1}, R_{L2}, R_R and negative phase rules are applied until no more rules are applicable. Then a decision rule is applied which focuses on one formula. One needs, however, to be careful on whether the focused formula's main connective is $\sharp^+$ or $\sharp^*$. If it is the former, then we have committed to one copy of the formula and therefore, it can be modified to be $\sharp^*$, while the latter does not change.

The number of rules in Fig. 2 simply reflects the different cases emerging due to the presence or not of formulas whose main connective is $\sharp^+$ or $\sharp^*$. For example, $\backslash_{L2}$ considers the case when the splitting of the context occurs exactly on a formula $\sharp^+ C$. In this case, the decision is to commit to use a copy of C in the right-premise, thus containing $\sharp^+ C$, while $\sharp^* C$ on the left-premise. The other rules follow the same reasoning.

Finally, notice that in the rules $I, !_R$ and ∇_R the context may contain formulas with main connective $\sharp^*$ in their conclusion, but not in their premise. This illustrates the lazy decision of how many copies of a formula are needed.

Theorem 4. *A sequent $\Gamma \rightarrow G$ is provable in* SLLM# *if and only if the sequent* $\cdot : \Uparrow\Gamma \rightarrow \Uparrow G$ *is provable in* SLLMF.

The proof of this theorem follows the same ideas as detailed in [26] and [12].

Corollary 1. *Let Γ, G be a list of formulas not containing $\sharp^*$ nor $\sharp^+$. A sequent $\Gamma \rightarrow G$ is provable in* SLLM *if and only if the sequent* $\cdot : \Uparrow\Gamma \rightarrow \Uparrow G$ *is provable in* SLLMF.

Remark 4. The focused proof system introduced above enables the use of more sophisticated search mechanisms. For example, lazy methods can reduce the non-determinism caused by the great number of introduction rules caused by managing $\sharp^*, \sharp^+$, *e.g.*, Fig. 2.

Positive Phase Rules

$$\frac{}{\cdot : [\Gamma_1^*, A, \Gamma_2^*] \to \Downarrow A}\; I \qquad \frac{}{\cdot : [\Gamma_1^*, \sharp^{+/*}A, \Gamma_2^*] \to \Downarrow A}\; I \qquad \frac{\Theta_1 : [\Gamma_1] \to \Downarrow F \quad \Theta_2 : [\Gamma_1] \to \Downarrow G}{\Theta_1, \Theta_2 : [\Gamma_1, \Gamma_2] \to \Downarrow F \cdot G}\; \cdot_R$$

$$\frac{\cdot : \Uparrow F \to \Uparrow G}{\cdot : [\Gamma_1^*, !F, \Gamma_2^*] \to \Downarrow !G}\; !_R \qquad \frac{\cdot : \Uparrow F \to \Uparrow G}{F : [\Gamma^*] \to \Downarrow \nabla G}\; \nabla_R$$

Negative Phase Rules

$$\frac{\Theta : [\Gamma_1], \Uparrow F, G, \Delta, [\Gamma_2] \to \mathcal{R}}{\Theta : [\Gamma_1], \Uparrow F \cdot G, \Delta, [\Gamma_2] \to \mathcal{R}}\; \cdot_L \qquad \frac{\Theta : \Uparrow F, [\Gamma] \to \Uparrow G}{\Theta : [\Gamma] \to \Uparrow F \backslash G}\; \backslash_R \qquad \frac{\Theta : [\Gamma], \Uparrow G \to \Uparrow F}{\Theta : [\Gamma] \to \Uparrow F / G}\; /_R$$

Decision and Reaction Rules

$$\frac{\Theta : [\Gamma_1], \Uparrow N_P^{\nabla}, [\Gamma_2] \to [G]}{\Theta : [\Gamma_1], \Downarrow N_P^{\nabla}, [\Gamma_2] \to [G]}\; R_L \qquad \frac{\Theta : [\Gamma] \to \Uparrow N_{NA}}{\Theta : [\Gamma] \to \Downarrow N_{NA}}\; R_R$$

$$\frac{\Theta : [\Gamma_1, N_P], \Uparrow \Delta, [\Gamma_2] \to \mathcal{R}}{\Theta : [\Gamma_1], \Uparrow N_P, \Delta, [\Gamma_2] \to \mathcal{R}}\; \Uparrow_{L1} \qquad \frac{\Theta, F : [\Gamma_1], \Uparrow \Delta, [\Gamma_2] \to \mathcal{R}}{\Theta : [\Gamma_1], \Uparrow \nabla F, \Delta, [\Gamma_2] \to \mathcal{R}}\; \Uparrow_{L2}$$

$$\frac{\Theta : [\Gamma] \to [N_N]}{\Theta : [\Gamma] \to \Uparrow N_N}\; R_R \qquad \frac{[\Gamma] \to \Downarrow G}{[\Gamma] \to [G]}\; D_R \qquad \frac{\Theta : [\Gamma_1], \Downarrow N_P, [\Gamma_2] \to [G]}{\Theta : [\Gamma_1, N_P, \Gamma_2] \to [G]}\; D_1$$

$$\frac{\Theta : [\Gamma_1], \Downarrow F, [\Gamma_2] \to [G]}{\Theta, F : [\Gamma_1, \Gamma_2] \to [G]}\; D_2 \qquad \frac{\Theta : [\Gamma_1, \sharp^+ F, \Gamma_2] \to [G]}{\Theta : [\Gamma_1, !F, \Gamma_2] \to [G]}\; D_3$$

$$\frac{\Theta : [\Gamma_1, \sharp^* F], \Downarrow F, [\Gamma_2] \to [G]}{\Theta : [\Gamma_1, \sharp^{+/*} F, \Gamma_2] \to [G]}\; D_4$$

Fig. 1. Focused proof system for SLLM#. N_N represent a non-negative formula. N_{NA} represent a non-atomic, non-negative formula. N_P represents a non-positive formula whose main connective is not ∇. N_P^{∇} represents a non-positive formula. We use $\mathcal{R}$ for both $\Uparrow G$ or $[N_N]$. We use $\sharp^{+/*}F$ for both $\sharp^+ F$ and $\sharp^* F$. We use Γ^* for a possibly empty list of formulas of the form $\sharp^* F$.

5 Complexity

In this section, we investigate the complexity of SLLM. In particular, we show that it is undecidable in general, by encoding Turing computations. This encoding also illustrate how the focused proof system SLLMF reduces non-determinism. We then identify decidable fragments for SLLM.

5.1 Encoding Turing Computations in SLLM

Any Turing instruction I is encoded by $!\nabla A_I$ with an appropriate A_I.
E.g., an instruction $I : q\xi \to q'\eta R$

if in state q looking at symbol ξ, replace it by η, move the tape head one cell to the right, and go into state q',

is encoded by $!\nabla A_i$ where

$$A_i = [(q_i \cdot \xi_i) \backslash (\eta_i \cdot q_i')]$$

$$\frac{\Theta_1 : [\Gamma_2], \rightarrow \Downarrow F \quad \Theta_2 : [\Gamma_1], \Downarrow G, [\Gamma_3] \rightarrow [H]}{\Theta_1, \Theta_2 : [\Gamma_1, \Gamma_2], \Downarrow F \setminus G, [\Gamma_3] \rightarrow [H]} \setminus_{L1}^{\star}$$

$$\frac{\Theta_1 : [\sharp^* C, \Gamma_2], \rightarrow \Downarrow F \quad \Theta_2 : [\Gamma_1, \sharp^+ C], \Downarrow G, [\Gamma_3] \rightarrow [H]}{\Theta_1, \Theta_2 : [\Gamma_1, \sharp^+ C, \Gamma_2], \Downarrow F \setminus G, [\Gamma_3] \rightarrow [H]} \setminus_{L2}$$

$$\frac{\Theta_1 : [\sharp^+ C, \Gamma_2], \rightarrow \Downarrow F \quad \Theta_2 : [\Gamma_1, \sharp^* C], \Downarrow G, [\Gamma_3] \rightarrow [H]}{\Theta_1, \Theta_2 : [\Gamma_1, \sharp^+ C, \Gamma_2], \Downarrow F \setminus G, [\Gamma_3] \rightarrow [H]} \setminus_{L3}$$

$$\frac{\Theta_1 : [\sharp^* C, \Gamma_2], \rightarrow \Downarrow F \quad \Theta_2 : [\Gamma_1, \sharp^* C], \Downarrow G, [\Gamma_3] \rightarrow [H]}{\Theta_1, \Theta_2 : [\Gamma_1, \sharp^* C, \Gamma_2], \Downarrow F \setminus G, [\Gamma_3] \rightarrow [H]} \setminus_{L4}$$

$$\frac{\Theta_1 : [\Gamma_2], \rightarrow \Downarrow G \quad \Theta_2 : [\Gamma_1], \Downarrow F, [\Gamma_3] \rightarrow [H]}{\Theta_1, \Theta_2 : [\Gamma_1], \Downarrow F / G, [\Gamma_2, \Gamma_3] \rightarrow [H]} /_{L1}^{\dagger}$$

$$\frac{\Theta_1 : [\Gamma_2, \sharp^* C], \rightarrow \Downarrow G \quad \Theta_2 : [\Gamma_1], \Downarrow F, [\sharp^+ C, \Gamma_3] \rightarrow [H]}{\Theta_1, \Theta_2 : [\Gamma_1], \Downarrow F / G, [\Gamma_2, \sharp^+ C, \Gamma_3] \rightarrow [H]} /_{L2}$$

$$\frac{\Theta_1 : [\Gamma_2, \sharp^+ C], \rightarrow \Downarrow G \quad \Theta_2 : [\Gamma_1], \Downarrow F, [\sharp^* C, \Gamma_3] \rightarrow [H]}{\Theta_1, \Theta_2 : [\Gamma_1], \Downarrow F / G, [\Gamma_2, \sharp^+ C, \Gamma_3] \rightarrow [H]} /_{L3}$$

$$\frac{\Theta_1 : [\Gamma_2, \sharp^* C], \rightarrow \Downarrow G \quad \Theta_2 : [\Gamma_1], \Downarrow F, [\sharp^* C, \Gamma_3] \rightarrow [H]}{\Theta_1, \Theta_2 : [\Gamma_1], \Downarrow F / G, [\Gamma_2, \sharp^* C, \Gamma_3] \rightarrow [H]} /_{L4}$$

Fig. 2. Focused left introduction rules for / and \ . The proviso $\star$ in $\setminus_{L1}$ states that the left-most formula of Γ_2 or the right-most formula of Γ_1 are not of the form $\sharp^* F$ nor $\sharp^+ F$. The proviso $\dagger$ in $/_{L1}$ is states that the right-most formula of Γ_2 or the left-most formula of Γ_3 are not of the form $\sharp^* F$ nor $\sharp^+ F$.

Let M lead from an initial configuration, represented as $B_1 \cdot q_1 \cdot \xi \cdot B_2$, to the final configuration q_0 (the tape is empty).

We demonstrate how focusing improves search with the encoding of Turing computations. For example, an instruction that write to the tape and moves to the right has the form: $A_i = [(q_i \cdot \xi_i) \setminus (\eta_i \cdot q_i')]$ while the Turing Machine (TM) configuration is encoded as in the sequent context: $[B_1, q_1, \xi, B_2]$ specifies A TM at state q_1 looking at the symbol ξ in the tape B_1, ξ, B_2.

Assuming that $A_1, \ldots, A_n$ is used at least once, $!\nabla A_1, \ldots, !\nabla A_n$ specifies the behavior of the TM. This becomes transparent with focusing. The following focused derivation illustrate how a copy of an instruction encoding, below A_1, can be made available to be used. Recall that the one has to look at the derivation from bottom-up.

$$\cfrac{\cfrac{\cfrac{A_1 : [\sharp^* \nabla A_1, \ldots, !\nabla A_n, B_1, q_1, \xi, B_2] \rightarrow [q_0]}{\cdot : \Downarrow \nabla A_1, [\sharp^* \nabla A_1, \ldots, !\nabla A_n, B_1, q_1, \xi, B_2] \rightarrow [q_0]} R_L, \Uparrow_2}{\cdot : [\sharp^+ \nabla A_1, \ldots, !\nabla A_n, B_1, q_1, \xi, B_2] \rightarrow [q_0]} D_4}{\cdot : [!\nabla A_1, \ldots, !\nabla A_n, B_1, q_1, \xi_1, B_2] \rightarrow [q_0]} D_3$$

Notice that A_1 is placed in the Θ context. This means that it can be moved at any place. Also notice that since one copy of A_1 is used, $\sharp^+\nabla A_1$ is replaced by $\sharp^*\nabla A_1$.

The following derivation continues from the premise of the derivation above by focusing on A_1, $\mathcal{A} = \sharp^*\nabla A_1, \ldots, !\nabla A_n$:

$$\dfrac{\dfrac{\dfrac{}{\cdot : [q_1] \rightarrow \Downarrow q_1}\ I \quad \dfrac{}{\cdot : [\eta_1] \rightarrow \Downarrow \eta_1}\ I}{\cdot : [q_1, \xi] \rightarrow \Downarrow q_1 \cdot \xi_1}\ \cdot_R \quad \dfrac{\dfrac{\cdot : [\mathcal{A}, B_1, \eta_1 \cdot q_1', B_2] \rightarrow [q_0]}{\cdot : [\mathcal{A}, B_1], \Uparrow \eta_1 \cdot q_1', [B_2] \rightarrow [q_0]}\ \cdot_L, 2\times \Uparrow_{L_1}}{\cdot : [\mathcal{A}, B_1], \Downarrow \eta_1 \cdot q_1', [B_2] \rightarrow [q_0]}\ R_L}{\dfrac{\cdot : [\mathcal{A}, B_1, q_1, \xi], \Downarrow(q_1 \cdot \xi_1) \backslash (\eta_1 \cdot q_1'), [B_2] \rightarrow [q_0]}{A_1 : [\mathcal{A}, B_1, q_1, \xi, B_2] \rightarrow [q_0]}\ D_4}\ \backslash_L$$

Notice that the resulting premise (at the right of the tree) specifies exactly the TM tape resulting from executing the instruction specified by A_1.

Moreover, notice that for the rule D_4, there are many options on where exactly to use A_1. However, it will only work if done as above. This is because otherwise it would not be possible to apply the initial rules as to the left of the derivation above. This reduces considerably the non-determinism involved for proof search.

Finally, once the final configuration q_0 (our Turing machine is responsible for garbage collection, so the configuration is just q_0, with no symbols on the tape) is reached, one finishes the proof:

$$\dfrac{\dfrac{}{\cdot : [\sharp^* A_1, \ldots, \sharp^* A_n, q_0] \rightarrow \Downarrow q_0}\ I}{\cdot : [\sharp^* A_1, \ldots, \sharp^* A_n, q_0] \rightarrow [q_0]}\ D_R$$

The focusing discipline guarantees that the structure of the proof described above is the only one available. Therefore our encoding soundly and faithfully encodes Turing computations, resulting in the following theorem. However, notice additionally, that due to the non-determinism due to the ! left introduction, the encoding of Turing computations is not on the level of derivations, but on the level of proofs, following the terminology in [24].

The absence of Weakening seems to reduce the expressive power of our system SLLM in the case where not all instructions might have been applied within a particular computation, see Remark 5. However, we are still able to get a strong positive statement.

Theorem 5. *We establish a strong correspondence between Turing computations and focused derivations in* SLLM.

Namely, given a subset $\{I_1, I_2, \ldots, I_m\}$ *of Turing instructions, the following two statements are equivalent:*

(a) The deterministic Turing machine M leads from an initial configuration, represented as $B_1 \cdot q_1 \cdot \xi \cdot B_2$, to the final configuration q_0 so that I_1, I_2, ..., I_m are only those instructions that have been actually *applied in the given Turing computation.*

(b) A sequent of the following specified form is derivable in SLLM.

$$!\nabla A_{I_1}, !\nabla A_{I_2}, \ldots, !\nabla A_{I_m}, B_1 \cdot q_1 \cdot \xi \cdot B_2 \vdash q_0 \tag{9}$$

Corollary 2. *The derivability problem for* SLLM *is undecidable.*

Proof. Assume the contrary: a decision algorithm α decides whether any sequent in SLLM is derivable or not. In particular, for any Turing machine M and any initial configuration of M, α decides whether any sequent of the form (9) is derivable or not, where $B_1 \cdot q_1 \cdot \xi \cdot B_2$ represents the initial configuration.

Then for each of the subsets $\{I_1, I_2, \ldots, I_m\}$ of the instructions of M, we apply α to the corresponding sequent of the form (9).

If all results are negative then we can conclude that M does not terminate on the initial configuration, represented as $B_1 \cdot q_1 \cdot \xi \cdot B_2$.

Otherwise, M does terminate.

Since the halting problem for Turing machines is undecidable, we conclude that α is impossible. □

Remark 5. The traditional approach to machine-based undecidability is to establish a one-to-one correspondence between machine computations and derivations in the system under consideration. Namely, given a Turing/Minsky machine M, we encode M as a formula $\mathtt{code}_M$, the 'product' of all C_I representing instructions I. The 'product' ranges over **all** M's instructions. Then we have to establish that, whatever initial configuration W we take, a sequent of the form

$$\mathtt{code}_M,\, W \vdash q_0 \tag{10}$$

is decidable in the logical system at hand iff M terminates on the initial configuration represented as W. Here one and the same $\mathtt{code}_M$ is used for each of the initials W.

Now, suppose that M terminates on each of the initial configurations W_1 and W_2 so that some instruction represented by $\widetilde{C}$ is used within the computation starting with W_1, but it is not used within the computation starting with W_2. Then, because of W_1, $\widetilde{C}$ takes an active part within $\mathtt{code}_M$. On the other hand, $\widetilde{C}$ becomes redundant and must be 'ignored' within the derivation for the sequent $\mathtt{code}_M, W_2 \vdash q_0$. To cope with the problem, it would be enough, e.g., to assume Weakening in the system under consideration, which is not a case here.

The *novelty* of our approach to the machine-based undecidability is to circumvent the problem caused by the absence of Weakening, by changing the order of quantifiers.

Instead of $\mathtt{code}_M$, *one and the same for all* W, we introduce *a finite number of candidates* of the form (9) with ranging over subsets $\{I_1, I_2, \ldots, I_m\}$ of the instructions of M.

According to Theorem 5, for any initial configuration W, there exists a sequent, say $\mathtt{code}_{M,W}, W \vdash q_0$, chosen from the finite set of candidates of the form (9) such that the terminated computation starting from W corresponds to an SLLM derivation for the chosen $\mathtt{code}_{M,W}, W \vdash q_0$. See also Corollary 2. □

5.2 Decidable Fragments: Syntactically Defined

An advantage of our approach is that, unlike the contraction rule, we can syntactically control the multiplexing to provide decidable fragments still sufficient for applications.

Theorem 6. *If we bound k in the multiplexing rule in the calculus* SLLM *with a fixed constant* k_0*, such a fragment becomes decidable.*

Proof Sketch. Each application of Multiplexing of the form:

$$\frac{\Gamma_1, \overbrace{A, A, \ldots, A}^{k \text{ times}}, \Gamma_2 \to C}{\Gamma_1, !A, \Gamma_2 \to C} \;!_L \; (1 \leq k \leq k_0)$$

multiplies the number of formulas with the factor k_0, which provides an upper bound on the size S of the sequents involved:

$$S = O(S_0 \cdot k_0^n)$$

here S_0 is the size of the input, and n is a bound on the nesting depth of the !-formulas. It suffices to apply a non-deterministic decision procedure but generally on the sequents of exponential size. □

Theorem 7. *In the case where we bound k in the multiplexing rule in the calculus* SLLM *with a fixed constant* k_0*, and, in addition, we bound the depth of nesting of* $!A$*, we get NP-completeness.*

Remark 6. In fact Theorem 7 gives NP-procedures for parsing complex and compound sentences in many practically important cases.

Remark 7. The strong lower bound is given by the following.

The !-free fragment that invokes only one implication, $(A \backslash B)$, and ∇A is still NP-complete.

6 Concluding Remarks

In this paper we have introduced SLLM, a non-commutative intuitionistic linear logic system with two subexponentials. One subexponential implements permutation and the other one obeys the multiplexing rule, which is a weaker, miniature version of contraction. Our system was inspired by *subexponentials* [23], *linear logic* [6], *light linear logic* [7], *soft linear logic* [16], and *easy linear logic* [10].

We have also provided a complete focused proof system for our calculus SLLM. We have illustrated the expressive power of the focused system by modelling computational processes.

The general aim is to develop more refined and efficient procedures for the miniature versions of non-commutative systems, e.g., Lambek calculus and its extensions, based on the multiplexing rule. We aim to ensure a reasonable balance between the expressive power of the formal systems in question and the complexity of their algorithmic implementation. The calculus SLLM, with multiplexing instead of contraction, provides simultaneously three properties: cut elimination, substitution, and Lambek's restriction.

One particular advantage of our system SLLM to systems in our previous work [12,15] is the fact that it naturally incorporates Lambek's non-emptiness restriction, which is incompatible with stronger systems involving contraction [15]. Lambek's non-emptiness restriction plays a crucial role in applications of substructural (Lambek-style) calculi in formal linguistics (type-logical grammars). Indeed, overcoming this impossibility result is one of our main motivations in looking for a system that would satisfy cut-elimination, substitution, and the Lambek's restriction. The new system proposed in this paper is our proposed solution to this subtle and challenging problem.

Moreover, there is no direct way of reducing the undecidability result in this paper, say, to the undecidability results from our previous papers [11,14,15] by a logical translation or representation of the logical systems. Since a number of those systems are also undecidable, there are of course Turing reductions both ways to the system in this paper. However, the Turing reductions factor through the new representation of Turing machines introduced in this paper. That is, the undecidability result in this paper is a new result.

Also the focused proof system proposed here has innovations to the paper [12]. For example, our proof system contains relevant subexponentials that do not allow contraction, something that was not addressed in [12]. Indeed such subexponentials have also been left out of other focused proof systems, e.g., in the papers [23,25].

Another advantage of this approach is that, unlike the contraction rule, we can syntactically control the multiplexing to provide feasible fragments still sufficient for linguistic applications.

As future work, we plan to investigate how to extend the systems proposed in this paper with additives. In particular, the proposed focused proof system using the introduced connectives $\sharp^*, \sharp^+$ may have to be extended in order to support additives. This is because it seems problematic, with these connectives, to provide that the same number of copies of a contractable formula are used in both premises when introducing an additive connective.

Financial Support. The work of Max Kanovich was partially supported by EPSRC Programme Grant EP/R006865/1: "Interface Reasoning for Interacting Systems (IRIS)." The work of Andre Scedrov and Stepan Kuznetsov was prepared within the framework of the HSE University Basic Research Program and partially funded by the Russian Academic Excellence Project '5–100'. The work of Stepan Kuznetsov was also partially supported by grant MK-430.2019.1 of the President of Russia, by the Young Russian Mathematics Award, and by the Russian Foundation for Basic Research grant 20-01-00435. Vivek Nigam's participation in this project has received funding from the European Union's

Horizon 2020 research and innovation programme under grant agreement No 830892. Vivek Nigam is also partially supported by CNPq grant 303909/2018-8.

References

1. Andreoli, J.-M.: Logic programming with focusing proofs in linear logic. J. Log. Comput. **2**(3), 297–347 (1992)
2. Barry, G., Hepple, M., Leslie, N., Morrill, G.: Proof figures and structural operators for categorial grammar. In: Proceedings of the Fifth Conference of the European Chapter of the Association for Computational Linguistics, Berlin (1991)
3. Carpenter, B.: Type-Logical Semantics. MIT Press, Cambridge (1998)
4. Cervesato, I., Pfenning, F.: A linear logic framework. In: Proceedings of the Eleventh Annual IEEE Symposium on Logic in Computer Science, New Brunswick, New Jersey, pp. 264–275. IEEE Computer Society Press, July 1996
5. Cervesato, I., Pfenning, F.: A linear logical framework. Inform. Comput. **179**(1), 19–75 (2002)
6. Girard, J.-Y.: Linear logic. Theor. Comput. Sci. **50**(1), 1–101 (1987)
7. Girard, J.-Y.: Light linear logic. Inform. Comput. **143**(2), 175–204 (1998)
8. Kanovich, M.I.: Horn programming in linear logic is NP-complete. In: Proceedings of the Seventh Annual IEEE Symposium on Logic in Computer Science , pp. 200–210 (1992)
9. Kanovich, M.I., Okada, M., Scedrov, A.: Phase semantics for light linear logic. Theor. Comput. Sci. **294**(3), 525–549 (2003)
10. Kanovich, M.I.: Light linear logics with controlled weakening: expressibility, confluent strong normalization. Ann. Pure Appl. Logic **163**(7), 854–874 (2012)
11. Kanovich, M., Kuznetsov, S., Nigam, V., Scedrov, A.: Subexponentials in non-commutative linear logic. Math. Struct. Comput. Sci. **29**(8), 1217–1249 (2019)
12. Kanovich, M., Kuznetsov, S., Nigam, V., Scedrov, A.: A logical framework with commutative and non-commutative subexponentials. In: Galmiche, D., Schulz, S., Sebastiani, R. (eds.) IJCAR 2018. LNCS (LNAI), vol. 10900, pp. 228–245. Springer, Cham (2018). https://doi.org/10.1007/978-3-319-94205-6_16
13. Kanovich, M., Kuznetsov, S., Scedrov, A.: L-models and R-models for Lambek calculus enriched with additives and the multiplicative unit. In: Iemhoff, R., Moortgat, M., de Queiroz, R. (eds.) WoLLIC 2019. LNCS, vol. 11541, pp. 373–391. Springer, Heidelberg (2019). https://doi.org/10.1007/978-3-662-59533-6_23
14. Kanovich, M., Kuznetsov, S., Scedrov, A.: Undecidability of the Lambek calculus with a relevant modality. In: Foret, A., Morrill, G., Muskens, R., Osswald, R., Pogodalla, S. (eds.) FG 2015-2016. LNCS, vol. 9804, pp. 240–256. Springer, Heidelberg (2016). https://doi.org/10.1007/978-3-662-53042-9_14
15. Kanovich, M., Kuznetsov, S., Scedrov, A.: Reconciling Lambek's restriction, cut-elimination, and substitution in the presence of exponential modalities. J. Logic Comput. **30**(1), 239–256 (2020)
16. Lafont, Y.: Soft linear logic and polynomial time. Theor. Comput. Sci. **318**(1–2), 163–180 (2004)
17. Lambek, J.: The mathematics of sentence structure. Amer. Math. Monthly **65**(3), 154–170 (1958)
18. Lincoln, P., Mitchell, J.C., Scedrov, A., Shankar, N.: Decision problems for propositional linear logic. Ann. Pure Appl. Logic **56**(1–3), 239–311 (1992)

19. Moot, R., Retoré, C.: The Logic of Categorial Grammars. A Deductive Account of Natural Language Syntax and Semantics. LNCS vol. 6850. Springer, Cham (2012). https://doi.org/10.1007/978-3-642-31555-8
20. Moortgat, M.: Constants of grammatical reasoning. In: Bouma, G., Hinrichs, E., Kruijff, G.-J., Oehrle, R. (eds.) Constraints and Resources in Natural Language Syntax and Semantics, pp. 195–219. CSLI Publications, Stanford (1999)
21. Morrill, G.: Categorial Grammar: Logical Syntax, Semantics, and Processing. Oxford University Press, Oxford (2011)
22. Morrill, G., Valentín, O.: Computational coverage of TLG: nonlinearity. In: Kanazawa, M., Moss, L., de Paiva, V. (eds.) Proceedings of NLCS 2015. Third Workshop on Natural Language and Computer Science, EPiC, vol. 32, pp. 51–63. EasyChair (2015)
23. Nigam, N., Miller, D.: Algorithmic specifications in linear logic with subexponentials. In: Proceedings of the 11th ACM Sigplan Conference on Principles and Practice of Declarative Programming, pp. 129–140 (2009)
24. Nigam, V., Miller, D.: A framework for proof systems. J. Automated Reason. **45**(2), 157–188 (2010)
25. Nigam, V., Pimentel, E., Reis, G.: An extended framework for specifying and reasoning about proof systems. J. Logic Comput. **26**(2), 539–576 (2016)
26. Saurin, A.: Une étude logique du contrôle. Ph.D. Thesis (2008)
27. Yetter, D.N.: Quantales and (noncommutative) linear logic. J. Symb. Log. **55**(1), 41–64 (1990)

Mechanised Modal Model Theory

Yiming Xu[1(✉)] and Michael Norrish[2,1(✉)]

[1] Australian National University, Canberra, Australia
u5943321@anu.edu.au
[2] Data61, CSIRO, Canberra, Australia
michael.norrish@data61.csiro.au

Abstract. In this paper, we discuss the mechanisation of some fundamental propositional modal model theory. The focus on models is novel: previous work in mechanisations of modal logic have centered on proof systems and applications in model-checking. We have mechanised a number of fundamental results from the first two chapters of a standard textbook (Blackburn *et al.* [1]). Among others, one important result, the Van Benthem characterisation theorem, characterises the connection between modal logic and first order logic. This latter captures the desired saturation property of ultraproduct models on countably incomplete ultrafilters.

1 Introduction

The theory of modal logic has long been a fruitful area when it comes to mechanisation. The proof systems are appealing, and the applications in model-checking are of clear real-world interest. It helps also that the subject domain (proof calculi and automata) are well-suited to "standard" theorem-proving technology (rule inductions and interesting data types).

There has been much less work on the model theory behind modal logic; indeed even in first order logic, most developments concern themselves only with model theory inasmuch as it is required to show completeness of an accompanying proof system. As our experience demonstrates, it is also clear that modern theorem-proving systems are not necessarily so well-suited to the mathematics behind model theory. Harrison [5] complained in 1998 that the very notion of validity is awkward to capture in HOL, and our own work shows up further failings in simple type theory.

Nonetheless, there is much interesting mathematics to be found even in the early chapters of a standard text such as Blackburn *et al.* [1]. The fact that mechanising only as far as [1, Chapter 2] requires what we believe to be the first mechanisation of the notion of ultraproduct (ultimately leading to Łoś's theorems and other results), is a strong suggestion that we are exploring novel mathematical ground for interactive theorem-proving systems.

Contributions. This paper presents the first mechanised proofs of a number of basic results from the first two chapters of Blackburn *et al.* [1] (e.g., bounded

N. Peltier and V. Sofronie-Stokkermans (Eds.): IJCAR 2020, LNAI 12166, pp. 518–533, 2020.
https://doi.org/10.1007/978-3-030-51074-9_30

morphisms, bisimulations and the finite model property *via* selection), as well as

- Two versions of Łoś's theorem on the saturation of ultraproduct models;
- modal equivalence as bisimilarity between ultrafilter extensions; and
- a close approximation of Van Benthem's Characterisation Theorem.

We also discuss where HOL's simple type theory lets us down: some standard results (including the best possible statement of Van Benthem's Characterisation Theorem) seem impossible to prove in our setting.

HOL4 Notation. All of our theorems have been pretty-printed to LaTeX from the HOL theory files. We hope that most of the basic syntax is easy to follow. In a few places we use CHOICE s to denote the arbitrary choice of an element from set s (appealing to the Axiom of Choice). The power-set of a set s is written $\mathcal{P}\ s$. In a number of places, we use HOL's "itself" type to allow us to explicitly mention a type *via* a term. The type α `itself` has just one value inhabiting it for any given choice of α; that value is written $(:\alpha)$.

Source Availability. Our HOL4 sources are available from GitHub at

https://github.com/u5943321/Modal-Logic

The sources build under HOL4 commit with SHA `03829d8986f`.

2 Syntax, Semantics and the Standard Translation

In our mechanisation, we consider the basic modal language, in which the only primitive modal operator is the '$\Diamond$'. A modal formula is either of form $\mathsf{V_m}\ p$, where p is of type `num`, enumerating all the possible variable symbols, a disjunction DISJ $\phi\ \psi$ (pretty-printed to $\phi \vee_{\mathsf{m}} \psi$ in most places), the falsity $\bot_{\mathsf{m}}$, a negation $\neg_{\mathsf{m}}\ \phi$, or, finally, of the form $\Diamond\phi$. We define a data type called `form` to represent the formulas of this modal language.

Definition 1. [1, Definition 1.9]

$$form\ =\ \mathsf{V_m}\ num \mid \mathsf{DISJ}\ form\ form \mid \bot_{\mathsf{m}} \mid \neg_{\mathsf{m}}\ form \mid \Diamond\ form$$

If we wanted to consider modal operators with any arity, we should change the last constructor of modal formulas so it takes two parameters: a natural number indexing the modal operator, and a list of modal formulas. This would in turn require a well-formedness predicate to be defined over formulas to make sure that modalities were applied to the right number of arguments.

A model where these formulas can be interpreted consists of a *frame* and a *valuation*, where a β `frame` is a β-set with a relation on it, and a model adds valuations for the variables present at each world:

Definition 2. [1, Definition 1.19]

$$\beta\ frame = \langle\!\langle \mathsf{world} : \beta \to bool;\ \mathsf{rel} : \beta \to \beta \to bool \rangle\!\rangle$$
$$\beta\ model = \langle\!\langle \mathsf{frame} : \beta\ frame;\ \mathsf{valt} : num \to \beta \to bool \rangle\!\rangle$$

In the rest of the paper, the field M.valt of a model M will be called the *valuation*, and M^W, M^R and M^V are used to denote the world set, the relation, and the valuation of M respectively. The interpretation of modal formulas on a model is given by the predicate *satisfaction*. We read '$M, w \Vdash \phi$' as 'ϕ is satisfied at the world w in M'.

Definition 3. [1, Definition 1.20]

$$\begin{aligned}
M, w \Vdash \mathsf{V_m}\, p &\stackrel{\mathsf{def}}{=} w \in M^W \land w \in M^V\, p \\
M, w \Vdash \bot_\mathsf{m} &\stackrel{\mathsf{def}}{=} \mathsf{F} \\
M, w \Vdash \neg_\mathsf{m}\, \phi &\stackrel{\mathsf{def}}{=} w \in M^W \land M, w \nVdash \phi \\
M, w \Vdash (\phi_1 \lor_\mathsf{m} \phi_2) &\stackrel{\mathsf{def}}{=} M, w \Vdash \phi_1 \lor M, w \Vdash \phi_2 \\
M, w \Vdash \Diamond\phi &\stackrel{\mathsf{def}}{=} w \in M^W \land \exists v.\, M^R\, w\, v \land v \in M^W \land M, v \Vdash \phi
\end{aligned}$$

By requiring $w \in M^W$ in various clauses above, we ensure that models' world sets must be non-empty if they are to satisfy any formulas.

Two worlds $w_1 \in {M_1}^W$ and $w_2 \in {M_2}^W$ are *modal equivalent* (written $M_1, w_1 \leftrightsquigarrow M_2, w_2$) if they satisfy the same set of modal formulas. If ϕ_1, ϕ_2 are modal formulas, we say they are *equivalent* over β models (written $\phi_1 \equiv_{(:\beta)} \phi_2$) if they are satisfied in the same worlds in every model:

Definition 4 (Notions of equivalence).

$$M, w \leftrightsquigarrow M', w' \stackrel{\mathsf{def}}{=} \forall \phi.\, M, w \Vdash \phi \iff M', w' \Vdash \phi$$

$$\begin{aligned}
&(\phi_1 : \mathit{form}) \equiv_{(:\beta)} (\phi_2 : \mathit{form}) \stackrel{\mathsf{def}}{=} \\
&\forall (M : \beta\ \mathtt{model})\ (w : \beta).\, M, w \Vdash \phi_1 \iff M, w \Vdash \phi_2
\end{aligned}$$

We cannot omit the type parameter $(:\beta)$ in the definition, as there would otherwise be a type, namely the type of the underlying set of the models we are talking about, that only appears on the right-hand side but not on the left-hand side of the definition. HOL forbids such definitions for soundness reasons. Also, HOL does not permit quantification over types, so it is impossible to write $\forall \mu.\, \phi_1 \equiv_\mu \phi_2$, with μ a type. Therefore, this definition is not exactly encoding the equivalence in the usual sense: when we mention equivalence of formulas in usual mathematical language, we are implicitly referring to the class of all models, but the constraint here bans us from talking about all models of all possible types at once.

A modal formula can be translated into a first-order formula via the *standard translation*. To mechanise this translation, we build on Harrison's construction of first-order logic [5]. The first-order connectives are decorated with an f. A first order model M is a set M.Dom with interpretation of function symbols M.Fun and predicate symbols M.Pred. A valuation σ of M is a function that maps all the natural numbers into the domain of M. If a first-order formula ϕ is satisfied in a first-order model M with σ a valuation assigning free variables of ϕ elements in the domain of M, we write $M, \sigma \vDash \phi$.

For a modal formula ϕ, $\mathsf{ST}_x\, \phi$ is the standard translation of ϕ using x as the only free variable that may occur:

Definition 5. [1, Definition 2.45 (Standard Translation)]

$$
\begin{aligned}
\mathsf{ST}_x\ (\mathsf{V_m}\ p) &\stackrel{\text{def}}{=} \mathsf{P_f}\ p\ (\mathsf{V_f}\ x) \\
\mathsf{ST}_x\ \bot_\mathsf{m} &\stackrel{\text{def}}{=} \bot_\mathsf{f} \\
\mathsf{ST}_x\ (\neg_\mathsf{m}\ \phi) &\stackrel{\text{def}}{=} \neg_\mathsf{f}\ (\mathsf{ST}_x\ \phi) \\
\mathsf{ST}_x\ (\phi \lor_\mathsf{m} \psi) &\stackrel{\text{def}}{=} \mathsf{ST}_x\ \phi \lor_\mathsf{f} \mathsf{ST}_x\ \psi \\
\mathsf{ST}_x\ (\Diamond\phi) &\stackrel{\text{def}}{=} \exists_\mathsf{f}\ (x + 1)\ (\mathsf{R_f}\ (\mathsf{V_f}\ x)\ (\mathsf{V_f}\ (x + 1)) \land_\mathsf{f} \mathsf{ST}_{x+1}\ \phi)
\end{aligned}
$$

As one would expect, we translate $\Diamond\phi$ into an existential formula. To ensure we use a fresh variable, we use $x + 1$ as our new variable symbol in this clause. The standard translation gives a first-order reformulation of satisfaction of modal formulas:

Proposition 1. [1, Theorem 2.47 (i)]

$$\vdash M, w \Vdash \phi \iff \mathsf{mm2folm}\ M, (\lambda\, n.\ w) \vDash \mathsf{ST}_x\ \phi$$

Here mm2folm is the function that turns a modal model into a first-order model, defined as:

$$
\begin{aligned}
&\mathsf{mm2folm}\ M \stackrel{\text{def}}{=} \\
&\quad \langle\!\langle \mathsf{Dom} := M^W;\ \mathsf{Fun} := (\lambda\, n\ l.\ \mathsf{CHOICE}\ M^W); \\
&\quad\ \ \mathsf{Pred} := \\
&\quad\quad (\lambda\, p\ zs. \\
&\quad\quad\quad \mathtt{case}\ zs\ \mathtt{of} \\
&\quad\quad\quad\ \ [\,] \Rightarrow \mathsf{F} \\
&\quad\quad\quad \mid [w_1] \Rightarrow w_1 \in M^W \land M^V\ p\ w_1 \\
&\quad\quad\quad \mid [w_1;\ w_2] \Rightarrow p = 0 \land M^R\ w_1\ w_2 \land w_1 \in M^W \land w_2 \in M^W \\
&\quad\quad\quad \mid w_1 :: w_2 :: w_3 :: ws \Rightarrow \mathsf{F})\rangle\!\rangle
\end{aligned}
$$

That is, the model obtained by converting a modal model M has domain M^W, maps every term $\mathsf{Fn_f}\ f\ l$ into an arbitrary world, maps each propositional letter to distinct predicates on worlds, and uses one binary predicate (the "0th predicate") to encode the frame relation.

3 Basic Results

We discuss some highlights of mechanised results from Blackburn *et al.* [1, §2.1–§2.3] below.

3.1 Tree-Like Property

A tree-like model is a model whose underlying frame is a tree. If Tr, a frame, is also a tree with root r, we write $\mathsf{tree}\ Tr\ r$:

Definition 6. [1, Definition 1.7]

$$\begin{array}{l}
\mathsf{tree}\ Tr\ r \stackrel{\mathsf{def}}{=} \\
\quad r \in Tr.\mathsf{world} \wedge (\forall\, w.\ w \in Tr.\mathsf{world} \Rightarrow Tr.\mathsf{rel}\,|_{Tr.\mathsf{world}}{}^{*}\ r\ w) \wedge \\
\quad (\forall\, w.\ w \in Tr.\mathsf{world} \Rightarrow \neg Tr.\mathsf{rel}\ w\ r) \wedge \\
\quad \forall\, w.\ w \in Tr.\mathsf{world} \wedge w \neq r \Rightarrow \exists!\, w_0.\ w_0 \in Tr.\mathsf{world} \wedge Tr.\mathsf{rel}\ w_0\ w
\end{array}$$

The tree-like property says each satisfiable modal formula can be satisfied in a tree-like model:

Proposition 2. [1, Proposition 2.15]

$$\begin{array}{l}
\vdash (M : \beta\ \mathit{model}), (w : \beta) \Vdash (\phi : \mathit{form}) \Rightarrow \\
\quad \exists (M' : \beta\ \mathit{list\ model})\ (r : \beta\ \mathit{list}).\ \mathsf{tree}\ M'.\mathsf{frame}\ r \wedge M', r \Vdash \phi
\end{array}$$

The world set of the tree-like model constructed from M is a set of lists of worlds in M (such lists are effectively paths from the root to various positions within the tree). Thus, passing to a tree-like model does not preserve the model type. The tree-like lemma is used to prove the finite model property via selection afterwards.

3.2 Bisimulation

Though apparently verbose, the definition of bisimulation in HOL is straightforward.

Definition 7. [1, Definition 2.16 (Bisimulations)]

$$\begin{array}{l}
M_1 \overset{Z}{\leftrightarrows} M_2 \stackrel{\mathsf{def}}{=} \\
\quad \forall\, w_1\ w_2. \\
\qquad w_1 \in M_1^W \wedge w_2 \in M_2^W \wedge Z\ w_1\ w_2 \Rightarrow \\
\qquad\quad (\forall\, p.\ M_1, w_1 \Vdash \mathsf{V_m}\ p \iff M_2, w_2 \Vdash \mathsf{V_m}\ p) \wedge \\
\qquad\quad (\forall\, v_1. \\
\qquad\qquad v_1 \in M_1^W \wedge M_1^R\ w_1\ v_1 \Rightarrow \\
\qquad\qquad\quad \exists\, v_2.\ v_2 \in M_2^W \wedge Z\ v_1\ v_2 \wedge M_2^R\ w_2\ v_2) \wedge \\
\qquad\quad \forall\, v_2. \\
\qquad\qquad v_2 \in M_2^W \wedge M_2^R\ w_2\ v_2 \Rightarrow \\
\qquad\qquad\quad \exists\, v_1.\ v_1 \in M_1^W \wedge Z\ v_1\ v_2 \wedge M_1^R\ w_1\ v_1
\end{array}$$

$$M, w \leftrightarrows M', w' \stackrel{\mathsf{def}}{=} \exists\, Z.\ M \overset{Z}{\leftrightarrows} M' \wedge w \in M^W \wedge w' \in M'^W \wedge Z\ w\ w'$$

It is trivial to prove by induction that bisimilar worlds are modal equivalent. As the most significant theorem on the basic theory of bisimulations, we proved the Hennessy-Milner theorem, which states that modal equivalence and bisimulation on *image-finite* models are the same thing. An image-finite model is a model where every world can only be related to finitely many worlds. In HOL, we get:

Theorem 1. [1, Theorem 2.24 (Hennessy-Milner Theorem)]

$$\vdash \text{image-finite } M_1 \wedge \text{image-finite } M_2 \wedge w_1 \in M_1^W \wedge w_2 \in M_2^W \Rightarrow (M_1, w_1 \leftrightsquigarrow M_2, w_2 \iff M_1, w_1 \leftrightarrows M_2, w_2)$$

Bisimulation is an interesting topic in modal logic. Three other significant theorems on bisimulations (including an approximation of Van Benthem Characterisation theorem) are discussed later.

3.3 Finite Model Property

There are two classical approaches to constructing finite models using model theory, namely via selection and via filtration. The complicated one is the former: Given $M_1, w_1 \Vdash \phi$, where ϕ has degree k, we can construct M_2, M_3, M_4 and M_5 consecutively, such that M_5 is the finite model we want, where:

- M_2 is the tree-like model obtained from Proposition 2 with root w_2 such that $M_2, w_2 \Vdash \phi$.
- M_3 is the restriction of M_2 to height k.
- M_4 is obtained from M_3 by modifying the valuation into $\lambda\, p\ v.$ `if` $p \in$ prop-letters ϕ `then` ${M_3}^V\ p\ v$ `else` F, where prop-letters ϕ is the set of all propositional letters used by ϕ.

The construction of M_5 requires a lemma:

Lemma 1. [1, Proposition 2.29]

$$\vdash \text{FINITE}\,(\Phi : \mathit{num} \to \mathtt{bool}) \wedge \text{INFINITE}\,\mathcal{U}(:\beta) \Rightarrow \forall (n : \mathit{num}).\ \text{FINITE}\ \{\ \phi \mid \text{DEG}\ \phi \le n \wedge \text{prop-letters}\ \phi \subseteq \Phi\ \} /{\equiv_{(:\beta)}}$$

The proof of Lemma 1 further relies on the following fact: Given a set A of modal-formulas that is finite up to equivalence, if we combine the elements of A using only connectives other than $\Diamond$, then we get only finitely many non-equivalent formulas. To show this, we prove that there is an injection from the set of equivalence classes of such combinations to a finite set. For the antecedent of Lemma 1, we require the assumption that the universe of β is infinite since we rely on the fact that two modal formulas $\Diamond\phi_1$ and $\Diamond\phi_2$ are equivalent if and only if ϕ_1 and ϕ_2 are equivalent. This would be easy to prove in set theory. However, in simple type theory, the proof of $\phi_1 \equiv_{(:\beta)} \phi_2$ iff $\Diamond\phi_1 \equiv_{(:\beta)} \Diamond\phi_2$ requires us (in the right-to-left direction) to be able to construct a model with a new world inserted, something only sure to be possible if the β universe is infinite. As the construction used Proposition 2, we change the type of the model by passing to a finite model via selection:

Theorem 2. [1, Theorem 2.34 (Finite model property, via selection)]

$$\vdash (M_1 : \beta\ \mathtt{model}), (w_1 : \beta) \Vdash (\phi : \mathtt{form}) \Rightarrow \exists (M : \beta\ \mathtt{list}\ \mathtt{model})\ (v : \beta\ \mathtt{list}).\ \text{FINITE}\ M^W \wedge v \in M^W \wedge M, v \Vdash \phi$$

We also mechanised the filtration approach, but omit the details for lack of space. The advantage of filtration is that the resulting finite model is over worlds of the same type as in the starting model.

All the results proved above can be captured using `num` models everywhere. If one takes β to be `num` (or any infinite type) in Theorem 2, one can also exploit the fact that numbers and lists of numbers have the same cardinality to derive a finite model result that preserves the "input type".

4 Mechanising Ultrafilters and Ultraproducts

A number of results in Blackburn *et al.* [1, §2.5–§2.7] rely on theorems about ultrafilters and ultraproducts.

4.1 Ultrafilters

Given a non-empty set J, a set $L \subseteq \mathcal{P}\, J$ is a *filter* if it contains J itself, is closed under binary intersection, and is closed upward.

Definition 8. [1, Definition A.12 (Filters)]

$$\begin{array}{l}\mathsf{filter}\ L\ J \stackrel{\text{def}}{=} \\ \quad J \neq \emptyset \wedge L \subseteq \mathcal{P}\, J \wedge J \in L \wedge \\ \quad (\forall X\ Y.\, X \in L \wedge Y \in L \Rightarrow X \cap Y \in L) \wedge \\ \quad \forall X\ Z.\, X \in L \wedge X \subseteq Z \wedge Z \subseteq J \Rightarrow Z \in L\end{array}$$

We call L a *proper filter* if L is not the whole power set. An *ultrafilter* is a filter U such that for every $X \subseteq J$, exactly one of X or $J \setminus X$ is in U. Intuitively, subsets $X \subseteq J$ in an ultrafilter U are considered as 'large' subsets of J.

The ultrafilter theorem states that every proper filter is contained in an ultrafilter:

Theorem 3. [1, Fact A.14, first half]

$$\vdash \mathsf{proper\text{-}filter}\ L\ J \Rightarrow \exists\, U.\ \mathsf{ultrafilter}\ U\ J \wedge L \subseteq U$$

(The proof uses Zorn's Lemma.)

A subset A of the power set on J has *finite intersection property* if once we take the intersection of a finite, nonempty family in A, the resultant set is nonempty.

Definition 9. [1, Definition A.13 (Finite Intersection Property)]

$$\begin{array}{l}\vdash \mathsf{FIP}\ A\ J \iff \\ \quad A \subseteq \mathcal{P}\, J \wedge \forall B.\, B \subseteq A \wedge \mathsf{FINITE}\ B \wedge B \neq \emptyset \Rightarrow \bigcap B \neq \emptyset\end{array}$$

As a corollary of ultrafilter theorem, a set with finite intersection property is contained in an ultrafilter.

4.2 Ultraproducts

The notion of ultraproducts is defined for sets, modal models, and first-order models.

Ultraproduct of Sets. A family of sets indexed by J is a function A^s in HOL. For $j \in J$, $A^s\ j$ is the set indexed by j. Given a family A^s indexed by a non-empty set J such that each $A^s\ j$ is non-empty, the ultraproduct of A^s is defined as a quotient of the cartesian product of the family.

Definition 10. [1, Page 495 (Cartesian product)]

$$\mathsf{Cart\text{-}prod}\ J\ A^s \stackrel{\mathsf{def}}{=} \{ f \mid \forall j.\ j \in J \Rightarrow f\ j \in A^s\ j \}$$

If U is an ultrafilter on J, for two functions f, g in the Cartesian product $\mathsf{Cart\text{-}prod}\ J\ A^s$, we say f and g are U-equivalent (notation: $f \sim_U^{A^s} g$) if the set $\{ j \mid j \in J \wedge f\ j = g\ j \}$ (where the values of f and g agree) is in U. The ultraproduct of A^s modulo U is the quotient of $\mathsf{Cart\text{-}prod}\ J\ A^s$ by $\sim_U^{A^s}$.

Definition 11. [1, Definition 2.69 (Ultraproduct of Sets)]

$$\mathsf{ultraproduct}\ U\ J\ A^s \stackrel{\mathsf{def}}{=} \mathsf{Cart\text{-}prod}\ J\ A^s / \sim_U^{A^s}$$

We write f_U to denote the equivalence class that f belongs to. In the case where $A^s\ j = A$ for all $j \in J$, the ultraproduct is called the ultrapower of A modulo U.

Ultraproduct for Modal Models. Given a family M^s of modal models indexed by J and an ultrafilter U on J, the ultraproduct model of M^s modulo U (notation: $\prod_U M^s$) is described as follows:

- The world set is the ultraproduct of world sets of M^s modulo U.
- Two equivalence classes f_U, g_U of functions are related in $\prod_U M^s$ iff there exist $f_0 \in f_U, g_0 \in g_U$, such that $\{ j \in J \mid (M^s\ j)^R\ (f_0\ j)\ (g_0\ j) \}$ is in U.
- A propositional letter p is satisfied at f_U in $\prod_U M^s$ iff there exists $f_0 \in f_U$ such that $\{ j \mid j \in J \wedge f_0\ j \in (M^s\ j)^V\ p \}$ is in U.

Definition 12. [1, Definition 2.70 (Ultraproduct of Modal Models)]

$$
\begin{aligned}
&\textstyle\prod_U M^s \stackrel{\mathsf{def}}{=}\\
&\quad \langle\!\langle \mathsf{frame} :=\\
&\qquad \langle\!\langle \mathsf{world} := \mathsf{ultraproduct}\ U\ J\ (\lambda j.\ (M^s\ j)^W);\\
&\qquad\ \mathsf{rel} :=\\
&\qquad\quad (\lambda f_U\ g_U.\\
&\qquad\qquad \exists f_0\ g_0.\\
&\qquad\qquad\quad f_0 \in f_U \wedge g_0 \in g_U \wedge\\
&\qquad\qquad\quad \{ j \mid j \in J \wedge (M^s\ j)^R\ (f_0\ j)\ (g_0\ j) \} \in U)\rangle\!\rangle;\\
&\quad\ \mathsf{valt} :=\\
&\qquad (\lambda p\ f_U.\ \exists f_0.\ f_0 \in f_U \wedge \{ j \mid j \in J \wedge f_0\ j \in (M^s\ j)^V\ p \} \in U)\rangle\!\rangle
\end{aligned}
$$

As $\sim_U^A$ is an equivalence relation, if one element in an equivalence class satisfies the required condition, then all the elements in the equivalence class will satisfy the condition. Therefore, if we replace all the existential quantifiers with universal quantifiers in the above definition, the construction is still valid, and will give the same model as the current definition.

The critical result we need about ultraproducts of modal models is a modal version of the fundamental theorem of ultraproducts, also known as Łoś's theorem.

Theorem 4 (Łoś's theorem, Modal version).

$$\begin{array}{l}\vdash \mathsf{ultrafilter}\ U\ J\ \wedge\ f_U\ \in\ (\prod_U M^{\mathsf{s}})^W\ \Rightarrow \\ \quad (\prod_U M^{\mathsf{s}}, f_U \Vdash \phi\ \iff \\ \qquad \exists f_0.\ f_0\ \in\ f_U\ \wedge\ \{\, j \mid j\ \in\ J\ \wedge\ M^{\mathsf{s}}\ j, f_0\ j \Vdash \phi \,\}\ \in\ U)\end{array}$$

According to our intuition about ultrafilters, we can gloss this theorem to mean that the ultraproduct of a family of modal models satisfies a modal formula if and only if 'most of' the models in the family satisfy the formula. Though it is possible to derive this result from Łoś's theorem of first-order models using the standard translation, our proof is direct, by structural induction on ϕ.

Ultraproducts for First-Order Models. Given a family M^{s} of first-order models indexed by J and an ultrafilter U on J, the ultraproduct model of M^{s} modulo U (notation: ${}^{\mathsf{f}}\prod_U M^{\mathsf{s}}$) is given by:

- The domain is the ultraproduct of the domains of M^{s} over U on J.
- A function named by symbol (natural number) n sends a list zs of equivalence classes to the equivalence class of a function that sends $j \in J$ to $(M^{\mathsf{s}}\ j).\mathsf{Fun}\ n\ l$, where the k-th member of the list l is a representative of the k-th member (which is an equivalence class) of zs.
- A predicate named by symbol p will hold for a list zs of equivalence classes if and only if when we have a list zr, where k-th member is a representative of the k-th member of zs, the set of elements $j \in J$ such that $(M^{\mathsf{s}}\ j).\mathsf{Pred}\ p\ zr$ is in U.

Definition 13. [1, Definition A.18 (Ultraproduct of First-Order Models)]

$$\begin{array}{l}{}^{\mathsf{f}}\prod_U M^{\mathsf{s}} \stackrel{\mathsf{def}}{=} \\ \quad \langle\!\langle \mathsf{Dom}\ :=\ \mathsf{ultraproduct}\ U\ J\ (\lambda j.\ (M^{\mathsf{s}}\ j).\mathsf{Dom}); \\ \quad\ \ \mathsf{Fun}\ := \\ \quad\quad (\lambda\, n\ zs. \\ \quad\qquad \{\, y \mid \\ \quad\qquad (\forall j.\ j\ \in\ J\ \Rightarrow\ y\ j\ \in\ (M^{\mathsf{s}}\ j).\mathsf{Dom})\ \wedge \\ \quad\qquad \{\, j \mid j\ \in\ J\ \wedge\ y\ j\ =\ \mathsf{Fun\text{-}component}\ M^{\mathsf{s}}\ n\ zs\ j \,\}\ \in\ U \,\}); \\ \quad\ \ \mathsf{Pred}\ :=\ (\lambda\, p\ zs.\ \{\, j \mid j\ \in\ J\ \wedge\ \mathsf{Pred\text{-}component}\ M^{\mathsf{s}}\ p\ zs\ j \,\}\ \in\ U)\rangle\!\rangle\end{array}$$

Here we fix the representative of each equivalence class f_U to be $\mathsf{CHOICE}\ f_U$. Therefore, as described above, the functions Fun-component and Pred-component are:

$$\mathsf{Fun\text{-}component}\ M^{\mathsf{s}}\ n\ fs\ i \stackrel{\mathsf{def}}{=} (M^{\mathsf{s}}\ i).\mathsf{Fun}\ n\ (\mathsf{MAP}\ (\lambda f_U.\ \mathsf{CHOICE}\ f_U\ i)\ fs)$$

$$\mathsf{Pred\text{-}component}\ M^{\mathsf{s}}\ p\ zs\ i \stackrel{\mathsf{def}}{=} (M^{\mathsf{s}}\ i).\mathsf{Pred}\ p\ (\mathsf{MAP}\ (\lambda f_U.\ \mathsf{CHOICE}\ f_U\ i)\ zs)$$

The semantic behavior of ultraproduct models is characterised by Łoś's theorem: for the ultraproduct of a family M^{s} of first-order models over an ultrafilter U on J, a formula ϕ is satisfied under a valuation σ if and only if the set indexing the models $M^{\mathsf{s}}\ j$ in the family where ϕ is true under the valuation $\lambda\, v.\ \mathsf{CHOICE}\ (\sigma\ v)\ j$ is in the ultrafilter U.

Theorem 5. [1, Theorem A.19 (Łoś's theorem)]

$$\begin{array}{l} \vdash \mathsf{ultrafilter}\ U\ J\ \wedge\ \mathsf{valuation}\ ({}^{\mathsf{f}}\Pi_U\ M^{\mathsf{s}})\ \sigma\ \wedge \\ \quad (\forall j.\ j\ \in\ J\ \Rightarrow\ \mathsf{wffm}\ (M^{\mathsf{s}}\ j))\ \Rightarrow \\ \quad\ ({}^{\mathsf{f}}\Pi_U\ M^{\mathsf{s}}, \sigma \vDash \phi\ \iff \\ \quad\quad \{\, j \mid j\ \in\ J\ \wedge\ M^{\mathsf{s}}\ j, (\lambda\, v.\ \mathsf{CHOICE}\ (\sigma\ v)\ j) \vDash \phi \,\}\ \in\ U) \end{array}$$

where wffm M means the functions of M never map a list out of the domain of M.

5 Ultrafilter Extensions

The first application of the theory of ultrafilters above is to construct the ultrafilter extension of a model, which has the nice property of being *modally saturated* (m-saturated hereafter). To define m-saturation, we give the following three definitions (the first two are called *finitely satisfiable*, *satisfiable*) consecutively:

Definition 14. [1, Definition 2.53]

$$\begin{array}{l} \mathsf{satisfiable\text{-}in}\ \Sigma\ X\ M \stackrel{\mathsf{def}}{=} \\ \quad X\ \subseteq\ M^W\ \wedge\ \exists\, w.\ w\ \in\ X\ \wedge\ \forall \phi.\ \phi\ \in\ \Sigma\ \Rightarrow\ M, w \Vdash \phi \end{array}$$

$$\mathsf{fin\text{-}satisfiable\text{-}in}\ \Sigma\ X\ M \stackrel{\mathsf{def}}{=} \forall S.\ S\ \subseteq\ \Sigma\ \wedge\ \mathsf{FINITE}\ S\ \Rightarrow\ \mathsf{satisfiable\text{-}in}\ S\ X\ M$$

$$\begin{array}{l} \mathsf{m\text{-}sat}\ M \stackrel{\mathsf{def}}{=} \\ \quad \forall\, w\ \Sigma. \\ \quad\quad w\ \in\ M^W\ \wedge\ \mathsf{fin\text{-}satisfiable\text{-}in}\ \Sigma\ \{\, v \mid v\ \in\ M^W\ \wedge\ M^R\ w\ v \,\}\ M\ \Rightarrow \\ \quad\quad\ \mathsf{satisfiable\text{-}in}\ \Sigma\ \{\, v \mid v\ \in\ M^W\ \wedge\ M^R\ w\ v \,\}\ M \end{array}$$

For m-saturated models, bisimulation and modal equivalence coincide:

Proposition 3. [1, Proposition 2.54]

$$\begin{array}{l} \vdash \mathsf{m\text{-}sat}\ M_1\ \wedge\ \mathsf{m\text{-}sat}\ M_2\ \wedge\ w_1\ \in\ M_1^W\ \wedge\ w_2\ \in\ M_2^W\ \Rightarrow \\ \quad (M_1, w_1 \leftrightsquigarrow M_2, w_2\ \iff\ M_1, w_1 \leftrightarrows M_2, w_2) \end{array}$$

Given a model M and a set X of worlds of M, the set of worlds that 'can see' X (notation: $M_{\Diamond}(X)$) is the set of worlds w of M such that there exists some $v \in X$ such that $M^R\ w\ v$. We define the ultrafilter extension ${}^{ue}M$ of M as:

- The world set is the set of all ultrafilters on M^W.
- Two ultrafilters u, v on M are related in the ultrafilter extension of M if for every $X \in v$, the set of worlds that can see X is in u.
- A propositional letter p to be satisfied at an ultrafilter v if and only if the set of worlds in M which satisfies p is in v.

In HOL:

Definition 15. [1, Definition 2.57 (Ultrafilter Extension)]

$$
\begin{array}{l}
{}^{ue}M \stackrel{\mathsf{def}}{=} \\
\quad \langle\!\langle \mathsf{frame} := \\
\quad\quad \langle\!\langle \mathsf{world} := \{\, u \mid \mathsf{ultrafilter}\ u\ M^W \,\}; \\
\quad\quad\quad \mathsf{rel} := \\
\quad\quad\quad\quad (\lambda\, u\ v. \\
\quad\quad\quad\quad\quad \mathsf{ultrafilter}\ u\ M^W \wedge \mathsf{ultrafilter}\ v\ M^W \wedge \\
\quad\quad\quad\quad\quad \forall X.\ X \in v \Rightarrow M_{\Diamond}(X) \in u)\rangle\!\rangle; \\
\quad\quad \mathsf{valt} := (\lambda\, p\ v.\ \mathsf{ultrafilter}\ v\ M^W \wedge \{\, w \mid w \in M^W \wedge M^V\ p\ w \,\} \in v)\rangle\!\rangle
\end{array}
$$

Using the ultrafilter theorem and some basic properties about ultrafilters, we derive:

Proposition 4. [1, Proposition 2.59 (i)]

$$
\begin{array}{l}
\vdash \mathsf{ultrafilter}\ u\ M^W \Rightarrow \\
\quad (\{\, w \mid w \in M^W \wedge M, w \Vdash \phi \,\} \in u \iff {}^{ue}M, u \Vdash \phi)
\end{array}
$$

In particular, every world $w \in M^W$ is embedded as the principal filter $\pi_w^{M^W}$ on M^W generated by w in the ultrafilter extension or M. Also, the above leads to the proof of the fact that the ultrafilter extension of every model is m-saturated. The m-saturatedness of ultrafilter extensions together with Proposition 3 immediately gives the central result about ultrafilter extension: bisimilarity of worlds in a model M can be characterised as bisimilarity in ${}^{ue}M$.

Theorem 6. [1, Proposition 2.62]

$$
\begin{array}{l}
\vdash w_1 \in M_1^W \wedge w_2 \in M_2^W \Rightarrow \\
\quad (M_1, w_1 \leftrightsquigarrow M_2, w_2 \iff {}^{ue}M_1, \pi_{w_1}^{M_1^W} \leftrightarrows {}^{ue}M_2, \pi_{w_2}^{M_2^W})
\end{array}
$$

6 Countable Saturatedness of Ultrapower Models

Given a first-order model M with no information about interpretation of its function symbols, we can expand the model M by adding an interpretation of some function symbols. For our purpose, we are only interested in adding the interpretation of finitely many nullary function symbols, also called *constants*. We write expand $M\ A\ f$ to denote the model that is the result of adding each element in A to M as a new constant. Further, the function f is a bijection between $\{0, \cdots, n-1\}$ and A, which is assumed to be finite, so that each nullary function symbol c will be interpreted as $f\ c$ in M'.

Definition 16. [1, Definition A.9 (Expansion)]

$$\begin{array}{l} \mathsf{expand}\ M\ A\ f \stackrel{\mathsf{def}}{=} \\ \quad \langle\!\langle \mathsf{Dom} := M.\mathsf{Dom}; \\ \quad\ \ \mathsf{Fun} := \\ \quad\ \ \ (\lambda\, c\, l.\ \mathtt{if}\ c < \mathsf{CARD}\ A \wedge l = [\,]\ \mathtt{then}\ f\ c\ \mathtt{else}\ \mathsf{CHOICE}\ M.\mathsf{Dom}); \\ \quad\ \ \mathsf{Pred} := M.\mathsf{Pred}\rangle\!\rangle \end{array}$$

As is apparent from the definition, the only difference between a model and its expansion is the interpretation of function symbols.

A set Σ of first-order formulas is called *consistent* with a model M if for every finite subset $\Sigma_0 \subseteq \Sigma$, there exists a valuation of M such that all elements of Σ_0 are satisfied, in this case, we write consistent $M\ \Sigma$. A set Γ of first-order formula is an *x-type* if for each formula in Γ, the only free variable that may contain is x. In this case, we write 'ftype $x\ \Gamma$' in HOL. If Γ is an x-type, when evaluating formulas in Γ, the valuations will only control where the only free variable x goes to. We say Γ is *realised* in M if there is an element w in the domain of M such that $M, (\lambda\, v.\ w) \vDash \phi$ for all $\phi \in \Gamma$. In this case, we write 'frealises $M\ x\ \Gamma$' in HOL. Let M be a model and n be a natural number. If for every $A \subseteq M.\mathsf{Dom}$ with $|A| < n$ and every $f : \mathbb{N} \to M.\mathsf{Dom}$, the model expand $M\ A\ f$ realises every x-type Γ that is consistent with expand $M\ A\ f$, then we say M is n-saturated. In HOL:

Definition 17. [1, Definition 2.63 (n-Saturated)]

$$\begin{array}{l} \mathsf{n\text{-}saturated}\ M\ n \stackrel{\mathsf{def}}{=} \\ \quad \forall\, A\ \Gamma\ x\ f. \\ \qquad \mathsf{IMAGE}\ f\ \mathcal{U}(:\mathit{num}) \subseteq M.\mathsf{Dom} \wedge \mathsf{FINITE}\ A \wedge \mathsf{CARD}\ A \le n\ \wedge \\ \qquad A \subseteq M.\mathsf{Dom} \wedge \mathsf{BIJ}\ f\ (\mathsf{count}\ (\mathsf{CARD}\ A))\ A\ \wedge \\ \qquad (\forall \phi.\ \phi \in \Gamma \Rightarrow \mathsf{form\text{-}functions}\ \phi \subseteq \{\, (c, 0) \mid c < \mathsf{CARD}\ A \,\})\ \wedge \\ \qquad \mathsf{ftype}\ x\ \Gamma \wedge \mathsf{consistent}\ (\mathsf{expand}\ M\ A\ f)\ \Gamma \Rightarrow \\ \qquad\ \ \mathsf{frealises}\ (\mathsf{expand}\ M\ A\ f)\ x\ \Gamma \end{array}$$

We say M is countably saturated if M is n-saturated for every natural number n. The ultimate goal is to prove a lemma to be used in the proof of Van Benthem characterisation theorem: For a family of non-empty models, their ultraproduct on a countably incomplete ultrafilter is *countably saturated.*

Lemma 2. [1, Lemma 2.73]

$$\begin{array}{l} \vdash (\forall j.\ j \in J \Rightarrow (M^{\mathsf{s}}\ j)^W \neq \emptyset) \wedge \mathsf{countably\text{-}incomplete}\ U\ J \Rightarrow \\ \quad \mathsf{countably\text{-}saturated}\ (\mathsf{mm2folm}\ (\prod_U M^{\mathsf{s}})) \end{array}$$

Here a countably incomplete ultrafilter is an ultrafilter that contains a countably infinite family that intersects to the empty set. We prove in HOL that such ultrafilters do exist using Theorem 3. The above theorem is not simply a direct consequence of Łoś's theorem: that result is about ultraproducts of first-order models, and it says nothing about expansion. But to prove Lemma 2, we must prove a statement for an expanded first-order model, and this first order model is itself obtained by converting a ultraproduct of modal models.

To deal with this issue, the key observation is that constants are nothing more than forcing some symbols to be sent to some points in a model under every valuation, hence rather than use nullary function symbols, we fix a set of variable letters, each corresponding to a function symbol, and only consider the valuations that send these variable letters to certain fixed points. With this idea, we can remove all the constants in a formula, and hence change our scope from an expanded model back to the unexpanded model. To get rid of the constants $\{0, \cdots, n-1\}$, we replace every $\mathsf{V_f}\ m$ with $\mathsf{V_f}\ (m + n)$, and replace every constant $\mathsf{Fn_f}\ c\ []$ by $\mathsf{V_f}\ c$. This operation is done by the function shift-form which takes a natural number (the number of constants we want to remove), and a first-order formula (where the only function symbols may appear are the constants $0, \cdots, n-1$). Since $0, \cdots, n-1$ in a shifted formula are now designed to be sent to fixed places $f\ 0, \cdots, f\ (n-1)$, it does not make sense to assign these variable symbols anywhere else. Therefore, to talk about evaluation of shifted formula, the first thing is to make sure that the valuations we are considering send the variables which actually denote constants to the right place. Hence we shift the valuations accordingly, and then prove that a formula is satisfied on an expanded model is satisfied under a valuation if and only if the shifted formula is satisfied under the shifted valuation. Also, we prove that 'taking the ultraproduct first-order model commutes with the convertion from modal to a first-order model on certain formulas', in the sense that the resulting models satisfies the same first-order formulas without function symbols. By putting these two results together, we prove Lemma 2 using the proof in Chang and Keisler [3].

7 Van Benthem's Characterisation Theorem

Note that the standard translation of any modal formula can only contain unary predicate symbols which correspond to propositional letters, one binary predicate symbol which corresponds to the relation, and no function symbols. A first-order formula which only uses these symbols is called an $\mathcal{L}^1_\tau$-formula. An $\mathcal{L}^1_\tau$-formula which contains only one free variable is called *invariant under bisimulation* if for all models M and N with $w \in M^W$ and $v \in N^W$, if there exists a bisimulation relation between M and N relating w and v, then ϕ holds at w if and only if it holds at v when both M and N are viewed as first-order models.

Definition 18. [1, Definition 2.67 (Invariant for Bisimulations)]

$$\begin{array}{l} \mathsf{invar4bisim}\ (x : \mathit{num})\ (:\alpha)\ (:\beta)\ (\phi : \mathtt{folform}) \stackrel{\text{def}}{=} \\ \quad \mathsf{FV}\ \phi \subseteq \{\, x \,\} \wedge \mathcal{L}^1_\tau\ \phi \wedge \\ \quad \forall (M : \alpha\ \mathtt{model})\ (N : \beta\ \mathtt{model})\ (v : \beta)\ (w : \alpha). \\ \qquad M, w \leftrightarrows N, v \Rightarrow \\ \qquad\quad (\mathsf{mm2folm}\ M, (\lambda\, (x : \mathit{num}).\ w) \vDash \phi \iff \\ \qquad\qquad \mathsf{mm2folm}\ N, (\lambda\, (x : \mathit{num}).\ v) \vDash \phi) \end{array}$$

Because of the same problem we met when defining equivalence of formulas, the type parameters are necessary here. However, although it is possible to prove theorems for different types α and β in the above definition, in the theorems to come, we will only consider the case where α and β are the same.

The Van Benthem characterisation theorem says an $\mathcal{L}^1_\tau$ formula with at most one free variable x is invariant under bisimulation precisely when it is equivalent to the standard translation of some modal formula at x. It is immediate from Proposition 1 that every such formula which is equivalent to a standard translation is invariant for bisimulation. We cannot prove it as an 'if and only if' statement, since according to the proofs in [1], we can only prove the two directions separately as:

Proposition 5. [1, Theorem 2.68, as two separate directions]

$$\vdash \mathsf{FV}\ \delta \subseteq \{\, x \,\} \wedge \mathcal{L}^1_\tau\ \delta \wedge \delta\ {}^{\mathsf{f}}\!\!\equiv_{(:\alpha)} \mathsf{ST}_x\ \phi \Rightarrow \mathsf{invar4bisim}\ x\ (:\alpha)\ (:\alpha)\ \delta$$

$$\begin{array}{l} \vdash \mathsf{INFINITE}\ \mathcal{U}(:\alpha) \wedge \\ \quad \mathsf{invar4bisim}\ x\ (:(\mathit{num} \to \alpha) \to \mathtt{bool})\ (:(\mathit{num} \to \alpha) \to \mathtt{bool})\ \delta \Rightarrow \\ \qquad \exists \phi.\ \delta\ {}^{\mathsf{f}}\!\!\equiv_{(:\alpha)} \mathsf{ST}_x\ \phi \end{array}$$

which cannot be put together into a double implication. To see the reason: given an $\mathcal{L}^1_\tau$-formula ϕ with no more then one free variable, by the second theorem above, if ϕ is invariant under bisimulation for models with $(\mathtt{num} \to \alpha) \to \mathtt{bool}$-worlds, then ϕ is equivalent to a standard translation on a model with α-worlds. However, if we want to prove the converse of this statement, we need to start with the assumption that ϕ is equivalent to a standard translation on models with α-worlds, and prove that ϕ is invariant for bisimulation for models with $(\mathtt{num} \to \alpha) \to \mathtt{bool}$-worlds. But by the first theorem above, we can only conclude ϕ is invariant for bisimulation for models of type α. The point is that it is not the fact that all our desired operations can be taken within a type. In particular, we cannot take ultraproducts of models and preserve cardinalities. The cardinality of the type universe of $(\mathtt{num} \to \alpha) \to \mathtt{bool}$ is too large to be embedded into α, so we cannot just fix the 'base type' to be α and get an 'if and only if' statement-we cannot derive ϕ is invariant for bisimulation for models with $(\mathtt{num} \to \alpha) \to \mathtt{bool}$-worlds from the fact that ϕ is invariant for bisimulation for models with α-worlds. If we could quantify over types (as we could in a theorem prover based on dependent type theory), then we can could define 'invariant under bisimulation for models of every type', and hence prove the original statement of Van Benthem characterisation theorem.

For the proof of the two theorems above, the first one is immediate from Proposition 1, and the second one requires another critical lemma saying 'modal equivalence between two worlds implies bisimilarity of the two worlds when embedded in some other models'. More precisely, if two worlds $w \in M^W$ and $v \in N^W$ are modal equivalent, then we can find an ultrafilter U on J such that in ultrapower models of M and N on U respectively, there is a bisimulation between the worlds corresponding to w and v.

Theorem 7. [1, Theorem 2.74, one direction]

$$\begin{array}{l} \vdash w \in M^W \wedge v \in N^W \wedge (\forall \phi.\, M, w \Vdash \phi \iff N, v \Vdash \phi) \Rightarrow \\ \quad \exists\, U\, J. \\ \qquad \textsf{ultrafilter}\ U\ J\ \wedge \\ \qquad \Pi_U\ (\lambda j.\, M), \{\, f \mid (\lambda j.\, w) \sim_U^{\textsf{worlds}\ (\lambda j.\, M)} f \,\} \leftrightarrows \Pi_U\ (\lambda j.\, N), \{\, g \mid \\ \qquad (\lambda j.\, v) \sim_U^{\textsf{worlds}\ (\lambda j.\, N)} g \,\} \end{array}$$

The proof of the above relies on Lemma 2.

8 Conclusion

To summarise, we have mechanised all of the results (appearing as propositions, lemmas and theorems) in the first two chapters in Blackburn *et al.* [1] that can be captured by the HOL logic, and which are about the basic modal language. The exceptions are:

- The result in Sect. 2.6 about 'definability', which requires a definition of the 'models closed under taking ultraproducts'. Simple type theory cannot capture such large sets.
- The result about 'safety' in Sect. 2.7 is a result about the PDL language, which has infinitely many modal operators. For the moment, we have restricted our attention to the basic modal language, with only $\Diamond$ (and the derived $\Box$).

The two characterisation theorems from Blackburn *et al.* [1], namely Theorem 2.68 (Van Benthem's Characterisation Theorem) and Theorem 2.78, are the only two mechanised theorem such that translating the 'if and only if' statements from set theory into simple theory does not yield an 'if and only if' statement. Blackburn *et al.*'s proof of Theorem 2.78 has the same pattern as Van Benthem's Characterisation Theorem (discussed earlier), and is less complicated.

For each of the mechanised definitions and results, we write the statement in HOL to be as close as possible to the original statement in [1]. We believe that this makes it as easy as possible for people who are interested in mechanising other results in [1] to continue with our work as a starting point. The work on ultraproducts up to Łoś's theorem is independent of our work on modal model theory, and should be generally useful in other model-theoretic applications.

8.1 Related Work

We believe that we are the first to mechanise the bulk of the results in this paper. Of course, much work has been done in this and similar areas. For example, de Wind's thesis [7] is a notable early mechanisation of modal logic, mainly focusing on proving the validity of modal formulas via natural deduction. Of similar vintage is Harrison's mechanisation of foundational results about first order model theory [5], in particular compactness. We used this mechanisation directly in our own work. A great deal of work has also been done in the mechanisation of first order *proof* theory, such as the recent pearl by Blanchette *et al.* [2], showing completeness in elegant fashion.

The connections between modal logic and process algebra are well-understood and there has been a great deal of mechanised work on the operational theory of such (co-)algebraic systems, starting at least as far back as Nesi [6]. Our proof of the Hennessy-Milner theorem (Theorem 1) is a gesture in this direction, but Van Benthem's theorem is much deeper and uses bisimulations as a tool to understanding the connection between modal and first order logics, rather than as a connection to process algebras.

Mechanised work with ultrafilters began with Fleuriot's use of them to mechanise non-standard analysis [4]. We are unaware of any previous mechanised use of ultraproducts or ultrapowers.

References

1. Blackburn, P., de Rijke, M., Venema, Y.: Modal Logic. Cambridge University Press, Cambridge (2001)
2. Blanchette, J.C., Popescu, A., Traytel, D.: Unified classical logic completeness. In: Demri, S., Kapur, D., Weidenbach, C. (eds.) IJCAR 2014. LNCS (LNAI), vol. 8562, pp. 46–60. Springer, Cham (2014). https://doi.org/10.1007/978-3-319-08587-6_4
3. Chang, C.C., Keisler, H.J.: Model Theory. North Holland, Amsterdam (1990)
4. Fleuriot, J.: A Combination of Geometry Theorem Proving and Nonstandard Analysis with Application to Newton's Principia. Springer, London (2001). https://doi.org/10.1007/978-0-85729-329-9
5. Harrison, J.: Formalizing basic first order model theory. In: Grundy, J., Newey, M. (eds.) TPHOLs 1998. LNCS, vol. 1479, pp. 153–170. Springer, Heidelberg (1998). https://doi.org/10.1007/BFb0055135
6. Nesi, M.: Mechanising a modal logic for value-passing agents in HOL. Electron. Notes Theor. Comput. Sci. **5**, 31–46 (1996). https://doi.org/10.1016/S1571-0661(05)80682-6
7. de Wind, P.: Modal Logic in Coq. Master's thesis, Vrije Universiteit (2001)

[illegible]

[illegible]

[illegible] them [illegible] previous [illegible]

[illegible]

[illegible]

Author Index

Affeldt, Reynald II-3
Aitken, Dave II-464
Allamigeon, Xavier II-185

Baader, Franz I-163, I-413
Baanen, Anne II-21
Baranowski, Marek I-13
Barbosa, Haniel I-141
Barnett, Lee A. I-32
Barrett, Clark I-218, I-238
Basin, David I-432
Baumgartner, Peter I-337
Bhayat, Ahmed I-259, I-278, II-361
Biere, Armin I-32
Blackburn, Patrick I-474
Blanchette, Jasmin I-316
Blazy, Sandrine II-324
Bohrer, Rose I-454
Bonacina, Maria Paola I-356
Brakensiek, Joshua I-48
Bray, Matt II-464
Bride, Hadrien II-369
Brown, Chad E. II-489

Cai, Cheng-Hao II-369
Calvanese, Diego I-181
Cerna, David I-32
Chew, Leroy I-66
Chlipala, Adam II-119
Chvalovský, Karel II-448
Clymo, Judith I-66
Cohen, Cyril II-3
Cohen, Liron I-375
Cruanes, Simon II-464
Czajka, Łukasz II-28

Dalmonte, Tiziano II-378
Dardinier, Thibault I-432
De Angelis, Emanuele I-83
de Moura, Leonardo II-167
de Vilhena, Paulo Emílio II-204
Delaware, Benjamin II-119
Dong, Jin Song II-369

Duarte, André II-388
Dutertre, Bruno I-103

Eßmann, Robin II-291

Fioravanti, Fabio I-83
Fontaine, Pascal I-238
From, Asta Halkjær I-474
Fürer, Basil II-58

Ghilardi, Silvio I-181
Gianola, Alessandro I-181
Girlando, Marianna II-398
Gleiss, Bernhard I-297, I-402
Gligoric, Milos II-97
Goertzel, Zarathustra Amadeus II-408
Gore, Rajeev II-369
Graham-Lengrand, Stéphane I-103
Gross, Jason II-119
Gunther, Emmanuel II-221
Gutiérrez, Raúl II-416, II-436
Guttmann, Walter II-236

Hague, Matthew I-122
Hales, Thomas II-254
Hashim, Mohammed II-480
Hausmann, Daniel I-482
He, Shaobo I-13
Heimes, Lukas I-432
Heule, Marijn I-48
Hóu, Zhé II-369

Ignatovich, Denis II-464

Jakubův, Jan II-448
Jovanović, Dejan I-103

Kagan, Elijah II-464
Kanishev, Kostya II-464
Kanovich, Max I-500
Kapur, Deepak I-163
Katz, Ricardo D. II-185
Kerjean, Marie II-3

Kirst, Dominik II-79
Kohlhase, Michael I-395
Korovin, Konstantin II-388
Kovács, Laura I-297
Krstić, Srđan I-432
Kuznetsov, Stepan I-500

Lammich, Peter II-307
Lampert, Timm I-201
Lange, Jane I-238
Larchey-Wendling, Dominique II-79
Larraz, Daniel I-141
Léchenet, Jean-Christophe II-324
Lechner, Mathias I-13
Li, Junyi Jessy II-97
Lin, Anthony W. I-122
Lochbihler, Andreas II-58
Lucas, Salvador II-416, II-436

Mackey, John I-48
Maclean, Ewen II-464
Mahboubi, Assia II-3
Mahony, Brendan II-369
Marić, Filip II-270
McCarthy, Jim II-369
Mometto, Nicola II-464
Montali, Marco I-181

Naeem, Zan II-480
Nakano, Anderson I-201
Narváez, David I-48
Nguyen, Thanh Son I-13
Nie, Pengyu II-97
Nigam, Vivek I-500
Nipkow, Tobias II-291, II-341
Norrish, Michael I-518
Nötzli, Andres I-218

Olivetti, Nicola II-378
Olšák, Miroslav II-448

Pagano, Miguel II-221
Palmskog, Karl II-97
Passmore, Grant II-464
Paulson, Lawrence C. II-204
Pease, Adam II-158, II-472
Pettorossi, Alberto I-83
Pichardie, David II-324
Piotrowski, Bartosz II-448
Piskac, Ruzica I-3
Pit-Claudel, Clément II-119
Platzer, André I-454
Pozzato, Gian Luca II-378
Proietti, Maurizio I-83

Rabe, Florian I-395
Rakamarić, Zvonimir I-13
Raszyk, Martin I-432
Rath, Jakob I-297
Rau, Martin II-341
Raya, Rodrigo II-254
Reger, Giles I-259, I-278, II-361
Reis, Giselle II-480
Reynolds, Andrew I-141, I-218
Ringeissen, Christophe I-238
Rivkin, Andrey I-181
Robillard, Simon I-316, II-291
Rouhling, Damien II-3
Rowe, Reuben N. S. I-375
Rümmer, Philipp I-122
Rydval, Jakub I-413

Sacerdoti Coen, Claudio I-395
Sakaguchi, Kazuhiko II-3, II-138
Sánchez Terraf, Pedro II-221
Scedrov, Andre I-500
Schaefer, Jan Frederik I-395
Schneider, Joshua I-432, II-58
Schröder, Lutz I-482
Schulz, Stephan II-158
Sheng, Ying I-238
Straßburger, Lutz II-398
Strub, Pierre-Yves II-185
Suda, Martin I-402, II-448

Tinelli, Cesare I-141, I-218
Tourret, Sophie I-316
Traytel, Dmitriy I-432, II-58

Ullrich, Sebastian II-167
Urban, Josef II-448, II-489

Villadsen, Jørgen I-474

Waldmann, Uwe I-316
Wang, Peng II-119
Winkler, Sarah I-356
Wu, Zhilin I-122

Xu, Yiming I-518

Zohar, Yoni I-238
Zombori, Zsolt II-489

Printed in the United States
by Baker & Taylor Publisher Services